Ocean Remote Sensing Technologies

Other volumes in this series:

Published by SciTech Publishing, an imprint of The Institution of Engineering and Technology, London, United Kingdom

The Institution of Engineering and Technology is registered as a Charity in England & Wales (no. 211014) and Scotland (no. SC038698).

© The Institution of Engineering and Technology 2022

First published 2021

The Institution of Engineering and Technology
Michael Faraday House
Six Hills Way, Stevenage
Herts, SG1 2AY, United Kingdom

www.theiet.org

British Library Cataloguing in Publication Data
A catalogue record for this product is available from the British Library

ISBN 978-1-83953-161-3 (hardback)
ISBN 978-1-83953-162-0 (PDF)

Typeset in India by Exeter Premedia Services Private Limited

Ocean Remote Sensing Technologies

High frequency, marine and GNSS-based radar

Edited by
Weimin Huang and Eric W. Gill

The Institution of Engineering and Technology

Contents

12 Observation of sea surface waves by noncoherent X-band marine radar   285

Zhongbiao Chen, Yijun He, and Weimin Huang

13 Wavelet-based methods to invert sea surfaces and bathymetries from X-band radar images   313

Pavel Chernyshov, Teodor Vrecica, and Yaron Toledo

14 Wave field reconstruction using orthogonal decomposition of Doppler velocities   333

Erin E Hackett

17 Wind parameter measurement using X-band marine radar images   401
Xinwei Chen, Weimin Huang, and Björn Lund

18 Introduction to remote sensing using GNSS signals of opportunity   425
Adriano Camps

About the Editors

Weimin Huang is a professor in the Department of Electrical and Computer Engineering at Memorial University of Newfoundland, Canada. He was previously a design engineer with Rutter Technologies, Canada. He has authored over 230 research papers. His research interests include the mapping of oceanic surface parameters via HF ground wave radar, X-band marine radar, and global navigation satellite systems. He is an associate editor for IEEE ACCESS, IEEE Canadian Journal of Electrical and Computer Engineering, and an editorial board member of Remote Sensing.

Eric W. Gill is an emeritus professor in the Department of Electrical and Computer Engineering at Memorial University of Newfoundland, Canada. He is a senior member of the IEEE. His research primarily deals with the scattering of high frequency (HF) radiation from time-varying random rough surfaces. He also works on the application of HF surface wave radar (HFSWR) operating in an ocean environment. He is an associate editor for IEEE Journal of Oceanic Engineering.

Preface

Covering more than seventy percent of the Earth's surface and containing roughly ninety seven percent of its water, the ocean plays an inestimable role in affecting global climate and marine transportation, exploration and development, as well as a myriad of other commercial and recreational enterprises. Unlocking the ocean's many intriguing mysteries has been the subject of human pursuits since the dawn of recorded history. With the proliferation of technological development over the last two centuries, it is not surprising that significant scientific advances have been made in enhancing the general body of knowledge associated with this enormous resource.

While a vast array of ocean instrumentation has been developed for research purposes, since the middle of the twentieth century remote sensing technologies have become increasingly important. Among this latter class of instruments, high frequency (HF) surface and skywave radar and microwave marine radar have been successfully implemented in gathering information on large tracts of the ocean surface. Since the 1990s, relatively new techniques for ocean observation involving signals from global navigation satellite systems (GNSS) have appeared. All three of these types of ocean radars are able to provide sea surface information associated with winds, waves, currents, and hard targets, and, in some cases, bathymetry with differing spatial and temporal resolutions.

Though many successful applications have been reported for each of the three classes of ocean radar technologies referred to above, there are plenty of questions that deserve further exploration. A basic objective of this book is to provide ocean researchers, including students, with a systematic introduction to the principles, state-of-the-art methods and applications of these radars and to explore ongoing challenges. In some cases, particularly for GNSS-based developments, recent outcomes are addressed.

This book includes 23 chapters that are organized into three parts, mainly according to the sensor types. The first covers work related to HF radar, the second focusses on microwave marine radar, and the third concentrates on GNSS-based radar. Each part consists of an introduction chapter that provides an overview of the corresponding sensor followed by chapters focussing on fundamental theory, specific application or advanced algorithm development. Each of the chapters is self-contained and readers should be aware that there may be across-chapter differences in symbols used for various parameters. Because the work is intended for a variety of readers, content may be selected according to interest and background.

The authors are researchers and developers from academic institutions, the private sector, and government agencies. We sincerely thank all authors for accepting the invitation to contribute their expertise to this book. Their patience and cooperation during the peer-review and revision processes are gratefully acknowledged. We also thank all reviewers (see the complete list of reviewers in the Appendix) for their time and energy to provide professional suggestions. With their valuable volunteer work and insightful comments, the quality of the book has been significantly improved. In particular, we would like to extend our thanks to Drs Stuart J. Anderson (University of Adelaide, Australia), Merrick C. Haller (Oregon State University, USA), Yonggang Liu (University of South Florida, USA) and Adriano Camps (Universitat Politècnica de Catalunya, Spain) for their constructive advice on the book development. The contributions of these individuals over a two-year editorial period, exacerbated by the COVID-19 pandemic, have made the work a reality.

Weimin Huang, Eric W. Gill
Editors

Chapter 1
HF radar in a maritime environment

Eric W. Gill[1] and Weimin Huang[1]

The general term, *remote sensing*, refers to the science of acquiring information about an object or phenomenon from a distance. This is in contrast to *in situ* measurement, which is accomplished by instruments located at or near the position of information gathering.

Perhaps, in the first instance, *remote sensing* conjures ideas of instruments being flown on aircraft or spacecraft, including satellites, to obtain data from the earth's surface, from objects on or near that surface, or from space itself. Furthermore, such instruments are generally understood to transmit some form of electromagnetic (EM) signal which, after reflecting or scattering from the distant 'object', is then received and analysed, either at the location of the transmission – i.e., in *monostatic* operation – or at some other receiver distant from both the transmitting (TX) source and the scattering object or surface – i.e., in *bistatic* operation. However, this notion of remote sensing is somewhat restrictive for at least two reasons. First, in addition to using EM signals, sound sources may be used to acoustically ascertain information from a distance; second, EM signal sources may be operated, not only from air or space, but from moving or stationary land- or ocean-based platforms. It is significant that early analyses associated with high-frequency (HF) radar were founded on prior developments in acoustics. In this chapter, an overview is given of the science, technology and techniques associated with the development of HF radar as an important instrument in the arsenal of tools available for remotely sensing the ocean surface and objects on or near it. This particular aspect of ocean sensing has matured to being generally referred to as *radiowave oceanography*, or, simply, *radio oceanography*.

1.1 HF radar as an ocean remote sensor – introduction

By the turn of the twentieth century, EM signals were being used to detect the presence of objects remote from the TX sources. Technological advances leading up to World War II permitted target detection and positioning by radar on a routine basis.

[1]Department of Electrical and Computer Engineering, Memorial University of Newfoundland, Canada

In the context of this present work, the target may be a portion of the ocean surface or an object, such as a ship, on the surface or an aircraft, which may or may not be close to the surface. It will be seen that the signal, received monostatically or bistatically after it scatters from the ocean or some other target, contains a wealth of information regarding the physical characteristics of that portion of the scattering surface or target. This is what makes radar in general such an important remote sensing tool, but the question may be asked, 'Why HF radar?'

1.1.1   *A few fundamentals*

In attempting to answer the question 'Why HF radar?' for ocean remote sensing, there are a few basic facts to consider. In this introductory material, little or no attempt is made to explain the 'why' of these facts, but more details are contained throughout the HF radar section of the book.

Radio waves occupy a portion of the EM spectrum ranging in wavelength from roughly 1 mm to 10 000 km, corresponding to a frequency range of 30 Hz to 300 GHz, with some texts including 3 Hz or less (extra low frequency) at the lower frequency end. The so-called *high-frequency* portion, or HF band, is a subset of the radio spectrum from 3 to 30 MHz (or wavelengths of 10 to 100 m). The name, 'high frequency' or, equivalently, 'short wave', of the band may be seen as a bit of a misnomer. The misnomer is a vestige of band nomenclature occurring at a time when those frequencies were indeed *high* compared to other technologically usable radio frequencies.

It is well-known physics that, generally, the detail that may be resolved by reflecting an EM signal from an object depends on the wavelength of the signal. Within the HF band, the signal wavelengths are of the same order of magnitude as the most energetic ocean gravity waves [1]. While some of the detail is addressed in Section 1.3 of this chapter, it is presently sufficient to observe that HF signals thus couple strongly with the most energetic ocean waves, resulting in the fact that the signals received after scattering may be analysed to obtain important features of oceanic phenomena. It will be seen throughout the HF radar section of the book that the information deduced directly from the HF scattered signal includes, for example, surface current velocities and directional wave fields. Other phenomena, such as surface wind fields and current shear, may be indirectly determined.

The most common type of radar used in coastal ocean applications is the HF surface wave radar (HFSWR). For these HFSWRs, when the source, that is, the signal launched from the radar TX antenna, is close to the surface of the ocean, the principal mode of propagation is a so-called *surface wave* [2]; hence the name 'surface wave radar'. The simplest form of a TX antenna used in most HFSWR ocean applications is a *vertical* thin-wire *monopole* or some combination of such monopoles. These monopoles usually have a length that is an appreciable fraction of the signal wavelength, generally not exceeding a quarter wavelength. The signal transmitted will have a so-called *vertical polarisation* meaning roughly, in this case, that the electric field vector of the EM wave oscillates perpendicular to the ocean surface, the latter being a highly conductive medium at HF frequencies. It may be noted that

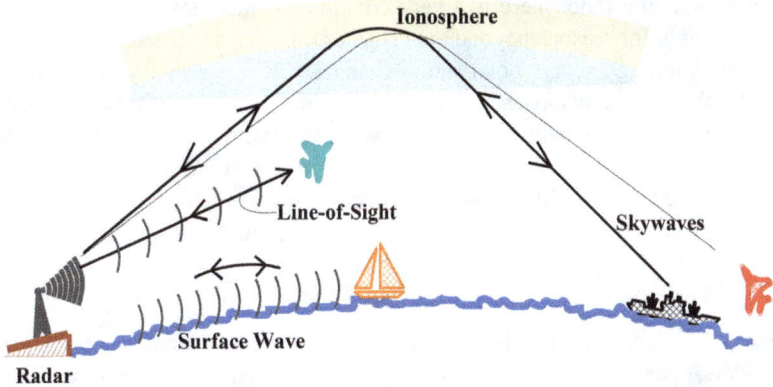

Figure 1.1   Pictorial representation of radar operating in various modes

horizontally polarised signals, for which the electric field vector is roughly parallel to the surface, do not propagate well over the ocean and are not useful for HFSWR. Thus, in HFSWR, both the TX and receiving (RX) antenna systems should be configured optimally for vertically polarised signals.

It is noteworthy that the high conductivity of the ocean surface at HF, as mentioned above, is an extremely important feature as it supports the surface wave propagation of the EM signal to large distances (see Figure 1.1). This explains why HF radar is highly useful in the ocean context, but not so in remotely sensing large bodies of fresh water, which have much lower conductivities. Also, not surprisingly, as a general rule, the lower the operating frequency, the greater the useful range of the HFSWR. Of course, there are several competing issues as, for example, for the lower frequencies, the spatial resolution will generally be degraded. Still, depending on a variety of radar operating parameters, including operating frequency, TX power and internal and external noise levels, and the ocean surface characteristics associated with the wave directional spectrum, HF radar can be used to collect data on some ocean parameters well beyond the line-of-sight horizon to distances greater than 200 km from the transmission site.

Also depicted in Figure 1.1 is the so-called *skywave* propagation mode (see Figure 3.1 for a more comprehensive illustration). For the purpose of this introductory material, in skywave mode the TX signal may be considered to be reflected from the ionosphere to some distant patch of ocean (typically, 1 000 to 4 000 km away) where it scatters in all directions. Some of that scattered signal may be received, via the ionosphere, at the original site – i.e., in monostatic operations – or it may be received bistatically via surface wave or the ionosphere at a remote site. For obvious reasons, radars operating in skywave mode are often referred to as *over-the-horizon* radars (OTHR), although this should not be confused with HFSWR which, as observed previously, may also be used to gather data from well beyond the line-of-sight horizon.

Of course, the ionosphere is a very complicated medium with stochastic EM properties, and, for reasons beyond the scope of this chapter, the signals interacting with it may undergo severe polarisation changes [3]. While in skywave mode, the TX and RX signals may have any polarisation, as noted in [3], for ocean applications involving combined skywave and surface wave modes, the RX antennas should be vertically polarised (meaning they should be configured to optimally receive vertically polarised signals). Some of the added complexities associated with OTHR as compared to HFSWR are considered in Chapter 3 of this book.

1.1.2   Common classes and properties of ocean-mapping HFSWR

By far, HF radars, whose primary purpose is the measurement of ocean surface parameters, operate in surface wave mode. HFSWRs themselves may be categorised into two main groups: (1) large, so-called phased-array radars that most commonly yield signal direction information via a process referred to as *beamforming* (BF) and (2) smaller, compact systems that provide direction via a process generally known as *direction finding* (DF). The commonalities and differences associated with these systems form the remainder of the discussion in this section. The requisite definitions of terms (including what is meant by BF and DF) either appear as required or are appropriately referenced. Since only the rudiments necessary for an understanding of the applications that follow in this and later chapters are broached here, the reader interested in much more comprehensive analyses is directed to such works as [4, 5].

1.1.2.1   Range

All HFSWRs, whether large or compact, used in ocean remote sensing, or for that matter any other application, must have characteristics that permit the user to identify the *location* of the phenomenon being interrogated, whether it is a surface current feature or some wave parameter. Obviously, in general, the more the precision associated with location, the more trustworthy and useful the information obtained.

As previously intimated, establishing location, say of a particular patch of ocean being studied, involves determining how far away it is from the radar site – i.e., the range (R) – as well as its direction from a reference, such as true north. In monostatic operation, the range may be trivially given as half the product of the elapsed two-way transmission time, Δt, of the signal from the radar transmitter to an ocean patch and back and the speed c of the EM wave. That is,

$$R = \frac{c\Delta t}{2}. \tag{1.1}$$

In practice, the exact manner in which a reflected signal is established as coming from a particular range will depend on the waveform of the transmitted signal. Figure 1.2 depicts two particular waveforms that have been or are currently being used in oceanographic HFSWR. Figure 1.2(a) illustrates a simple voltage (V) versus time (t) plot for *pulse* transmission, the invention of which dates back to World War II. In the most basic implementation of this signal, the transmitter is 'on' for a brief period, perhaps a few microseconds, during which it may transmit a

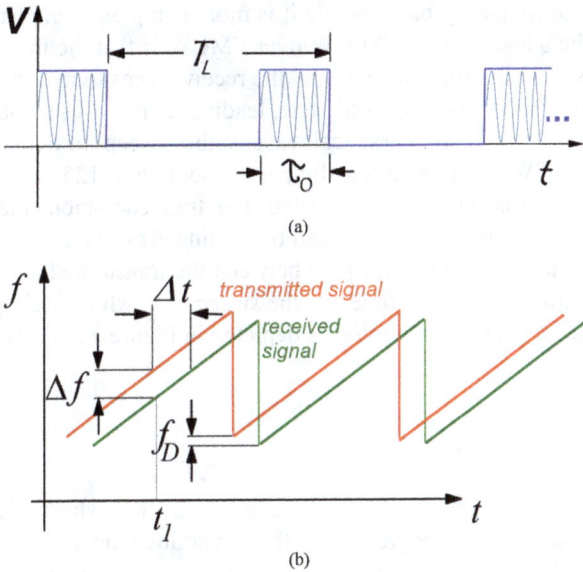

Figure 1.2   Plot of (a) voltage V versus time (t) for a pulse transmission and
(b) frequency (f) versus time (t) for an FMCW transmission

sinusoidal waveform. It is then shut off for generally a much longer time, say a few hundred microseconds, before this 'on–off' or *gating* process is repeated. Thus, the term *pulse* radar aptly fits the behaviour, and, with reference to Figure 1.2(a), the *pulse width* is τ_0 while the *pulse period* is T_L. In a simple implementation of pulse radar, during the time that the transmitted signal is off, the radar receiver 'listens' for the reflection coming from 'targets' located at increasingly distant ranges until the signal is no longer strong enough to be detected. The process is then repeated, sometimes for several 10 s of minutes, so that target information may be obtained by averaging the data from many pulses. As already emphasised, in purely oceanographic applications, the 'target' is actually an ocean surface patch, the size of which is discussed later in this section. For pulse radar, it is intuitively obvious how (1.1) may be used to determine the range.

Although the so-called *pulse* radar is perhaps the most conceptually simple, in modern, relatively low-cost, low-power HFSWR used for oceanographic measurement, the transmitted signal is generally a frequency-modulated continuous wave (FMCW) [6] or a frequency-modulated interrupted continuous wave (FMICW) [7]. The general form of an FMCW signal is depicted in Figure 1.2(b). As the name implies, FMCW means that there is no on–off gating of the frequency-modulated transmission as it is *swept* over a specified range of frequencies or *bandwidth*. Care must be taken not to overload the receiver from a direct transmitter signal since, in this mode, both operate simultaneously. This limits the TX power and, consequently, the range. In FMICW operation, the FM transmitter signal is 'off' or *interrupted* for part of the time that the receiver is 'on' as the signal is swept over the frequency band. While it is more complex to implement in hardware, one of the advantages of FMICW over FMCW is that the transmitter power may be increased during the 'on' state and the receiver sensitivity may be increased for the portions of the transmitter 'off' state, leading to increases in maximum radar range. For regulatory and other practical reasons, the 'sweep' of frequencies in both FMCW and FMICW is limited, typically being about 25 to 125 kHz in extent (see [8] for details), and is repeated as depicted. For frequency-modulated radars, the range measurement may be accomplished by writing Δt of (1.1) as the quotient of the absolute value of the difference $|\Delta f|$ between the transmitted and received frequencies at a particular point in time and the *sweep rate*, which is simply the slope, $\frac{df}{dt}$, of the 'frequency versus time' plot, as depicted in Figure 1.2(b). That is,

$$R = \frac{c|\Delta f|}{2\left(\dfrac{df}{dt}\right)}.$$

(1.2)

It is important to notice that in oceanographic HFSWRs, whether large or compact, the received signal is referenced to the transmitted signal through a process referred to as *coherent detection* and, in the FMCW case, being one of the simplest to illustrate, the Δf of (1.2) and the Doppler shift f_D in Figure 1.2(b) are the outcomes of this process. The exact mechanics of accomplishing this coherence is well explained, for example, in [9]. In fact, as will be seen in Section 1.3, this coherent detection also permits the development of Doppler spectra, each frequency of which corresponds to a velocity component of the scatterer (for example, waves travelling on the ocean may be the 'scatterer').

1.1.2.2   Direction

It is common in both BF and DF ocean-sensing HFSWRs for the TX antenna systems to flood the ocean surface with incident radiation equally, or nearly so, in all directions at a given range (see Figure 1.3). The spatial shape of, for example, the power distribution radiated from or received by an antenna or an array of antennas is referred to as the *radiation* or *beam pattern* (see IEEE standards [10] for a more general definition). For example, in the horizontal plane, which is of primary interest for the surface wave mode, the theoretical beam pattern of a typical vertical monopole is a circle, and such antennas, referred to as being *omnidirectional*, are often used in HFSWR applications (see, for example [7]), though they may not be the most efficient in radar deployments when a substantial amount of the radiation is wasted over land. Another commonly used TX system consists of a four-element array in which monopoles appear at the corners of an appropriately sized rectangular area (see, for example [11]). In either case, the intention is to floodlight the ocean surface in front of the radar (see Figure 1.3). Of course, the TX signal scatters from the random ocean surface in random directions, and a portion of that radiation will be acquired at the radar RX site which, as previously indicated, may or may not be co-located with the transmitter. Determining the direction, relative to a fixed reference, from which the RX antennas receive the scattered signal is obviously essential in the final

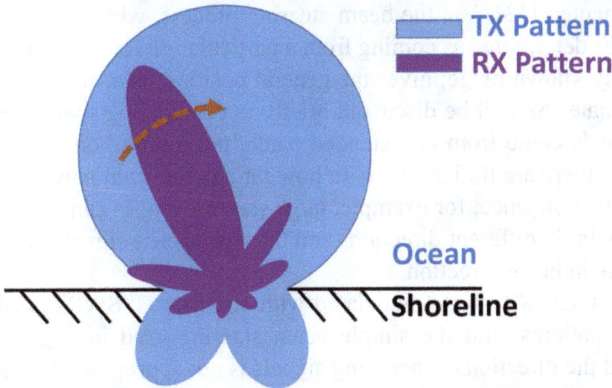

Figure 1.3   Illustration of a broad TX beam pattern floodlighting the ocean while a narrow RX beam is scanned, as indicated by the arrow, to receive the scattered radiation. See also Figure 1.4.

assignment of proper locations to the ocean parameters derived from the signals. The process of determining this direction is what fundamentally distinguishes BF and DF systems. In passing, it may be noted that *azimuth* refers to an angle between 0° and 360° measured clockwise from true north, while *bearing* is simply the same direction as referenced to one of the cardinal directions.

In BF systems, the RX beam pattern, generally consisting of a relatively large amplitude but narrow main lobe and smaller sidelobes (see Figures 1.3 and 1.4), is 'steered' through a wide angle, up to about 120°, by appropriately phasing and adding the received signals from each of the antennas in the RX array (see [12], for example). In simple ocean-sensing HFSWR systems, the RX antenna array often consists of a line of elements spaced approximately a half-wavelength apart. Other arrangements, suiting the array to topographical or regulatory constraints, have also

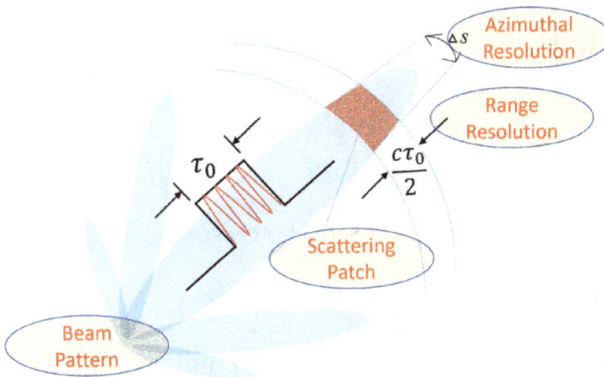

Figure 1.4   Typical RX beam pattern indicating scattering patch resolution. Symbols are defined in the text.

been implemented [13]. Via the beam steering process, which occurs in software, the signals are determined as coming from a particular direction and this, along with the previously known range, gives the general position from which these scattered signals originate. As will be discussed briefly in considering spatial resolution, the received signals come from an extended patch, not a point, on the ocean surface. As intimated, there are limitations as to how far a main beam may be steered off its broadside direction since, for example, large steering angles can result in large sidelobes that point in different directions and thus may sense signals not coming from the desired main beam direction.

The RX antenna systems associated with compact HFSWRs necessarily have broad beam patterns, and the simple beam steering used in large, phased-array radars to find the direction of incoming signals is not appropriate for these compact radars. A variety of DF algorithms have been used with compact coastal HF radars, and Barrick and Lipa [14] provide a succinct description of the historical development. The DF algorithm currently implemented in the CODAR SeaSonde [7] finds its basis in the Multiple Signal Classification (MUSIC) routine initially proposed by Schmidt [15]. Excellent elaborations on the application of MUSIC to compact coastal radars are found in [14, 16]. Without resorting to the intricacies of the DF analysis found therein, the basic ideas used to 'assign direction' to signals carrying information regarding, in the case of ocean-sensing HFSWR, ocean surface currents, may be briefly summarised as follows: (1) for an N-element RX antenna system, the incoming, coherently received signals are converted to the spectral domain (see, for example Figures 1.5–1.7); (2) a covariance matrix, the size of which depends on N, is formed from the results of (1) for each Doppler frequency (it is these Doppler frequencies that correspond to components of the ocean surface current velocities heading towards or away from the radar – the so-called *radial* current velocity); (3) under satisfactory signal-to-noise conditions, an eigenfunction analysis of the

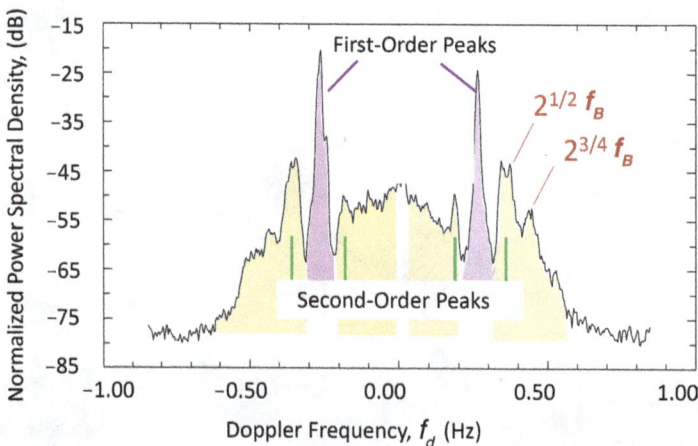

Figure 1.5   *Depiction of an actual pulsed HFSWR Doppler spectrum from a 6.75 MHz radar*

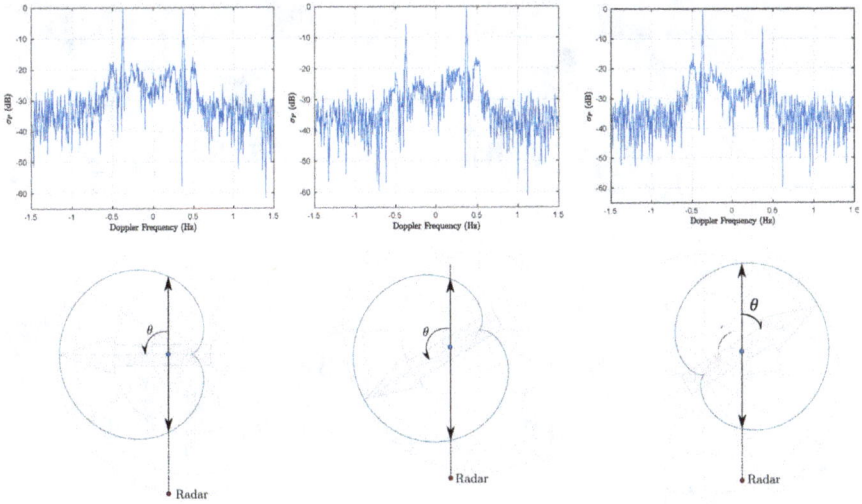

Figure 1.6   *The spectral effect of changing wind direction $\bar{\theta}$ relative to the radar look direction*

spectral covariance matrix permits the separation of the signal eigenvectors from the noise eigenvectors; (4) next, incorporating information on the directional characteristics of the RX elements using a so-called *mode* vector [15] or *signal model vector* [14], the MUSIC algorithm is used to find the direction at which this vector is orthogonal to the set of noise eigenvectors, this direction being the correct one associated with the particular signal (Doppler) component being considered. MUSIC permits $N - 1$ possible directions for each signal component (in the present context, each radial velocity component), and other statistical means may be used to find the direction(s) that best fit the data.

Comparative analyses of both BF and DF have been conducted by several research teams, and a sample of these may be found in [14, 17–19]. In particular, the paper by Laws *et al.* [17] provides a clear description of the basic ideas behind how MUSIC assigns directions of arrival (DOA) to signals associated with specific Doppler frequencies, and thus, for example, specific current velocity components, in the radar spectrum.

1.1.2.3   Range and azimuthal resolutions

Since, at this point of the discussion, a portion of the ocean *surface* is being considered, it is important to establish the size of the smallest surface patch, at a given range, that may be unambiguously distinguished from other such patches. The degree of this importance is commensurate with the rapidity with which the ocean surface features change with location. Thus, having established a means of determining the range and direction of an ocean patch being observed, it is essential to know both its radial (i.e., for monostatic operation, along the line from the radar to

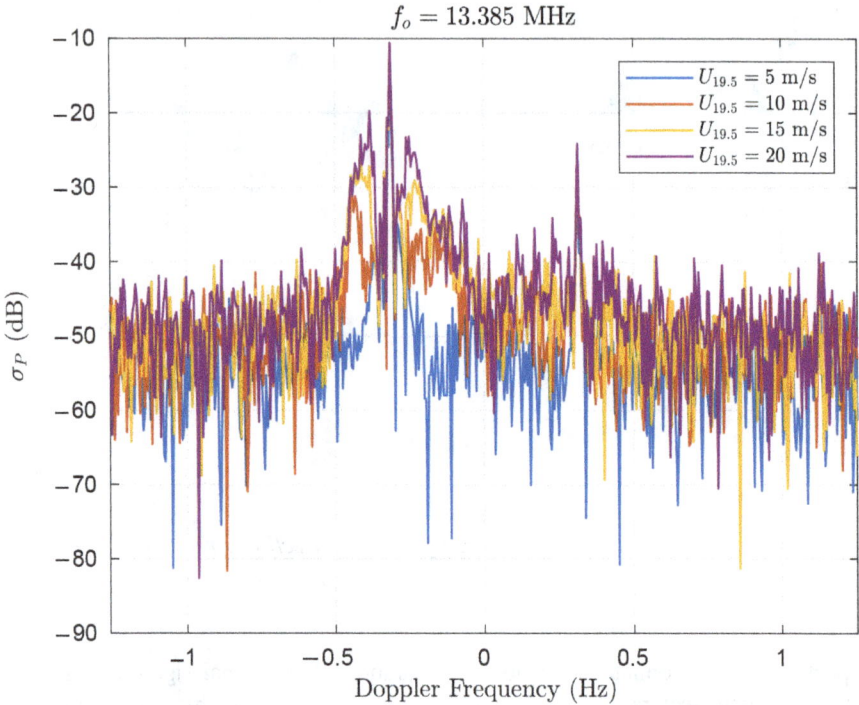

*Figure 1.7   The effect that changing wind speed (or sea state) has on the RCS.
$U_{19.5}$ is identical to U in the text.*

the scattering patch) and angular extents (see Figure 1.4). That is, the *range resolution* and *azimuthal resolution* must be addressed.

In the case of pulse radars, the range resolution ΔR may be easily determined to be (see, for example [20])

$$\Delta R = \frac{c\tau_0}{2}. \qquad (1.3)$$

Equation (1.3) may be written equivalently, for practical purposes, in terms of the signal bandwidth B as

$$\Delta R = \frac{c}{2B}. \qquad (1.4)$$

Equation (1.4) also suitably expresses the range resolution for FM radar systems, whether they are of the larger, phased-array or compact type. Further elucidation of the ideas behind using $B = 1/\tau_0$ may be found throughout the literature on Fourier analysis as applied to signal theory.

The angular resolution associated with a particular azimuth or bearing as it pertains to BF systems is a straightforward concept, at least when simple beam patterns are assumed for the RX array. Again referencing Figure 1.4, the beamwidth is generally taken to mean (1) the angle between the nulls of the main lobe or (2) the angle between the points at which the power associated with angular positions on the main lobe is reduced to half of its maximum value. The latter of these definitions is commonly used in the HF ocean radar community. Using θ_{BW} to represent the half-power beamwidth, the azimuthal resolution Δs may thus be simply defined as

$$\Delta s = R\theta_{BW}. \tag{1.5}$$

For the case of a radar operating with a nominal wavelength of λ_0 and having a linear receive array of $N+1$ elements spaced d_s apart and scanned at an angle θ_0 from the broadside direction (the direction 90° from the array axis),

$$\theta_{BW} = \frac{2.65\lambda_0}{\pi(N+1)d_s \cos\theta_0} \tag{1.6}$$

(see, for example [21, 22]), the azimuthal resolution becomes

$$\Delta s = \frac{2.65R\lambda_0}{(N+1)\pi d_s \cos\theta_0}. \tag{1.7}$$

Using the range resolution of (1.4) along with this azimuthal resolution, the area A of a patch of ocean, which may be unambiguously distinguished from other such patches, may be approximated as

$$A \approx \Delta R\Delta s = \frac{1.325Rc\lambda_0}{\pi B(N+1)d_s \cos\theta_0}. \tag{1.8}$$

While it is obvious from the foregoing that angular resolution in a BF radar system is largely a function of RX antenna size, as pointed out by Laws *et al.* [17], the DOA of the received signals in DF systems can, in principle, be completely resolved. For practical purposes, however, the angular resolution is set by a parameter in the MUSIC processing scheme. Even so, as again discussed in [17], the limitation in angular resolution is generally determined by the frequency resolution of the spectrum of the segment of received data under consideration. These and other issues that may affect the overall quality of results for ocean features such as surface currents obtained by DF radars are addressed by those authors and have more recently been discussed in [23].

1.1.2.4 Summary of BF and DF systems

Both BF and DF coastal radar systems can be used to much profit in obtaining ocean parameters with reasonable resolutions over vast areas of the ocean surface. The phased-array systems, usually operated in BF mode, are suitable when large tracts of land are available for deployment. However, there are many instances, especially along highly populated coastlines, where deployment footprints must be kept to a minimum and, in such places, the compact DF systems have proven to be the instrument of choice, yielding good results. There has been some indication that DF can outperform BF in, for example, rapidly changing current regimes. Several studies (see, for example [17–19]) suggest that there is no single set of circumstances in which one of these modes of operation always outweighs the other.

1.2   A brief historical perspective on relevant theory and technology

Having considered a variety of possible HF radar characteristics that have caused HFSWR and OTHR to emerge as important ocean sensors, an attempt is next made to paint a broad picture of the path of the basic science and resulting technologies that have led to the present state of the art. To this end, the authors have opted in this section to provide a brief summary of theoretical developments in the scattering of signals from rough surfaces that have had a direct impact on how ocean information may be deduced from the radars. Additionally, a high-level sketch of the development of the related technologies is considered. The background theory, the technology and experimentation have been the topics of hundreds of publications, but the purpose here is simply to provide the interested reader with enough details to trace some of the more relevant information directly related to the discussion at hand. A more specific consideration of the application of scattering theory to the development of HF *radar cross sections* (RCSs) of the ocean surface, which in a broad sense in the present context is a measure of the ability of a surface patch of ocean to reflect radar signals in the direction of the receiver, is deferred to Section 1.3. As will be seen there, the RCS properties will be governed by, among other parameters, the roughness and directional characteristics of the ocean surface features as well as various radar operating parameters such as frequency and waveform.

1.2.1   *Relevant propagation and scattering theory*

It is interesting to note that one of the most important developments in the scattering of signals from rough surfaces appeared in the work of Lord Rayleigh (John W. Strutt) in his classic work, *The Theory of Sound* [24]. Being, amongst other things, one of the premier scientists of his day, it is perhaps not surprising that the techniques Rayleigh employed in relation to *acoustical* signals found their way into the analyses of those addressing ideas associated with the propagation and scattering of EM signals. Since Hertz's work [25] in the late 1800s, there have been a plethora of theoretical analyses and applications associated with EM radiation in the context of signals travelling on or near the earth's surface. An excellent summary, which tackles some of the growing pains of the research associated with the propagation of *ground waves* – i.e., waves that would exist in the vicinity of the surface of the earth in the absence of an ionosphere [2] – and covers about a century's worth of developments, was carried out by Wait [26]. It may be noted that in the strict sense of IEEE standard definitions, a *surface wave* forms a component of the total EM radiation of a signal propagating in the presence of ground (i.e., the *ground wave*) [2]. Names such as Sommerfeld, Heaviside, Norton, Tesla, Bremmer, van der Pol, Rice and Wait himself, among others, some of whom appear later in this work, were instrumental in establishing the fundamental science on which present-day applications are founded (see [26] for an extensive list of important players and their seminal research publications).

Once the general theory of *surface wave* of propagation over flat, smooth, ideally conducting half-spaces had gained a firm footing, it is perhaps not too surprising that a good deal of effort was expended on extensions to curved, rough, highly conducting (but not perfectly so) surfaces and stratified media. From the very large number of publications along these lines, for the purpose of this work, we distil the list to a few examples involving the rough, highly conducting earth's surface. Of course, as may be found from any electromagnetics text, the term *highly conducting* involves not only the EM properties of the surface medium, but also the frequency of the radiation travelling on that surface. As intimated earlier, at HF frequencies, the surface of the ocean is seen to be highly conducting and a vertically polarised wave launched from a source close to the ocean may be consequently capable of travelling as a surface wave well beyond the line-of-sight horizon.

In 1951, Rice [27] used the so-called perturbation theory developed by Lord Rayleigh [24] in his acoustic investigations to explore the scattering of EM radiation from *non-time-varying*, two-dimensional, rough, perfectly conducting surfaces. From 1955, when Crombie [28] made the first attempt at explaining the mechanism of the scattering of EM signals from the ocean, the potential for using HF radar for ocean parameter measurements became apparent, as did the necessity of developing more detailed models for the scattering of EM radiation from the ocean surface, which is *stochastic in both time and space*. Subsequently, several analyses, such as those in 1971 by Wait [29] and, in particular, Barrick [30, 31], who closely followed the Rice technique, laid a solid foundation for further theoretical and experimental pursuits.

Historically, in the context of EM scattering from the ocean, models were generally developed under the assumption of plane wave scattering. Typically, scattered field derivations imposed both small-height and small-slope restrictions on the scattering surface. The former of these restrictions implies that for a surface deviation of ξ and a transmitted signal of nominal wavenumber k_0 ($= 2\pi/\lambda_0$), then $k_0\xi \ll 1$, while the latter restriction requires the gradient of the surface to be small compared to unity ($|\vec{\nabla}\xi| \ll 1$). Later, electric field derivations were conducted for a general source located just above the rough surface with particular applications being conducted for both pulsed [32] and FMCW dipole sources [33]. As is discussed in Section 1.3.1, scattering models developed with specific sources reflect reality somewhat better than assuming an ideal plane wave source.

In addition to developments in the context of *ground wave propagation*, by the 1920s, studies were beginning to validate the notion of reflection of radio waves from the ionosphere, and earlier experimental work that had suggested such *skywave propagation* began to be established upon a more theoretical foundation. Headrick and Anderson in Chapter 20 of the *Radar Handbook* [5] provide an extensive list of theoretical and experimental publications related to skywave radar developments and succinctly summarise, where appropriate, their relevance to radio oceanography. It is seen from [5] that much of the theory associated with surface wave propagation and scatter is transferable to the skywave case. For an extensive theoretical treatment of skywave propagation, the reader is referred to [34]. More recent studies

[35–39] have seen the development of new scattering models for the case of iono-spheric/ocean, mixed-path propagation and scattering.

Again, as will be seen in Section 1.3.1, in order to properly model the interaction of the EM signal with the scattering (ocean) surface, the interaction of the ocean waves themselves also needs to be considered. While the results are important to the ideas contained in the EM scattering analysis, the attendant hydrodynamical analyses are outside the present context. However, no historical account of HFSWR, regardless of its brevity, should go without referencing the seminal, related work by Klaus Hasselmann on the ocean gravity wave spectrum [40] and its role in HF scattering problems [41]. Further mention of this is reserved for Section 1.3.1.

1.2.2   Technological advances

While many advances in HF and other radar technology were directly linked to military initiatives (see, for example, *The Invention that Changed the World* [42]), by the 1970s, HFSWR systems, whose primary purpose was for oceanographic applications, were being developed. Teague *et al.* [43] provide a summary of various such systems that appeared over the course of the past three decades of the twentieth century.

From the North American perspective, much of the persisting, ocean-focused HFSWR development finds its genesis in the work headed by Donald Barrick working, at the time, at the National Oceanographic and Atmospheric Administration of the United States Department of Commerce. A brief description of one such pulsed phased-array radar is described in [44] and the design of a more compact system, specifically intended for surface current measurement, appeared in [45]. Various forms of these pulsed systems, generally referred to as Coastal Ocean Dynamics Applications Radars (CODAR), appeared in the literature, and, in particular, one having a four-element square RX array operating in tandem with a Yagi TX antenna was reported in [46]. Besides the early American technological advances leading to the pulsed radars reported in [44] and [45], as early as 1973, Barrick [9] had suggested the relevance of FMCW waveforms for radar operation. With the formation of CODAR Ocean Sensors Limited in Monterey, California in 1984, the technology enjoyed significant evolution. Presently, the CODAR SeaSonde, the most compact form of which consists of a monopole antenna for transmission sitting atop a dome loopstick antenna assembly having two orthogonal elements, is the state of the art in DF HFSWR. The three-channel combination of the monopole and loopstick assembly forms the receive array, and the waveform used in this DF system is FMICW.

In the late 1970s, work led by Pierre Broche of the University of Toulon, France, resulted in the development and deployment of a dual-frequency, phased-array HF radar in the western Mediterranean [47]. While there does not appear to have been a commercial outcome of these systems, the Toulon radar group continues to be actively involved in HF radar deployments (see, for example [48, 49]).

During the 1980s, Ramsay Shearman from the University of Birmingham, United Kingdom, led extensive efforts in the development and implementation of an FMICW radar operating in both the lower and mid HF sub-bands [50]. The TX antenna was chosen to be a log-periodic dipole array while the RX system consisted of two nested 15-element vertical loop arrays, the latter corresponding to the two operational sub-bands. The initiative was carried out with collaboration between Shearman's university group and Neptune Radar Limited. Refinements of this BF system, referred to as Pisces, and its application to wave, current and wind measurements appear in [51]. Further discussion on this system appears in Chapter 6.

As noted by Teague *et al.* [43], during the 1980s, Marex Limited, England, using foundational work conducted by CODAR, the NOAA Wave Propagation Laboratory and researchers out of Stanford University, developed the so-called Ocean Surface Current Radar. This phased-array system was used over a couple of decades to measure waves as well as currents (see, for example, Chapter 6).

Beside the efforts in the United States and Britain, by the late 1990s, a team from the University of Hamburg had designed the so-called Wellen Radar (WERA) [6]. The hardware development of this HFSWR phased-array system, one form of which consists of a four-element monopole TX array and a linear RX array having up to 16 monopoles, continues at the German company, Helzel Messtechnik GmbH. A more compact version of the WERA system, consisting of a four-element square array, has also been designed to operate in DF mode [52].

In addition to the American and European technological achievements, Chinese researchers have also been very active in advancing HFSWR techniques and developing their own systems. The Experimental HF Surface Over-the-Horizon Radar (EHFR) test, which was led by the Harbin Institute of Technology, was conducted in late 1980s [53]. The focus of EHFR is on detection and tracking of beyond line-of-sight ships and low-altitude aeroplanes [54]. Later, their interest was expanded to shipborne HF radar for both ship tracking and oceanographic applications [55, 56]. Since the 1980s, Wuhan University has been significantly investing in the development of experimental and commercial HF Ocean State Monitoring and Analyzing Radar (OSMAR) systems such as OSMAR2000 [57], OSMAR071 [58] and portable OSMAR-S [59]. Most recently, HF radar operating in hybrid sky–surface wave mode has attracted tremendous interest among many Chinese groups [60–62].

As intimated by Anderson in Chapter 3 of this book, the use of skywave radar for the purpose of ocean remote sensing did not evolve as a result of OTHR development for that dedicated purpose. However, as is evidenced by material found in references [6–11] of that chapter, it became clear early in the evolution of skywave radar that ocean parameter measurement by that means would be a possibility. Extensive efforts in developing the Australian Jindalee Operational Radar Network, including oceanographic applications, are documented by Anderson in [63]. In addition to obtaining knowledge of the ocean surface itself, it is readily apparent that a fundamental understanding of ocean scattering processes is integral to the development of schemes for the suppression of so-called ocean clutter in order to enhance the detection capabilities of OTHR for hard targets such as aircraft and ships. The interested reader is directed to Fabrizio's text [64] as a modern, comprehensive account of various aspects of this technology.

1.3   RCSs of the ocean

The IEEE Standards [2] defines an RCS as *the projected area required to intercept and isotropically radiate the same power that a scatterer (target) scatters towards the receiver*. As usual, for present purposes, the target is a patch of ocean surface and it is common to normalise the RCS to the patch area. It should also be noted that the RCS of an ocean patch is not generally the same as the geometric area but is a quantity that depends on the signal properties and the surface roughness along with its spatial distribution. The importance of the RCS lies in the fact that it provides the relationship between Doppler spectra developed coherently from the radar return signal and the directional spectrum of the ocean, the latter containing the oceanographic quantities of interest. In this section, for the purpose of illustrating RCS derivation, a procedure for developing an *HFSWR cross section* of the ocean is outlined, along with final results being provided for a few cases, such as for monostatic and bistatic operation from land or a moving platform. While the exact forms of the RCSs supplied depend on a variety of assumptions, in all but one case reported here, three constraints are common: (1) the surface is assumed to be highly conducting; (2) surface slopes are assumed to be small; and (3) wave heights are taken to be small compared to the wavelength of the transmitted signal. Elaborations of these assumptions and rudimentary interpretation of the models are provided. However, other works are dedicated to detailed theoretical derivations, and the interested reader is directed to such analyses and models as are found in, for example, [30–32, 65–67]. The application of the models forms the content of subsequent chapters.

1.3.1   *A technique for developing an RCS of the ocean*

As noted in Section 1.2.1, the first, best known and most widely used RCSs of the ocean surface were developed under the assumption of a plane wave source [31, 65]. Because the technology has advanced significantly, especially with the advent of software-defined radio [68] in which waveforms can be rather easily constructed appropriate to a particular purpose, the authors have chosen to present an RCS development technique that may be used for any waveform. As referenced above, the intricate details of the derivations may be sourced elsewhere.

Figure 1.8 illustrates the rough surface, initially taken to be time-invariant, which forms the boundary between the upper and lower half-spaces and from which a source field produced by a current density located near the origin at $z = 0^+$ is scattered. With reference to the figure, a Heaviside function may be defined in terms of the half-spaces as

$$h(z - \xi(x,y)) = \begin{cases} 1, & z > \xi(x,y) \\ 0, & z \le \xi(x,y) \end{cases} \tag{1.9}$$

from which the EM parameters for the whole space may be defined as follows:
Permittivity: $\varepsilon = \varepsilon_0 h + \varepsilon_1(1 - h)$.
Conductivity: $\sigma = (1 - h)\sigma_0$.
Conduction current density: $\vec{J}_c = \sigma_0(1 - h)\vec{E}$.

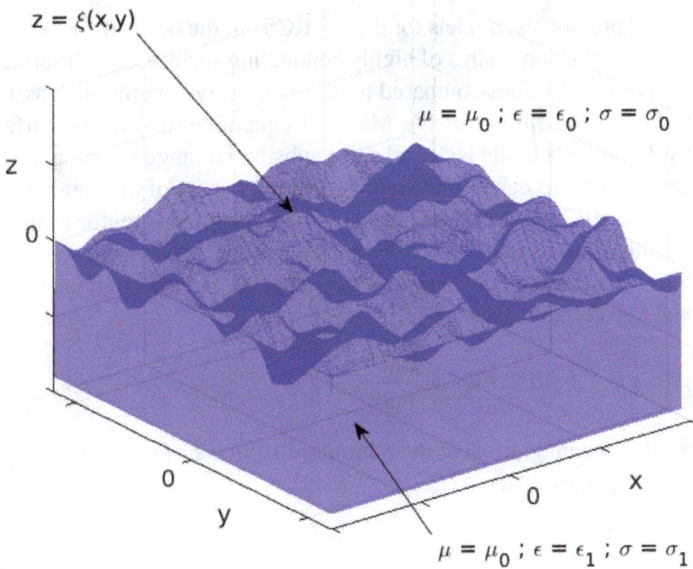

Figure 1.8 *A randomly rough scattering surface, $\xi(x, y)$. The various μs, ϵs and σs are defined in the accompanying text.*

Permeability: $\mu = \mu_0$, the permeability of free space.

Here, $\vec{E}$ is the electric field intensity. These parameters may be substituted into the Maxwell equations (see any elementary electromagnetics text) written for a general current density $\vec{J} = \vec{J}_c + \vec{J}_s$, where $\vec{J}_s$ is the source current density to give

$$\nabla \times \vec{E} = -j\omega\mu_0\vec{H} \tag{1.10}$$

$$\nabla \times \vec{H} = j\omega\vec{D} + \vec{J} \tag{1.11}$$

$$\nabla \cdot \vec{D} = \rho_{sc} \tag{1.12}$$

$$\nabla \cdot \vec{H} = \nabla \cdot \vec{B} = 0. \tag{1.13}$$

with $\vec{D} = \varepsilon\vec{E}$ and $\vec{B} = \mu_0\vec{H}$, $\vec{H}$ being the magnetic field intensity and ρ_{sc} being the surface charge density. The initial aim, with a view to eventually determining the RCS of the surface, is to solve (1.10)–(1.13) for the electric field in the top half-space shown in Figure 1.8. While, here, in this most basic of scattering possibilities, the equations have not been set up to include a second scattering body (such as the ionosphere), a portion of their solution may indeed be shown to be a *surface wave*. Furthermore, the general ideas behind RCS development, which this presentation is intended to depict, can be redeveloped for more complicated scattering scenarios/geometries.

1.3.1.1   Constraints revisited

As emphasised previously, models for the HF RCSs of the ocean have been obtained historically under the constraints of highly conducting surfaces, small surface slopes and small surface deviations compared to the signal wavelength. Such assumptions greatly facilitate the solution of the Maxwell equations to give the surface field. Fortunately, for signals in the HF band and a substantial range of ocean surface conditions, the assumptions can be shown to be valid. The case of the highly conducting ocean surface at HF frequencies has been well documented (see, for example [30]).

The validity of the small-slope assumption may be addressed using the work of Phillips [69], who gives the *mean square surface slope* in terms of wind speed U and gravitational acceleration g as

$$\overline{|\nabla_{xy}\xi(x,y)|^2} = 0.46 \times 10^{-2} \ln \frac{2\pi U^2}{0.3g}. \tag{1.14}$$

It transpires that the magnitude of the normal ($|\vec{n}|$) to the surface shown in Figure 1.8 is related to $\overline{|\nabla_{xy}\xi(x,y)|^2}$ via (see [32])

$$|\vec{n}|^2 = 1 + \overline{|\nabla_{xy}\xi(x,y)|^2} \approx 1, \tag{1.15}$$

the approximation being valid for small surface slopes. To illustrate the requirement, it may be observed from Phillips' model in (1.14) that for a wind speed of $U = 15$ m/s, for example, $\overline{|\nabla_{xy}\xi(x,y)|^2} \approx 0.028$, which produces less than 1.5% error in the magnitude of the surface normal, a critical feature in the electric field, and subsequent RCS, derivations [32, 67].

The small-height assumption, historically manifesting itself in the perturbation theories of Rice [27] and Barrick [30] limits the sea states that may be legitimately measured using HFSWR. A common rule of thumb such as appears in [70] limits the maximum measurable significant wave height, H_s (see Chapter 6 of this book for an elaboration on H_s) for a radar signal wavelength of λ_0 as

$$H_s < 2\frac{\lambda_0}{\pi}. \tag{1.16}$$

Clearly, for the HF band (3–30 MHz), the maximum H_s which can be measured, depending on the operating frequency, must lie between about 6.37 and 63.7 m. For normal sea states – i.e., not extremely high or extremely low – it should be pointed out that there are other factors that affect the inversion of radar data to provide useful ocean parameter estimations and the reader is directed to Chapters 6 and 7, where such factors are discussed. At this point, it is sufficient that the reader remains cognisant of the fact that any assumptions made in the theory will also impact the final models for the RCS.

1.3.1.2   The first-order received electric field

Next, consideration is given to radiation from a *dipole source* (T) at the origin (see Figure 1.9) being received at R, after scattering from a point (x_1, y_1) on the surface. On the basis of the three underlying assumptions, Walsh and Gill [32] show that the

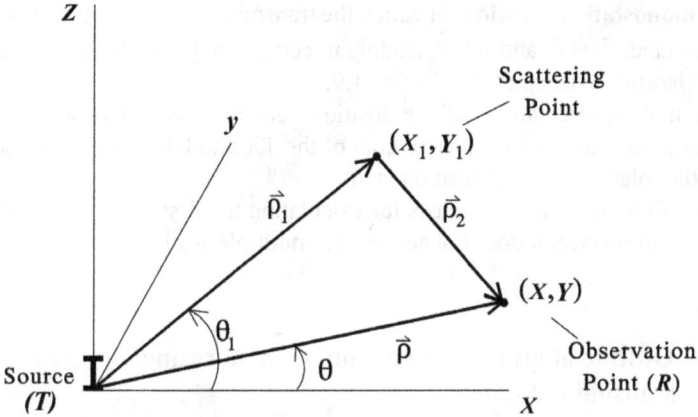

Figure 1.9   *The so-called first-order scattering geometry for a signal sent from T and arriving at R after a single scatter on the rough surface at point (x_1, y_1)*

normal, frequency-domain surface field (i.e., the vertically polarised field), $(E_{0_n}^+)_1$, received after a *single* scatter (from each surface point), if the source is a vertical dipole of length $\Delta \ell$ carrying a continuous current I of wavenumber k and radian frequency ω, and situated near the water surface, is given by

$$(E_{0_n}^+)_1 \approx -jkC_0 \frac{1}{(2\pi)^2} \int_{y_1} \int_{x_1} \left[\hat{\rho}_1 \cdot \nabla_{x_1 y_1}(\xi(x_1, y_1)) \right] F(\rho_1) F(\rho_2) \cdot \frac{e^{-jk(\rho_1 + \rho_2)}}{\rho_1 \rho_2} dx_1 dy_1$$

(1.17)

where $C_0 = \dfrac{I\Delta \ell k^2}{j\omega\varepsilon_0}$ and F is the Sommerfeld surface attenuation function [71]. The displacement vectors, $\vec{\rho}$, $\vec{\rho}_1$, $\vec{\rho}_2$, are shown in the accompanying figure and their respective magnitudes appear in this field equation. The frequency dependence is understood but has been notationally suppressed in the quantities $(E_{0_n}^+)_1$, I, and F.

Equation (1.17) forms the prototype for the development of an ocean surface RCS at HF. However, depending on the applications intended, there are important modifications and elaborations required, including those associated with the following facts:

1. The **dipole source** will not generally have a continuous sinusoid as the current excitation. Thus, the expression for I will depend on the transmission waveform.
2. The **surface** ξ must be specified and, in the case of the ocean, it must be time-varying. Furthermore, if the surface slope and/or height restrictions are removed, the scattering theory itself will need to be reworked to provide new forms of the scattered field expressions.
3. The **directional features imposed by the TX and RX antennas** in practical radar deployments will eventually need to be accounted for.

4. For **monostatic operation**, in which the transmitter and receiver are essentially co-located, $\vec{\rho}_1 = \vec{\rho}_2$ and $\vec{\rho} = \vec{0}$, leading to certain simplifications not available in the **bistatic case** depicted in Figure 1.9.

5. Dynamic representations of the position vectors $\vec{\rho}$, $\vec{\rho}_1$ and $\vec{\rho}_2$ will need to be introduced if one or the other or both of the TX and RX systems are located on floating platforms rather than on land.

6. While (1.17) provides the basis for calculating the first-order HFSWR RCS of the ocean surface, it does not account for multiple scattering.

1.3.1.3   Outline of an RCS derivation incorporating a pulsed-sinusoidal dipole

Despite the above realities, it is still possible to get a general idea of RCS development by considering a special case of this first-order field, one of the simplest being that resulting from a pulsed dipole source. In this case, with reference to Figure 1.2(a), the expression for the time-domain pulsed current $i(t)$ with magnitude I_0 and angular frequency ω_0 may be written using Heaviside functions h as

$$i(t) = I_0 e^{j\omega_0 t}[h(t) - h(t - \tau_0)] \tag{1.18}$$

or, by Fourier transformation, in its equivalent frequency-domain form as

$$I(\omega) = I_0 \tau_0 \mathrm{Sa}\left[(\omega - \omega_0)\frac{\tau_0}{2}\right] \tag{1.19}$$

where the so-called sampling function

$$\mathrm{Sa}(x) = \frac{\sin(x)}{x}. \tag{1.20}$$

Having considered a particular source current form, it is next required that the randomly rough surface be specified. Figure 1.10., which is a modification of a depiction in [72], shows the complex ocean surface wind-wave field as a sum of simple, single-direction waves having differing propagation directions and wavelengths. It is common to assume that the ocean surface may be represented as a three-dimensional Fourier series in (x, y, t). That is,

$$\xi(\vec{\rho}, t) = \sum_{\vec{K}, \omega} P_{\vec{K}, \omega}\, e^{j\vec{K}\cdot\vec{\rho}}\, e^{j\omega t} \tag{1.21}$$

where $\vec{\rho} = x\hat{x} + y\hat{y}$ is the surface position vector, $\vec{K}$ and ω are, respectively, the wave vector and angular frequency of the scattering components (i.e., the ocean surface waves) and $P_{\vec{K}, \omega}$ are random Fourier coefficients.

In addition to the small-height, small-slope and good conductor restrictions placed on the scattering surface, it is equally important to be aware of other assumptions placed on the wind-wave field in arriving at fairly concise and usable forms of the cross sections. It has been typical to assume the following for the wave field:

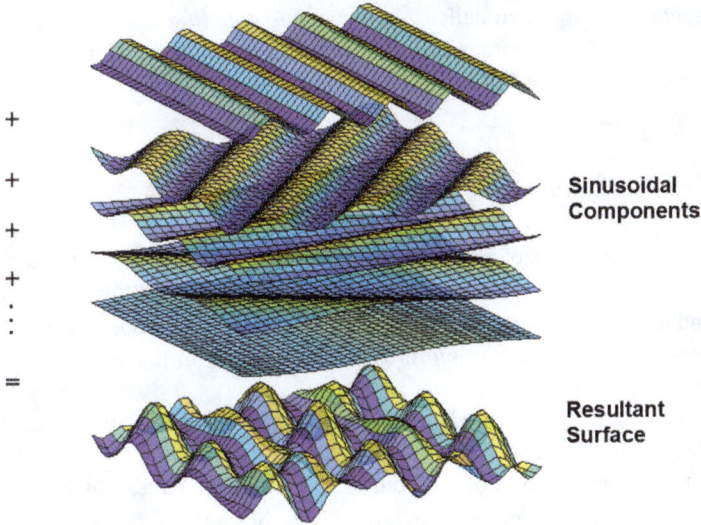

Figure 1.10   *Depiction of constituent waves adding to give a resultant surface (simulated ocean wind-wave field)*

- It is taken to be a *zero-mean Gaussian process*. Since it is a *real, zero-mean surface*, $P_{-\vec{K},-\omega} = P^*_{\vec{K},\omega}$, where * indicates conjugation.
- It is considered to be *stationary* in time and *homogeneous* in space. Consequently, the surface statistics are invariant to (1) the addition of a constant vector, $\Delta\vec{\rho}$, to all position vectors (space points) and (2) the addition of a constant time, τ, to all time points. A further important consequence is that the autocorrelation of the surface function depends only on $\Delta\vec{\rho}$ and τ.
- The Fourier coefficients are *uncorrelated* random variables.

As discussed by Hasselmann [40], and others since then, the hydrodynamic equations describing the ocean surface are nonlinear. A so-called second-order gravity wave, resulting from the interaction of two first-order waves may be accounted for in the Fourier notation as follows.

- First, the surface displacement may be written to second order as

$$\xi(\vec{\rho}, t) = {}_1\xi(\vec{\rho}, t) + {}_2\xi(\vec{\rho}, t);$$ (1.22)

- Then, using the Fourier series form for the first- and second-order displacements,

$${}_1\xi(\vec{\rho}, t) = \sum_{\vec{K},\omega} {}_1P_{\vec{K},\omega}\, e^{j\vec{K}\cdot\vec{\rho}}\, e^{j\omega t}$$ (1.23)

where $\vec{K}$ is associated with a first-order gravity wave;

$${}_2\xi(\vec{\rho}, t) = \sum_{\vec{K},\omega} {}_2P_{\vec{K},\omega}\, e^{j\vec{K}\cdot\vec{\rho}}\, e^{j\omega t},$$ (1.24)

where $\vec{K}$ is associated with a second-order gravity wave.

• Fortunately, using a so-called *hydrodynamic coupling coefficient* Γ_H (Chapter 6, (6.8)), the second-order Fourier coefficients may be written in terms of the first-order coefficients as

$$_2P_{\vec{K},\omega} = \sum_{\substack{\vec{K}_1+\vec{K}_2=\vec{K} \\ \omega_1+\omega_2=\omega}} \Gamma_H \, {}_1P_{\vec{K}_1,\omega_1} \, {}_1P_{\vec{K}_2,\omega_2} \tag{1.25}$$

where the subscripted $\vec{K}$s and ωs represent, respectively, the wave vectors and radian frequency of the two first-order waves constituting the second-order $\vec{K}$ and ω. It is noteworthy that while the first-order waves obey the gravity wave *linear dispersion relationship* provided in Chapter 6, the second-order waves do not.

The importance of describing the surface as above is that the autocorrelation of the electric field, taken as an important step in determining the RCS, essentially involves the autocorrelation of the Fourier surface coefficients, which conveniently and explicitly brings the power spectral density, $S(\vec{K}, \omega)$, of the ocean surface into the cross-section integrals (see, for example [67]). It is the determination of this latter quantity – essentially, the directional ocean wave spectrum – which constitutes one of the prime interests in the use of HFSWR as an ocean remote sensor. Chapters 6 and 7 elaborate the ideas involved in using the RCS in the so-called inversion process for determining $S(\vec{K}, \omega)$ from the radar return.

Having specified a particular current source from (1.18) and a form for the scattering surface ξ from (1.22), in principle, the field in (1.16) may be deduced. Considering the two orders of the surface, there should result two components of the first-order electric field in (1.16). With a view to carrying out time-domain autocorrelations of the field as required for the RCS development under consideration, it is convenient to *inverse Fourier transform* (1.17) to the time domain (see, for example [67], for details). The field then takes the general form

$$\left(E_{0n}^+\right)_1(t_0, t) = \left(E_{0n}^+\right)_{11}(t_0, t) + \left(E_{0n}^+\right)_{12}(t_0, t) \tag{1.26}$$

in which $\left(E_{0n}^+\right)_{11}(t_0, t)$ accounts for a *single scatter* from *first-order gravity waves* while $\left(E_{0n}^+\right)_{12}(t_0, t)$ accounts for a *single scatter* from *second-order gravity waves*, the latter arising from wave–wave interactions. It is important to note that t_0 fixes the *range* from which the field is received while t is the general time variable that allows for the *temporal variation of the surface*. While the surface changes from one pulse to the next, it is *assumed to be time-invariant for each individual measurement*. The reader is cautioned that (1.21) does not account for multiple scattering. However, the *form* of the second term is identical to that for the field arising as a result of two scatters before reception, the only difference lying in the the nature of the coupling coefficient for the two situations. For this reason, since the intent here is simply to show the ideas that constitute an RCS derivation, the first-order portion represented by $\left(E_{0n}^+\right)_{11}(t_0, t)$, will be pursued. Results from other components of the scattered field follow analogously.

In determining the portion of the RCS corresponding to $\left(E_{0n}^{+}\right)_{11}(t_0, t)$, the next step is to calculate the corresponding autocorrelation, which for present purposes is defined in normalised form as

$$\mathscr{R}_{11}(\tau) = \frac{A_r}{2\eta_0} < \left(E_{0n}^{+}\right)_{11}(t_0, t + \tau)\left(E_{0n}^{+}\right)_{11}^{*}(t_0, t) > . \tag{1.27}$$

Here A_r and η_0 are, respectively, the free space aperture of the RX antenna and the intrinsic impedance of free space. For a signal wavelength of λ_0 and an RX antenna with a gain G_r, $A_r = (\lambda_0^2/4\pi)G_r$. The normalisation conveniently makes $\mathscr{R}_{11}(0)$ equivalent to the average power received from a particular surface patch.

In the process of moving towards an RCS of the ocean surface, it is necessary to calculate the power spectral density $\mathscr{P}_{11}$, defined as the power per unit frequency in a signal as a function of frequency, by Fourier-transforming the autocorrelation. Here, then,

$$\mathscr{P}_{11}(\omega_d) = \mathscr{F}[\mathscr{R}_{11}(\tau)], \tag{1.28}$$

where ω_d, the radian *Doppler frequency*, is the frequency-domain transform variable for the time domain τ. Physically, this interpretation is obvious as during the time shift evident in (1.27), the moving surface imposes a Doppler frequency shift on the signal.

Carrying out the operations in (1.27) and (1.28), the first-order power spectral density for *monostatic reception* when the source is a pulsed dipole can be readily shown from the bistatic analysis found in [67] to be

$$\mathscr{P}_{11}(\omega_d) = \frac{A_r \eta_0 k_0^4 \left|I_0 \Delta\ell\right|^2 (\Delta\rho)^2}{2\sqrt{g}\,(2\pi)^2 (\rho_0)^3} |F(\rho_0, \omega_0)|^4 \sum_{m=\pm 1} \int_{-\pi}^{\pi} S_1(m\vec{K})K^{\frac{5}{2}} \\ \cdot \mathrm{Sa}^2\left[\tfrac{\Delta\rho}{2}\left(K - 2k_0\right)\right] d\theta_{\vec{K}}. \tag{1.29}$$

In this expression, ρ_0 can be considered as the distance to the centre of the scattering patch of width $\Delta\rho$ (which is exactly the range resolution given by ΔR of (1.2)). $S_1(m\vec{K})$ is the spectrum of first-order ocean waves of wavenumber K and direction $\theta_{\vec{K}}$, while k_0 is the wavenumber of the transmitted signal corresponding to the angular frequency ω_0 of (1.18). Further elaboration of properties appearing in this equation follows the cross-section expression in (1.32).

As in [67], with a view to obtaining the first-order RCS, it is convenient to write the power spectral density per scattering area A in differential form as

$$\frac{d\mathscr{P}_{11}(\omega_d)}{dA} = \frac{A_r \eta_0 k_0^4 \left|I_0 \Delta\ell\right|^2 (\Delta\rho)}{2\sqrt{g}(2\pi)^2(\rho_0)^4} |F(\rho_0, \omega_0)|^4 \\ \cdot \sum_{m=\pm 1} S_1(m\vec{K})K^{\frac{5}{2}}\mathrm{Sa}^2\left[\tfrac{\Delta\rho}{2}\left(K - 2k_0\right)\right]. \tag{1.30}$$

With (1.30) in place, the stage is set for comparing this result from an analysis of the scattered field with that given by the so-called *radar range equation*, a necessary step in arriving at the final RCS of the ocean surface. In the literature, the radar range equation appears in several different, but essentially equivalent, forms, depending on the application. In a basic sense, it involves the relationship between the received power, the transmitted power, the distance to the 'target', the wavelength of the transmitted signal, the gains of the TX and RX antennas and the RCS (see, for example [20]). For the situation at hand, involving a dipole TX antenna, it may be shown that a suitable form of the radar range equation is given by

$$\frac{d\mathscr{P}_{11}(\omega_d)}{dA} = \frac{A_r \eta_0 k_0^2 \left|I_0 \Delta \ell\right|^2 \left|F(\rho_0, \omega_0)\right|^4}{16(2\pi)^3 \rho_0^4} \sigma_{11}(\omega_d). \tag{1.31}$$

The quantity of interest is the first-order Doppler RCS of the ocean surface, symbolised as $\sigma_{11}(\omega_d)$. Clearly, equating the right-hand sides of (1.30) and (1.31) directly gives

$$\sigma_{11}(\omega_d) = 2^4 \pi k_0^2 \Delta \rho \sum_{m=\pm 1} S_1(m\vec{K}) \frac{K^{\frac{5}{2}}}{\sqrt{g}} \, \mathrm{Sa}^2\left[\frac{\Delta \rho}{2}\left(K - 2k_0\right)\right]. \tag{1.32}$$

Before writing down the corresponding expressions for plane wave and FMCW waveforms, a few comments on this pulsed-waveform solution are in order.

- This is cross section per area and its unit is $\frac{\mathrm{m}^2}{\mathrm{m}^2}$ (radian/second)$^{-1}$ or simply (radian/second)$^{-1}$.
- The components of the surface, the spectrum of which is $S_1(m\vec{K})$, are of *small height* and *small slope* and the surface is assumed to be a *good conductor*, *homogeneous* and *stationary*.
- The sampling function, and thus the cross section, peaks when its argument is 0, that is, when $K = 2k_0$. An ocean wave represented by this special K has a wavelength that is one-half that of the transmitted signal. Historically, Crombie conjectured [28] that first-order scatter gave evidence of constructive interference of the scattered signal occurring from ocean waves whose wavelength is $\lambda_0/2$ and which travel directly towards ($m = -1$, $\omega_d > 0$) or directly away ($m = 1$, $\omega_d < 0$) from the radar. The first impression was that this was analogous to scatter from layers of a crystal lattice and the term 'Bragg scatter' was adopted and has remained the choice of terminology for EM scattering from the ocean surface. However, as pointed out by others (see, for example [73]), and as is intrinsic to the analysis outlined here, the HFSWR signal scatters from the whole wave, not only from wave crests. Still, the literature ubiquitously refers to the first-order peaks in the RCS of ocean scatter as *Bragg peaks*. It may also be noted that there is a natural spread in the Bragg region decreasing in amplitude in accordance with the sampling squared function (i.e., there is not, in reality, just a single Bragg wave).
- The presence of $\Delta \rho$, the radial width (resolution) of the scattering patch, ensures that as the *patch increases in width* the sampling squared function and therefore the *Bragg Doppler regions become narrower*. Therefore, there is a *trade-off* between the *surface resolution* and the *spread of the Bragg peaks* in the Doppler radar spectrum.

1.3.2 Other cross-section results

Having presented a particular technique for obtaining an RCS of the ocean for a special case of the transmitted waveform being a pulsed sinusoid, it should be emphasised that there have been several valuable analyses conducted. Some of those outcomes are summarised here and in subsequent chapters. Graphical depictions of the RCSs are reserved until Section 1.3.3.

1.3.2.1 A second-order solution for patch scatter – pulsed dipole source

It has been emphasised that (1.32) represents a model for the case of a single scatter of incident radiation from *first-order gravity waves* on a patch of ocean surface whose size is to be determined by the range and azimuthal resolutions of the radar. However, it does not include the portion of the RCS arising from the field for a *single scatter*, represented by $\left(E_{0n}^+\right)_{12}(t_0, t)$ in (1.26), from *second-order gravity waves*. The derivation of this so-called *hydrodynamic second-order* RCS (even though *electromagnetically* this is still a first-order phenomenon) is completely analogous to that for $\sigma_{11}(\omega_d)$, but incorporates the second-order wave field cast as $_2\xi(\vec{\rho}, t)$ in (1.22). Symbolised here as $\sigma_{12}(\omega_d)$, to emphasise that it comes from $\left(E_{0n}^+\right)_{12}(t_0, t)$, the monostatic result, readily following from the bistatic form in [67], is

$$
\sigma_{12}(\omega_d) = 2^3 \pi k_0^2 \Delta\rho \sum_{m_1=\pm 1} \sum_{m_2=\pm 1} \int_0^\infty \int_{-\pi}^\pi \int_0^\infty \left\{ S_1(m_1\vec{K}_1) S_1(m_2\vec{K}_2) \left|\Gamma_H\right|^2 K^2 \right.
$$
$$
\left. \cdot \mathrm{Sa}^2\left[\frac{\Delta\rho}{2}(K - 2k_0)\right] \cdot \delta\left(\omega_d + m_1\sqrt{gK_1} + m_2\sqrt{gK_2}\right) K_1 \right\} dK_1 d\theta_{\vec{K}_1} dK.
$$

(1.33)

As indicated in the definition of the second-order Fourier surface coefficients following (1.22), $\vec{K}_1$, $\vec{K}_2$, ω_1 and ω_2 are associated with two first-order ocean waves, and $\vec{K}_1 + \vec{K}_2 = \vec{K}$ while $\omega_1 + \omega_2 = \omega$, where $\vec{K}$ and ω are second-order quantities.

In accounting for a *double scattering* on the same patch of ocean (i.e., within the same $\Delta\rho$), the conditions on the $\vec{K}_1$, $\vec{K}_2$, ω_1 and ω_2 of the two first-order ocean waves involved in the process, as dictated by a Dirac delta function identical to that in (1.33), remain the same. The difference in the scattering model from that in (1.33) is accounted for by a so-called *electromagnetic coupling coefficient*, Γ_E. Various forms of this have been developed (for example [32, 74]). Replacing Γ_H above by a total coupling coefficient, say Γ, incorporating both Γ_H and Γ_E gives the full second-order patch cross section, say $\sigma_2(\omega_d)$. The exact way as to how the two coupling coefficients should be combined lies in the various definitions and approaches used to develop the scattering analysis. One such combination appears in Chapter 6. In the notation of this chapter, the total RCS for a patch of ocean remote from the radar, to second order in scatter, may be written as

$$
\sigma(\omega_d) = \sigma_{11}(\omega_d) + \sigma_{12}(\omega_d) + \sigma_{21}(\omega_d)
$$

(1.34)

where $\sigma_{21}(\omega_d)$ represents the EM double scatter. From the discussion above,

$$\sigma(\omega_d) = \sigma_{11}(\omega_d) + \sigma_2(\omega_d) \tag{1.35}$$

where

$$\sigma_2(\omega_d) = \sigma_{12}(\omega_d) + \sigma_{21}(\omega_d). \tag{1.36}$$

1.3.2.2   An RCS assuming a plane wave source

As noted previously, the first, and still most widely used, HFSWR ocean surface RCS models assumed a plane wave source [31, 65]. Strictly speaking, such a source dictates that the scattering patch width $\Delta\rho$ becomes infinite. Since a real $\Delta\rho$ is typically tens or even hundreds of wavelengths in extent, the plane wave assumption, while masking some details, is often sufficient.

It is well known that the sampling function appearing in power spectral density (1.29) may be used to define a Dirac delta function [75], according to the relationship

$$\lim_{M \to \infty} M\mathrm{Sa}^2[Mx] = \pi\delta(x). \tag{1.37}$$

To arrive at (1.29) from (1.28), it is required that a dK integral be completed. Moving a $\Delta\rho$ inside that particular integral and applying the plane wave limit lead to

$$\lim_{\Delta\rho \to \infty} \Delta\rho\mathrm{Sa}^2\left[\frac{\Delta\rho}{2}(K - 2k_0)\right] = \pi\delta\left(\frac{K - 2k_0}{2}\right) = 2\pi\delta(K - 2k_0), \tag{1.38}$$

which allows immediate evaluation of the K integral. Continuing the procedure as in Section 1.3.1 leads directly to the first-order HFSW RCS of the ocean surface as

$$\sigma_{11}(\omega_d) = 2^6\pi^2 k_0^4 \sum_{m=\pm 1} S_1(2m\vec{k}_0)\delta(\omega_d + m\sqrt{2gk_0}). \tag{1.39}$$

When interpreted in terms of Barrick's notations and normalisations, this RCS is the same as his. Two features are immediately noticeable, namely, that the dependence on patch width (which, in reality, depends on the signal pulse width or, in the frequency domain, on bandwidth) is lost and, as a consequence, so is the spread in the first-order Doppler spectrum.

Applying the same analysis to the second order power spectral densities yields a plane wave result of

$$\sigma_2(\omega_d) \approx 2^6\pi^2 k_0^4 \sum_{m_1=\pm 1}\sum_{m_2=\pm 1} \int_{-\pi}^{\pi}\int_0^\infty \left\{ S_1(m_1\vec{K}_1)S_1(m_2\vec{K}_2) \right.$$
$$\left. |\Gamma_P|^2\,\delta\left(\omega_d + m_1\sqrt{gK_1} + m_2\sqrt{gK_2}\right)K_1 \right\} dK_1\,d\theta_{\vec{K}_1}, \tag{1.40}$$

where Γ_P represents the combined hydrodynamic and EM coupling coefficients. The relationships between the wave vectors and angular frequencies remain unchanged from the finite-patch-width case (see Chapter 6, where notations are slightly altered, the models are essentially equivalent).

1.3.2.3   An RCS assuming an FMCW source

For a given required average power, the usually very large peak powers associated with pulsed sources can be avoided by using FM waveforms. This is important in modern, low-power coastal installations. While the plane wave RCS solutions of [31, 65] have been implemented in ocean measurement applications, in 2011, RCSs developed using an FMCW source appeared in the literature [33].

The 'shape' of the FM waveform, as the name implies, indicates a sinusoid of varying frequency – Figure 1.11, for illustration purposes only, depicts a linear frequency modulation. The frequency may be 'swept' up or down or both. In complex exponential form, the current, whose magnitude is I_0 and whose centre angular frequency is ω_0, may be written for one sweep as

$$i(t) = I_0 e^{j(\omega_0 t + \alpha \pi t^2)} \cdot \left\{ h\left[t + \frac{T_r}{2}\right] - h\left[t - \frac{T_r}{2}\right] \right\} \tag{1.41}$$

where α is the sweep rate in Hz/s and T_r is the sweep interval. Using this expression in the time-domain version of the electric field equations and following the general procedure outlined after (1.17) lead to FMCW RCSs of the ocean surface. The FMCW first-order model, corresponding to the pulsed case appearing in (1.31), is given in [33] as

$$\sigma_1\left(\omega_d\right) = 16\pi k_0^2 \sum_{m=\pm1} S_1\left(m\vec{K}\right) \frac{K^{2.5}}{\sqrt{g}} \left[\Delta\rho\mathrm{Sm}^2\left(K, k_B, \Delta_r\right)\right]. \tag{1.42}$$

An interpretation of this RCS is considerably more intricate than the pulsed and plane wave counterparts. First, it is noted that the Sm function is defined as

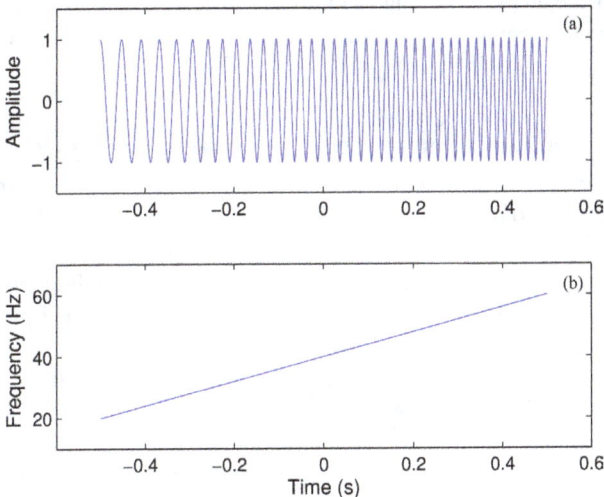

Figure 1.11   *A single FMCW sweep showing (a) the waveform and (b) the linearity of the sweep*

$$\text{Sm}\left(K,\ k_B,\ \Delta_r\right) = \frac{1}{\pi}\left\{Si\left[\left(K - 2k_0 + k_B\right)\Delta_r\right] - Si\left[\left(K - 2k_0 - k_B\right)\Delta_r\right]\right\} \quad (1.43)$$

where

$$Si(x) = \int_0^x \frac{\sin(t)}{t}\,dt. \quad (1.44)$$

A complication here, while not explicitly visible in the final cross section, is that a sampling function appearing in the intermediate analysis leading to the form of (1.42) indicates that spectral energy from one 'range cell' actually consists of some energy, which 'spills over' into it from other cells. Here, Δr is really a parameter that the user employs to set the extent of this *spillover* effect. However, it is common in practice to assume that the sampling function in question is narrow enough so that $\Delta r = \Delta \rho/2$ may be employed and that $c/2B$ is the range resolution. The parameter k_B is defined in [33] as $k_B = (2\pi B)/c$. As for the first-order pulsed-radar case, $\omega_d = -m\sqrt{gK}$, where the direction of $\vec{K}$ is the radar look direction (for narrow beam reception).

Carrying out analysis completely analogous to the first-order, the second-order FMCW HF RCS of the ocean surface becomes

$$\sigma_2\left(\omega_d\right) = 2^3\pi k_0^2 \sum_{m_1=\pm 1} \sum_{m_2=\pm 1} \int_0^\infty \int_{-\pi}^\pi \int_0^\infty S_1\left(m_1\vec{K}_1\right) S_1\left(m_2\vec{K}_2\right) \left|\Gamma_P\right|^2 K^2$$
$$\delta\left(\omega_d + m_1\sqrt{gK_1} + m_2\sqrt{gK_2}\right)\left[\Delta\rho\text{Sm}^2\left(K,\ k_B,\ \Delta_r\right)\right] K_1\,dK_1\,d\theta_{\vec{K}_1}\,dK$$
$$(1.45)$$

where the various Ks and ωs may be interpreted as in (1.33) and Γ_P is the total coupling coefficient incorporating both Γ_H and Γ_E. As for the FMCW first-order case, the details of the analysis leading to (1.45) appear in [33]. If Δr, or, equivalently, $\Delta\rho$ is allowed to become unbounded, then (1.33) reduces to the plane wave case for second order appearing in (1.40).

1.3.2.4   A summary of more recent cross-section analyses
The analysis techniques similar to those already referenced in this section have more recently been applied to other scattering situations. These include the following:

1. Analytical results for an RCS of the ocean in which the small-height assumption has been removed [76–78]. Much of this work, while verified in a qualitative manner, awaits or is still being subjected to quantitative investigation, both by the authors and by others in the field.
2. The case in which one or the other, or both the TX antenna and RX antenna are situated on a moving ocean structure [79–87].
3. The case where the TX signal is reflected by the ionosphere before returning to the RX site via propagation along the ocean surface [35–39].
4. Incorporation of *swell* (not only wind waves) into the pulsed-radar ocean RCS [88].

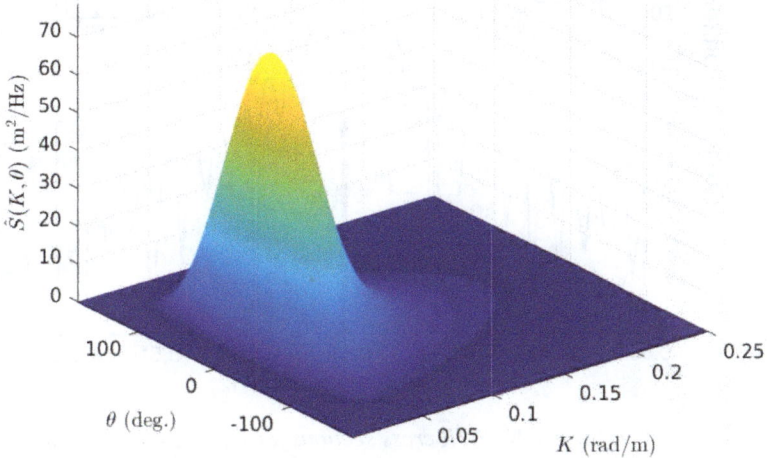

Figure 1.12 Depiction of a directional ocean spectrum for a wind speed of $U=5m/s$, $\bar{\theta}=0°$ and $s = 2$

1.3.3 RCS depictions and discussion

There are two basic questions that may be posed in attempting to acquire insights into how the RCS may be used to determine ocean surface information: (1) What parameters associated with the ocean conditions affect the RCS? and (2) How do the radar operating parameters affect the RCS for a given set of ocean conditions? First, then, for the purpose of illustration, a form of the ocean directional spectrum $S(\vec{K})$ must be chosen. As a starting point, a typical directional ocean spectrum, used purely for illustration – *with no claim as to its appropriateness universally* – may consist of the product of one form of the non-directional Pierson–Moskowitz (PM) spectrum [89, 90] and a cosine-based directional factor. Here, the result takes the form

$$
S_1(m\vec{K}) = \left[\frac{\alpha_{PM}}{4K^2} e^{\left(\frac{-0.74g^2}{K^2 U^4}\right)}\right] \cdot \left[\frac{4}{3\pi} \cos^{2s}\left(\frac{\theta_{\vec{K}} + \frac{(1-m)\pi}{2} - \bar{\theta}}{2}\right)\right] \quad (1.46)
$$

where α_{PM} is a dimensionless parameter whose value is 0.0081, U is the wind speed measured 19.5 m above the mean ocean surface, $\bar{\theta}$ is the overall mean direction of wave energy propagation (generally taken to be the wind direction); s is the so-called spread parameter and, in the RCS depictions of this section, a value of $s = 2$ is used. An example is illustrated in Figure 1.12, and further discussion of such a wave

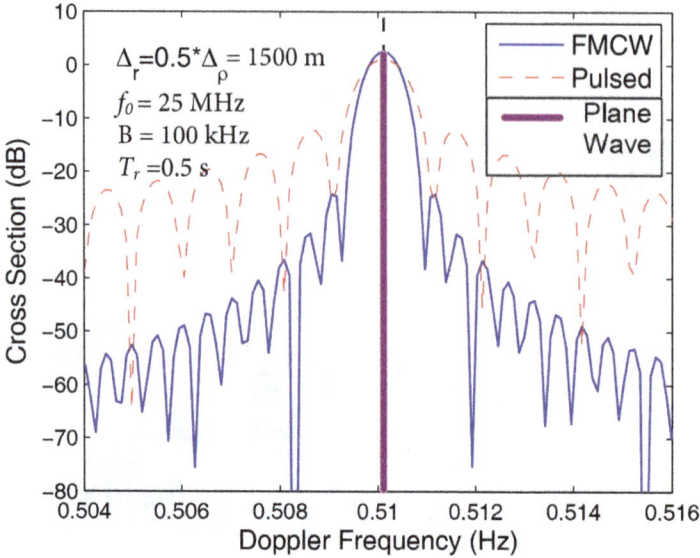

Figure 1.13   *First-order HFSWR cross sections for a pulsed, FMCW and plane wave source. Doppler spectrum from a 25 MHz radar.*

spectrum appears in Chapter 6. Because of the difference in the surface expansion used by Pierson in [89] as compared to that used in the RCS depicted subsequently, it transpires that the PM non-directional spectrum appearing in the first set of square brackets of (1.46) must be multiplied by 0.5.

To illustrate in an unencumbered fashion the RCSs for the cases of pulsed, plane wave and FMCW sources, Figure 1.13. depicts only the first-order cases presented in (1.32), (1.39) and (1.42). For each case, $\omega_d = -m\sqrt{gK}$ and since $m = \mp 1$, there will be *positive* and *negative* Doppler spectral regions, corresponding to scatter from first-order waves travelling, respectively, *towards* or *away from* the radar under the assumption of monostatic operation. Of course, the Doppler frequency in hertz is simply $f_d = \omega_d/(2\pi)$. As discussed following (1.32), the spectral peaks occur when the wavenumber for the *approaching* and *receding* waves is $2k_0$ (i.e., the water wavelength is half the transmission wavelength) and the so-called Bragg Doppler frequencies in hertz will occur at

$$\pm f_B = \pm\omega_B/(2\pi) = \pm\sqrt{g/(\pi\lambda_0)}. \tag{1.47}$$

As is discussed in Chapters 2 and 4, establishing the shift in the location of this peak from its theoretical value is essential in determining ocean surface currents from HFSWR spectra. The accompanying figure depicts only the positive Doppler regions, corresponding to $m = -1$, for the case of a TX signal of centre frequency $f_0 = 25$ MHz, bandwidth (for the pulsed and FMCW cases) of $B = 100$ kHz and, in the FMCW case, particular values of $\Delta_r = \Delta\rho/2 = 1\,500$ m and sweep interval of $T_r = 0.5$ s. The Doppler continuum observed for the non-zero bandwidth cases as compared to the plane wave result is seen to be more realistic when compared to an actual Doppler spectrum such as appears in Figure 1.5. The major difference between the first-order pulsed and FMCW spectral results is the lower sidelobe level evident in the latter.

It transpires that examination of the second-order plane wave RCS model in (1.33), with Γ_H replaced by Γ_P, or the plane wave model of (1.40) will give insight as to the utility of HFSWR in ocean wave and wind parameter estimation. The wave vectors, $\vec{K}_1$ and $\vec{K}_2$, associated with both hydrodynamic and EM second-order processes, for the case of monostatic operation, require that

$$\vec{K}_1 + \vec{K}_2 = \vec{K} = 2k_0\hat{k}_0 \tag{1.48}$$

where $\hat{k}_0$ is the unit vector in the radar look direction. Many such pairs of $\vec{K}_1$ and $\vec{K}_2$ satisfy this condition for *each* Doppler frequency, f_d, and the integration at a given f_d (or, equivalently, radian Doppler frequency ω_d) gives the magnitude of the cross section at that position. Thus, there is a continuum to the second-order scattering. In fact, in actual HFSWR spectra, the continuum consists of even higher orders of scatter [32], but traditionally only first and second orders have been used for oceanographic purposes.

Figure 1.5 shows the entire first- and second-order Doppler continuum for a 6.75 MHz pulsed HFSWR. In the second-order regions, there are two important relative maxima. One of these, occurring when $\vec{K}_1 \perp \vec{K}_2$, is due to the EM coupling coefficient Γ_E becoming maximum. It can be shown that the Doppler frequency at which this happens is given by

$$f_d = \pm 2^{\frac{3}{4}} f_B. \tag{1.49}$$

The phenomenon is referred to as the second-order electromagnetic corner reflector. Whether it is observed in field data will depend on both the state of the ocean surface as well as the noise level. The other special relative maximum in the second-order return can be shown to occur when $|\vec{K}_1| = |\vec{K}_2| = k_0$ while meeting the condition for $\vec{K}_1 + \vec{K}_2 = 2k_0\hat{k}_0$. This particular second-order wave has a speed of $\sqrt{2}v_B$ (v_B being the Bragg wave speed) and produces a peak at

$$f_d = \pm\sqrt{2}f_B. \tag{1.50}$$

This is a hydrodynamic effect and is often visible in the Doppler spectra.

Considering (1.33), it is obvious that the overall shape of the Doppler spectrum will be governed by sea state and, thus, by *wind velocity*, as well as by radar operating parameters including *frequency* and *bandwidth*. While not evident in the cross sections themselves, the Doppler spectrum will also depend on such factors as *transmitted power*, *gains* of the TX and RX antennas, which will be especially important in complicated ocean current regimes, the *conductivity* of the surface (for less saline water, the attenuation at HF will be higher) and *noise levels* (ionospheric or otherwise). Assuming that the system is designed to address this latter list, attention is now turned to the ways in which ocean properties or surface winds manifest themselves in the RCS. In this chapter, a qualitative analysis is provided, while in subsequent chapters, or in other work on inversion techniques, details of parameter extraction are addressed.

It has been intimated that *ocean currents* cause shifts in the location of the so-called first-order Bragg peaks from their theoretical values given by (1.47). Various techniques have been developed to determine ocean currents from these shifts, and several excellent references appear at the end of Chapter 2.

It is evident that at HF frequencies, the signals do not interact with the atmosphere immediately above the ocean in any significant way. Thus, if HFSWR is to be used to obtain surface wind information, this can only be accomplished through observing the effect that the wind has on the ocean surface. Figure 1.6 gives a general idea of this effect. The leftmost frame in the figure indicates a surface wind blowing perpendicular to the direction in which the radar 'looks' or is steered (see the discussion on *beamforming* in Section 1.1.2.). In the middle frame, there is a component of the wind towards the radar receiver, and the rightmost case depicts a Doppler spectrum obtained when there is a component of the wind away from the radar receiver. A $\cos^4$ directional distribution is assumed in developing these simulated Doppler spectra. Evidently, then, when the wind drives the waves generally towards the radar, there is an increase in power associated with the positive Doppler frequencies as compared to the negative. Of course, the opposite is true when the wind drives the waves generally away from the radar. Techniques have been reported that use the ratio of the positive to negative Doppler powers to estimate the *wind direction*. An elaboration of these ideas appears in Chapter 8, and a list of appropriate references is provided. It is interesting to note that one of the earliest examinations of the extraction of wind direction from HF radar data involves skywave operation [91].

As is true for wind direction, *wind speed* also may be estimated from HFSWR Doppler spectra only by using relationships between this property and sea conditions. Again, a variety of methods have been reported in the literature [92–94] and others appear in Chapter 8. Figure 1.7 shows the effect that changing wind speed, and therefore sea state, has on the RCS. Clearly, there is an increase in the second-order continuum as sea state increases. Not surprisingly, this continuum plays a significant role in the inversion process, which leads to the *extraction of the ocean directional spectrum* and related parameters as discussed in Chapter 6 as long as the ocean conditions meet the requirement dictated by (1.16).

1.3.3.1   Summary

Over the past forty years, there has been substantial maturity realised in both the theoretical and practical aspects associated with the use of HF radar as an ocean remote sensing tool. The resulting body of literature is vast, consisting of hundreds of peer-reviewed documents, some containing intricate fundamental scattering analyses and others manifesting creative signal processing and spectral inversion techniques for various implementations of the technology. Still, the saga of HFSWR is unfinished and research continues to advance both the theory and the technology. Much of the remainder of this book, composed by individuals who have invested entire careers in this or related fields, is a testament to these facts.

Acknowledgement

The authors wish to thank Dr Reza Shahidi for constructive discussions on parts of the text involving direction finding and Mr Michael Royle and Dr Murilo Silva for their contributions to Figures 1.8 and 1.10 and Figures 1.6, 1.7 and 1.12, respectively. The work was also supported by Natural Sciences and Engineering Council (NSERC) Discovery Grants to each of the authors.

References

[1]   Kinsman B. *Wind Waves*. New York: Dover Publications Inc; 1984.
[2]   IEEE, 211-2018, IEEE standard definitions of terms for radio wave propagation. New York; 2019.
[3]   Barrick D.E. 'Status of HF radars for wave-height directional spectral measurements in measuring ocean waves'. Symposium and Workshop on Wave Measurement Technology April 1981; Washington, DC, USA; 1982. pp. 112–8.
[4]   Skolnik M.I. 'Introduction to Radar Systems.' in Plains W. (ed.). New York. NY: McGraw-Hill Education; 2002.
[5]   Headrick J.M., Anderson S.J. 'HF over-the-horizon radar' in Skolnik M.I. (ed.). *Radar Handbook*. New York, NY: McGraw Hill; 2008. pp. 20.1–20.83.
[6]   Gurgel K.W., Antonischki G., Essen H.H., *et al.* 'Wellen Radar (WERA): a new ground-wave HF radar for ocean remote sensing'. *Coastal Engineering.* 1999;**37**:219–34.
[7]   Barrick D.E. *The seasonde: continuous surface current mapping and wave monitoring* [online]. 2020. Available from http://www.codar.com/SeaSonde. shtml [Accessed 5 Apr 2021].
[8]   Technical and operational characteristics of oceanographic radars operating in sub-bands within the frequency range 3-50 MHz, recommendation ITU-R M.1874-1. Geneva; 2013.
[9]   Barrick D.E. FM/CW radar signals and digital processing. 8ERL 283-WPL 26. Boulder, CO, USA: National Oceanic and Atmospheric Administration; 1973.
[10] IEEE, 145-2013, IEEE standard for definitions of terms for antennas. New York; 2014.
[11] Helzel T. *WERA ocean radar* [online]. 2020. Available from https://helzel-messtechnik.de/de/9307-wera-features [Accessed 5 Apr 2021].
[12] Balanis C.A. *Antenna Theory, Analysis and Design*. 3rd Edn. Hoboken, New Jersey: John Wiley and Sons Inc; 2005.

[13]   Heron M.L., Prytz A., Searson S. 'The Australian coastal ocean radar network (ACORN)'. Proceedings of IEEE/OES 9th Working Conference on Current Measurement Technology. Charlston; SC, USA; 2008.

[14]   Barrick D.E., Lipa B. 'Evolution of bearing determination in HF current mapping radars'. *Oceanography*. 1997;**10**(2):72–5.

[15]   Schmidt R.O. 'Multiple emitter location and signal parameter estimation'. *IEEE Transactions on Antennas and Propagation*. 1986;**34**:276–80.

[16]   Barrick D.E., Lipa B. Radar angle determination with MUSIC direction finding. US Patent Office; 1999. US Patent 5,990,834. Available from https://patents.google.com/patent/US5990834.

[17]   Laws K.E., Fernandez D.M., Paduan J.D. 'Simulation-based evaluations of HF radar ocean current algorithms'. *IEEE Journal of Oceanic Engineering*. 2000;**25**:481–91.

[18]   Heron M.L. 'Comparisons of different HF ocean surface-wave radar technologies'. Proceedings of IEEE Current, Waves and Turbulence Conference; St. Petersburg, FL, USA; 2015.

[19]   Wang W., Gill E.W. 'Evaluation of beamforming and direction finding for a phased array HF ocean current radar'. *Journal of Atmospheric Oceanic Technology*. 2016;**33**(12):2599–613.

[20]   Barton D.K. *Modern Radar Systems Analysis*. Norwood, MA: Artech House Inc; 1987.

[21]   Frank J., Richards J.D. 'Phased array radar antennas' in Skolnik M.I. (ed.). *Radar Handbook*. New York, NY: McGraw Hill; 2008. pp. 13.1–13.74.

[22]   Collin R.E. *Antennas and Radio Wave Propagation*. New York: McGraw-Hill Book Company; 1985.

[23]   Heron M.L., Wyatt L.R., Atwater D.P. 'The Australian coastal ocean radar network: lessons learned in the establishment phase'. Proceedings of IEEE Oceans Conference; Yeosu, Korea; 2012.

[24]   Strutt J.W. *The Theory of Sound*. **2**. New York: Dover; 1945.

[25]   Hertz H. *Electric Waves*. London: Macmillan and Company; 1893.

[26]   Wait J.R. 'The ancient and modern history of em ground-wave propagation'. *IEEE Antennas and Propagation Magazine*. 1998;**40**(5):7–24.

[27]   Rice S.O. 'Reflection of electromagnetic waves from a slightly rough surface' in Kline K. (ed.). *Theory of Electromagnetic Waves*. **4**. New York: Interscience 08; 1951. pp. 351–78.

[28]   Crombie D.D. 'Doppler spectrum of sea echo at 13.56 Mc./s'. *Nature*. 1955;**175**:681–2.

[29]   Wait J.R. 'Perturbation analysis for reflection from two-dimensional periodic sea waves'. *Radio Science*. 1971;**6**:387–91.

[30]   Barrick D. 'Theory of HF and VHF propagation across the rough sea (Parts 1 and 2)'. *Radio Science*. 1971;**6**:517–33.

[31]   Barrick D.E. 'First-order theory and analysis of MF/HF/VHF scatter from the sea'. *IEEE Transactions on Antennas and Propagation*. 1972;**20**:2–10.

[32] Walsh J., Gill E.W. 'An analysis of the scattering of high frequency electromagnetic radiation from rough surfaces with application to pulse radar operating in backscatter mode'. *Radio Science*. 2000;**35**(6):1337–59.

[33] Walsh J., Zhang J., Gill E.W. 'High frequency radar cross section of the ocean surface for an FMCW waveform'. *IEEE Journal of Oceanic Engineering*. 2011;**36**(4):615–26.

[34] Budden K.G. *The Propagation of Radio Waves: the Theory of Radio Waves of Low Power in the Ionosphere and Magnetosphere*. Cambridge, UK: Cambridge University Press; 1988.

[35] Walsh J., Gill E.W., Huang W., Chen S. 'On the development of a high frequency radar cross section model for mixed path ionosphere-ocean propagation'. *IEEE transactions on antennas and propagation*. 2015;**63**(6):2655–64.

[36] Chen S., Huang W., Gill E.W. 'A vertical reflection ionospheric clutter model for HF radar used in coastal remote sensing'. *IEEE Antennas Wireless Propagation Letters*. 2015;**14**:1689–93.

[37] Chen S., Gill E.W., Huang W. 'A first-order HF radar cross section model for mixed-path ionosphere-ocean propagation with an FMCW source'. *IEEE Journal of Oceanic Engineering*. 2016;**41**(4):982–92.

[38] Chen S., Gill E.W., Huang W. 'A high frequency surface wave radar ionospheric clutter model for mixed-path propagation with second-order sea scattering'. *IEEE Transactions on Antennas and Propagation*. 2016;**64**(12):5373–81.

[39] Chen S., Huang W., Gill E.W. 'First-order bistatic high frequency radar power for mixed-path ionosphere-ocean propagation'. *IEEE Geoscience and Remote Sensing Letters*. 2016;**13**(12):1940–4.

[40] Hasselmann K. 'On the nonlinear energy transfer in a gravity wave spectrum, Part 1. General theory'. *Journal of Fluid Mechanics*. 1962;**12**:481–500.

[41] Hasselmann K. 'Determination of ocean wave spectra from Doppler return from sea surface'. *Nature: Physical Science*. 1971;**229**:16–17.

[42] Boderi R. *The Invention that Changed the World*. New York: Touchstone; 1997.

[43] Teague C., Vesecky J., Fernandez D. 'HF radar instruments, past to present'. *Oceanography*. 1997;**10**(2):40–4.

[44] Barrick D.E., Headrick J.M., Bogle R.W., *et al.* 'Remote sensing of sea backscatter at HF: interpretation and utilization of the echo'. *Proceedings of the IEEE*. 1974;**62**(6):673–80.

[45] Barrick D.E., Evans M.W. Implementation of coastal currentmapping HF radar system, progress report No. 1. ERL WPL47. U.S. Department of Commerce: National Oceanic and Atmospheric Administration; 1976.

[46] Jeans P., Donnelly R. 'Four-element CODAR beam forming'. *IEEE Journal of Oceanic Engineering*. 1986;**11**(2):296–303.

[47] Broche P. 'Estimation du spectre directionnel des vagues par radar decametrique coherent'. AGARD Conference Proceedings (Special Topics in HF Propagation), 263; Washington, DC, USA; 1979. pp. 31–1–31–12.

[48]   Grosdidier S., Forget P., Barbin Y., Guérin C.A. 'HF bistatic ocean Doppler spectra: simulation versus experimentation'. *IEEE Transactions on Geoscience and Remote Sensing.* 2014;**52**(4):2138–48.

[49]   Domps B., Dumas D., Guérin C.A., Marmain J. 'High-frequency radar ocean current mapping at rapid scale with autoregressive modeling'. *IEEE Journal of Oceanic Engineering.* 2021;**46**(3):891–9.

[50]   Shearman E.D.R., Burrows G.D., Moorhead M.D. 'An FMICW ground-wave radar for remote sensing of ocean waves and current'. Proceedings of IEEE International Conference Radar; London, UK; 1987.

[51]   Shearman E.D.R., Moorhead M.D. 'Pisces: a coastal ground-wave HF radar for current, wind and wave mapping to 200 KM ranges'. Proceedings of International Geoscience and Remote Sensing Symposium (IGARSS'88); Edinburgh, UK; 1988.

[52]   Helzel T., Kniephoff M., Petersen L. 'Accuracy and reliability of ocean radar WERA in beam forming or direction finding mode'. Proceedings of IEEE Current, Waves and Turbulence Conference; Monterey, CA, USA; 2011.

[53]   Liu Y. 'Target detection and tracking with a high frequency ground wave over-the-horizon radar'. Proceedings of the International Radar Conference; Ann Arbor, USA; 1996. pp. 29–33.

[54]   Liu Y., Xu R., Zhang N. 'Progress in HFSWR research at Harbin Institute of Technology'. Proceedings of the International Radar Conference; Adelaide, AU; 2003. pp. 29–33.

[55]   Xie J., Yuan Y., Liu Y. 'Suppression of sea clutter with orthogonal weighting for target detection in shipborne HFSWR'. *IEE Proceedings – Radar, Sonar and Navigation.* 2002;**149**(1):39–44.

[56]   Xie J., Yao G., Sun M., *et al.* 'Measuring ocean surface wind field using shipborne high-frequency surface wave radar'. *IEEE Transactions on Geoscience and Remote Sensing.* 2018;**56**(6):3389–97.

[57]   Wu S., Yang Z., Wen B., *et al.* 'Test of HF ground wave radar OSMAR2000 at the Eastern China Sea'. Proceedings of the MTS/IEEE Oceans Conference; Honolulu, USA; 2001. pp. 646–8.

[58]   Wu X., Li L., Shao Y., *et al.* 'Experimental determination of significant waveheight by OSMAR071: comparison with results from buoy'. *Wuhan University Journal of Natural Sciences.* 2009;**14**(6):499–504.

[59]   Zhou H., Wen B., Wu S., *et al.* 'Sea states observation with a portable HFSWR during the 16th Asian games sailing competition'. *Chinese Journal of Radio Science.* 2012;**27**(2):293–300.

[60]   Zhao Z., Wan X., Zhang D., *et al.* 'An experimental study of HF passive bistatic radar via hybrid sky-surface wave mode'. *IEEE Transactions on Antennas and Propagation.* 2013;**61**(1):415–24.

[61]   Li Y., Wei Y., Xu R., *et al.* 'Simulation analysis and experimentation study on sea clutter spectrum for high-frequency hybrid sky-surface wave propagation mode'. *IET Radar, Sonar, and Navigation.* 2014;**8**(8):917–30.

[62] Li M., Zhang L., Wu X., *et al.* 'Ocean surface current extraction scheme with high-frequency distributed hybrid sky-surface wave radar system'. *IEEE Transactions on Geosciences and Remote Sensing.* 2018;**56**(8):4678–90.

[63] Anderson S. 'Remote sensing applications of HF skywave radar: the Australian experience'. *Turkish Journal of Electrical Engineering and Computer Science.* 2010;**18**(3):339–72.

[64] Fabrizio G.A. *High Frequency Over-the-Horizon Radar: Fundamental Principles, Signal Processing and Practical Applications.* New York: McGraw-Hill Education; 2013.

[65] Barrick D. 'Remote sensing of sea state by radar' in Derr V.E. (ed.). *Remote Sensing of the Troposphere.* Washington, DC: U.S. Government Printing Office; 1972. pp. 1–46.

[66] Gill E.W., Walsh J. 'Bistatic form of the electric field equations for the scattering of vertically polarized high frequency ground wave radiation from slightly rough, good conducting surfaces'. *Radio Science.* 2000;**35**(06):1323–35.

[67] Gill E.W., Walsh J. 'High-frequency bistatic cross sections of the ocean surface'. *Radio Science.* 2001;**36**(06):1459–75.

[68] El-Darymli K., Hansen N., Dawe B., *et al.* 'Design and implementation of a high-frequency software-defined radar for coastal ocean applications'. *IEEE Aerospace Electronic Systems Magazine.* 2018;**33**(3):14–21.

[69] Phillips O.M. *The Dynamics of the Upper Ocean.* Cambridge, England: Cambridge University Press; 1977.

[70] Gurgel K., Essen H., Schlick T. 'An empirical method to derive ocean waves from second-order Bragg scattering: prospects and limitations'. *IEEE Journal of Oceanic Engineering.* 2006;**31**(4):1–8.

[71] Sommerfeld A. 'The propagation of waves in wireless telegraphy'. *Annals of Physics.* 1909;**28**:665–736.

[72] Shearman E.D.R. 'Radio science and oceanography'. *Radio Science.* 1983;**18**(3):299–320.

[73] Naylor G.R.S., Robson R.E. 'Interpretation of backscattered HF radio waves from the sea'. *Australian Journal of Physics.* 1986;**39**:395–9.

[74] Weber B.L., Barrick D.E. 'On the nonlinear theory for gravity waves on the ocean's surface. Part I: derivations'. *Journal of Physical Oceanography.* 1977;**7**(1):3–10.

[75] Lathi B.P. *Random Signals and Communication Theory.* Scranton, Penn: International Textbook Company; 1968.

[76] Silva M., Huang W., Gill E.W. 'Bistatic high-frequency radar cross-section of the ocean surface with arbitrary wave heights'. *Remote Sensing.* 2020;**12**(4):667.

[77] Silva M., Huang W., Gill E.W. 'High-frequency radar cross-section of the ocean surface with arbitrary roughness scales: a generalized functions approach'. *IEEE Transactions on Antennas and Propagation.* 2021;**69**(3):1643–57.

[78] Silva M., Huang W., Gill E.W. 'High-frequency radar cross-section of the ocean surface with arbitrary roughness scales: higher-order corrections and general form'. *IEEE Transactions on Antennas and Propagation.* 2021.

[79]   Walsh J., Huang W., Gill E.W. 'The first-order high frequency radar ocean sur-
         face cross section for an antenna on a floating platform'. *IEEE Transactions
         on Antennas and Propagation.* 2010;**58**(9):2994–3003.
[80]   Walsh J., Huang W., Gill E.W. 'The second-order high frequency radar ocean
         surface cross section for an antenna on a floating platform'. *IEEE Transactions
         on Antennas and Propagation.* 2012;**60**(10):4804–13.
[81]   Walsh J., Huang W., Gill E.W. 'The second-order high frequency radar ocean
         surface foot-scatter cross section for an antenna on a floating platform'. *IEEE
         Transactions on Antennas and Propagation.* 2013;**61**(11):5833–8.
[82]   Ma Y., Gill E.W., Huang W. 'First-Order bistatic high-frequency radar ocean
         surface cross section for an antenna on a floating platform'. *IET Radar, Sonar
         and Navigation.* 2016;**10**(6):1136–44.
[83]   Ma Y., Huang W., Gill E.W. 'The second-order bistatic high frequency radar
         ocean surface cross section for an antenna on a floating platform'. *Canadian
         Journal of Remote Sensing.* 2016;**42**(4):332–43.
[84]   Ma Y., Huang W., Gill E.W. 'Bistatic high frequency radar ocean surface
         cross section for an FMCW source with an antenna on a floating platform'.
         International Journal of Antennas and Propagation. 2016;**2016**(1):1–9.
[85]   Gill E.W., Ma Y., Huang W. 'Motion compensation for high frequency sur-
         face wave radar on a floating platform'. *IET Radar, Sonar and Navigation.*
         2018;**12**(1):37–45.
[86]   Ma Y., Huang W., Gill E.W. 'High frequency radar ocean surface cross sec-
         tion incorporating a dual-frequency platform motion model'. *IEEE Journal of
         Oceanic Engineering.* 2018;**43**(1):195–204.
[87]   Ma Y., Gill E.W., Huang W. 'Bistatic high frequency radar ocean surface
         cross section incorporating a dual-frequency platform motion model'. *IEEE
         Journal of Oceanic Engineering.* 2018;**43**(1):205–10.
[88]   Shen C., Gill E.W., Huang W. 'HF radar cross sections of swell contami-
         nated seas for a pulsed waveform'. *IET Radar, Sonar and Navigation.*
         2014;**8**(4):382–95.
[89]   Pierson W.J. 'Wind generated gravity waves'. *Advances in Geophysics.*
         1955;**2**:93–178.
[90]   Pierson W.J., Moskowitz L. 'A proposed spectral form for fully developed
         wind seas based on the similarity theory of S. A. Kitaigorodskii'. *Journal of
         Geophysical Research.* 1964;**69**(24):5181–90.
[91]   Long A.E., Trizna D.B. 'Mapping of North Atlantic winds by HF radar sea
         backscatter interpretation'. *IEEE Transactions on Antennas and Propagation.*
         1973;**21**(5):680–5.
[92]   Green D., Gill E.W., Huang W. 'An inversion method for extraction of wind
         speed from high-frequency ground-wave radar oceanic backscatter'. *IEEE
         Transactions on Geoscience and Remote Sensing.* 2009;**47**(10):3338–45.
[93]   Gaffard C.J.P. 'Remote sensing of wind speed at sea surface level using HF
         skywave echoes from decametric waves'. *Geophysical Research Letters.*
         1990;**17**(5):615–18.

[94] Maresca J.W., Barnum J.R. 'Measurement of oceanic wind speed from HF sea scatter by skywave radar'. *IEEE Transactions on Antennas and Propagation.* 1977;**25**(6):132–6.

Chapter 2

Oceanographic applications of high-frequency (HF) radar backscatter

Jeffrey D Paduan[1]

The collection of high-frequency (HF) radar backscatter observations has many potential uses for management and scientific studies in coastal waters. This is because HF backscatter observations contain information related to the interaction of the electromagnetic waves and the ocean waves. Radiowaves in the HF portion of the electromagnetic (EM) spectrum have physical wavelengths that match those of wind-driven gravity waves on the ocean surface. Because of these facts, there is the potential to extract environmental information about the ocean, and about the atmosphere that is forcing it, by examining details of the HF backscatter. Indeed, this has been shown to be a viable remote sensing method to infer ocean surface currents and, to a lesser extent, ocean surface wave conditions and overwater wind conditions [1].

The nature of HF backscatter and the theory for how to use it to extract environmental parameters are discussed in detail throughout this book. In this chapter, the goal is to review potential applications of these environmental parameters with a particular focus on the relevant space and timescales. When does it make sense to use observations from one or more coastal HF radar stations and what are the pitfalls and limitations of these data?

2.1 Factors influencing HF backscatter

The relevant time and space scales for which HF backscatter observations become useful environmental indices result from a specific set of coincidences and practical limitations. The most important of those cases are reviewed here and their cumulative effects are used to define the relevant scales.

2.1.1 The electromagnetic spectrum and the speed of light

High-frequency signals represent a portion of the EM spectrum as highlighted in Figure 2.1. Many other portions of the EM spectrum are used for many different applications, such as communications (e.g., UHF and VHF), navigation (e.g.,

[1]Naval Postgraduate School, USA

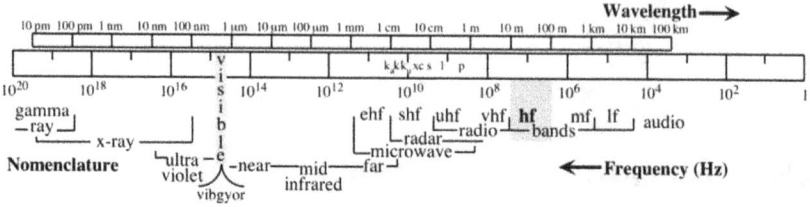

Figure 2.1   *Designated regions of the electromagnetic spectrum highlighting the HF band with wavelengths of a few meters to a few tens of meters and frequencies from around 30 MHz down to around 3 MHz (from [2])*

microwave), or medical diagnostics (e.g., X-ray). So why do oceanographers transmit in the HF band to extract information about coastal ocean conditions? The answer lies in the coincidence of the physical scale of HF EM waves and gravity waves on the ocean surface. In the case of the EM waves, their wavelength, λ, is controlled by the speed of light according to:

$$\lambda = c/f \tag{2.1}$$

where c is the speed of light and f is the frequency. For example, a 10-MHz EM wave in the middle of the HF band has a wavelength of 30 m. By contrast, a 10-GHz EM wave, which is at the high end of the microwave band, has a wavelength of 0.03 m or 3 cm.

Because of the differing physical scales of EM waves for different frequency bands, the waves interact with different physical phenomena. In the case of HF frequencies, the EM waves have the same physical scale as ocean surface gravity waves. In the case of microwave frequencies, the EM waves have the same physical scale as capillary waves on the ocean surface. The EM waves can be expected to have maximum interactions with those corrugations on the sea surface with similar physical scales. In the case of backscattered information, i.e., EM energy reflected from the surface, a predominate effect is the phenomenon of Bragg resonance that occurs when corrugations on the reflecting surface have a physical wavelength that is exactly half as long as the transmitted EM wavelength [3]. (In Chapter 1 of this book, the wave–wave interactions that control the precise details of the reflections are reviewed.)

2.1.2   Factors related to the use of HF transmissions

Knowing how Bragg resonance is related to physical wavelengths and how EM wavelengths relate to the speed of light, the scientist or engineer is able to select which physical waves will most influence the backscattered signal by selecting the frequency of the EM transmission. Given that ability, the choice of frequency then falls to other factors, including practical constraints on the size of the equipment, propagation performance of the EM waves, and, very importantly, what is known about the resonant reflectors. The HF band has been selected for coastal ocean observations because two of these three factors are favorable in that band. This is despite the fact that the third factor, practical constraints on the size of the equipment, is not favorable in the HF band.

Starting with the unfavorable factor, it is the case that antennae needed to broadcast and receive HF frequencies are most effective when they have physical spacing that is similar to the EM wavelength. That is, with HF wavelengths on the order of 10 m, the antenna array size should be multiples of that length or on the order of 100 m. Antenna arrays of this size can be and are constructed, but they are not as convenient as, e.g., arrays configured for microwave frequencies that can be constructed at scales of 1 m or less. This means that microwave-based instruments may be suitable for shipboard or satellite-borne applications whereas HF-based instruments are not.

Propagation of EM transmissions in the HF band is, however, more favorable than in many other bands due to the phenomenon of ground wave propagation [3]. When moving over the surface of a perfect conductor, EM transmissions in the HF band can propagate without loss of energy. Seawater happens to be a near-perfect conductor, meaning that these transmissions can propagate great distances with very little loss of energy. Hence when the HF transmitters and receivers are placed very near to seawater (i.e., within a few wavelengths), the energy can travel out and back along the sea surface with less loss of energy into the underlying media when compared with non-conducting media. This groundwave path is distinct from and generally much longer than the line-of-site path that limits, e.g., microwave radar transmissions.

2.1.2.1   The well-established theory of surface gravity waves

Waves on the sea surface are governed by different physical properties depending on their wavelength [4]. At the smaller scales of order 1 cm, capillary waves are dominated by surface tension. At intermediate scales of order 1 m, including those that are resonant with HF transmissions, ocean waves (or wind waves) are dominated by gravity. And at the largest scales of order 100 km, a combination of gravity and earth's rotation (i.e., Coriolis acceleration) dominate.

HF transmissions are used to measure ocean surface currents, despite the requirement for relatively large antenna arrays, because a great deal is known about the behavior of the reflecting surface, i.e., the behavior of ocean surface gravity waves. In particular, the phase speed of ocean surface gravity waves in deep water is related to their wavelength according to [5]:

$$c^2 = \frac{g}{k} = \frac{g\lambda}{2\pi},$$ (2.2)

where g is gravitational acceleration and k is the wavenumber. This relationship is critical because it allows the operator of HF-based observing systems to know, a priori, the phase speed of the Bragg-resonant reflectors, c_B, which is given by:

$$c_B^2 = \frac{g\lambda_R}{4\pi},$$ (2.3)

where λ_R is the wavelength of the HF radar transmission.

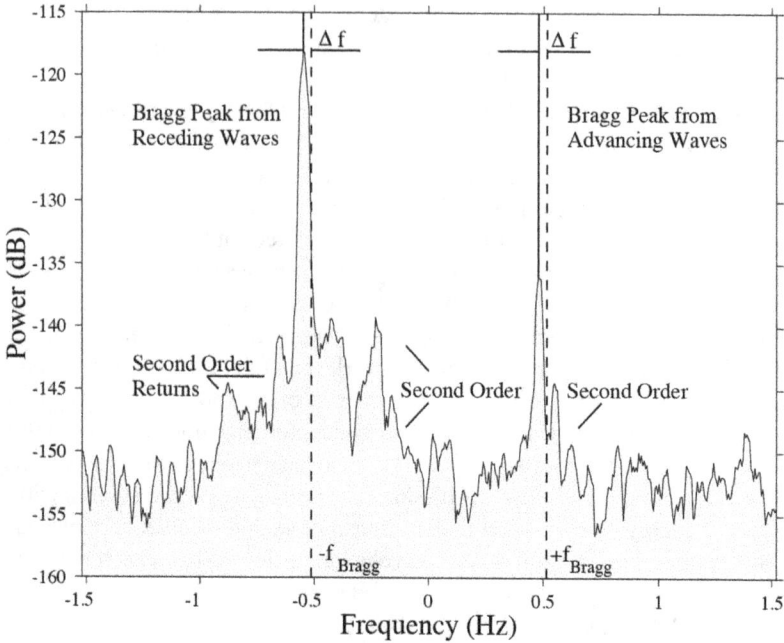

Figure 2.2   *Example Doppler backscatter spectrum from an HF radar system broadcasting at 25.4 MHz illustrating first-order peaks due to approaching and receding waves and their respective Doppler shifts along with a much smaller Doppler offset, Δf, attributed to the underlying current (from [2])*

2.1.2.2   Reflector speed and the Doppler backscatter spectrum

The importance of accurate prediction of the reflector speed to the process of estimating ocean surface currents from HF backscatter observations cannot be overstated. Without the ability to predict the reflector speed, and to predict it with great accuracy, the measurement would not be possible. This fact is illustrated by the classic example of a Doppler backscatter spectrum shown in Figure 2.2. It is clear from this example that the Doppler frequency shifts due to movement of the surface waves are on the order of ten times larger than the Doppler frequency shift due to the underlying current. Another way to express this is to note that an uncertainty of only 10 per cent in the specification of the surface wave speed would be enough to completely negate the ability of the measurement to detect the underlying current.

Hence, the remote measurement of ocean surface current using Doppler backscatter observations is restricted to HF frequencies because those frequencies are resonant with deep-water ocean surface waves and, in turn, the phase speed of those waves is accurately predicted as a function of their wavelength. At higher frequencies, above the VHF band, the EM transmissions are resonant with waves governed by a mixture of gravity and surface tension effects and the phase speed cannot be accurately predicted[a]. At even higher frequencies in the microwave band, the EM transmissions are resonant with capillary waves, which are dominated by surface tension effects. In that case, capillary wave phase speeds have been shown empirically to be correlated with overwater wind speeds. That correlation forms the basis for measuring winds from satellite scatterometers.

2.1.3 Impacts of noise and averaging

Once the frequency of the EM transmissions has been chosen, it then becomes possible to determine practical limits on the resolution of the measurements. The absolute lower limit on, e.g., the spatial resolution is set by the wavelength of the EM transmissions. Information is not available on scales smaller than about two times the wavelength. In practice, however, measurements over many more than two wavelengths are required in order for the constructive interference inherent in the Bragg resonance phenomenon to produce backscattered energy levels that exceed background levels from all other reflections. Those background levels and their manifestations across the Doppler spectrum fluctuate with wind and wave conditions. This can be seen in the level of the noise floor in the outer portions of the spectrum in Figure 2.2 and in the second-order reflections that appear around the first order, Bragg-resonant reflections in that spectrum. In addition to these effects, which are related to the physical shape of the sea surface, the background noise levels may be increased by the presence of radiowave interference from both internal and external sources.

The resulting practical limits on HF radar observations of ocean surface currents are such that the reflecting patch on the ocean surface needs to encompass around 100 wavelengths [6]. For example, the range cell average size for a 10-m wavelength Bragg wave should be, at least, 1 km. Larger cells will produce stronger returns with a built-in trade-off that assumes homogeneity of ocean surface currents over a larger area. Commonly used HF band systems operate between 5 and 25 MHz with range cell sizes from 10 km down to 2 km, respectively, and, in real-world situations, minimum range cell sizes are most often controlled by external noise levels or maximum available broadcast bandwidth.

While spatial resolution is affected by native wavelengths, surface wave conditions, and external noise, temporal resolution is also affected by these factors. It is necessary to dwell on a reflecting patch for some amount of time and to subsequently average those results over some number of segments in order to produce a stable Doppler backscatter spectrum (i.e., frequency resolution is determined by the length of the time series and probability analyses predict the mean conditions based on the number of replicate samples). It is not clear a priori what are the relevant independence timescales that would determine the degrees of freedom in the spectrum and help predict what dwell and averaging times are needed. Experimentation shows that dwell times of several minutes and averaging times between 20 min and

---

[a]It should be noted that the techniques described here can be used for frequencies in the VHF band, particularly for sheltered ocean locations, extending to about 100 MHz.

3 hours are required when paired with the typical patch sizes listed here. In practice, the majority of HF radar systems deployed to measure ocean surface currents are configured to output results every 1 hour with the underlying time series collected over periods ranging from 4 to 20 minutes.

2.1.4   Relevant time and space scales

It is helpful to assemble the temporal and spatial resolutions available from HF radar observations of ocean surface currents and to pair them with the range of interesting processes in the coastal ocean and with the practical necessities involved in maintaining a network of field sites. The collection of ocean surface current mapping data from a network of HF radar systems represents a unique capability. Two-dimensional mapping is combined with relatively high temporal sampling. The position of instruments along the coast on land, as opposed to in the water, provides for the possibility to maintain observations over extended periods with reliability approaching that of operational systems.

The results of these factors create two primary regimes of applicability for HF radar-derived surface currents. The first regime involves real-time applications in support of emergencies, such as oil spill response or a search-and-rescue event. Beyond these practical applications, contributions to understanding oceanographic and ecosystem processes require much longer data sets that encompass many realizations of the processes. The general space–time range of applicability for HF radar-derived surface currents is presented in Figure 2.3. The full range of possible scales is set by the HF radar system's space and time resolution, which are typically around 2 km and 1 hour, respectively, paired with the extent of instrumented coastline and the length of the measurements. As Figure 2.3 suggests, however, the intermediate

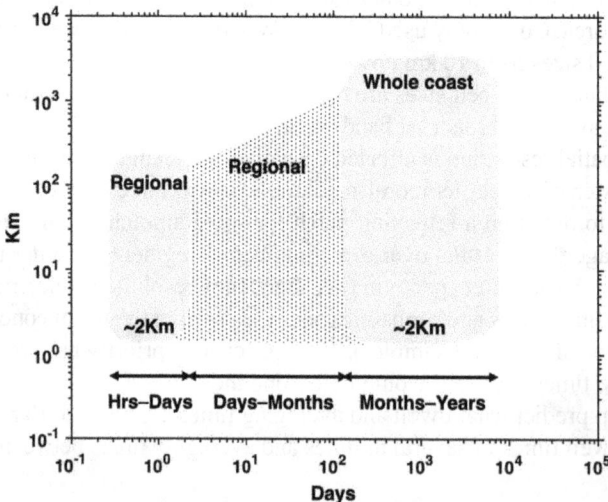

Figure 2.3   Space and timescales of applicability for surface current mapping data from commonly configured HF radar systems

scales are accessible from HF radar measurements but are somewhat less applicable due to the fact that the critical phenomena cannot be sampled repeatedly.

Despite the practical gap in the timescales of interest for HF radar-derived surface current mapping data, the observations are uniquely useful. Other systems capable of providing mapping data, such as airborne or satellite observations, are generally not able to support the temporal resolution, immediate results, and sustained observations available through use of HF radar systems. When temporal resolution is high, such as through moored observations, (horizontal) spatial coverage is generally limited.

The potential space–time coverage of large networks of HF radar systems includes a wide range of important oceanographic processes. This is illustrated by the many processes listed in Figure 2.4, which fall within the nominal capabilities of HF radar observations. Processes such as internal tides, mesoscale fronts and eddies, and coastally trapped waves are well resolved by HF radar. These scales, together with the surface-only nature of the measurements discussed in the next section, should prioritize the applications of HF radar-derived surface current maps.

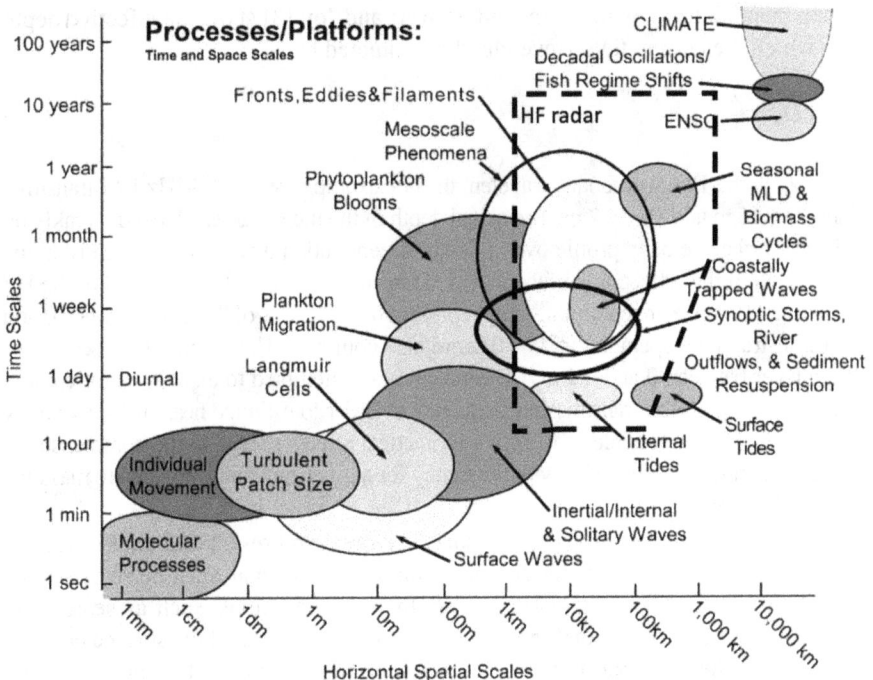

Figure 2.4   *Space and timescales of important oceanographic processes in relation to the nominal coverage envelope for a network of HF radar systems as approximated in Figure 2.3 and outlined here (dashed line; after [7])*

Monitoring and studying of internal tides are, e.g., well suited to HF radar observations while study of ocean turbulence is not.

2.1.5   *Depths observed by HF radar*

While the strength of HF radar systems lies in their ability to collect two-dimensional mapping data, it is always important to keep in mind what depth(s) are being observed. For that information, we return, briefly, to the chain of events that support remote sensing of ocean surface currents using HF backscatter observations. Recall that EM transmissions in the HF band are resonantly backscattered through Bragg scattering over a patch of surface gravity waves whose wavelengths are exactly half the length of the outgoing EM wave. Movement of the Bragg-resonant reflectors causes positive and negative Doppler shifts for waves approaching and receding from the receiver, respectively, and deep-water gravity wave theory provides for a priori prediction of those shifts (Figure 2.2). Any additional Doppler offset, Δf, is attributed to the underlying current field. But what governs interactions between that field and the reflectors?

Interactions between the current field and surface gravity waves are presumed to be controlled by the particle motions of the gravity waves. Those motions exhibit an exponential decay with depth, and Stewart and Joy [8] show the effective depth over which the current field is weighted is estimated by:

$$D_e = \frac{\lambda_B}{4\pi},$$
(2.4)

where λ_B is the Bragg-resonant wavelength. For example, with 10 MHz EM transmissions, $\lambda_B = 15$ m and $D_e = 1.2$ m. The actual depth of the measurement depends weakly on the near-surface velocity profile over a depth of penetration that, in turn, depends on the wavelength of the reflecting surface waves. Transmissions near the lower end of the HF band around 5 MHz are common. They represent observations of the layer between 0 and 2.4 m. Instruments operating at 25 MHz are also common. They represent observations of the layer between 0 and 0.5 m. Researchers have attempted to exploit the differences in effective depth with transmission frequency in order to estimate near-surface velocity shear [8, 9]. While possible in theory, the practical effects related to the commensurate changes in scattering patch size with frequency for a given antenna array tend to mask the near-surface velocity shear.

In effect, HF radar-derived ocean surface currents represent estimates of current in the upper ~1 m of the water column. This is a factor that must be remembered when interpreting HF radar observations. In many situations, such as search and rescue support or tracking drifting larvae, direct observations of the surface currents are useful. When comparing with other measurements or model output, however, the shallow depth of the HF radar observations may complicate the interpretation.

Finally, it should be noted that HF radar observations of surface current include, at least, some portion of the Stokes Drift. That portion is assumed to result from the particle motion of all waves present with wavelengths equal to or greater than λ_B [10, 11]. This factor provides one more complication when comparing HF radar-derived surface currents with, e.g., moored current meter observations because the Eulerian current meter observations are not expected to include contributions from Stokes Drift.

2.2 Real-time applications of HF radar backscatter

Networks of HF radar systems have been deployed along large stretches of coastline in many parts of the world. In virtually all cases, these networks have been justified as quasi-operational monitoring systems in support of real-time applications [12, 13]. Those applications focus, primarily, on safety of persons or the environment in support of more efficient search-and-rescue operations or containment of hazardous spills. To be effective, an emergency response environmental monitoring system like this should be working everywhere all the time. This makes real-time applications of HF radar-derived surface current maps relatively costly compared with science-focused data collection activities that can tolerate some gaps. The ability to maintain the instruments from the shoreline with only an occasional need to put to sea for the collection of calibration data means that large networks of HF radar systems can be, at least, feasible as a supplier of real-time response data. While there are no definitive answers to the question of what constitutes real-time availability and how much of a network must report in that time frame to be considered "operational", the international community is developing standards based on the growing set of common experiences [14].

2.2.1 Considerations of real-time applications

What makes real-time operations of HF radar networks challenging? It is not the intent here to provide a comprehensive answer. But it is helpful to list some of the most common challenges. First, as the descriptive term *network* implies, the desired products require coordination among a number of independent instruments. In the case of HF radar systems, the instruments are often deployed at remote locations along the coastline where the supply of power and communications can be interrupted without the ability to mount an immediate response. This means that spatial and temporal gaps in the surface current maps may be created due to external factors. Other external factors, such as bursts of radiowave interference, ionospheric reflections, or corrosion of individual antenna components, can also lead to gaps in the data or, even worse, errors in the reported data that may not get flagged. As a result, many real-time applications of HF radar measurements require a range of automated processing to fill gaps and quality control procedures to flag potentially bad data [14]. This is not to imply that real-time observations cannot be supported as part of operational networks. Rather, it is intended to illustrate challenges related to operating those networks.

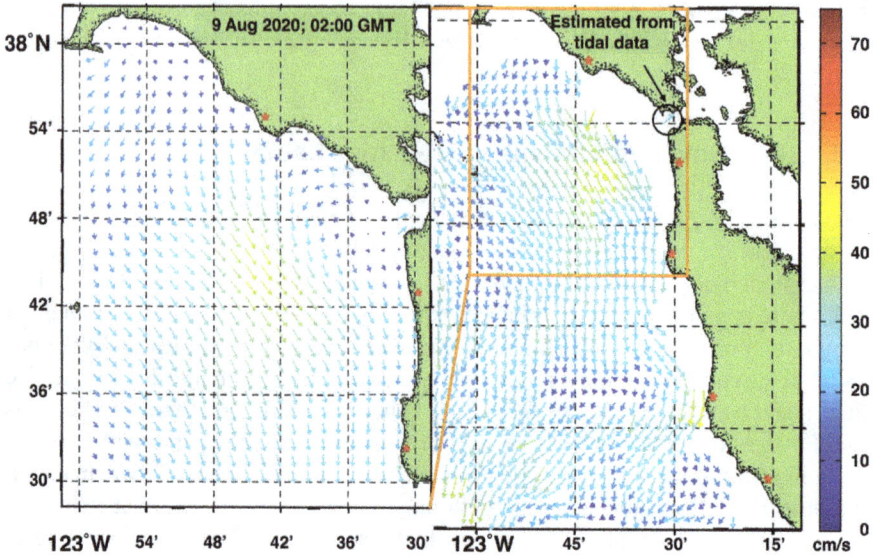

Figure 2.5   *Right: vector surface current mapping results from five individual HF radar installations offshore San Francisco (* symbols) and estimated from historical tidal observations (circle) and Left: results of open modal analysis spatial smoothing in the subregion of the Gulf of the Farallones*

2.2.2   Examples of real-time applications

An example of a real-time processing stream applied to HF radar observations off-shore San Francisco is provided in Figure 2.5. In this case, radial current observations (not shown) were collected from five individual HF radar sites and combined to form the hourly vector map shown here. Despite the fact that radial current observations were collected from multiple sites, the resulting surface current map contains a number of coverage gaps, particularly near the coastline. In addition to those spatial gaps, the domain includes the important and dynamic ocean region offshore of the Golden Gate, which is the entrance to San Francisco Bay. The HF radar network geometry in this case does not allow for measurements near the Golden Gate, yet experience and historical observations indicate this to be a region of very strong currents. This is particularly true of the variable tidal currents that regularly flow into or out from San Francisco Bay.

For real-time applications, the gappy observations in the subregion of the Gulf of the Farallones were filled using the open modal analysis (OMA) gap-filling technique [15]. In this particular domain, a tidal model representation of current in the Golden Gate was added to the HF radar-derived surface current observations prior to application of the OMA technique. This has a stark impact on the estimated surface currents in the area offshore of the Golden Gate. These estimates are almost certain to be more realistic than those produced by the alternative of assuming a closed boundary at that location.

The advantage of spatially smoothing the real-time observations using a procedure like OMA is the ability of the resulting uniform data to support follow-on analyses, such as trajectory computations. (This is not meant to imply that gap-filling techniques do not themselves introduce errors into the subsequent computations but rather to point out that, as a practical matter, many computations require uniformly distributed data.) In the Gulf of the Farallones observing region offshore San Francisco, these techniques were effectively employed to support the environmental cleanup in response to a major oil spill that occurred in San Francisco Bay. On the morning of Wednesday, November 7, 2007, the cargo vessel *Costco Busan* struck a tower of the western span of the Bay Bridge while operating in a heavy fog. For a couple of days, spilled oil contaminated shorelines within San Francisco Bay. It was known, however, that significant amounts of drifting oil would soon leave through the Golden Gate putting at risk beaches along the open ocean and possibly even the extremely sensitive Farallon Islands, which lie about 30 km offshore. HF radar observations of ocean surface currents were used to guide responders to the most at-risk beaches by computing estimated trajectories for simulated particles initiating from positions immediately offshore of the Golden Gate. The summary trajectories in Figure 2.6 illustrate all of the estimated particle motions during the week following the spill. During the event, these results clearly showed that the at-risk beaches were those to the north of the Golden Gate and, thankfully, that the Farallon Islands were unlikely to receive a direct delivery of oil-laden water.

In addition to improving the response to hazardous material spills in the coastal ocean, real-time surface current mapping data from HF radar systems can be used to speed up life-saving rescue operations. In any rescue operation, time is critical. This is particularly true for open ocean search and rescue because the ability of ships or aircraft to cover wide swaths of ocean is limited. The ability of victims to survive in the water is also limited. Real-time data from HF radar support this process by *increasing* the understanding of drift patterns and rates and, thereby, *decreasing* the size of the initial search area. Studies by U.S. Coast Guard, together with university researchers, demonstrate that including HF radar-derived surface current mapping data in their search and rescue planning can decrease the initial search area by a factor of four compared with model-only planning tools [16].

2.3 Example of an intermediate-scale observation

While phenomena with timescales of days to months were described as less applicable in Section 2.1.4, HF radar are uniquely able to capture them. Features such as mesoscale eddies and, in particular, submesoscale eddies can be identified in HF radar-derived surface current maps. Figure 2.7, which is reproduced from [17], provides a beautiful example of a submesoscale frontal eddy on the inshore side of the East Australian Current. In this example, the cold-core feature is also seen in the MODIS sea surface temperature data. With its all-weather capability, however,

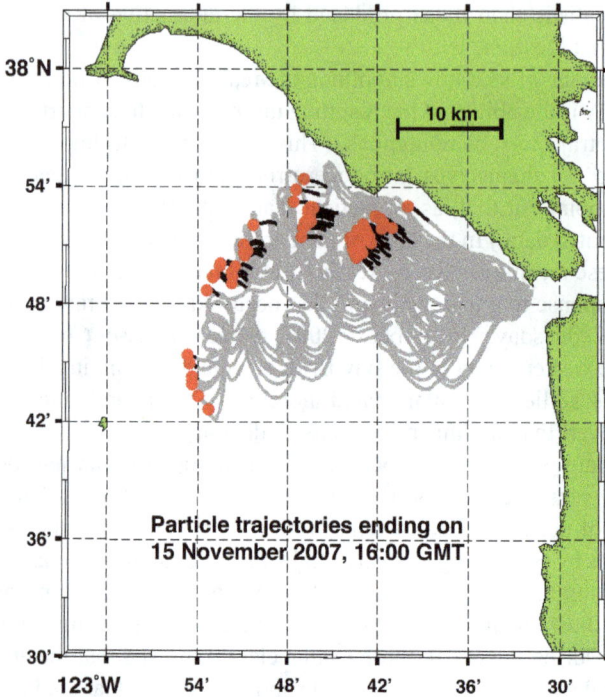

Figure 2.6   *Surface trajectories and endpoints for November 15, 2007, at 16:00*
GMT (red symbols) for clusters of simulated particles released at the
easternmost location near the mouth of San Francisco Bay every 24
hours for the days prior to that time beginning November 8, 2007 at
16:00 GMT. Black lines represent the 24 hours prior to the endpoint
time. Gray lines are the trajectories during the preceding days.

the HF radar network has the capability of identifying such features even in the
presence of cloud cover. This example is also impressive in that it shows how the
satellite-based geostrophic current estimates from altimetry completely miss the
submesoscale feature due to their relatively low resolution. Note that the HF radar-
derived surface current map in this example includes only every sixth vector. The
high resolution of the observations allows for an estimate of the surface divergence
field, which is a computation that will be revisited in the next section. As dramatic
as the individual example of Figure 2.7 is, it can be argued that the process insights
are more significant in the descriptions of the 40 cyclonic eddies and 16 anticyclonic
eddies that were objectively identified in the full 297-day HF radar record collected
by these authors [17]. It is this distinction between an example of what is possible
and a statistical description of what is likely that argues for the applicability gap in
Figure 2.3.

It should be noted here that Figure 2.7 is useful and important in that it derives
from set of WERA phased array HF radar instruments. Those instruments differ

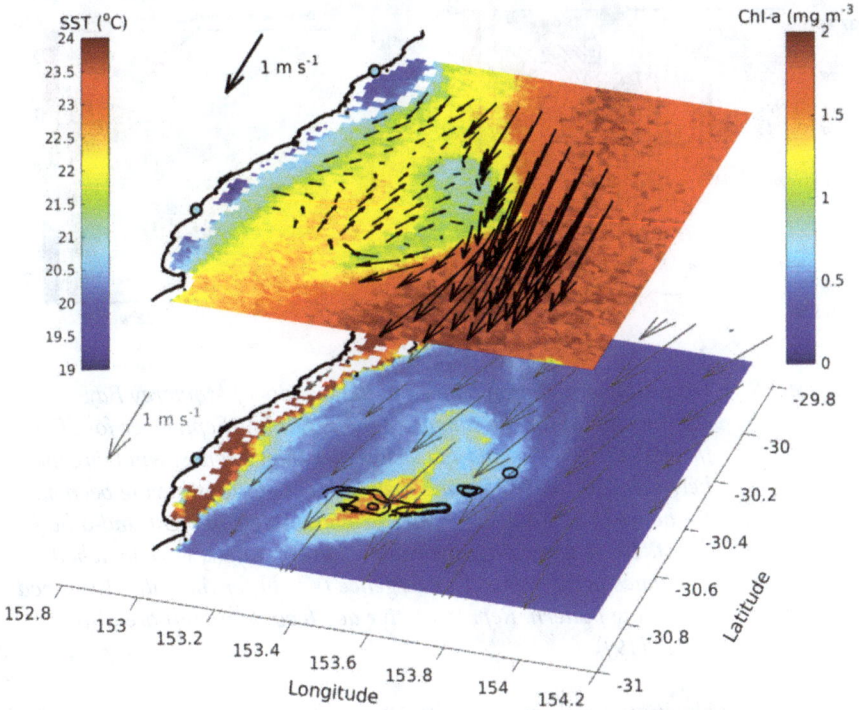

Figure 2.7   Top view of MODIS SST (top section, left colorbar) and ocean
              color (bottom, right colorbar) remote-sensed images on September
              29, 2013, showing the signature of a frontal eddy. Velocity vectors
              show the surface currents measured by HF radars on September 29,
              2013, 08:00 (plot every sixth grid point, top, black) and geostrophic
              current from altimetry (bottom, gray). Black contours overlaid on
              the chlorophyll-a concentrations (bottom) show positive surface
              divergence calculated from the HF radar velocities (contours of
              0.2|f|, 0.3|f|, 0.4|f|, 0.5|f|, increasing to maximum in the center). Blue
              dots indicate the location of the two HF Radar systems (from [17]).

from the SeaSonde direction-finding HF radar instruments that provided the other
examples in this chapter in the way look direction (or azimuth angle) is determined.
The advantages and disadvantages of the two genres of HF radar systems are discussed in many other publications (e.g., [2]) and elsewhere in this book.

2.4   Process studies using HF radar backscatter

Deeper understanding of physical processes at the ocean surface may be supported by observations from HF radar systems as a result of the extremely near-
surface nature of the observations. Progress here can be seen in the work related to

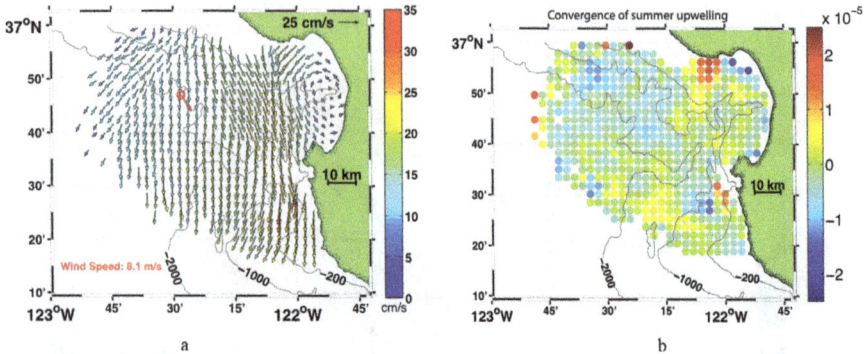

Figure 2.8   Upwelling circulation pattern in the vicinity of Monterey Bay, California, (a) for the summer months of May–September for 2006 through 2010 with wind speeds greater than 5.0 m/s, wind direction between 280° and 325°, and these joint conditions having been met for between 204 and 216 hours (a spin up time of eight-and-a-half days). These upwelling-favorable wind conditions were matched 127 times in 12 events. Convergence (s^{-1}; b) for this fully developed upwelling pattern. Representative depth contours (m) are also shown (after [19]).

near-surface velocity shear observations [9, 18] and the work related to Stokes Drift characterization [10, 11]. Potential exists to push this topic further. Any operator of HF radar systems for ocean surface current observations will note that the currents respond extremely quickly to changes in wind forcing.[b] While the observations of rapid response to wind are plentiful, commensurate observations of near-surface stratification are not. Future experiments with detailed measurements of density in the upper few meters of the water column over time combined with HF radar observations should be instructive.

As mentioned throughout this chapter, HF radar observations are most impactful when their ability to *map* ocean surface parameters is exploited. This is the backbone of real-time products in support of hazardous material spill mitigation and search and rescue operations. It is also the basis for advanced learning about physical and bio-chemical processes near the air–ocean interface.

The power of ocean surface current mapping over an extended time period was wielded in the work of [19], where the repeating coastal upwelling circulation pattern in the area of Monterey Bay, California, was exposed through conditional averaging (Figure 2.8). In this case, the conditional test data were independent observations of overwater wind speeds and directions. The fidelity of the resulting map of upwelling circulation is greatly improved when observations are selected at times when wind

---

[b]This is based on personal observations and the knowledge that the measurement is tracking a very thin layer of the upper ocean. Nonetheless, a quantitative assessment of how quickly the measurement responds to wind is warranted.

forcing was upwelling favorably as opposed to any predetermined, fixed averaging period. The results shown in Figure 2.8 are derived from the later stage, fully developed periods of multiple upwelling events, which have remarkable event-to-event consistency. By varying the "spin-up" time with upwelling-favorable winds prior to averaging, the study was also able to show that the upwelling circulation pattern in this location reaches equilibrium (i.e., becomes fully developed) after about three days of continuous, upwelling-favorable winds. Recently, this approach was repeated for the upwelling circulation offshore Lisbon, Portugal, where fully developed upwelling circulation occurs after five-and-a-half days of upwelling-favorable winds [20].

Well-characterized, stable ocean circulation patterns also provide an opportunity to compute and visualize one of the more important but difficult-to-observe dynamical properties of the flow field: the horizontal divergence. The divergence field associated with the fully developed upwelling circulation in the area of Monterey Bay is shown in Figure 2.8. (Here, the results are presented in terms of the horizontal convergence so that upwelling areas appear blue and downwelling areas appear red.) The results provide evidence for the expected upwelling region, which in this case appears just offshore of the strong jet crossing the mouth of Monterey Bay. Also striking is the strong correlation in scale and, approximate, locations of the contrasting upwelling and downwelling regions derived directly from these HF radar observations with earlier measurements from an aircraft and a high-resolution atmospheric circulation model, showing the distribution of wind stress curl for this same location [21]. That suggests that small-scale wind structure, driven by local coastal shape and orographic effects, may directly control the locations and scale of coastal upwelling.

2.5 Conclusions

Environmental remote sensing techniques arise through a combination of physics and luck. In the cases discussed here in which HF Doppler backscatter information is used to detect ocean surface currents, luck is involved in the fact that the combination of Bragg-resonant backscatter occurs at scales that can be broadcast and monitored and for which the behavior of the reflector (i.e., ocean surface gravity waves over deep water) is very precisely known. This match of scale and physics is coincidence, not a priori destiny. In fact, engineers and experimentalists would surely prefer to build their instruments with higher broadcast frequencies and smaller wavelengths than the relatively large HF wavelengths. That would make the physical size of the necessary antenna array smaller and the ability to steer the measurement better. That option is not available, however, for this same type of Doppler-shift computation used for HF wavelengths because the physics governing smaller waves on the ocean surface do not provide for a deterministic expression for their propagation speed as a function of their wavelength. The movement (i.e., Doppler shift) of the smaller waves cannot be removed from the measurements and, therefore, the additional Doppler shift due to the movement of the underlying ocean current cannot be determined using the higher frequencies.

So, the frequency of a remote sensing system for monitoring ocean surface currents is set by the range of broadcast wavelengths that are commensurate with the wavelengths of wind waves in deep water. Given the transmission speed (i.e., the speed of light), this turns out to be the HF radiowave band. Given that reality, other factors then come into play to determine realistic space and timescales that can be monitored using HF backscatter observations. These factors include how much power and bandwidth can be legally broadcast by the instrument and the resulting signal-to-noise performance at each range. In the case of HF radar instruments, the practical ranges in space and time happen to overlap with those associated with a large number of important oceanographic processes (Figure 2.4). It is also the case that other observing methods based in the water, in the air, or from space are not able to match the resolution, range, and persistence capabilities of HF radar systems. This motivates both the present-day and expanding use of HF radar measurements.

While the spatial scales of HF radar measurements are generally set by the particular broadcast characteristics of the instrument, the temporal scales, particularly the duration, are often affected by resources available to deploy and maintain the instruments. This means that oceanographic applications of HF radar measurements tend to come in two types: (1) near real-time applications, which are dominated by practical tasks related to mitigating hazardous spills or search and rescue operations and (2) very much longer-term observations for which repeated process sampling can be used to study the underlying forcing of ocean currents and the impacts they have on coastal ecosystems. In many locations, resources needed to deploy a network of HF radar systems have been justified through the desire to support near real-time applications. Where that support has been successful and data record lengths have grown from days to months and years, the ability to study the full suite of oceanographic processes within the covered space and timescales has been unleashed.

As more results based on long-term observations from HF radar systems are produced, the argument for these systems being both uniquely able to monitor ocean processes and being cost effective over time is strengthened. The insights provided here around coastal upwelling circulation represent a small subset of what has been learned already from long-term observations of coastal ocean surface currents from HF radar networks around the world. And, while this chapter focused on oceanographic applications, it is also true that simultaneous observations of ocean surface currents and contributions to vessel tracking systems can be made using the same HF radar network [22, 23]. As these methods become better established, networks of coastal HF radar systems can be seen and operated as a win-win-win set of observations for coastal managers, ocean scientists, and security professionals.

References

[1] Paduan J.D., Washburn L. 'High-frequency radar observations of ocean surface currents'. *Annual Review of Marine Science.* 2013;**5**:115–36.

[2] Paduan J., Graber H. 'Introduction to high-frequency radar: reality and myth'. *Oceanography.* 1997;**10**(2):36–9.

[3] Hansen P. 'Measurements of basic transmission loss for HF ground wave propagation over seawater'. *Radio Science.* 1977;**12**(3):397–404.

[4] Toffoli A., Biner-Gregersen E.M. 'Types of ocean surface waves, wave classification'. *Encyclopedia of Maritime and Offshore Engineering.* John Wiley & Sons, Ltd.; 2017.

[5] Kinsman B. *Wind Waves: Their Generation and Propagation on the Ocean Surface.* New York: Dover Publications; 1965.

[6] Tyler G.L., Teague C.C., Stewart R.H., Peterson A.M., Munk W.H., Joy J.W. 'Wave directional spectra from synthetic aperture observations of radio scatter'. *Deep Sea Research and Oceanographic Abstracts.* 1974;**21**(12):989–1016.

[7] Dickey T.D. 'Emerging ocean observations for interdisciplinary data assimilation systems'. *Journal of Marine Systems.* 2003;**40–41**:5–48. [The Use of Data Assimilation in Coupled Hydrodynamic, Ecological and Bio-geochemical Models of the Ocean. Selected papers from the 33rd International Liege Colloquium on Ocean Dynamics, held in Liege, Belgium on May 7–11th, 2001].

[8] Stewart R.H., Joy J.W. 'HF radar measurement of surface current'. *Deep Sea Research.* 1974;**21**:1039–49.

[9] Teague C.C., Vesecky J.F., Hallock Z.R. 'Multifrequency HF radar observations of the vertical structure of near-surface currents during COPE-3'. IEEE 1999 International Geoscience and Remote Sensing Symposium, IGARSS'99; 1999.

[10] Ardhuin F., Marié L., Rascle N., *et al.* 'Observation and estimation of Lagrangian, Stokes, and Eulerian currents induced by wind and waves at the sea surface'. *Journal of Physical Oceanography.* 2009;**39**(11):2820–38.

[11] Chavanne C. 'Do high-frequency radars measure the wave-induced stokes drift?' *Journal of Atmospheric and Oceanic Technology.* 2018;**35**(5):1023–31.

[12] Paduan J.D., Kosro P.M., Glenn S.M. 'A national coastal ocean surface current mapping system for the United States'. *Marine Technology Society Journal.* 2004;**38**(2):102–8.

[13] Roarty H., Cook T., Hazard L., *et al.* 'The global high frequency radar network'. *Frontiers in Marine Science.* 2019;**6**(164).

[14] Carlo Mantovani C., Corgnati L.P., Horstmann J., *et al.* 'Best practices on high frequency radar deployment and operation for ocean current measurement'. *Frontiers in Marine Science.* 2020;**7**(00210).

[15] Kaplan D.M., Lekien F. 'Spatial interpolation and filtering of surface current data based on open-boundary modal analysis'. *Journal of Geophysical Research.* 2007;**112**(C12).

[16] O'Donnell J., Ullman D., Spaulding M., *et al.* 'Integration of coastal ocean dynamics application radar (CODAR) and short-term predictive system (STPS) surface current estimates into the search and rescue optimal planning system (SAROPS)'. *Report No. CG-D-01-2006. U.S. Coast Guard Research & Development Center 1082 Shennecossett Road Groton, CT.* 2005.

[17] Schaeffer A., Gramoulle A., Roughan M., *et al.* 'Characterizing frontal eddies along the East Australian current from HF radar observations'. *Journal of Geophysical Research: Oceans.* 2017;**122**(5):3964–80.

[18] Meadows L.A. 'High frequency radar measurements of friction velocity in the marine boundary layer'. *Ph.D. Dissertation, University of Michigan, Ann Arbor.* 2002:125.

[19] Paduan J.D., Cook M.S., Tapia V.M. 'Patterns of upwelling and relaxation around Monterey Bay based on long-term observations of surface currents from high frequency radar'. *Deep Sea Research Part II: Topical Studies in Oceanography.* 2018;**151**(2):129–36.

[20] Tavares A.C.G. '2020: analysis of surface current patterns to the SW of Lisbon using high-frequency radar data'. *M.S. Thesis, Naval Postgraduate School.* Monterey, CA; September 2020. pp. 97.

[21] Ramp S.R., Davis R.E., Leonard N.E., *et al.* 'Preparing to predict: the second autonomous ocean sampling network (AOSN-II) experiment in the Monterey Bay'. *Deep Sea Research Part II: Topical Studies in Oceanography.* 2009;**56**(3-5):68–86.

[22] Chuang L.Z.H., Chung Y.-J., Tang S.T. 'A simple ship echo identification procedure with SeaSonde HF radar'. *IEEE Geoscience and Remote Sensing Letters.* 2015;**12**:2491–5.

[23] Braca P., Grasso R., Bryan K., *et al.* 'Maritime surveillance with multiple over-the-horizon HFSW radars: an overview of recent experimentation'. *IEEE A&E Systems Magazine.* 2015;**150004**.

Chapter 3

Symbiosis of remote sensing and ocean surveillance missions of HF skywave radar

Stuart J Anderson[1]

HF over-the-horizon radar (OTHR) systems are widely used for ocean surveillance, that is to say, for detecting and tracking ships and aircraft over wide areas extending far beyond the range of shore-based microwave radars [1–3]. For ranges less than about 400 km, the surface wave mode of propagation can be exploited, while, for ranges beyond 1 000 km, skywave illumination is generally available. The signal transfer characteristics of the skywave channel are highly variable, necessitating sophisticated frequency management, flexible choice of waveform, and adaptive signal processing, but the potential coverage is vast – the one-hop zone for a single skywave radar may exceed 5×10^6 km^2 in area. Moreover, the earth–ionosphere waveguide supports multi-hop propagation, and it is by this mechanism that large ships have been detected at ranges in excess of 6 000 km [4].

While aircraft detection was generally the primary task for skywave OTHR systems until the 1990s, it is now the case that several national systems accord comparable priority to ship detection, with missions addressing problems such as drug smuggling, illegal immigration, sovereignty of disputed waters, fishing zone protection, piracy and ship routing. The associated objective of classifying or even identifying the observed vessels remains a work in progress, with various approaches having been considered (see [5] for a survey) but no robust solution yet implemented in operational systems.

A far more widespread application of HF radar technology is the remote sensing of the ocean surface, presently implemented in perhaps 500 HF surface wave radar (HFSWR) systems. There are far fewer instances of remote sensing missions carried out with skywave radars, yet some meaningful measurements of ocean surface parameters were performed with skywave radars even before they were demonstrated with HFSWR systems. The first such mapping of oceanic wind fields took place in 1972 [6] using the NRL MADRE radar, and independently later in the 1970s by a group at SRI, with the WARF radar [7, 8], and at DSTO in Australia, with the Jindalee Stage A radar [9]. The wind field observations were followed by measurements of wave

[1]University of Adelaide, Australia

height [10, 11], mid-ocean currents [12, 13] and, subsequently, estimates of, first, the non-directional wave spectrum [14], and then the directional wave spectrum [15]. By 1983, the Australian Jindalee Stage B radar was delivering daily oceanic wind maps over more than 10^6 km^2 of the Indian Ocean [16]. In the 1980s, mapping of wind, waves and currents excited by tropical cyclones yielded unique insights into the formation and evolution of tropical cyclones [17–19]. All this information is extracted from the spatial and temporal properties of the echoes from the dynamic ocean surface, that is, the sea clutter, with the information retrieval techniques varying according to the dimensionality of the radar measurements.

Beyond these established observables, numerous more ambitious measurement capabilities have been explored, addressing phenomena such as internal waves [20], rainfall [21], surfactants [22], sea ice [23], tropical convective cells [24] and tsunamis [25]. Theoretical models for the associated radar signatures have been developed, and most of these phenomena have been demonstrated experimentally, but, at the present time, these advanced capabilities have not been integrated into the standard missions of today's HF skywave radars.

It might be thought that ship/aircraft detection and remote sensing missions are so different in nature, with completely independent objectives, serving disjoint user communities and motivating different radar designs and signal processing, that the prospects for mutual benefit are slim. Indeed, it is a fact that almost all HF skywave radars are specifically tailored to one or other of these missions[a]. Yet, as we discuss in this chapter, performance of the former task can be enormously enhanced by utilising information provided by the latter; moreover, there are specialised observations where the reverse holds.

The distinctive characteristics of skywave OTHR oblige us to present our analysis in a structured framework that factorises the radar observation process, thereby enabling us to explore how coupling between propagation and scattering mechanisms provides new avenues for signal interpretation and performance enhancement. The natural choice of framework is the radar process model formalism reported in [28, 29] specifically for OTHR investigations; a brief summary is presented in Section 3.1.1. In the subsequent sections, we review the remote sensing and surveillance missions and then proceed to examine some of the ways in which information about the oceanic environment can enhance and extend radar performance. We show that the primary ocean surveillance missions of target detection, location and classification are all refashioned when remote sensing information is exploited, opening new vistas for signal analysis and interpretation. Moreover, to consummate the integration of surveillance and remote sensing missions, new strategies for radar parameter selection and resource allocation must be implemented. One notable consequence of the new regime is that traditional guidelines for frequency and waveform selection must be replaced with more sophisticated criteria.

---

[a]The only exception known to the author is the Jindalee Stage B radar whose design integrated the ship detection requirements with those needed to optimise remote sensing [26, 27].

Beyond the primary missions, the unified approach has spin-off for electronic protection and tactical communications, though we shall not dwell on these. However, we do look briefly at ways in which the radar can exploit remote sensing data to help friendly assets optimise their situational awareness, survivability, disposition and course planning. Finally, we consider the reverse flow of information, where the surveillance output might be used to enhance the quality and diversity of remote sensing products.

3.1 Modelling the radar observation process

The analysis and interpretation of radar echoes is usually predicated on the assumption that we have detailed knowledge of the signal incident on the target and some information about the target's response to that excitation as manifested in the form of the scattered field in the vicinity of the target, with the incident and scattered fields related through the target's polarisation scattering matrix (see e.g. [30]),

$$\vec{E}^{scat} \equiv \begin{bmatrix} E_\alpha^{scat} \\ E_\beta^{scat} \end{bmatrix} = \begin{bmatrix} S_{\alpha\alpha} & S_{\alpha\beta} \\ S_{\beta\alpha} & S_{\beta\beta} \end{bmatrix} \begin{bmatrix} E_\alpha^{inc} \\ E_\beta^{inc} \end{bmatrix} \equiv \tilde{S}(\alpha, \beta)\, \vec{E}^{inc} \tag{3.1}$$

Here $\{\alpha, \beta\}$ denotes an orthonormal basis for the transverse field. The term *target* is interpreted here to include both discrete scatterers and distributed scatterers such as the ocean surface; the conventional radar cross section (RCS) and surface scattering coefficient matrices can each be expressed as

$$\tilde{\sigma}(\alpha, \beta) = \tilde{S}^*(\alpha, \beta) \circ \tilde{S}(\alpha, \beta) \tag{3.2}$$

where $\circ$ denotes the Hadamard product.

With HF skywave radar, we have access to neither $\vec{E}^{inc}$ nor $\vec{E}^{scat}$. Our analysis must proceed on the basis of only the following information:

- exact knowledge of the chosen waveform entering the power amplifier, en route to the transmitting antenna feeders and radiating structure;
- reasonably good models for the directional characteristics of the transmitting and receiving antennas;
- a satisfactory model for the geomagnetic field;
- equations that describe the skywave propagation of radio waves for specified atmospheric and ionospheric conditions;
- approximate knowledge of the ionosphere electron density structure and hence of the general form of the propagation paths, though not the time-dependent amplitude, phase and polarisation modulations they impose on transiting signals;
- an expression for the Doppler spectrum of radio waves scattered from a specified ocean surface as represented by a directional wave spectrum, subject to certain limits on its validity;
- the radar output in the form of digital times series data delivered to the signal processor; ideally these data sample the radio wave field incident on the

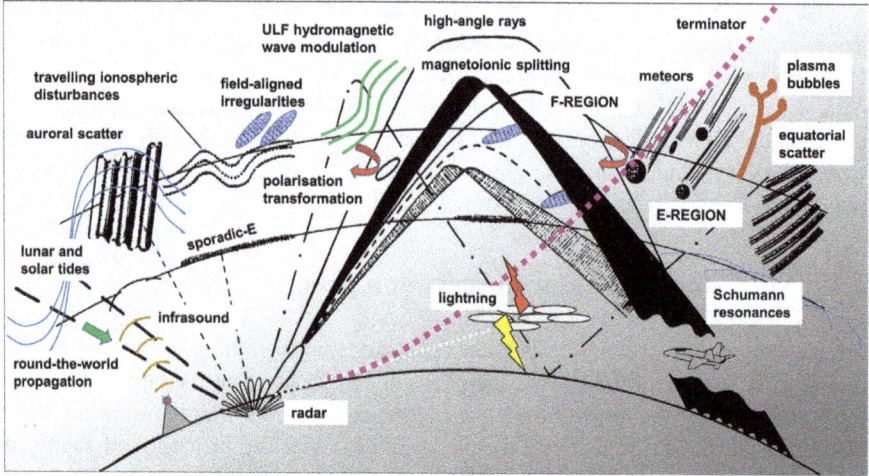

Figure 3.1   A schematic representation of the skywave propagation environment

receiving antenna but, in reality, they are also subject to time-dependent gain and phase changes accumulated during passage through the antenna structure, cabling and any analog stages of the receiver prior to digitisation.

The relative severity of the gaps in our knowledge may vary from one situation to another but, almost always, the greatest uncertainties arise from the complexity of skywave propagation, where reliance on the ionosphere exposes the radar to a host of geophysical phenomena.

Figure 3.1 illustrates some of the effects and disturbances that complicate life for skywave OTHR. This environment, along with scattering mechanisms and noise distributions that differ substantially from those encountered with microwave radar, forces us to abandon the conventional radar equation as a modelling tool; instead we employ a process model that can be adapted to incorporate as much or as little of the physical phenomenology as a given task might require.

3.1.1   The radar process model

The mathematical description of the radar observation process can be formulated as a concatenation of operators that represent the progressive transformation of the selected waveform into an electromagnetic wave that undergoes propagation and scattering before being acquired by a receiving complex where it is converted into a digital signal form, finally undergoing processing aimed at extracting the desired information.

$$s = \sum_{n_B=1}^{N} \tilde{R} \left[\prod_{j=1}^{n_B} \tilde{M}_{s(j)}^{s(j+1)} \tilde{S}(j) \right] \tilde{M}_T^{s(1)} \tilde{T} w$$

$$+ \sum_{l=1}^{N_j} \sum_{m_B=1}^{M} \tilde{R} \left[\prod_{k=1}^{m_B} \tilde{M}_{s(j)}^{s(j+1)} \tilde{S}(j) \right] \tilde{M}_N^{s(1)} n_l \; + \; m \tag{3.3}$$

where

w represents the selected waveform,

$\tilde{T}$ represents the transmitting complex, including amplifiers and antennas,

$\tilde{M}_T^{s(1)}$ represents the propagation from transmitter to the first ground scattering zone,

$\tilde{S}(j)$ represents all the scattering processes in the jth scattering zone,

$\tilde{M}_{s(j)}^{s(j+1)}$ represents propagation from jth scattering region to the $(j+1)$th zone,

$S(n_B)$ and $S(m_B)$ represent the receiver location,

N_j represents the number of external noise sources, interferers or jammers,

$\tilde{M}_N^{s(1)}$ represents propagation from a noise source to its first ground scattering region,

n_B, m_B represent the number of bounces (hops),

N, M represent the upper bounds on the number of bounces,

m represents the internal noise,

N_j represents the number of jammers/interferers,

$\tilde{R}$ represents the receiving complex, including antennas and receivers,

s represents the signal delivered to the processing stage.

In order to implement a computer model of the observation process, so that signal analysis and interpretation can proceed, each stage of the chain of events, i.e. each operator, is assumed to be characterised by a set of physical parameters, some known exactly, some known approximately and some unknown. The goal of the radar observation is to estimate the parameters that specify the physical phenomena of interest via their contributions to $\tilde{S}$, such as the presence or absence of an object (detection), its geographical position and altitude (location) and the nature of the object (classification or geophysical parameter measurement). Not surprisingly, superior radar performance can be achieved when parameters are estimated jointly, not individually.

The extent to which these goals can be accomplished depends on both the environmental conditions and the fidelity of the physical models employed by the radar. Accordingly, the first step in optimising skywave radar performance is to minimise the uncertainty in our knowledge of the prevailing values of the parameters that define the observation process. We achieve this by calibration.

3.1.2   *Calibration*

In the present context, calibration refers to procedures invoked to ensure that radar measurements are not corrupted through the assignment of erroneous values to the parameters that specify the radar process model. Calibration tasks that are common to many types of radar include:

- minimising corruption of the radiated waveform during transmission by sampling and pre-distortion;
- refining models of the absolute gain and phase patterns for transmit and receive antennas;
- estimation of the receiving antenna array manifold in the presence of environmental coupling effects and element placement errors;
- quantification of the coupling between nominally orthogonal channels in polarimetric radars;
- estimation of gain and phase errors that accumulate through the receiver chain from the base of each antenna, through the associated cabling and receivers, up to analog-to-digital conversion prior to signal processing.

These tend to be essentially constant or very slowly varying relative to the radar coherent integration times and the timescales of variations in the dynamical behaviour of targets and environment; so, in many radars they are performed periodically as part of system maintenance or on an opportunity basis. The exception is the receiver chain calibration, which is obviously frequency-dependent, so it may be measured for each waveform/frequency combination or interpolated from a look-up table that is updated according to some schedule.

For these quasistatic tasks, radar-derived remote sensing information might seem irrelevant to OTHR, but that is not quite the case. As an example, remote sensing yields the estimates of the spatially varying scattering coefficient of the sea surface, which can be used under very plausible assumptions to provide information on the far-field antenna gain product $G_T G_R \cong \tilde{T}^H \tilde{R}^H \tilde{R} \tilde{T}$, a difficult quantity to measure by traditional techniques as the near-field may extend to more than 1 000 km. But, by and large, these quasistatic calibration tasks derive no great benefit from access to remote sensing measurements.

However, given the composite structure of the radar process model, it is not unreasonable to extend our concept of calibration from the aforementioned hardware stages, represented by $\tilde{T}$ and $\tilde{R}$, to include the determination of the dynamic parameters of $\tilde{M}$ that define the propagation channel. We know that these can be highly dynamic, necessitating correction or demodulation on timescales commensurate with the signal coherent integration time. A number of these parameters play central roles in OTHR, and significant resources are assigned to their determination. Perhaps the best-known dynamic calibration operations are those that deal with:

- coordinate registration
- time-varying channel propagation loss
- ionospheric Doppler shift correction
- time-varying channel phase path desmearing
- correcting for wavefront distortion
- optimising receiver channel gain in the presence of non-linearity.

In carrying out these tasks, remote sensing information can be of substantial value, both for basic first-order correction techniques and for more sophisticated algorithms that make use of higher order effects. We shall return to this subject later in the chapter.

3.1.3   Sea clutter modelling I: the direct problem

For the present analysis, we need not explore the technical details associated with radar scattering from discrete targets, but we do need to review the mechanism of scattering from the sea surface as it plays a central role in what follows. There are two widely used theoretical models that describe HF scattering from the ocean surface. Both take the form of perturbation expansions, one, originally developed by Barrick [31], building on the Rice theory for scattering from static rough surfaces [32], and one developed by Walsh, Srivastava, Gill and colleagues (see, e.g. [33–35]) based on the Walsh theory for scattering from static surfaces [36]. Quite a few generalisations of each of these have appeared over the years [37] but only those based on the Barrick theory have been developed for skywave OTHR applications. In both approaches, the solution for the Doppler (power) spectral density of the scattered field takes the form of an expansion in orders of surface roughness, with only even powers represented explicitly so it is expressible in terms of the directional ocean wave spectrum. The modelling used in this chapter employs the model reported in [38] which handles arbitrary bistatic geometries and polarisation states.

The general expression for the Doppler spectrum of the scattered field in $(\vec{k}, \omega, \pi)$ space, to second order, then takes the form

$$
\begin{aligned}
D(\vec{k}, \omega, \pi) = &\, F_0(\vec{k}_{inc}, \omega_0, \pi_0, \pi)\delta(\vec{k}_{scat} - \vec{k}_{inc} + 2\vec{k}_{inc} \cdot \hat{n}\,\hat{n})\delta(\omega - \omega_0) \\
&+ \int d\vec{\kappa}_1\, F_1(\vec{k}_{scat}, \pi, \vec{k}_{inc}, \pi_0, \vec{\kappa}_1) S(\vec{\kappa}_1)\delta(\vec{k}_{scat} - \vec{k}_{inc} + \vec{\kappa}_1)\delta(\omega - \omega_0 + \Omega(\vec{\kappa}_1)) \\
&+ \iint d\vec{\kappa}_1\, d\vec{\kappa}_2\, F_2(\vec{k}_{scat}, \pi, \vec{k}_{inc}, \pi_0, \vec{\kappa}_1, \vec{\kappa}_2) S(\vec{\kappa}_1) S(\vec{\kappa}_2)\delta(\vec{k}_{scat} - \vec{k}_{inc} + \vec{\kappa}_1 + \vec{\kappa}_2) \\
&\quad\times \delta(\omega - \omega_0 + \Omega(\vec{\kappa}_2) + \Omega(\vec{\kappa}_1))
\end{aligned}
\tag{3.4}
$$

In this equation, the wave vectors of the incident and scattered radio waves are denoted by $\vec{k}_{inc}$ and $\vec{k}_{scat}$, with angular frequencies and polarisations (ω_0, π_0) and (ω, π) respectively. $S(\vec{\kappa})$ represents the sea directional wave spectrum, while the dispersion relation appears as $\Omega(\vec{\kappa})$; it is this function that determines the contours of integration that yield the Doppler power spectral density. Expressions for the kernel functions F_1 and F_2 can be found in the cited literature; the Fresnel reflection coefficient F_0 is a function of the water temperature and salinity.

The art of ocean surveillance with OTHR, for it is an art as well as a science, cannot be mastered without a deep understanding of the implications of this relationship.

3.1.4   Sea clutter modelling II: the inverse problem

For some purposes it suffices to be able to model the clutter Doppler spectrum for a given directional wave spectrum but, for many remote sensing tasks, the goal is to invert the relationship, retrieving the directional wave spectrum from the radar measurements. This problem is not so easily accomplished – in practice it takes the form of a non-linear Fredholm integral equation of the first kind, derived from (3.4). Equations of this class are notoriously difficult to solve. Even in the linear case, they are, in general, ill-posed; in particular, they suffer from numerical instability, as well as non-uniqueness. Thus, in order to make progress, researchers have resorted to various approximations and parameterisations.

A categorisation of possible approaches was described in [39]. The earliest attempt, by Lipa [40] in 1977, assumed separability of the wave spectrum, that is, the angular distribution was assumed independent of wavenumber. Subsequent approaches by Barrick and Lipa [41] and Howell and Walsh [42] used low-rank parameterisations of the angular distribution, while Wyatt used physical arguments to simplify the integral and render it amenable to solution by regularisation [43]. A parametric approach was adopted by Hisaki [44], and also by Hashimoto and Tokuda [45], combined with a conventional discretisation scheme. Some of the limitations of these techniques were discussed in [46, 47].

All the inversion work referred to above addresses the case of monostatic, HFSWR radars and carried the analysis to second order. Recently the bistatic HFSWR case was considered by Hardman *et al.* [48]. Inversion of quasi-monostatic skywave radar spectra has been carried out since the 1980s, initially by fitting a seven-parameter wave spectrum model to the radar data [16]; later a more sophisticated approach was developed though no details have been released.

3.2   Characteristics of OTHR radar missions

As in almost all radar systems, the detection and subsequent tracking of a discrete target at HF frequencies require the amplitude of the target echo to exceed that of the sum from competing sources of energy – clutter, external noise, interference, jamming and internal noise – as observed in the signal analysis domain. Typically, that domain consists of some subset of the attributes azimuth, elevation, range, Doppler and polarisation, though other parametrisations exist, and it is the goal of the radar operator to maximise the signal-to-(interference+noise+clutter) ratio, the SINCR. The standard tools for this task are: (i) wide-sweep sounders, which deliver maps of received signal power as a function of range (strictly speaking, group range or delay) and radar frequency, and (ii) narrow-sweep sounders which trade the radar frequency dimension for Doppler frequency, thus emulating the radar itself. Optionally the maps are normalised by the received noise power at each frequency, yielding the sub-clutter visibility (SCV) which is more important for many practical purposes [49].

For aircraft, whose velocities typically yield Doppler shifts exceeding the frequency band occupied by sea or land clutter, target echoes compete most of the time against the external noise background, so conventional thinking suggests that the detailed spectral characteristics of the clutter are largely irrelevant and SNR is the metric of concern. (As we shall see, this is a flawed argument.) Nevertheless, it is standard practice to perform the aircraft detection mission by following two steps:

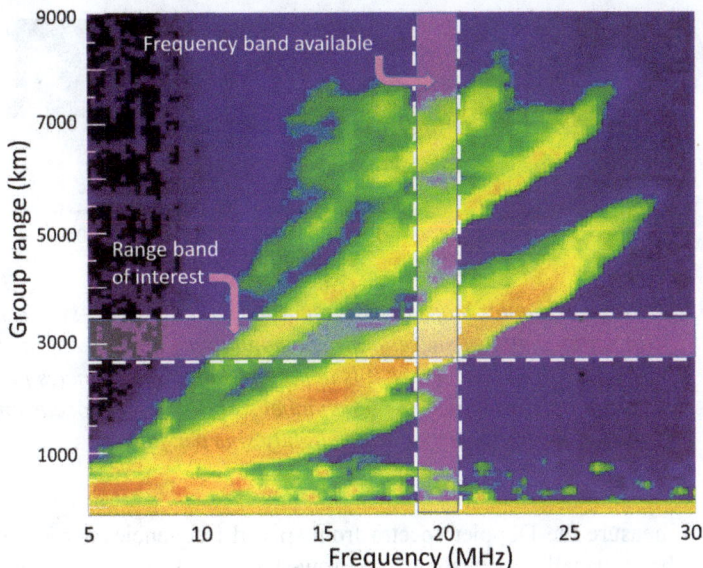

Figure 3.2   *A backscatter ionogram showing the distribution of echo power as a function of radar frequency and group range. The logic governing conventional frequency selection is indicated by the shaded regions.*

i.   Examine a wide-sweep backscatter ionogram and find the broad band of frequencies that deliver strong returns from the range extent of interest. An example of a wide-sweep backscatter ionogram is shown in Figure 3.2.

ii.   Check spectral occupancy in that frequency band and select a clear channel of bandwidth adequate to accommodate the chosen waveform without interfering with any other users of the HF spectrum.

In the case of ship detection, it is almost always sea clutter that constitutes the dominant component of the background in which the target echoes are immersed, so the form of the Doppler spectrum is of central importance insofar as it determines the signal-to-clutter ratio (SCR). The dependence of that Doppler signature on radar frequency, waveform and scattering geometry can be pronounced, complicating frequency selection even under benign conditions, but the observed clutter spectrum is also distorted by ionospheric propagation effects whose obscuring power can dominate the intrinsic spectral masking. Standard practice here involves the following three steps:

i.   examine a wide-sweep backscatter ionogram and locate the full band of frequencies that deliver a reasonably strong return from the range band of interest;

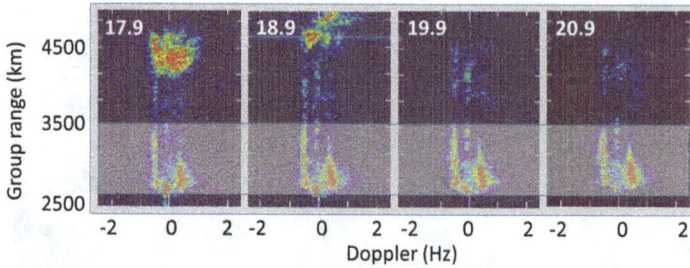

Figure 3.3   *Range–Doppler maps of SCV recorded with a narrow-sweep sounder in rapid temporal succession. The nominal range band of interest is shaded; from these maps it is apparent that all frequencies provide reasonable illumination and channel stability on this occasion, with the lowest frequency offering slightly sharper focus.*

  ii.   measure the Doppler spectra from spaced frequencies across that band, usually by means of a narrow-sweep backscatter sounder or 'mini-radar', and select the frequency that shows minimal Doppler smearing as manifested by the first-order Bragg peaks. An example of a family of narrow-sweep backscatter soundings is shown in Figure 3.3;

 iii.   check spectral occupancy around that frequency and select a clear channel of bandwidth adequate to accommodate the chosen waveform without interfering with any other users of the HF spectrum.

The situation for remote sensing of the ocean surface is different again. Here the desired information is encoded in the detailed global form of the Doppler spectrum, so priority in frequency selection goes to jointly maximising (i) the sensitivity of the intrinsic spectrum to the particular environmental parameter of interest, and (ii) the quality of the actual echo that has been subjected to the distortion of the prevailing ionosphere. As of today, there is no standard procedure for remote sensing with skywave radar, but the principles on which one should be based are reasonably well understood and new tools are being developed to facilitate implementation. (That is not to say that experienced operators cannot achieve high levels of performance today, but the automation of the procedure is a work in progress.)

Another important distinction between these radar missions lies in their respective demands on radar resources. Despite the vast coverage afforded by skywave propagation, the sheer number of candidate surveillance tasks at any given time means that there is severe pressure on the time line of the radar. Aircraft detection is normally undertaken with coherent integration times (CITs) of order 10^0 s, whereas ship detection missions require CITs of order 10^1–10^2 s. Remote sensing tasks vary in their demands but can exceed the upper end of this range. The requirements of long coherent integration times for ship detection and remote sensing conflict with the need for high revisit rates on aircraft targets, leading to awkward trade-offs [12].

3.3 Remote sensing information for enhanced surveillance

It should come as no surprise to learn that access to detailed, synoptic, real-time, remote sensing information about waves and currents can contribute some measure of improved performance to the primary surveillance missions, especially ship detection where the coupling between target and environment is so direct. What is far less obvious is the fact that clever exploitation of radar-derived remote sensing information can open the door to entirely new radar capabilities. In this section, we shall describe a number of topics where remotely sensed environmental information enhances radar surveillance capabilities, grouping the applications under the headings Detection, Location, Classification, Resource Management and Tactical Intelligence.

3.3.1 Detection

3.3.1.1 The problem of kinematic blind speeds

The motion of a ship or aircraft imposes a Doppler shift that varies with scattering geometry:

$$\omega_D = (\vec{k}_{\text{scat}} - \vec{k}_{\text{inc}}) \cdot \vec{V} \tag{3.5}$$

where $\vec{V}$ is the platform velocity with $\vec{k}_{\text{inc}}$ and $\vec{k}_{\text{scat}}$ as before. The sea clutter background against which a ship is to be detected has the characteristic form described earlier, with a complex dependence on radar frequency and sea state. For example, the two dominant peaks associated with first-order Bragg resonance possess Doppler shifts that vary with the square root of the frequency. The main operational consequence of this behaviour as far as detection is concerned is that the SCR of the ship echo varies strongly across the Doppler domain. As successful detection requires that the SCR exceeds some threshold, the Doppler domain is partitioned into a number of regions, in each of which the radar is either blind or able to detect the target. In the 1970s, Barnum [50] modelled this effect using a simple parametric model to represent the sea state and assuming a target RCS independent of aspect; variations on this approach have been used by others to ensure radar design adequacy.

A modern implementation is illustrated in Figures 3.4 and 3.5 which overlay high fidelity computations of the RCS of a representative vessel – the Fremantle Class Patrol Boat (see Figure 3.6) – on the sea clutter spectrum computed for a representative OTHR at a range of 2 000 km. Here we have shown the results for vertically polarised incident and scattered fields. Results are presented for four radar frequencies (4, 8, 16 and 24 MHz) and for two ship speeds – cruising (12 kn) and fast (24 kn) so as to sample a full operational envelope. The RCS curve has been evaluated over a full 360° at the indicated speed; as the figures show, the vessel's RCS is almost symmetric about its centre plane except for one or two minor departures. In each case treated, there are blind speed bands where the SCR detectability condition is not met. An important feature of these plots is that the abscissa is radial velocity, not Doppler frequency, making it easy to gain an intuitive feel for the implications.

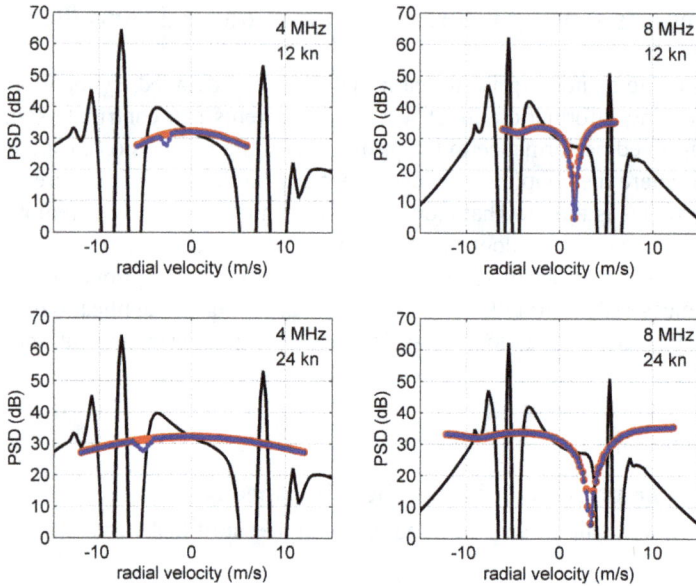

Figure 3.4   Doppler spectra and RCS traces for a Fremantle Class Patrol Boat at 12 kn and 24 kn as a function of radial velocity (relative to the radar direction). (a) and (c) show the results for a radar frequency of 4 MHz, (b) and (d) for 8 MHz.

Clearly it would be enormously helpful if the radar operator were equipped with knowledge of the blind speed bands for targets of interest over the region being surveyed and hence able to choose a radar frequency (or set of frequencies) that maximises SCR over the Doppler bands of interest.

It might be thought that one would need to measure the sea clutter at many frequencies to explore this variation and thereby find by trial and error the best frequency for detection. This is not the case. If we employ one of the inversion techniques developed for remote sensing, the data collected at a single frequency yield an estimate of the sea directional wave spectrum, to which the scattering theory can then be applied to compute the spectrum that would be observed at each of any number of other radar frequencies.

The situation is not always straightforward. The deep nulls in the clutter Doppler spectrum afford narrow windows of enhanced detectability, but one can control the position of these windows only by varying the radar frequency, which brings propagation factors into play. This can lead to highly non-intuitive guidelines for frequency selection [51].

3.3.1.2   Detection via multiple scatter

When modelling the scattering of radio waves by targets of interest, it is all too easy to overlook the electromagnetic coupling between targets and their environment. A

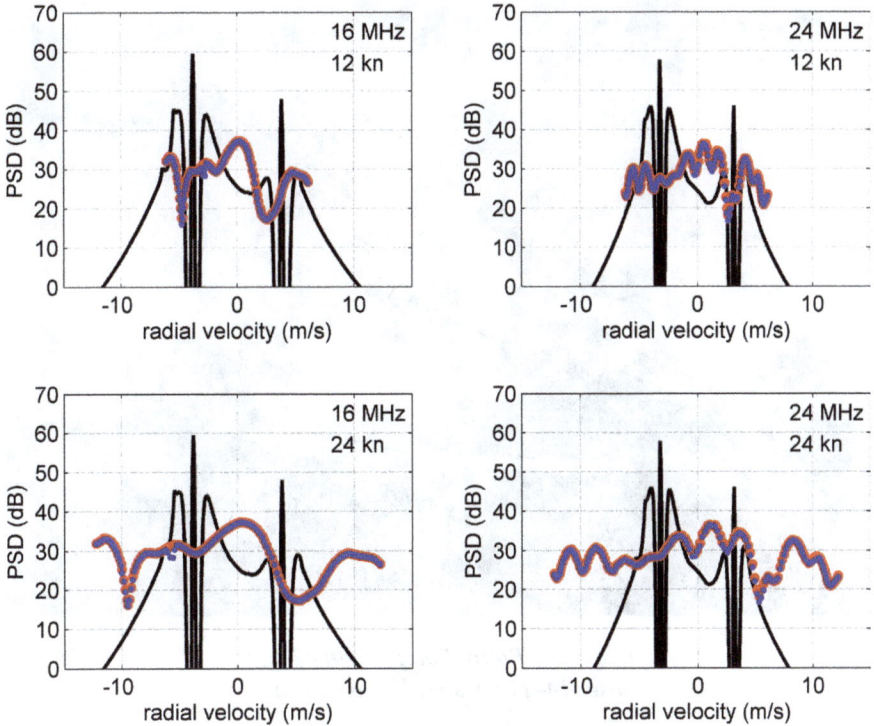

Figure 3.5   *Doppler spectra for a Fremantle Class Patrol Boat at 12 kn and 24
              kn as a function of radial velocity (relative to the radar direction).
              (a) and (c) show the results for a radar frequency of 16 MHz, (b) and
              (d) for 24 MHz.*

ship at sea is embedded in a time-varying ocean wave field that modulates the electromagnetic field incident on the ship. Similarly, an aircraft at altitude is illuminated
by both a descending skywave signal and a diffuse field of waves reflected upwards
from the ocean surface. In each instance, the scattered field can return to the radar
via both direct and ocean surface reflection paths. Thus the Doppler structure of the
composite echo at the radar receiver bears the imprint of the prevailing sea surface
modulations. As we shall see in a later section, this mechanism has a number of
valuable attributes, providing uniquely powerful insights into target parameters. In
the context of detection, we need only point out that even if the direct echo from the
target is buried under strong clutter, the spread Doppler of the diffuse multiple scatter contributions from around the target can extend well beyond the clutter band and
hence support detection. This mechanism is potentially useful in the case of a smart
opponent wishing to avoid detection by choosing a course–speed combination that
places the platform direct skin echo inside the clutter peak.

Of course, the magnitude of the multiply scattered field may be tens of dB
lower than that received via the direct–direct signal path; this may explain why little

*Figure 3.6   A Fremantle Class Patrol Boat; length 42 m, beam 7 m, draught 2 m,
          speed (cruise) 14–17 kn, speed (max) 30 kn*

attention has been paid to this sensing modality. Yet, as technology advances, this excuse wears thin. Figure 3.7 shows the Doppler spectrum of a waveform transmitted from a ship at sea to a shore-based receiver. On this occasion the sea was extremely calm, Beaufort scale 1 conditions, but even here the multiple scattering contributions are pronounced. Two spectra from the same range (time delay) cell are superimposed on the common axis, one for a moving ship, one for a stationary ship. The scattering theory enables us to predict the former from the latter, given the ship course and speed. To obtain the two-way (radar observation) case, one need only convolve the spectrum with itself. Moreover, given just the data, one can in principle invert the echo to retrieve the course and speed of the ship without needing to establish a track.

In the case of an aircraft flying over the ocean, the skywave radar echo includes multiple scatter contributions from everywhere on the sea surface within the aircraft visual horizon [52]. This information provides many unique benefits, some discussed later in the chapter; all of them are dependent on having access to the output of the remote sensing mission.

3.3.1.3   Detection exploiting non-stationarity of the ocean wave field
In predicting the performance of HF radars tasked with ship detection, it is common practice to employ a spectral description of the expected sea conditions and then apply the relevant HF scattering theory to compute the clutter background against

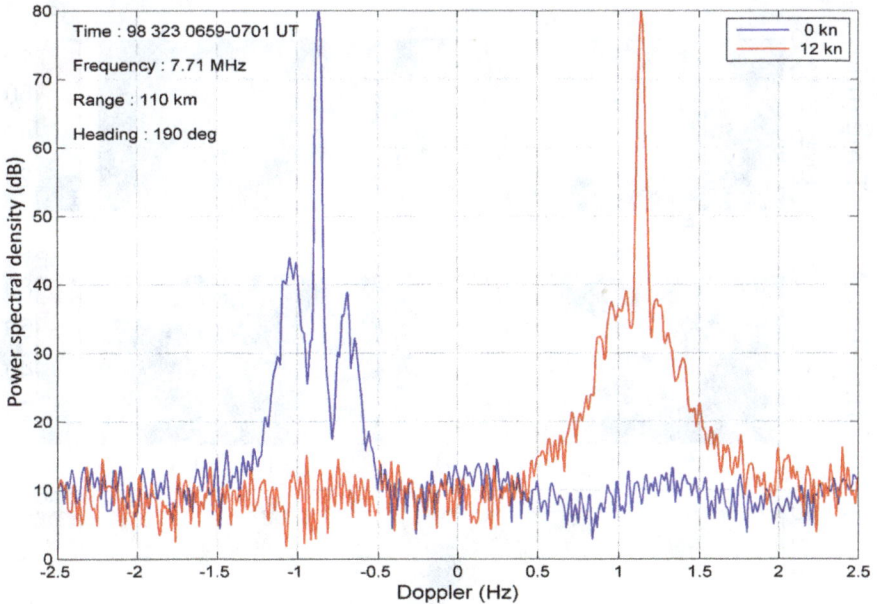

Figure 3.7   Doppler-spread clutter resulting from multiple scatter on the one-way path from a shipborne transmitter to a shore-based radar. The prevailing sea conditions determine the shape of the Doppler sidebands. For clarity, the two spectra – stationary ship and moving ship – have been displaced in Doppler.

which detection is to be essayed, as described in the discussion earlier on blind speeds. This is typically approached through the adoption of a low-rank parametric wave spectrum model such as the Pierson-Moskowitz, JONSWAP or Elfouhaily form, along with an assumed directional spreading function. To the extent that temporal variation is accommodated, it is generally accepted that the evolution is adiabatic, that is, the state evolves on the manifold defined by the adopted parametric model.

Theoretical and experimental studies of the temporal variation of the wave spectrum have shown that, under sufficiently rapidly changing wind stress, different components of the wave spectrum adjust at different rates, skewing the directional spectrum. In situ measurements have revealed that such changes are more the rule than the exception. This has motivated the modelling of the HF sea clutter spectrum evolution for non-adiabatic states [53]. The results reveal that there are very significant implications for ship detection, including gains or losses of the order of 5–10 dB in SCR. An example illustrating these variations is presented in Figure 3.8.

The relevance of this finding to predicting ship detection performance in a statistical sense is obvious; such calculations are essential for radar design and siting. Of even greater potential value is the exploitation of the remote sensing functionality

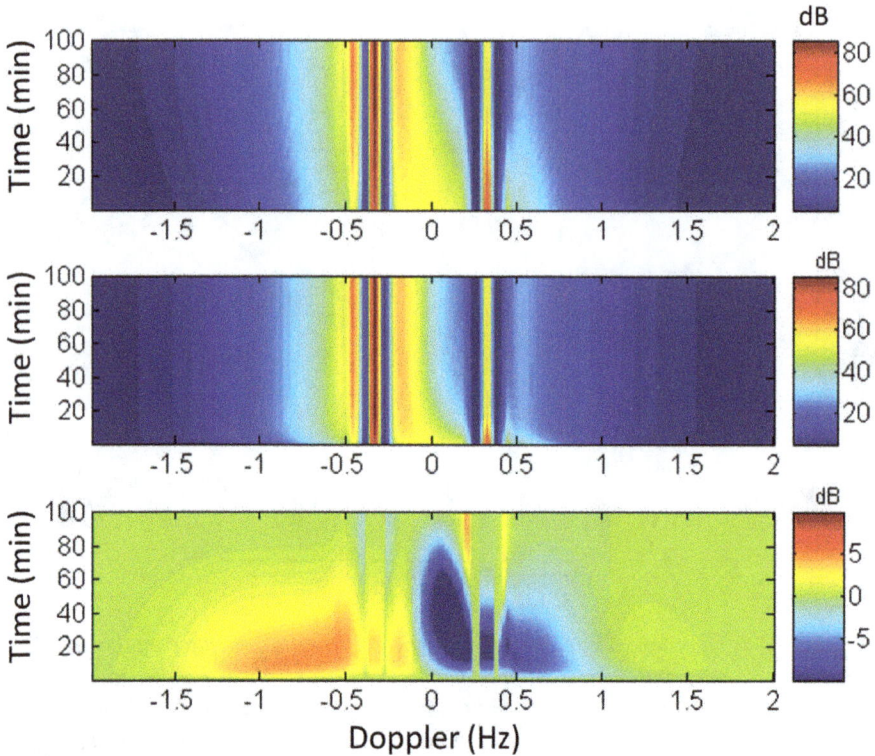

Figure 3.8   Modelled Doppler spectra from a patch of sea experiencing a change in wind direction of 120°, occurring over a span of 100 minutes. The top panel shows the spectra that would result under the assumption of adiabaticity, the middle panel when non-adiabatic behaviour follows a law derived from wave staff array measurements, and the bottom panel, the difference between the two.

of OTHR to identify windows of enhanced ship detectability in real time, and couple these opportunities into the radar mission scheduling process.

3.3.1.4   Detection exploiting modulational instability of ocean waves

The classical model of the ocean surface structure is that of a superposition of weakly interacting Stokes waves. This model was formulated in 1847 and survived for over a century before it was found that such waves are subject to a modulation sideband instability [54]. Traditional HF radar modelling assumes the Stokes wave description, as the instability was long regarded as an exotic phenomenon, observed in the laboratory under highly constrained conditions but never observed in the open sea. A unique experiment in 1995 [29] showed that the instability can indeed be observed with HF radar when the wave field angular spread is narrow. The results revealed that the waves responsible for the strong Bragg peaks in sea clutter undergo

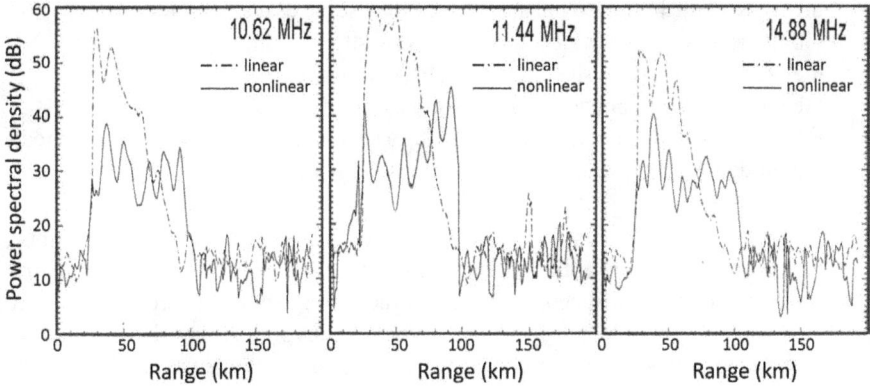

Figure 3.9   Measurements of Fermi-Pasta-Ulam recurrence associated with modulational instability of surface gravity waves

amplitude modulation of about 10 dB, exactly in accord with the corresponding theory, and moreover, that the associated phenomenon of Fermi-Pasta-Ulam occurrence could be observed over many cycles. Figure 3.9 shows several examples. In these images, the time-averaged echo power is shown by a dashed line, indicating the expected decrease with range from the radar. The solid line plots the intensity of the Bragg-resonant second harmonic of the Stokes waves of half the wavenumber of those responsible for the Bragg peaks, normalised by the intensity of the latter. Thus the solid line is an indicator of the power spectral density of those longer waves, revealing that their Bragg peaks are experiencing modulation.

What it means for ship detection is that, when such conditions prevail, spatial patterns of enhanced detectability are forming and moving across the sea surface, offering the prospect of exploitation by astute radar frequency management informed by remote sensing observations. Again, the practical exploitation of these windows of opportunity relies on real-time access to the remote sensing products of the radar.

3.3.1.5   Detection of ship and submarine wakes

The distinctive wave pattern set up in the wake of a moving ship has long been considered as a potential avenue for detection by OTH radars operating in the HF band. Although intuition suggests that the skin echo from a ship is likely to be orders of magnitude larger than that from the wake, interest in the prospect of wake echo detection is driven by quite a number of considerations:

- the assumption that the skin echo dominates may not hold, especially if the ship adopts low-observable technologies;
- the ship may constrain its velocity vector so as to conceal its conventional skin echo under the sea clutter peaks;
- the ship may be a submarine, in which case there will be no skin echo;
- the wake echo may contain information that assists target classification;

- the wake contribution to the sea surface may serve as a known input signal providing access to the ambient sea state via multiple scatter;
- the wake echo can resolve the left-right wave directional ambiguity;
- the environmental impact of the wake may be of interest;
- HF radar may offer unique advantages as a tool for confirming marine engineering models employed in ship design.

Early experiments focused on resonant scattering from the Kelvin wake alone, but the limitations of this approach are apparent – the need to satisfy a geometric constraint involving ship speed, bearing and radar frequency. Accordingly, a more sophisticated scattering model has been developed, involving both the wake structure and the ambient sea [55]. Under this model, we can compute the received signal from a double scatter process in which the radar signal experiences two consecutive 'bounces', one from the wake and one from the ocean waves.

The extension of the expression (3.4) for the sea clutter Doppler spectrum to predict the spectrum in the presence of a ship wake is straightforward. As the wake components are statistically independent of the ambient wave field, at least to second order in the perturbation parameter, the power spectral density of the composite surface can be written as the sum of the individual power spectral densities

$$S(\vec{\kappa}) = S_{wave}(\vec{\kappa}) + S_{wake}(\vec{\kappa}) \tag{3.6}$$

Substituting this expression into (3.4) yields

$$
\begin{aligned}
\tilde{D}(\vec{k}_{scat}, \vec{k}_{inc}; \omega) = & \\
& \int d\vec{\kappa}_1 F_1(\vec{k}_{scat}, \vec{k}_{inc}, \vec{\kappa}_1) \left[S_{wave}(\vec{\kappa}_1) \right] \\
& + \int d\vec{\kappa}_1 F_1(\vec{k}_{scat}, \vec{k}_{inc}, \vec{\kappa}_1) \left[S_{wake}(\vec{\kappa}_1) \right] \\
& + \iint d\vec{\kappa}_1 d\vec{\kappa}_2 F_2(\vec{k}_{scat}, \vec{k}_{inc}, \vec{\kappa}_1, \vec{\kappa}_2) \left[S_{wave}(\vec{\kappa}_1) S_{wave}(\vec{\kappa}_2) \right] \\
& + \iint d\vec{\kappa}_1 d\vec{\kappa}_2 F_2(\vec{k}_{scat}, \vec{k}_{inc}, \vec{\kappa}_1, \vec{\kappa}_2) \left[S_{wake}(\vec{\kappa}_1) S_{wake}(\vec{\kappa}_2) \right] \\
& + \iint d\vec{\kappa}_1 d\vec{\kappa}_2 F_2(\vec{k}_{scat}, \vec{k}_{inc}, \vec{\kappa}_1, \vec{\kappa}_2) \left[S_{wave}(\vec{\kappa}_1) S_{wake}(\vec{\kappa}_2) \right] \\
& + \iint d\vec{\kappa}_1 d\vec{\kappa}_2 F_2(\vec{k}_{scat}, \vec{k}_{inc}, \vec{\kappa}_1, \vec{\kappa}_2) \left[S_{wake}(\vec{\kappa}_1) S_{wave}(\vec{\kappa}_2) \right]
\end{aligned}
\tag{3.7}
$$

where we have suppressed the common polarisation arguments for readability. The terms on the right hand side have simple interpretations. The first represents the familiar first-order Bragg scatter, while the second yields the contribution of any wake energy which happens to have precisely the right wavelength and direction of travel to satisfy the Bragg resonance. Next we have the wave–wave term, which is the familiar second-order sea clutter spectrum of (3.4), the basis for most remote sensing applications of HF radar and the distribution that sets the clutter threshold for ship skin echo detection. Following that, we have a term describing double scatter from the wake. As the wake spectral components are obliged to lie on (or very close to) the characteristic curve, it is not generally possible to satisfy the Phillips resonance constraints when only a single ship wake is present, so this term makes no contribution. The final two terms, which are mathematically equivalent, describe wave-wake scattering, and it is these contributions that describe the ship wake echo. Accordingly, the solution for the wake Doppler spectrum is a linear integral equation in the unknown, $S_{wave}(\vec{\kappa})$, where the kernel includes the a priori unknown directional wave spectrum of the ambient sea as a factor. But, as we know, the estimation of the directional wave spectrum of the prevailing sea is the central task of the remote sensing mission of HF radars. Thus by monitoring the directional wave spectrum via remote sensing techniques, the radar has all the information it needs to construct a matched filter for wake detection.

As with the multiple scatter mechanism described earlier, the resultant Doppler spectrum will be distributed across the Doppler domain, and may exceed the detection threshold over parts of that domain. Figure 3.10 shows an instance of this for a merchant ship. The ship course and speed are unchanged throughout, but the wind direction is stepped over 120° in 60° increments. The clutter spectrum, shown in black, changes as expected, but so does the wake signature (shown in red) as it is determined jointly by the ship parameters and the sea conditions.

Thus, by performing the remote sensing mission and obtaining the directional wave spectrum, one can achieve quite a number of important surveillance goals.

3.3.1.6   External noise rejection and skywave signal desmearing

The global propagation of HF radio waves from a host of natural or man-made sources delivers a spatially and temporally non-uniform additive noise field incident onto the receiving array. An important task of the signal processing is to achieve high suppression of the unwanted additive signal components without simultaneously corrupting the desired information contained in the echoes from targets and the ocean surface. This can be dealt with quite effectively by means of space–time adaptive signal processing (STAP) techniques [56–58], provided that sufficient data are available with which to construct the STAP filter. Figure 3.11 illustrates the efficacy of this approach. That proviso is not always met, so the question arises, how might one make best use of the available radar returns, given the clutter field inhomogeneity? One approach is to exploit contemporaneous remote sensing information generated by the radar. We recall that the radar echoes from first-order Bragg-resonant waves provide an immediate map – the familiar 'wind direction map' – of the directional distribution of one spectral component, which can reasonably be taken as indicative of ocean wave field spatial homogeneity. As illustrated by the inferred wind map example shown in Figure 3.12, a range of spatial scales can be observed.

This information can enable optimum partitioning of the range–azimuth space for the estimation of the spatial covariance functions used to construct the filter. It might seem that detailed knowledge of the 'clutter subspace' is irrelevant to estimation of the additive noise field to be suppressed, but it is not always possible to find a clutter-free region that can be identified as the noise subspace. The reason is that,

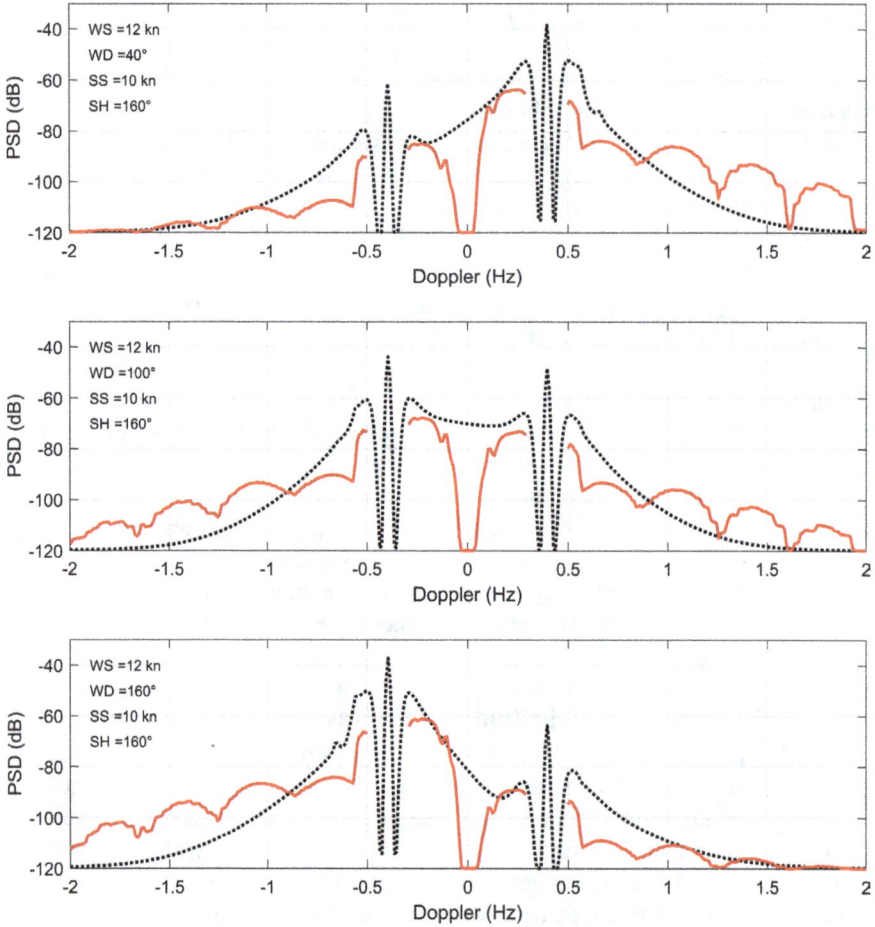

Figure 3.10   Wake spectra for a merchant ship travelling at 10 kn. The ship heading is constant but the wind direction steps from 40° to 160°. The black line shows the sea clutter spectrum and the red line shows the wake echo spectrum.

for high dynamic range radars, the sea clutter Doppler support can extend to many times the Bragg frequency, thus occupying a lot of the domain needed to mine for samples with which to estimate the noise covariance matrix. In general, the spread of the clutter energy is asymmetric, being highly dependent on the sea state, so ideally the choice of domain for noise field estimation should adjust accordingly. On those occasions when this consideration is relevant, we might also note that knowledge of the directional wave spectrum can serve in the corresponding role to specify temporal sample support because the coherence of the ocean wave field is limited by non-linear wave interactions, themselves dependent on the full directional wave spectrum.

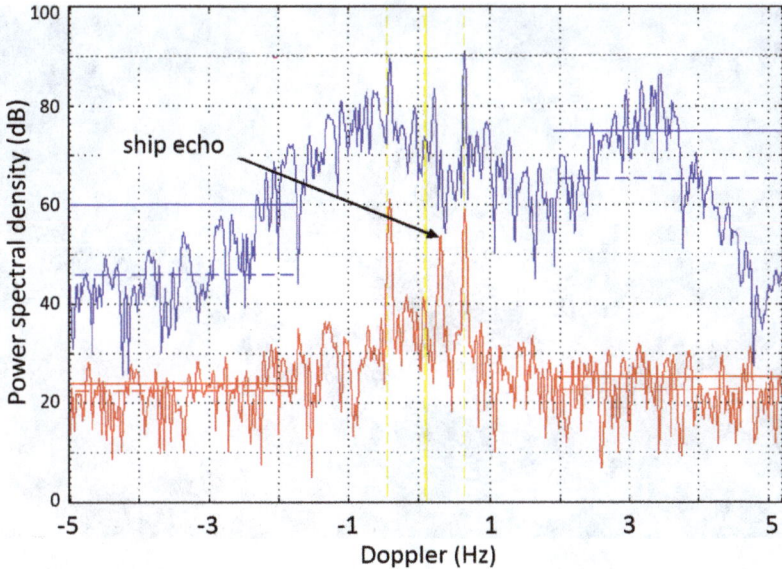

Figure 3.11   *An example of effective STAP rejection of additive time-varying noise achieved by achieving adequate sample support. The noise-corrupted spectrum is displaced by 30 dB along the vertical axis for clarity.*

Another class of skywave signal corruption arises from multiplicative distortion, a common form of which is modulation of the ionospheric plasma by magnetohydrodynamic waves from above, transferring the field pulsations onto the radar signal, described in Figure 3.13.

When applying a desmearing technique to estimate and remove phase fluctuations arising from ionospheric modulation, knowledge of the ocean directional wave spectrum can inform the choice of signal subspace, based on the expected rank of the autoregressive model used to represent it. Combining the desmearing methodology with STAP rejection of additive interference can yield striking benefits [59].

3.3.2   Location

The mapping of the standard radar measurement coordinates – time delay, direction of arrival and Doppler frequency – onto the required physical coordinates – range, bearing and velocity – plays a crucial role in skywave radar where the relationships are non-trivial, even for monostatic radars. For many applications, imprecise geolocation greatly reduces the value of successful detection as it may preclude the initiation of an operational response, such as interception or over-the-horizon targeting. The situation is further aggravated by the inability of conventional skywave radar systems to determine aircraft altitude reliably and with any precision, even when they are fitted with waveforms and processing designed to estimate altitude. In the absence of meaningful altitude information, options

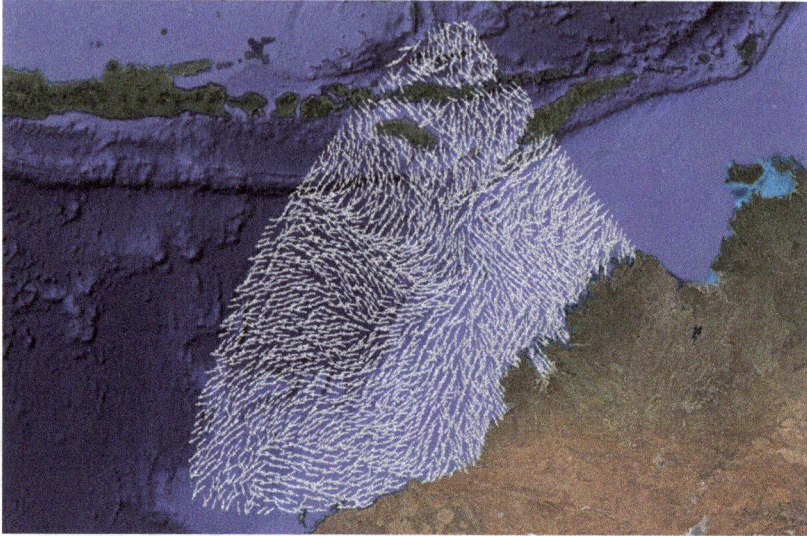

Figure 3.12   A radar map of wind direction as inferred from the Bragg ratio of short gravity waves, illustrating the diverse spatial scales of wind/ wave field homogeneity

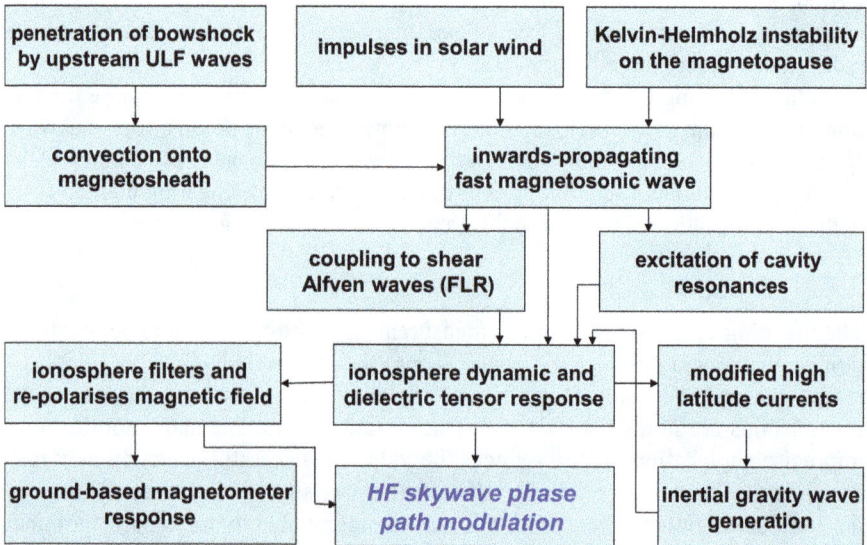

Figure 3.13   The physical mechanism whereby magnetospheric disturbances modulate HF skywave radar signals

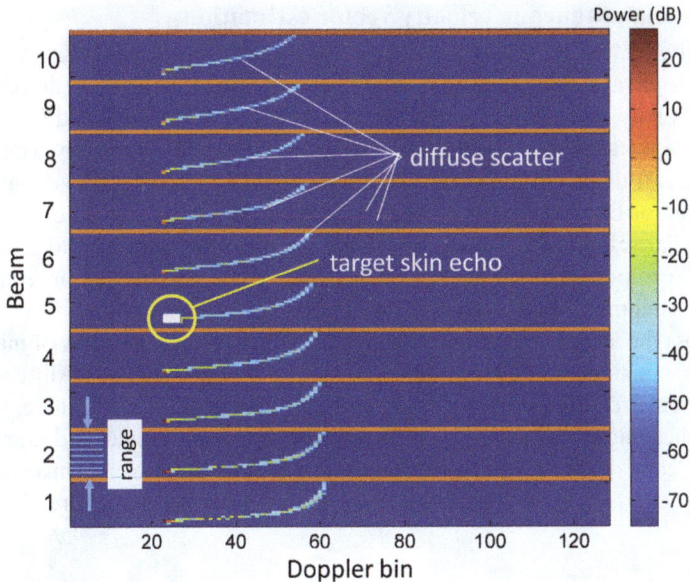

Figure 3.14   The skywave radar echo structure for an aircraft over the ocean. The skin echo is localised in range, beam and Doppler, but the diffuse scatter is widely distributed in a form dependent on the directional wave spectrum.

for response are even more limited. It follows that any technique that could provide good estimates of aircraft altitude would add enormously to the utility of detections.

3.3.2.1   Altitude estimation

Existing techniques for altitude estimation rely on the classical four-path model, allowing for specular ground reflections of descending waves as well as direct illumination of the target. We have already mentioned in our discussion of detection that diffuse scatter from surface roughness spreads target echoes in the Doppler domain. This is not the only effect – the diffuse scatter can occur from any and all parts of the sea surface within the visible horizon as viewed from the aircraft. This 'side scatter' spreads an otherwise discrete echo across all three processing dimensions – range, bearing and (radial) velocity, as illustrated in Figure 3.14. It has been shown by experiment and through theoretical analysis [52] that the spread echo encodes altitude information in an easily retrievable form, with both accuracy and precision to better than 1 000 metres. This is an extraordinary capability, and the only requirement for the retrieval is knowledge of the directional wave spectrum as generated by the remote sensing mission.

3.3.2.2   Instantaneous velocity vector estimation

Amazingly, that is not the end of the largesse of diffuse scatter. With OTHR we are almost always concerned with tracking, not just instantaneous detections, so 'location' really extends to establishing a trajectory in space-time. Although subject to variance introduced by ionospheric motions, we might reasonably assert that a single aircraft detection based on just one or two seconds of observations affords an estimate of the component of velocity normal to the ellipse of constant range, that is, radial velocity for monostatic radar configurations. However, to estimate the tangential component of velocity and thus complete the velocity vector requires that we establish a track and maintain it for long enough for the echo to move appreciably across the azimuth domain. This may take many minutes, orders of magnitude greater than a single look. But when we examine the distribution of diffuse scatter contributions, for the same target, at the same geolocation and altitude, but with different headings, the patterns of diffuse scatter are quite distinct and immediately yield the corresponding total velocity vector. This is a major advance in OTHR capability, courtesy of the remote sensing mission that provides the directional wave spectrum.

3.3.2.3   Coordinate registration

Land-sea boundary mapping has been a standard coordinate registration technique since 1983 [27]. The same idea – spatial variations in clutter spectral density – need not be discarded in the case of the open sea, because mesoscale weather systems forming, evolving and moving across the ocean surface write their own signatures on the directional wave spectrum and hence on the HF radar Doppler spectrum. Properly implemented, the remote sensing mission can monitor the spatial distribution of $S(\vec{\kappa})$ and infer the prevailing meteorological conditions. But how does this help CR? The answer is that there are many spaceborne sensors that map clouds, winds, sea surface temperature and waves, without the complications of skywave propagation. While the spatial resolution of some older sensors is barely adequate to aid CR, others offer resolution of the order $\sim10^0$ m, far better than required. Thus, given a satellite image of the clouds or wave height or sea surface temperature in an area of interest, one could segment the image according to some threshold criteria, then correlate this map against one showing selected features in the HF radar signature. As an illustration of the feasibility of meteorological CR, it has already been established experimentally that mesoscale tropical convective cells present strong, sharp HF radar signatures [24]. Thus the practical measure of usefulness reduces to the frequency of occurrence of meteorological features that introduce distinguishable imprints on the sea surface.

3.3.3   Target classification

The utility of knowledge about the presence of ships or aircraft within the zone of interest is extremely limited unless there is some additional information about the *types* of vessel or aircraft involved. Does this echo indicate a cruise ship or is it a cruiser? Is that echo from a civil airliner or a military transport? To answer such questions, we need to develop and apply techniques of target classification.

In the past, a variety of methods have been proposed for extracting class and even identity of HF radar scatterers, and some have been implemented in operational systems. A fairly comprehensive review can be found in [5]. Yet, with the exception of targets whose intrinsic signatures have characteristic distributions in the Doppler domain, such as helicopters, the performance of conventional OTHR target classification techniques falls far short of operational requirements. We shall see in what follows that remote sensing information can ameliorate this state of affairs.

3.3.3.1 Classification based on intrinsic scattering characteristics

The radar scattering behaviour of an obstacle is a function of aspect, frequency and incident polarisation, even in the simplest cases. In an anechoic chamber, with scale models, it may be possible to explore all these domains but, with operational OTHR, the accessible measurement space is extremely limited and such observations as are possible are subject to complicating factors. Specifically,

- the characteristic scattering properties of ships and aircraft are strongly frequency-dependent, but the range of frequencies F that can be used to interrogate a target at a given range is limited by the propagation conditions, as shown in Figure 3.2, may be quite narrow. Typically the fractional bandwidth – the ratio of F to the nominal carrier frequency F – satisfies $\frac{\Delta F}{F} \lesssim 0.15$;
- the elevation angle of illumination at the target cannot be measured so it must be estimated from a real-time ionospheric model and coordinate registration information;
- the skywave propagation channel experiences a variety of signal attenuation, modulation, focussing and defocussing mechanisms that make it very difficult to obtain absolute measurements of RCS parameters;
- the characteristic scattering properties of ships and aircraft are strongly polarisation-dependent;
- the propagation channel transforms the polarisation state of signals during skywave propagation, so the state of the field incident on the target is a priori unknown. Similarly, the scattered field leaving the target experiences unknown degrees of repolarisation and depolarisation on the path to the radar receiver.

Thus it might be thought that there is little prospect of assigning a target to some class based only on the magnitude of the target echo. A modest capability is still achievable through the analysis of statistical distributions [5], at least as far as discriminating between a small number of distinct types, but there remains the challenge of ensuring that the sample measurements are sufficiently uniformly distributed over the polarisation domain. In order to accumulate high quality statistical information, the process model still needs to be calibrated, and here again, inversion of the sea clutter plays the central role.

The exploitation of contemporaneous remote sensing information offers solutions to some of these challenging 'facts of life' of OTHR:

- the absolute scattering coefficient of the sea can be ascertained from the regional time history of the clutter Doppler spectrum, even without introducing the standard parametric wave spectrum models. This calibration reference enables us to compensate for attenuation and focussing, so the composite scalar return from the target can be converted to absolute radar cross section as projected by the prevailing polarisation transformation onto the receiving antenna;
- some information about the polarisation state of the signal incident on the sea around the target may be extracted, in principle, either from the received signal augmented by a separate receiving channel with orthogonal polarisation, or by varying the radar frequency slightly and making use of the known structure of the sea surface scattering matrix and the associated Doppler spectra [60]. Figure 3.15 presents modelled results for the polarisation state in the illuminated zone of a skywave radar for the situation where the dominant mechanism of polarisation transformation is Faraday rotation. An experimentally measured example of the resulting clutter modulation is shown superimposed, aligned with the azimuth and range bracket at which it was recorded. We can see that the ranges at which enhanced echo strength occurs correspond to the same polarisation phase. The agreement is excellent, and from the physics of HF scattering from the sea surface, we know that these bands correspond to vertical polarisation.

3.3.3.2   Classification based on multiple scattering

The diffuse scatter effect discussed in Section 3.3.2 is equally relevant to target classification. Of the complete bistatic scattering pattern defined on the angular coordinates, $\sigma(\theta_{scat}, \phi_{scat}, \theta_{inc}, \phi_{inc}; \ldots)$, only a tiny part is manifested in the direct and surface specular target echo. If we consider the additional contributions from multiple scatter, where the rough sea surface serves as a diffuser of the incident and scattered field, the entire azimuthal arc, 360°, contributes to both incident and scattered fields, while the elevation domain contributes from 180° (nadir) out to some angle close to 90° (corresponding to the horizon). We do not have direct access to the functional dependence because the pattern is convolved with the surface scattering coefficient, but the resulting signal is spread across time delay (range) and Doppler, as illustrated earlier in Figure 3.13, so once again the remote sensing information provides the kernel in a linear integral equation for the target scattering pattern over most of its domain.

3.3.3.3   Classification based on target–environment coupling

While we customarily think of radar signatures in terms of the intrinsic electromagnetic properties of the scatterers, whether or not diffuse scatter is involved, the mechanical coupling between target and environment introduces new signature components. The most immediate of these may be categorised as (i) effects of the target on the

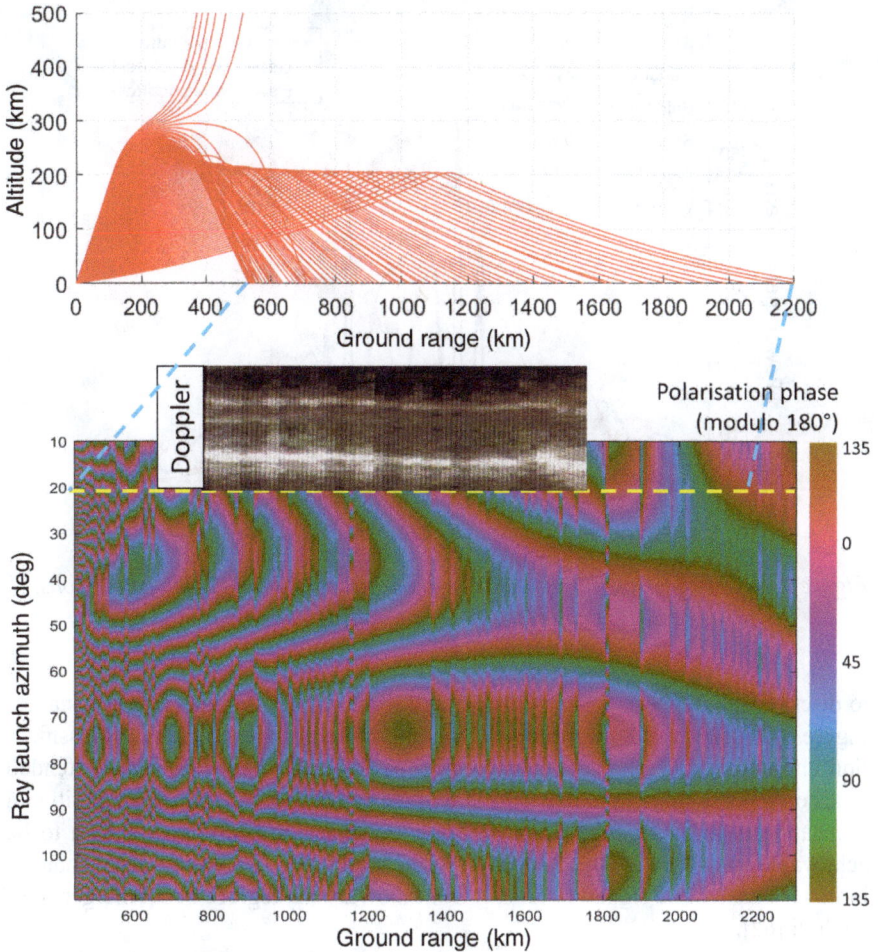

Figure 3.15 The polarisation map predicted for a particular radar, frequency and ionospheric environment, overlaid with experimental data showing the intensity modulation of the sea clutter for a particular azimuth; the modelled ray fan for that azimuth is plotted at the top

environment and (ii) effects of the environment on the target. The archetypes of these mechanisms are, for aircraft, wing-tip vortex generation and response to turbulence, respectively, and for ships, generation of ship wakes and dynamical response to wave forcing respectively.

3.3.3.4   Ship wakes

The exploitation of wake echoes for detection has been discussed in Section 3.3.1; they can also offer a powerful means of discrimination and classification. The study reported in [55] demonstrated that the wake echo contains sufficient information

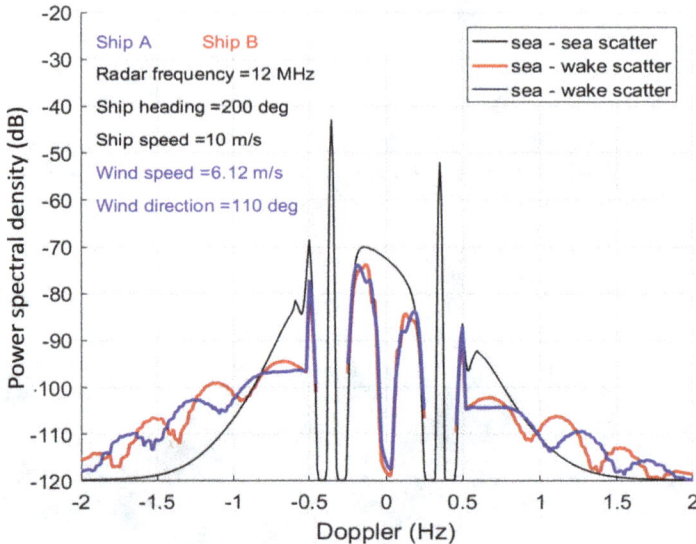

Figure 3.16   Wake Doppler spectra for two frigates of comparable dimensions, computed for the same sea and sailing conditions

to discriminate between two vessels of very similar size – frigates in that instance. Figure 3.16 shows another example. Thus the more general problem of classification into multiple classes seems eminently feasible. The theory has been extended to accommodate more than one ship in a given resolution cell, where an additional term appears [61]. An even more ambitious goal has recently been shown to be achievable on occasion – the estimation of the loading of a given ship from changes in its HF radar wake signature arising from the varying degrees of submergence of the hull [62].

3.3.3.5   Ship classification from dynamical response to wave forcing

The motion of a ship in rough seas is classically formulated in terms of inertial reaction to hydrodynamic forces on the hull, with characteristic response moments corresponding to six degrees of freedom – pitch, heave, roll, surge, sway and yaw. More sophisticated models allow for ship flexural motions, but these do not concern us here as they have magnitudes too small to be observable at HF.

The observed response cross-spectra in a complex sea are governed by the product of the ship inertial response matrix and the excitation – the directional wave frequency spectrum, transformed into the ship's frame of reference (the encounter frequency domain) via the relation

$$\omega_{enc} = \omega - \frac{\omega^2 U}{g}\cos(\chi) \tag{3.8}$$

where U is the ship speed and χ is the encounter angle, with $\chi = 0$ corresponding to a following sea. Thus to extract information about the ship's motion and hence its

Table 3.1   Typical motion parameters for various ship classes

Ship type	Roll period (s)	Pitch period (s)
Trawler	6 8	3 4
Patrol boat	6 7	
Ice breaker	8 12	3 5
Frigate	9 13	6 9
Tanker	9 15	7 11
20 kT container ship	16 19	7 9
70 kT container ship	23 28	
Cruise liner	20 28	10 12

dynamic radar signature, we need to know the directional wave spectrum. That, of course, is exactly the output of the remote sensing mission.

In low sea states it is often valid to treat the degrees of freedom as dynamically independent. Experiments with HF radar have shown that roll usually imposes the largest modulation of the scattering amplitude and hence the most obvious feature in the measured Doppler signature. Natural roll and pitch periods of different ship classes are presented in Table 3.1. An example of measured motion spectra for a research vessel are presented in Figure 3.17 [63].

We can see that expected roll periods are shorter than a typical coherent integration time for an HF radar tasked with ship detection, and as a consequence a time-frequency analysis of a ship echo acquired over 30–60 s, say, reveals several roll cycles. An example is shown in Figure 3.18.

From (3.8) it follows that the amplitude of the roll motion, and hence the magnitude of the modulation of the scattered radio wave field, depends on the heading relative to the prevailing sea. An example of this roll response dependence is shown in Figure 3.19 [64].

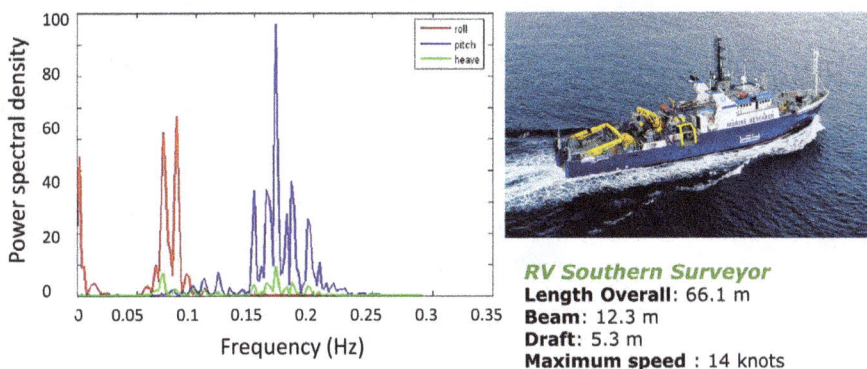

RV Southern Surveyor
Length Overall: 66.1 m
Beam: 12.3 m
Draft: 5.3 m
Maximum speed : 14 knots

Figure 3.17   An example of boat motion spectra, here observed in a sea with a low swell and sea state 2

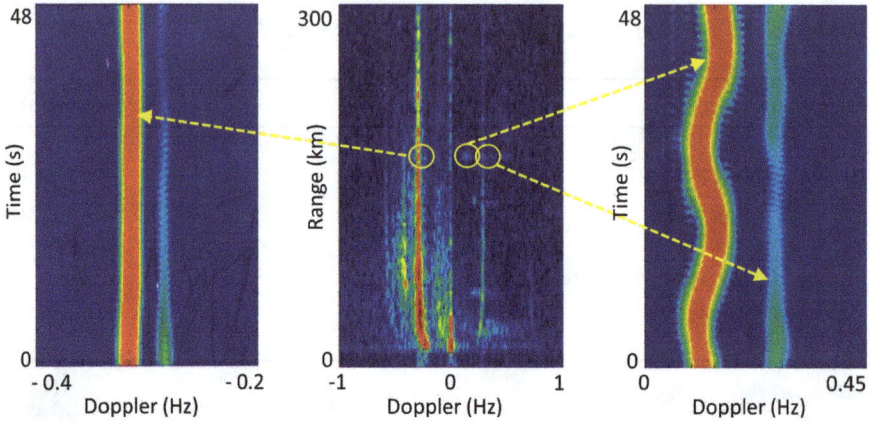

Figure 3.18　*A range-Doppler map, recorded with an HFSWR, not a skywave radar, but showing a good example of a roll signature modulation of the radar signal. The side panels show time histories of the data from the indicated range bins, clutter on the left, ship on the right. The period is consistent with a medium size container ship.*

3.3.4　Resource management

One of the defining characteristics of HF skywave radar is the intense pressure on the time line. The vast coverage, the sheer variety of candidate remote sensing missions, the need for long coherent integration times in many cases and the temporal variation of geophysical parameters, necessitating frequent updates, all these conspire to make time the resource of greatest scarcity to the radar operator. It thus

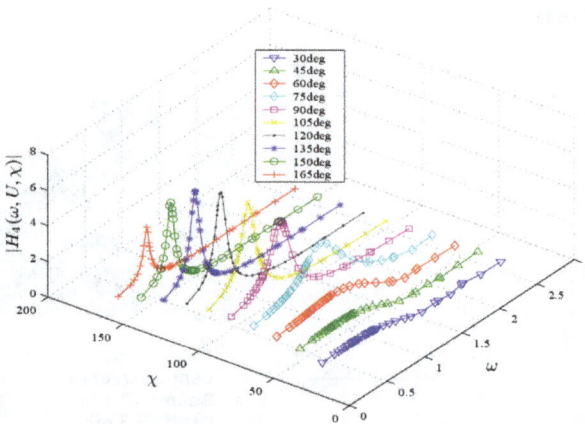

Figure 3.19　*The roll frequency response function of a naval vessel at 15 kn computed via linear potential theory for different encounter angles [63]*

becomes of paramount importance to ensure that missions are conducted wisely, preferably when conditions are especially favourable for the current mission but, more crucially, not when the chances of success of the selected mission are low or non-existent while another task on the 'to-do' list has far higher likelihood of success.

As noted earlier, the selection of radar parameters is driven by the mission objectives as well as the environmental conditions, both oceanographic and ionospheric. Only by possessing a detailed knowledge of the wave directional spectrum can one plan a sensible ship echo detection strategy, or a wake echo interpretation strategy, for another example. This statement holds irrespective of signal contamination during propagation. Access to that pre-requisite information can be threatened by disturbed propagation conditions, but fortunately the remote sensing mission has a built-in resilience. As explained in Section 3.3.1, the wave spectrum is accessible in principle at any radar frequency, though there are some parameter sensitivity issues (see for example [47]). Thus, only one frequency offering good quality propagation is needed to service all candidate radar frequencies, for each of which the prevailing channel transfer function can be applied to yield the expected clutter map and guide resource allocation.

3.3.5 Tactical intelligence

Most naval operations are heavily dependent on sea state and wind conditions. Shipboard meteorological and oceanographic observations are limited in detail, accuracy and range, so when knowledge of the directional wave spectrum over the entire area within range of the ship's sensors and systems is needed, external sources must be consulted. Moreover, apart from the diverse applications to surface vessels, the performance of sensors on airborne platforms, such as radars on long-range maritime patrol aircraft, varies strongly with aspect relative to prevailing seas. Real-time optimisation of flight patterns thus relies on region-scale knowledge of conditions, knowledge that HF skywave radar can provide. In addition, the surveillance and tracking function of the radar means that the environmental information can be registered precisely to the location of cooperating platforms, avoiding the need for absolute coordinate registration with the attendant errors.

When communicated to the platform, the information can be used to correct for bias in RF sensors [65], plan engagements, maintain optimum routes, exploit clutter for concealment against adversarial OTHR systems and may other purposes.

It may seem fanciful to think that the intent of a non-cooperative vessel can be deduced by the radar, but we should recall that evidence can accumulate from radar observations of point of earliest detection, speed, course, existence of significant locations further along the axis of sailing, observations of rendezvous at sea and so on. To these we should add data on sea state, ocean currents, wind speed and direction, and other environmental information that can be measured by the radar in its remote sensing mission because the prevailing conditions govern the feasibility of many marine operations. For example, ship-to-ship materiel transfers and aircraft launches from carriers are both undertaken by adjusting ship heading to align with the most favourable combination of wind and waves, while container ships in high seas must adopt the combinations of course and speed that avoid the risk of cargo movement.

3.4 Summary

The largely independent historical development of remote sensing and ocean sur-veillance missions for HF skywave radar has deprived some national OTHR systems of potential gains in performance and access to new capabilities. The benefits to per-formance in standard surveillance tasks derive mainly from information that enables the radar operator to optimise parameters such as frequency, bandwidth, coherent integration time, scan pattern, signal processing settings and mission scheduling. Traditionally, real-time selection of these parameters has been driven overwhelm-ingly by general mission type and ionospheric propagation considerations but, as the examples in this chapter illustrate, optimal settings should take joint account of iono-spheric and oceanographic conditions, along with an understanding of the full spec-trum of possible observables afforded by the prevailing environment. These include some signatures for which theoretical models have only recently been formulated. The radar itself can provide most or all of the necessary supporting information via its remote sensing mission. Gains in detectability, positional accuracy, target clas-sification, resource allocation and event analysis have been identified, including:

- absolute calibration of target RCS and estimation of propagation losses
- improved coordinate registration over the open sea
- interpretation of ship dynamic signatures
- detection of ship wakes, even when the ship echo is concealed in sea clutter
- possible detection of submarine wakes under favourable conditions
- robust aircraft altitude estimation
- instantaneous determination of the full velocity vector of detected targets
- resistance to some common techniques of electronic attack
- ancillary information to support event analysis.

The extent to which these capabilities have been implemented in operational OTHR systems around the world cannot be known with certainty, but informed opinion inclines to the view that many opportunities remain unexploited.

References

[1]   Headrick J.M., Skolnik M.I. 'Over-the-horizon radar in the HF band'. *Proceedings of the IEEE*. 1974;**62**(7):664–73.
[2]   Headrick J.M., Anderson S.J. 'HF over-the-horizon radar' in Skolnik M.I. (ed.). *Radar Handbook*. McGraw-Hill; 2008.
[3]   Fabrizio G.A. '*High Frequency Over-The-Horizon Radar*'. McGraw-Hill; 2013.

[4] Anderson S.J. 'Some recent results obtained using the JINDALEE radar in SHIP mode'. *Australian Defence Science and Technology Organisation, JINDALEE Working Paper*. 1985.

[5] Anderson S.J. 'Target classification, recognition and identification with HF radar'. Proceedings of the NATO Research and Technology Agency Sensors and Electronics Technology Panel Symposium. Paper presented at the RTO SET Symposium on 'Target Identification and Recognition Using RF Systems', held in Oslo, Norway, 11-13 October 2004, and published in RTO-MP-SET-080.2004.

[6] Long A., Trizna D. 'Mapping of North Atlantic winds by HF radar sea back-scatter interpretation'. *IEEE Transactions on Antennas and Propagation*. 1973;**21**(5):680–5.

[7] Stewart R.H., Barnum J.R. 'Radio measurements of oceanic winds at long ranges: an evaluation'. *Radio Science*. 1975;**10**:853–7.

[8] Maresca J.W. Jr., Carlson C.T. 'Tracking and monitoring hurricanes by HF skywave radar over the Gulf of Mexico'. *Technical Report 1, Contract F49620-76-C-0023, SRI International*. Menlo Park, CA, USA; 1977.

[9] Anderson S.J. 'The remote sensing of tropical meteorological and oceano-graphic phenomena by over-the-horizon radar'. Proceedings of the 1st International Conference on Tropical Cyclones. Perth; Australia; Dec 1979.

[10] Maresca J.W., Georges T.M. 'Measuring RMS wave height and the sca-lar ocean wave spectrum with HF skywave radar'. *Journal of Geophysical Research*. 1980;**85**(C5):2759–71.

[11] Barrick D.E. 'Status of HF radars for wave-height directional spectral measurements'. Proceedings of a Symposium and Workshop on Wave-Measurement Technology, April 1981; Washington DC, National Academy Press; 1982. pp. 112–7.

[12] Maresca J.W., Jr., Carlson C.T. 'Longshore currents on the fringe of Hurricane Anita'. *Journal of Geophysical Research*. 1980;**8 5**(C3):1640–1.

[13] Trizna D.B. 'Mapping ocean currents using over-the-horizon HF radar'. *International Journal of Remote Sensing*. 1982;**3**(3):295–309.

[14] Lipa B.J., Barrick D.E., Maresca J.W. 'HF radar measurements of long ocean waves'. *Journal of Geophysical Research*. 1981;**86**(C5):4089–102.

[15] Trizna D.B., Headrick J.M., Bogle R.W., Moore J.C. 'Directional sea spec-trum determination using HF Doppler radar techniques'. *IEEE Transactions on Antennas and Propagation*. 1977;**AP-25**(1):4–11.

[16] Anderson S.J. 'Remote sensing applications of HF skywave radar: The Australian experience'. *Elektrik, Turkish Journal of Electronic Engineering and Computer Science*. 2010;**18**(3):339–72.

[17] Keenan T.D., Anderson S.J. 'Some examples of surface wind field analysis based on Jindalee skywave radar data'. *Australian Meteorological Magazine*. 1987;**35**(4):153–61.

[18] Maresca J.W., Carlson C.T. 'High-frequency skywave radar measurements of Hurricane Anita'. *Science*. 1980;**209**(4462):1189–96.

[19]   Anderson S.J., Abramovich Y.I., Skinner A.I. 'Robust mapping of tropical cyclone wave fields using HF skywave radar'. IEE Conference Publication 490, Proceedings of RADAR 2002; Edinburgh, UK; Oct 2002. pp. 47–50.

[20]   Anderson S.J. 'HF radar detection of internal waves in the ocean'. Proceedings of the IEEE International Radar Conference; Lille, France; October 2014.

[21]   Anderson S.J. 'HF radar signatures of rain-modified ocean surfaces'. Proceedings of the Asia-Pacific Microwave Conference; Adelaide, Australia; August 1992.

[22]   Anderson S.J. 'HF radar detection and tracking of oil spills in the marine environment'. Paper presented at the IEEE International Geophysics and Remote Sensing Symposium, IGARSS'97; Singapore; August 1997.

[23]   Anderson S.J. 'Monitoring the marginal ice zone with HF radar'. Proceedings of the IEEE US International Radar Conference; Seattle; May 2017.

[24]   Anderson S.J. 'HF radar signatures of ocean surface geometry and dynamics arising from localized disturbances'. Proceedings of the 2015 IEEE US Radar Conference; Alexandria, USA; May 2015.

[25]   Anderson S.J. 'HF skywave radar performance in the tsunami detection and measurement role' in Marner N M. (ed.). *The Tsunami Threat – Research and Technology*. Intech; 2011. pp. 41–66.

[26]   Anderson S.J. 'Design implications of a ship detection requirement for Jindalee Stage B'. *Weapons Research Establishment Technical Memorandum WRE-TM-B95 (AP)*. Adelaide, Australia; 1976.

[27]   Anderson S. 'Remote sensing with the JINDALEE skywave radar'. *IEEE Journal of Oceanic Engineering*. 1986;**11**(2):158–63.

[28]   Anderson S.J. 'The challenge of signal processing for HF over-the-horizon radar'. Proceedings of the Workshop on Signal Processing and Applications, WOSPA'93; Brisbane, Australia; Dec 1993.

[29]   Anderson S. 'OTH radar phenomenology: signal interpretation and target characterization at HF'. *IEEE Aerospace and Electronic Systems Magazine*. 2017;**32**(12):4–16.

[30]   Ulaby F.T., Elachi C. (eds.). *Radar Polarimetry for Geoscience Applications*. Basingstoke, Hampshire: Taylor and Francis; 1990.

[31]   Barrick D.E. 'Remote sensing of sea state by radar' in Derr V.E. (ed.), *Remote Sensing of the Troposphere*. Washington DC: US Government Printing Office; 1972.

[32]   Rice S.O. 'Reflection of electromagnetic waves from slightly rough surfaces'. *Communications on pure and applied mathematics*. 1951;**4**(2–3):351–78.

[33]   Srivastava S., Walsh J. 'An analysis of the second-order Doppler return from the ocean surface'. *IEEE Journal of Oceanic Engineering*. 1985;**OE10**(4):443–5.

[34]   Gill E.W., Walsh J. 'High-frequency bistatic cross sections of the ocean surface'. *Radio Science*. 2001;**36**(6):1459–75.

[35]   Gill E.W., Huang W., Walsh J. 'On the development of a second-order bistatic radar cross section of the ocean surface: a high frequency results for a finite scattering patch'. *IEEE Journal of Oceanic Engineering*. 2004;**OE31**(4):740–50.

[36] Walsh J. *On the theory of electromagnetic propagation across a rough surface and calculations in the VHF region.* CanadaTechnical Report N00232, prepared for DREA; 1980.

[37] Anderson S. 'Bistatic and stereoscopic configurations for HF radar'. *Remote Sensing*;12(4):689.

[38] Anderson S.J., Anderson W.C. 'Bistatic HF scattering from the ocean surface and its application to remote sensing of seastate'. Proceedings of the 1987 IEEE APS International Symposium; Blacksburg, VA, USA; Jun 1987.

[39] Anderson S.J. 'The extraction of wind and sea state parameters from HF skywave radar echoes'. IREECON International Digest, IREECON-83; Sydney; September 1983. pp. 654–6.

[40] Lipa B. 'Derivation of directional ocean-wave spectra by integral inversion of second-order radar echoes'. *Radio Science.* 1977;**12**(3):425–34.

[41] Lipa B., Barrick D. 'Methods for the extraction of long-period oceanwave parameters from narrow beam HF radar sea echo'. *Radio Science.* 1980;**15**(4):843–53.

[42] Wyatt L.R. 'A relaxation method for integral inversion applied to HF radar measurement of the ocean wave directional spectrum'. *International Journal of Remote Sensing.* 1990;**11**(8):1481–94.

[43] Howell R., Walsh J. 'Measurement of ocean wave spectra using narrow-beam HF radar'. *IEEE Journal of Oceanic Engineering.* 1993;**18**(3):296–305.

[44] Hisaki Y. 'Nonlinear inversion of the integral equation to estimate ocean wave spectra from HF radar'. *Radio Science.* 1996;**31**(1):25–39.

[45] Hashimoto N., Tokuda M. 'A Bayesian approach for estimation of directional wave spectra with HF radar'. *Coastal Engineering Journal.* 1999;**41**(2):137–49.

[46] Wyatt L.R. 'Progress in the interpretation of HF sea echo: HF radar as a remote sensing tool'. *IEE Proceedings F Radar and Signal Processing.* 1990;**137**(2):139–47.

[47] Wyatt L.R. 'Limits to the inversion of HF radar backscatter for ocean wave measurement'. *Journal of Atmospheric and Oceanic Technology.* 2000;**17**(2):1651–66.

[48] Hardman R.L., Wyatt L.R., Engleback C.C. 'Measuring the directional ocean spectrum from simulated bistatic HF radar data'. *Remote Sensing.* 2020;**12**(2):313–42.

[49] Earl G.F., Ward B.D. 'The frequency management system of the Jindalee over-the-horizon backscatter HF radar'. *Radio Science.* 1987;**22**(2):275–91.

[50] Barnum J.R. 'Theoretical limitation of the sea on the detection of low-Doppler targets by OTH radar'. Menlo Park, CA: SRI International; 1980.

[51] Anderson S.J. 'Cognitive HF radar'. *Journal of Engineering.* 2019;**20**:6772–6.

[52] Anderson S.J. 'Estimation of target class and altitude in OTHR via diffuse surface scatter'. Proceedings of the IET International Radar Conference; Hangzhou, China; Oct 2015.

[53] Anderson S.J. 'On the sensitivity of HFSWR wave height and directional wave spectrum estimation to non-stationarity of forcing winds'. Paper

presented at the IUGG General Assembly; IAPSO Session, Melbourne; Jul 2011.

[54]   Benjamin T.B., Feir J.E. 'The disintegration of wave trains on deep water Part 1. theory'. *Journal of Fluid Mechanics.* 1967;**27**(3):417–30.

[55]   Anderson S.J. 'HF radar signatures of ship and submarine wakes'. *Journal of Engineering.* 2019;**21**:7512–20.

[56]   Abramovich Y.I., Spencer N.K., Anderson S.J., Gorokhov A.Y. 'Stochastic-constraints method in nonstationary hot-clutter cancellation. I. Fundamentals and supervised training applications'. *IEEE Transactions on Aerospace and Electronic Systems.* 1998;**34**(4):1271–92.

[57]   Abramovich Y.I., Spencer N.K., Anderson S.J. 'Stochastic-constraints method in nonstationary hot-clutter cancellation. II. Unsupervised training applications'. *IEEE Transactions on Aerospace and Electronic Systems.* 2000;**36**(1):132–50.

[58]   Abramovich Y.I., Spencer N.K., Johnson B.A. 'Band-inverse TVAR covariance matrix estimation for adaptive detection'. *IEEE Transactions on Aerospace and Electronic Systems.* 2010;**46**(1):375–96.

[59]   Anderson S.J., Abramovich Y.I. 'A unified approach to detection, classification, and correction of ionospheric distortion in HF sky wave radar systems'. *Radio Science.* 1998;**33**(4):1055–67.

[60]   Anderson S.J. 'Skywave channel characterisation for polarimetric OTH radar'. Proceedings of the IEEE US Radar Conference; Philadelphia, USA; May 2016.

[61]   Anderson S.J. 'HF radar scattering from a sea surface perturbed by multiple ship wakes'. Proceedings of the International Conference on Electromagnetics in Advanced Applications; Granada, Spain; Sep 2019.

[62]   Anderson S.J. 'An OTH Plimsoll line? Remote measurement of ship loading with HF Radar'. Proceedings of the IEEE International Radar Conference; Toulon, France; Sep 2019.

[63]   Perez T., Control S.M. *Advances in Industrial Control.* London: Springer-Verlag; 2005.

[64]   Anderson S.J. 'Motion compensation techniques for ship-borne SAR at HF'. Proceedings of CEOS 2005, the Committee on Earth Observation Satellites Workshop on SAR; Adelaide; Sep 2005.

[65]   Anderson S.J. 'Scattering effects on DOA estimation and geolocation from HF surface wave intercepts'. Proceedings of the HF/DF Symposium; South-West Research Institute, San Antonio; May 2008.

Chapter 4

Sea surface current mapping with HF radar – a primer

Clifford R Merz[1], Yonggang Liu[1], and Robert H Weisberg[1]

Shore-based oceanographic high frequency (HF) radars are frequently used to remotely sense and map coastal sea surface currents. This chapter begins with a review of the development and utilization of HF radar sea-echo interactions and their relationship in the determination of the radial component of the sea surface current and vector coverage map, followed by a brief discussion of recent ongoing HF radar observations on the West Florida Shelf (WFS). Reported are HF radar performance and its complicated relationships with environmental factors.

4.1 Introduction

Ocean currents are the continuous movement of seawater. In the coastal ocean, the currents are what deliver nutrients from the deep-ocean and the estuaries to the shelf, thereby fueling primary productivity and initiating the complex biological and chemical interactions resulting in the shelf ecology [1]. Commercial and recreational fisheries depend on this as do blooms of harmful algae [2]. Similarly, safe and efficient maritime commerce and missing boater search and rescue operations depend on our ability to specify currents, sea level, and sea state on the basis of the ocean and atmosphere interactions.

Two different methods of current measurement are commonly in use: Lagrangian and Eulerian. The Lagrangian method tracks the path followed by the moving fluid and consists of placing a floating object in the water and allowing it to drift away from its initial position. This method has progressed from early efforts using floating coconuts, drifter bottles, a timer, and known distance traveled to highly advanced instruments equipped with internal position logging, near- and real-time satellite transmission [3], and satellite-tracked surface drifters [4]. In the Eulerian method, a monitoring instrument is placed at a fixed location and the speed and direction of the moving fluid are measured with respect to that fixed location. Trends can be determined by analysis of the time series generated at that given point. Some

[1]College of Marine Science, University of South Florida, United States

examples of Eulerian measurements include: cable-attached electro-mechanical current meters, which measure the velocity at a single depth; bottom, mounted upward-looking and buoy-mounted downward-looking Acoustic Doppler Current Profilers (ADCP), which measure the velocity throughout the water column [5]. These in situ current observations are made at a single point at a time, which are very limited in spatial coverage.

In contrast to those traditional single-point current measurements, shore-based HF radars can map surface currents with large spatial coverages. They have become an important component of coastal ocean observing systems [6]. HF radar-measured currents have been found in many oceanographic applications [7–9] as well as for assimilation in numerical ocean circulation models [10–12] because of their ability to remotely sense and map surface currents over large horizontal extents and at high sample resolution. Radiated by a vertically polarized shore-based antenna, transmitted HF electromagnetic (EM) radio waves (3–30 MHz) interact with ocean waves, resulting in Doppler shifted echoes backscattered by sea-surface gravity waves possessing a wavelength equal to one-half of the transmitted EM radar wavelength [13, 14]. The Doppler effect (also referred to as Doppler shift) is the change in frequency of a propagating wave in relation to an observer moving relative to the wave source and is named after Austrian physicist Christian Doppler who first described the phenomenon in 1842. Radar and sonar are but two of many successful examples where direct application of the Doppler principle is applied to measure the relative speed of a moving target.

As discussed in Chapter 1 [15], HF radar EM waves can propagate via two operational modes: (1) ground wave and (2) sky wave. Here, we consider the effects associated with ground wave propagation only to map sea surface currents using the concept of Bragg scattering at near-grazing incidence [16–20]. This chapter is written to serve as both a fundamental chapter on HF radar surface current mapping estimation and a companion to Chapter 5 [21].

4.2   Theory behind radial and vector current derivation from HF radar Doppler spectrum

Barrick [22] reported that HF radar sea echoes were first observed on air-defense nets around the English Channel during World War II, where the resulting "clutter" periodically imposed limitations to the detection of aircraft. Based on HF radar experimental observations at 13.5 MHz, Crombie [16] was the first to correctly deduce the physical mechanism producing sea scatter and identify the distinctive features of sea-echo Doppler spectra. Crombie observed that the echo Doppler spectrum contained two dominant peaks symmetrically located about the transmitted EM radio carrier frequency.

Crombie reasoned that since there were two dominant peaks in the Doppler echo, the source of the originating scatter must be from two targets moving at a given velocity and that these "targets" were ocean wave trains with radial wave phase velocity components along the radar look direction. He further calculated the wavelength of the ocean wave train seen by the HF radar using the deep-water gravity-wave dispersion relationship. Finally upon equating these two relationships, the wavelength of the backscattering ocean wave train, λ_{Ocean}, was found to equal one-half of the transmitted EM radar frequency wavelength, λ_{EM}.

The "resonant" effect often mentioned during HF radar operational discussions shares similar characteristics with Bragg backscatter, the same backscatter diffraction-grating mechanism associated with Bragg's law, which is widely applied to X-ray crystallography and laser holography [23, 24]. The λ_{Ocean} backscattered signal is in phase with the backscattered signal from the next λ_{Ocean} wave of the wave train. With the result being an "amplified" signal strength when all the in-phase backscattered signals that reach the shore base receiver are summed together [25].

An important property of the HF sea echo is its random Gaussian nature. Since the heights of the Bragg-scattering waves are random variables, the sea echo must also be a random variable [26]. Through constructive interference of these backscattered EM waves, an energy peak can be detected in the HF radar spectra, from which

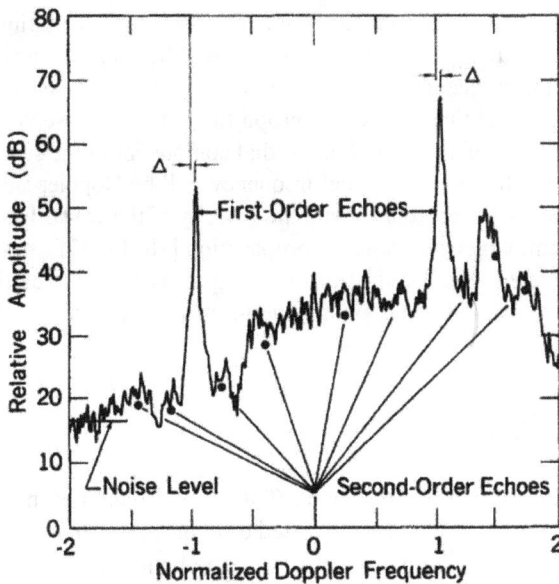

Figure 4.1   *Example surface-wave sea-echo Doppler spectrum from a 13.4 MHz radar (Reprinted by permission from Springer Nature, Boundary-Layer Meteorology, HF Radio Oceanography – A Review, Barrick, D.E., Copyright 1978). The Doppler frequency units of the abscissa are normalized such that 0 corresponds to the transmitted EM carrier frequency (λ_{EM}) position and ±1 refers to the predicted positions of the first-order Bragg echo peaks. Δ is the normalized frequency shift of the record due to the presence of a moving sea-surface current [22, 26].*

the ocean variance density spectrum can be determined (Figure 4.1). The first-order spectral maxima region (observed peak) is dependent solely on the Bragg resonant scattering effect and the surrounding second-order spectral return regions are caused by double scattering or non-linearities in the wave field [27, 28]. Barrick and Peake [29] and Barrick [23, 30] showed that to first and second order, the *average* sea-echo Doppler spectrum is related to the *average* sea wave height directional spectrum evaluated at the required first- and second-order Bragg wavenumbers [26]. While examining a sinusoidal sea wave train, Wait [31] related the strength of the signal voltage at the Doppler-echo peak to the height of the Bragg-resonant wave train. Even though over 50 years have passed since some of these transformational contributions occurred, it's important to not let the many technical advances that have occurred since reduce these observation-based achievements to a mere footnote or passing reference. It is for this reason that the historical detail and albeit time-dated Figure 4.1 is included herein.

Although all of the sea wave trains on the sea surface interact with the transmitted EM radar wave, as noted above, the waves that can produce strong backscatter energy toward the radar are those that have wavelengths equal to half wavelengths of the EM waves, $\lambda_{Ocean} = \lambda_{EM}/2$, and are moving either toward (resulting in a positive Doppler shift in frequency [+ peak]) or away from the radar (resulting in a negative Doppler shift in frequency [− peak]). In the absence of any sea surface current, the Doppler positions of these peaks are proportional to the phase velocity of these waves. In the presence of any underlying radial component of the surface current at the point of scatter, the received signal frequency will be Doppler shifted from the transmitted frequency in accordance with the velocity of the sea surface current field within which the moving ocean wave is propagating [18, 19, 32].

The backscattered Doppler shift of the Bragg peak from the carrier frequency due to Bragg resonant scattering along a non-moving sea surface is given by the expression:

$$f_{BP} = \pm \sqrt{\left(\frac{gf}{\pi c_o}\right)} \tag{4.1}$$

where g is the acceleration due to gravity (9.8 m/s^2), f is the transmitted EM radar carrier frequency, and c_o is the vacuum speed of light (3.0 × 10^8 m/s) [33].

In the presence of non-zero sea surface current, both Bragg peaks are shifted in the same direction by a frequency amount Δ given by

$$\Delta = 2V_r f/c_o \tag{4.2}$$

where V_r is the radial speed of the sea surface current along the look direction of the radar. By measuring Δ, the radial component (V_r) of the sea surface current can be calculated [34].

A single HF radar only measures the radial component of the sea surface current by analyzing the Doppler spectra of the received backscattered signal. When two or more spatially separated HF radar sites are used, the calculation of a two-dimensional current vector surface coverage map can be made. In order for this to

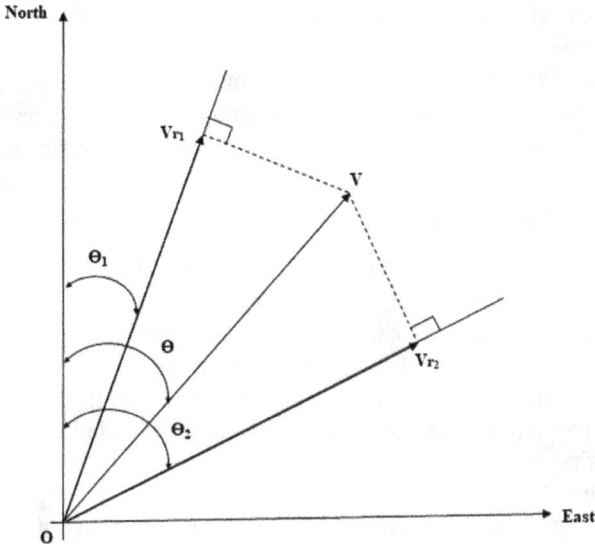

Figure 4.2   *Cartesian geometry for vectorization of the radial components of the sea surface current velocities measured by a two-site HF radar system. Location O is the observed offshore sea surface within the overlap area. Source: Adapted from [35].*

happen, the two sites must be adequately spaced and must overlook the same area of ocean from two different angles. The geometry for vectorization of the radial components of the sea surface current velocities measured by a two-site (master/slave) HF radar system in the overlap scattering area is shown in Figure 4.2 [35].

The vector current in the HF radar overlap observation area can be calculated from current radial components measured by two radars using the following expressions [33, 35]:

$$u = (V_{r1}\cos\theta_2 - V_{r2}\cos\theta_1)/\sin(\theta_1 - \theta_2) \tag{4.3}$$
$$v = (-V_{r1}\sin\theta_2 + V_{r2}\sin\theta_1)/\sin(\theta_1 - \theta_2) \tag{4.4}$$

where u and v are the eastward (or across-shelf) component and northward (or along-shelf) component of current, V_{r1} and V_{r2} are the radial components of the current vector measured by radars 1 (master) and 2 (slave), and θ_1 and θ_2 are referenced clockwise from the north are the beam directions for HF radars 1 and 2, respectively. The vector current shown in Figure 4.2 is $\mathbf{V} = u + i\,v$. Due to the different values of observation direction and range, radars located at different site locations can have different backscattered echo intensities for the same portion of observed sea surface scattering area.

4.3   Factors affecting current measurements

HF radar system current measurements depend upon many factors including the type of HF radar system used affecting radial current azimuthal resolution, range resolution, geometrical dilution of precision, and signal propagation and sea state.

4.3.1   HF radar system types

Currently, two types of oceanographic HF radars are in frequent use worldwide: direction-finding HF radars such as the commercially available SeaSonde developed by CODAR Ocean Sensors, Ltd, and phased array HF radars, such as the commercially available WERA developed by the University of Hamburg and manufactured by Helzel Messtechnik GmbH (Helzel). Other HF radar systems comprising these two types found in the technical literature include COSRAD (James Cook University), LERA (University of Hawaii at Manoa), and OSMAR (Wuhan University, China).

The CODAR SeaSonde direction-finding HF radar utilizes a compact directional receive antenna consisting of three collocated antenna elements (two orthogonally mounted loops and a vertical dipole or monopole element) in a single antenna housing. A single, omnidirectional antenna is used to transmit a pulsed, gated frequency modulated continuous wave (FMCW) waveform. The WERA phased array HF radar utilizes two separate, simultaneously operated antenna arrays: one for receiving and the other for transmitting FMCW chirps. WERAs utilize a 1–4 element array to transmit and a 12–16 element linear array to receive and record the backscattered radio wave signals. By virtue of the increased number of antennas, it can operate both in a direction-finding and in a directional beam-forming mode.

All HF radars resolve the target range and radial current speed to a high degree; however, system limitations exist in the ability to accurately determine the bearing of the specific surface scattering point. The Bragg region is taken to be a region of spread around the Bragg peaks whose energy and position are caused by Bragg-resonant waves and the underlying current velocity. Although the external factors affecting the strength of the received Bragg backscattered sea-echo signal may be the same, the mechanics of the reception and signal processing is different depending upon the HF radar system type used – phased array or direction-finding.

Phased array HF radars (such as the WERA) use multiple receive antennas and resolve signal arrival bearings using beamforming techniques. After fast Fourier transformation of the received signal, the time delays between antenna elements become phase shifts. The weighted signals at all the elements are summed in software to sweep a narrow beam across a series of desired azimuthal directions in which signals from particular arrival directions sum constructively, whereas signals for other arrival directions sum destructively, ultimately forming a steered beam across the radar footprint [34, 36, 37]. On the other hand, in direction-finding HF radars (such as the CODAR SeaSonde), signal arrival bearings are resolved in software by fitting the data from all three elements of the compact receive antenna to either a measured or modeled antenna response function consisting of relative phase and amplitude versus bearing with the direction of arrival (DOA) determined using the Multiple Signal Classification (MUSIC) algorithm [38–41]. The Doppler shift (or radial current velocity) is calculated for each frequency bin in the region followed by a determination of the ultimate DOA for that radial current velocity [42]. While MUSIC has advantages, attention must be paid during operational times of low signal-to-noise ratio (SNR) (the general SNR cut-off threshold is 6 dB) and if variations occur in the measured receive antenna beam pattern in order to minimize observational errors [43].

In both system types, the antenna pattern distortion and DOA algorithms will affect the azimuth accuracy and resolution of current radial component and vector accuracies. It should also be noted that the coherent integration (or dwell) time used for Fourier transformation will affect the Doppler resolution of the spectrum and, thus, the current speed accuracy.

4.3.2   Range resolution

As discussed in Chapter 1 [15], range resolution is an important parameter affecting spatial resolution of HF radar current measurements. The relationship between FMCW bandwidth and range cell resolution (cell depth) for some typical settings is presented in Table 4.1 [44].

The maximum bandwidth available, and therefore the range resolution used, is usually set by the local or regional frequency licensing authority, for example, the Federal Communications Commission (FCC) in the United States of America.

4.3.3   Geometrical dilution of precision

The HF radar site-to-site separation distance and transmit frequency control the size and shape of the domain where currents can be resolved through the geometry of the intersection angles of the radial component of the sea surface currents. This influence of the geometry on the surface current measurement errors is known as the Geometrical Dilution of Precision (GDOP). GDOP describes the reduction of accuracy caused solely by geometry and does not take into account the effects of reduced SNR that might be expected to decrease the accuracy of the current estimates in the farthest range bins [45].

Table 4.1   Relationship between bandwidth and range cell resolution (depth) for a FMCW chirp

Bandwidth (kHz)	Range Resolution (km)
25	6.0
50	3.0
100	1.5
150	1.0
250	0.6

Chapman *et al.* [45] derived the GDOP for the Cartesian current components based on the radar's mean look direction and the half angle between the intersecting beams. The GDOP is defined as the variance ratio of (σ_u/σ) and (σ_v/σ), where σ is the total root mean square (RMS) variance of current differences [34]. When the azimuthal difference between the two HF radar beams is a right angle (90°), the current vector measurement error is at its minimum value. When the azimuthal difference becomes very different from a right angle, the aspect ratio of the contour changes from unity and the measurement error increases [35] with major variations occurring within the radar's near and far fields. For each grid point, the X- and Y- components of GDOP are calculated depending upon the angular positions relative to each other, with radial velocity data from GDOP regions that exceed specified input criteria excluded from vector current estimation.

4.3.4   Signal propagation and sea state

For a given radar system, the broadness of the Doppler spectrum and the noise level are dependent on the sea state [46]. Regardless of the HF radar system used, successful current measurement depends upon the quality and magnitude of the received signal relative to noise level. The extent of the offshore working range is ultimately determined by the SNR, which, in turn, depends on the sea surface roughness (or sea state), the radar's operating frequency and corresponding propagation loss, and the background noise level at the receive antenna.

Depending upon the frequency, EM waves can be affected by parameters relevant to both the sea and the atmosphere; two such possible sources include:

1.   Variations in the viscosity and complex relative dielectric constant (ϵ) of the seawater. The attenuation of HF propagation is dependent on the imaginary part of ϵ, frequency, and conductivity, where the conductivity is a function of the salinity and temperature [37, 47].
2.   The atmospheric radio refractivity (N), which is dependent on variations in the atmospheric humidity, temperature, and air pressure [48, 49]. However, it is well-known these effects are considered minimal within the HF band.

The backscattered signal can be strongly affected by the sea state. A well-defined statistic to denote the characteristic height of the random waves in a sea state is the significant wave height (H_s), defined as the mean wave height, trough to crest, of the highest 1/3 of the waves measured.

As reported by Maresca *et al.* [50], for high sea states, the backscattered power is larger than for smaller sea states. Liu *et al.* [51] reported that the radial offshore coverage on the WFS decreased rapidly once H_s dropped below 1 m and decreased to about 60% of its peak value when H_s decreased to 0.5 m. However, the HF radar data quality remained satisfactory until H_s decreased below 0.3 m corresponding to an RMS wind speed of <3 m/s.

4.4 HF radar current observations on the West Florida Shelf

The University of South Florida (USF) Coastal Ocean Monitoring and Prediction System (COMPS) HF radar network consists of six operational sites: four long-range CODAR SeaSonde direction-finding systems operating at 4.9 MHz and two 12-element WERA phased array systems operating between 12.275 and 13.20 MHz, all overlooking an array of moored instrumentation (surface meteorology and water column ADCP velocity, temperature, and salinity), which together comprise a unique HF radar testbed along the WFS and lower Florida Keys region. The combined HF radar network provides real-time sea surface current measurements on the WFS with footprints of the network coverage shown in Figure 4.3. Detailed operational specifications and performance measurements covering these specific sites along with detailed discussion of the combined USF HF radar network, over-all system layout, remote site design, and general COMPS program can be found in [52–55]. The CODAR radial current velocity (radials) and vector totals are

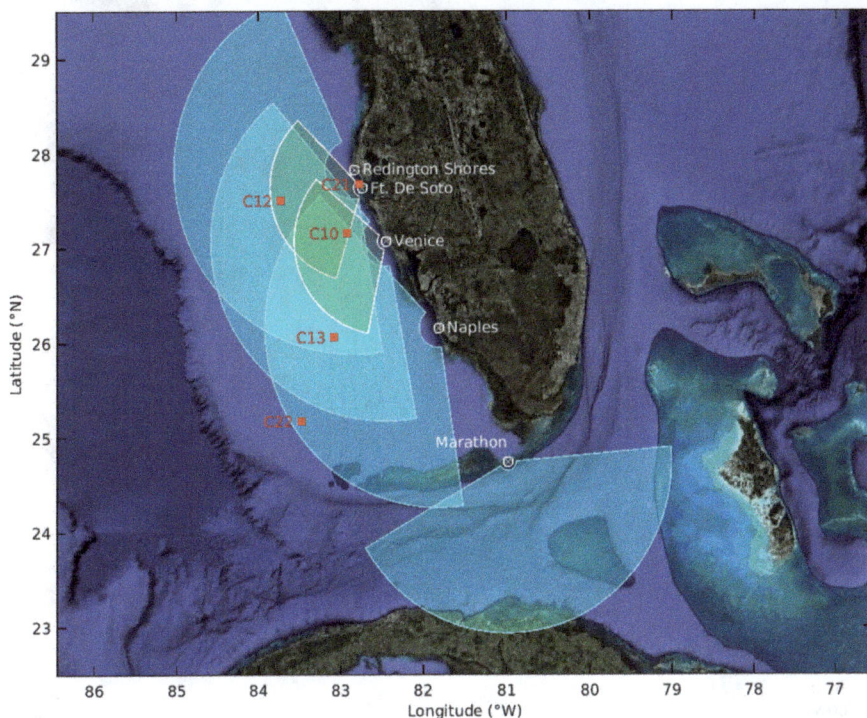

Figure 4.3   *Theoretical coverages of the USF COMPS HF radars on the WFS (Redington Shores, Ft. De Soto, Venice, and Naples, Florida) and in the Florida Keys (Marathon, Florida) overlooking the Straits of Florida. Also shown are the locations of the real-time moorings (C10, C12, C13, C21, and C22) on the WFS.*

processed using standard CODAR software with a nominal range and bearing of 5.8 km and 5°, respectively, and a transmit bandwidth of 25.734 kHz. WERA radial current velocity (radials) and vector totals are processed using standard WERA software with a nominal range and bearing of 1.5 km and 10°, respectively, but at broadside (90°) changing to 1.5 km and 20° at $+/- 60°$. The WERA systems provide the ability to set the measurement bandwidth or utilize an adaptive noise reduction algorithm to dynamically adapt the HF radar center transmit frequency and measurement bandwidth to locally varying Radio Frequency Interference (RFI) conditions.

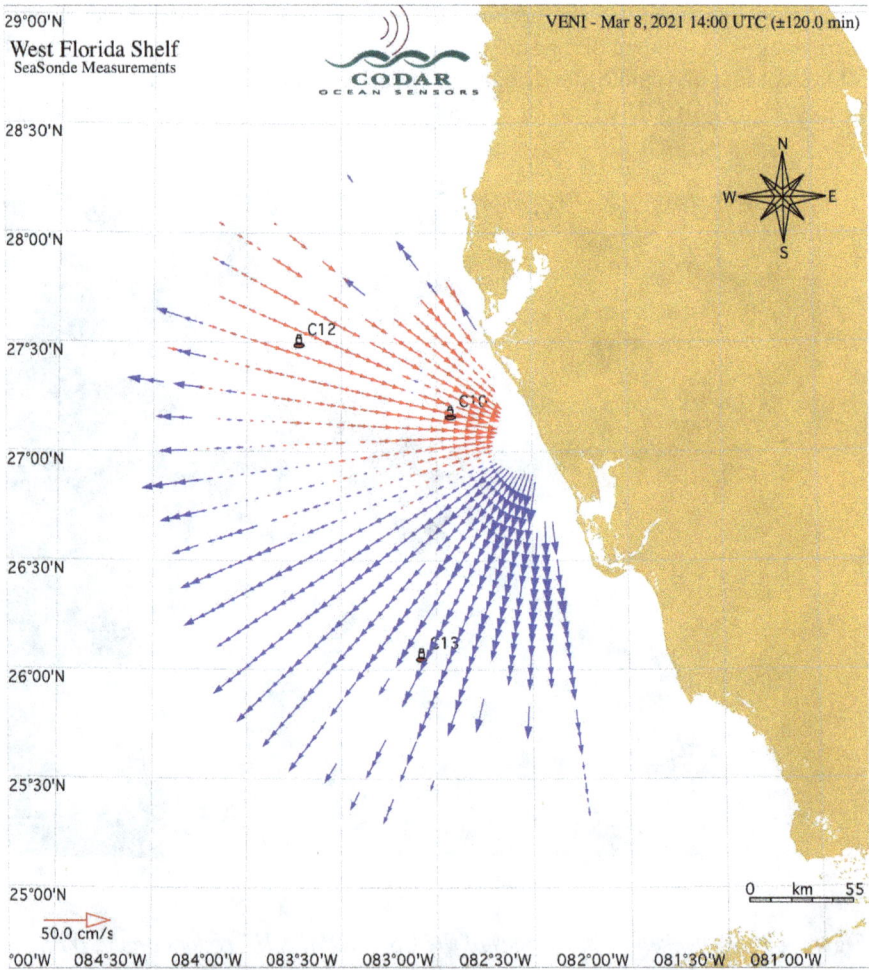

Figure 4.4   Example of measured CODAR sea surface radial component velocity field from the USF COMPS Venice Site. Dual colors signify radial vectors coming either toward the radar (red) or moving away (blue).

As previously mentioned, a single radar site can only measure the radial velocity component of the sea surface current. Thus, radial currents from two or more sites are required to calculate two-dimensional vector surface currents. Figure 4.4 presents an example of a measured sea surface radial component velocity field from the USF COMPS CODAR Venice HF radar site. Figures 4.5 and 4.6 show typical measured sea surface vector current field maps using radial velocities from USF COMPS CODAR and WERA HF radar sites, respectively.

Hourly data from each remote HF radar site are pulled via scripting to a central processing station located at the USF College of Marine Science in St. Petersburg, Florida, where the data are processed and web served in near real-time through the COMPS website http://comps.marine.usf.edu, the Southeast Coastal Ocean Regional Association (SECOORA) website http://secoora.org, and the U.S. Integrated Ocean Observing System (IOOS) National HF Radar Network's Coastal Observing Research and Development Center (CORDC) website http://cordc.ucsd. edu/projects/mapping/stats, where they are ultimately integrated into ocean models for various uses such as improved boater safety, supporting U.S Coast Guard Search

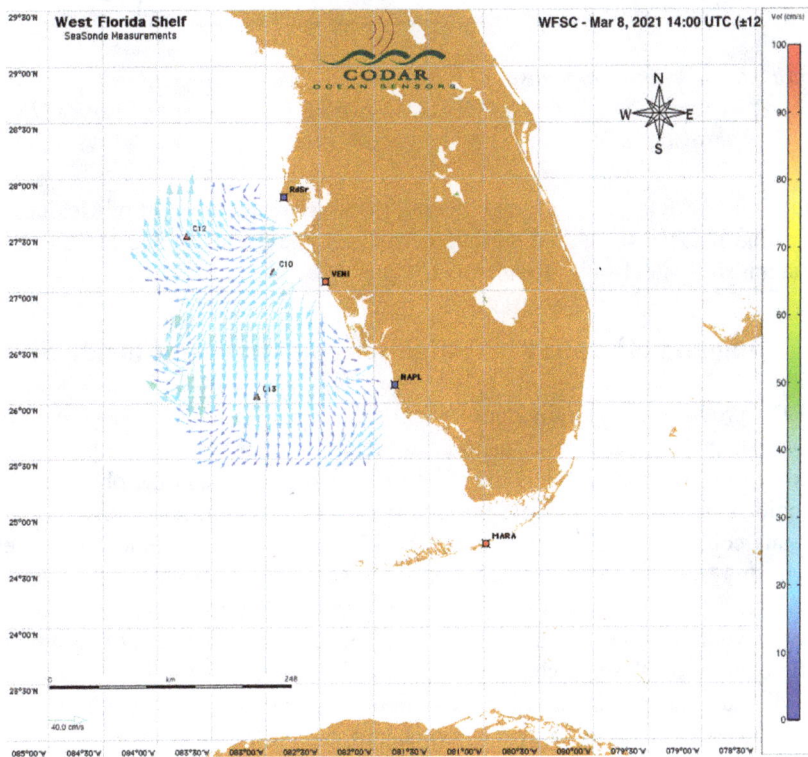

Figure 4.5   *Example of measured CODAR sea surface current field using radial velocities from the USF COMPS WFS Redington Shores, Venice, and Naples HF radar sites.*

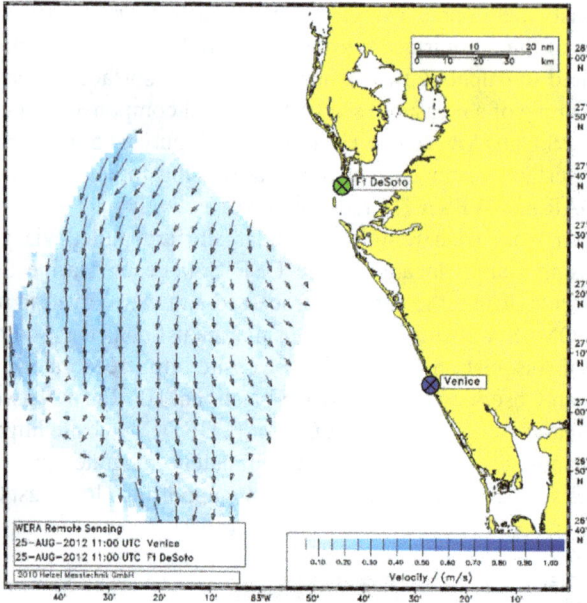

Figure 4.6   Example of measured WERA sea surface current field using radial velocities from the USF COMPS WFS Ft. De Soto and Venice HF radar sites

and Rescue (SAR) operations, and oil spill tracking within the Gulf of Mexico. An evaluation of observed radial surface currents in the Straits of Florida using the Marathon HFR site data is provided in Chapter 5 [21].

4.5   Ongoing HF radar investigations on the West Florida Shelf

CORDC provides useful individual HFR site diagnostic pages with specific performance parameters and time series for near real-time monitoring. One particularly useful operational parameter is the offshore working range and the observance of periodic variations. One possible cause of these offshore range variations is the low sea state conditions typically found on the WFS during various times of the year, which can result in low backscatter. Reduced backscatter does not necessarily produce higher current measurement errors, but it reduces the SNR and corresponding offshore working range. The SNR can be reduced further by increases of the ambient background noise through changes in the local electromagnetic RFI environment (i.e., diurnal ionospheric variations, weather-related lightning, local radio stations, near-shore manufacturing centers, airports, and ship traffic) [46].

4.5.1 An event of offshore working range drop

Observance of the CORDC offshore working range time series data during the November/December 2020 time period revealed, to varying degrees, a several-day range level drop at all the five WFS HF radar sites. According to a previous study on HF radar data returns on the WFS, the 4.9 MHz CODAR system's data return was closely related with the conditions of sea state, while the ~12.7 MHz WERA system's performance did not exhibit an obvious sea state relationship [55]. Thus, we focus herein on the WERA system for the remainder of this chapter. This low data return event is analyzed using the coastal ocean observations available during those days. Real-time data from the Ft. De Soto HF radar and C12 air–sea interaction buoy were selected for discussion because of their relative proximity and the C12's near-broadside location within the Ft. De Soto HF radar offshore coverage area (Figure 4.3).

Figure 4.7 presents the 31-day (November 11–December 11, 2020) offshore working range real-time data measured by the Ft. De Soto WERA HF radar operating between 13.1 and 13.2 MHz. A drop in offshore range is observed during the 4-day (November 26–29, 2020) time period.

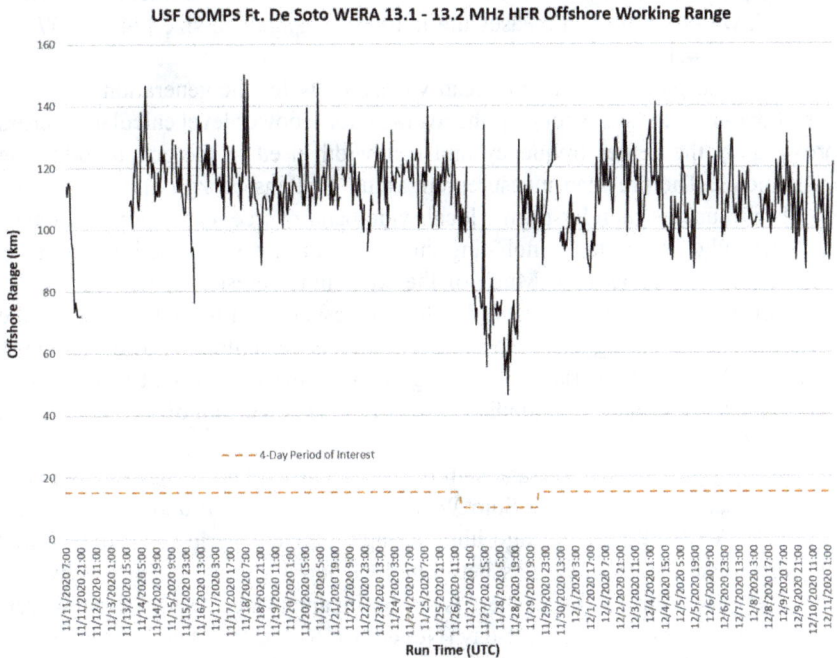

Figure 4.7   *USF COMPS Ft. De Soto WERA HF radar offshore working range covering the 31-day period between November 11 to December 11, 2020. Real-time data downloaded from CORDC's FDS site diagnostic page. Highlighted 4-day period of interest covers November 26 to 29, 2020.*

As mentioned, the extent of the offshore working range is ultimately determined by the SNR, which, in turn, depends upon several environmental variables including the background noise level at the receive antenna, the radar's operating frequency and corresponding propagation loss, and sea surface roughness. Investigations into several of these variables were conducted and are presented herein while other variables are currently under examination.

4.5.2   *Average background noise and RFI effect*

Both of the COMPS WFS WERA HF radar systems utilize the "Listen-Before-Talk" (LBT) adaptive noise reduction algorithm to dynamically adapt the HF radar center transmit frequency and measurement bandwidth to varying local RFI conditions. Performing like a spectrum analyzer using 32 linear frequency chirps with the transmitter turned off, prior to each full acquisition, a one-minute pre-scan measurement is made across the entire anticipated frequency band. Real-time analysis of the pre-scan data reveals regions of varying external noise with the quietest allowable bandwidth determined and the corresponding mid span transmit frequency selected for subsequent use in the following full measurement. If the frequency pre-scan detects too much external noise, the measurement bandwidth is reduced, which, in turn, increases the individual range cell size [44, 56, 57] as shown in Table 4.1.

Storage and plotting of the pre-scan values allow for the generation of a time series of one-minute "snapshots" of the average noise power level calculated across approximately the center frequency and bandwidth used in the actual follow-on measurements. The pre-scan measurement value contains background noise (e.g., atmospheric such as thunderstorms, local wide-band sources) and varying signals from other radio transmitters, including those very far away reflected by the ionosphere, which appear as RFI. Much of the RFI clutter present in the pre-scan is reduced/mitigated during subsequent WERA processing so it is not the actual value the WERA "sees" during later processing. However, examination at this stage is useful in looking at the variation and magnitude of the RFI present as compared to the characteristic shape and median value of the average sum of the background noise and RFI data.

Figure 4.8 presents the average sum of the background noise and RFI pre-scan data as measured by the Ft. De Soto WERA HF radar system under the application of WERA's LBT adaptive algorithm during the 31-day period. The transmitted operational central frequency was between 13.1 and 13.2 MHz with bandwidths varying between 25 to 100 kHz. There is no significant change observed in the average background noise during the 4-day (November 26–29, 2020) period as opposed to the entire 30-day period. This indicates that the drop in offshore range is related to a reduction in backscattered signal and not an increase in local background noise.

The distance between the Venice and Ft. De Soto WERA sites shown in Figure 4.3 is ~68.5 km. As expected, graphical results for the Venice WERA HF radar pre-scan data reveal similar results as that of presented for Ft. De Soto in Figure 4.8, with many of the same RFI signals but differing slightly in strength due

USF COMPS Ft. De Soto WERA 13.1 - 13.2 MHz Frequency Band Var Tx and 25 - 100 kHz BW

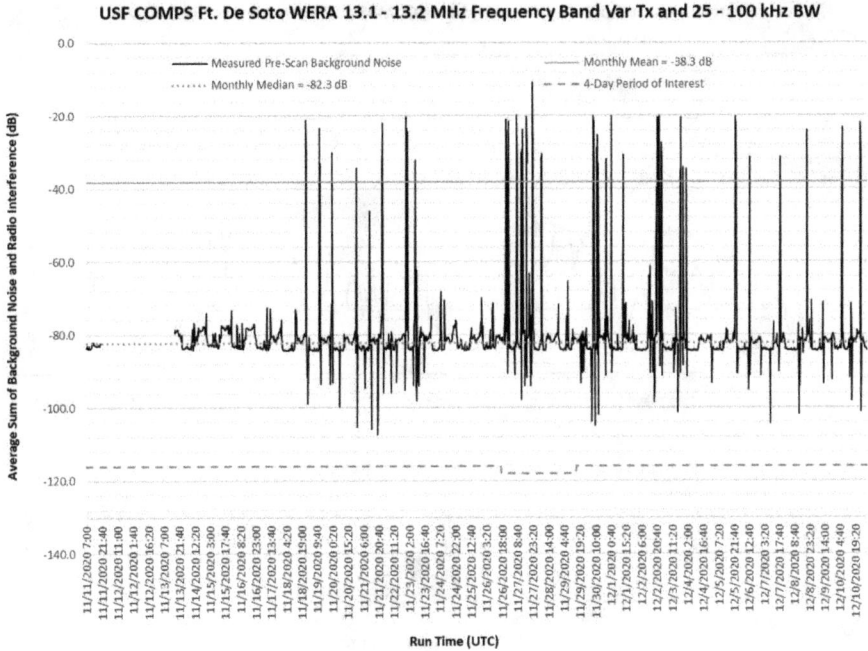

Figure 4.8   *USF COMPS Ft. De Soto WERA HF Radar average sum of background noise and radio interference during the 31-day period November 11 to December 11, 2020. Highlighted 4-day period of interest covers November 26 to 29, 2020.*

Table 4.2   *Pre-scan summary of the median value of the average sum of measured background noise and radio frequency interference*

Days/Date	Ft. De Soto WERA (dB)	Venice WERA (dB)
31-day (Nov 11–Dec 11, 2020)	−82.3	−82.7
4-day (Nov 26–Nov 29, 2020)	−82.0	−82.3

to the separation distance. Table 4.2 summarizes the background noise levels for both the Ft. De Soto and Venice WERA HF radar sites.

4.5.3   Atmospheric radio refractivity (N) effect

Figure 4.9 presents the computed atmospheric radio refractivity *(N)* for the period November 11 through December 11, 2020, using data obtained from the COMPS C12 air/sea interaction buoy. The value of *N* is computed according to the ITU-R [48] as:

$$N = \frac{77.6}{T}\left(P + 4\,810\frac{E}{T}\right) \tag{4.5}$$

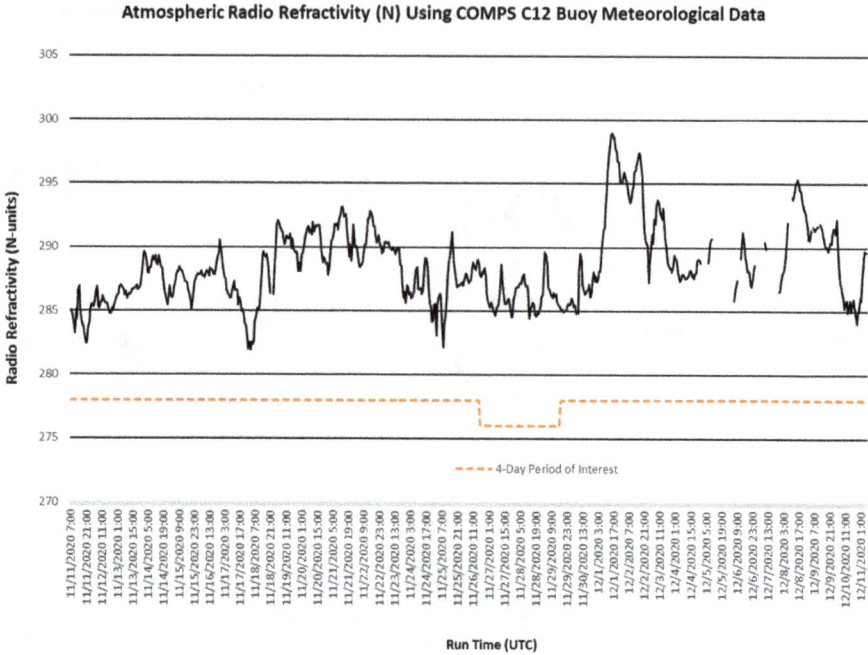

Figure 4.9   *USF COMPS C12 meteorological buoy computed atmospheric radio refractivity covering the 31-day period November 11 to December 11, 2020. Highlighted 4-day period of interest covers November 26 – 29, 2020.*

where P = atmospheric pressure (hPa), E = water vapor pressure (hPa), and T = absolute temperature (K). The relationship between water vapor pressure (E) and relative humidity (H) is given by:

$$E = (H)(Es)/100 \tag{4.6}$$
$$Es = a \exp(bt/(t + c)) \tag{4.7}$$

where H = relative humidity (%), t = temperature (°C), Es = saturation vapor pressure (hPa) at the temperature t (°C), and the coefficients a, b, c for water are a = 6.1121, b = 17.502, c = 240.97 (valid between −20° to +50° with a ±0.20% accuracy).

As expected, no significant change was observed in the atmospheric radio refractivity values during November 26–29, 2020, thus, confirming and quantifying prior well-known expressed comments that the strength of the first-order echo returns is not dependent upon air temperature, relative humidity, or barometric pressure within the HF frequency band.

4.5.4   Wind speed effect

Figure 4.10 presents the wind speed for the same 31-day period using wind data converted to 10 m height obtained from the COMPS C12 air/sea interaction buoy. There is a similar drop in the wind speed during the same 4-day time period. Superimposed on Figure 4.10 is the computed wind speed RMS during the 4-day low wind (~3.4 m/s) event and the remaining 27 days (9.0 m/s). While the computed 4-day 3.4 m/s wind speed RMS compares favorably to the <3 m/s wind speed RMS previously observed on the WFS by Liu *et al.* [53] and briefly discussed in Section 4.3.4, close examination of Figure 4.10 reveals other periods in the record of equally low wind speed RMS values without a corresponding drop in the offshore working range of the Ft. De Soto WERA HF radar (Figure 4.7).

Although a telling clue in itself, the wind speed is but a contributor to the underlying cause. Wave height is affected by wind speed, wind duration (or how long the wind blows), and fetch (the distance over water that the wind blows in a single direction). If the wind speed is low, only small waves appear, regardless of wind duration or fetch. Weakening winds may result in the reduction of the sea state and corresponding scattering strength of the rough sea surface which, in turn, increases the backscattered signal's propagation loss. Work is continuing on examining the interaction of energy loss and backscattering strength in terms of sea state and seawater physical conditions (e.g., water temperature and salinity) as they may relate to attenuation of the EM wave propagation along the sea surface.

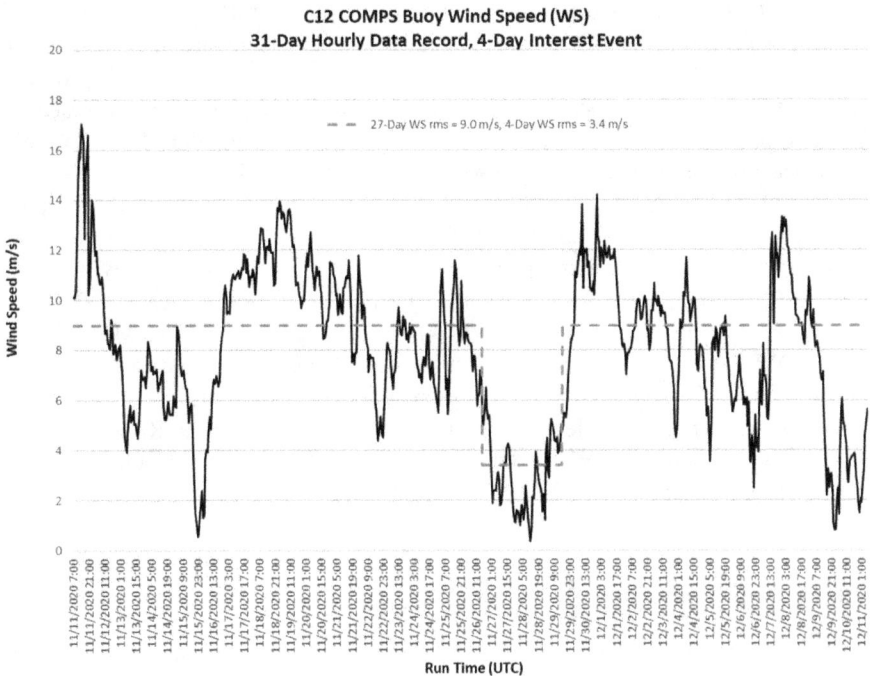

Figure 4.10   *USF COMPS C12 meteorological buoy wind speed at 10m height covering the 31-day period November 11 to December 11, 2020. Highlighted 4-day period of interest covers November 26–29, 2020.*

4.6   Summary

Sea surface current measurements using oceanographic HF radar systems are reviewed followed by a brief discussion of recent ongoing HF radar observations on the WFS. Overall HF radar system performance of current measurement depends upon many external factors including oceanic conditions, type of HF radar system used, SNR of the signal received at the shore-based radar antennas, frequency resolution of the Doppler spectrum, and the accurate identification of the Bragg peaks. Challenges of HF radar current observation on the WFS are often manifested in low SNR, reduced data returns, and offshore range. A careful review of these factors along with a thorough evaluation of the unique environmental characteristics surrounding the site location under consideration for use is required in order to achieve a successful measurement outcome.

Acknowledgment

Support was provided by NOAA via the U.S. IOOS Office through the Southeast Coastal Ocean Observing Regional Association (SECOORA); NOAA Award Number NA16NOS0120028, SECOORA Subnumber: IOOS.16 (028) USF. BW.OBS-HFR.1. This work was supported by the Gulf Research Program of the National Academies of Sciences, Engineering, and Medicine under the Grant Agreement number 2000009917. The content is solely the responsibility of the authors and does not necessarily represent the official views of the Gulf Research Program or the National Academies of Sciences, Engineering, and Medicine. This work was also supported by the University of South Florida's College of Marine Science via the Coastal Ocean Monitoring and Prediction System (COMPS) program. USF seagoing buoy activity is attributed to J. Law with J. Donovan providing real-time data management competency. USF HF Radar Network activity is attributed to C. Merz. We thank H. Parikh (CODAR Ocean Sensors, Ltd., CODAR), K-W Gurgel (Univ. of Hamburg, WERA), and L. Petersen (Helzel Messtechnik GmbH, WERA) for their technical discussions and suggestions and to L. Petersen (Helzel Messtechnik GmbH, WERA) for creating the custom script used to retrieve the pre-scan average noise level of the LBT selected band from a .RAW_ascii file and append it to a log file. We also thank J. Chen, M. Otero, and T. Cook (CORDC) for HF radar diagnostic page data downloads.

References

[1]   Weisberg R.H., Zheng L., Liu Y. 'Basic tenets for coastal ocean ecosystems monitoring: a West Florida perspective'. in *Coastal Ocean Observing Systems*; London, UK: Elsevier (Academic Press); 2015. pp. 40–57.
[2]   Weisberg R.H., Liu Y., Lembke C., Hu C., Hubbard K., Garrett M. 'The Coastal Ocean Circulation Influence on the 2018 West Florida Shelf *K*.

brevis red tide bloom'. *Journal of Geophysical Research: Oceans.* 2019;**124**(4):2501–12.

[3] Merz C.R., Weisberg R.H., Smith P., Law J., Cole R.D., Donovan J. 'Development of a self-contained, satellite based, surface buoy position tracking device'. MTS/IEEE Oceans 2007 Conference Proceedings; 2007.

[4] Liu Y., Weisberg R.H., Hu C., Kovach C., Riethmüller R. 'Evolution of the loop current system during the deepwater horizon oil spill event as observed with drifters and satellites'. in *Monitoring and Modeling the Deepwater Horizon oil Spill: A Record-Breaking Enterprise. Geophysical Monograph Series*; 2011. pp. 91–101.

[5] Weisberg R.H., Liu Y., Mayer D.A. 'Mean circulation on the West Florida continental shelf observed with long-term moorings'. *Geophysical Research Letters.* 2009;**36**:L19610.

[6] Liu Y., Kerkering H., Weisberg R.H. (eds.). *Coastal Ocean Observing Systems.* **461**. London: Academic Press; 2015.

[7] Abascal A.J., Castanedo S., Medina R., Losada I.J., Álvarez-Fanjul E. 'Application of HF radar currents to oil spill modelling'. *Marine Pollution Bulletin.* 2009;**58**(2):238–48.

[8] Zelenke B.C., Moline M.A., Jones B.H., Ramp S.R., Crawford G.B., Largier J.L. 'Evaluating connectivity between marine protected areas using CODAR high-frequency radar'. Proceeding at OCEANS MTS/IEEE, Biloxi, MS 2009; 2009.

[9] Roarty H., Glenn S., Kohut J., *et al.* 'Operation and application of a regional high-frequency radar network in the mid-Atlantic Bight'. *Marine Technology Society Journal.* 2010;**44**(6):133–45.

[10] Lewis J.K., Shulman I., Blumberg A.F. 'Assimilation of CODAR observations into ocean models'. *Continental Shelf Research.* 1998;**18**:541–59.

[11] Paduan J.D., Shulman I. 'HF radar data assimilation in the Monterey Bay area'. *Journal of Geophysical Research.* 2004;**109**(C07S09).

[12] Barth A., Alvera-Azcárate A., Weisberg R.H. 'Assimilation of high-frequency radar currents in a nested model of the West Florida shelf'. *Journal of Geophysical Research.* 2008;**113**(C08033).

[13] Barrick D.E., Evans M.W., Weber B.L. 'Ocean surface currents mapped by radar'. *Science.* 1977;**198**(4313):138–44.

[14] Lipa B., Nyden B. 'Directional wave information from the SeaSonde'. *IEEE Journal of Oceanic Engineering.* 2005;**30**(1):221–31.

[15] Gill E.W., Huang W. 'HF radar in a marine environment' in Huang W., Gill E.W. (eds.). *Ocean Remote Sensing Technologies: High Frequency, Marine and GNSS-Based Radar.* IET; 2021.

[16] Crombie D.D. 'Doppler spectrum of sea echo at 13.56 Mc./s'. *Nature.* 1955;**175**(4459):681–2.

[17] Stewart R.H., Joy J.W. 'HF radio measurements of surface currents'. *Deep Sea Research and Oceanographic Abstracts.* 1974;**21**(12):1039–49.

[18] Barrick D.E. 'Extraction of wave parameters from measured HF radar sea-echo Doppler spectra'. *Radio Science.* 1977;**12**(3):415–24.

[19]   Barrick D.E. 'The ocean waveheight nondirectional spectrum from inversion of the HF sea-echo Doppler spectrum'. *Remote Sensing of Environment.* 1977b;**6**(3):201–27.

[20]   Paduan J., Graber H. 'Introduction to high-frequency radar: reality and myth'. *Oceanography.* 1997;**10**(2):36–9.

[21]   Liu Y., Merz C.R., Weisberg R.H., *et al.* 'An initial evaluation of high-frequency radar radial currents in the Straits of Florida in comparison with altimetry and model products'. *Ocean Remote Sensing Technologies: High Frequency, Marine and GNSS-Based Radar.* IET; 2021.

[22]   Barrick D.E. 'HF radio oceanography – a review'. *Boundary-Layer Meteorology.* 1978;**13**(1–4):23–43.

[23]   Barrick D. 'First-Order theory and analysis of MF/HF/VHF scatter from the sea'. *IEEE Transactions on Antennas and Propagation.* 1972a;**20**(1):2–10.

[24]   Wang W., Gill E.W. 'High-resolution spectral estimation of HF radar data for current measurement applications'. *Journal of Atmospheric and Oceanic Technology.* 2015;**32**(8):1515–25.

[25]   High Frequency Radar (HF Radar). 1999.Rutgers University Instutiute of Marine and Coastal Sciences. Available from https://web.archive.org/web/20210309192121/https://marine.rutgers.edu/cool/education/class/josh/hf_radar.html [Accessed 9 Apr 2021].

[26]   Barrick D., Snider J. 'The statistics of HF sea-echo Doppler spectra'. *IEEE Transactions on Antennas and Propagation.* 1977;**AP-25**(1):19–28.

[27]   Barros F.F.C. Validation and quality assessment of HF radar wave measurements in the algarve shore. 2019.Master's Thesis, Geophysical Sciences (Meteorology and Oceanography). Available from http://hdl.handle.net/10451/40479.

[28]   Kirincich A. 'Remote sensing of the surface wind field over the coastal ocean via direct calibration of HF radar backscatter power'. *Journal of Atmospheric and Oceanic Technology.* 2016;**33**(7):1377–92.

[29]   Barrick D.E., Peake W.H. 'A review of scattering from surfaces with different roughness scales'. *Radio Science.* 1968;**3**(8):865–8.

[30]   Barrick D.E., Derr V.E. 'Remote sensing of the troposphere' in Derr V.E. (ed.). *Remote Sensing of Sea State by Radar.* Washington, DC: U.S. Government Printing Office; 1972.

[31]   Wait J.R. 'Theory of HF ground wave backscatter from sea waves'. *Journal of Geophysical Research.* 1966;**71**(20):4839–42.

[32]   Saviano S., Kalampokis A., Zambianchi E., Uttieri M. 'A year-long assessment of wave measurements retrieved from an HF radar network in the Gulf of Naples (Tyrrhenian Sea, Western Mediterranean sea)'. *Journal of Operational Oceanography.* 2019;**12**(1):1–15.

[33]   Shay L.K., Cook T.M., Peters H., *et al.* 'Very high-frequency radar mapping of surface currents'. *IEEE Journal of Oceanic Engineering.* 2002;**27**(2):155–69.

[34]   Martinez-Pedraja J., Shay L.K., Haus B.K., Whelan C. 'Interoperability of SeaSondes and Wellen radars in mapping radial surface currents'. *Journal of Atmospheric and Oceanic Technology.* 2013;**30**(11):2662–75.

[35] Nadai A., Kuroiwa H., Mizutori M., Sakai S. 'Measurement of ocean surface currents by the CRL HF ocean surface radar of FMCW type. Part 2 current vector'. *Journal of Oceanography*. 1999;**55**(1):13–30.

[36] Gurgel K.-W., Antonischki G., Essen H.-H., Schlick T. 'Wellen radar (WERA): a new ground-wave HF radar for ocean remote sensing'. *Coastal Engineering*. 1999a;**37**(3–4):219–34.

[37] Gurgel K.-W., Essen H.-H., Kingsley S.P. 'High-frequency radars: physical limitations and recent developments'. *Coastal Engineering*. 1999a;**37**(3–4):201–18.

[38] Schmidt R. 'Multiple emitter location and signal parameter estimation'. *IEEE Transactions on Antennas and Propagation*. 1986;**34**(3):276–80.

[39] Laws K.E., Fernandez D.M., Paduan J.D. 'Simulation-based evaluations of HF radar ocean current algorithms'. *IEEE Journal of Oceanic Engineering*. 2000;**25**(4):481–91.

[40] Barrick D., Lipa B. 'Evolution of bearing determination in HF current mapping radars'. *Oceanography*. 1997;**10**(2):72–5.

[41] de Paolo T., Terrill E. 'Skill assessment of resolving ocean surface current structure using compact-antenna-style HF radar and the music direction-finding algorithm'. *Journal of Atmospheric and Oceanic Technology*. 2007;**24**(7):1277–300.

[42] Huang W., Gill E.W. 'HF surface wave radar'. *Wiley Encyclopedia of Electrical and Electronics Engineering. https://doi.org/10.1002/047134608X. W8376.* 2019:1–11.

[43] Emery B., Washburn L. 'Uncertainty estimates for SeaSonde HF radar ocean current observations'. *Journal of Atmospheric and Oceanic Technology*. 2019;**36**(2):231–47.

[44] Merz C.R., Liu Y., Gurgel K.-W., Petersen L., Weisberg R.H. 'Effect of radio frequency interference (RFI) noise energy on WERA performance using the "Listen Before Talk" adaptive noise procedure on the West Florida Shelf' in Liu Y., Kerkering H., Weisberg R.H. (eds.). *Coastal Ocean Observing Systems*. London: Elsevier; 2015. pp. 229–47.

[45] Chapman R.D., Shay L.K., Graber H.C., *et al.* 'On the accuracy of HF radar surface current measurements: intercomparisons with ship-based sensors'. *Journal of Geophysical Research: Oceans*. 1997;**102**(C8):18737–48.

[46] Essen H.-H., Gurgel K.-W., Schlick T. 'On the accuracy of current measurements by means of HF radar'. *IEEE Journal of Oceanic Engineering*. 2000;**25**(4):472–80.

[47] Tiuri M., Sihvola A., Nyfors E., Hallikaiken M. 'The complex dielectric constant of snow at microwave frequencies'. *IEEE Journal of Oceanic Engineering*. 1984;**OE-9**(5):377–82.

[48] The radio refractive index: its formula and refractivity data. Recommendation ITU-R P.453-9. International Télécommunication Union, Radiocommunication Sector of ITU, P series, Radiowave propagation; 2003.

[49] Tamosiunaite M., Tamosiunas S., Zilinskas M., Tamosiuniene M. 'Atmospheric attenuation due to humidity'. IntechOpen; 2011.

[50]   Maresca S., Braca P., Grasso R., Horstmann J. 'The impact of sea state on HF surface-wave radar ship detection and tracking performances'. Proceedings of the OCEANS'15 MTS/IEEE Conference; *Genova*; 2015.

[51]   Liu Y., Weisberg R.H., Merz C.R., Lichtenwalner S., Kirkpatrick G.J. 'HF radar performance in a low-energy environment: CODAR SeaSonde experience on the West Florida Shelf'. *Journal of Atmospheric and Oceanic Technology.* 2010;**27**(10):1689–710.

[52]   Liu Y., Weisberg R.H., Merz C.R. 'Assessment of CODAR SeaSonde and WERA HF radars in mapping surface currents on the West Florida Shelf'. *Journal of Atmospheric and Oceanic Technology.* 2014;**31**(6):1363–82.

[53]   Merz C.R., Weisberg R.H., Liu Y. 'Evolution of the USF/CMS CODAR and WERA HF radar network'. Oceans'12 MTS/IEEE Conference Proceedings – Xplore; *Hampton Roads, VA*; 2012.

[54]   Merz C.R. 'An Overview of the Coastal ocean Monitoring and Prediction System (COMPS)'. Oceans'01 MTS/IEEE Conference Proceedings – IEEE Xplore; *Honolulu, HI*; 2001. pp. 1183–7.

[55]   Liu Y., Merz C.R., Weisberg R.H., Venkatesan R., Tandon A. 'Data return aspects of CODAR and WERA high frequency radars in mapping currents' in Venkatesan R., Tandon A., D'Asaro E., Atmanand M. (eds.). *Observing the Oceans in Real Time.* Springer; 2017. pp. 227–41.

[56]   Gurgel K.-W., Barbin Y., Schlick T. Radio frequency interference suppression techniques in FMCW modulated HF radars. OCEANS 2007 – Europe; Aberdeen, UK; 2007. pp. 1–4.

[57]   Cosoli S. 'Implementation of the listen-before-talk mode for SeaSonde high-frequency ocean radars'. *Journal of Marine Science and Engineering.* 2020;**8**(57):57–13.

Chapter 5

An initial evaluation of high-frequency radar radial currents in the Straits of Florida in comparison with altimetry and model products

Yonggang Liu[1], Clifford R. Merz[1], Robert H. Weisberg[1], Lynn K. Shay[2], Scott M. Glenn[3], and Michael J. Smith[3]

A long-range (4.9 MHz) CODAR SeaSonde was deployed at Marathon, Florida (along the Florida Keys chain), in December 2019 to observe surface currents in the Straits of Florida. An analysis of the initial seven months of High-Frequency Radar (HFR) data provides an opportunity to assess the CODAR performance in this area of complex ocean current dynamics and to compare these HFR radial data with the surface geostrophic currents derived from the along-track and gridded sea surface heights from satellite altimetry, and with surface currents from the data assimilative Gulf of Mexico HYbrid Coordinate Ocean Model (HYCOM). The HFR and the along-track altimetry-derived radial velocity components agree to within a root-mean-square difference (RMSD) of about 21 cm/s in a region of strong currents (the Florida Current, 100~200 cm/s). The agreement with the gridded altimetry product varies within the HFR footprint with an RMSD range of 16.2–61.2 cm/s and mean value of 34.1 cm/s when spatially averaged over the HFR domain. Lesser agreement is found with the HYCOM output, wherein the RMSD range is 15.2–82.3 cm/s and the mean value is 39.9 cm/s. The largest RMSD values are generally within the Florida Current frontal regions where frequent mesoscale eddy activity occurs. These findings have important implications for users of both altimetry data products and data assimilative numerical ocean circulation models.

5.1   Introduction

Shore-based High-Frequency Radars (HFRs) measure surface currents over ranges up to 200 km from the coastline [1]. Their relatively large spatial coverages make them an important component of coastal ocean observing systems [2]. HFR

[1]College of Marine Science, University of South Florida, United States
[2]Rosenstiel School of Marine Science, University of Miami, United States
[3]Department of Marine and Coastal Sciences, Rutgers University, United States

measured current (and wave) data are widely used in various oceanographic applications, e.g. [3–5], as well as for assimilation in numerical ocean circulation models, e.g. [6–9].

Another important data source for assimilation into ocean circulation models is satellite altimetry, e.g. [10, 11]. By repeatedly measuring sea surface height (SSH) over the global ocean, satellite altimeters find their applications traditionally in large scale ocean circulation and climate research, e.g. [12, 13], as well as in regional oceanography, e.g. [14–16]. Progress in along-track post-processing and re-tracking has also led to satellite altimetry applications in coastal ocean regions [17, 18]. Two levels of satellite altimetry SSH data are widely used in oceanography: along-track and merged/gridded products. The along-track SSH data are generated by individual altimeters with high resolution along the satellite ground tracks. However, the repeat cycles differ among altimetry satellites. For example, the main altimetry satellites, the Jason series, have a repeat cycle of about nine days. Other satellites presently in orbit have longer repeat cycles. Multiple satellite along-track SSH data are merged through optimal interpolation to generate gridded products. Along-track altimetry products are preferred by the modeling community for data assimilation, while merged/gridded products, in particular altimetry-derived surface current fields, are widely used by the non-modeling community as a source of ocean circulation information. Several altimetry-derived surface current products are provided by different research groups, e.g., the Archiving, Validation and Interpretation of Satellite Oceanographic Data (AVISO+) gridded product [19], the Ocean Surface Currents Analyses Real-time (OSCAR) [20, 21], the Geostrophic and Ekman Current Observatory (GEKCO) [22], and the GlobCurrent [23]. These data products have been validated in different ways with other data sets [24–27], and some of these data products have been shown to perform better than data assimilative numerical models for simulating surface drifter trajectories in the Gulf of Mexico (GoM) [26]. Given that the quality of such data products may vary widely with location and time, it is important to assess the accuracy of both the data and model products before using these in oceanographic applications [28].

The GoM Loop Current (LC) system is the main dynamical feature of the GoM. The LC enters the Gulf through the Yucatan Channel, exits through the Straits of Florida as the Florida Current, and then continues farther north as the Gulf Stream. Beginning with the seminal work in 1972 by Reid [29], the LC's evolution within the GoM through its complicated intrusion, retraction, and eddy separation cycles have been active research topics [15, 30–32]. The LC not only plays an important role in air–sea interactions [33] but also interacts with surrounding continental slopes and shelves [34, 35] and influences coastal ecosystems [36, 37]. In particular, the West Florida Shelf (WFS) region near the Dry Tortugas is referred to as the "pressure point" because LC/shelf interactions there can set the entire shelf into motion [35]. By doing so, it may also serve as an anchor, impeding the penetration of the LC back into the GoM until the LC separates from the pressure point [16].

Recently, the National Academies of Sciences, Engineering, and Medicine (NASEM) implemented the Understanding Gulf of Mexico System (UGOS), as part of their Gulf Research Program (GRP), to advance our knowledge on this

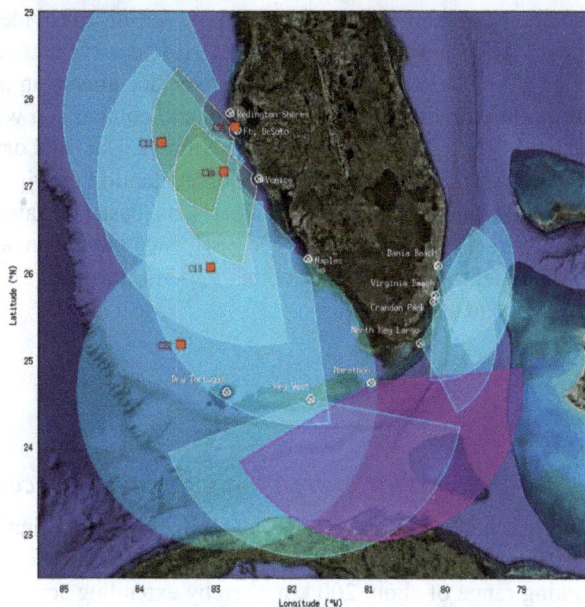

Figure 5.1   *Theoretical coverages of the HFRs in the Florida Keys (the Dry Tortugas/Fort Jefferson National Park, and the cities of Key West and Marathon, Florida), on the WFS (Redington Shores, Ft. De Soto, Venice, and Naples, Florida), and on the Florida east coast (North Key Largo, Virginia Beach, Crandon Park, and Dania Beach, Florida). Also shown are the locations of the real-time moorings (C10, C12, C13, C21, and C22) on the WFS. Buoy C22 is also called the Pressure Point Mooring.*

decades-long science conundrum of what controls the LC evolution. Part of UGOS is the deployment of a set of three HFRs along the Florida Keys at the Dry Tortugas/ Fort Jefferson National Park, and within the Florida sites of Key West and Marathon, respectively (Figure 5.1). While the Dry Tortugas and Key West HFRs are still awaiting site permits, the Marathon CODAR SeaSonde was successfully installed at the Curry Hammock State Park (24° 44.417' N, 80° 59.000' W) on Crawl Key and has been in operation since mid-December 2019. The radial velocity component data collected thus far provide an opportunity for comparison with other data and model products. It is further noted that UGOS will eventually also have HFRs overlooking the LC inflow region and that the planned Dry Tortugas and Key West HFRs will also include the southern portion of the WFS and, in particular, the region currently covered by an USF installed and maintained oceanographic/meteorologi- cal buoy C22 (the Pressure Point Mooring), which is also funded by NASEM UGOS (Figure 5.1). The outflow region HFRs will further help to fill the gaps between the five-site HFR network on the WFS [38–40] and the four-site HFR network along the Miami coast [41, 42].

Basic information about HFR as an ocean remote sensor is provided by Gill and Huang in Chapter 1 [43], and the mechanisms of HFR current mapping are introduced by Merz *et al.* in Chapter 4 [44]. This chapter documents an application of the HFR in current measurement, its performance, and comparison with other data sets. The remainder of this chapter is arranged as follows: Data and data processing are described in Section 5.2. Velocity comparison/evaluation metrics are provided in Section 5.3. Radial velocity component comparisons, based on satellite altimetry products, both along-track and gridded/merged SSH, and output from an operational ocean circulation model, are given in Sections 5.4–5.6, respectively. Section 5.7 provides a discussion and summary.

5.2   Data sets

5.2.1   *High-frequency radar current data and post-processing*

The Marathon HFR is a CODAR SeaSonde that operates at a nominal transmit frequency of 4.9 MHz with the intended purpose of observing surface currents out to an offshore working range of about 200 km, thereby extending across the Straits of Florida to Cuba (Figure 5.1) with nominal range and bearing resolutions of 5.8 km and 5°, respectively. Data collection started on 28 December 2019 using a measured nearshore "walking" antenna pattern followed on 12 February 2020 with a measured 1-km offshore "boat" antenna pattern, which provided additional angular coverage to the east. For a consistent comparison, we use the 7-month interval from 12 February 2020 to 10 September 2020.

The quality of radial current data is dependent on the quality of the Doppler spectra produced from the backscatter of radio waves off the ocean surface. At times, the HFR receiver may receive unwanted background noise in the form of man-made radio interference (RF) or naturally occurring ionospheric echoes which may be processed by the real-time radial current processing software. Human-in-the-loop data evaluation is necessary to ensure that good data are not discarded and bad data are not distributed [45]. To accomplish this task, we developed an HFR post-processing approach with five major and distinct steps. The first-step is to examine the system diagnostics as the first check to identify time periods requiring special attention. The second step is to plot the spectra to ensure that the first order lines, which are used to inform the real-time software as to where to process target echoes inside the Doppler spectra into radial current vectors, are adjusted correctly and to identify any sources of outside interference that may affect data quality. The third is to recalculate the radial current vectors with the best antenna patterns and first-order line settings to produce the best radial current vectors based on steps one and two. The fourth step is to apply the full suite of "required" and "recommended" Quality Assurance/Quality Control of Real-Time Oceanographic Data (QARTOD) [45] radial tests to the reprocessed radials, flagging the vectors for each of the tests that are not passed. As a final check, the radial currents before and after post-processing are plotted and visually inspected.

The data were post-processed starting with a thorough review of the site's radial diagnostics. Useful diagnostics include sea echo amplitudes and phases, signal-to-noise, noise floor, and average radial bearing. A step change in sea echo phases and amplitudes or a change in average radial bearing with no coinciding change in antenna bearing can signal an error in the configuration that will need to be corrected in reprocessing. Data from stations reporting low signal-to-noise and/or high background noise diagnostics for long periods of time are often indicative of equipment failures and/or environmental conditions and were excluded from processing. The diagnostics were used to identify possible events that may require a new antenna pattern measurement or equipment checks, but none were found.

Radar cross-spectra were plotted and first-order lines were adjusted when necessary in order to better define the areas around the Bragg scattering peak, which were to be processed into radial currents. Maximum velocity was raised from 150 cm/s to 200 cm/s in order to view the higher speeds of the Florida Current. At this point, interference may be observed both as RF interference from an unknown man-made source in the form of vertical striping and as ionospheric echoes which appear as a band of range cells with strong returns across a broad range of Doppler frequencies. The times when interference occurred, along with their corresponding ranges and bearings, were noted in a database. These data were inserted into the appropriate radial files as an operator flag in order so that end users have the option to filter out. The diagnostics along with the first-order line settings notes were used to inform the processing of spectra into radials using the appropriate configuration settings and files.

For directional calibration, a measured antenna pattern [46] generated by combining ship-measured and shore-based walking patterns was judged to be the best and applicable to the entire data set since no changes in the diagnostics were identified.

Reprocessed radial currents were then tested based on Version 1.0 of the QARTOD Manual for Real-Time Quality Control of High Frequency Radar Surface Current Data by U.S. Integrated Ocean Observing System (2016) [45]. The QARTOD radial tests that have been applied are listed in Table 5.1 along with threshold values that were chosen to implement the tests.

The syntax is a collection of tests that ensures proper formatting and existence of fields within a radial file. The radial file is tested for proper parsing and content, file format, site code, appropriate time stamp, time zone, site coordinates, antenna pattern type, and internally consistent row/column specifications. For an example, the test requires that the following metadata be present in the file: file type LLUV, site code, timestamp, site coordinates, antenna pattern type, and time zone. Other requirements include: (1) the file name timestamp must match the timestamp reported within the file, (2) radial data tables (Longitude, Latitude, U, V) must not be empty, (3) radial data table columns stated must match the number of columns reported for each row (4) site location must be within the following ranges: $-180 \leq$ Longitude ≤ 180 and $-90 \leq$ Latitude ≤ 90, and (5) time zone must be Greenwich Mean Time. If any of the tests fails, the radial file is not created to community standards and the entire file is flagged as rejected [45].

Table 5.1   Quality control flags of the HFR radial currents

Test Code	Test Name	Suspect Flag	Fail Flag
QC06	Syntax	N/A	See text
QC07	Max Threshold	N/A	If RSPD > RSPDMAX, RSPDMAX = 200 cm/s
QC08	Valid Location	N/A	If FLOC = 128
QC09	Radial Count	RCNT >= RC_MIN and RCNT <= RC_LOW	If RCNT < RC_MIN, RCLOW = 270 RCMIN = 90
QC10	Spatial Median	N/A	If VELO – median(VELO-neighboring) > CURLIM, RCLIM = 2.1 range cells ANGLIM = 10 degrees CURLIM = 50 cm/s
QC11	Temporal Gradient	GRADWARN <= GRADIENT < GRADFAIL	GRADIENT >= GRADFAIL, GRADWARN = 25 cm/s GRADFAIL = 32 cm/s

RF interference may cause strong radar echoes in the spectra in areas near the Bragg peaks where radial currents are processed from. These strong echoes show as high velocities in the radial current maps. A maximum threshold test was written to ensure that a radial current speed is not unrealistically high. If a radial speed exceeds the maximum reasonable surface radial velocity as defined by the HFR operator, then this speed realization will be flagged [45].

HFR surface velocity data are measured from a stationary land-based remote sensor and are placed on a fixed grid created based on range and bearing from the receiver location. Sometimes, the software may place velocity data over land. A valid location test was implemented with a high-resolution coastline to determine whether a point near a distant coast is a valid ocean point or a land point. All points that do not lay over water are flagged [45].

Low radial counts indicate poor radial map coverage. The radial count test rejects radial files that contain less than a minimum number of radial vectors (RCMIN). The RCMIN threshold is site specific and dependent on the number of radial grid cells that are available given 40 range cells, 5° of bearing resolution, and omitting any cells that are invalid (e.g. over or behind land). RCMIN is defined as 10 per cent of the available radial grid cells "rounded" to the nearest 25. RCLOW is defined as 30 per cent of the available radial grid cells "rounded" to the nearest 25.

The spatial median filter is used to reduce outlier velocities. For each separate radial vector, the median of all velocities within a 10-km radius is computed. If the difference between the source vector and the median velocity is greater than 50 cm/s, the vector is discarded. If the difference is less than the threshold, the median velocity is used [45].

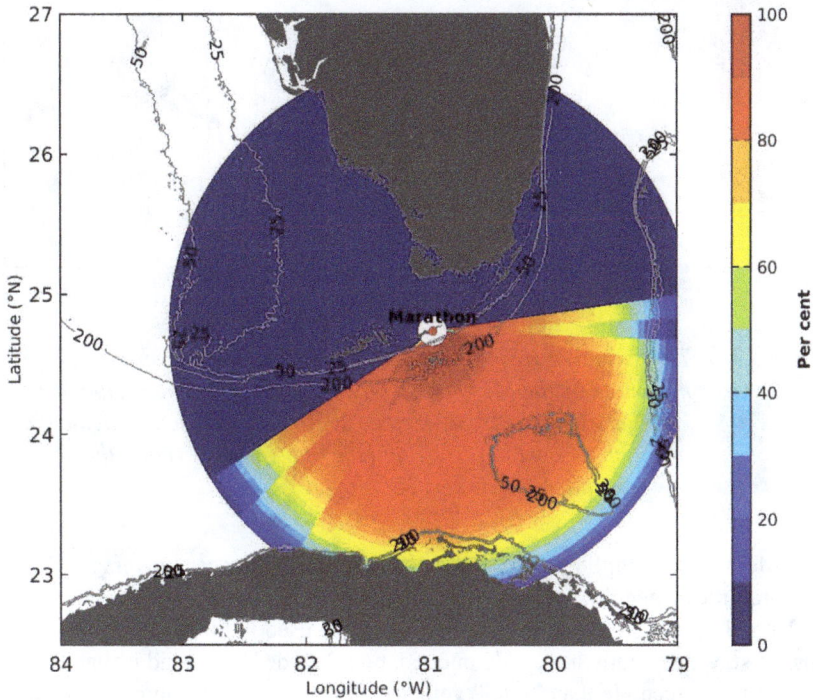

Figure 5.2   Spatial coverage of HFR radial data return (per cent coverage) as measured by the Marathon CODAR SeaSonde during the time period of F12 February 2020 to 10 September 2020. Bathymetry units in m.

The temporal gradient test checks for satisfactory temporal rate of change of radial components. The test determines whether changes between successive radial velocity measurements at a specific radial source point are within an acceptable range [45].

The flags for each radial current vector are appended to the file, saved in standard CODAR tabular separated variable formats, and served via United States National Ocean and Atmospheric Administration (NOAA) Environmental Research Division's Data Access Program (ERDDAP). The HFRadarPy toolbox is available on GitHub (https://github.com/rucool/HFRadarPy).

Data returns are shown in two forms: The first is a spatial map of data return by radial sector (Figure 5.2). For each sector, the total number of valid radial velocity component data points is divided by the record length and is shown as per cent. Higher data returns occur near the site origin and lower data returns sometimes occur towards the outermost ranges. Certain bearing directions also show reduced data returns. These findings are similar to those found with the CODAR SeaSonde on the WFS [40, 47]. The second form is a time series of data returns (Figure 5.3), defined as the number of sectors returning valid data each time normalized by the maximum number of sectors. In this way, we see the wax and wane of HFR coverage over

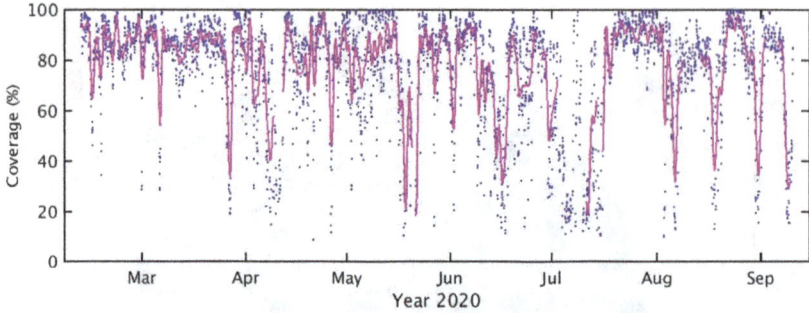

Figure 5.3   *Time series of HFR radial data return (per cent coverage) as measured by the Marathon CODAR during the time period of 12 February 2020 to 10 September 2020. The blue dots indicate the original per cent coverage and the magenta line shows the 36-hour low-pass filtered time series.*

time. Most of the sampling interval shows more than 80 per cent coverage, although there are times when the coverage decreases rapidly to 40 per cent.

Note that only radial velocity components are used in this analysis for two reasons. First, we presently have only one radar deployed. The second is that the radial data are more accurate than "total" velocity data that may be computed within the overlap regions of radial current data obtained from two or more radars (Figure 5.1). Moreover, the accuracy of the "total" velocity vector is subject to the geometrical dilution of precision (GDOP) [48] owing to the angle at which the radial velocity components intersect. So, radial velocity components themselves may be more useful for the purpose of velocity comparison with other products [25, 40, 47]. For the same reason, the radial velocities were preferred over the total velocities in data assimilation into a coastal ocean circulation model [8].

5.2.2   Satellite altimetry-derived current products

Post-processed along-track altimetry data would be preferred for current comparison [25, 49, 50]; however, such products are updated in delayed mode and are not available until at least half a year later. Therefore, we opt here to compare the recent Marathon HFR data with near real-time altimetry data. Jason-3 is the main altimetry satellite mission that is currently in operation in the TOPEX/Poseidon, Jason-1 and Jason-2 series. Jason-3's 1 Hz along-track data are used in this analysis.

Even though several gridded satellite altimetry-derived current products are available, they are primarily based on the same sources of data. Liu *et al.* (2014) [26] compared major three altimetry-derived current products (OSCAR [51], AVISO+ [19, 52], GEKCO [22]) and found that they have about the same performance in simulating the drifter trajectories in the GoM region during summer 2010. Thus, we choose to use the AVISO+ product that has been used in a series of our altimetry data applications [15, 16, 35]. The gridded altimetry SSH data set is a global product with a horizontal solution of 1/4° and daily timestamps.

The near real-time along-track and merged/gridded altimetry, data are produced by AVISO+ (www.aviso.altimetry.fr) with a support from National Centre for Space Studies (CNES), France, and served via the EU's Copernicus Marine Environment Monitoring Service (CMEMS) (marine.copernicus.eu/).

5.2.3   *Numerical model output*

We use the surface currents output from the HYCOM + NCODA Gulf of Mexico 1/25° Analysis (GOMu0.04/expt_90.1m000). This is a data assimilative, operational model [11, 53–56] covering the entire GoM region including the Straits of Florida. The output hourly time series are publicly available through the HYCOM Consortium (hycom.org).

5.3   Evaluation metrics

A correlation coefficient (*CC*) is often used to quantify the agreement (co-linearity) between two time series. Mean squared error (*MSE*) is an alternative, commonly used measure of accuracy in numerical ocean modeling, e.g., Liu *et al.* [57],

$$MSE = \left\langle \left(v_m - v_o\right)^2 \right\rangle \tag{5.1}$$

where v_m and v_o are time series of the modeled (or altimetry-derived) and HFR-observed velocities, respectively, and $\langle\,\rangle$ denotes a mean. The *MSE* can also be written as

$$MSE = MB^2 + SDE^2 + CCE^2 \tag{5.2}$$

i.e., the *MSE* is composed of three parts: the mean bias,

$$MB = \langle v_m \rangle - \langle v_0 \rangle \tag{5.3}$$

the standard deviation error (SDE),

$$SDE = S_m - S_o, \tag{5.4}$$

and the cross-correlation error,

$$CCE = [2S_m S_o(1 - CC)]^{1/2}, \tag{5.5}$$

where $\langle v_m \rangle$ and $\langle v_o \rangle$ are the respective mean velocities; S_m and S_o are the respective standard deviations. The square root of *MSE,* i.e., *RMSE,* has the same unit as the variable (velocity). In our case, it does not necessarily mean "error", rather it measures the difference between the two types of velocities and is called root mean squared difference (*RMSD*). Similarly, SDE and CCE should be, respectively, called standard deviation difference (SDD) and correlation coefficient difference (CCD) instead.

Based on the *MSE*, a quantitative model skill was presented by Willmott [58],

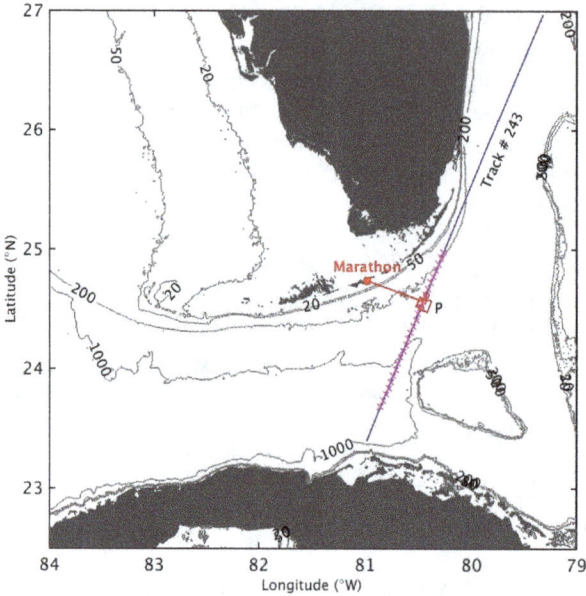

Figure 5.4   Locations of the HFR site origin (Marathon, Florida) and Jason-3 altimeter satellite ground track # 243. The magenta crosses along the track indicate the locations of the geostrophic velocity estimates.

$$S = 1 - MSE / \left\langle \left(\left| v_m - \langle v_o \rangle \right| + \left| v_o - \langle v_o \rangle \right| \right)^2 \right\rangle \tag{5.6}$$

The highest value, $S = 1$, means perfect agreement between model and observation, while the lowest value, $S = 0$, indicates complete disagreement. The non-dimensional Willmott skill score is widely used in performance evaluation of numerical ocean circulation models [57]. It will be used to quantify the agreement between the modeled (altimetry-derived) and HFR-observed velocities.

5.4   Comparison with geostrophic currents derived from along-track altimetry

The Jason-3 altimeter has two ground tracks overlapping the Marathon HFR footprint. Only one of these tracks (#243) has a point that is in the HFR coverage domain and has a perpendicular intersection with an HFR radial (Figure 5.4). It is at this particular location (Point P in Figure 5.4) that both the HFR radial velocity component and the along-track altimetry-derived surface geostrophic velocity component are in the same direction and thus can be compared without further rotation. Point P is 56.8 km from the HFR site origin, with a bearing of 340° (0 = East, increases anticlockwise), and it corresponds to the radial sector with a bearing of 338° and a range of 58.3 km.

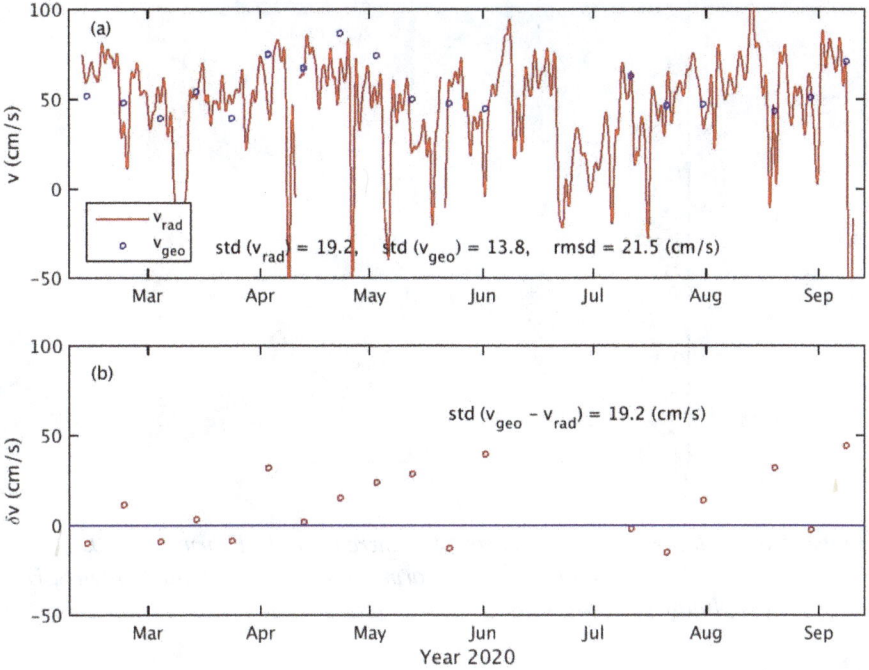

Figure 5.5   *Comparison of velocity components perpendicular to the Jason-3 altimeter satellite ground track #243 (Point P in Figure 5.3): altimetry-derived surface geostrophic current (blue open circles) vs. 36 h low-pass filtered HFR radial current components (a) and their differences (b).*

The "filtered" version of the SSH anomaly is already corrected for atmospheric effects and tides. Adding a mean dynamic topography (MDT_CNES-CLS13) [23] transforms the SSH anomaly to an absolute dynamic height. Following previous work of comparing along-track altimetry-derived currents with HFR radial currents on the WFS [25], we compute surface geostrophic currents that are perpendicular to the altimeter ground track 243 using the following relationship:

$$v_g = \frac{g}{f} \frac{\Delta h}{\Delta x} \tag{5.7}$$

where g is acceleration of gravity, f is Coriolis parameter, and Δh and Δx are the along-track sea level difference and distance between two adjacent points, respectively. This includes the application of a compact, optimal difference operator to minimize the white noise of the SSH measurements when computing $\Delta h/\Delta x$ by weighted smoothing along-track SSH, using a proposed optimum filter [59], with an error correction to the published code [25].

Both the HFR radial current speeds and the along-track altimetry-derived surface geostrophic current speeds are shown in Figure 5.5. The comparison looks

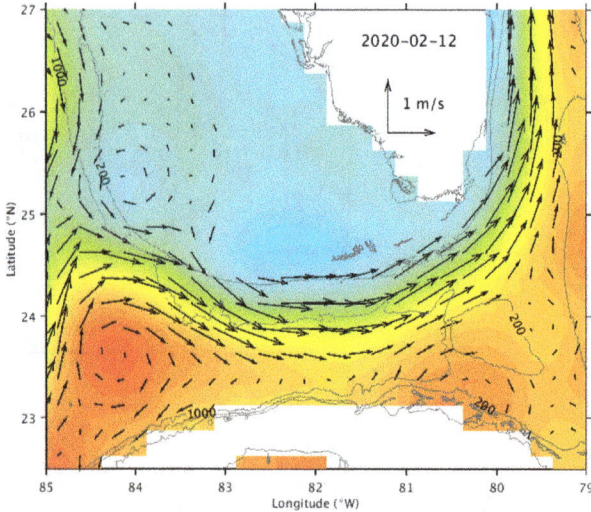

*Figure 5.6   Altimetry-derived geostrophic currents on 12 February 2020.
Color coded are SSH with warm colors corresponding to high SSH.
Bathymetry units in m.*

encouraging. The HFR data are hourly time series, while the altimetry-derived currents are sampled every nine days approximately. Statistics are made on common data timestamps only, i.e., when both data are available. There are 18 such data points during the 7 month period. The mean current speeds of HFR and altimetry-derived current radial speeds are 45.3 cm/s and 55.8 cm/s, respectively, with standard deviations of 13.8 cm/s and 19.2 cm/s, respectively. Their RMSD is 21.5 cm/s. The difference between the two time series range from −15.1 cm/s to 44.4 cm/s, with mean difference of 10.6 cm/s and standard deviation of 19.2 cm/s.

5.5   Comparison with geostrophic currents derived from gridded altimetry

The altimetry-derived surface geostrophic currents must first be projected onto the HFR radial directions so that the two data sets can be compared. This is implemented through three steps: (1) for each radial sector, the hourly HFR radial data are daily averaged; (2) for each day, the altimetry-derived surface geostrophic currents are bi-linearly interpolated onto the central locations of the radial sectors; and (3) for each day and each radial sector, the altimetry-derived current vector is projected towards the HFR radial direction to get a radial component. This is implemented through rotating the total velocity vector by an angle that is determined by the HFR bearing of the sector. As an example for visual verification, the altimetry-derived surface geostrophic currents on 12 February 2020 are shown in Figure 5.6, and their corresponding radial currents as determined through the above procedure

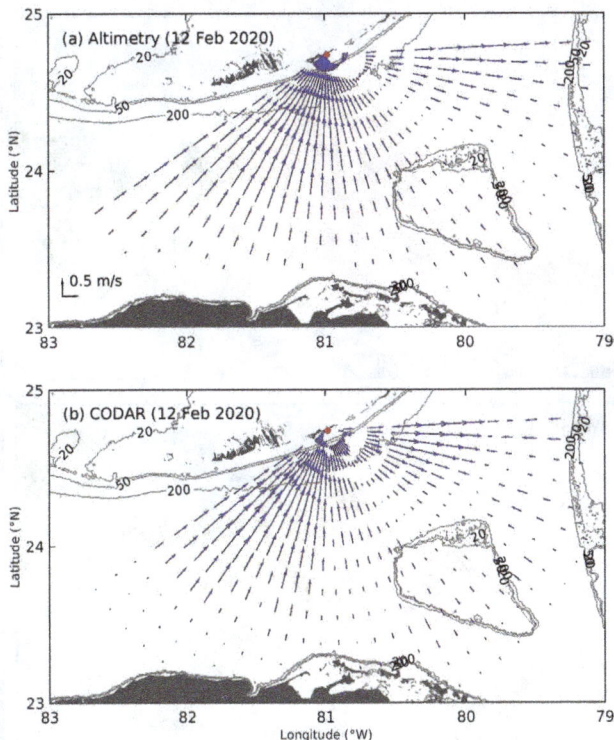

Figure 5.7   *Comparison of (a) the surface geostrophic radial currents derived from altimetry and (b) the daily averaged radial currents measured by the HFR on 12 February 2020. Bathymetry units in m.*

are shown in Figure 5.7a. The observed daily averaged HFR radial currents are shown in Figure 5.7b. The visual comparison between the two radial current maps is reasonably acceptable. As the Florida Current flows eastward in the Straits of Florida, the radial currents on the western part of the HFR footprint are in the directions towards the HFR site origin and those on the eastern part are in the direction away from the HFR site origin. The strongest radial currents also generally correspond well with locations of the center of the Florida Current on that day. All these indicate successful interpolation and projection of the geostrophic currents onto the radial sectors.

Note that the altimetry product is continuous in time, while the daily averaged HFR data may still have gaps. The following quantitative and visual comparisons are made only when valid records from the two data sets are commonly available, i.e., only on the sectors and dates when both radial data are available.

The record-length mean radial currents and corresponding mean bias are shown in Figure 5.8. Both data sets agree well within the main Florida Current flow regions of the HFR footprint, with large positive values

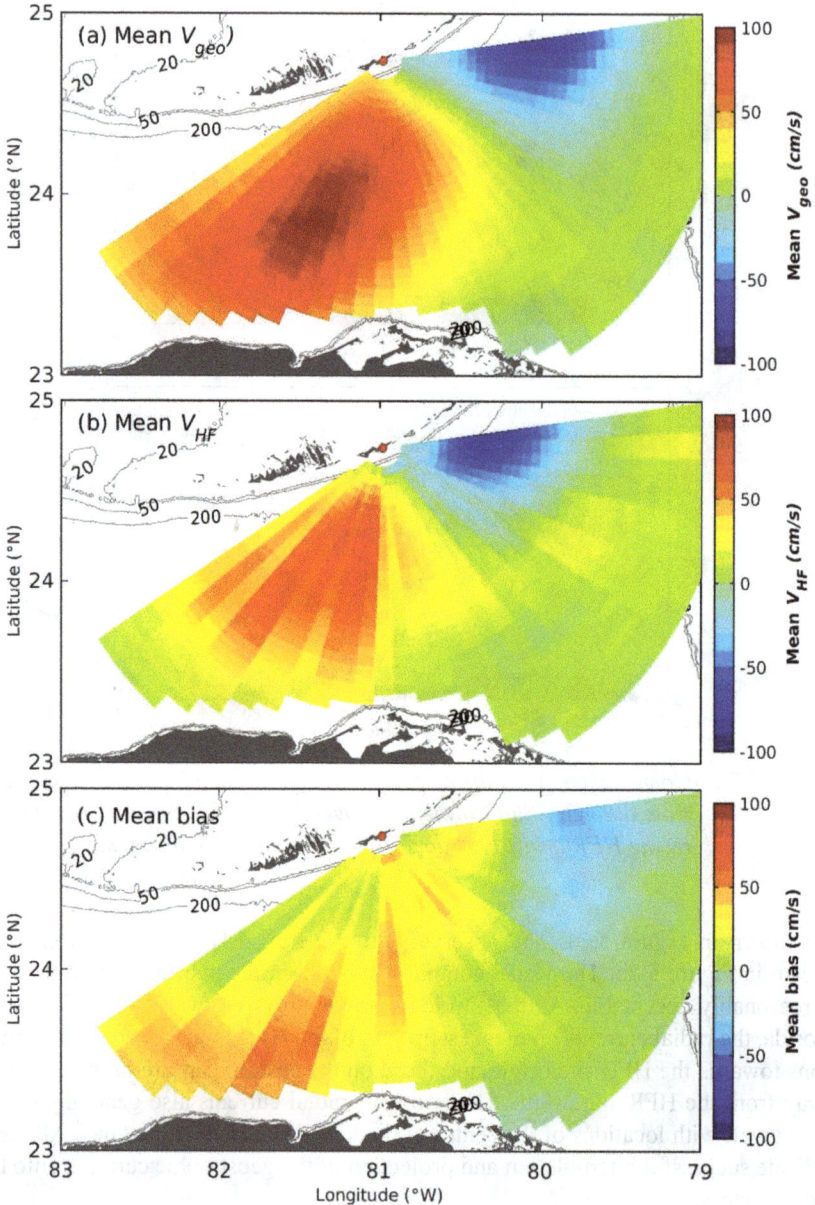

Figure 5.8   *Record-long mean radial currents from (a) altimetry-derived
geostrophic velocity and (b) HFR, and their difference (c) mean bias.*

(indicating radial currents towards the site origin) on the western sideand those
of the HFR data and negative values (indicating radial currents away from
the site origin) on the eastern side. The minimum and maximum values of the

record-long mean altimetry-derived radial currents are −87.1 cm/s and 93.2 cm/s, respectively, and those of the HFR data are −87.0 cm/s and 69.8 cm/s, respectively. On the eastern part of the domain, the two data show almost the same magnitude of the outflow component (−87 cm/s). However, the altimetry-derived mean currents are generally stronger than the HFR-observed mean currents with a mean bias of about 20–40 cm/s in the strong current areas. Large mean bias is also seen along the outer range of the HFR foot map, which may be due to the data quality issues. The mean bias is 10.7 cm/s within the entire HFR radial domain.

The standard deviations of the two data sets are shown in Figure 5.9. The difference between the two standard deviations is shown as the SDE in Figure 5.9c. In general, the standard deviations show more differences than agreements. The altimetry-derived radial currents generally have smaller standard deviations (maximum 30.0 cm/s, mean 13.0 cm/s) than the HFR-observed radial currents (maximum 58.7 cm/s, mean 24.3 cm/s). The large standard deviations in the HFR data are located in several bearing directions that correspond to the low data return rates. It may also be the case that the altimetry product underestimates the standard deviations of the currents, while the HFR data may overestimate the standard deviations because of low data availability and quality issues within certain radial sectors.

The RMSD is an overall quantitative measure of the differences that include the mean bias, the standard deviation error, and the cross-correlation error. The RMSD ranges from 16.2 cm/s to 61.2 cm/s in the HFR radial domain, with a mean value of 34.1 cm/s across all radial sectors (Figure 5.10a). Large RMSD values are seen in certain bands corresponding to low data return sectors.

The HFR domain area mean of the Willmott skill score is 0.37, with a highest score of 0.73 occurring on the western side within 50 km range of the site origin (Figure 5.10b). The region of the highest skill score corresponds with that of the lowest RMSD.

5.6 Comparison with data assimilative model output

The GoM HYCOM assimilates altimetry data in its operational analysis, but unlike the altimetry data, the HYCOM dynamics allow for ageostrophic motions as also exist within the HFR domain. Thus it is instructive to also compare the model-simulated surface currents with HFR observations. Consistent with the gridded altimetry product occurring with a daily time interval, the GoM HYCOM currents are first daily averaged for comparison. Following a similar procedure, the GoM HYCOM surface currents are interpolated onto HFR radial sector locations and further projected to the radial directions. The model output on same day (12 February 2012) as in Section 5.4 is used as an example to demonstrate the outcome of this procedure. The GoM HYCOM surface currents before and after interpolation/projection are shown in Figures 5.11 and 5.12a, respectively, with the corresponding daily averaged HFR radial data shown in Figure 5.12b. The model-derived radial currents generally agree well with HFR observations. The different pattern near the site origin on the western side is due to the presence of the flow curvature, which

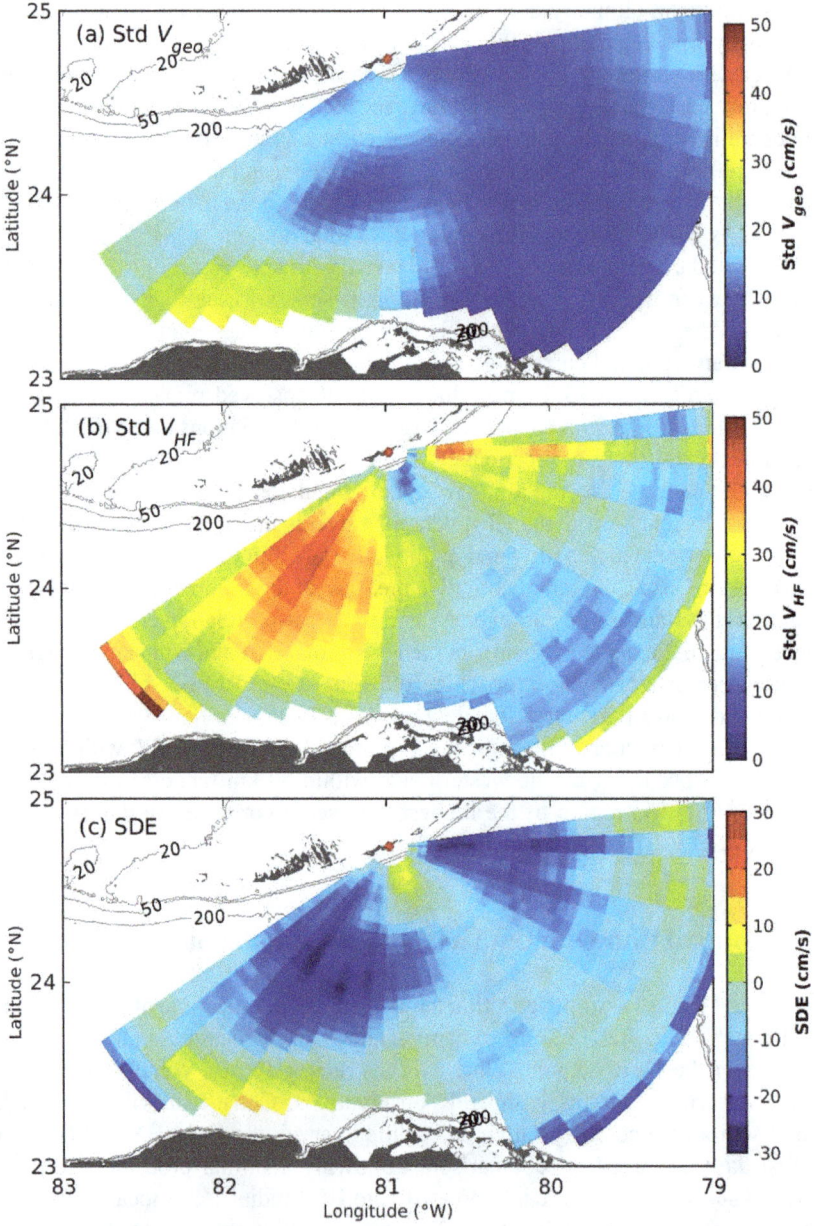

Figure 5.9   Standard deviations of the radial currents from (a) altimetry-derived geostrophic velocity, (b) HFR, and their difference (c) standard deviation error (SDE).

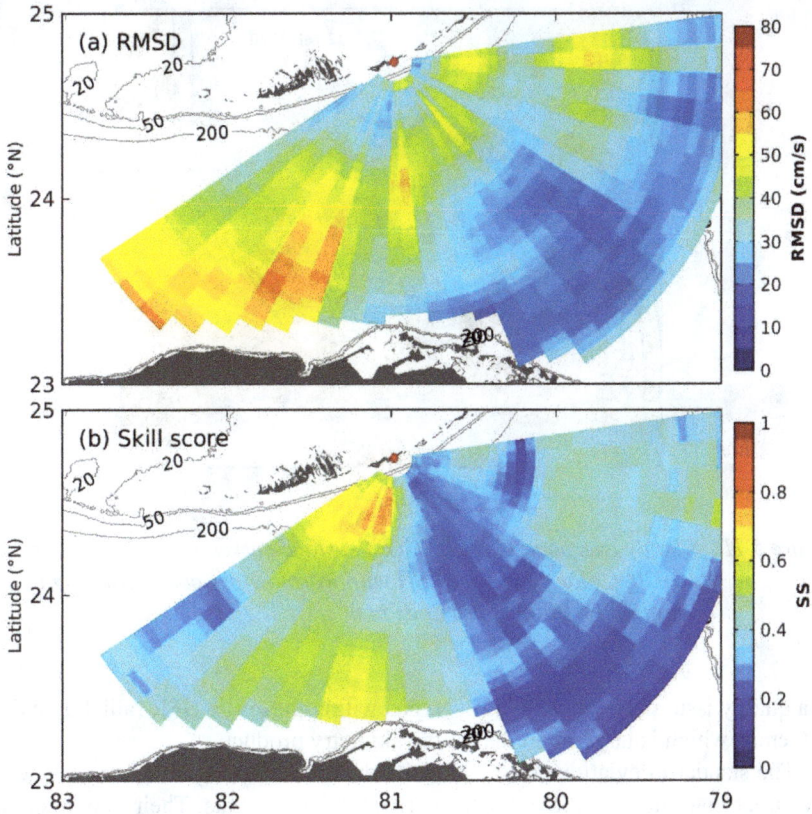

Figure 5.10   *Evaluation of the altimetry-derived surface geostrophic velocity against HFR data: (a) RMSD, and (b) Willmott skill score*

may possibly be associated with a mesoscale eddy [60–62]. Such small-scale surface variability between the Florida Current and the coastal region was observed in previous observations [63–65].

The temporal mean model and HFR radial currents, and the mean bias are shown in Figure 5.13. Both data sets show the main Florida current components to be large and positive on the western side of the HFR radial coverage domain (indicating radial currents towards the site origin) and negative (indicating radial currents away from the site origin) values on the eastern side. The minimum and maximum values of the record-long mean model-derived radial currents are −99.5 cm/s and 99.6 cm/s, respectively. These values show a stronger mean modeled Florida Current than the altimetry-derived currents. Similar to the altimetry-derived mean currents, the model currents are generally stronger than the HFR-observed mean currents with a mean bias of about 20–60 cm/s in the strong current areas. Large mean bias is also seen along the outer range of the HFR domain, which may be due to the HFR

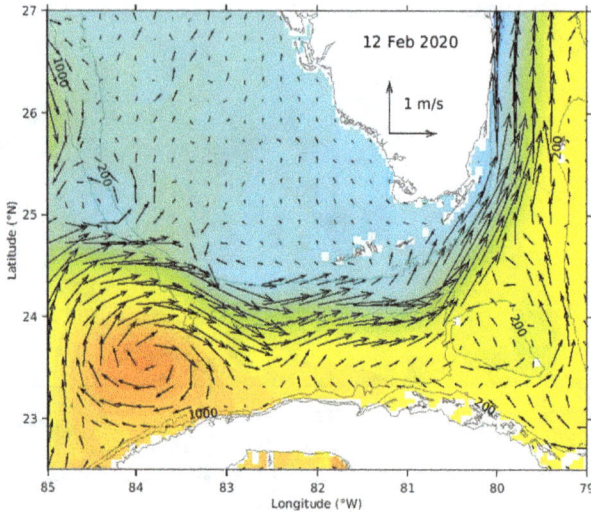

Figure 5.11   *Daily averaged GoM HYCOM surface currents on 12 February*
2020. Color coded are SSH with warm colors corresponding to
high SSH. Bathymetry units in m.

data quality issues. The mean bias averaged within the entire HFR radial domain is
17.0 cm/s, which is larger than that of the altimetry product.

The standard deviations of the model radial currents (Figure 5.14a) are generally larger than those of the altimetry-derived radial currents. Their maximum and
mean values are 43.0 cm/s and 22.5 cm/s, respectively. The mean standard deviation value is close to that of HFR radial currents (24.1 cm/s). The RMSD values
of the model radial currents are larger than those of the altimetry-derived currents,
with a range of 15.2–82.3 cm/s and a mean value of 39.9 cm/s for all the radial sectors (Figure 5.15a). The Willmott skill scores of the GoM HYCOM have an area
mean value of 0.37 (same as that of the altimetry-derived product), with the highest
score of 0.76 similarly within the 50-km range of the site origin on the west side
(Figure 5.15b).

5.7   Summary and discussion

The radial currents measured by the Marathon long-range CODAR HFR SeaSonde,
which overlooks the Straits of Florida to Cuba, were used to evaluate the along-
track and merged/gridded altimetry-derived geostrophic current products (AVISO+)
and the surface currents output from the data assimilative ocean circulation model
(GoM HYCOM). Similar comparisons between along-track altimetry-derived and
HFR radial currents were previously made on the WFS [25], but this is the first
time that CODAR SeaSonde HFR-observed radial currents for the Florida Current

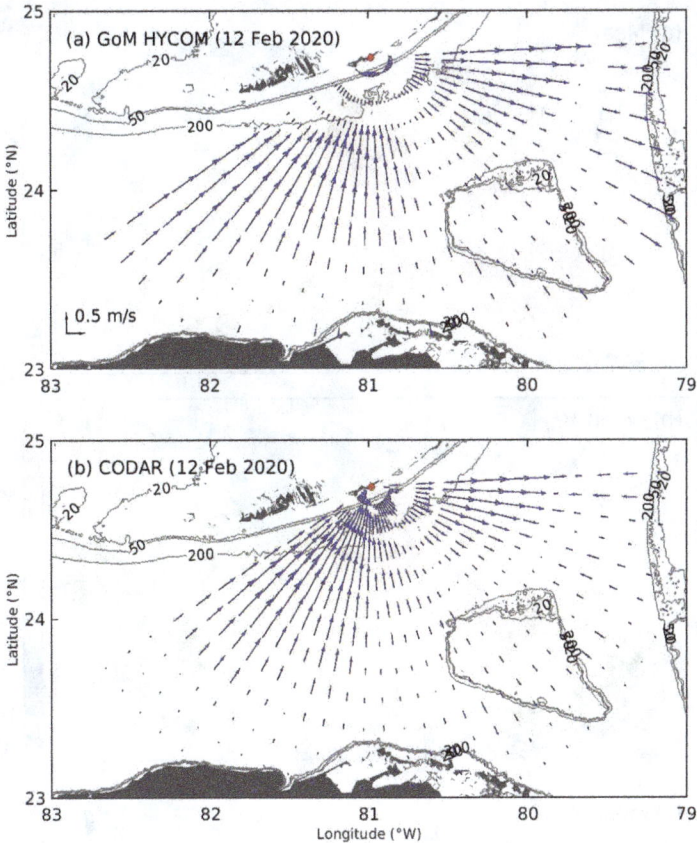

Figure 5.12   *Comparison of (a) the daily averaged radial currents extracted from the GoM HYCOM and (b) the daily averaged radial currents measured by the HFR on 12 February 2020. Bathymetry units in m.*

region have been evaluated against merged/gridded altimetry products and numerical model outputs.

It was found that among the three current products, the along-track altimetry-derived radial speeds agreed the best with the HFR radial currents, with the RMSD of about 21 cm/s. This value is larger than those obtained from similar comparisons on the WFS (8–11 cm/s) [25]. However, considering the much stronger currents in the Straits of Florida than on the WFS (100–200 cm/s vs. 10–30 cm/s) and that HFR's accuracy could vary with environmental conditions, this larger RMSD value is reasonable. Also, the altimetry-derived geostrophic currents lack ageostrophic influences that are part of this dynamically complex region with frequent mesoscale and submescoscale eddies [41, 62, 66] where non-linear dynamics might also be in play [67]. The currents in the Straits of Florida have large across-strait variability with the relative vorticity at least in the same order as the planetary vorticity [68].

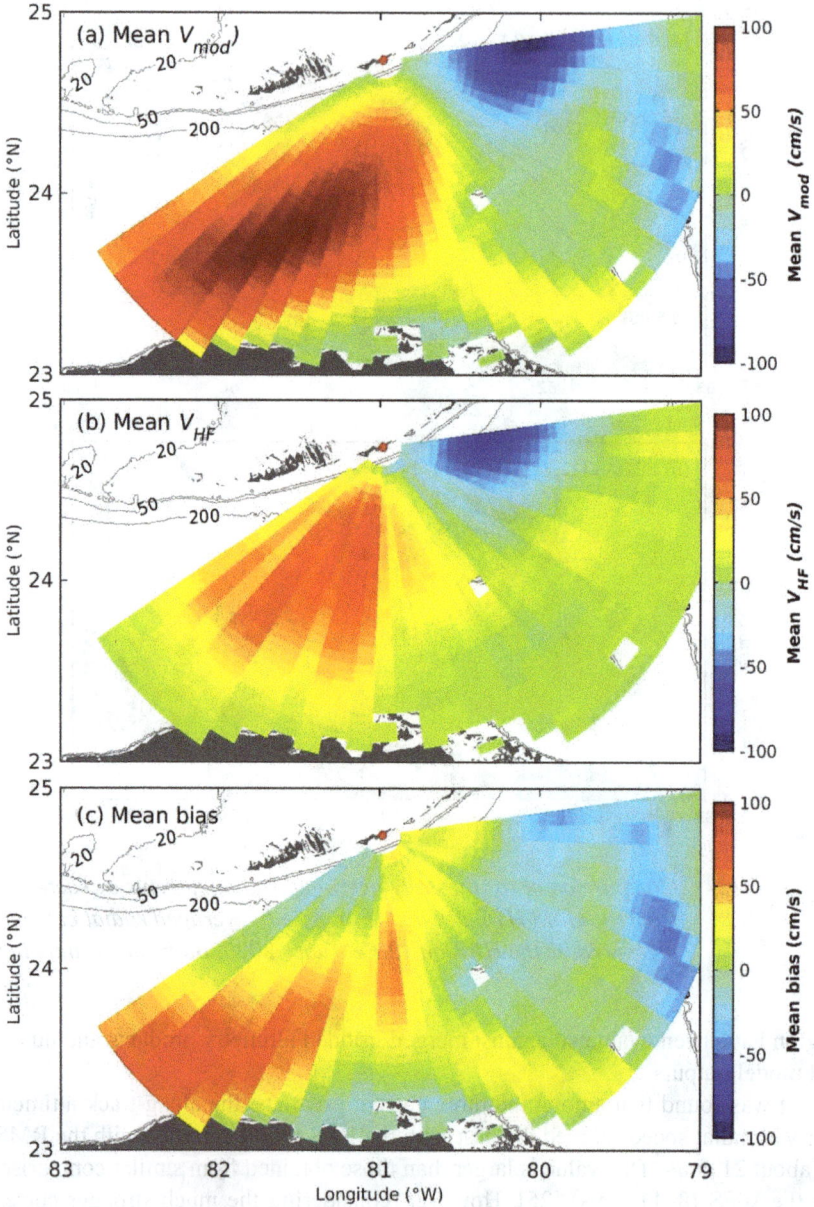

Figure 5.13   Record-long mean radial currents from (a) GoM HYCOM and (b) HFR, and their difference (c) mean bias

The agreement between the merged/gridded altimetry product and the HFR radial currents was reduced with the RMSD of 34.1 cm/s. The RMSD varied among the HFR radial sectors with a range of 16.2–61.2 cm/s. This was mainly because

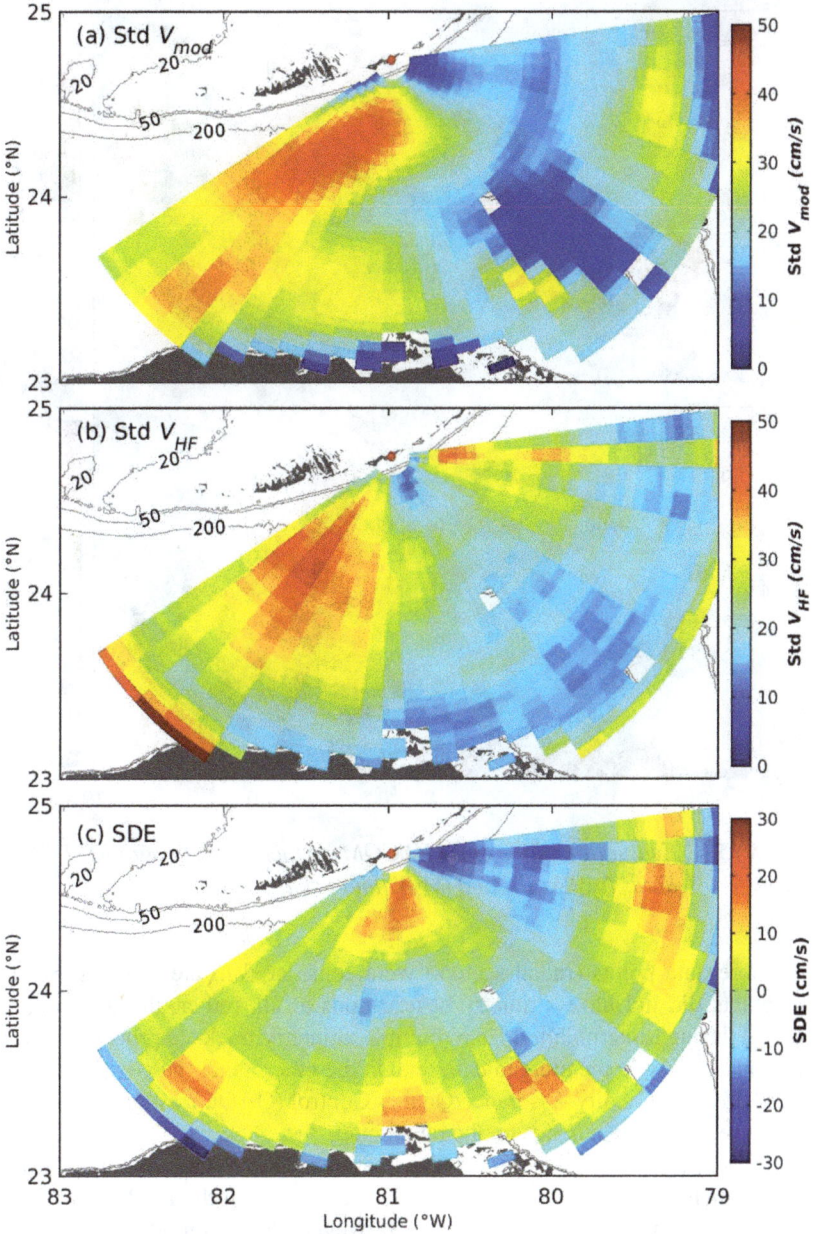

Figure 5.14   Standard deviations of the radial currents from (a) GoM HYCOM and (b) HFR, and their difference (c) standard deviation error (SDE)

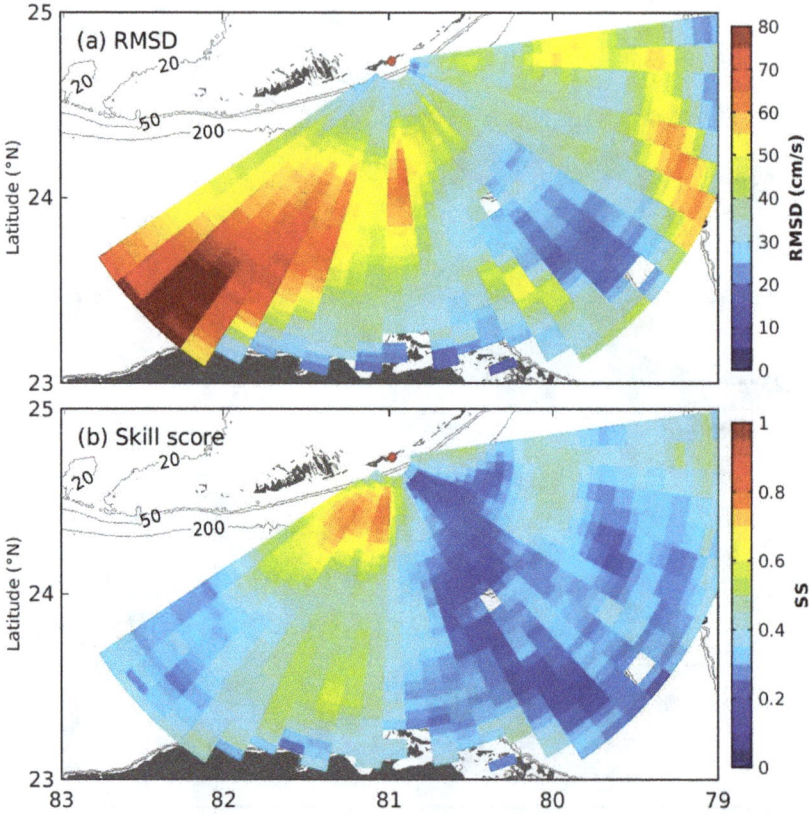

Figure 5.15   Evaluation of the GoM HYCOM surface velocity against HFR data: (a) RMSD, and (b) Willmott skill score

many more data points (radial sectors) were considered in the analysis. Some of these points did not have altimetry observations at all, rather, the data might be produced through interpolation from multiple altimetry tracks, a process that could induce uncertainty in the merged data set. On the other hand, there is still room for further HFR data Quality Assurance/Quality Control (QAQC), as large RMSD values were seen in certain bands on the radial map.

The least agreement was found with the numerical model output in terms of RMSD, which varied in a range of 15.2–82.3 cm/s among the HFR radial sectors. The spatial mean value was 39.9 cm/s. However, the Willmott skill scores did show the GoM HYCOM and the AVISO+ merged/gridded altimetry product had about the same performance when they were compared with the HFR radial data. A previous study also showed that the altimetry products outperformed the data assimilative ocean circulation models in terms of the skills in simulating the drifter trajectories in the GoM [26]. Larger RMSD values are found in the Florida Current frontal areas with frequent mesoscale eddy activities. These findings have important implications to the altimetry data users, particularly, numerical ocean modeling community with interest in data assimilation.

The Marathon HFR is just the first of three lower Florida Keys long-range CODAR SeaSonde HFRs funded by the NASEM UGOS GRP. With the other two HFRs anticipated to be installed soon, the overall lower Keys HFR additions will measure the surface currents in the Straits of Florida and reduce the regions of missing HFR coverage gaps between the west and east Florida coasts [38–42, 64, 65, 68, 69]. By increasing real-time observations for assimilation into numerical circulation models in the vicinity of the Pressure Point Mooring [16, 35] via HFR observations overlapping with those from COMPS/SECOORA WFS HFR observations we will greatly increase the ability for models to more properly account for the circulation in this very high gradient, spatially and temporally variable region, while at the same time addressing the surface circulation of the GoM LC system outflow region. Given that flows driven by the LC and its associated eddies also play important roles in the water properties and ecological influences for Florida Bay and the Florida Keys reef track, these observations will also be of benefit studies pertaining to the Everglades and the various protected areas associated with the Dry Tortugas National Park and the Florida Keys.

Acknowledgment

This work was supported by the Gulf Research Program of the National Academies of Sciences, Engineering, and Medicine under the Grant Agreement number 2000009917. The content is solely the responsibility of the authors and does not necessarily represent the official views of the Gulf Research Program or the National Academies of Sciences, Engineering, and Medicine.

Support was also provided by the University of South Florida's College of Marine Science via the Coastal Ocean Monitoring and Prediction System (COMPS) program with public outreach support provided under purchase order to the Southeast Coastal Ocean Observing Regional Association (SECOORA).

References

[1]   Barrick D.E., Evans M.W., Weber B.L. 'Ocean surface currents mapped by radar'. *Science*. 1977;**198**(4313):138–44.
[2]   Liu Y., Kerkering H., Wesiberg R.H. (eds.). *Coastal Ocean Observing Systems*. **461**. London: Academic Press; 2015.
[3]   Abascal A.J., Castanedo S., Medina R., Losada I.J., Alvarez-Fanjul E. 'Application of HF radar currents to oil spill modelling'. *Marine Pollution Bulletin*. 2009;**58**(2):238–48.
[4]   Zelenke B.C., Moline M.A., Jones B.H., *et al.* 'Evaluating connectivity between marine protected areas using CODAR high-frequency radar'. Proceeding at OCEANS MTS/IEEE; Biloxi, MS; 2009.

[5]   Roarty H., Glenn S., Kohut J., *et al.* 'Operation and application of a regional high-frequency radar network in the mid-Atlantic Bight'. *Marine Technology Society Journal.* 2010;**44**(6):133–45.

[6]   Lewis J.K., Shulman I., Blumberg A.F. 'Assimilation of CODAR observations into ocean models'. *Continental Shelf Research.* 1998;**18**:541–59.

[7]   Paduan J.D., Shulman I. 'HF radar data assimilation in the Monterey Bay area'. *Journal of Geophysical Research.* 2004;**109**:C07S09.

[8]   Barth A., Alvera-Azcárate A., Weisberg R.H. 'Assimilation of high-frequency radar currents in a nested model of the West Florida shelf'. *Journal of Geophysical Research.* 2008;**113**(C8):C08033.

[9]   Ren L., Hartnett M. 'Hindcasting and forecasting of surface flow fields through assimilating high frequency remotely sensing radar data'. *Remote Sensing.* 2017;**9**(9):932.

[10] Wilkin J.L., Bowen M.M., Emery W.J. 'Mapping mesoscale currents by optimal interpolation of satellite radiometer and altimeter data'. *Ocean Dynamics.* 2002;**52**(3):95–103.

[11] Cummings J.A. 'Operational multivariate ocean data assimilation'. *Quarterly Journal of the Royal Meteorological Society.* 2005;**131**(613):3583–604.

[12] Willis J.K., Roemmich D., Cornuelle B. 'Interannual variability in upper ocean heat content, temperature, and thermosteric expansion on global scales'. *Journal of Geophysical Research.* 2004;**109**(C12):C12036.

[13] Chambers D.P., Merrifield M.A., Nerem R.S. 'Is there a 60-year oscillation in global mean sea level?' *Geophysical Research Letters.* 2012;**39**(18):L18607.

[14] Liu Y., Weisberg R.H., Yuan Y. 'Patterns of upper layer circulation variability in the South China Sea from satellite altimetry using the self-organizing map'. *Acta Oceanologica Sinica.* 2008;**27**(Supp.):129–44.

[15] Liu Y., Weisberg R.H., Vignudelli S., Mitchum G.T. 'Patterns of the loop current system and regions of sea surface height variability in the eastern Gulf of Mexico revealed by the self-organizing maps'. *Journal of Geophysical Research: Oceans.* 2016;**121**(4):2347–66.

[16] Weisberg R.H., Liu Y. 'On the Loop Current penetration into the Gulf of Mexico'. *Journal of Geophysical Research: Oceans.* 2017;**122**(12):9679–94.

[17] Vignudelli S., Vignudelli S., Kostianoy A.G., Cipollini P., Benveniate J. (eds.). *Coastal Altimetry.* Berlin: Springer; 2011.

[18] Birol F., Fuller N., Lyard F., *et al.* 'Coastal applications from nadir altimetry: example of the X-TRACK regional products'. *Advances in Space Research.* 2017;**59**(4):936–53.

[19] Pascual A., Faugère Y., Larnicol G., Le Traon P.-Y. 'Improved description of the ocean mesoscale variability by combining four satellite altimeters'. *Geophysical Research Letters.* 2006;**33**(2):L02611.

[20] Bonjean F., Lagerloef G.S.E. 'Diagnostic model and analysis of the surface currents in the tropical Pacific ocean'. *Journal of Physical Oceanography.* 2002;**32**(10):2938–54.

[21] Dohan K., Maximenko N. 'Monitoring ocean currents with satellite sensors'. *Oceanography.* 2010;**23**(4):94–103.

[22] Sudre J., Morrow R.A. 'Global surface currents: a high-resolution product for investigating ocean dynamics'. *Ocean Dynamics*. 2008;**58**(2):101–18.

[23] Rio M.-H., Mulet S., Picot N. 'Beyond GOCE for the ocean circulation estimate: synergetic use of altimetry, gravimetry, and in situ data provides new insight into geostrophic and Ekman currents'. *Geophysical Research Letters*. 2014;**41**(24):8918–25.

[24] Johnson E.S., Bonjean F., Lagerloef G.S.E., Gunn J.T., Mitchum G.T. 'Validation and error analysis of Oscar sea surface currents'. *Journal of Atmospheric and Oceanic Technology*. 2007;**24**(4):688–701.

[25] Liu Y., Weisberg R.H., Vignudelli S., Roblou L., Merz C.R. 'Comparison of the X-TRACK altimetry estimated currents with moored ADCP and HF radar observations on the West Florida shelf'. *Advances in Space Research*. 2012;**50**(8):1085–98.

[26] Liu Y., Weisberg R.H., Vignudelli S., Mitchum G.T. 'Evaluation of altimetry-derived surface current products using Lagrangian drifter trajectories in the eastern Gulf of Mexico'. *Journal of Geophysical Research: Oceans*. 2014;**119**(5):2827–42.

[27] Feng H., Vandemark D., Levin J., Wilkin J. 'Examining the accuracy of GlobCurrent upper ocean velocity data products on the northwestern Atlantic shelf'. *Remote Sensing*. 2018;**10**(8):1205.

[28] Emery B., Washburn L. 'Uncertainty estimates for SeaSonde HF radar ocean current observations'. *Journal of Atmospheric and Oceanic Technology*. 2019;**36**(2):231–47.

[29] Reid R.O. ' A simple dynamic model of the loop current' in Reid J.L., Capurro L.R.A. (eds.). *Contributions to the Physical Oceanography of the Gulf of Mexico*. **2**. College Station, TX: Texas A&M: University Oceanography Studies; 1972. pp. 157–9.

[30] Sturges W., Lugo-Fernandez A. (eds.). *Circulation in the Gulf of Mexico: Observations and Models*. **161**. Washington, DC: AGU; 2005.

[31] Alvera-Azcárate A., Barth A., Weisberg R.H. 'The surface circulation of the Caribbean sea and the Gulf of Mexico as inferred from satellite altimetry'. *Journal of Physical Oceanography*. 2009;**39**(3):640–57.

[32] Yang Y., Weisberg R.H., Liu Y., San Liang X, Liang X.S. 'Instabilities and multiscale interactions underlying the Loop Current eddy shedding in the Gulf of Mexico'. *Journal of Physical Oceanography*. 2020;**50**(5):1289–317.

[33] Shay L.K., Uhlhorn E.W. 'Loop Current response to hurricanes Isidore and Lili'. *Monthly Weather Review*. 2008;**136**(9):3248–74.

[34] He R., Weisberg R.H. 'A Loop Current intrusion case study on the West Florida Shelf'. *Journal of Physical Oceanography*. 2003;**33**(2):465–77.

[35] Liu Y., Weisberg R.H., Lenes J.M., Zheng L., Hubbard K., Walsh J.J. 'Offshore forcing on the "pressure point" of the West Florida Shelf: Anomalous upwelling and its influence on harmful algal blooms'. *Journal of Geophysical Research: Oceans*. 2016;**121**(8):5501–15.

[36]   Weisberg R.H., Zheng L., Liu Y., Lembke C., Lenes J.M., Walsh J.J. 'Why no red tide was observed on the West Florida continental shelf in 2010'. *Harmful Algae*. 2014;**38**(6):119–26.

[37]   Weisberg R.H., Liu Y., Lembke C., Hu C., Hubbard K., Garrett M. 'The coastal ocean circulation influence on the 2018 West Florida Shelf K . brevis redtide bloom'. *Journal of Geophysical Research: Oceans*. 2019;**124**(4):2501–12.

[38]   Merz C.R., Weisberg R.H., Liu Y. 'Evolution of the USF/CMS CODAR and WERA HF radar network'. Proceedings of Oceans'12 MTS/IEEE Conference. Hampton Roads; VA; 2012.

[39]   Liu Y., Weisberg R.H., Shay L.K. 'Current patterns on the West Florida Shelf from joint self-organizing map analyses of HF radar and ADCP data'. *Journal of Atmospheric and Oceanic Technology*. 2007;**24**(4):702–12.

[40]   Liu Y., Weisberg R.H., Merz C.R., Lichtenwalner S., Kirkpatrick G.J. 'HF radar performance in a low-energy environment: CODAR SeaSonde experience on the West Florida Shelf'. *Journal of Atmospheric and Oceanic Technology*. 2010;**27**(10):1689–710.

[41]   Shay L.K., Cook T.M., An P.E. 'Submesoscale coastal ocean flows detected by very high frequency radar and autonomous underwater vehicles'. *Journal of Atmospheric and Oceanic Technology*. 2003;**20**(11):1583–99.

[42]   Archer M.R., Shay L.K., Jaimes B. 'Observing frontal instabilities of the Florida Current using high frequency radar' in Liu Y., Kerkering H., Weisberg R.H. (eds.). *Coastal Ocean Observing Systems*. London: Academic Press (Elsevier); 2015. pp. 179–208.

[43]   Gill E.W., Huang W. 'HF radar in a marine environment' in Huang W., Gill E.W. (eds.). *Ocean Remote Sensing Technologies: High Frequency, Marine and GNSS-Based Radar*. IET; 2021.

[44]   Merz C.R., Liu Y., Weisberg R.H. 'Sea surface current mapping with HF radar – a primer' in Huang W., Gill E.W. (eds.). *Ocean Remote Sensing Technologies: High Frequency, Marine and GNSS-Based Radar*. IET; 2021.

[45]   U.S Integrated Ocean Observing System. *Manual for real-time quality control of high frequency radar surface current data: a guide to quality control and quality assurance for high frequency radar surface current observations [online]*. version 1.0. Silver spring, MD: U.S. department of commerce, national oceanic and atmospheric administration, integrated ocean observing system. 2016. Available from https://cdn.ioos.noaa.gov/media/2017/12/HFR_QARTOD_Manual_05_26_16.pdf [Accessed 21 Jan 2021].

[46]   Kohut J.T., Glenn S.M. 'Improving HF radar surface current measurements with measured antenna beam patterns'. *Journal of Atmospheric and Oceanic Technology*. 2003;**20**(9):1303–16.

[47]   Liu Y., Weisberg R.H., Merz C.R. 'Assessment of CODAR SeaSonde and WERA HF radars in mapping surface currents on the West Florida Shelf'. *Journal of Atmospheric and Oceanic Technology*. 2014;**31**(6):1363–82.

[48]   Chapman R.D., Shay L.K., Graber H.C., *et al.* 'On the accuracy of HF radar surface current measurements: Intercomparisons with ship-based sensors'. *Journal of Geophysical Research: Oceans*. 1997;**102**(C8):18737–48.

[49] Pascual A., Lana A., Troupin C., *et al.* 'Assessing SARAL/AltiKa data in the coastal zone: comparisons with HF radar observations'. *Marine Geodesy.* 2015;**38**(1):260–76.

[50] Idris N.H., Deng X., Idris N.H. 'Comparison of retracked coastal altimetry sea levels against high frequency radar on the continental shelf of the great barrier reef, Australia'. *Journal of Applied Remote Sensing.* 2017;**11**(3):032403.

[51] Lagerloef G.S.E., Mitchum G.T., Lukas R.B., Niiler P.P. 'Tropical Pacific near-surface currents estimated from altimeter, wind, and drifter data'. *Journal of Geophysical Research: Oceans.* 1999;**104**(C10):23313–26.

[52] Le Traon P.-Y., Antoine D., Bentamy A., *et al.* 'Use of satellite observations for operational oceanography: recent achievements and future prospects'. *Journal of Operational Oceanography.* 2015;**8**(1):s12–27.

[53] Chassignet E., Hurlburt H., Metzger E.J., *et al.* 'US GODAE: Global ocean prediction with the HYbrid Coordinate Ocean Model (HYCOM)'. *Oceanography.* 2009;**22**(2):64–75.

[54] Cummings J.A., Smedstad O.M. 'Variational data assimilation for the global ocean' in Xu L., Park S.K. (eds.). *Data Assimilation for Atmospheric, Oceanic and Hydrologic Applications.* **13**. Berlin Heidelberg: Springer-Verlag; 2013. pp. 303–43.

[55] Helber R.W., Townsend T.L., Barron C.N. 'Validation test report for the improved synthetic ocean profile (ISOP) system, Part I: synthetic profile methods and algorithm'. NRL Memo. Report, NRL/MR/7320—13-9364; 2013.

[56] Hogan T., Liu M., Ridout J., *et al.* 'The navy global environmental model'. *Oceanography.* 2014;**27**(3):116–25.

[57] Liu Y., MacCready P., Hickey B.M., Dever E.P., Kosro P.M., Banas N.S. 'Evaluation of a coastal ocean circulation model for the Columbia river plume in summer 2004'. *Journal of Geophysical Research.* 2009;**114**(3):C00B4.

[58] Willmott C.J. 'On the validation of models'. *Physical Geography.* 1981;**2**:184–94.

[59] Powell B.S., Leben R.R. 'An optimal filter for geostrophic mesoscale currents from along-track satellite altimetry'. *Journal of Atmospheric and Oceanic Technology.* 2004;**21**(10):1633–42.

[60] Shay L.K., Lee T.N., Williams E.J., Graber H.C., Rooth C.G.H. 'Effects of low-frequency current variability on near-inertial submesoscale vortices'. *Journal of Geophysical Research: Oceans.* 1998;**103**(C9):18691–714.

[61] Kourafalou V.H., Kang H. 'Florida current meandering and evolution of cyclonic eddies along the Florida keys reef tract: are they interconnected?' *Journal of Geophysical Research: Oceans.* 2012;**117**(C5):C05028.

[62] Zhang Y., Hu C., Liu Y., Weisberg R.H., Kourafalou V.H. 'Submesoscale and mesoscale eddies in the Florida Straits: observations from satellite ocean color measurements'. *Geophysical Research Letters.* 2019;**46**(22):13262–70.

[63] Haus B.K., Graber H.C., Shay L.K., Cook T.M. 'Alongshelf variability of a coastal buoyancy current during the relaxation of downwelling favorable winds'. *Journal of Coastal Research.* 2003;**19**(2):409–20.

[64] Parks A.B., Shay L.K., Johns W.E., Martinez-Pedraja J., Gurgel K.-W. 'HF radar observations of small-scale surface current variability in the Straits of Florida'. *Journal of Geophysical Research: Oceans*. 2009;**114**(C8):C08002.

[65] Martinez-Pedraja J., Shay L.K., Haus B.K., Whelan C. 'Interoperability of SeaSondes and Wellen radars in mapping radial surface currents'. *Journal of Atmospheric and Oceanic Technology*. 2013;**30**(11):2662–75.

[66] Lee T.N., Leaman K., Williams E., *et al.* 'Florida Current meanders and gyre formation in the southern Straits of Florida'. *Journal of Geophysical Research*. 1995;**100**(C5):8607–20.

[67] Hiron L., Cruz B.J., Shay L.K. 'Evidence of Loop Current frontal eddy intensification through local linear and nonlinear interactions with the Loop Current'. *Journal of Geophysical Research: Oceans*. 2020;**125**(e2019 JC015533).

[68] Peters H., Shay L.K., Mariano A.J., Cook T.M. 'Current variability on a narrow shelf with large ambient vorticity'. *Journal of Geophysical Research*. 2002;**107**(C8):3087.

[69] Liu Y., Merz C.R., Weisberg R.H., O'Loughlin B.K., Subramanian V. 'Data return aspects of CODAR and WERA high frequency radars in mapping currents' in Venkatesan R., Tandon A., D'Asaro E., Atmanand M.A. (eds.). *Observing the Oceans in Real Time*. Berlin: Springer; 2017. pp. 227–41.

Chapter 6

Ocean wave measurement

Lucy R. Wyatt[1]

The Doppler spectrum provides a measurement of the magnitude and frequency of the received radar signal scattered from ocean waves. We therefore need to understand something about ocean waves and how they interact with electromagnetic waves in order to understand how we can use the Doppler spectrum to extract quantitative oceanographic information. Therefore, we begin this chapter with an introduction to ocean waves.

6.1 Introduction to ocean waves

Readers are referred to Kinsman [1], Komen *et al.* [2], Tucker and Pitt [3], Janssen [4], Holthuijsen [5] and Hauser *et al.* [6] (amongst others) for more information about the theory, modelling and/or measurement of ocean waves. An interesting account of the early work on ocean wave science can be seen on this video [7]. Here we focus on those aspects that are important for high-frequency (HF) radar measurement. The simplest (albeit approximate) way to think of the ocean surface wave field is as a collection of different sinusoidal waves with different amplitudes, wavelengths, periods and propagation directions, which add together to provide a typical ocean surface. These waves are generated by the wind either locally, where they are called wind waves, or have propagated into the area of observation from a distant storm, where they are called swell. Swell is usually narrowband in both wavelength and direction and so, in the absence of any significant local wind, has a simple sinusoidal form (see Figure 6.1). Wind waves are usually more complicated with a much wider range of amplitudes, wavelengths and propagation directions (see Figure 6.2).

Both of these can be present at the same time. The most general description of such a surface is the two-dimensional (2D) directional wave spectrum, $S(\vec{k})$, where $\vec{k}$ is the vector wavenumber; $\vec{k} = k(\sin\theta, \cos\theta)$, where $\lambda = \frac{2\pi}{k}$ is the wavelength and θ is the propagation direction measured from due north. $S(\vec{k})$, which has units of m^4, is defined such that $\rho g \iint_{\vec{k}} S(\vec{k})d\vec{k} = E$, the total energy per unit area of the sea

[1]School of Mathematics and Statistics, University of Sheffield, UK

Figure 6.1   Swell waves

Figure 6.2   Wind waves

surface (ρ is the water density and g the is gravity). This energy is usually expressed in terms of a significant wave height, $H_s = 4 \iint S(\vec{k})d\vec{k}$.

Ocean waves are dispersive, i.e. their propagation speeds depend on their wavelength with long waves propagating faster than short waves. Longer waves can propagate faster than the wind that generated them and hence can arrive at the coast as swell before the following storm. In terms of frequency, ω in $rads^{-1}$, wavenumber k in m^{-1} and water depth d in m, the dispersion relationship is written as $\omega = \sqrt{gk \tanh kd}$. Often wave measurements are expressed in terms of frequency, f, in Hz ($\omega = 2\pi f$), instead of wavenumber in which case the directional spectrum is written as $S(f, \theta)$, which is related to $S(\vec{k})$ by $S(f, \theta) = \frac{dk}{df}kS(\vec{k})$ (assuming no current). Integrating this over all directions, we obtain the frequency spectrum, $E(f) = \int_\theta S(f, \theta)d\theta$. The 0th moment of this spectrum, $m_0 = \int_f E(f)df$, is also related to significant wave height, $H_s = 4\sqrt{m_0}$, which is the most common wave measurement parameter. Another parameter commonly used in describing ocean waves is the mean period, which is sometimes defined in terms of the first (T_1) and sometimes in terms of the second moment of the spectrum (in the latter case, this parameter is sometimes referred to as the zero-crossing period, T_z), i.e. where $m_n = \int_f E(f)f^n df$ we define $T_1 = \frac{m_0}{m_1}$ and $T_z = \sqrt{\frac{m_0}{m_2}}$. Wave power applications refer to energy period,

$T_e = \frac{m_{-1}}{m_0}$. Because this focuses on the lower, energy containing frequencies, it is a useful parameter for radar comparisons where there is an HF cut-off in the measurement (Section 6.3). The directional parameters most commonly used are mean direction, Th_m, and directional spreading, which are obtained from the directional spectrum (see Tucker and Pitt [3], for example). Period, T_p, direction, Th_p, and spread parameters at the spectral peak are also sometimes referred to. It is also sometimes useful to look at the variation in mean direction over frequency given

by $D(f) = \tan^{-1} \dfrac{\int_\theta S(f, \theta) \sin(\theta)d\theta}{\int_\theta S(f, \theta) \cos(\theta)d\theta}$ and the variation of energy with direction given by

$E(\theta) = \displaystyle\int_f S(f, \theta)df$ although these can be misleading in bimodal seas.

Figure 6.3 shows an example of $E(f), D(f)$ and $S(f, \theta)$. The example is taken from a trial of the Pisces HF radar in the Celtic Sea (between England, Wales and Ireland) and compares the radar and buoy measurements at the same time and location. On this occasion, both show swell with a peak frequency of about 0.075 Hz and a wind sea with a peak at about 0.15 Hz in roughly the same direction. More details of this trial are given in Section 6.4. Wave buoys do not in fact measure $S(f, \theta)$; statistical methods have to be used based on the measured Fourier coefficients [6]. In the example shown here, a maximum likelihood method [8] has been used.

The way in which the spectrum varies with wind speed and fetch (the distance over which the wind has been blowing) has been the subject of many studies and theories. If the wind has been blowing with constant speed and direction for a long time over a long distance, the waves are described as fully developed and are often

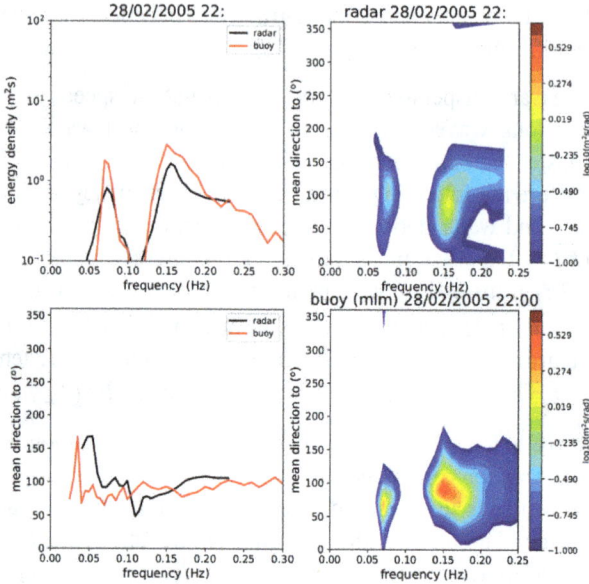

Figure 6.3   Examples of wave spectra from the Pisces trial in the Celtic Sea. Top
left: E(f) buoy in red, radar in black. Bottom left: D(f). On the right:
S(f, θ) for the radar above and buoy below.

modelled with a Pierson–Moskowitz spectrum [9]. If the fetch or time is short, a
Jonswap spectrum [10] is the model often used.

6.2   Waves in the Doppler spectrum

6.2.1   First order

With a single radar in a monostatic configuration, the first-order Bragg peaks are
backscatter from ocean waves with half the radio wavelength propagating towards
(to give the positive Doppler peak) and away from (negative) the radar site. These
first-order Bragg waves are generally short compared to the waves that are of most
interest in applications. This can be seen in Figure 6.4, which shows Pierson–
Moskowitz spectra for various significant wave heights. The ocean wave frequencies corresponding to the Bragg waves for a range of radio frequencies are shown
and it is clear that, at 10 MHz for example, the Bragg wave is far from the peak in
the spectrum above wave heights of 0.5 m or so. Hence the first-order information is
not so useful for wave measurement applications.

However, because they are short waves that generally respond quickly to and
align with changing wind direction, this information can be used to estimate wind
direction if a model of short-wave directionality is assumed. Such models are of
the form $S(f_{BW}, \theta) = E(f_{BW})g(\theta)$, where $g(\theta)$ usually takes either a $\cos^{2s} \frac{(\theta - \theta_w)}{2}$ [11]
or a $\operatorname{sech}^2(\beta(\theta - \theta_w))$ [12] form. The parameter s or β measures the spread of the

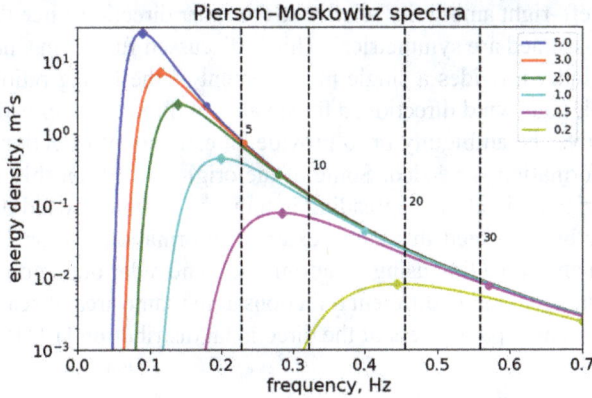

Figure 6.4   *Pierson–Moskowitz ocean wave spectra for the significant wave heights indicated. Vertical dashed lines show the Bragg-matched linear wave frequencies for the radio frequencies indicated.*

directional distribution about the short-wave (or wind) direction, θ_w. The two first-order Bragg wave amplitudes correspond to two points on the directional distribution, 180° apart along the radar look direction, as seen in Figure 6.5. So measuring the ratio of those peaks determines the wind direction for a given spreading parameter

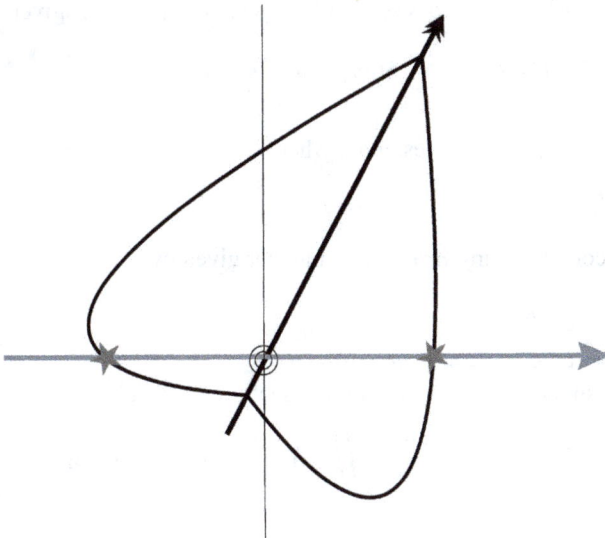

Figure 6.5   *Short-wave-directional spreading model. Wind direction is the black arrow, radar direction the grey arrow and grey asterisks mark amplitudes proportional to the first-order peaks.*

albeit with a left–right ambiguity relative to the radar direction since the directional distributions assumed are symmetrical. This is discussed further in Chapter 7.

A single radar provides a single measurement of the Bragg ratio; so to determine the ambiguous wind direction, a fixed value of directional spreading has to be used. To remove the ambiguity or to provide an estimate of directional spreading, additional information is needed. Some of the original work on this, e.g. [13–15], included a variable directional spreading in their estimation of wind direction from a single radar but obtained this using external information. Heron [16] extracted both direction and spreading using data from the same radar but on different beams. Using two radars looking in different directions at the same area of sea makes it possible to measure both parameters of the directional distribution [17, 18].

6.2.2   Second order

It is the second-order continuum surrounding the first-order peaks that provides wave measurements. Barrick and Weber [19] showed that this could be explained by a combination of second-order hydrodynamic interactions and second-order electromagnetic scattering giving rise to what have become known as Barrick's equations. Similar equations were obtained by Walsh [20, 21], Hisaki [22] and Hardman *et al.* [23]; see also Chapter 1 ((1.31) and (1.32)). Here we adopt the notation of Hardman *et al.* and write the equations for the power spectrum of the first- and second-order scatter for a bistatic radar system as

$$\sigma(\omega) = \sigma_1(\omega) + \sigma_2(\omega) \tag{6.1}$$

where the first-order radar cross section for a bistatic angle φ_{bi} is given by

$$\sigma_1(\omega) = 2^5 \pi k_0^4 \cos^4 \varphi_{bi} \sum_{m=\pm 1} S(m\vec{k}_B)\delta(\omega - m\omega_B), \tag{6.2}$$

defined at the Bragg frequencies, $\pm\omega_B$, where

$$\omega_B = \sqrt{2gk_0 \cos \varphi_{bi} \tanh(2k_0 d \cos \varphi_{bi})}, \tag{6.3}$$

and k_B is the corresponding Bragg wavenumber given by

$$k_B = 2k_0 \cos \varphi_{bi}, \tag{6.4}$$

where k_0 is the radio wavenumber.

The corresponding second-order cross section is given by

$$\sigma^{(2)}(\omega) = 2^5 \pi k_0^4 \cos^4 \varphi_{bi} \sum_{m,m'=\pm 1} \iint |\Gamma_E - i\Gamma_H|^2 S(m\vec{k}_1)S(m'\vec{k}_2)$$
$$\times \delta(\omega - m\omega_1 - m'\omega_2)\, dp\, dq, \tag{6.5}$$

for wavevector pairs $\vec{k}_1$ and $\vec{k}_2$ (with associated angular frequencies ω_1 and ω_2) such that

$$\vec{k}_1 + \vec{k}_2 = \vec{k}_B. \tag{6.6}$$

Explicitly, $\vec{k}_1 = (p - k_0, q)$ and $\vec{k}_2 = (-k_0 \cos(2\varphi_{bi}) - p, k_0 \sin(2\varphi_{bi}) - q)$, where p is aligned with the direction from the transmitter to the scattering patch. In (6.5), the four combinations of m and m' denote the four second-order sidebands either side of the two first-order Bragg peaks, e.g. $m = m' = 1$ is the sideband to the right of the positive Bragg peak and $m = -1$, $m' = 1$ is the sideband to the left of that peak.

Γ_E is the electromagnetic coupling coefficient given by

$$\Gamma_E = \frac{1}{2^2 \cos^2 \varphi_{bi}} \left(\frac{a_1}{b_1 - k_0 \Delta} + \frac{a_2}{b_2 - k_0 \Delta} \right) \tag{6.7}$$

where

$$\Delta = 0.011 - 0.012i \tag{6.8}$$

is the normalised surface impedance derived by Barrick [24] and

$$a_1 = -k_{1x}(\vec{k}_2 \cdot \vec{a}) - 2 \cos^2 \varphi_{bi} \left(-k_2^2 + 2k_0(\vec{k}_2 \cdot \vec{a}) \right) \tag{6.9}$$

$$a_2 = -k_{2x}(\vec{k}_1 \cdot \vec{a}) - 2 \cos^2 \varphi_{bi} \left(-k_1^2 + 2k_0(\vec{k}_1 \cdot \vec{a}) \right) \tag{6.10}$$

$$b_1 = \sqrt{-k_2^2 + 2k_0(\vec{k}_2 \cdot \vec{a})} \tag{6.11}$$

and

$$b_2 = \sqrt{-k_1^2 + 2k_0(\vec{k}_1 \cdot \vec{a})} \tag{6.12}$$

(noting that both b_1 and b_2 can be real or imaginary depending on the argument), where $\vec{a}$ is a unit vector in the direction of the receiver from the scattering patch $\vec{a} = (- \cos(2\varphi_{bi}), \sin(2\varphi_{bi}))$.

Γ_H is the hydrodynamic coupling coefficient given by

$$\Gamma_H = \frac{1}{2} \left\{ k_1 \tanh(k_1 d) + k_2 \tanh(k_2 d) + \frac{\omega}{g} \frac{(m\omega_1^3 \operatorname{csch}^2(k_1 d) + m'\omega_2^3 \operatorname{csch}^2(k_2 d))}{(\omega^2 - \omega_B^2)} \right.$$
$$\left. + \frac{(k_1 k_2 \tanh(k_1 d) \tanh(k_2 d) - \vec{k}_1 \cdot \vec{k}_2)}{mm' \sqrt{k_1 k_2 \tanh(k_1 d) \tanh(k_2 d)}} \left(\frac{\omega^2 + \omega_B^2}{\omega^2 - \omega_B^2} \right) \right\} \tag{6.13}$$

The hydrodynamic interactions are the most important as far as wave measurement is concerned because they generally provide larger contributions to the radar signal. Work by Creamer *et al.* [25] and Janssen [4, 26] has shown that the Barrick and Weber hydrodynamic analysis omitted a contribution from the interaction of first- and third-order waves. This term becomes increasingly important in higher seas and is essentially a wave height dependent scaling for the first-order amplitude. We will come back to this point later on.

It is important to look at the different components of the second-order integral in order to understand the limitations inherent in the data available. This is illustrated in Figure 6.6. On the left of this figure are two simulated backscatter (i.e. $\varphi_{bi} = 0$) Doppler spectra using a Pierson–Moskowitz ocean wave spectrum for a wind speed

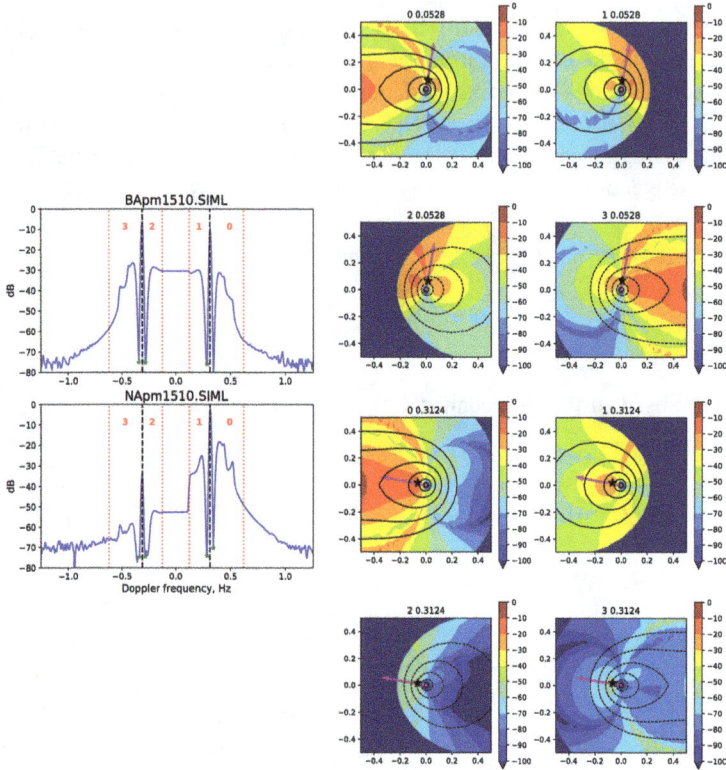

*Figure 6.6   The left panel shows simulated Doppler spectra for a Pierson–
Moskowitz spectrum with a 15 ms⁻¹ wind speed at a direction relative
to true north of 10°. The dashed black lines show the first-order
peaks. The upper four panels on the right show contributions to the
integral giving the spectrum on the left upper panel between the
dotted red lines surrounding each first-order peak with the lower four
panels on the right corresponding to the lower spectrum on the left.
The four panels in each set are numbered 0 to 3 corresponding to
the regions shown by the labels on the Doppler spectrum plots. The
panels are oriented along the radar direction, the wind direction is
shown with a magenta arrow and the wavenumber of the maximum
of the input spectrum is shown with ∗. The axes are the x— and y—
components of the long wavenumber vector $\vec{k}$. The black lines are the
integration contours for Doppler frequencies at 0.1 to 0.6 × Bragg
frequency from the corresponding first-order peak. The colours show
the magnitude (in dB) of the integrand at each vector wavenumber
relative to the maximum for that Doppler spectrum (shown above
each figure).*

of 15 ms^{-1} with a unimodal directional distribution around a wind direction of 10°. In the upper case the bearing from the radar relative to due north is −70° and in the lower case 160°. Therefore the upper spectrum is nearly perpendicular (80°) to the wind direction and the lower spectrum is nearly along (170°) the wind direction. The right-hand panel shows the contributions to the integral at each Doppler frequency from each long wavenumber vector $\vec{k}$ for the different regions of the Doppler spectrum identified with the numbers 0–3. These plots are aligned along the radar direction, and the axes are the $x-$ and $y-$ components of $\vec{k}$. The wind direction relative to the radar is shown with a magenta arrow. Close to the Bragg lines, the Doppler contours (selected ones are shown in black) are nearly circular and thus contain contributions from very similar long wavenumbers, but further away, there are contributions from wider ranges of wavenumbers. In addition, in the lower set of four plots, it can be seen that the maximum in the ocean wave spectrum is roughly aligned with the maximum contributions to the integral at the corresponding Doppler frequency whereas that is not the case in the upper set of plots. Thus, unlike the hypothesis of Hasselman [27], there is no 1–1 correspondence between Doppler frequency and ocean wavenumber in all cases. When the radar and waves are roughly aligned with the wind, as in the lower case, there is an approximate but not perfect 1–1 correspondence. In the more general case of mixed wind sea and swell in different directions, the relationship between the ocean wave spectrum and resulting Doppler spectrum is even more complex.

Another important property of the integration that needs to be considered is the symmetry. This is illustrated in Figure 6.7. In this case, the wind direction is 210° and is thus again 80° to the radar direction for the upper Doppler spectrum, which is nearly identical to the spectrum in Figure 6.6. Any differences are related to the very low level noise added to the spectra. The contributions to the integral for this Doppler spectrum, shown in the upper four panels, are the mirror image about the x-axis (radar direction) to those in Figure 6.6. It is therefore not possible, using a single-radar Doppler spectrum, to uniquely determine the wave direction. It has a left–right ambiguity with respect to the radar direction as is also the case for short-wave direction determined from the first-order Bragg lines (see Section 6.2.1).

The way the integral is constructed for a second radar looking from a different direction at the same wave field (the lower panels in Figures 6.6 and 6.7) is very different. This can be exploited by combining the information from both sites to avoid the amplitude and direction ambiguities, as will be discussed in Section 6.3.

The bistatic forms of the equations have been given here for completeness. However, monostatic radar systems are used for the examples given in the rest of this chapter.

6.3 Inversion

The radar measurement is not directly the backscatter cross-section spectrum given by (6.1) but is this multiplied by factors such as antenna gain and propagation loss

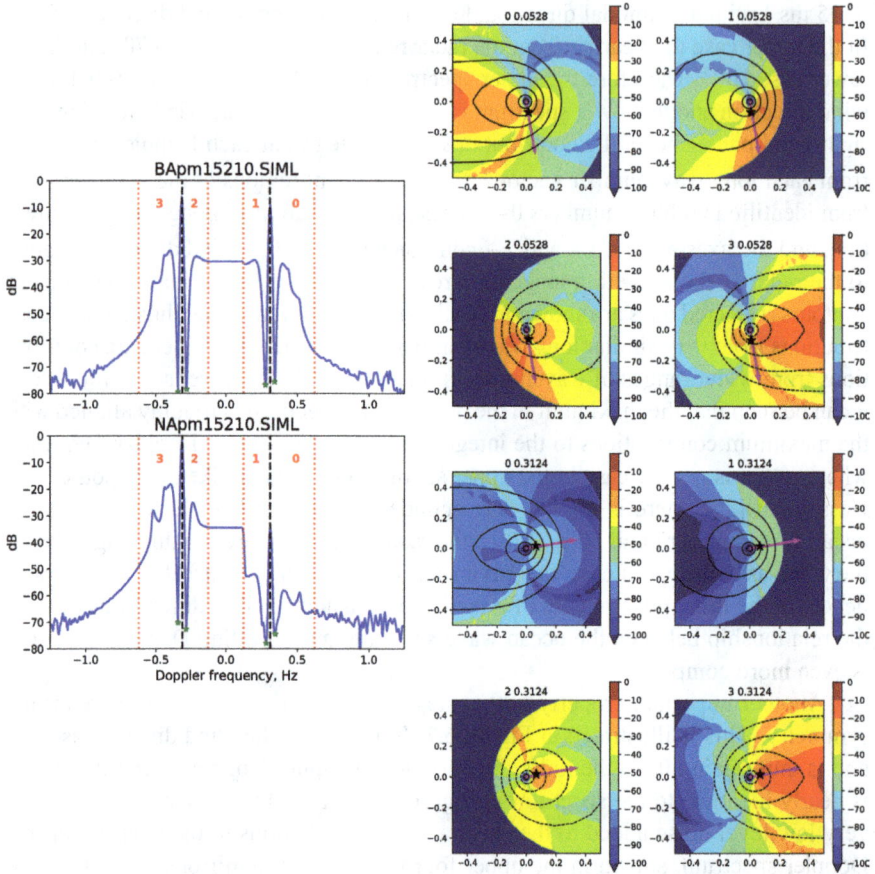

Figure 6.7   *The left panel shows simulated Doppler spectra for a Pierson–Moskowitz spectrum with a 15 ms^{-1} wind speed at a direction relative to true north of 210°. The details are described in the caption for Figure 6.6.*

(see discussion around (1.26) in Chapter 1). These factors multiply both first- and second-order terms, i.e.

$$\sigma_M(\omega) = R(\sigma_1(\omega) + \sigma_2(\omega)) \tag{6.14}$$

where $\sigma_M(\omega)$ is the measured power spectrum and R contains all the radar- and site-specific terms. To remove this factor, we can integrate the measured first-order radar cross section, σ_{1M}, over, say, the positive Bragg peak (in practice, the larger of the two peaks) to get

$$\int_\omega \sigma_{1M}(\omega)d\omega = R \times 2^5 \pi k_0^4 \cos^4 \varphi_{bi} S(\vec{k_B}) \tag{6.15}$$

This provides a value of R that can then be used to remove these factors from the second-order part of the measured spectrum. So we have

$$\frac{\sigma_{2M}(\omega)}{\int_\omega \sigma_{1M}(\omega)d\omega} = \frac{\sum_{m,m'=\pm 1} \iint |\Gamma_E - i\Gamma_H|^2 S(m\vec{k}_1)S(m'\vec{k}_2)\delta(\omega - m\omega_1 - m'\omega_2)dpdq}{S(\vec{k}_B)} \tag{6.16}$$

The left-hand side of the equation contains the measured radar Doppler spectrum, and to obtain the ocean wave directional spectrum $S(\vec{k})$, we need to invert this equation. It is convenient to non-dimensionalise all the variables, e.g. the Doppler frequency using the Bragg frequency, $\eta = \frac{\omega}{\omega_b}$, and wavenumbers using the monostatic Bragg wavenumber, $2k_0$, with other terms also non-dimensionalised appropriately [28], so that the inversion process is radar frequency independent. For the rest of this chapter, the non-dimensionalised left-hand side of (6.16) will be denoted by $\sigma_{2N}(\eta)$.

The equation is a non-linear integral equation for $S(\vec{k})$, and careful numerical methods are needed to obtain a reliable solution.

6.3.1 Approximations and empirical methods

Simplified approaches have therefore been sought to obtain wave measurements without full inversion. Some authors have sought approximations to (6.16) based on the work of Barrick [29]. In this work, Barrick linearised the equations and developed a weighting function $w(\eta)$ to provide a closed form relationship between the ocean wave frequency spectrum and the Doppler spectrum of the form $E(\omega_b|\eta - 1|) = \dfrac{4\sigma_{2M}(\omega_b\eta)/w(\eta)}{k_0^2 \int_\omega \sigma_{1M}(\omega_b\eta)d(\omega_b\eta)}$. This expression implies a 1–1 correspondence between the Doppler and ocean wave spectra, and we have already argued that this is not the case in Section 6.2. Barrick [29] removed the directional dependence by averaging the coupling coefficient over direction, noting that this did involve some approximations. However, he did not account for directional differences in the limits of the averaging integral that were noted by Wyatt [30], where Barrick's weighting function was shown to correspond to the integral for waves travelling perpendicular to the radar. The resulting wave height calculations using the weighting function that is given in [29] do show a directional dependence. It is not surprising therefore that in subsequent work using this approach (e.g. [31, 32]), the weighting function had to be recalibrated for local conditions.

Others have used empirical methods to derive relationships between (possibly) integrated properties of the Doppler spectrum and buoy data (e.g. [33–36]). However, as pointed out in Section 6.2, the radar response to a particular wave spectrum depends on the angle between the radar beam and the wave direction. Thus empirical coefficients relating, for example, significant wave height to an integral of the Doppler spectrum will only be valid for particular radar beam/wave direction combinations. This can be demonstrated using the simple case presented in Figure 6.6. Taking the integral over the second-order sidebands surrounding the larger first-order peaks, as is often done since the sidebands around the smaller first-order peaks are often close to the noise level, the value for upper spectrum is 0.05

and that for the lower spectrum is 0.13. Using all the sidebands, the corresponding figures are 0.15 and 0.13. These are two different numbers that correspond to the same significant wave height. This motivated Wyatt [34, 35] to find two relationships, one for waves propagating roughly along the radar beam and one for those perpendicular. A similar argument applies to methods that involve coefficients relating the ocean wave frequency spectrum to the Doppler spectrum [36], i.e. assuming there is a 1–1 relationship between Doppler frequency and ocean wave frequency. There is no such relationship as was made clear in Section 6.2 and in [30].

However such approaches can sometimes be, and have been, usefully applied to data from a single-radar system if there is a dominant wave direction and the method is calibrated with a buoy (for example), but there will be wave propagation directions that will not provide accurate measurements. A second radar (with a different calibration) can be used to reduce the directional errors.

Note that changing water depth cannot easily be taken into account with such approaches.

6.3.2 Integral inversion

To invert the equation, it must first be discretised. The power spectrum of the back-scattered spectrum is already discretised since it has been obtained using a fast Fourier transform (FFT) on digitised data so that we can write the left-hand side of the normalised equation as: $\sigma_{2N}(\eta_i) = \sigma_i$, where $i = 1, \ldots, n$ and n is the number of frequency bins used.

The integral on the right-hand side of (6.16) must also be discretised. There are a number of ways of doing this. For the purpose of describing the problems associated with the solution of this system of equations, here we will adopt one simple approach and discretise in angle. Thus the numerator of the right-hand side is written as

$$\sum_{j=1}^{m} K(\eta_i, \theta_j) S(k_{ij}, \theta_{ij}) S(k'_{ij}, \theta'_{ij}) \delta\theta_j \qquad (6.17)$$

where k'_{ij}, θ'_{ij} and k_{ij} are all functions of θ_j and η_i obtained by solving the second-order Bragg constraint (6.6) and $K(\eta_i, \theta_j)$ includes the discretisation of all other terms in (6.16). Close to the first-order Bragg lines, k'_{ij} is a high wavenumber and k_{ij} a low wavenumber as can be seen in Figure 6.6. The wavenumber k'_{ij} is not shown in those plots but is obtained from the Bragg constraint. Far away from the first-order lines, both k'_{ij} and k_{ij} are high wavenumbers. In most of the methods we will discuss, high wavenumbers are assumed to be wind-driven and are modelled in some way. For the moment, we will limit attention to an inversion close to the first-order lines and model the k' spectrum with a suitable wind-wave model. This process is referred to as a linearisation of the integral equation. The equation can thus be simplified to read: $\sigma_i = \sum_{j=1}^{m} K'(\eta_i, \theta_j) S(k_{ij}\theta_{ij})$ for $i = 1, \ldots, n$, where $K'(\eta_i, \theta_j)$ contains all the known or modelled terms including the first-order Bragg term in the denominator. To make further progress, we need to make some assumption about the form of the long wave directional spectrum, $S(k, \theta)$.

For the moment, we will assume that it is separable, i.e. $S(k, \theta) = E(k)\Theta(\theta)$, and that $\Theta(\theta)$ can also be modelled. We now need to pose the problem in such a way that a solution can be found for $E(k)$. To do this, we discretise k as k_κ, where $\kappa = 1, \ldots, l$, and hence write $E(k)$ as $\vec{a} = (a_1, \ldots, a_l)$. The whole equation can now be written in the form $\sigma = A\vec{a}$, where A is an $n \times l$ matrix absorbing K' and Θ and containing non-zero terms along rows only at the entries corresponding to the set of wavenumbers, k_{ij}, that contribute to a particular Doppler frequency η_i (see Figure 6.6). We have now a system of overdetermined (since $n > l$) linear algebraic equations. The solution of such a problem involves well-established numerical methods, some of which are discussed in more detail below. At the basis of these is the standard least squares problem, i.e. a solution must be found that minimises the quantity $\sum_i (\sigma_i - A_{ik}a_k)^2$ with respect to the unknowns a_k. This problem has solution $\vec{a} = (A^T A)^{-1} A^T \sigma$ (see for example, Noble and Daniel [37]). Unfortunately, for our problem, $A^T A$ has a number of small eigenvalues and hence a very unstable inverse. Small errors in σ (which is a quantity derived by FFT from a measured time series and is therefore never known exactly) or small changes in the discretisation parameters and hence the matrix A can lead to a very different solution for $\vec{a}$.

Standard least square methods are not therefore appropriate and a number of different methods have been developed, which tackle this intrinsic ill-conditioning in different ways. These include regularisation [38–40], singular value decomposition [41], constrained iteration [42] (more details in Section 6.3.3) [43, 44], Bayesian methods [45], optimisation [46] (more details in Chapter 7) and neural networks [47], and the reader is referred to those papers for the details. In addition to the inversion approach, the way in which the directional spectrum is represented in the numerical procedure differs between these methods with some [40] solving for a limited number of Fourier coefficients of the spectrum (thus providing data in the form provided by directional buoys) and others solving for the discretised spectrum on a regular [44] or irregular [42, 43] grid. The spectrum on a grid provides the most general solution and Fourier coefficients can be extracted from this, using standard methods [6, 48, 49]. Although formally one can solve the equation for data from a single-radar system, it is more usual to use data from two (or more) radars since this removes the amplitude and directional ambiguities referred to earlier. Other methods to constrain the solution have been suggested [50] including use of neighbouring Doppler spectrum measurements or measuring with more than one frequency at the same time and location, but these have not yet reached the accuracy obtainable with a second radar. As discussed in Chapter 7, Hisaki [51, 52] has demonstrated that single-radar solutions are possible when constrained with wave models.

In the above, it was noted that the problem can be linearised, i.e. $S(\vec{k_2})$ is modelled and a solution found for $S(\vec{k_1})$, as long as the inversion is limited to a small range of Doppler frequencies close to the first-order peaks. The range used determines the range of ocean wave frequencies for which a full linearised inversion is possible [53, 54], and this ocean wave frequency range depends on radio frequency. For low radio frequencies, the upper limit is lower than that at high radio frequencies, e.g. at 10 MHz it is about 0.22 Hz whereas at 25 MHz it is about 0.35 Hz, the exact limit also depending on the wind direction (see [54] for further details). Attempts have been made to solve the non-linear inversion problem [46] (see Chapter 7) [43], in order to extend the range of ocean wave frequencies, needed in particular at low radio frequencies in lower seas. However, the inversion at high wavenumbers only covers a small range of directions and this limits the value of the method. Also, using a wider range of Doppler frequencies is not usually feasible because of the noise levels.

For broad beam radar systems, such as the CODAR SeaSonde, the second-order backscatter at a particular range is described in (6.5) convolved with the antenna beam pattern [28]. This makes the inversion more difficult. Progress either requires assumptions about spatial homogeneity of the wave field and no finite depth effects [55] or assumes that any spatial variations are only due to refraction as the waves move into shallow water and so the equations can be solved for the deep sea spectrum that propagated into the measurement region. Lipa *et al.* [28, 55] describe both a full inversion for five Fourier coefficients and a model-fitting approach assuming a Pierson–Moskowitz spectrum with a cardioid directional distribution, the latter being implemented in the SeaSonde operational software.

6.3.3   The constrained iteration method

Although a number of different empirical approaches and methods for integral inversion have been developed, very few have been tested using good quality radar data; so it is difficult to determine the best method. In Section 6.4, examples using a constrained iteration method [42, 43, 44] are presented; so a little more detail on that method is given here. This is based on the work of Chahine [56] and Twomey [57] modified to account for the extra dimension in the HF radar problem where we are seeking a 2D solution, $S(f, \theta)$, from a one-dimensional measurement, $\sigma(\omega)$. As noted above, the method has been applied to the full non-linear inversion problem, but here we assume that the equations have been linearised by assuming a wind-wave model for the short waves in the spectrum. The inversion is therefore applied only to normalised Doppler frequencies within ± 0.6 from the first-order Bragg peak.

The wind-wave model used is a Pierson–Moskowitz model using an empirical estimate of wave height [35] to determine peak frequency. A directional distribution of the form $\mathrm{sech}^2(\beta(\theta - \theta_w))$ [12], where β and wind direction θ_w are determined from the Doppler spectra using a maximum likelihood method [17, 18, 58], is needed to obtain a wind wave $S(f, \theta)$, which is transformed to $S(\vec{k})$ and used for $S(m'\vec{k}_2)$ in (6.11). This $S(f, \theta)$ is used to initialise the spectrum at all frequencies and in particular $S(m\vec{k}_1)$, the spectrum we are seeking.

In order to discretise $S(m\vec{k}_1)$, it is expressed as a linear combination of radial basis functions, centred on the discretised wavenumber vectors [44], the coefficients of which are then modified during the iteration. The inversion proceeds by integrating (6.5) with the initialising spectrum in its radial basis form to obtain the corresponding Doppler spectrum. Changes are then made to the coefficients using the difference between the integrated and measured spectra and the value of the coupling coefficient at each wavevector relative to the maximum along the Doppler contour (Section 6.2.2) upon which it lies (see [44] for details). Provided radar data quality is good, this approach converges and good quality wave data are found. Convergence is measured using the mean of the difference between measured and integrated Doppler spectra relative to the mean of the measured spectrum across all Doppler frequencies used. Values ≤ 0.3 usually indicate good convergence.

Section 6.2.2 referred to the work of Creamer [25] and Janssen [4], which identified a second-order interaction term missing from the Barrick and Weber [19] analysis. The impact of this can be approximated by applying a wave height dependent scaling to the first-order Bragg peaks, and this is implemented at the initialisation stage in the method discussed here. Further work is needed to embed the additional term at each stage of the iteration process, but this approach has increased accuracy at higher wave heights.

6.4 Examples and validations

In this section, examples of data obtained using the [42–44] method are presented. This method has been tested on several radar systems (Pisces, Ocean Surface Current Radar (OSCR), WERA and the University of Hawaii generic HF radar) at different radio frequencies (from 5 to 27 MHz) and under different oceanographic conditions in many parts of the world. It has probably had the most exhaustive evaluation of all methods and has revealed many of the limitations described in Section 6.5. The comparisons have also demonstrated that when radar data quality is good, directional spectra, and therefore parameters, such as significant wave height, peak period and direction, derived therefrom, can be extracted with good accuracy.

There are three main steps to establish the validity of the radar measurements. Results from all of these follow. See also discussions on wave validation in [6] and [48]. Comparisons of simple wave parameters like wave height are essential but also comparisons of other parameters, e.g. mean and peak direction, Fourier coefficients, the frequency spectrum and other parameters broken down over narrower frequency bands.

- Time series comparisons with other measurement systems, most commonly but not exclusively wave buoys, to look at overall agreement as well as responses to changing wind conditions, to mixed wind-sea and swell events, extreme events.
- Statistical comparison. Many studies present root-mean-square (rms) differences and/or correlation coefficients. These are helpful but do not always tell the full story. Combining them through a Taylor diagram [59] can be more helpful especially if the results from different studies or with different methods need to be compared. More detailed statistical evaluations give greater insight and enable more confident assertions of accuracy.
- Time series of temporal maps assessing reasonableness of the spatio-temporal development. Possibly comparing with wave models. Once statistical validation has been done, demonstrating reasonable and useful spatio-temporal variability is key to convincing users of the value of HF radar for wave monitoring.

6.4.1   Time series

Figure 6.8 shows time series comparisons using three different radar systems operating at different frequencies, or frequency ranges, in seven different experiments. Netherlands-UK Radar Wavebuoy Experiment Comparison 2 (NURWEC2) was the first trial of the Pisces dual radar system originally developed at the University of Birmingham [60] and commercialised by Neptune Radar Ltd. Data collection was manual, so infrequent, and good-quality Doppler spectra for inversion were hand-selected, hence the sparse but good comparison seen in the figure. Pisces was next used to collect wave data in the Celtic Sea from late 2003 to mid-2005 and operated automatically over a range of frequencies to minimise external interference [61]. Four months of these data are included in the Figure 6.8. Unfortunately, the buoy disappeared in December 2004 and was not replaced until mid-January after the period of strong storms seen in the radar data. Wave model data during this period (not shown) are consistent with the radar measurements. The radar tends to overestimate wave height in low seas when using the lower radio frequencies in its range, but generally good agreement is seen. The OSCR is no longer available but was used for some short-wave measurement trials in the 1990s, one being at Holderness on the north-east coast of England [62] and shown in the Figure 6.8. The other four time series used the WERA radar first developed during the EU Scawvex project [63] and now deployed at a number of locations around the world. Shown here are results from the EU EuroROSE project in Norway and Spain [64], both very dynamic regions, and revealing an upper wave height of about 5 m for radar measurements at 27 MHz. A WERA system was deployed in Liverpool Bay for about seven years, and four months of data are shown here [54]. Waves are generally low in the bay, and results were rather noisy in these conditions. Finally shown in the figure is one month of data from the University of Plymouth (UoP) radar system located on the north coast of Cornwall in south-west England. This has been collecting data intermittently since 2011, and Lopez and Conley [65] include a more extensive validation of the method at this site. In all cases shown, when measurement is possible, the radar picks up the temporal variability seen in the buoy data.

Some quality control measures have been applied to these data before plotting as follows. Radar-specific criteria are (1) second-order signal to noise > 15 dB; (2) inversion residual < 0.3; (3) first-order Bragg frequency > peak in the Pierson–Moskowitz spectrum. The inversion residual measures the convergence of the constrained iteration method. The third criterion is needed to ensure that the shorter waves contributing to the Doppler spectrum, which are of the order of the first-order Bragg frequency, are wind waves. The Pierson–Moskowitz peak frequency, f_m, is related to significant wave height by $f_m^2 \approx \frac{0.04}{H_s}$ [66]. The first-order Bragg frequency, f_B, is given by $(f_B)^2 = \frac{2gk_0}{(2\pi)^2} \approx 0.5k_0$ (for a monostatic radar); so $f_B = f_m$ when $k_0 H_s \approx 0.08$. The Pierson–Moskowitz spectrum is obtained in the limit of a fully developed sea. In practice, the peak frequency in the wind-wave part of the spectrum is likely to be higher than this; so we set a low wave height threshold of $k_0 H_s = 0.12$ for radar wave height measurement. This threshold is set to a higher value of 0.32 when comparing period and direction parameters as discussed in [54]. The buoy measured H_s is used

Pisces NURWEC2 6-17MHz

Pisces Celtic Sea 6-13MHz

OSCR Holderness 25.4MHz

WERA EuroROSE Norway 27.65MHz

WERA EuroROSE Spain 27.65MHz

WERA Liverpool Bay 12.5-13.5MHz

WERA UoP Cornwall 12-13MHz

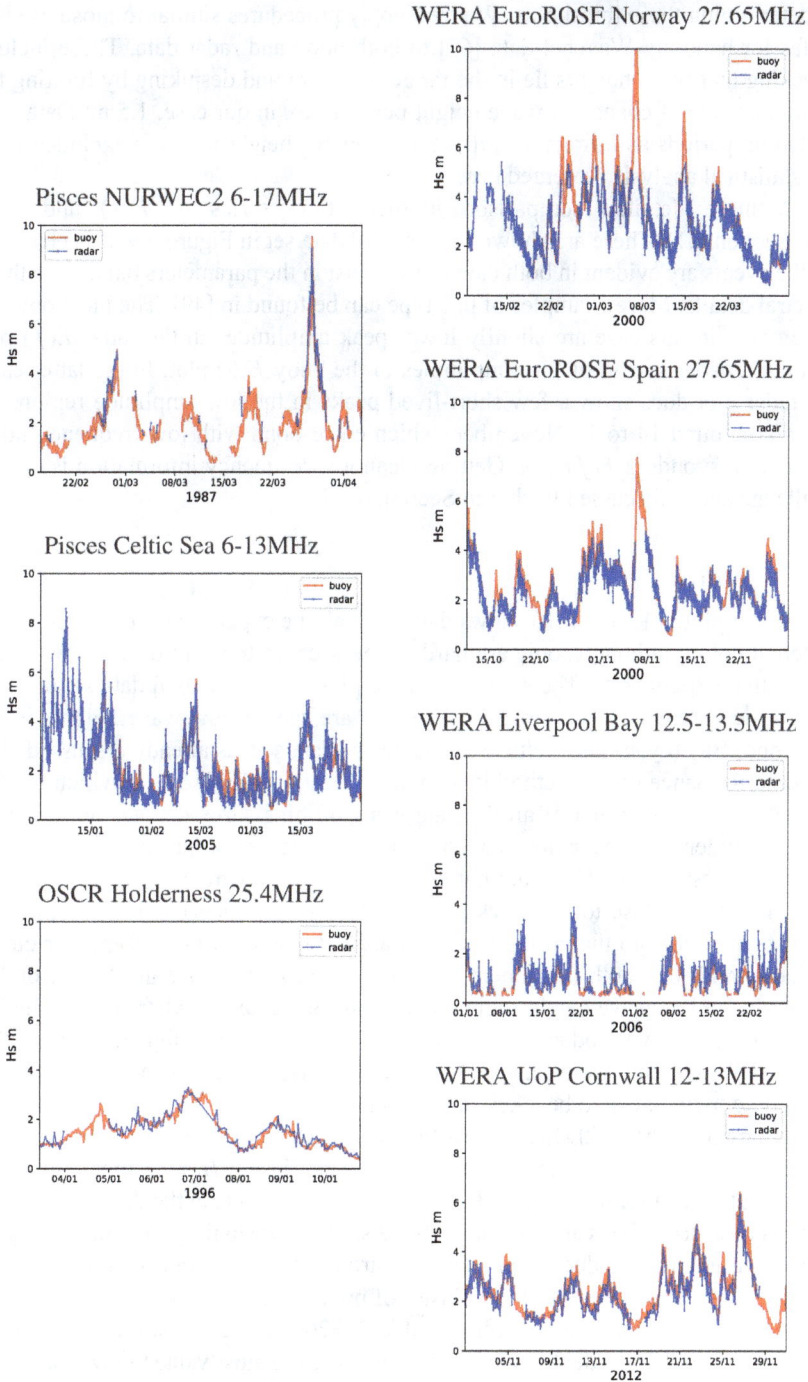

Figure 6.8   Time series of significant wave height: buoy in red, radar in blue

in determining these thresholds. We then apply procedures similar to those used by Cefas for handling Wavenet data [67] to both buoy and radar data. These include checking that wave heights lie in the range $0 - 20$ m and despiking by limiting the maximum rate of change in wave height per hour to, in our case, 1.5 m. Data sets, including periods and directions, that fail the wave height tests are excluded from the statistical analysis presented here.

A more detailed comparison involves time series of $E(f)$ and $E(\theta)$ (see Section 6.1). These are shown for the UoP data set in Figure 6.9. It is clear that all the events are evident in both data sets not just in the parameters but also in these spectral details. More examples of this type can be found in [49]. The most obvious differences in this case are slightly lower peak amplitudes in the radar $E(f)$ plot and the broader spread of low amplitudes in the buoy $E(\theta)$ plot. In the latter case, the radar plot does show a few short-lived peaks in the low-amplitude region, for example around 14 to 16 November, which correspond with low-frequency noise in the corresponding $E(f)$ plot. Getting clean low-frequency information is quite a challenge and is discussed further in Section 6.5.1.

6.4.2   Statistics

The scatterplot in Figure 6.10 shows data from all the experiments, each one with a different colour although some are hard to see because they are overlaid with data from other experiments. The statistics on the plot are for the total data set of 5 518 paired measurements. The largest differences are seen at low wave heights in the Liverpool Bay (cyan) and Celtic Sea (magenta) data sets as already discussed. The good performance of the method is seen more clearly in Figure 6.11, which is a 2D histogram or density plot of all the data using 40 bins. Grey denotes no measurements; the densities are colour-coded as shown. The performance of the different data sets in estimating H_s is summarised in the Taylor diagram in Figure 6.12. The goal is to be as close to the black square (perfect agreement) as possible, preferably staying close to the normalised standard deviation (std) of 1.0 quarter circle, which implies that both instruments measure the same variance in H_s. The outlier is clearly the low wave height regime Liverpool Bay data set, where using a higher frequency may have produced better results. In evaluating this figure, the facts that the NURWEC2 data were hand-selected and the Holderness data covered a very short time period need to be taken into account.

Figure 6.13 shows 2D histograms for some of the spectral parameters discussed in Section 6.1. Although there is more scatter than in the H_s measurements, good agreement is seen for most cases. There is a cluster of points in the T_p plot where the buoy is measuring 4–6 s and the radar 10–12 s. These are mainly from the Liverpool Bay radar and may be due to ship signals introducing low-frequency noise into the directional spectra. There is also a cluster of points in the Th_p plot with buoy directions of 120–150° and radar directions of 270–320°. In these cases, these are from the Norwegian WERA deployment and again these are mostly due to low-frequency noise in the spectra possibly arising because the data were not averaged sufficiently [68]. However, such differences in T_p and Th_p can also arise when data quality is

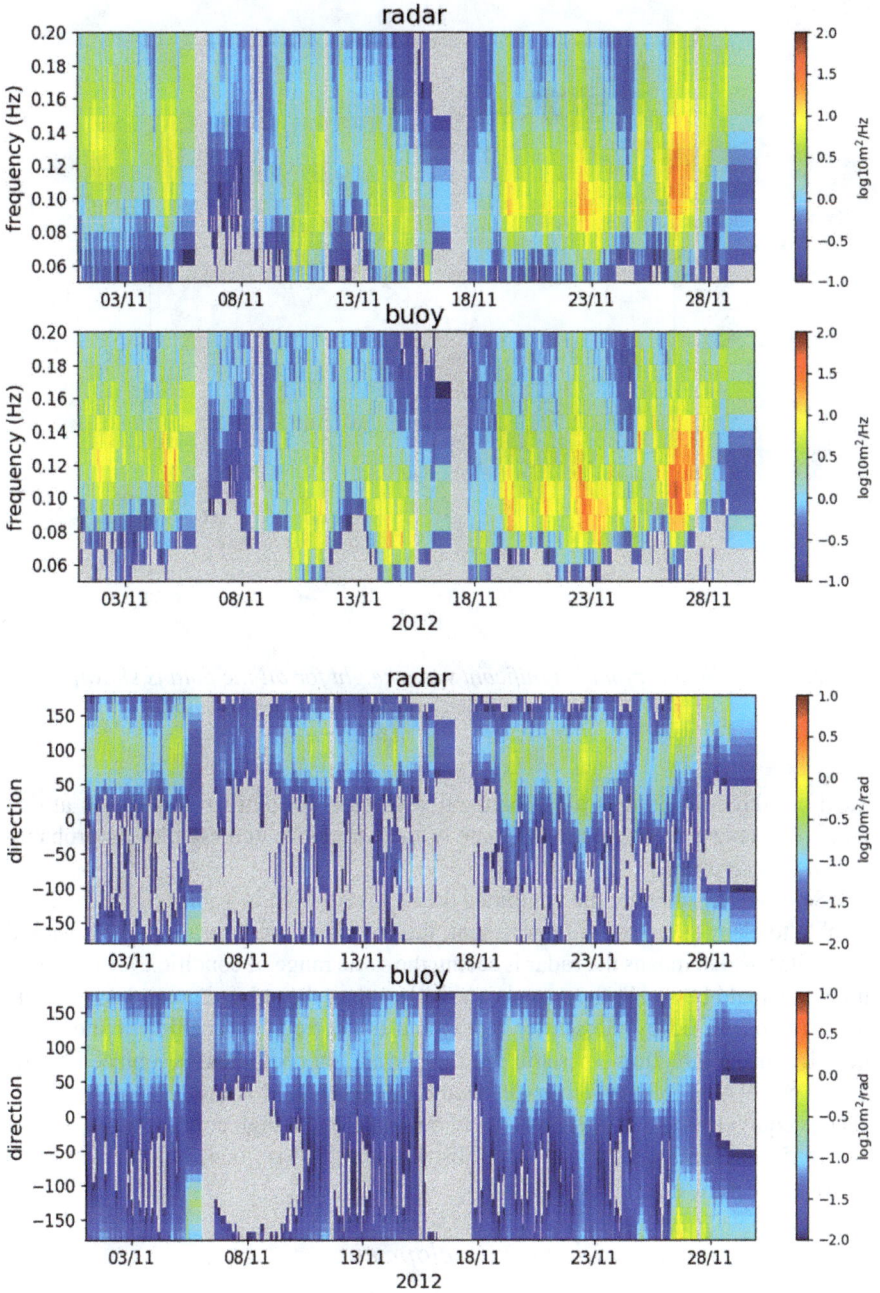

*Figure 6.9 Time series of UoP E(f) (upper pair) and E(θ). In each pair, the
radar measurement is above and buoy below. Grey indicates no
measurement or measurement below the low-amplitude threshold.*

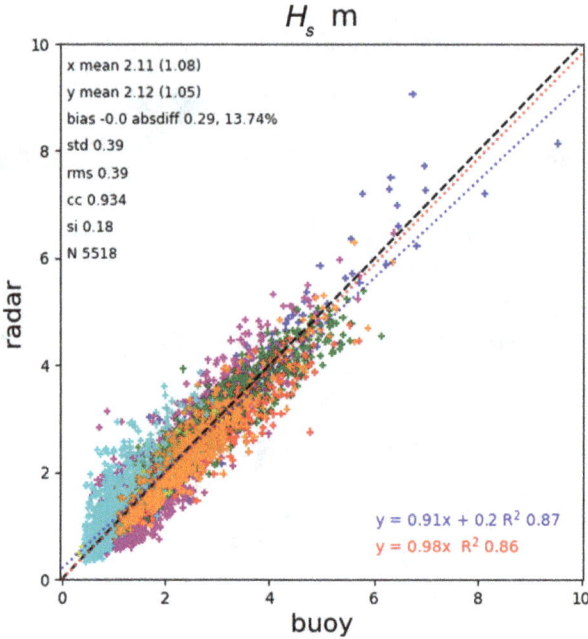

Figure 6.10   Scatterplot of significant wave height for all the data is shown in Figure 6.8. Each data set is shown in a different colour.

good but directional spectra are bimodal with peaks of similar amplitudes at different frequencies and directions. Some of the variability in these plots is probably linked to such cases.

Selected statistics for the combined data sets are given in Tables 6.1 and 6.2. To be able to claim good overall agreement, the buoy and radar means and std should be similar, which means the radar is seeing the same range of conditions as the buoy, the bias should be small, the rms should be less than the std and much less than the mean (as measured by the scatter index) and the correlation coefficient should be near 1. For the directional parameters, the correlation coefficient is replaced by the vector correlation [69], which again should be near 1. In this case, the mean difference is equivalent to the bias and should be small, and a high concentration parameter (> 5 represents a tightly grouped distribution) [70] is good. The figures in the tables are therefore very encouraging.

6.4.3   Spatio-temporal wave development

The previous two sections demonstrate that, when data quality is good, the radar measured spectra and parameters compare well with those of a wave buoy. Those comparisons are at a single position in the field of view of the radar. Having established credibility, in this section we look at how the waves vary across the radar coverage region, now looking only at the radar data.

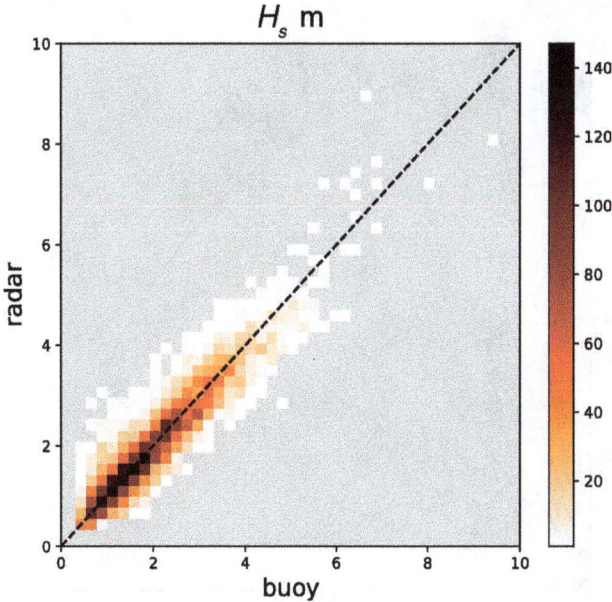

Figure 6.11   *Density plot of significant wave height for all the data is shown in Figure 6.8*

Figures 6.14–6.18 show the wave field responding to a change in wind direction. The parameter maps are taken from the Seaview data viewer for this data set that can be accessed from the Seaview Sensing website (www.seaviewsensing.com). There are wave spectra at every point where there are H_s and T_p values, and it is not possible to display all of these on a map. Cells have been selected to illustrate the spatial and temporal variation in the directional spectra. All spectra are scaled relative to the same maximum (10 m²/Hz), and the levels shown are logarithmic so that it is easier to see the lower amplitudes in the spectra. The spectral map also shows wind direction with the magenta arrow and surface current with the black arrow at these selected cells. This example starts with mostly swell propagating against the wind and finishes with a wind-dominated wave field. During the transition, the spectrum and peak period maps show that the swell persists but the wind sea increases in amplitude; so at some locations, the peak is in the swell and at others in the wind sea.

6.5   Sources of error and limitations

There are several sources of error in HF radar wave measurements. These relate to the quality of the radar data itself, the sampling and averaging processes needed to obtain the radar Doppler spectrum, the scattering model used (e.g. (6.2) and (6.5)) and errors associated with the numerical method used in the inversion or in the application of empirical methods.

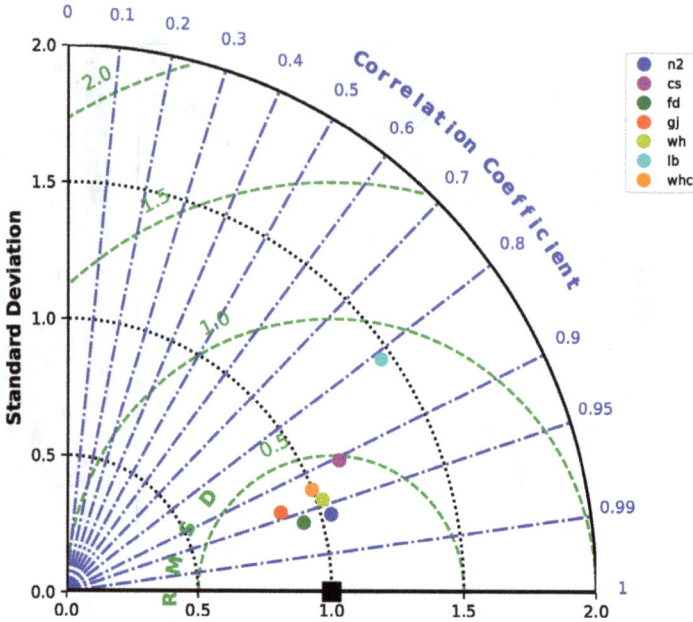

Figure 6.12   Taylor diagram of significant wave height for the seven data sets colour-coded is shown in Figure 6.10 and labelled n2 – NURWEC2, cs – Celtic Sea, fd – Norway, gj – Spain, wh – Holderness, lb – Liverpool Bay, whc – UoP

6.5.1   Radar data quality

As has already been said, the radar operating frequency determines the range of ocean wave frequencies that can be measured with a full inversion. This frequency therefore needs to be selected to suit the conditions likely to be encountered in a particular location and in many cases having more than one frequency will allow for a greater range of conditions. However, different radio frequencies have different radio frequency interference susceptibilities and ionospheric propagation characteristics and these can sometimes conflict with the oceanographic requirements. Wave measurements are made using the second-order part of the spectrum and are thus particularly susceptible to contamination by interference and other noise sources, e.g. scatter from targets, usually ships, moving at similar speeds to the ocean waves. Whilst efforts are being made to identify and remove these unwanted signals (see [71–75]), these are not yet available for operational applications and, as a result, radar wave measurements tend to be noisier than those made with other instruments such as wave buoys. These errors can sometimes be identified by looking for large changes in wave height over periods of less than one hour or from cell to cell spatially since such large changes are not usually encountered in nature. This despiking (with a 3 to 6 m threshold, so higher than the one used in this chapter) is also used by Cefas in their QA/QC procedures for handling data for Wavenet [67]. The National

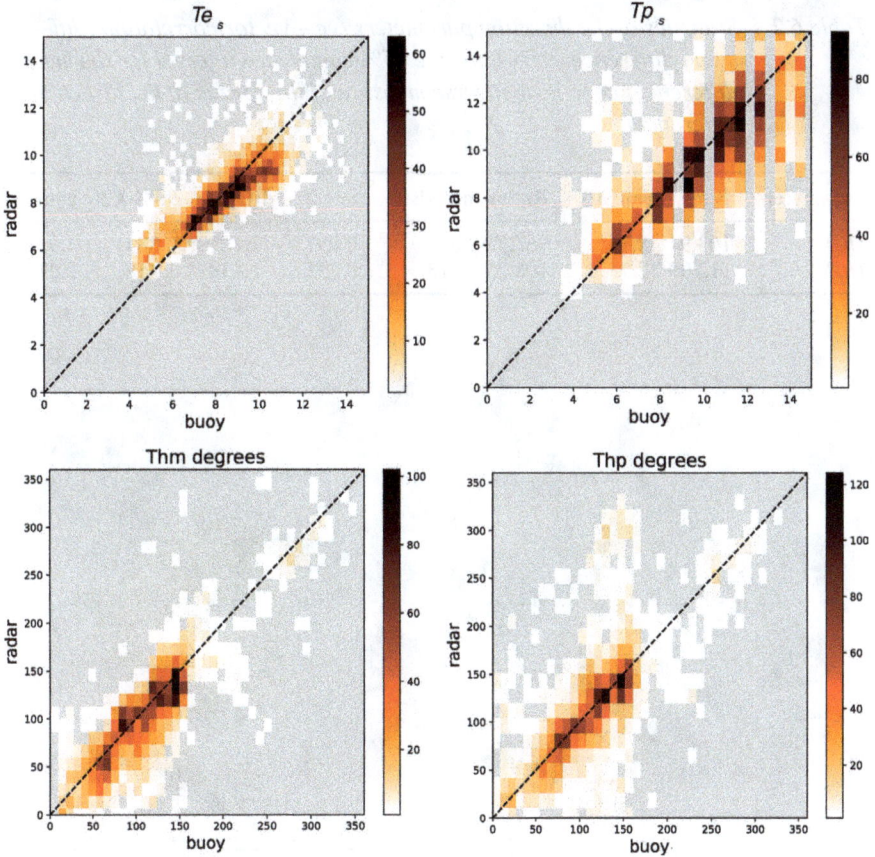

Figure 6.13   *Density plots of (top) energy, T_e, and peak period, T_p; (bottom) mean, Th_m, and peak direction, Th_p, for all the data shown in Figure 6.8*

Oceanic and Atmospheric Administration (National Data Buoy Center) uses an std method to despike, removing data that are more than 3 std from the mean with an algorithm that passes through the data three times and also carries out a number of

Table 6.1   *Statistics of the magnitude parameters (rms – root-mean-square difference; SI – scatter index; cc – correlation coefficient)*

Parameter	Buoy mean/std	Radar mean/std	Bias	rms	SI	cc
H_s, m	2.112/1.083	2.117/1.051	−0.005	0.39	0.185	0.934
T_e, s	8.409/1.798	8.417/1.415	−0.008	1.063	0.126	0.807
T_p, s	10.304/2.793	10.132/2.714	0.172	2.155	0.209	0.696

Table 6.2   *Statistics of the direction parameters (vc – vector correlation; diff – mean difference; 95% CI is the 95% confidence interval for the mean difference; conc is an inverse measure of the spread of the directional difference)*

Parameter	Buoy mean/std	Radar mean/std	vc	diff	95% CI	conc
$Th_m°$	103.91/48.1	96.27/52.07	0.892	−7.1	0.884	4.935
$Th_p°$	112.09/47.75	108.25/65.13	0.731	−1.539	1.458	2.217

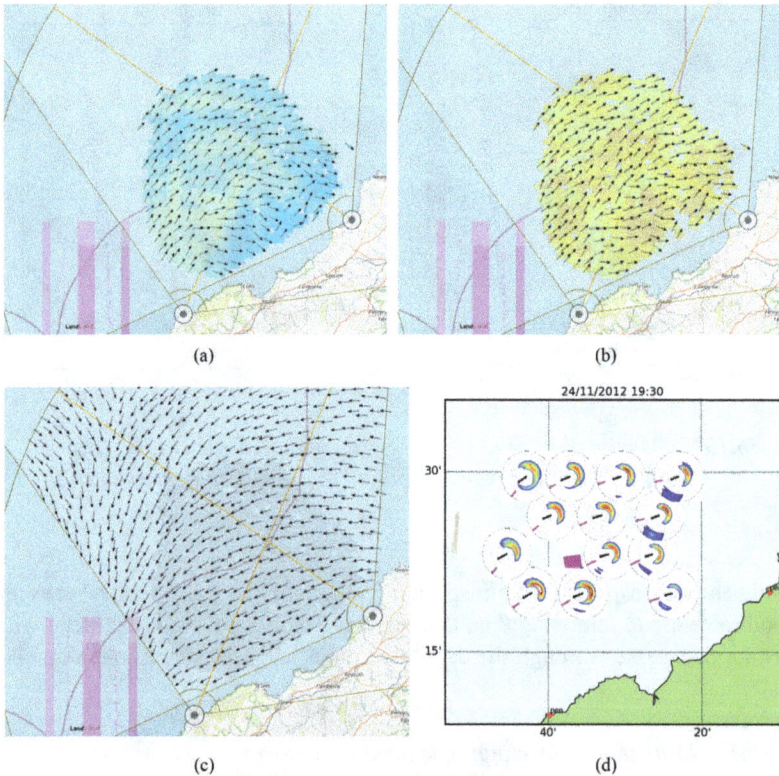

Figure 6.14   *Maps of (a) significant wave height and mean direction (m), (b) peak period and direction (s), (c) wind direction, (d) selected directional spectra from the UoP radar on 24/11/2012, 19:30. The radar sites are shown in red on the spectra map, which also shows the location of the wavehub in magenta partially covered by one of the spectra.*

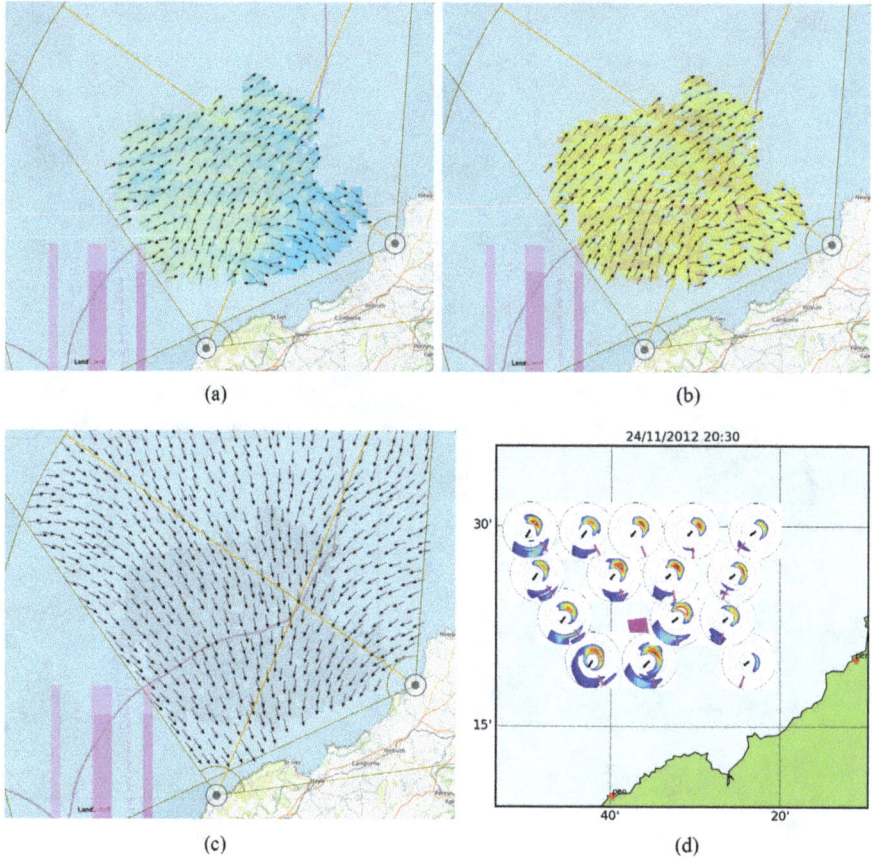

(a)

(b)

(c)

(d)

Figure 6.15 *As Figure 6.14 on 24/11/2012, 20:30*

other checks including a time continuity algorithm also based on the std of expected measurements [76].

Antenna sidelobes, due to poorly calibrated or failing antennas, can also lead to spurious wave measurements, particularly contributions to the low-frequency part of the spectrum. First-order Bragg peaks along a sidelobe are poorly suppressed and appear in the second-order part of the spectrum if the radial currents are different along the main lobe and sidelobes [54, 77]. One way to spot the presence of this problem is to follow the first-order peaks in the Doppler spectra at a fixed range (the range at which the wave height is suspiciously high) varying the azimuth. Where the wave height anomaly occurs, it should be possible to see an apparent second-order peak aligned with a first-order peak from another direction, as shown in [77]. In-cell current variability can also lead to similar problems.

In the presence of such effects and other factors influencing data quality, determining a robust boundary between first and second order is difficult but is key to obtaining a good wave measurement. Some of these issues severely limited wave

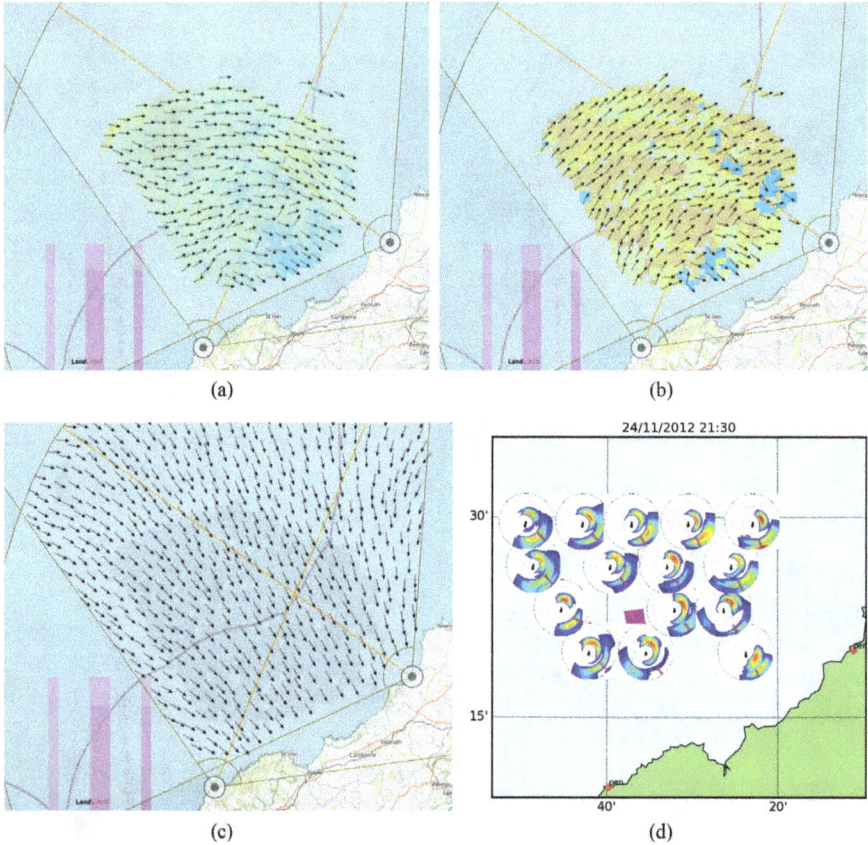

Figure 6.16   As Figure 6.14 on 24/11/2012, 21:30

measurements in a recent French experiment, HYD2M [78]. Solving these problems, either with signal processing methods or by ensuring better data quality, remains an area of active research.

6.5.2   Averaging

Ocean waves are stochastic in nature, and averaging is required to reduce the variance in the estimated wave spectrum whether measured with a wave buoy or an HF radar. Building on the work of Barrick and Snider [79], Wyatt *et al.* [68] show that averaging periods of 15–20 minutes are the minimum requirement for HF radar to approach the sampling variability of buoy measurements. An alternative is to average in space with the assumption that spatial homogeneity prevails [80].

6.5.3   The scattering model

The success of HF radar wave measurements to date demonstrates that the Barrick and Weber [19] hydrodynamic analysis provides a good theoretical basis. However,

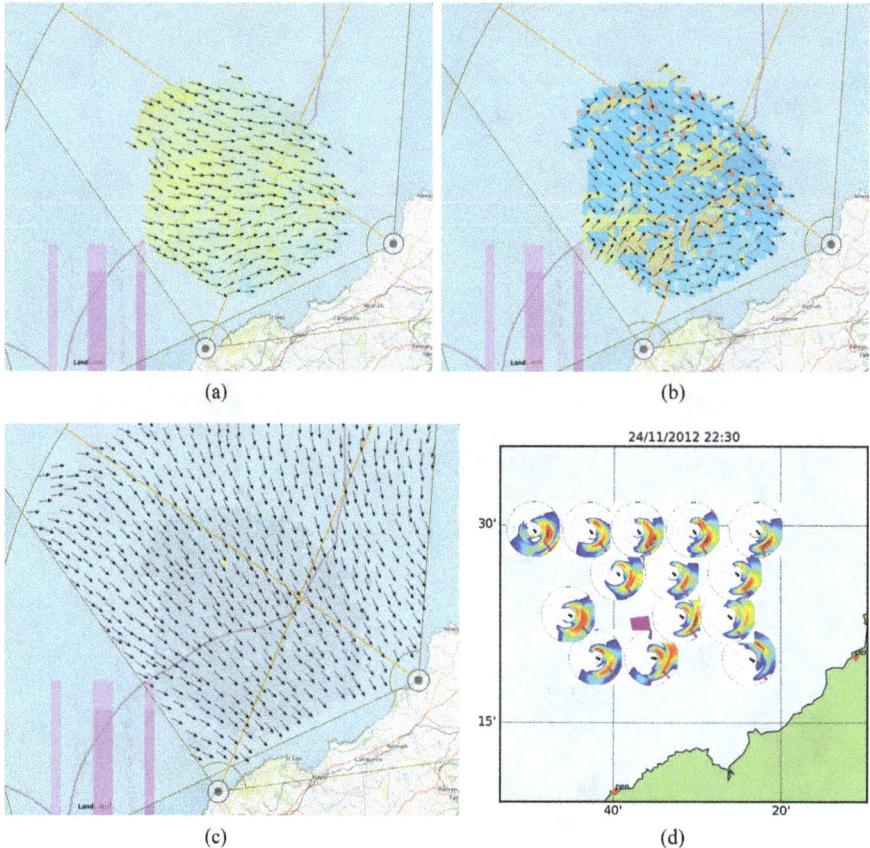

Figure 6.17   As Figure 6.14 on 24/11/2012, 22:30

there is evidence that measured Doppler spectra depart from this theory in higher seas particularly at higher radio frequencies [64, 81]. Work by Creamer *et al.* [25] and Janssen [26] has shown that there is a contribution from the interaction of first- and third-order waves missing from the Barrick and Weber analysis. This term becomes increasingly important in higher seas and is essentially a wave height dependent scaling for the first-order amplitude. Wyatt *et al.* [54] show that a simple approximation to this theory is sufficient to improve the accuracy of wave height estimation in higher seas, but no work has yet been done on a full inversion that includes this extra term. Hisaki [82] and Silva *et al.* [83] have also looked at this issue, but their approaches have not yet been validated with measured radar data.

6.5.4   Numerical methods

The success or otherwise of the numerical methods that have been adopted either for a full inversion or using empirical methods has only really been judged by direct comparisons with other measurement systems as discussed earlier. Sova [84] looked

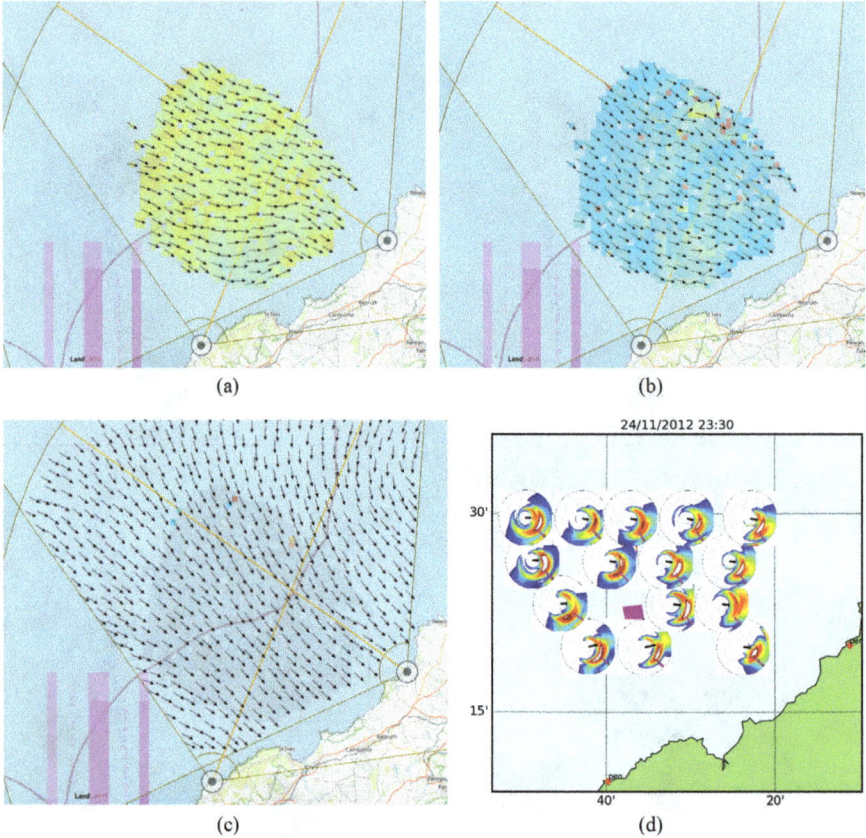

Figure 6.18   As Figure 6.14 on 24/11/2012, 23:30

at the propagation of sampling variability in the radar Doppler spectra through the constrained iteration inversion method to estimate the variance in the output wave spectra and derived parameters. This analysis used simulated radar spectra; so the 'true' wave spectrum was known. The evidence from measurements is that this estimate of variance is small compared to actual differences between radar and buoy measurements (part 2, Section 5.2 of Hauser *et al.* [6]). This may mean that the use of simulated data and its sampling variability is not adequately reflecting the errors in the inversion for real, and usually much more complicated, wave conditions or may mean that the inversion errors are small compared with the errors associated with radar data quality.

6.6   Summary

In this chapter, we have reviewed the theoretical basis for extracting wave information from HF radar Doppler spectra. We have explained why there are limitations in obtaining such data accurately from single-radar measurements. Additional information in the form of e.g. measurements from a second radar, using different radio frequencies and constraining with a wave model (as discussed in Chapter 7) is needed to remove directional ambiguities inherent in the scattering process. In the examples presented in this chapter, data from two radars have been used to obtain the results. Wave measurements are much more dependent on radar data quality than is the case for surface current measurement with these systems, but the results presented here, and in some of the references mentioned, show that the constrained iterative inversion method can provide accurate wave measurements when radar data quality is good enough. At the moment, poor quality data are usually identified and either removed before wave processing or flagged after processing. This of course reduces the amount of data available with an associated reduction in spatial and temporal continuity in the data sets. The challenge is to improve the quality of the radar data with signal processing methods that still allow for near-real-time monitoring of the coastal ocean.

Acknowledgements

The NURWEC2 data were collected by the University of Birmingham and the buoy data were provided by the Netherlands Rijskwaterstaat. The Pisces Celtic Sea data were provided by Neptune Radar and collected during a project funded by DEFRA and the MetOffice. The buoy data for this experiment have been provided by Cefas. The OSCR data were collected by the University of Sheffield during the Proudman Oceanographic Laboratory (now NOCL) Holderness project. The Norwegian data were obtained during the EU-funded EuroROSE project, and we thank all the EuroROSE team for their contributions. We are grateful to the radar team at POL, and the support team at Neptune Radar, for providing the Liverpool Bay radar data. The corresponding buoy data were obtained from the Cefas Wavenet website. The UoP radar and buoy data were provided by Daniel Conley and his team.

References

[1]   Kinsman B. *Wind Waves: Their Generation and Propagation on the Ocean Surface.* Prentice Hall; 1965.

[2]   Komen G.J., Cavaleri L., Donelan M. (eds.). *Dynamics and Modelling of Ocean Waves.* Cambridge University Press; 1994.

[3]   Tucker M.J., Pitt E.G. *Waves in Ocean Engineering.* Elsevier; 2001.

[4]   Janssen P. *The Interaction of Ocean Waves and Wind.* Cambridge University Press; 2004.

[5]   Holthuijsen L.H. *Waves in Oceanic and Coastal Waters.* Cambridge University Press; 2007.

[6]   Hauser D., Kahma K., Krogstad H.E. (eds.). *Measuring and Analyzing the Directional Spectra of Ocean Waves.* Luxembourg: Luxembourg Office for Official Publications of the European Communities; 2005.

[7]   The Unconquered Ocean Waves [online]. Available from https://www.you-tube.com/watch?v=-PJ9RNODzM0 [Accessed 29 Mar 2021].

[8]   Hashimoto N. 'Analysis of the directional wave spectrum from field data'. *Advances in Coastal and Ocean Engineering. World Scientific*; 1997.

[9]   Pierson W.J., Moskowitz L. 'A proposed spectral form for fully developed wind seas based on the similarity theory of S. A. Kitaigorodskii'. *Journal of Geophysical Research*. 1964;**69**(24):5181–90.

[10]   Hasselmann K., Barnett T.P., Bouws E., *et al.* 'Measurement of wind-wave growth and swell decay during the joint North sea wave project (JONSWAP)'. *Deutsche Hydrographische Zeitschrift*. 1973;**A**(8):12, 00.

[11]   Longuet-Higgins M.S., Cartwright D.E., Smith N.D. *Observations of the Directional Spectrum of Sea Waves Using the Motions of a Flaoting Buoy*. Prentice Hall: Ocean Wave Spectra; 1963.

[12]   Donelan M.A., Hamilton J., Hui W.H. 'Directional spectra of wind-generated waves'. *Philosophical Transactions of the Royal Society of London. Series A*. 1985;**315**:509–62.

[13]   Tyler G.L., Teague C.C., Stewart R.H., *et al.* 'Wave directional spectra from synthetic aperture observations of radio scatter'. *Deep Sea Research*. 1974;**21**:989–1016.

[14]   Stewart R.H., Barnum J.R. 'Radio measurements of oceanic winds at long ranges: an evaluation'. *Radio Science*. 1975;**10**:853–7.

[15]   Heron M.L., Dexter P.E., McGann B.T. 'Parameters of the air-sea interface by high-frequency ground-wave Doppler radar'. *Australian Journal of Marine and Freshwater Research*. 1985;**36**(5):655–70.

[16]   Heron M.L. 'Directional spreading of short wavelength fetch-limited wind waves'. *Journal of Physical Oceanography*. 1987;**17**:281–5.

[17]   Wyatt L.R., Ledgard L.J., Anderson C.W. 'Maximum likelihood estimation of the directional distribution of 0.53Hz ocean waves'. *Journal of Atmospheric and Oceanic Technology*. 1997;**14**:591–603.

[18]   Wyatt L. 'A comparison of scatterometer and HF radar wind direction measurements'. *Journal of Operational Oceanography*. 2018;**11**:54–63.

[19]   Barrick D.E., Weber B.L. 'On the nonlinear theory for gravity waves on the ocean's surface. Part II: Interpretation and applications'. *Journal of Physical Oceanography*. 1977;7:11–21.

[20]   Srivastava S.K., Walsh J. 'An analytical model for the HF backscattered Doppler spectrum for the ocean surface'. *IEEE Journal of Oceanic Engineering*. 1986;**11**:293–5.

[21]   Walsh J., Gill E.W. 'An analysis of the scattering of high-frequency electromagnetic radiation from rough surfaces with application to pulse radar operating in backscatter mode'. *Radio Science*. 2000;**35**:1337–59.

[22]   Hisaki Y., Tokuda M. 'VHF and HF sea echo Doppler spectrum for a finite illuminated area'. *Radio Science*. 2001;**36**(3):425–40.

[23]   Hardman R.L., Wyatt L.R., Engleback C.C. 'Measuring the directional ocean spectrum from simulated bistatic HF radar data'. *Remote Sensing*. 2020;**12**.

[24] Barrick D.E. 'Theory of HF and VHF propagation across the rough sea, 1, the effective surface impedance for a slightly rough highly conducting medium at grazing incidence'. *Radio Science*. 1971;**6**(5):517–26.

[25] Creamer D.B., Henyey F., Schult R., Wright J. 'Improved linear representation of ocean surface waves'. *Journal of Fluid Mechanics*. 1989;**205**:135–61.

[26] Janssen P. 'On some consequences of the canonical transformation in the Hamiltonian theory of water waves'. *Journal of Fluid Mechanics*. 2009;**637**:1–44.

[27] Hasselmann K. 'Determination of ocean wave spectra from measured HF radar sea-echo Doppler spectra'. *Nature: Physical Science*. 1971;**229**:681–2.

[28] Lipa B.J., Barrick D.E. 'Extraction of sea state from HF radar sea echo: mathematical theory and modelling'. *Radio Science*. 1986;**21**:81–100.

[29] Barrick D.E. 'Extraction of wave parameters from measured HF radar sea-echo Doppler spectra'. *Radio Science*. 1977;**12**:415–24.

[30] Wyatt L.R. On Barrick's inversion procedure; University of Birmingham, Dept of Electronic and Electrical Engineering, Deptl Memo 487. 1981. Available from https://www.researchgate.net/publication/279878594_On_Barrick's_Inversion_Procedure.

[31] Heron S.F., Heron M.L. 'A comparison of algorithms for extracting significant wave height from HF radar ocean backscatter spectra'. *Journal of Atmospheric and Oceanic Technology*. 1998;**15**:1157–63.

[32] Ramos R.J., Graber H.C., Haus B. 'Observation of wave energy evolution in coastal areas using HF radar'. *Journal of Atmospheric and Oceanic Technology*. 2009;**26**:1891–190.

[33] Maresca J.W., Georges T.M. 'Measuring rms waveheight and the scalar ocean wave spectrum with HF skywave radar'. *Journal of Geophysical Research-C*. 1980;**85**:2759–71.

[34] Wyatt L.R. 'Significant waveheight measurement with HF radar'. *International Journal of Remote Sensing*. 1988;**9**:1087–95.

[35] Wyatt L.R. 'An evaluation of wave parameters measured using a single HF radar system'. *Canadian Journal of Remote Sensing*. 2002;**28**:205–18.

[36] Gurgel K.W., Essen H.H., Schlick T. 'An empirical method to derive ocean waves from second-order Bragg scattering: prospects and limitations'. *IEEE Journal of Oceanic Engineering*. 2006;**31**:804–11.

[37] Noble B., Daniel J.W. *Applied linear algebra*. New Jersey: Prentice Hall; 1977.

[38] Lipa B.J. 'Derivation of directional ocean-wave spectra by inversion of second order radar echoes'. *Radio Science*. 1977;**12**:425–34.

[39] Lipa B.J. 'Inversion of second-order radar echoes from the sea'. *Journal of Geophysical Research-C*. 1978;**83**:959–62.

[40] Lipa B.J., Barrick D.E. 'Methods for the extraction of long-period ocean wave parameters from narrow beam HF radar sea echo'. *Radio Science*. 1980;**15**:843–53.

[41] Howell R., Walsh J. 'Measurement of ocean wave spectra using narrow beam HF radar'. *IEEE Journal of Oceanic Engineering*. 1993;**18**:296–305.

[42] Wyatt L.R. 'A relaxation method for integral inversion applied to HF radar measurement of the ocean wave directional spectrum'. *International Journal of Remote Sensing*. 1990;**11**:1481–94.

[43] Wyatt L.R. 'Limits to the inversion of HF radar backscatter for ocean wave measurement'. *Journal of Atmospheric and Oceanic Technology*. 2000;**17**:1651–66.

[44] Green J.J., Wyatt L.R. 'Row-action inversion of the Barrick-Weber equations'. *Journal of Atmospheric and Oceanic Technology*. 2006;**23**:501–10.

[45] Hashimoto N N., Tokuda M. 'A Bayesian approach for estimating directional spectra with HF radar'. *Coastal Engineering Journal*. 1999;**41**:137–49.

[46] Hisaki Y. 'Nonlinear inversion of the integral equation to estimate ocean wave spectra from HF radar'. *Radio Science*. 1996;**31**:25–39.

[47] Hardman R.L., Wyatt L.R. 'Inversion of HF radar Doppler spectra using a neural network'. *Journal of Marine Science and Engineering*. 2019;**7**(8):255–72.

[48] Krogstad H.E., Wolf J., Thompson S.P., *et al.* 'Methods for the intercomparison of wave measurements'. *Coastal Engineering*. 1999;**37**:235–58.

[49] Wyatt L. 'Measuring the ocean wave directional spectrum 'First Five' with HF radar'. *Ocean Dynamics*. 2019;**69**:123–44.

[50] Wyatt L.R., Moorhead M.D., Fairley I.A. 'Developments in metocean HF radar technology, applications and accuracy'. Proceedings of the ASME 2019 38th International Conference on Ocean, Offshore and Arctic Engineering OMAE 2009 June 9–14, 2019; Glasgow, Scotland; 2019.

[51] Hisaki Y. 'Ocean wave directional spectra estimation from an HF ocean radar with a single antenna array: observation'. *Journal of Geophys Researech*. 2005;**110**:11004.

[52] Hisaki Y. 'Ocean wave directional spectra estimation from an HF ocean radar with a single antenna array: methodology'. *Journal of Atmospheric and Oceanic Technology*. 2006;**23**:268–86.

[53] Barrick D.E. 'The ocean waveheight nondirectional spectrum from inversion of the HF sea-echo Doppler spectrum'. *Remote Sensing of Environment*. 1977;**6**:201–27.

[54] Wyatt L.R., Green J.J., Middleditch A. 'HF radar data quality requirements for wave measurement'. *Coastal Engineering*. 2011;**58**:327–36.

[55] Lipa B.J., Nyden B. 'Directional wave information from the SeaSonde'. *IEEE Journal of Oceanic Engineering*. 2006;**30**:221–31.

[56] Chanine M.T. 'Determination of the temperature profile in an atmosphere from its outgoing radiance'. *Journal of the Optical Society of America*. 1968;**58**:1634–7.

[57] Twomey S. *Introduction to the Mathematics of Inversion in Remote Sensing and Indirect Measurements*. Dover; 1996.

[58] Wyatt L.R. 'Shortwave direction and spreading measured with HF radar'. *Journal of Atmospheric and Oceanic Technology*. 2012;**29**:286–99.

[59] Taylor K.E. 'Summarizing multiple aspects of model performance in a single diagram'. *Journal of Geophysical Research: Atmospheres*. 2001;**106**(D7):7183–92.

[60] Wyatt L.R., Venn J., Moorhead M.D., Burrows G., Ponsford A., Van Heteren J. 'HF radar measurements of ocean wave parameters during NURWEC'. *IEEE journal of oceanic engineering*. 1986;**OE-11**:219–34.

[61] Wyatt L.R., Green J.J., Middleditch A., *et al.* 'Operational wave, current and wind measurements with the pisces HF radar'. *IEEE Journal of Oceanic Engineering*. 2006;**31**:819–34.

[62] Wyatt L.R., Thompson S.P., Burton R.R. 'Evaluation of HF radar wave measurement'. *Coastal Engineering*. 1999;**37**:259–82.

[63] Gurgel K.W., Antonischki G., Essen H.H., Schlick T. 'Wellen radar (WERA): a new ground-wave HF radar for ocean remote sensing'. *Coastal Engineering*. 1999;**37**:219–34.

[64] Wyatt L.R., Green J.J., Gurgel K.W., *et al.* 'Validation and intercomparisons of wave measurements and models during the EuroROSE experiments'. *Coastal Engineering*. 2003;**48**:1–28.

[65] Lopez G., Conley D.C. 'Comparison of HF radar fields of directional wave spectra against in situ measurements at multiple locations'. *Journal of Marine Science and Engineering*. 2019;**7**(8):271.

[66] Tucker M.J. *Waves in Ocean Engineering Measurement, Analysis, Interpretation*. Ellis Horwood; 1991.

[67] CEFAS. QA/QC procedure [online]. Available from https://www.cefas.co.uk/data-and-publications/wavenet/qa-qc-procedure/ [Accessed 4 Feb 2020].

[68] Wyatt L.R., Green J.J., Middleditch A. 'Signal sampling impacts on HF radar wave measurement'. *Journal of Atmospheric and Oceanic Technology*. 2009;**26**:793–805.

[69] Kundu P.K. 'Ekman veering observed near the ocean bottom'. *Journal of Physical Oceanography*. 1976;**6**:238–42.

[70] Bowers J.A., Morton I.D., Mould G.I. 'Directional statistics of the wind and waves'. *Applied Ocean Research*. 2000;**22**:13–30.

[71] Abramovich Y.I., Spencer N.K., Anderson S.J., Gorokhov A. 'Stochastic-constraints method in nonstationary hot-clutter cancellation-Part I:Fundamentals and supervised training applications'. *IEEE Transactions Aerospace and Electronic Systems*. 1998;**34**:20.

[72] Abramovich Y.I., Spencer N.K., Anderson S.J. 'Stochastic-constraints method in nonstationary hot-clutter cancellation-Part II:Unsupervised training applications'. *IEEE Transactions Aerospace and Electronic Systems*. 2000;**36**:132–50.

[73] Gurgel K.W., Barbin Y., Schlick T. 'Radio frequency interference suppression techniques in FMCW modulated HF radars'. Proceedings of Oceans 2007 – Europe; 2007. pp. 538–41.

[74] Zhou H., Wen B., Wu S. 'Dense radio frequency interference suppression in HF radars'. *IEEE Signal Processing Letters*. 2005;**12**(5):361–36.

[75] Wang W., Wyatt L.R. 'Radio frequency interference cancellation for sea-state remote sensing by HF'. *IET Radar, Sonar and Navigation*. 2011;**5**(4):405–15.

[76] NOAA. 'Handbook of automated data quality control checks and procedures'. 2009.

[77]   Wyatt L.R., Liakhovetski G., Graber H.C., Haus B.K. 'Factors affecting the accuracy of SHOWEX HF radar wave measurements'. *Journal of Atmospheric and Oceanic Technology*. 2005;**22**(7):847–59.

[78]   Lopez G., Bennis A.C., Barbin Y., *et al.* 'Hydrodynamics of Alderney Race: HF Radar Wave Measurements'. 7th International Conference on Ocean Energy 2018; Cherbourg, France; 2018.

[79]   Barrick D.E., Snider J.B. 'The statistics of HF sea-echo Doppler spectra'. *IEEE Transactions on Antennas and Propagation*. 1977;**AP-25**:19–28.

[80]   Wyatt L.R., Jaffrés J.B.D., Heron M.L. 'Spatial averaging of HF radar data for wave measurement applications'. *Journal of Atmospheric and Oceanic Technology*. 2013;**30**:2216–24.

[81]   Wyatt L.R. 'High order nonlinearities in HF radar backscatter from the ocean surface'. *IEE Proceedings–Radar, Sonar and Navigation*. 1995;**142**:293–300.

[82]   Hisaki Y. 'Correction of amplitudes of Bragg lines in the sea echo Doppler spectrum of an ocean radar'. *Journal of Atmospheric and Oceanic Technology*. 1999;**16**:1416–33.

[83]   Silva M.T., Huang W., Gill E.W. 'Bistatic high-frequency radar cross-section of the ocean surface with arbitrary wave heights'. *Remote Sensing*. 2020;**12**:667.

[84]   Sova M.G. The sampling variability and the validation of high frequency wave measurements of the sea surface[PhD thesis]. University of Sheffield; 1995.

Chapter 7

A non-linear method to estimate the wave directional spectrum by HF radar

Yukiharu Hisaki[1]

7.1   Introduction

High-frequency (HF) ocean radar radiates HF radio waves and can measure surface currents and ocean wave spectra by analyzing backscattered signals from the ocean. There are four types of wave estimation methods by HF ocean radars: Barrick's approximation method [1], parameter-fitting method [2], linear inversion method [3], and non-linear inversion method. The first three methods have been discussed in Chapter 6. This chapter introduces the non-linear inversion method to estimate wave spectrum from the HF radar. The non-linear inversion method is the one to obtain the wave spectra without approximating the integral equation of the relationship between the Doppler spectra and the wave spectra to a linear integral equation with respect to the wave spectra.

   As will be described later, Doppler frequencies are related to the frequencies and directions of the wave. This method uses not only integral equations but also other constraints on the wave spectrum. It is possible to estimate the spectral values in a wider wave number plane than other methods. The non-linear inversion method presented here was proposed by [4]. Then, it was extended to a method applicable to both single and dual radars.

7.2   Equations of radar cross sections

The mathematical formulation of HF radio wave scattering from the sea surface was described in Chapters 1 and 6, but it is briefly described to explain the inversion method.

   A Doppler spectra $P(\omega_D)$ is estimated by radiating HF radio waves and analyzing backscattered signals from the ocean, where ω_D is the radian Doppler frequency. We can obtain Doppler spectra $P(\omega_D)$ at radial grids on the polar coordinate with

[1]Department of Physics and Earth Sciences, University of the Ryukyus, Japan

origin at the radar position. The wave spectra at the radial grid points are estimated from Doppler spectra at the grid points.

A Doppler spectrum $P(\omega_D)$ is proportional to the radar cross section (RCS)

$$P(\omega_D) = P_1(\omega_D) + P_2(\omega_D) \tag{7.1}$$

$$P(\omega_D) = A\sigma(\omega_D) \tag{7.2}$$

$$P_n(\omega_D) = A\sigma_n(\omega_D) \qquad n = 1, 2 \tag{7.3}$$

where $\sigma(\omega_D)$ is the RCS, A is an unknown factor, and $P_1(\omega_D)$ and $P_2(\omega_D)$ are the first- and second-order scattering. The equations of the first-order RCS ($\sigma_1(\omega_D)$) and the second-order RCS ($\sigma_2(\omega_D)$) are given in Chapter 6. An ocean wave spectrum ($S(\mathbf{k})$) for wave number vector $\mathbf{k}$ is estimated by inverting (7.1)–(7.3).

To simplify the treatment here, we have normalized physical values by radar parameters. Here, $\mathbf{k}_0$ is the radio wavenumber vector:

$$\omega_B = [2gk_0 \tanh(2k_0 d)]^{\frac{1}{2}}, \tag{7.4}$$

is the Bragg radian frequency, g is the gravitational acceleration, d is the water depth, k_0 is the magnitude of $\mathbf{k}_0$, and the Bragg wave number vector is $-2\mathbf{k}_0$.

The subscript N represents such a normalization. In the normalized form, for example, $\mathbf{k}_N = \mathbf{k}/(2k_0)$ and $S_N(\mathbf{k}_N) = (2k_0)^4 S(\mathbf{k})$. The normalized first-order and second-order RCSs $\sigma_{1,2N}(\omega_{DN}) = \omega_B \sigma_{1,2}(\omega_D)$ for normalized Doppler frequency $\omega_{DN} = \omega_D/\omega_B$ for deep water are expressed as follows:

$$\sigma_{1N}(\omega_{DN}) = 4\pi \sum_{m_2 = \pm 1} S_N(-m_2 \mathbf{n}_N)\delta(\omega_{DN} - m_2), \tag{7.5}$$

$$\sigma_{2N}(\omega_{DN}) = 4\pi \sum_{m_1 = \pm 1} \sum_{m_2 = \pm 1} \int_{-\infty}^{+\infty} \int_{-\infty}^{+\infty} |\Gamma_{EN}^s - i\Gamma_{HN}^s|^2 \tag{7.6}$$
$$S_N(m_1 \mathbf{k}_{1N}) S_N(m_2 \mathbf{k}_{2N})$$
$$\delta(\omega_{DN} - m_1 k_{1N}^{\frac{1}{2}} - m_2 k_{2N}^{\frac{1}{2}}) dp_N dq_N,$$

where $\mathbf{n}_N$ is a unit vector defined as $\mathbf{n}_N = 2\mathbf{k}_0/(2k_0)$. The normalized hydrodynamic coupling coefficient for deep water is expressed as follows:

$$\Gamma_{HN}^s = \frac{1}{2}\left[k_{1N} + k_{2N} + \frac{(k_{1N}k_{2N} - \mathbf{k}_{1N} \cdot \mathbf{k}_{2N})}{m_1 m_2 (k_{1N}k_{2N})^{\frac{1}{2}}} \left(\frac{1 + \omega_{DN}^2}{1 - \omega_{DN}^2}\right) \right]. \tag{7.7}$$

Moreover, the normalized electromagnetic coupling coefficient is

$$\Gamma_{EN}^s = \frac{1}{2}\left[\frac{(\mathbf{k}_{1N} \cdot \mathbf{n}_N)(\mathbf{k}_{2N} \cdot \mathbf{n}_N) - 2\mathbf{k}_{1N} \cdot \mathbf{k}_{2N}}{(\mathbf{k}_{1N} \cdot \mathbf{k}_{2N})^{\frac{1}{2}} - \Delta/2} \right]. \tag{7.8}$$

Here, $\mathbf{k}_{nN}$ and ω_{nN} are expressed as follows for $\kappa_N = (p_N, q_N)$;

$$\mathbf{k}_{nN} = -\frac{1}{2}\mathbf{n}_N + (3 - 2n)\kappa_N, \tag{7.9}$$

$$\omega_{nN} = k_{nN}^{\frac{1}{2}}, \tag{7.10}$$

and wave numbers $k_{nN} = |\mathbf{k}_{nN}|$ ($n = 1, 2$). The radian frequency-directional spectrum is defined as

$$G(\omega, \theta)C_g = S(\mathbf{k}), \tag{7.11}$$

where $\mathbf{k} = (k\cos\theta, k\sin\theta)$, θ is the wave propagation direction, $\omega = gk^{1/2}$ is a radial wave frequency, and $C_g = \partial\omega/\partial k = g/(2\omega)$ is a group velocity. The normalized frequency-directional spectrum is

$$G_N(\omega_N, \theta) = (2k_0)^2\omega_B G(\omega, \theta). \tag{7.12}$$

7.3 Discretization of the integral equation

The relationship between the frequency and direction of the wavenumber vector $m_1\mathbf{k}_{1N}$ and the wavenumber vector $m_2\mathbf{k}_{2N}$ in (7.6) for deep water is written as follows:

$$\omega_{DN} = m_1\omega_{1N} + m_2[\omega_{1N}^4 + 2m_1\omega_{1N}^2\cos(\theta_1 - \phi_b) + 1]^{\frac{1}{4}}, \tag{7.13}$$

$$\omega_{DN} = m_1[\omega_{2N}^4 + 2m_2\omega_{2N}^2\cos(\theta_2 - \phi_b) + 1]^{\frac{1}{4}} + m_2\omega_{2N}, \tag{7.14}$$

where ϕ_b is the radar beam direction and (ω_{nN}, θ_n) is the frequency and direction of the wavenumber vector $m_n\mathbf{k}_{nN}$ ($n = 1, 2$) and $k_{1N} \leq k_{2N}$.

Figure 7.1 shows the relationship between (ω_{nN}, θ_n) and positive Doppler frequencies for the case of $\phi_b = 0$ ($n = 1, 2$). This figure is symmetrical with respect to $\theta_n = 0$. Figure 7.1 is shown only for positive Doppler frequencies. For negative Doppler frequencies, the figure is symmetrical to Figure 7.1 with respect to $\theta_n = 90°$. The integration of (7.6) is performed along the contour line for a given Doppler frequency. Figure 7.1 also shows an example of the discretization on the wave frequency-direction plane. The dotted lines and black points show the discretization. The wave spectrum is estimated on the grid in Figure 7.1.

For example, if the Doppler frequency range of the Doppler spectrum for inversion is in the range of $0.6 \leq \omega_{DN} \leq 0.8$, the spectral values of the grid points between and near the red and blue contours in Figure 7.1 will be obtained. For the numerical integration of (7.6), wave numbers (frequencies) k_{1N} (ω_{1N}) and k_{2N} (ω_{2N}) and θ_2 are evaluated for a given Doppler frequency (ω_{DN}) and θ_1. A spectral value at the quadrature point on the Doppler frequency contour is expressed by spectral values on the the grid points by the bilinear interpolation. This procedure is described in [4].

7.4 Other constraints

The signal-to-noise (SN) ratios of Doppler spectra are critical for wave estimation, and high SN ratios of Doppler spectra from two radars are required. However, it is not often the case that the SN ratio of both Doppler spectra is high enough to estimate the wave spectrum.

By adding constraints on wave spectra, the wave spectra can be estimated even in a region where only one radar can measure Doppler spectra. The constraints

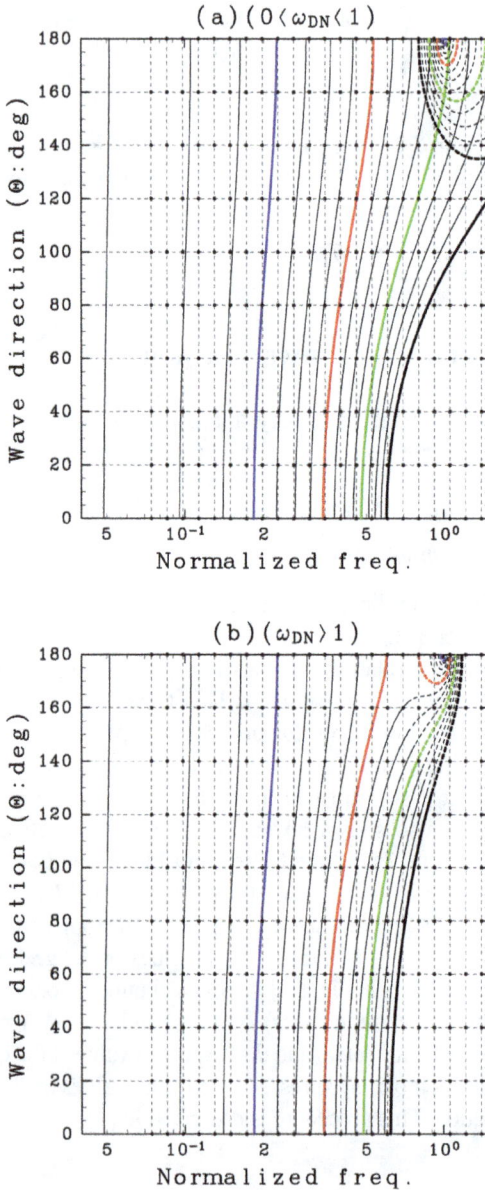

Figure 7.1   Relationship between (ω_{nN}, θ_n) (n = 1, 2) and Doppler frequencies (ω_{DN}) for $0 < \omega_{DN} < 1$. Solid line: (ω_{1N}, θ_1) (7.13). Dashed line: (ω_{2N}, θ_2) (7.14). Blue: $\omega_{DN} = 0.8$. Red: $\omega_{DN} = 0.6$, Green: $\omega_{DN} = 0.4$. Contour interval= 0.05. The dotted lines and black points are grid lines and points of the discretization on the wave frequency-direction plane. (b) Same as (a) but for $\omega_{DN} > 1$. Blue: $\omega_{DN} = 1.2$. Red: $\omega_{DN} = 1.4$, Green:$\omega_{DN} = 1.6$.

between ocean wave spectra and Doppler spectra that are applied here are as follows: (c1) the relationship between the first-order scattering and the ocean wave spectrum (7.5), (c2) the relationship between the second-order scattering and the ocean wave spectrum (7.6), (c3) the energy balance equation, (c4) the two-dimensional continuity equation of the wind, (c5) the smooth spectral values in the frequency-direction plane, and (c6) a small propagation term in the energy balance equation.

The constraint (c3) is written as

$$\mathbf{C}_{gN} \cdot \nabla_N G_N(\omega_N, \theta, \mathbf{x}_N) - S_{tN} = 0, \tag{7.15}$$

where $G_N(\omega_N, \theta, \mathbf{x}_N) = G_N(\omega_N, \theta)$ is the wave spectrum at the position $\mathbf{x}_N = (x_N, y_N)$, $\mathbf{C}_{gN} = \partial \omega_N / \partial \mathbf{k}_N$ is the normalized group velocity vector of waves for wave frequency ω_N and wavenumber vector $\mathbf{k}_N$, ∇_N denotes the normalized horizontal gradient, and S_{tN} is the total source function [5]. The total energy source function S_{tN} for deep water is written as

$$S_{tN} = S_{inN} + S_{dsN} + S_{nlN}, \tag{7.16}$$

and the effect of bottom friction is neglected. The parameterizations of S_{inN} (wind input source function), S_{dsN} (dissipation source function), and S_{nlN} (non-linear interaction source function) are almost same as those in WAM cycle-3 [6], because the derivatives of S_{tN} by wave spectral values can be calculated from analytical equations.

The source function S_{inN} in (7.15) and (7.16) is dependent on not only the wave spectrum but also the wind vector. The unknowns to be estimated are spectral values and wind vectors. The constraint (c4) is the continuity equation of wind vectors. The continuity equation of the non-divergent sea surface wind vector $\mathbf{u}_N = (u_N \cos\theta_w, u_N \sin\theta_w)$ is written as

$$\nabla_N \cdot \mathbf{u}_N = 0, \tag{7.17}$$

where u_N is the normalized sea surface wind speeds and θ_w is the wind direction.

The regularization constraint (c5) in the frequency-direction (spectral) plane is expressed as

$$\begin{aligned}
&\log(G_N(k_f + 1, l_d)) + \log(G_N(k_f - 1, l_d)) \\
&+ \log(G_N(k_f, l_d - 1)) + \log(G_N(k_f, l_d + 1)) \\
&- 4\log(G_N(k_f, l_d)) = 0 \quad \text{for} \quad 1 < k_f < M_f \quad \text{and} \quad 1 \le l_d \le M_d, \tag{7.18}
\end{aligned}$$

$$\text{or} \quad \begin{aligned}
&\log(G_N(k_f, l_d - 1)) + \log(G_N(k_f, l_d + 1)) \\
&- 2\log(G_N(k_f, l_d)) = 0 \quad \text{for} \quad k_f = 1, M_f
\end{aligned}$$

where k_f is the wave frequency index number, M_f is the number of frequencies, l_d is the wave direction index number, and M_d is the number of directions. Here, $G_N(k_f, 0) = G_N(k_f, M_d)$ and $G_N(k_f, M_d + 1) = G_N(k_f, 1)$. The constraint (c6) is written as

$$\mathbf{C}_{gN} \cdot \nabla_N G_N(\omega_N, \theta) = 0 \tag{7.19}$$

which is used as a regularization constraint in the horizontal $(x_N - y_N)$ plane.

These constraints are written as $F_K(\mathbf{s}) = 0 \, (K = 1, \ldots, N_t)$, where N_u-dimensional vector $\mathbf{s}$ denotes unknowns $(\log(G_N(\omega_N, \theta, \mathbf{x}_N)), \log(u_N), \theta_w)$, N_u is the number of unknowns, K is the index number of the constraints (c1)–(c6), and N_t is the number of constraints. The unknowns are estimated by seeking the values to minimize the objective function as

$$U(\mathbf{s}) = \frac{1}{2} \sum_{K=1}^{N_t} [\lambda_w(M, K)F_K(\mathbf{s})]^2, \tag{7.20}$$

where $\lambda_w(M, K)$ is the weight and the index M ($M = 1, \ldots, 6$) denotes the type of constraint. The weights of constraints (c1) and (c2) are proportional to the number of the Doppler spectra for the wave estimation in the grid cell [7].

This method estimates wave spectra in all of the grid points simultaneously. The number of unknowns is $N_u = N_g N_s + 2N_g$, where N_g is the number of grid points and $N_s = M_f N_d$ is the number of total spectral values of the wave spectrum. The order of N_u is 10^4 for high spatial and spectral resolution, which requires large computer memory. As a result, spatial and spectral resolutions are low, which is the drawback of the method. In the case of coarse spatial resolution, there are multiple Doppler spectra in a grid cell. These Doppler spectra in a grid cell are spatially averaged to reduce the computation for inversion. The procedure is described in [7].

7.5   Algorithm

The minimization problem of (7.20) is solved iteratively, for which the algorithm is written as

$$\mathbf{d}_m = -\mathsf{H}^{(m)}\nabla U = -\mathsf{H}^{(m)} \mathsf{J}_f^T \mathbf{f}, \tag{7.21}$$

$$\mathbf{s}^{(m+1)} = \mathbf{s}^{(m)} + \alpha_m \mathbf{d}_m, \tag{7.22}$$

where m indicates the step number and α_m is a positive constant that has been adjusted such that $U(\mathbf{s}^{(m+1)}) < U(\mathbf{s}^{(m)})$. The vector $\mathbf{s}^{(m)} = (s_1^{(m)}, \cdots, s_{N_u}^{(m)})$ is $\mathbf{s}$ for the m th step. The vector $\mathbf{f}$ is $\mathbf{f} = (f_K) = (\lambda_w(M, K)F_K)$ ($K = 1, \ldots, N_t$), and J_f is the Jacobian matrix defined as

$$\mathsf{J}_f(K, L) = \frac{\partial f_K}{\partial s_L^{(m)}}$$
$$(K = 1, \ldots, N_t) \tag{7.23}$$
$$(L = 1, \ldots, N_u).$$

The $N_u \times N_u$ matrix $\mathsf{H}^{(m)}$ is positive definite. The matrix $\mathsf{H}^{(m)}$ is a diagonal matrix and it is to save the computer memory such as

$$\mathsf{H}^{(m)} = [\mathrm{diag}(\mathsf{J}_f^T \mathsf{J}_f)]^{-1} \tag{7.24}$$

or

$$\mathsf{H}^{(m)} = [\mathrm{diag}(\mathsf{J}_f^T \mathsf{J}_f) + \mathsf{I}]^{-1}, \tag{7.25}$$

or $\mathsf{H}^{(m)} = \mathsf{I}$, where I is the unit matrix. The derivative of F_K with respect to $\mathbf{s}$ can be calculated from analytic form. The derivatives such as $\partial\sigma_{nN}(\omega_{DN})/\partial G(\omega_N, \theta)$ ($n = 1, 2$), $\partial S_{iN}/\partial G(\omega_N, \theta)$, $\partial S_{inN}/\partial u_{nN}$, and $\partial S_{inN}/\partial\theta_w$ on the four-dimensional grid

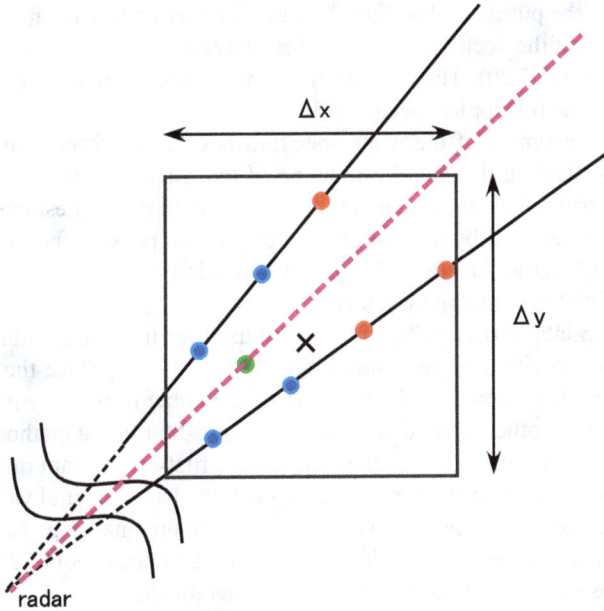

Figure 7.2   Schematic illustration of averaging Doppler spectra

points $(\omega_N, \theta, x_N, y_N)$ are computed. The derivatives with respect to s of the regularization constraints and wind continuity equation are also computed.

7.6   Procedure for wave spectrum estimation

The procedure for wave spectrum estimation is summarized as follows: We set grid points on which wave spectra are estimated for the area covered by the HF ocean radars. The Doppler spectra in each grid cells are obtained for each HF radars. In the observation, Doppler spectra are obtained on polar coordinate grid points centered on the radar. If the grid spacing is greater than the radar resolution, the Doppler spectrum should be averaged in some way, which requires the selection of high SN Doppler spectra [7].

Figure 7.2 shows a schematic illustration of the Doppler spectra averaging. The rectangle is the grid cell and the lengths of the side (Δx and Δy) indicate the spatial resolution of the estimated wave spectra. The symbol × is the grid point at the center of the rectangle. Ocean waves in rectangular cells are assumed to be statistically homogeneous. The Doppler spectra are sampled on the radial grid points indicated by the red and blue circles in Figure 7.2.

The Doppler spectra on the blue circles are selected for wave estimation, while the Doppler spectra on the red circles are discarded due to low SN ratios. The Doppler spectra on blue circles are averaged. The green circle in Figure 7.2 shows the average position of the blue circles and the average radar beam direction is indicated by the purple dashed line. This beam direction is used in (7.5) and (7.6) as the direction of the vector $\mathbf{n}_N$. The number of averaged Doppler spectra is related with the weight of (7.20). The Doppler frequency ranges used for the inversion are determined for each Doppler spectra.

The initial estimate of the wave spectrum is obtained from Doppler spectra. There are two steps in the initial estimation of the wave spectra. The initial wave spectra at all grid points are assumed to be the same. In the first estimate, the wave spectrum is represented by several parameters. Those parameters are obtained to minimize the objective function (7.20), which consists of only constraints (c1) and (c2). The Monte Carlo method is adopted here.

The constraints of (c1), (c2), and (c5) are used for the initial estimation in the next step. The algorithm is the same as in (7.21) and (7.23). Since the initial wave spectrum is not dependent on grid points and the number of unknowns is not large, it is possible to use other methods such as Levenberg-Marquardt method. The initial wind direction is estimated from the ratio of the first-order scattering. The wind velocity is estimated from the wave spectrum obtained as the initial value.

The wave spectrum depends on the grid points in the final step. The constraints (c1)–(c6) are used in the final step. The algorithm is an iteration of (7.21) and (7.23) and terminates with a predetermined number of iterations.

7.7   Example of wave estimation and issues to be addressed

The method can estimate wave directional spectra for both multiple radar case and single radar case. In the case of single radar, the grid on which wave spectrum is estimated is regular grid [7] or radial grid with the origin at the radar position [8]. In the case of multiple radar, the grids are regular grids as explained in the previous section. In a region not covered by radars, it is possible to estimate the wave spectrum on that grid, but the accuracy of the estimated wave parameters is low in that region.

The accuracy of the wave parameters depends primarily on the SN ratio of the Doppler spectra, which is true for any method. In any method, the accuracy of the wave parameter as a function of wave frequency (e.g., spectral levels and spectral mean wave period) is the best in the peak wave frequency band and decreases with increasing frequency. This is because the second-order Doppler peak is the most robust to the noise in the second-order Doppler spectrum. The difference with other methods is that the accuracy of the wave direction as a function of the wave frequency is higher at higher frequencies. This is due to the regularization constraint in the spectral plane (c5). The inversion method uses first-order scattering. The wave direction in the HF band is close to the wave direction at the Bragg wave frequency. The wave direction at the Bragg wave frequency is estimated from the first-order scattering, which is robust to the noise [9].

Figure 7.3 shows examples of wave directional spectra at 26° N, 128° E from HF radar for the same observation as those in [7] and [10]. The radio frequency was 24.515 MHz, and the Bragg frequency was $\omega_B/(2\pi) = 0.505$ Hz. The weights $\lambda_w(M, K)$ are given in [7]. The wave spectrum from ERA5 [11, 12] is also shown

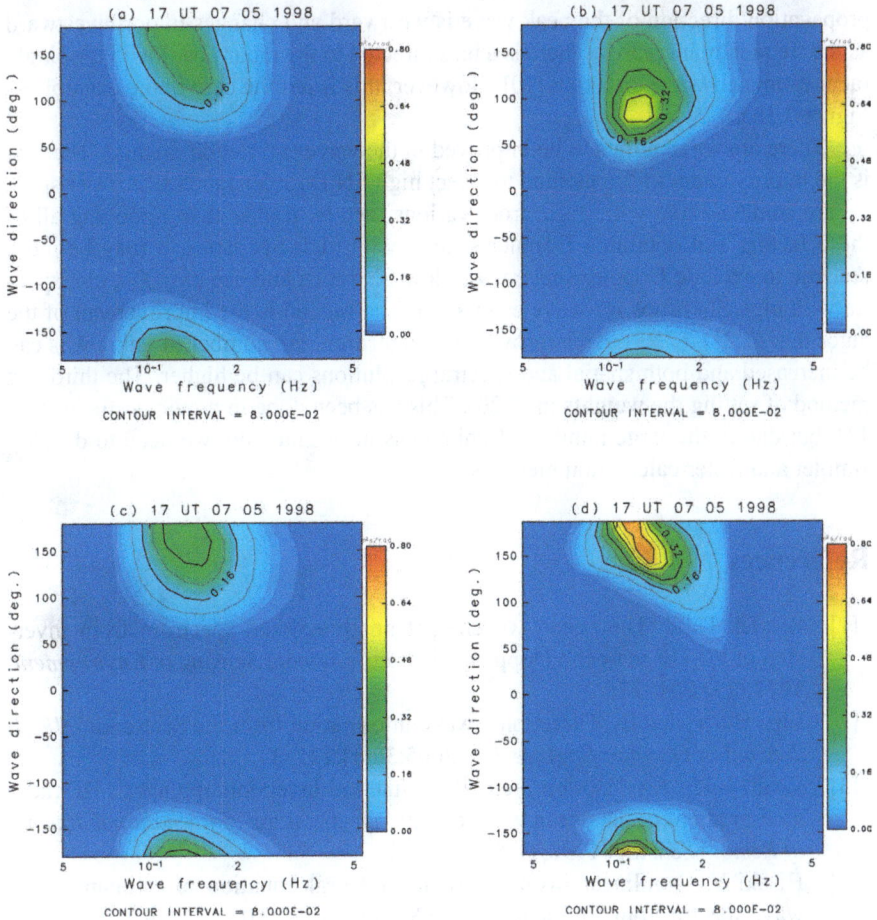

Figure 7.3   Examples of wave spectra. (a) HF radar estimated wave spectrum.
(b) Same as (a) but estimated from one of the single radars. The
direction from the average position in the grid cell (green circle in
Figure 7.2) to the radar is $\phi_b + \pi = 158°$. (c) Same as (b) but from
another radar. $\phi_b + \pi = 120°$. (d) ERA5 wave spectrum. The 0° of the
vertical axis means eastward propagation and the counter clockwise
is positive.

in Figure 7.3d. The ERA5 wave spectra were obtained by assimilating satellite
altimeter data into the ocean wave model. The frequency resolution (ratio of frequency) is 1.15 and direction resolution is 20° for HF radar. The spectral resolutions
of the ERA5 wave data are 1.1 and 15°, and the maximum spectral value of ERA5
wave spectrum is larger than those of HF radar. The wave heights in this examples are 1.27 m, 1.05 m, 1.18 m, and 1.06 m for Figure 7.3a,b,c,d, respectively. The

propagation direction of the peak wave is westward and changes to northwestward as the frequency increases. There is a bias directed to the radar position in the single radar estimated wave direction [10]. However, this is not the case in the example of Figure 7.3b,c.

There are some points to be improved in the wave estimation method. The first is the improvement of the method to select high SN Doppler spectra. Doppler spectra are contaminated with noise from various factors. Rather than removing all of these factors and obtaining Doppler spectra with high SN ratios, it may be more realistic to exclude Doppler spectra with low SN ratios and use only Doppler spectra with high SN ratios for wave estimation. The second is the improvement of the algorithm (7.21)–(7.23). By improving the algorithm, the number of unknowns can be increased and both spatial and spectral resolutions can be higher. The third is a method of setting the weights in (7.20). This has been done in previous studies [13, 14], but due to the large number of unknowns in our method, we need to develop simpler and faster calculation methods.

References

[1]   Barrick D.E. 'The ocean waveheight nondirectional spectrum from inversion of the HF sea-echo Doppler spectrum'. *Remote Sensing of Environment.* 1977;**6**(3):201–27.

[2]   Lipa B., Nyden B. 'Directional wave information from the SeaSonde'. *IEEE Journal of Oceanic Engineering.* 2005;**30**(1):221–31.

[3]   Wyatt L.R. 'A relaxation method for integral inversion applied to HF radar measurement of the ocean wave directional spectrum'. *International Journal of Remote Sensing.* 1990;**11**(8):1481–94.

[4]   Hisaki Y. 'Nonlinear inversion of the integral equation to estimate ocean wave spectra from HF radar'. *Radio Science.* 1996;**31**(1):25–39.

[5]   Hisaki Y. 'Ocean wave directional spectra estimation from an HF ocean radar with a single antenna array: methodology'. *Journal of Atmospheric and Oceanic Technology.* 2006;**23**(2):268–86.

[6]   WAMDI Group. 'The WAM model—a third generation ocean wave prediction model'. *Journal of Physical Oceanography.* 1988;**18**(12):1775–810.

[7]   Hisaki Y. 'Development of HF radar inversion algorithm for spectrum estimation (HIAS)'. *Journal of Geophysical Research: Oceans.* 2015;**120**(3):1725–40.

[8]   Hisaki Y. 'Ocean wave directional spectra estimation from an HF ocean radar with a single antenna array: observation'. *Journal of Geophysical Research.* 2005;**110**(C11).

[9]   Hisaki Y. 'Inter-comparison of wave data obtained from single high-frequency radar, *in situ* observation, and model prediction'. *International Journal of Remote Sensing.* 2014;**35**(10):3459–81.

[10]   Hisaki Y. 'Ocean wave parameters and spectrum estimated from single and dual high-frequency radar systems'. *Ocean Dynamics.* 2016;**66**(9):1065–85.

[11] Copernicus Climate Change Service. ERA5: fifth generation of ECMWF atmospheric reanalyses of the global climate *[online]*. 2017. Available from https://cds.climate.copernicus.eu/cdsapp#!/home [Accessed 23 Sept 2021].

[12] Hersbach H., Bell B., Berrisford P., *et al.* 'The ERA5 global reanalysis'. *Quarterly Journal of the Royal Meteorological Society.* 2020;**146**(730):1999–2049.

[13] Hashimoto N., Tokuda M. 'A Bayesian approach for estimation of directional wave spectra with HF radar'. *Coastal Engineering Journal.* 1999;**41**(02):137–49.

[14] Hashimoto N., Wyatt L.R., Kojima S. 'Verification of a Bayesian method for estimating directional spectra from HF radar surface backscatter'. *Coastal Engineering Journal.* 2003;**45**(02):255–74.

Chapter 8

HF radar observation of nearshore winds

Brian Emery[1] and Anthony Kirincich[2]

8.1 Introduction

Following Ekman's 1905 mathematical description of the influence of wind stress on the ocean [1], measurement of the wind field itself has been a critical part of any effort to understand the movement of ocean currents. This is particularly true in the coastal zone, where winds and waves interact with the coastal boundary to drive spatially and temporally complex currents. In addition to understanding and predicting coastal flows, observations of nearshore surface winds are fundamental to fulfilling both scientific needs (e.g. circulation, mixing, biological productivity, larval transport) and societal needs (e.g. shipping, wind power).

Nearshore wind observations, either in situ or remote, typically have high temporal resolution or high spatial resolution but not both. Buoy-based observations, such as those accessible via the National Data Buoy Center (NDBC), provide time series of wind products from most coastal areas but lack the spatially relevant resolution for observing many – if not most – of the critical small-scale circulation processes. Recent developments have improved satellite scatterometer capabilities to within 10–15 km from shore (e.g. [2, 3]), and similarly, Synthetic Aperture Radar (SAR) achieves sub-km resolution (0.5–1 km) up to 1–3 km from shore with root mean square (RMS) errors of 1.4–1.8 m s^{-1} [4, 5]. While planned future missions such as the Waves and Currents Mission [6] would further advance these techniques and provide maps of winds with high spatial resolution, satellite-based observations sample at periods of 12 hours or greater, limiting their scientific utility.

In contrast, land-based high frequency (HF) radar systems routinely provide both high spatial and high temporal resolution observations of surface currents in the coastal ocean in all weather conditions. The fundamental signal observed by the radar system results from the presence of relatively short ocean waves that respond quickly to changes in wind speed and direction. In the nearshore, wind observations from these systems would fill an important niche between satellite observations, which encounter difficulties close to landmasses, and moored observations,

[1]Marine Science Institute, University of California Santa Barbara, USA
[2]Woods Hole Oceanographic Institution, USA

for which spatially dense deployments are cost prohibitive. The spatial and temporal coverage possible, with timescales of tens of minutes and spatial scales of 2–6 km, matches the resolution needed to advance the present understanding and modeling of coastal ocean circulation [7–9].

The objective of this chapter is to describe the principles and methods for extracting wind speed and direction from the signals received by oceanographic HF radars. We begin with a review of early efforts – including the use of over-the-horizon (OTH) radars and phased array ground wave HF radars– most of which utilized the second-order scatter [10, 11], before reviewing the more recent use of first-order scatter to examine wind speed and direction [12–14]. As an example of this methodology, the empirical model used in [14] to extract wind from the first-order scatter is detailed. Finally, we discuss theoretical factors that may have inhibited previous efforts and provide suggestions for future work.

8.2   Background

It has long been recognized that oceanographic HF radar has a unique potential to observe surface winds. This recognition stems from the theoretical work of Barrick [15], who demonstrated that observed increases in backscattered signal power at certain Doppler shifts were indeed due to Bragg resonance as hypothesized by Crombie [16]. Barrick showed that the signal power returned was a function of the power spectrum of the ocean surface wave heights at the Bragg resonant frequency. In this section, we lay out the theoretical background and develop the framework underlying the discussion of methods for obtaining winds from HF radar.

In its simplest form, the returned power observed by a radar P_r can be expressed as the product of the power transmitted (P_t), path losses (F_t, F_r), and antenna gains (G_t, G_r), e.g. [17]

$$P_r = \frac{P_t G_t G_r F_t^2 F_r^2}{(4\pi)^3 r^4} \sigma \lambda^2. \tag{8.1}$$

Equation (8.1) can be used to estimate power received, P_r, from an object at range r with radar cross section σ using a radar with a transmission wavelength of λ. Partitioning (8.1) into terms representing the transmitted power density impinging on the distant object, the fraction of this power that is backscattered and impinging on the receiver, and the effective area of the receive antenna gives

$$P_r = \frac{P_t G_t F_t^2}{4\pi r^2} \frac{F_r^2 \sigma}{4\pi r^2} \frac{G_r \lambda^2}{4\pi}. \tag{8.2}$$

The $1/4\pi r^2$ in each of the first two terms on the right-hand side accounts for divergence or spreading of the signal with distance, while the F^2 terms account for along-path attenuation. For an incremental patch area dA (following [15]), (8.2) becomes

$$dP_r = \frac{P_t G_t F_t^2}{4\pi r^2} \frac{F_r^2 \sigma^0 dA}{4\pi r^2} \frac{G_r \lambda^2}{4\pi}, \tag{8.3}$$

where σ^0 is the radar cross section per unit area, dA gives the area of the scattering patch, and dP_r is the increment of received power. The relation in (8.3) assumes plane waves, rather than swept radio waves, but provides a useful framework for understanding the factors affecting the radar measurements and their scales.

Equation (8.3) also illustrates the role of the radar cross-section, σ^0, on the received power. Theoretically, σ^0 of the ocean is a function of the power spectrum of ocean surface wave heights [15] and hence provides a direct connection between the observed radar power and the ocean wave energy. Considering just the largest component of the radar cross section, the first-order return due to a resonant interaction with ocean waves with wavelength $\lambda_B = \lambda/2$, i.e. Bragg waves, σ^0 can be defined as [15]

$$\sigma^0 = 2^6 \pi k_0{}^4 S(\pm 2k_0), \tag{8.4}$$

given the radar wave number $k_0 = 2\pi/\lambda$, and S the spectral power of the ocean waves at the Bragg wave number, $2k_0$. $S(k,\theta)$ quantifies the surface roughness, such that increased $S(k,\theta)$ increases σ^0, resulting in more signal being returned to the radar. Combining (8.3) and (8.4), it can be seen that the equation for dP_r incorporates the spectral power of the ocean Bragg wave heights $S(k,\theta)$, which itself depends on wind speed and direction.

Thus, the observation of wind by HF radar assumes a relationship between the presence of waves, their heights or energies, and the wind speed and direction. With respect to dP_r for first-order scattering, only the Bragg waves propagating in the radar look direction ($S(\pm 2k_0, \pm \theta)$) contributes. Calculations of both dP_r and σ^0, as well as the calculation of wind direction thus depend on the spreading of the ocean wave spectrum around the primary wind direction. Initial studies of applications of HF radar to oceanography recognized the potential for wind sensing, but generally achieved mixed results, as we discuss in the remainder of Section 8.2.

8.2.1 Early studies

Some of the earliest applications of HF radar to observing winds employed OTH radars using sky-wave propagation. For example, the scientific use of defense OTH radar [18] and the use of the Wide Aperture Research Facility (WARF) built by the Stanford Research Institute [19] investigated wind speed and direction estimates at near-synoptic spatial scales. The sky-wave propagation used in these very large, high-powered phased array radars added additional complexity compared to modern coastal radars that rely on ground-wave propagation. With OTH radars, the specific ionospheric layer providing the bounce can vary on short timescales and may consist of multiple layers. Scattering from multiple layers results in the spreading of the Doppler signals into adjacent frequency bins, adding uncertainty and error to the inverted observations. At the time of these studies, several methods for extracting wind speed were hypothesized, while the methods for extracting wind direction were essentially the same as those used by later ground-wave radar studies.

OTH HF radars attained their best results when mapping wind directions. Typical results spanned large areas of the ocean surface and found differences with in situ observations (e.g. from ships within the coverage area) of 16°–33° for direction [19, 20]. WARF was capable of 15 km spatial resolution at 1 000–3 000 km range and could observe wind direction changes, for example, on either side of a cold front [21]. Unfortunately, the earliest studies typically used only a small number of radar observations (e.g. a few days), and even fewer in situ comparisons for validation, i.e. $N = 44$ radar measurements [18], $N = 24$ [19], and $N = 25$ [22]. The use of small data sets in early studies motivated a later, expanded investigation involving multiple OTH radars [20] (from both U.S. coasts), and significantly more in situ validation data. Using 12 days of OTH data, resulting in $N = 1\,900$ comparison data points (1 509 from a data-assimilating model, 231 from ships, and 160 from buoys) the authors found RMS differences of 33° [20]. Using only "high-quality" radar data, an RMS difference of 24° was obtained compared with buoys. Little difference was found in the results when parsing the data by observed wind speed, suggesting that the form of the directional spreading, though not accounted for, was not a significant influence on the empirical results. For comparison, [20] noted that an evaluation of NDBC mooring wind observations [23] had found wind direction RMS differences of 21° for buoys 39 km apart, suggesting that some of the RMS differences likely resulted from comparing the spatially averaged HF radar data with point measures of winds.

Compared with wind direction, obtaining wind speeds with OTH radar was less obvious theoretically, and empirical methods to do so only achieved moderate success. Methods hypothesized at the time include relating wind speed to the ratios of the beam-formed first- and second-order powers; the ratio of the first-order power to the zero Doppler power; the total power of both first-order peaks above a minimum between them [24]; or the width of the first-order peak at the point 10 dB down from the peak value [19]. The latter method essentially included the second-order scatter due to the low resolution of the Doppler processing used by the OTH radars. Typical bin widths of 0.08 Hz [19] or 0.078 Hz [22] for Doppler processing resulted from the short coherent integration times, used as a consequence of the temporally dynamic ionosphere. (Note that typical, modern coastal HF radars with a 2-Hz sweep and 512-point FFT length would have a Doppler bin width of 0.0039 Hz, or higher frequency resolution by a factor of 20.) Using the 10 dB down first-order width yielded a standard deviation of ± 4 m s^{-1} when compared with $N = 24$ in situ observations [19] and later ± 2.4 m s^{-1} [22] with $N = 25$ for two days of observations from WARF. This latter study also found a lower bound on observable wind speed to be about 5 m s^{-1}, below which the contribution from second-order scattering was too low to produce an effect. The results of much of these efforts for obtaining wind speed are summarized (from [25]): "although Bragg-line width appeared to be correlated with surface-wind speed, empirical formula expressing the observed relation proved inconsistent with subsequent measurements. Furthermore, Bragg-line width responds as much to ionospheric effects as to the state of the sea." This conclusion was followed up with the later suggestion that "wind speed estimates are better obtained from HF radar spectra by first estimating the directional wind-wave spectrum from the second-order echoes ... and then computing wind speed from a wind-wave prediction model" [26]. The use of OTH HF radar for ocean remote sensing transitioned briefly to relocatable OTH radar; however, the research use of OTH ceased due to the promise of satellite-based wind sensors (at least from the perspective of the U.S. Navy, the main provider of research funding at the time [27]).

8.2.2 Wind direction via wave spreading models

A relevant assumption made by each of the above methods for obtaining wind direction is that the direction of the Bragg resonant ocean waves is tightly coupled to that of the local winds. At the time of early OTH-based studies, it was assumed that the Bragg waves were locally generated, finding equilibrium with the wind in a relatively short time period, on the order of 30 min [20]. However, the prevailing oceanographic theory at the time suggested that a wave of frequency ω would respond to a change in the wind direction in roughly $T \sim 4e4/\omega$, which predicts that a 5 m wave would obtain equilibrium with the wind in 3.5 hours. A later analysis using several beams from a 1 kW beamforming radar confirmed the assumption of a tighter coupling, finding that ocean waves with wavelengths of 5 m did not take multiple hours to equilibrate with the local winds, but would adjust in a matter of minutes [28]. These findings used the first-order scatter and revealed spatial differences in the time of arrival of a cold front, along with unexpectedly high spatial variability between the nearshore and offshore wind directions. Thus, HF radar observations supported the assertion that the Bragg waves can be used to infer the wind direction and confirmed the assumptions of the early OTH studies.

In most HF radar studies, including those described above, the wind direction is obtained from ratios of the spectral power of the approaching and receding Bragg lines (i.e. the first-order return) from the same range and azimuth, following [18]. Essentially, this method assumes that wave heights follow an angular distribution with the maximum aligned with the wind direction. Since the beam of a radar look direction samples waves from two sides of this directional distribution, the wind direction can be estimated from the ratio of the amplitudes of the Doppler peaks (Figure 8.1). However, as illustrated by Figure 8.1, a 180° ambiguity must be resolved (e.g. by use of external data, more than one radar, or more than one radar look direction). Defining the heights of the first-order peaks associated with the approaching ($+f_B$) and receding ($-f_B$) Bragg waves, as P^+ and P^-, [18] defines the Bragg power ratio (in dB):

$$\zeta = 10 \log \left(\frac{P^+}{P^-} \right). \tag{8.5}$$

While an initial effort based on OTH observations obtained an empirical relation between ζ and wind direction (θ_w in degrees) [20],

$$|\theta_w| = 3.75\zeta + 90, \tag{8.6}$$

obtaining the wind direction from ζ typically assumes knowledge of, or a model for, the directional distribution of the ocean waves at the Bragg frequency.

Directional spreading models for wind-driven surface gravity waves take several forms, usually more complex than (8.6), with some including spreading variability

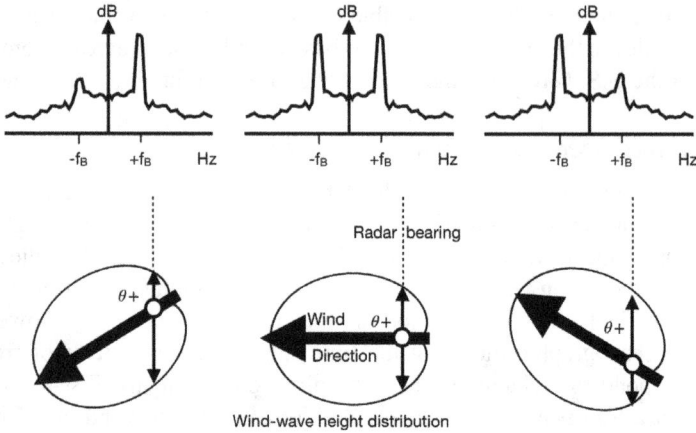

Figure 8.1   Example directional distributions of wave heights for different wind directions (bottom), and how HF radar sampling of these (along a look direction given by the dashed line from a location at the bottom of the figure), results in different peak values for first-order spectra. (shown along the top of the figure). Illustration based on figure 1 of [29].

that depends on the wind speed. In early works, it was assumed that the ocean wave directional spectrum had a functional form of $\cos^s(\theta/2)$ [19, 30] (based on the results of [31]), with the value of s varying with wave frequency. Defining the directional spectrum as $G(\theta) = \cos^s(\theta/2)$, where θ is the angle between the radar look direction and the wind direction, the Bragg power ratio in theory observes

$$\zeta = \frac{G(\theta)}{G(\theta + \pi)}. \tag{8.7}$$

Solving for θ using the $\cos^s()$ model obtains

$$\theta = 2\arctan(\zeta^{1/s}), \tag{8.8}$$

applicable when θ is "greater than a few degrees" [19]. Following [31], this model represents the directional part of the ocean wave spectrum, with the scalar part defined as $S(k)$.

More generally, for an assumed spreading model $G(\theta_w)$ sampled by a radar with look direction θ_r, the wind direction can be found as the angle, θ_w, that minimizes

$$\zeta - \frac{G(\theta_w - \theta_r)}{G(\theta_w - \theta_r + \pi)}. \tag{8.9}$$

This form can then be expanded for two radars by finding the θ that minimizes (in a least square sense)

$$\left[\zeta_1 - \frac{G(\theta_w - \theta_{r1})}{G(\theta_w - \theta_{r1} + \pi)}\right]^2 + \left[\zeta_2 - \frac{G(\theta_w - \theta_{r2})}{G(\theta_w - \theta_{r2} + \pi)}\right]^2, \tag{8.10}$$

for look directions θ_{r1} and θ_{r2} for radars 1 and 2, respectively.

Within the $\cos^s$ model, varying s changes the form of the directional spread. Initial models for $G(\theta)$ assumed frequency-dependent values for s [30], and later studies with HF radar also suggest that s varies with wind speed (e.g. [28]). Experimental results suggest s decreases from 4 for wind speeds up to 7 m s^{-1}, down to 1 at about 14 m s^{-1}, and less than 1 for greater wind speeds [19]. Low wind speeds and larger values for s result in a narrower directional spread for waves in the model, while higher winds and smaller s resulting in a broader directional wave field. Using data from different look directions, [28] assessed the validity of the $\cos^s$ model, and estimated s as ranging from 1 to 5 through time. Spreading models based on the $\cos^s$ require differences between the positive and negative Bragg peaks that are greater than 20 dB when the waves are aligned with the radar look direction ($< \pm 15°$), which were rarely observed in these early studies. A later work reported good agreement between measured wind directions when using the $\cos^s$ model [32], where HF radar-based wind directions derived from a 27 MHz WERA with 16-element linear array found 42° RMS differences when compared with an onshore wind sensor. While large, this measure of the difference appears to be influenced by a relatively few values of more than 90°.

Several publications investigating beam-forming observations with in situ wave and wind measurements have applied a different model, the $G(\theta) = \text{sech}^2(\beta\theta)$ model for directional spectra [33]. Improved matches with observations were found using $G(\theta) = \text{sech}^2(\beta\theta)$, with $\beta = 0.8$ based on the best fit with buoy observations and a 27 MHz WERA radar system deployed off the coast of Norway [34, 35]. Additionally, the application of maximum likelihood methods to both the first- and second-order spectra found better agreement than with the $\cos^s$ model, in part because the sech^2 model is not limited to wind directions outside of an envelope 15° from the radar look direction [34]. A subsequent analysis, again applying the use of second-order scatter, used this model, finding wind directions differences of 30–50° from an onshore anemometer [11]. More recent work [12, 36] using this model suggests that it gives the most realistic upwind/downwind ratio conditions for 5–20 m ocean waves (as in [35]). In this application, $G(\theta) = 0.5\beta\text{sech}^2(\beta\theta)$, where β is a directional spreading parameter that varies with wind speed, fetch, and wave age via a complex, empirically derived relationship.

There are several important caveats to the application of these directional spreading models for obtaining wind direction with HF radar data. First, OTH studies as well as those using WERA radars involved the use of beamforming, in which the radar is electronically "steered" into the look direction. These have the benefits of almost always having both positive and negative Doppler peaks for wind direction analysis. The more prevalent direction-finding (DF) systems, which use signal processing methods to assign observed signal power from a given Doppler bin to an azimuthal direction, may not assign the signal from both Doppler peaks to the same patch of the ocean surface. Thus, DF systems have an additional data requirement for simultaneous observations from both positive and negative Bragg peaks to compute wind direction. Given that the error in the DF increases with decreasing signal-to-noise ratio (SNR) [37], the likelihood of collocated observations from both Doppler peaks may also decrease with SNR. Second, precise observations of the directional distribution of surface gravity waves with wavelengths as small as 6 m in varying sea states are quite challenging for in situ observational systems, adding uncertainty to comparison data. Combining this with the uncertainty in the HF radar measurements makes the validation of the HF radar-derived spreading functions a difficult, and active area of research. Finally, there are important limits to the radar scattering theory that inhibit accurate estimates of wind direction, or other parameters for that matter, as significant wave heights get large. According to Gurgel [35], the Barrick-Weber theory [38, 39] breaks down at $k_0H_s < 4$, where $k_0 = 2\pi/\lambda$ is the radar wave number. This would suggest limits of $H_s \sim 6$ m for 25 MHz, or ~ 16 m for 13 MHz, above which the theory is no longer valid because of the wave height magnitudes.

8.2.3   Wind speed

An important step in the development of wind speed extraction was given by [40], which combined the closed-form inversion methods of [41, 42], along with empirical relationships between wind speed and sea state parameters to produce wind speed estimates. Essentially, the closed form estimates [41, 42] enabled the calculation of wave height (H_s) and peak frequency (f_p) from Doppler spectra. The method uses the first- and second-order scatter, though not the entire Doppler spectrum, to infer the wave parameters, thus relating the observed Doppler spectra to the power spectrum of the ocean waves. The characteristics of the wavefield were then used to estimate the generating wind field using empirical relationships developed for wave forecasting, known as the Sverdrup-Munk-Bretschneider curves (c.f. [43]), which typically take as input the 10-m wind speed and fetch to predict the significant wave height and dominant period. The results of the study were qualitatively promising, showing wind vectors along with a meteorological analysis map (using the methods of [19] for wind direction); however, quantitative comparisons to observations from a single ship were not significant. A later application of these methods [44], though promising, also lacked substantial comparison data. The lack of further application of these methods suggests the likely possibility that approximations in both the inversions of [41, 42] and the use of empirical relationships between wind speed, fetch, and the wave field could result in substantial errors. For example, assuming a 10% error in the wave parameters produces wind uncertainties of ± 2.7–4 m s^{-1} [40].

An application of the Multi-frequency Coastal Radar (MCR) was able to infer wind speed along with direction using a statistical relationship between buoy-based wind observations and radar-observed current shears [45, 46]. The MCR operated intermittently at frequencies from 4 to 22 MHz, enabling the observation of currents at several depths quasi-simultaneously. Using these observations along with buoy observations, the authors created a partial least squares model for wind speed, and then applied the model to predict winds for comparison on a reserved data set. Though not directly modeled, the use of currents to infer winds assumes continuity of stress across the air–sea interface, that is, $\rho_{air}u^2_{air}(z = 0) = \rho_{water}u^2_{water}(Z = 0)$

given the air density (ρ_{air}), wind speed (u_{air}), water density (ρ_{water}), and current speed (u_{water}). Applying the assumption from the Monin-Obukhov Theory of a logarithmic profile of near-surface currents, the wind speed may be estimated from current measurements at several depths. While noting the limitations of using the currents to infer the winds, the MCR study obtained standard errors of prediction of 1 m s^{-1} for wind speed (with −0.4 m s^{-1} bias and $r^2 = 0.8$) [46]. The study cited the use of methods from [18] for wind direction. The technique found large errors for winds below 3 m s^{-1}, which is near the 3.2 m s^{-1} phase speed of the resonant Bragg waves at 21.8 MHz, hinting at a likely limitation of the technique. An earlier study using the same radar [29] similarly found improved wind direction estimates above 5 m s^{-1}, though these were attributed to low SNR at times of low wind speed. Even controlling for times of sea-breeze conditions, numerous other factors affecting surface current shear tended to obscure the influence of wind speed and a direct relationship between the two, reducing the potential for inverting wind speed from current shear observations with radar. A subsequent analysis with a different multi-frequency radar found significantly non-exponential current shears, preventing the application of this method [27] and calling into doubt the assumptions for estimating wind speed via current shear. Note, however, that the MCR obtains currents from the first-order scatter, such that winds inferred from the currents ultimately result from the first-order backscattered signals.

8.3   Winds from second-order wave estimates

The potential for the combined first- and second-order returns of ground wave radars to provide estimates of the surface gravity wave spectrum drove a significant effort to extract these characteristics (c.f. [47–49]). As most extractions allow for a directional estimate of the wind-driven wave field, wave extractions often include an estimate of the wind speed and direction based on these observations. Wave extraction methods are covered elsewhere in this book in detail, and as they are not primarily focused on the wind, they are not covered here. However, we review what can be considered the most successful methods, since these may have important application as part of a complete wind observation method. It has been suggested, for example, that the use of second-order methods may find application for wind speeds above the saturation value [12] .

A demonstration of a successful application of the use of second-order for obtaining wind speed and direction from a ground-wave HF radar is provided by Wyatt *et al.* [34], who apply the methods of Gaffard *et al.* [50] along with wave inversion methods [10, 49]. This method uses the relative power of the second-order spectra to the first order as an indicator of wind speed, accounting for radar look direction relative to the wind direction. Using several data sets obtained from beamforming radars, it was shown that previous methods [40] can overestimate wind speed due to the influence of swell on the second-order determination. Later improvements to the second-order wave inversions [51] used advanced inversion methods with models such as the Pierson–Moskowitz type ocean wave spectrum [52] to simulate the second-order sidebands within the radar Doppler spectrum. While these more robust techniques led to improved results when the ocean waves were in equilibrium with the wind field, swell contamination remains a problem [12].

The use of second-order scatter to diagnose the wind speed is a potentially powerful tool in HF studies, as wind-wave theory suggests that the peak frequency is related directly to wind speed. This approach is also independent of the problems with Bragg wave saturation described below. However, a number of limitations exist. A key limitation is that the second order is, by definition, much lower in SNR than first-order scattering, and thus the calculation suffers from increased noise and uncertainty as well as reduced range relative to HF radar observations extracted from the first order only. Second, the full-wave spectrum will adjust more slowly to the surface winds than the higher frequency waves themselves, making wind estimates from the second-order wave inversion more representative of the slowly evolving wind field rather than the near-real-time winds (i.e. hours rather than tens of minutes as described above [28]). Finally, while winds from second order are considered by many to be an operational product, for the reason stated here as well as the confounding issue of non-locally generated swell, winds estimated via this method are not routinely utilized for research or operational data needs.

8.4   Winds from first order

Previous studies [18, 20, 29, 53, 54] used signals from the first-order region to obtain *wind direction*; however, diagnosing wind speed from the first-order returns was thought to be nonviable for a number of reasons. Primary among these is uncertainty in the absolute backscattered power received, resulting from propagation losses, spreading, and attenuation, which obscures the relationship between wind speed and first-order signal power. An additional reason is that a given radar frequency has a limited range of observable wind speeds, as shown in Table 8.1. The limited range of observable wind speeds results from either the lack of wind-wave equilibrium conditions at low winds or the saturation of the Bragg wave power at high winds [12]. However, signals in the first-order peaks often occur at levels 20 dB above the

Table 8.1   *Estimated wind speed observational range, based on a numerical model of the ocean wave response to wind inputs (WAve Model (WAM); based on [12]).*

Transmit Frequency (MHz)	Wind (ms^{-1})	
	Min.	Max.
5	5.7	17.8
13	3.8	10.8
25	2.8	9.4
42	1.4	7.5

second- and higher-order scatter and thus can theoretically provide greater spatial coverage and higher-accuracy wind inversions. In part, due to these potential advantages [12] and [14] describe models for inverting *wind speed and direction* from the first-order power.

Shen *et al.* [12] presented a theoretical outline relating wind speed to the received power, which was then solved using artificial neural networks. This investigation took the novel approach of using WAM to provide ocean wave spectra, along with 1 year of wind data as inputs, rather than using a simple model of wave and wind relationship that assumes fully developed or equilibrium (stationary) conditions. This approach to understanding the relationship between wind speed and Bragg wave spectral power is illustrated in Figure 8.2, produced using co-located wave and wind observations, rather than model output. The figure illustrates the relationship between the energy in the Bragg waves of different radar frequencies against wind speed and shows the usable range of wind speeds for each radar frequency. Theory suggests the lower limits result from the wind speed being lower than the ocean wave phase speed. Upper limits result from saturation as further wind input results in breaking, rather than greater wave heights (Table 8.1). The scatter in Figure 8.2 likely results from several factors, including the time delay between wind forcing and wave energy (1-hour averages were used to compare winds to wave power) and the use of "local" wind speed in comparison with potentially non-locally generated waves, which is particularly apparent for the lower wind speed comparisons of the waves sampled by 5 MHz radars. Still, the figure suggests, with data, a potentially linear relationship between the ocean wave spectral power and wind speed.

Using training data sets, [12] developed an artificial neural network-based solution for the wind speed and direction based on two separate forms of the radar data: (1) the absolute, received first-order signal power directly from the radars, and (2) differences between the left and right first-order signal powers cast within a wind speed-dependent directional distribution model ($\text{sech}^2(\beta\theta)$ as given above). The advantage of the second method over the first is that it, in theory, should not suffer from the wind speed limits imposed by the upper and lower bounds of wind-wave equilibrium and additionally not depend on unknown transmission and/or scattering losses inherent in the use of the absolute received first-order power. However, an advantage of the first method is that it is not dependent on a complex spreading model. Computing the neural network solutions for two, short-term data sets, [12] reported RMS differences for the validation data sets that were the same as the training data sets. However, the RMS differences for the training data set were significant, generally, 1.5–2.9 m s^{-1} with directional differences of around 20°.

Following the work of [12], [14] developed a model-based approach that relied on solving for empirical constants that could be calibrated with substantial in situ observations. The model-based approach combined the theoretical developments of previous works [12, 35] into a physical model that could be solved, in the least squares sense, for an arbitrary number of sites at each location. Additionally, novel aspect of this study was to utilize wind observations from autonomous surface vehicles (ASVs). ASVs provided low-cost, mobile wind observations which enabled calibrated extractions of surface winds from existing HF radars at the spatial

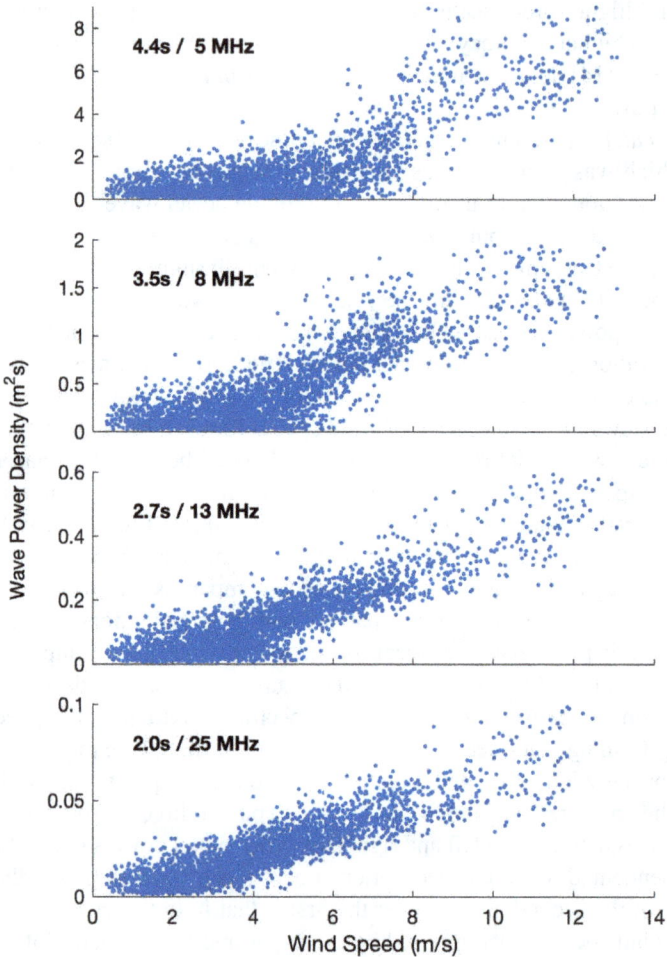

Figure 8.2   *Wind speed dependence of wave energy density levels for the Bragg
wave periods of typical radars operating at 5, 8, 13, and 25 MHz.
Hourly averaged wave power density was estimated from waveheight spectra calculated using 6 months of surface elevation
observations from a moored, near-surface upward-looking ADCP
in 40 m of water on the New England Shelf in 2019. Wind speeds
were observed with a nearby sonic anemometer measuring at 2 m
above mean sea level on a surface buoy co-located with the ADCP
mooring. Data were collected by, as part of the U.S. NSF-funded,
Submesoscale Dynamics of the Shelf project (SDOTS), led by
Kirincich.*

resolution (2–6 km) and ranges (up to 150 km) of the radars themselves. By calibrating the model with high spatial and temporal resolution wind observations, [14] realized a reproducible in situ method that could be applied to future sites for an operational sampling of the wind field. As in [12], RMS differences were again similar for both the training data set and the non-trained observations with RMS differences of 1.2–1.5 m s^{-1}. Importantly, variations in the calibration constants over time did not appear to significantly affect the wind extraction results, suggesting that a spatially varying but temporally constant set of calibration parameters might be sufficient for extracting the wind speed and direction to the accuracy of the model itself. Before discussing some of the key challenges for future advancement of HF radar-based wind sensing, the model and approach used by [14] are detailed below.

The model developed in [14] assumes that the power P received by a radar from the approaching and receding Bragg waves (at frequency $\pm f_B$) is given by $P(\pm f_B) = \kappa + Ew + Ea$, where κ is a radar-dependent reference power for each radar site as a function of range and bearing. κ includes system, site, and range-dependent effects on the signal power and can be represented empirically by the mean signal power in the radar coverage area over time. Ew and Ea represent deviations from this reference power due to the wind, accounting for the competing effects of wind wave growth (Ew) and attenuation (Ea) with range and increased surface roughness. Models of wave energy growth from wind suggest Ew has the form of $Ew = W_{fact}(U_{10}/c)^2$, where c is the phase speed of the Bragg wave, U_{10} is the wind speed at 10 m height, and W_{fact} is an empirically determined coefficient. Ea is also defined as a range-dependent function of U_{10}, given by $Ea = -(r/r_{max})(U_{10}/R_{fact})^3$, where r is the range from the radar, r_{max} is the maximum range, and R_{fact} is a second empirically determined coefficient. Combining these terms, we have

$$P(\pm f_B) = \kappa + \left(W_{fact}\left(\frac{U_{10}}{c}\right)^2 - \frac{r}{r_{max}}\left(\frac{U_{10}}{R_{fact}}\right)^3 \right)\text{sech}^2(\beta\theta), \qquad (8.11)$$

where a final factor multiplying the wind-dependent signal powers, $\text{sech}^2(\beta\theta)$, models the wave directional spread [33], as discussed in Section 8.2.2. As written, the method assumes that Bragg waves, averaged over radar sampling interval, are in equilibrium with the mean winds over the same time period.

The last component of (8.11), defined previously as $G(\theta)$, models the wave spreading around the wind direction and accounts for the wind direction relative to the radar look direction. Observations suggest that $G(\theta)$ varies with the wave number of the peak in ocean wave power, which is accounted for in the directional spreading parameter β [12, 55, 56], defined empirically as

$$\beta = 2.28(f_B/f_p)^{-0.65}. \qquad (8.12)$$

Here f_p is the peak wind-wave frequency, f_B is the Bragg wave frequency, and (8.12) is valid over the range $0.97 < f_B/f_p \le 2.56$. Given that f_p depends on the wind speed (U_{10}) and fetch (d), β itself will also vary with location as well as wind speed and radar frequency. Estimating f_p using

$$f_p = \frac{11}{\pi}[g^2/U_{10}d]^{\frac{1}{3}}, \qquad (8.13)$$

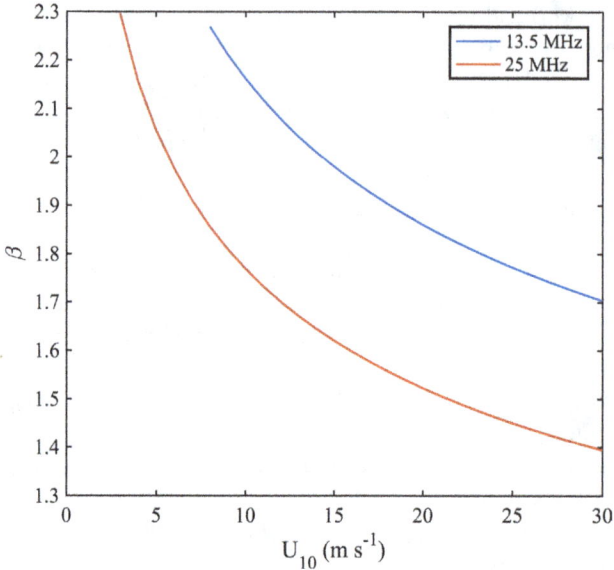

Figure 8.3   *Typical values of β for a range of wind speeds given by* U_{10}, *and at two HF transmit frequencies.*

it can be shown that $β$ ranges from 2.28 to ~1.1 as f_B/f_p ranges from 0.97 to 3 (Figure 8.3). Effectively, $β$ accounts for the increase in wave energy in crosswind directions at greater wind speeds.

Equations (8.11)–(8.13) define a fairly complex relationship between wind speed, direction, and received radar signal power. The empirical calibration technique, similar to those applied to satellite-based observations, was able to determine the spatially varying and site-specific coefficients $κ$, W_{fact}, and R_{fact} with relatively small amounts of in situ wind observations (<30 independent samples) [14]. Operationally, the model is first run backward to estimate, via least squares, the model coefficients for patches of the ocean surface given signal power observations from the approaching and receding Bragg waves, typically from two or more radars, along with spatially and temporally coincident in situ wind observations. $κ$ is determined from, for example, long-term averages of the radar power in range and bearing [14]. The approaching and receding waves at each location provide two values of the signal power, $P(±f_B)$, which along with the known wind speeds and directions can be used to compute values for the two constants W_{fact} and R_{fact}, with $β$ computed based on estimated or observed values for f_p.

After the site-specific coefficients are determined, the model is run forward to solve for wind speed and direction, given W_{fact} and R_{fact}, with f_p assumed to follow (8.13). To employ the model, a parameter space is formed, $Θ = \{U_{10}, θ, F\}$, for performing the minimization of the following:

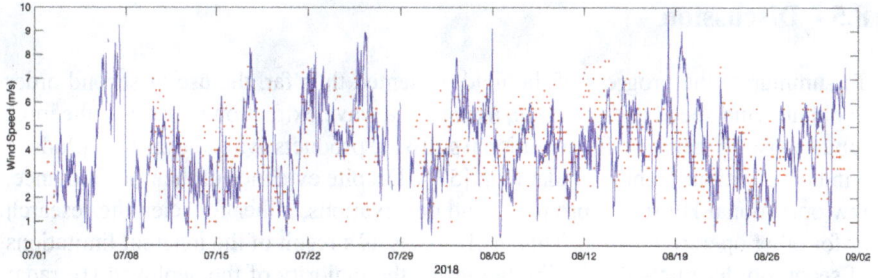

Figure 8.4   *Wind speed observations from a moored wind sensor (blue) and HF*
radar-based extractions (red dots) using the methodology of [14].
The time series have an RMS difference of 1.89 m s⁻¹.

$$\hat{\Theta}_i = arg\ min\left\{\kappa + \left(W_{fact}\left(\frac{U_{10}}{c}\right)^2 - \frac{r}{r_{max}}\left(\frac{U_{10}}{R_{fact}}\right)^3\right)\ sech^2(\beta\theta) - P(\pm f_B)\right\}. \quad (8.14)$$

The search is performed to determine the set $\hat{\Theta}_i$ for each spatial and temporal grid
point, with an arbitrary number of observations, though typically from at least two
HF radar sites.

   Applying this extraction method directly to first-order radar returns from a combined radar and buoy data set, yielded an RMS wind speed difference of 1.89 m s⁻¹
over the two-month comparison time series (Figure 8.4). This analysis defined the
first quarter of the time series as a training data set, which was used to estimate
calibration coefficients. While the calibration appears to be stable over the training
period, significant variability exists in the radar coefficients as estimated using the
training data set (Figure 8.5), which raises a number of issues that will be discussed
in Section 8.5.2.

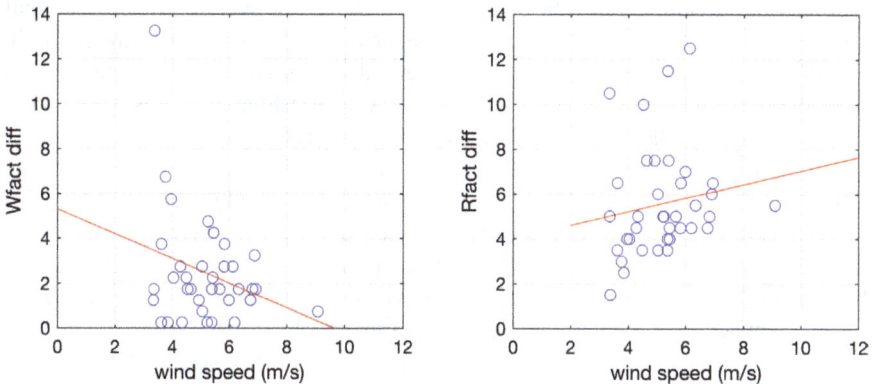

Figure 8.5 W_{fact} and R_{fact} *parameter estimates from the SDOTS training data set,*
with linear regressions to wind speed.

8.5 Discussion

To summarize the progress of the field presented thus far, the use of second order to obtain wind speed and direction, as part of a wave extraction method, is the most robust existing method with operational software products available that can produce wind observations from HF radar (e.g. [57]). Despite evidence of their performance, few operational HF radars produce wind observations, either for scientific research or for other operational uses. This likely exists as a result of the inherent limitations of second-order methods, the limitations of the majority of the deployed HF radar network (e.g. SeaSondes), and the fairly recent need for higher resolution observations of nearshore wind within the oceanographic community. As oceanography has transitioned from a data-sparse to data-rich science, modelers and oceanographers have driven resolutions, both spatial and temporal, to smaller scales. To date, coastal ocean wind observations have not kept pace. For example, the understanding of the role of across-shelf wind stress variation on upwelling and relaxation requires wind observations on hourly timescales and less than tens of kilometer spatial scales. It is likely that the limitations of second-order-based methods will prevent their use on the majority of deployed HF radars. Given these limitations, the potential for HF radars to produce wind observations over the same coverage area, and on the same spatial and temporal resolution as currents, would enable significant advances in coastal oceanography. While it remains to be seen if first- or second-order methods will find wide application, a number of problem areas must be resolved for any method to fulfill this potential. The remainder of this chapter examines the issues preventing further development and widespread use of wind measurements as a radar product, including limitations of second-order methods, measurement noise, and signal propagation losses.

8.5.1 Trade off between first- and second-order wind sensing

The most successful second-order wind extractions have utilized beamforming systems that are able to focus the receive arrays along relatively narrow directional pathways. This method for obtaining the radar look direction allows for spatially varying wind extractions over the area observed with second-order scatter. In this case, the theoretical beamwidth is well known, with most radars of this type having half power beamwidths of less than 10°, such that the spatial scale of the observations is well known (both for winds and currents). Large receive arrays of 8–16 elements also improve the characterization of the second-order spectra by increasing the relative SNRs compared with arrays of fewer antennas. Most DF systems such as the three-element SeaSonde use receive antennas optimized for small footprint and thus have theoretical beamwidths that are much wider than beamforming systems. As a result, compact DF systems estimate wave parameters by utilizing the second-order scatter from the entire angular coverage area of a range cell. Despite the increased noise in second-order spectra compared to beamforming systems, these compact systems may allow for a second-order-based wind estimate as an operational product. However, it is unlikely that additional advancements in second-order wave and wind sensing will occur for the SeaSonde systems, due in part to the inherent SNR of their second-order spectra. If true, second-order wind sensing will continue to be operational only for a narrow subset of deployed HF radars.

Meanwhile, first-order wind sensing has the potential to realize wind observations over the full range of the radar system, using the same DF methods and covering the same spatial area as currents. Thus, a first-order-based method is independent of the radar type or antenna array design. Exploratory investigations [12, 14] suggest that first-order winds are a complex data product, dependent on in situ calibrations or neural network approaches, that will nominally require multiple radars, with wind estimates produced by a centralized analysis effort independent of the radars themselves. As shown above (Figure 8.2, Table 8.1), multiple radar frequencies might also be necessary to improve first-order wind extractions for the widest range of winds possible. In summary, first-order winds hold promise for more widespread application but are not a data product that can be provided by an individual radar based on data collected by that radar alone.

8.5.2 Further radar noise issues

Any method to extract wind observations from HF radar signals will suffer from the inherent noise that exists within radar observations. In comparison to in situ point sensors of currents, winds, or waves, HF radar data are noisy due to a variety of factors that cannot be completely minimized. Similar to that for current estimates from radars, spectral noise and uncertainty are important contributors of noise in the wind observations, as they depend on multiple Fourier transforms to isolate the signal in space, and distinguish it from other signals and/or background noise. Low SNRs contribute directly to directional uncertainty for DF radars [37], while the spatial smoothing inherent in the beamwidth, and/or side lobe contributions, causes similar issues with beam-forming radars. While the resulting wind observations are essentially independent of currents, both observations would rely on the same DF or beamforming methods to locate the observations in bearing relative to the radar. Finally, measurement differences between point sensors and the wider spatially averaged radar estimates are a significant driver of the nominal 6–8 cm s^{-1} RMS differences reported for current comparisons (e.g. [58, 59]) which cannot be ignored for winds.

Wind extractions also depend on matching the observed powers to a model of the directional wave field. Methods using second-order spectra require the full-wave field, while methods using first-order spectra require a model for the directional spreading of Bragg waves only. Received signal powers vary over a narrow range, measured in dB, and thus a direct dependence of power on wave energy, while potentially linear, will also be weak and noisy. Figure 8.6 shows the dependence of HF radar signal power and wave energy, using data from the 2018 SDOTS observational campaign, led by Kirincich. The figure shows buoy measured wind speeds (from Figures 8.2 and 8.4), compared with the radar backscatter power (normalized by κ) only during times when the wind was directed toward the radar. Bin-averaged normalized power varies from 8 dB to 15 dB as

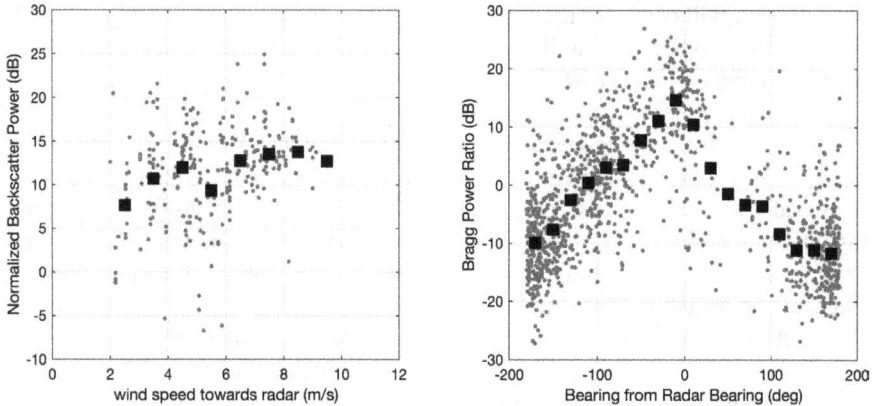

Figure 8.6   *A comparison of the (left) normalized radar power, with bin averages (black), as a function of wind speed when the wind was directly toward the radar, and (right) the Bragg power ratio as a function of observed wind direction relative to the radar look direction, for radar observations at the range and bearing of a moored anemometer (data from the 2018 SDOTS observational campaign).*

wind speed increases over the range $2-10$ m s^{-1}, with considerable scatter in the observed power.

Using the same SDOTS dataset, Figure 8.6 compares the observed Bragg power ratio to the bearing of the wind relative to the bearing of the radar from the buoy, illustrating both that the bearing-mean Bragg ratio varies from -10 dB to $+15$ dB, and that scatter about the bearing-mean is considerable.

These comparisons are admittedly simplistic views of the relationship between radar returns and winds, as they do not account for the number of factors that could cause scatter about a slowly varying mean trend (i.e. speed and/or fetch-dependent directional spreading). However, they illustrate the issue of radar noise as a critical component of wind extractions. While the scatter shown here, using Bragg wave returns only, directly pertains to wind sensing using the first-order methods, similar scatter contributes to noise in the results from methods using second-order spectra. The comparisons in Figures 8.4 and 8.6 suggest that any method to robustly observe surface winds via HF radars must be able to reduce this scatter or otherwise account for it. For example, it is not yet known how much of the scatter in Figure 8.6 is due to, for example, DF errors or spectral uncertainty, as opposed to wind speed dependence of the directional spreading or variable propagation losses. The first two are not likely easily resolved, as extractions of surface currents suffer the same problems, but the second two are open for additional research efforts. A fraction of the scatter shown here is almost certainly related to short time-scale variations in the propagation losses, which are discussed in the next section.

The results of uncertainty or noise in the received powers, bearing, and/or wind model application can be seen in the variability of the first-order wind extraction coefficients computed here for the SDOTS data set (Figure 8.5). Not only were the W_{fact} and R_{fact} parameters, as estimated, related to the wind speed itself, they also varied across the training data set, with W_{fact} having more potential variability than R_{fact}. Both the relation to the wind speed and the scatter around this relationship reveal the limitations of the model given by (8.11)–(8.13). That said, using the exact coefficients for each instance within the training data set, rather that the location mean, to predict the wind from the radar powers still resulted in RMS differences that were greater than 1 m s^{-1}, suggesting that the model itself needs further improvement.

8.5.3 Propagation losses

Theoretical understanding, as stated in (8.3), suggests a simple relationship between wind and HF radar observations. However, just as the wind-wave relationship described above, which is embedded within σ^0, is complex and challenging to model correctly, additional complexity resides in the attenuation factors (F), and transmit and receive gains. The attenuation factors F_t and F_r account for propagation losses occurring along the outbound and inbound paths due to the processes including absorption, diffraction, and shadowing [17]. Furthermore, transmit and receive antenna gains $(G_t$ and $G_r)$ include directional gains, resulting from both their intended design, and any differences due to antenna pattern distortions and differential overland losses, such as the variation in propagation over land at different bearings. Uncertainties in each of these can be the same scale as power changes resulting from the wind. It is likely that uncertainty in initial estimates for these terms and their variations over time and space prevented earlier works from achieving more significant contributions focused on first order, Bragg wave-based wind extractions.

For example, the attenuation factor F represents a significant complicating factor in (8.3). Defined as the normalized path attenuation, F is a non-dimensional function of the permittivity and conductivity of both the propagation medium (i.e. air) and the surface over which the radio wave is propagating (i.e. the salty ocean). Essentially, the electrical properties of these two mediums cause an impedance of the ground wave – quantified by F. Observations suggest a difference between computed and observed attenuation. One study observed increased attenuation (F) compared with theoretical values for frequencies less than 20 MHz, with decreased attenuation above 20 MHz [60], with the differences attributed to atmospheric conditions [61]. F must also account for varying attenuation that results from along path variation in surface roughness due to the sea state. An early example demonstrating the apparent increased attenuation to higher sea state [62] used published graphs [63] to show increased path loss with wind speed at both 7 MHz and 14 MHz. These observations indicate a possible source of variability that would be difficult to constrain without measurements in addition to HF radar observations.

The derivation of σ^0 for first-order scattering [15] also has dependencies on the scattering medium. As defined by [15], (8.4) holds for a slightly rough surface, where "slightly rough" is defined as having a "small" wave height in terms of λ, and "small" surface slopes. Furthermore, the patch size must be large with respect to λ,

Table 8.2   Published values for σ^0 (definitions vary between publications).

Reference	σ^0	Notes
Headrick *et al.* [64]	−29 dB	Chesapeake Bay, 10 MHz
Barrick [15]	−17 dB	Theoretical, same value as [64]
Teague *et al.* [65]	−29.7, −32.4, −38.1	1.95 MHz

but small with respect to range (r). Table 8.2 gives published values of σ^0 based on observations and theoretical calculations. Converting out of dBm, a typical value for σ^0 is on the order of 0.03, which in consideration of (8.3) illustrates the small fraction of transmitted power that is backscattered by the sea surface. These values are also illustrative of the wide range of values found for the radar cross-section, independent of the other terms in (8.3), and the departures from theoretical estimates. As described above for OTH radar-based efforts, a key problem area for constraining the signal loss components of the received signal power was the lack of comparative data for these types of experiments.

8.5.4   Future directions

Finally, the impediments to widespread operational application of wind extraction methods using coastal HF radars described here will require additional effort to solve. Much of the present focus is centered on extractions from the first-order Bragg wave returns, due to their longer range and potentially higher spatial resolutions. The neural network approach used by [12] could be of value in teasing apart the functional relationships that link wind to Bragg wave energy, wave spreading, and radar returns. The neural network-based results of [12] in part motivated empirical approaches to relating HF radar observations to wind speed and direction. Future efforts in these areas may lead to improved wind products based on present understanding; however, future investigations of wind extraction from HF radar may also lead to improved understanding of the ocean wave field and wind-wave generation physics.

A factor discussed in [19] but often ignored is that most spreading models for wind direction are themselves dependent on wind speed (e.g. (8.12)–(8.13)). A potential alternative methodology would forgo the use of signal power as in [14], in favor of using the spreading model itself to estimate speed along with direction. This approach was reviewed by [12], finding similar errors as methods using signal power directly. As this requires the relative difference in power between incoming and outgoing waves only, potential electromagnetic effects might be minimized in favor of additional complexity in the model inversion. This extraction method would benefit from improvements in inversion techniques, known priors, or the use of estimates of the full ocean wave power spectrum.

Though these approaches show promise, additional effort is needed to improve the wind extractions and realize the potential benefit to studies of the coastal ocean. Observations of nearshore wind stress and wind stress curl should be accurate to 1 m s^{-1} over 10 km if they are to reveal a new understanding of coastal ocean dynamics. RMS differences obtained by [14] (near 2 m s^{-1}; Figure 8.4) are similar to those of remote sensing observations such as scatterometer- or SAR-based extractions. The magnitude of the RMS differences suggests that the model described by (8.11)–(8.13) remains overly simplistic in key areas. Meanwhile, there are numerous potential optimizations for the equations defined above. For example, significant variability in the calibration coefficients (W_{fact} and R_{fact}) was observed, both in space and time (Figure 8.5). Sources for this variability, and in particular the wind speed dependence of R_{fact}, may include fetch limitations and wave age, radar scattering issues, or other radar processing issues. Constraining variability in the calibration coefficients may be sufficient to reduce RMS differences to scientifically important levels.

8.6   Summary

In this chapter, we have provided a review of methods, theory, remaining issues, and future prospects for obtaining coastal ocean surface wind observations with HF radar. All HF radar-based methods rely on the fact that the HF radar scatters from waves, and thus any radar-based wind product is based on an assumed relationship between the resonant waves and the wind. Most published methods to date either utilize the second-order Bragg scatter to obtain the ocean wave field, before inferring the generating wind field from there, or assume a relationship between the first and second-order scatter that depends on wind speed. Necessarily, these methods assume forms for the directional wave field. No single method for obtaining winds from HF radar has found widespread application. This may stem from the fact that the use of HF radar to measure waves has found wider application only recently. Furthermore, most investigations of wind extractions from HF radar have relied on relatively small data sets, which has limited their utility. While there are exceptions (e.g. [20, 66]), the full employment of larger data sets with accurate in situ wind observations has yet to be applied to this problem. Meanwhile, the need for wind observations on small spatial and temporal scales has only recently become urgent within coastal oceanography [7, 8, 67]. As model resolutions have improved, the need for accurate, high-resolution forcing has become more acute. HF radar has the potential to address this need, and as this chapter argues, the potential is still unrealized.

Acknowledgements

The authors acknowledge the support of the U.S. National Science Foundation under Grant OCE-1923465 (Emery) and OCE-1923927 (Kirincich). Any opinions, findings, and conclusions or recommendations expressed in this material are those of the authors and do not necessarily reflect the views of the National Science Foundation.

References

[1]   Ekman V.W. 'On the influence of the earth's rotation on ocean-currents'. **2**. Arkiv för matematik, astronomi och fysik; 1905. pp. 1–53.

[2]   Plagge A.M., Vandemark D.C., Long D.G. 'Coastal validation of ultra-high resolution wind vector retrieval from QuikSCAT in the Gulf of Maine'. *IEEE Geoscience and Remote Sensing Letters*. 2009;**6**(3):413–17.

[3]   Verhoef A., Portabella M., Stoffelen A. 'High-resolution ASCAT scatterometer winds near the coast'. *IEEE Transactions on Geoscience and Remote Sensing*. 2012;**50**(7):2481–7.

[4]   Ahsbahs T., Badger M., Karagali I., Larsén XG. 'Validation of Sentinel-1A SAR coastal wind speeds against scanning LiDAR'. *Remote Sensing*. 2017;**9**(6):552.

[5]   Lu Y., Zhang B., Perrie W., Mouche A.A., Li X., Wang H. 'A C-band geophysical model function for determining coastal wind speed using synthetic aperture radar'. *IEEE Journal of Selected Topics in Applied Earth Observations and Remote Sensing*. 2018;**11**(7):2417–28.

[6]   Rodríguez E., Wineteer A., Perkovic-Martin D., *et al.* Estimating ocean vector winds and currents using a Ka-Band pencil-beam Doppler Scatterometer'. *Remote Sensing*. 2018;**10**(4):576.

[7]   Capet X., Colas F., McWilliams J., Penven P., Marchesiello P. 'Eddies in eastern boundary subtropical upwelling systems'. *Ocean Modeling in an Eddying Regime, Geophysical Monograph Series*. 2008;**177**:131–47.

[8]   Colas F., Capet X., McWilliams J.C., Li Z. 'Mesoscale eddy buoyancy flux and eddy-induced circulation in eastern boundary currents'. *Journal of Physical Oceanography*. 2013;**43**(6):1073–95.

[9]   Kirincich A. 'The occurrence, drivers, and implications of submesoscale eddies on the Martha's Vineyard inner shelf'. *Journal of Physical Oceanography*. 2016;**46**(9):JPO–D–15–0191.1):2645:62.

[10] Wyatt L.R. 'Limits to the inversion of HF radar backscatter for ocean wave measurement'. *Journal of Atmospheric and Oceanic Technology*. 2000;**17**(12):1651–66.

[11] Wyatt L.R. 'Shortwave direction and spreading measured with HF radar'. *Journal of Atmospheric and Oceanic Technology*. 2012;**29**(2):286–99.

[12] Shen W., Gurgel K.W., Voulgaris G., Thomas S., Detlef S. 'Wind-speed inversion from HF radar first-order Backscatter signal'. *Ocean dynamics*. 2012;**62**(1):105–21.

[13] Kirincich A. 'Toward real-time, remote observations of the coastal wind resource using high-frequency radar'. *Marine Technology Society Journal*. 2013;**47**(4):206–17.

[14] Kirincich A.R. 'Remote sensing of the surface wind field over the coastal ocean via direct calibration of HF radar backscatter power'. *Journal of Atmospheric and Oceanic Technology*. 2016;**33**(7):1377–92.

[15] Barrick D.E. 'First-order theory and analysis of MF/HF/VHF scatter from the sea'. *IEEE Transactions on Antennas and Propagation.* 1972;**20**(1):2–10.

[16] Crombie D.D. 'Doppler spectrum of sea echo at 13.56 Mc./s'. *Nature.* 1955;**175**(4459):681–2.

[17] Skolnik M.I. *Radar Handbook.* **3**; 1990.

[18] Long A., Trizna D. 'Mapping of North Atlantic winds by HF radar sea backscatter interpretation'. *IEEE Transactions on Antennas and Propagation.* 1973;**21**(5):680–5.

[19] Stewart R.H., Barnum J.R. 'Radio measurements of oceanic winds at long ranges: an evaluation'. *Radio Science.* 1975;**10**(10):853–7.

[20] Harlan J.A., Georges T.M. 'An empirical relation between ocean-surface wind direction and the Bragg line ratio of HF radar sea echo spectra'. *Journal of Geophysical Research.* 1994;**99**(C4):7971–8.

[21] Barnum J., Maresca J., Serebreny S. 'High-resolution mapping of oceanic wind fields with skywave radar'. *IEEE Journal of Oceanic Engineering.* 1977;**2**(1):128–32.

[22] Maresca J., Barnum J. 'Measurement of oceanic wind speed from HF sea scatter by skywave radar'. *IEEE Transactions on Antennas and Propagation.* 1977;**25**(1):132–6.

[23] Gilhousen D.B. 'A field evaluation of NDBC moored buoy winds'. *Journal of Atmospheric and Oceanic Technology.* 1987;**4**(1):94–104.

[24] Ahearn J.L., Curley S.R., Headrick J.M., Trizna D.B. 'Tests of remote skywave measurement of ocean surface conditions'. *Proceedings of the IEEE.* 1974;**62**(6):681–7.

[25] Georges T. 'Progress toward a practical skywave sea-state radar'. *IEEE Transactions on Antennas and Propagation.* 1980;**28**(6):751–61.

[26] Maresca J., Barnum J. 'Estimating wind speed from HF skywave radar sea backscatter'. *IEEE Transactions on Antennas and Propagation.* 1982;**30**(5):846–52.

[27] Trizna D., Xu L. 'Target classification and remote sensing of ocean current shear using a dual-use multifrequency HF radar'. *IEEE Journal of Oceanic Engineering.* 2006;**31**(4):904–18.

[28] Heron M., Rose R. 'On the application of HF ocean radar to the observation of temporal and spatial changes in wind direction'. *IEEE Journal of Oceanic Engineering.* 1986;**11**(2):210–18.

[29] Paduan J.D., Delgado R., Vesecky J.F., Fernandez Y., Daida J.M., Teague C.C. 'Mapping coastal winds with HF radar'. Proceedings of the IEEE Sixth Working Conference on Current Measurement. IEEE; 1999. pp. 28–32.

[30] Tyler G.L., Teague C.C., Stewart R.H. 'Wave directional spectra from synthetic aperture observations of radio scatter. in: deep sea research and oceanographic Abstracts'. Deep-Sea Research and Oceanographic Abstracts; 1974. pp. 989–1016.

[31] Longuet-Higgins M.S., Cartwright D.E., Smith N.D. 'Observations of the directional spectrum of sea waves using the motions of a floating buoy'. *Ocean wave spectra.* Prentice-Hall, Inc.; 1961.

[32] Essen H.H., Gurgel K.W., Schlick T. 'Measurement of ocean wave height and direction by means of HF radar: an empirical approach'. *Deutsche Hydrografische Zeitschrift*. 1999;**51**(4):369–83.

[33] Donelan M.A., Hamilton J., Hui W. 'Directional spectra of wind-generated ocean waves'. *Philosophical Transactions of the Royal Society A*. 1985;**315**(1534):509–62.

[34] Wyatt L.R. 'HF radar wind measurement – present capabilities and future prospects'. Proceedings of the Radiowave Oceanography 1st International Workshop; Timberline, Oregon, April 9-12, 2001; 2001. pp. 83–95.

[35] Gurgel K.W., Essen H.H., Schlick T. 'An empirical method to derive ocean waves from second-order Bragg scattering: prospects and limitations'. *IEEE Journal of Oceanic Engineering*. 2006;**31**(4):804–11.

[36] Shen W., Gurgel K.W. 'Wind direction inversion from narrow-beam HF radar backscatter signals in low and high wind conditions at different radar frequencies'. *Remote Sensing*. 2018;**10**(9):1480.

[37] Emery B.M., Washburn L. 'Uncertainty estimates for SeaSonde HF radar ocean current observations'. *Journal of Atmospheric and Oceanic Technology*. 2019;**36**(2):231–47.

[38] Weber B.L., Barrick D.E. 'On the nonlinear theory for gravity waves on the ocean's surface Part I: Derivations'. *Journal of Physical Oceanography*. 1977;**7**(1):3–10.

[39] Barrick D., Weber B. 'On the nonlinear theory for gravity waves on the ocean's surface. Part II: Interpretation and applications'. *Journal of Physical Oceanography*. 1977;**7**(1):11–21.

[40] Dexter P.E., Theodoridis S. 'Surface wind speed extraction from HF sky wave radar Doppler spectra'. *Radio Science*. 1982;**17**(3):643–52.

[41] Barrick D.E. 'The ocean waveheight nondirectional spectrum from inversion of the HF sea-echo Doppler spectrum'. *Remote Sensing of Environment*. 1977;**6**(3):201–27.

[42] Barrick D.E., Snider J.B. 'The statistics of HF sea-echo Doppler spectra'. *IEEE Journal of Oceanic Engineering*. 1977;**2**(1):19–28.

[43] Sverdrup H.U., Munk W.H. *Wind, sea and swell: theory of relations for forecasting (No. 303)*. Hydrographic Office; 1947.

[44] Huang W., Wu S., Gill E.W., Hou J., Wen B. 'HF radar wave and wind measurement over the eastern China Sea'. *IEEE Transactions on Geoscience and Remote Sensing*. 2002;**40**(9):1950–5.

[45] Drake J., Vesecky J., Laws K., Teague C., Ludwig F., Paduan J. 'Vector wind field measurements using multifrequency HF radar'. Proceedings of the IEEE/OES Seventh Working Conference on Current Measurement Technology, 2003; 2003. pp. 88–91.

[46] Vesecky J.F., Drake J.A., Laws K., Ludwig F.L., Teague C.C., Meadows L.A. 'Using multifrequency HF radar to estimate ocean wind fields'. IGARSS 2004. 2004 IEEE International Geoscience and Remote Sensing Symposium. vol. 2. IEEE; 2004. pp. 1167–70.

[47]   Barrick D.E. 'Theory of HF and VHF propagation across the rough sea, 1, the effective surface impedance for a slightly rough highly conducting medium at grazing incidence'. *Radio Science.* 1971;**5**:517–26.

[48]   Barrick D.E. 'Extraction of wave parameters from measured HF radar sea-echo Doppler spectra'. *Radio Science.* 1977;**12**(3):415–24.

[49]   Wyatt L., Ledgard L., Anderson C. 'Maximum-likelihood estimation of the directional distribution of 0.53-Hz ocean waves'. *Journal of Atmospheric and Oceanic Technology.* 1997;**14**(3):591–603.

[50]   Gaffard C., Parent J. 'Remote sensing of wind speed at sea surface level using HF skywave echoes from decametric waves'. *Geophysical Research Letters.* 1990;**17**(5):615–18.

[51]   Green D., Gill E.W., Huang W. 'An inversion method for extraction of wind speed from high-frequency ground-wave radar oceanic backscatter'. *IEEE Transactions on Geoscience and Remote Sensing.* 2009;**47**(10):3338–46.

[52]   Pierson W.J., Moskowitz L. 'A proposed spectral form for fully developed wind seas based on the similarity theory of S. A. Kitaigorodskii'. *Journal of Geophysical Research.* 1964;**69**(24):5181–90.

[53]   Georges T., Harlan J., Meyer L., Peer R.G. 'Tracking Hurricane Claudette with the US air force over-the-horizon radar'. *Journal of Atmospheric and Oceanic Technology.* 1993;**10**(4):441–51.

[54]   Lipa B., Barrick D., Alonso-Martirena A., Fernandes M.J., Ferrer M., Nyden B. 'Brahan project high frequency radar ocean measurements: currents, winds, waves and their interactions'. *Remote sensing.* 2014;**6**(12):12094–117.

[55]   Banner M., Jones I.S., Trinder J. 'Wavenumber spectra of short gravity waves'. *Journal of Fluid Mechanics.* 1989;**198**:321–44.

[56]   Banner M.L. 'Equilibrium spectra of wind waves'. *Journal of Physical Oceanography.* 1990;**20**(7):966–84.

[57]   Wyatt L.R. 'A comparison of scatterometer and HF radar wind direction measurements'. *Journal of Operational Oceanography.* 2018;**11**(1):54–63.

[58]   Graber H.C., Haus B.K., Chapman R.D., Shay L.K. 'HF radar comparisons with moored estimates of current speed and direction: expected differences and implications'. *Journal of Geophysical Research.* 1997;**102**(C8):18749–66.

[59]   Ohlmann C., White P., Washburn L., Emery B., Terrill E., Otero M. 'Interpretation of coastal HF radar-derived surface currents with high-resolution drifter data'. *Journal of Atmospheric and Oceanic Technology.* 2007;**24**(4):666–80.

[60]   Hansen P. 'Measurements of basic transmission loss for HF ground wave propagation over seawater'. *Radio Science.* 1977;**12**(3):397–404.

[61]   Pappert R., Goodhart C. 'A numerical study of tropospheric ducting at HF'. *Radio Science.* 1979;**14**(5):803–13.

[62]   Forget P., Broche P., De Maistre J. 'Attenuation with distance and wind speed of HF surface waves over the ocean'. *Radio Science.* 1982;**17**(3):599–610.

[63]   Barrick D.E. 'Theory of HF and VHF propagation across the rough sea, 2, application to HF and VHF propagation above the sea'. *Radio Science.* 1971;**6**(5):527–33.

[64]   Headrick J.M., Rohlfs D.C., Ward E.W., Boyd F.E. HF radar as a fleet sensor. Washington, DC: Naval Research Lab; 1970.

[65]   Teague C.C., Tyler G.L., Stewart R.H. 'The radar cross section of the sea at 1.95 MHz: comparison of in-situ and radar determinations'. *Radio Science.* 1975;**10**(10):847–52.

[66]   Vesecky J.F., Drake J., Teague C.C., Ludwig F.L., Davidson K.L., Paduan J.D. 'Measurement of wind speed and direction using multifrequency HF radar'. 2002 IEEE International Geoscience and Remote Sensing Symposium, 2002. IGARSS'02. vol. 3. IEEE; 2002. pp. 1899–901.

[67]   Capet X.J., Marchesiello P., McWilliams J.C. 'Upwelling response to coastal wind profiles'. *Geophysical Research Letters.* 2004;**31**(13):1–4.

Chapter 9

HF radar in tsunami detection

ML Heron[1]

The primary product of high-frequency (HF) ocean radar systems is the mapping of surface currents in coastal waters, and some systems produce maps of significant wave heights and directional spectra of wind waves. Wind speed can be derived as a secondary parameter that depends on *a priori* knowledge of the link between wind speed and significant wave height and is not a primary observation of HF ocean radar. In the palette of capability, the detection of tsunamis by HF radar is a subset of surface current observations because the radar measures the to-and-fro surges in the surface velocity as a tsunami propagates through a radar pixel.

Substantial archives of HF ocean radar data have been established where the main parameter is hourly values of surface current vectors on a rectangular grid. Archived data have been widely used for general circulation studies, seasonal variability of coastal currents, and location of recurrent mesoscale eddies [1–4]. For some of these large-scale applications, the spatial resolution of up to 20 km and temporal low-pass filtering of up to 3 hours are sufficient, but for tsunami detection, finer resolution in time and space is required. HF radars can be optimised for the fine spatial resolution and a temporal resolution of a few minutes that are needed for tsunami detection. To reduce costs and provide flexibility, one of the challenges that has been faced by developers of HF radars is to be able to simultaneously meet the requirements of seemingly conflicting applications of different scales from one installation. For example, a method has been developed for phased-array radars to operate coherently for 1 hour with samples of time-series being taken out at short intervals of less than one minute for high temporal resolution applications.

Most of the installations of HF ocean radars around the world have been for research purposes. However, as the technology has matured, there has been a steady increase in installations for routine management of ports, marine reserves and other coastal assets. This includes identifying high-current streams, sub-mesoscale eddies and tidal features. These installations have provided data for the planning of a wide range of hazard mitigation like the accumulation of debris, minimising erosion and pollution control. The development of tsunami detection by HF radar has grown from early observations of large events in coastal currents that were observed on

[1]Marine Geophysics Laboratory and Physical Sciences, James Cook University, Australia

systems designed for general circulation and then refined to detection algorithms that are tailored to the characteristics of tsunamis which can give robust early-warning alerts even for small non-hazardous tsunamis.

The application of real-time data to the mitigation of hazards can be seen in the routine management of ports and other coastal assets, operational warnings for excessive wave heights, tsunami, storm surges and rip currents and Lagrangian Tracking of surface drift in search-and-rescue operations.

9.1   The underlying physics

It is shown in Chapter 1 (1.27) that the radar cross section for a monochromatic radar has two dominant first-order peaks with positive and negative frequencies as shown in Figure 1.5 of Chapter 1. These two frequencies are determined by the radar frequency and the phase velocity of the two scattering waves on the sea surface. In deep water, defined by $d > \lambda/4\pi$ [5], the phase velocity of gravity waves is fixed by its wavelength (1.37 in Chapter 1). If the surface waves are superposed on a current with radial component v_r then the first-order peaks occur at frequencies δf_D given by

$$\delta f_D = \pm\sqrt{\frac{g}{\pi\lambda}} - \frac{2v_r}{\lambda} \tag{9.1}$$

The two spectral lines predicted by (9.1) are shown as the two main peaks in Figure 9.1. The frequency offset δf is equivalent to the second term in (9.1) and is due to the surface current. Note that the offset is in the same direction for both the

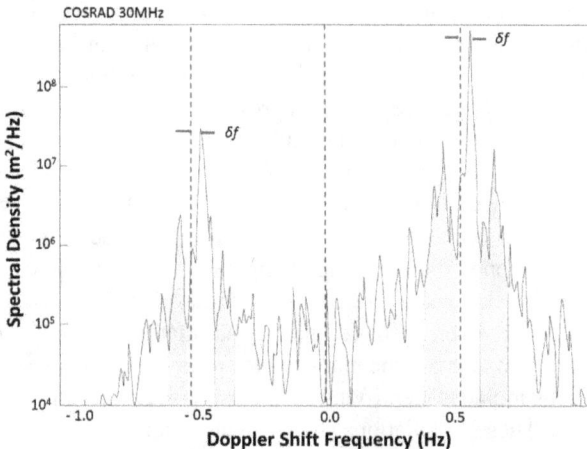

Figure 9.1   *Typical Doppler shift spectrum from a wide-aperture HF radar. The two dominant first-order peaks are shifted from the no-current positions (red dashed lines) by an amount δf given by the second term in (9.1). The shaded regions indicate the second-order parts of the spectrum that are used to extract wave heights.*

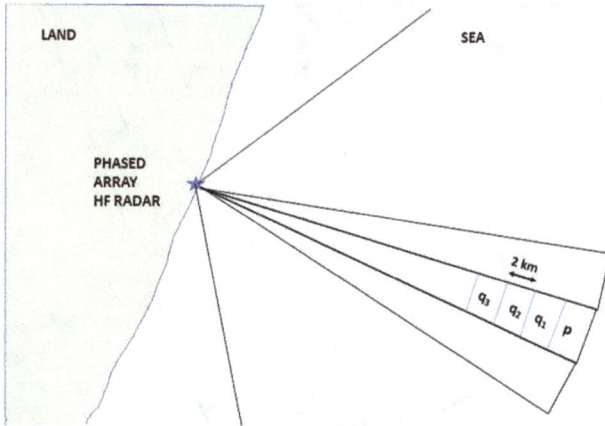

Figure 9.2 *Typical spatial sampling by a phased-array HF radar showing a 20 degree window with independent pixels p, q*

positive and negative first-order peaks. The second-order energy, shown shaded in Figure 9.1, is discussed in Chapter 1.

The main genres of High Frequency Surface Wave Radar (HFSWR) are introduced in Chapter 1 and here the key parameters for tsunami detection are summarised.

Figure 9.2 illustrates the raw scanning pattern of a typical phased-array radar. Sea echoes from independent pixels (e.g., *p*, *q* in Figure 9.2) are analysed to produce observed parameters. To reduce random fluctuations, the data from independent pixels are sometimes averaged in space and time. Raw analysis records the time series (and therefore the Doppler shift spectrum) in range/azimuth cells in a polar grid. Alternatively, by control of range and azimuth, the radar can populate a rectangular grid with radial components of velocity. The spatial resolution is determined by the width of the beam, and the range resolution is determined by the bandwidth at the operating frequency. For example, a typical phased-array radar with 16 antenna elements has an azimuthal resolution of about 2 km ((1.7) of Chapter 1) at 20 km range, and a radar with an operating bandwidth of 75 kHz has a range step of 2 km. This gives a spatial resolution of 2 km × 2 km at a range of approximately 20 km. To get 2-dimensional (2-D) maps of surface current velocity vectors, a second radar station is needed at a location that samples each target grid point from a different direction.

Two types of direction-finding radar systems are described in Chapter 1. One uses phase detection on vertically polarised antennas to determine the angle of arrival of sea echoes. The other uses antenna patterns of power on loop antennas to determine the azimuth from which it came. This is illustrated in Figure 9.3, where each pair of the red arcs represents a target annulus which is populated with radial components of velocity. The data fall naturally onto polar coordinates, and to transform to a Cartesian grid, the radial components that lie within an averaging circle are combined to get the averaged value at the grid point. The spatial resolution here is usually set by the radius of the averaging circle. If the averaging circle is set too

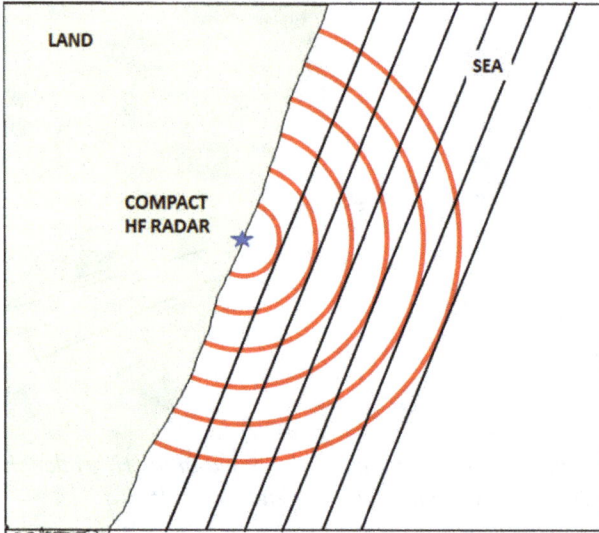

Figure 9.3   *The red arcs show the ring-shaped target cells for compact radar*
systems. The q-factor analysis averages radial current observations
in the rectangular slabs defined by the black lines.

small then the spatial resolution is limited by the range interval (2 km for an operating bandwidth of 75 kHz) and the uncertainty in extracting the azimuth for each
Doppler shift element in the first-order spread spectrum. Typical values for the averaging circle radius and standard error in the azimuth estimate (before averaging) are
up to 10 km and up to 18 degrees, respectively [6–8].

The spatial and temporal resolutions of an HF radar have important consequences
in applications to mitigate marine hazards. The International Telecommunications
Union (ITU) has allocated specific frequency bands for HF ocean radar systems at
a range of frequencies between 3 MHz and 50 MHz for different geographic zones
[1]. The allocated frequency bands are rather narrow for HF radar because the range
resolution is inversely proportional to the operating bandwidth. For example, at an
operating frequency of 5.25 MHz, the allocated band is 25 kHz; if a radar used the
whole band, the range resolution would be 6 km; the corresponding figures for the
39 MHz band are 500 kHz and 300 m. Usually, radar installations are designed to
optimise selected capabilities – at the expense of others. For example, by selecting
a higher operating frequency it will often be possible to be allocated a wider bandwidth by the national regulatory authority. This can lead to improved spatial resolution but the maximum operating range will be decreased at the higher frequency.
Spatial resolution is also adversely affected when spatial averaging is used to reduce
random fluctuations in data. Also, for tsunami detection, time samples have to be
tailored to record the shortest of tsunami waves (about 10 minutes period). This
impacts the accuracy of the surface current velocity components that are derived.

The integrity of HF radar technology depends on correct evaluation of the uncertainties in the final products and the verisimilitude in specifying temporal and spatial resolution [7].

9.2   Observation of surface currents

Most HF radar installations are designed to produce archived data for large-scale coastal currents. The temporal resolution of these data ranges from 10 minutes to 3 hours and these are usually archived as hourly values. Integration over longer time periods (3 hours) and larger areas (300 km^2) improves the signal-to-noise ratio and gives longer operating ranges. These data are most useful in identifying seasonal events like upwelling and convergence, general circulation and associated mesoscale eddies. For hazard mitigation, these measurements can be used for identifying pollution transport and concentration spots for marine debris. The large-scale data are not well conditioned for detecting tsunamis.

For the measurement of currents closer to the coast, a temporal resolution given by 1 hour or less gives data that will resolve tidal flows. This reveals coastal eddies and the impacts of flow, including pollution from the land and coastal industries. Coastal engineering works, like sewage outfalls and marine reserve management, require knowledge of tidal currents in their planning. Lagrangian tracking of parcels of surface water for search-and-rescue operations requires good accuracy in time and space. Again, these scales are not well conditioned for tsunami detection.

For detection of tsunamis approaching the coast, good resolution in the measurement of surface currents is required, and at the same time, the operating range needs to be maximised in order to detect a tsunami as early as possible.

9.3   Tsunami characteristics

In the spectrum of sea surface disturbances, tsunamis fall between long swell waves and tidal oscillations. Tsunami waves typically have periods of 10–40 minutes and wavelengths of tens of kilometres in coastal waters to hundreds of kilometres in the deep ocean.

Traditionally, tsunamis have been associated with seismic activity and the Japanese origin of the word "tsunami" indicates the impact of the phenomenon. Some (but not all) earthquakes with offshore epicentres generate tsunamis by an upthrust (or slump) of a strip of the ocean floor that produces a pulse of elevated water and, from that, a propagating wave. Seismic waves propagate much more quickly than tsunami waves, and seismic reports of earthquakes give a first alert to Tsunami Early Warning Systems (TEWS). Information on significant seismic events is rapidly disseminated around the globe, and pre-calculated models are used to estimate the potential time of arrival of possible tsunamis at specific locations [9].

While seismically generated tsunamis have generally had the most severe impacts on coasts, there are several other genesis mechanisms. Subsea landslides have generated tsunamis without any corresponding seismic signals. These tsunamis can be generated on the sloping edge of the continental shelf, especially in regions where there is an accumulation of sediment. A 1998 earthquake near the north coast of Papua New Guinea generated a moderate tsunami, but a delayed submarine slump produced a second tsunami that inundated 4 villages with the loss of over 1 600 people.

Sudden changes in atmospheric pressure or front lines of wind events can produce tsunami-like waves usually in shallow waters [10]. Storm setup on the coast and seiches in a harbour are hazards, and the associated currents are detectable by HF radar, but they are not propagating waves. The most likely meteorological genesis of tsunami-like waves is the Proudman resonance [11], where the atmospheric disturbance is travelling at the same speed as shallow-water gravity waves. Coupling between the boundary layers transfers energy from the atmosphere to the ocean. Atmospherically generated tsunamis have come to be known as meteotsunamis. Meteotsunamis tend to recur in specific areas [10] and while generally not as damaging as the bigger seismically generated tsunamis, they have been recorded at up to 3–4 m in the western Mediterranean [12] and other places.

Reports of tsunami amplitudes on tide gauges at the coast range from centimetres to tens of metres depending on bathymetry and coastal topography, which can induce funnelling. Many meteotsunami events have been reported in the Mediterranean Sea, where coastal impacts are felt to be severe because they are compared with a small regular tidal range in the order of 0.5 m. Much of the published work is focused on bays and inlets where funnelling and seiche mechanisms produce amplification of tsunami amplitudes [13].

The detection and early warning of tsunamis comes in three phases. The first is the seismic alert of a significant subsea earthquake. The second phase is the observation in the open ocean of cyclical water elevation of a propagating tsunami wave. The third phase includes observations near the coast. It is in the third phase that HF radar has a significant role in providing confirmation to coastal communities of magnitude and timing of an imminent tsunami up to one hour before impact. For most meteotsunamis, the first and second phases are bypassed and warnings depend solely on observations near the coast. This raises the importance of HF radar monitoring.

The Deep-ocean Assessment and Reporting (DART) buoy system has been developed by NOAA in USA and deployed in many open ocean locations, beginning in 2000 [14, 15]. The DART buoys measure pressure in 15-second samples with a resolution of 1 mm of sea water. A cubic polynomial is fitted to the past 3 hours of observations, and if the next two 15-second samples vary from the background by >3 cm (in the North Pacific) then the system goes into rapid reporting mode, which is to send data every minute through a link to a surface buoy and thence via a GOES satellite to NOAA's Tsunami Warning Centers. The DART buoys form an excellent complement to the seismic data to confirm whether a tsunami has been generated. Computer modelling, based on seismic warnings and DART buoy observations, is used to estimate arrival times in coastal waters. The role of the HF radar is to provide information on the amplitude and confirmation of the timing of tsunamis when they arrive in coastal waters and approach the coast. Details of the tsunami up to 40–60 minutes before impact are provided by HF radars.

In deep ocean water, tsunamis propagate at speeds up to 200 ms^{-1} and on a shallow continental shelf, typically around 20 ms^{-1}.

The development of HF radar methodology for tsunami detection took a big step forward following the mega-tsunami originating in Tohuku, Japan, in 2011, which was detected on HF radars of all genres right around the rim of the Pacific Ocean. Although none of the operating radars were configured for tsunami detection, post-analysis of the data from many sites showed conclusively that this technology could be used to detect tsunamis and had the potential to contribute to the mitigation of impacts. In subsequent years, it has been mostly meteotsunamis that have enabled improvement of systems with the first automatic alert for a tsunami-like disturbance from HF radar in 2015 [16].

9.3.1 Physics of tsunamis

Tsunamis originating from underwater earthquakes or landslides are normally manifest as a pulse of elevated water in the near field that propagates out as a wave with a wavelength of the same scale as the originating pulse. If a tsunami impacts the coast in its near field (i.e., within a few wavelengths of its source) then there will usually be a single flooding surge onto the coast. When a tsunami wave propagates away from its source, it will develop into a short train of gravity waves. Tsunamis are technically shallow-water gravity waves, even in the deep oceans, because their wavelengths are much larger than the water depth.

The phase velocity of a shallow-water gravity wave is

$$v_p = \sqrt{gd} \qquad (9.2)$$

where g is gravitational acceleration and d is water depth. As a tsunami wave propagates from the deep ocean into shallower water, for example, on a continental shelf, the phase velocity is reduced, the wavelength is shortened, the amplitude increases and the maximum horizontal velocity of orbiting particles increases. This makes the tsunami easier to detect by HF ocean radar as it approaches the coast. The design requirement for an HF radar for tsunami detection is to confirm an alert as far offshore as possible to maximise the warning time before impact on the coast.

If the water depth is slowly changing as the wave propagates from depth, D, to depth, d, then the way the amplitude of the wave changes is given by [17] as

$$a(d) = a(D) \left(\frac{D}{d}\right)^{1/4} . \qquad (9.3)$$

This is a simplified theory, and Lipa *et al.* [18] have shown that it may over-estimate the increase in wave amplitude by up to 25% over a very steep gradient in the bathymetry. Here we retain the simplified theory and adopt the caveat of gentle gradients, noting that other phenomena such as refractive focusing are also being ignored.

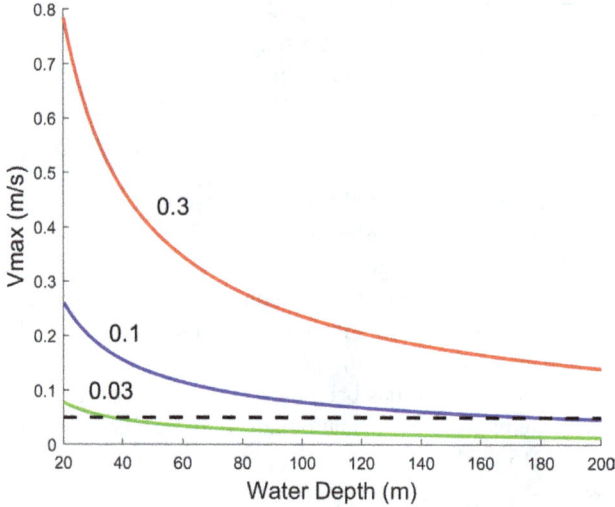

Figure 9.4   Tsunami maximum particle velocity as a function of water depth for deep ocean amplitudes of 0.03 m, 0.10 m and 0.3 m. The horizontal dashed line is a reference at $v_m = 0.05\ ms^{-1}$ for evaluating the performance of HF radars.

Combining the above equations, the maximum particle velocity depends on depth as

$$v_m\left(d\right) = v_m\left(D\right)\left(\frac{D}{d}\right)^{3/4}. \tag{9.4}$$

As a case study, note that the tsunami generated in north Sumatra in 2004 had a maximum amplitude of 0.5 m and wavelength over 200 km when observed in the southern Bay of Bengal in water about 4000 m deep by an altimeter on the Jason-1 satellite [19]. As this tsunami moved into shallower water, the amplitude increased according to (9.2) and it resulted in destructive impacts in many communities around the Indian Ocean. Reports of coastal inundation to depths up to 10 m were reported in Sri Lanka, India and Bangladesh. The changes in the maximum horizontal orbital velocity, v_m, are depicted by the red line in Figure 9.4. Guided by these data, we conclude that any tsunami with amplitude greater than 0.1 m in deep water (blue line in Figure 9.4) would be significantly hazardous. The dashed black line in Figure 9.4 shows the 0.05 ms^{-1} level of resolution that would have to be achieved by an HF ocean radar if it were to detect a threshold hazardous tsunami at a water depth of 200 m. The green line corresponds to the threshold for a DART buoy to go into rapid reporting mode. To match this, the HF radar would have to have a resolution of 0.01 ms^{-1} in a water depth of 200 m.

9.4   HF ocean radar detection of tsunamis

The feature of a tsunami that an HF ocean radar can sense is the surge velocity of surface water as the wave passes a target cell of the radar [20]. Kinsman [5] explains that the maximum horizontal velocity of orbiting particles as a shallow-water gravity wave propagates past a fixed point is given by (9.2). The threshold amplitude to trigger rapid mode in a DART buoy is 3 cm, and in deep water, a tsunami with this amplitude will have a maximum surge velocity of 0.0015 ms^{-1}. As the tsunami propagates into shallower coastal water at a depth of 200 m, the maximum surge velocity will increase to about 0.015 ms^{-1}. The challenge in using HF radars to detect tsunamis at this threshold is to measure small currents like this in sampling times of 2–4 minutes at ranges as far distant from the coast as possible in order to provide warnings as early as possible before impact on the coastline. The contrast with the normal configurations of HF ocean radars for observing coastal surface currents driven by tides and winds is demonstrated in Table 9.1.

The time resolution of 2–4 minutes is required for detecting a tsunami wave with a period of 10 minutes. But the irony is that in order to obtain a finer resolution of surface current, a longer time series is needed (9.5) and the operating range is extended by increasing the integration time for each sample. These are competing requirements, and some trade-offs and special processing are needed to configure a successful tsunami detection HF radar. This optimisation is currently a high priority area of development in HF ocean radar.

For a shallow-water shelf (<200 m) that extends beyond 100 km from the coast, HF radars are limited by the range at which they can achieve detection of maximum particle velocities of about 0.05 ms^{-1}. When the shelf is narrow (<100 km), the limit is set by a water depth of about 200 m as shown for the 0.1 m (deep water) case in Figure 9.4. If a resolution of 0.05 ms^{-1} can be achieved by the radar, then all the large tsunamis depicted in Figure 9.4 could be detected; otherwise smaller tsunamis would be detected only in water depths less than 200 m. In parallel with these conditions is the challenge of detecting velocity changes of about 0.05 ms^{-1} from a time-series window of 2–4 minutes. The threshold velocity that can be resolved by

Table 9.1   Parameters for an HF ocean radar designed to detect a typical approaching tsunami compared with the parameters used for general-purpose HF ocean radars

	Tsunami HF radar	General purpose coastal HF radar
Minimum resolved surface current	0.015 ms^{-1}	0.05 ms^{-1}
Time resolution	2–4 minutes	10–180 minutes
Spatial resolution	5 km	3–20 km
Operating range	Maximise	50–150 km

traditional Fourier analysis depends on the operating frequency and the length of the time-series window given by

$$V_{th} = \frac{\lambda}{2T} \qquad (9.5)$$

where λ is the radar wavelength and T is the length of the time-series window; but this can be considered a worst case because peak detection methods are used on the bundle of frequencies that contribute to the first-order peaks (see the last part of Chapter 1.3.1). In Figure 9.5, the reference level of 0.05 ms^{-1} is illustrated by the black dashed line and it indicates that the radar frequency should be above about 12 MHz for a 4-minute time-series window. In conflict with this is the need to detect the tsunami as early as possible (at a long range) to maximise the warning time; and HF radars achieve longer ranges when the operating frequency is low. The lowest frequency bands allocated by ITU are around 4–5 MHz, but these frequencies are poorly conditioned for the resolution of the velocity threshold.

Gurgel *et al.* [21] showed by simulation that a time-series window of about 2 minutes (512 samples) retained the fluctuation in surface current due to a tsunami with a period of about 10 minutes, while a 9-minute time-series window smoothed out the fluctuations. This confirmed the Nyquist sampling theorem for tsunami detection. Similarly, a tsunami wave with a 10-minute period in water of 20 m deep has a wavelength of about 8 km. Therefore spatial averaging in the radial direction should be limited to less than about 4 km.

The challenge for HF radars in tsunami detection is seen in the metrics in Table 9.1. The detection system should take independent time samples in less than 4 minutes, should average over not more than 4 km in the radial direction, should

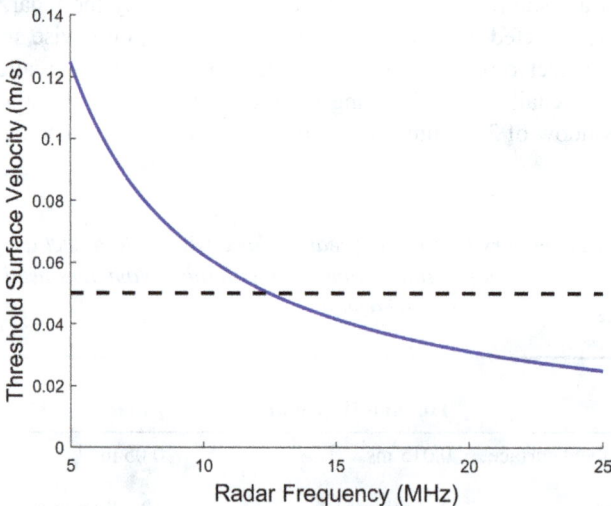

Figure 9.5   *Maximum horizontal orbital velocities smaller than 0.05 ms^{-1} require radar frequencies above about 12 MHz*

operate at about 12 MHz or more, and the operating range should be as long as possible by reducing noise in the signals. These challenges are addressed in different ways by the two main genres of HF radar and these are outlined in the next section.

9.4.1 Crossed-loop HF radar systems

Lipa *et al.* [18] have described an approach used widely by crossed-loop HF radar systems. This genre of HF radar uses directional patterns of essentially the amplitude response of orthogonally mounted loop antennas to determine the azimuth angle of arrival of echoes from the sea surface. A target range annulus is set by timing in the radar system illustrated by the red lines in Figure 9.3, and discrete Doppler shift elements (that define radial current components) have direction-of-arrival determined by the receiver antennas. The normal setting for observation of surface currents at long ranges (typically 150 km) is 20-minute time-series sampling to determine the azimuth and range of radial component measurements. In order to maximise the operating range, these radars typically operate at 4–5 MHz and the 20-minute samples are averaged over 3 hours for general circulation applications. At an operating frequency of 13 MHz, the typical operating range for a crossed-loop radar for coastal circulation is about 75 km [22].

For tsunami detection, the assumption is made that the bathymetric contours are parallel to the coastline so the tsunami will approach directly towards the coast at all points. The radial components in bands parallel to the coast (illustrated by black lines in Figure 9.3) are converted to up-slope velocities and the values in each band can be averaged to improve the capability to detect the tsunami signal. This procedure retains the spatial averaging of about the same scale as in the normal radar settings for general circulation applications.

Temporal averaging is reduced to 2–4 minutes in order to achieve the surface current resolution. The resulting surface current values are therefore more noisy than those obtained after long integration times for circulation applications [23].

The procedure proposed by [23] and [18] for tsunami detection is based on three distinguishing features that are observed in HF radar records of previous tsunami events. They are that velocities in neighbouring range bands are highly correlated; that the velocity fluctuations of the tsunami wave are clearly above the background surface current velocities; and that the velocity is consistently increasing or decreasing over three adjacent range bands. For crossed-loop systems, each site is calibrated from an extended data set for the site, where a typical range of surface currents is identified. Then the average on-shore current components, v_{os}, in each spatial band, and in each 4-minute time sample, is used to modify an empirical index of the presence of a tsunami, which are called the q-factors. The q-factor is altered according to the change of v_{os} between successive time steps and according to the correlation of v_{os} values in adjacent range bands [24].

A recent example of the application of further q-factor analysis is given by Lipa *et al.* [18] for the meteotsunami on the US East Coast on 13 June 2013 as a re-analysis of data from the crossed-loop radar at Brant Beach, operating at about 13.5 MHz. Data are shown for range bands out to 22 km where the perturbation in

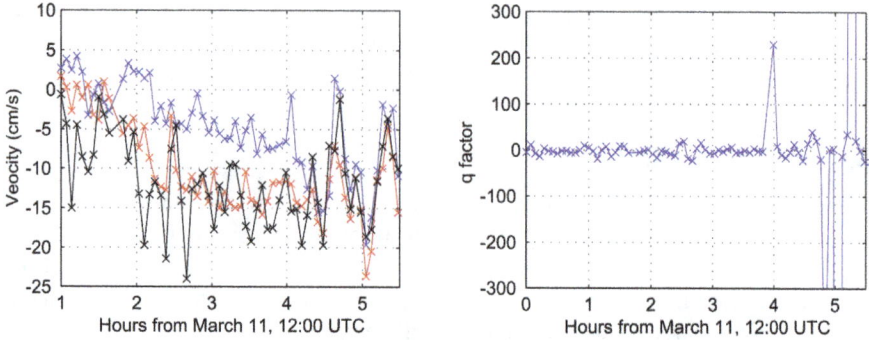

Figure 9.6   *SeaSonde time series of onshore velocity components and q-factors at Point Estero (a) blue: 8–10 km offshore; red: 10–12 km offshore and black: 12–14 km offshore; (b) q-factor for the 8–14 km interval. (From [24]).*

elevation is approximately 5 cm. At this range, the water is about 30 m deep so the range limit is set by v_{OS} resolution rather than by deep water. This meteotsunami registered a fluctuation of about 0.48 m on a tide gauge at Atlantic City about 35 km to the south. This was a small tsunami that did not have a hazardous impact, and it serves as a good demonstration of the ability of the HF radar to detect even a weak tsunami.

A second example, also taken from Lipa *et al.* [24], is the re-processed data from the 13 MHz crossed-loop radar station at Point Estero, California, on 11 March 2012 after the Tohoku earthquake off Sendai, Japan. The HF radar measured the largest fluctuation of ±13 cms^{-1} at 6–8 km offshore. Figure 9.6 shows that there was a clear tsunami signal in the 8–10 km range band at about 4 hours on the abscissa, but this first tsunami peak was missed by the q-factor metric ($q = 200$ when the threshold was $q = 500$) because there were noisy responses in the 10–12 km and 12–14 km bands. This is consistent with the depth dropping to 100 fathoms (180 m) about 9 km from the shore. The tide gauge at Port San Luis, some 40 km to the south, had a 2 m fluctuation. This was a small but hazardous tsunami, and the range for detection was limited by rapidly deepening water off the continental shelf.

Details are given by Lipa *et al.* [24] for observations of the tsunami from the Tohoku earthquake at an HF radar (code YSH2) at Yaquina Head South, Oregon, on the west coast of the USA at 12 MHz. The q-factor metric again missed the first fluctuation of the tsunami but was robust for subsequent fluctuations. The tsunami observations were clear at a range of 14 km. With the shelf width of nearly 40 km at this site, it appeared that the range limitation was due to the loss of quality signal-to-noise ratio rather than the reduced surface current fluctuations in deep water.

These results, and many others using re-analysis of raw data, demonstrate the potential of HF radar to detect an approaching tsunami in water shallower than about

200 m. The challenge is to increase the detection range for the q-factor metric and produce real-time warnings with minimal false alarms.

9.4.2 Phased-array HF radars

The phased-array genre of HF radar steers a beam by software to scan across the azimuth dimension by applying phase shifts to individual antenna elements, with effectively a time-based scan to determine ranges. For general circulation applications, a complete scan of all ranges and azimuths is carried out in typically 5–20 minutes and radial components of surface currents are transformed from the natural polar coordinates onto a rectangular grid by interpolation between the nearest four points without further averaging or smoothing. At an operating frequency of 13 MHz, the typical operating range for a beam-forming radar for coastal circulation is about 120 km [22]. As with all HF radars, the challenge is to meet the sampling window, surface current resolution and maximum range required for tsunami detection.

Gurgel *et al.* [21] suggested a probability approach to tsunami detection whereby surface currents from 132-second (512 samples) time series at each grid point were given a probability of being an anomalous fluctuation compared with background noise and other currents driven by tides and oceanographic phenomena. Ordered Statistics sorting was then applied to the probabilities at the grid points and these were plotted on a 2-D map. The maps can be examined either manually or by a pattern recognition method to determine whether an alert should be issued. The 132-second window is slid forward in steps (retaining coherence) to produce a new map every 33 seconds. The decision on issuing an alert, and avoidance of false alarms, is enhanced by comparing a new map with preceding ones.

This approach meets the short time sampling window requirement and the spatial integration to reduce noise is achieved by cluster analysis in the 2-D mapping of probabilities. The choice of 132-second time-series windows means that an operating frequency of about 23 MHz is required in order to detect a surface current threshold of 0.05 ms^{-1}. There is a capacity to increase the time-series window to about 4 minutes, in which case, the threshold operating frequency can be about 12.5 MHz and the operating range of the radar is increased. This procedure of plotting probabilities and cluster analysis increases the effective spatial averaging to more than the normal radar settings for general circulation applications.

Dzvonkovskaya [25] has further developed the probability approach and produced a robust algorithm that has been implemented in several phased-array radar sites. Signal-to-noise quality is enhanced by detrending the actual measurements of surface current at each grid point over the preceding 45–60-minute period to remove the strong background currents produced by tides, synoptic winds and large-scale circulation. The Gurgel *et al.* [21] approach is adopted by using a running time-series window over a 132-second interval with output issued every 33 seconds. The sign of the residual surface current (after detrending) at each independent grid point is a binary value that is assumed to have a Bernoulli probability distribution. Laplace's Law of Succession is applied to samples at 33-second time intervals at each grid point. The probabilities are sorted by an Ordered Statistics method and

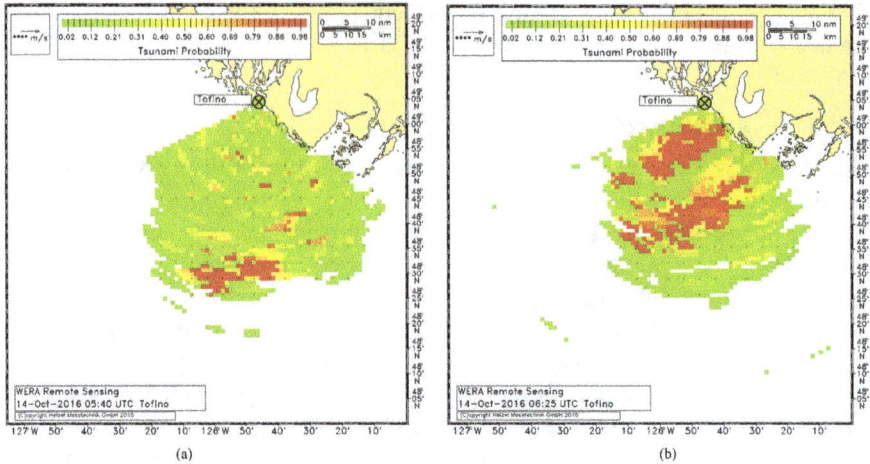

Figure 9.7   Tsunami probability plotted at grid points for a WERA phased-array HF radar at Tofino, BC, Canada, in real-time on 14 October 2016. (a) 0540h UTC a few minutes after the alert was given; (b) at 0625h UT when the second wave was confirmed. The first impact at the site of the radar was at 0636h (from [16]).

plotted on a 2-D map like that shown in Figure 9.7. The final alert decision is derived from the probability maps by a pattern recognition algorithm.

This probability analysis was developed using observations made at the Rumena, Chile, HF radar station for the Tohoku event in 2011 and has been installed on several HF radar systems. The tsunami detection algorithm described by Dzvonkovskaya [25] was installed at Tofino on Victoria Island, Canada, in 2015, operating in real time.

On 14 October 2016, the WERA HF phased-array radar at Tofino, operating at 13.5 MHz, issued an alert in real-time for a tsunami-like disturbance approaching the coast. This alert was the first time that any HF radar had detected such a disturbance in real-time and marks a significant milestone in the development of HF radar technology for tsunami detection. Figure 9.7 shows (a) the probability plots near the time of the first detection and (b) later when a second band had become clear.

The disturbance triggered a real-time warning from the current Tofino WERA system at a range of about 60 km, approximately 45 minutes before it arrived at the site of the radar station. The crest of the disturbance was oblique to the coast and arrived at a point about 20 km to the southeast of the radar about 25 minutes after the first alert. The impact point on the coast then travelled north-west and arrived at the radar some 20 minutes later. The distance to the 200 m depth contour directly off Tofino is about 50 km, but the distance to the 200 m contour in the direction of the disturbance is over 150 km. This means that the detection range was limited by the noise of the radar system rather than by the detection point being in deepwater. The tide gauge at Tofino registered a maximum elevation of 18 cm, which means that this was not a hazardous event. This real-time detection and alert for a minor event serves as a proof-of-concept for the phased-array HF radar to provide early detection and real-time alerts for hazardous events.

Grilli *et al.* [26, 27] have suggested an analysis of data from phased-array HF radars to improve tsunami detection at longer ranges by taking advantage of the progression of a tsunami as it approaches the coast. The surge in surface velocity due to the tsunami at range cell p at time t, $V_p(t)$, will, to a good approximation, be seen at a later time $t+\Delta t$ at range cells q as shown in Figure 9.2.

The delay is $\Delta t = \int_{rp}^{rq} V_g^{-1} dr$, where rp and rq are the ranges to cells p and q, and $V_g = \sqrt{gd(r)}$ is the phase speed of the tsunami wave in the water of depth $d(r)$.

The time correlation algorithm (TCA) uses a cross-correlation between the time series of the radar signal at cell p with the signal at cell q, delayed by Δt, and a second-order phase correction. The advantage of the TCA is that calculations are made in the time domain and the length of the time series is not limited to 2–4 minutes, imposed by the need for spectral resolution in the Doppler shifts. The background surface currents due to tides and other large-scale phenomena are removed using data from the radar in the recent past (e.g., 1 hour). The process can be extended to q_1, q_2, q_3, etc. as shown in Figure 9.2.

The TCA method was applied by Guérin *et al.* [28, 29] to the raw data from the WERA radar at Tofino on 14 October 2016 and showed that TCA processing could have issued a confirmed alert at 0549UT, within a few minutes of the first alert issued in real-time by the automatic probability analysis (Figure 9.7a). The TCA method has shown potential but needs further development to become a production option for phased-array HF radar systems.

9.5 Definition of a hazardous tsunami

Historically, tsunamis have been characterised by depth of inundation at points on the coast [30]. The metric of coastal inundation is useful in comparing different events at the same location, but a global standard metric for tsunami amplitude needs to be separate from local bathymetry and coastline topography, preferably in deep water away from continental shelves. For example, the tsunami from the Sumatra-Andaman earthquake in 2004 was observed in deep water by the Jason-1 altimeter [19] to have an amplitude of about 50 cm in water depth 4000 m. This single data point establishes the tsunami as severely hazardous. In this extreme case, the availability of these data in real-time would have been very helpful for any TEWS across the Indian Ocean. It is clear that the water elevation in the deep ocean is a global metric, and it is this that prompted the development of the DART buoy systems. The threshold for escalating to rapid response mode of the DART buoys in the North Pacific is set at 0.03 m. Anecdotal reports of coastal inundation from the Sumatra-Andaman Tsunami were up to 10 m. Assuming all scaling is linear, a 0.03 m tsunami in open water would lead to coastal inundation of around 0.6 m which is suggested here as a useful provisional threshold for defining a hazardous tsunami. Of course, with focusing and amplification, a 0.6 m surge could result in significant damage at some locations, and that has to be addressed at local TEWS levels.

A tsunami of deep water magnitude 0.03 m has a maximum surge velocity of 0.01 ms^{-1} at a depth of 200 m, as discussed in Section 9.1.1 and Figure 9.4. This is a challenge for HF radar technology.

9.6   Discussion and summary

9.6.1   Oblique tsunamis

Meteotsunamis are normally generated in shallow water because here the phase velocity of the tsunami wave is more likely to match that of a meteorological event. Also, because meteotsunamis tend to be generated locally, they are less likely to encounter sustained bathymetry gradients and are less likely to refract into directly cross contour directions than their seismically generated cousins. This indicates that meteotsunamis may approach the coast obliquely.

The probability analysis of phased-array radar data provides a 2-D map of the probability of tsunami conditions on a cell-by-cell basis. There is no assumption about the direction of approach of a tsunami; in fact, the only real-time alert, on Vancouver Island, was oblique to the coast. The TCA analysis method carries out analysis along radials from the radar station and has the potential to sense a tsunami's direction of approach.

The current q-factor analysis is based on the assumption that an approaching tsunami will always be up the maximum bathymetric gradient. This needs to be modified in order for this method to reliably detect meteotsunamis.

9.6.2   Maximising the alert period

All three of the analyses discussed in this chapter depend on spatial correlations of disturbances in surface currents to identify a tsunami. This is possible because the spatial scales of tsunamis are much greater than the spatial resolution of the HF radars. Ultimately, however, it is the signal-to-noise ratio for distant echoes that determines the maximum detection range and, therefore, the period between the alert and the coastal impact. The elapsed time between detection and impact is critical and Figure 9.8 shows that detection at a range of 75 km gives about a 40-minute warning for typical continental shelf waters. Detection is at the distance to the 200 m depth contour or the detection range of the radar, whichever is shorter.

One of the strengths of the probability analysis is that the 2-D plot has an independent measure of probability at each grid point. Figure 9.7 shows two of these plots and it is clear in them that there is redundancy in the spatial resolution. The processing step to combine the 2-D probability map into a single alert index is critical in extending the range at which an alert may be issued.

The TCA analysis as applied to the Tofino data achieved a similar alert distance from the radar as the probability method. This analysis is attractive because of its use of coherent raw signals in adjacent cells along a radar radial. It may become

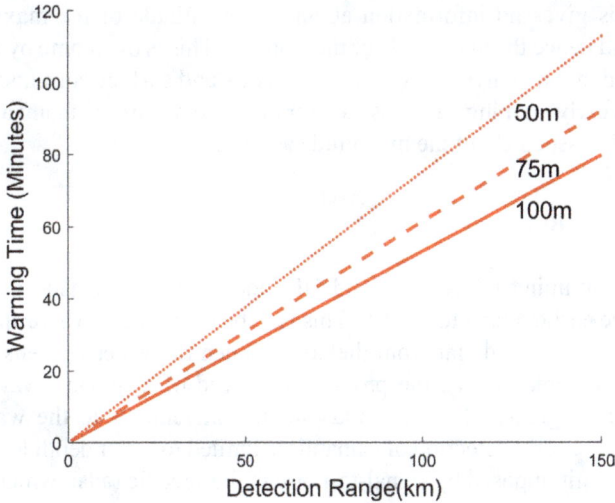

Figure 9.8   *Time between detection and impact depends on the depth shown for 100 m (solid), 75 m (dashed) and 50 m (dotted)*

more robust in the next few years but needs further development to see it installed and operating in real-time.

The q-factor analysis has been successful in validating the ability of the crossed-loop systems to observe tsunamis in simulations and reprocessed field data. The correlation of velocity disturbances over 3 adjacent 2 km bands is a significant limitation on the spatial resolution for detection near to shore (within about 20 km) but is not so serious for longer ranges. A significant improvement is needed in the q-factor method to achieve longer ranges and, therefore, longer warning times.

9.6.3   Achieving surface current resolution

The time-series sampling of tsunami signals for the probability analysis and the q-factor method has to be done in less than 4 minutes in order to satisfy the Nyquist Sampling Criterion for a tsunami with a period as short as 8 minutes.

The probability analysis method addresses this by running the radar coherently for 55 minutes (and then pausing for station ID and housekeeping) [21]. Every 33 seconds, a running 132-second time series is extracted and analysed. One advantage of this scheme is that time-series samples can be taken off for general circulation applications (10 minutes) and for significant wave height applications (20 minutes). In this way, an installation can be serving several purposes. There is a wealth of methods for surface current estimation that avoid the time-frequency trade-off of the classical spectral analysis (9.4), for example, methods based on autoregressive-moving-average modelling [31, 32] or a probabilistic approach using a parametric model [28].

The TCA method avoids limitations of surface current resolution because the analysis is carried out in the time domain. One of the consequences is that the

TCA analysis gives no information about the magnitude of the maximum surge velocity – and hence the magnitude of the tsunami. This is overcome by carrying out a frequency domain analysis to get Doppler shifts and surface currents. The case is made that an early warning can be issued for the approach of a tsunami and a follow-up alert can be issued about the magnitude at a later time [27].

9.7   Conclusion

There is a continuing effort to extend HF radar detection of tsunamis to longer ranges to give earlier alerts to TEWS. This is to be encouraged where shallow water (less than 200 m) extends far from the coast. When the water deepens (e.g., at the edge of a continental shelf), the propagation speed of a tsunami wave is greater and detection at greater distances adds smaller increments to the warning time. Realistically, the early detection of tsunamis is limited to water depth less than about 200 m or the limit imposed by signal-to-noise in the specific radar, whichever comes first. All systems need to be more robust at longer ranges.

During the five years following the first real-time automatic alert issued by the phased-array radar at Tofino, there have been several HF radar installations that are optimised for tsunami detection. To our knowledge at the time of writing, there has been only one other real-time alert for a tsunami-like disturbance, and that was issued, also by the Tofino phased-array radar, on 5 January 2020 (Manman Wang, personal communication) and research is in progress to identify its properties. The next decade will produce more validation data, but this technology is now poised to make significant contributions to tsunami mitigation in the locations where the proven operational systems are installed.

The phased-array systems have the potential to be improved by the time correlation analysis methodology.

Some further development of the q-factor method is needed to detect locally generated tsunamis that may not have experienced refraction at a shelf edge, to be propagating up the bathymetric gradients. Operational range remains a challenge for these systems.

As HF radar becomes more widely used for tsunami detection, statistical reports of false alerts and missed alerts will contribute to fine-tuning of detection algorithms.

Acknowledgements

We are most grateful for comments on the draft manuscript by Charles-Antoine Guérin and Klaus-Werner Gurgel during preparation. Commercial products SeaSonde and WERA are mentioned. For technical details, the reader is referred to the manufacturers' web sites.

References

[1] Roarty H., Cook T., Hazard L., *et al.* 'The global high frequency radar network'. *Frontiers in Marine Science.* 2019;**6**:164.

[2] Rubio A., Mader J., Corgnati L., *et al.* 'HF radar activity in European coastal seas: next steps toward a pan-European HF radar network'. *Frontiers in Marine Science.* 2017;**4**(8):00008.

[3] Parks A.B., Shay L.K., Johns W.E., Martinez-Pedraja J., Gurgel K.-W. 'HF radar observations of small-scale surface current variability in the Straits of Florida'. *Journal of Geophysical Research.* 2009;**114**(C8):C08002.

[4] Mao Y., Luick J.L. 'Circulation in the southern great barrier reef studied through an integration of multiple remote sensing and in situ measurements'. *Journal of Geophysical Research: Oceans.* 2014;**119**(3):1621–43.

[5] Kinsman B. *Wind Waves.* NJ, USA: Prentice-Hall, Englewood Cliffs; 1965.

[6] Mantovani C., Corgnati L., Horstmann J., *et al.* 'Best practices on high frequency radar deployment and operation for ocean current measurement'. *Frontiers in Marine Science.* 2020;**7**:210.

[7] Heron M.L., Atwater D.P. 'Temporal and spatial resolution of HF ocean radars'. *Ocean Science Journal.* 2013;**48**(1):99–103.

[8] Atwater D.P., Heron M.L., Heron M.L. 'High-frequency radar two-station baseline bisector comparisons of radial'. OCEANS Sydney IEEE Xplore; 2010.

[9] Titov V.V., Gonzalez F.I. 'Implementation and testing of the method of splitting tsunami (most) model'. **11**. NOAA Technical Memorandum ERL PMEL-112; 1997.

[10] Monserrat S., Vilibić I., Rabinovich A.B. 'Meteotsunamis: atmospherically induced destructive ocean waves in the tsunami frequency band'. *Natural Hazards and Earth System Sciences.* 2006;**6**:1035–51.

[11] Proudman J. 'The effects on the sea of changes in atmospheric pressure'. *Geophysical Journal International.* 1929;**2**:197–209.

[12] Rabinovich A.B., Monserrat S. 'Generation of meteorological tsunamis (large amplitude seiches) near the Balearic and Kuril islands'. *Natural Hazards.* 1998;**18**(1):27–55.

[13] Rabinovich A.B., Monserrat S. 'Meteorological tsunamis near the balearic and kuril islands: descriptive and statistical analysis'. *Natural Hazards.* 1996;**13**(1):55–90.

[14] Milburn H.B., Nakamura A.I., Nakamura A.I., Gonzalez F.I., Gonzalez F.I. Real-time tsunami reporting from the deep ocean. Proceedings of the MTS/IEEE Oceans Conference; 1991.

[15] Meinig C., Stalin S.E., Nakamura A.I., González F., Milburn H.G. 'Technology developments in real-time tsunami measuring, monitoring and forecasting'. Oceans 2005 MTS/IEEE; Washington, September; 2005. pp. 19–23.

[16] Dzvonkovskaya A., Petersen L., Insua T.L. 'Real-time capability of meteotsunami detection by WERA ocean radar system'. *2017 18th International Radar Symposium (IRS)*; Prague; 2017. pp. 1–10.

[17] Green G. 'On the motion of waves in a canal of variable depth and width'. *Transactions of the Cambridge Philosphical Society*. 1838;**6**:457–62.

[18] Lipa B., Barrick D., Isaacson J. 'Coastal Tsunami Warning with Deployed HF Radar Systems'. *Tsunami*. in Mohammad Moktari (Ed.), InTech; 2016. Available from http://www.intechopen.com/books/tsunami/coastal-tsunami-warning-with-deployed-hf-radar-systems.

[19] Gower J. 'The 26 December 2004 tsunami measured by satellite altimetry'. *International Journal of Remote Sensing*. 2007;**28**(13–14):2897–913.

[20] Barrick D.E. 'A coastal radar system for tsunami warning'. *Remote Sensing of Environment*. 1979;**8**(4):353–8.

[21] Gurgel K.-W., Dzvonkovskaya A., Pohlmann T., Schlick T., Gill E.W. 'Simulation and detection of tsunami signatures in ocean surface currents measured by HF radar'. *Ocean Dynamics*. 2011;**61**(10):1495–507.

[22] Heron M., Dzvonkovskaya A., Helzel T. 'HF radar optimised for tsunami monitoring, OCEANS 2015 Genova'. IEEE Xplore; 2015.

[23] Lipa B., Parikh H., Barrick D., Roarty H., Glenn S. 'High-frequency radar observations of the June 2013 US East Coast meteotsunami'. *Natural Hazards*. 2014;**74**(1):109–22.

[24] Lipa B., Isaacson J., Nyden B., Barrick D. 'Tsunami arrival detection with high frequency (HF) radar'. *Remote Sensing*. 2012;**4**(10):1448–61.

[25] Dzvonkovskaya A. 'HF surface wave radar for tsunami alerting: from system concept and simulations to integration into early warning systems'. *IEEE Aerospace and Electronic Systems Magazine*. 2018;**33**(3):48–58.

[26] Grilli S.T., Grosdidier S., Guérin C.-A. 'Tsunami detection by high-frequency radar beyond the continental shelf: 1. algorithms and validation on idealized case studies'. *Pure and applied geophysics*. 2015;**173**:3895–934.

[27] Grilli S.T., Guérin C.-A., Shelby M., *et al.* 'Tsunami detection by high frequency radar beyond the continental shelf: II. extension of time correlation algorithm and validation on realistic case studies'. *Pure and Applied Geophysics*. 2017;**174**(8):3003–28.

[28] Guérin C.-A., Grilli S.T. 'A probabilistic method for the estimation of ocean surface currents from short time series of HF radar data'. *Ocean Modelling*. 2018;**121**(1):105–16.

[29] Guérin C.-A., Grilli S.T., Moran P., Grilli A.R., Insua T.L. 'Tsunami detection by high-frequency radar in British Columbia: performance assessment of the time-correlation algorithm for synthetic and real events'. *Ocean Dynamics*. 2018;**68**(4-5):423–38.

[30] Tinti S., Maramai A., Graziani L. 'A new version of the european tsunami catalogue: updating and revision'. *Natural Hazards and Earth System Sciences*. 2001;**1**(4):255–62.

[31] Khan R.H. 'Ocean-clutter model for high-frequency radar'. *IEEE Journal of Oceanic Engineering*. 1991;**16**(2):181–8.

[32]   Martin R.J., Kearney M.J. 'Remote sea current sensing using HF radar: an autoregressive approach'. *IEEE Journal of Oceanic Engineering*. 1997;**22**(1):151–5.

Chapter 10

High-frequency surface wave radar for target detection

Anthony Miles Ponsford[1] and Peter Moo[2]

10.1 Introduction to high-frequency surface wave radar basics

In today's resource-driven economy, maritime nations are claiming economic borders that extend out to 200 nautical mile (nm) exclusive economic zone (EEZ). However, claiming this large economic area places responsibilities on the parent nation to exercise jurisdiction through surveillance and enforcement.

The requirement to see beyond the horizon has been a long-time goal of maritime security forces. Today, this is largely dependent on cooperative vessels voluntarily communicating their intentions to local shore-side authorities as well as vessel sightings reported by patrollers. Recent advances in radar technology provide maritime nations with the ability to provide more systematic surveillance of both cooperative and non-cooperative vessels.

This chapter presents a high-level view of land-based high-frequency surface wave radar (HFSWR) used to provide persistent, active surveillance of surface vessels throughout the EEZ.

HF radar systems have been in operation for many decades and, as illustrated in Figure 10.1, can be divided into those radars that operate using the skywave mode of propagation and those that use the surface wave mode of propagation. Both systems share common propagation modes and issues with their communication counterparts. Of note is that system performance is ultimately limited by external noise and that at frequencies below about 10 MHz this external noise increases during the night-time resulting in degradation in system performance.

Skywave radars provide coverage over vast areas of the earth and operate at frequencies determined by the prevailing ionospheric conditions and the range to the zone of interest. Thus, the entire HF band, 3 to 30 MHz, may be exploited by a single radar at different times. In general, skywave radars are designed for detection and tracking of air-breathing targets, such as aircraft, over thousands of kilometres but with a minimum range

[1]Maerospace Co: Radar Advisory Board Member and Consultant, Canada
[2]Leader of the Wide Area Surveillance Radar Group in the Radar Sensing and Exploitation Section, Defence R&D, Canada

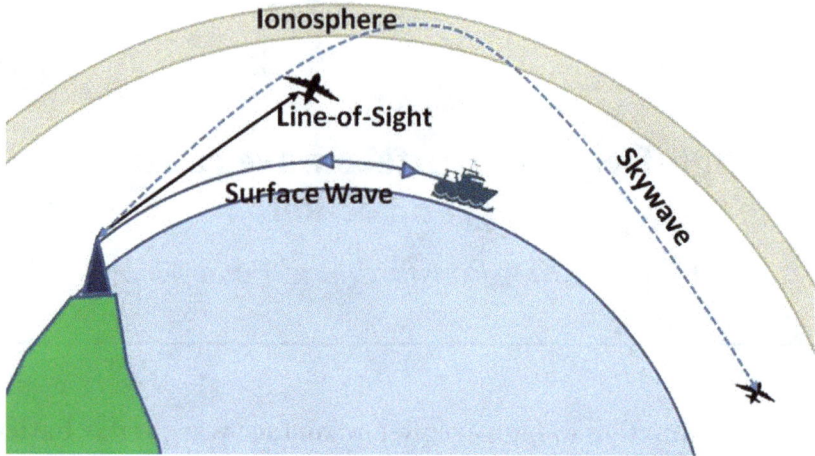

Figure 10.1   HF radar modes of operation: skywave, surface wave and line-of-sight

in the order of 1 000 km. These radars also have the capability of detecting larger ships over similar ranges.

HFSWR systems fall into the following broad categories:

• Systems designed for coastal monitoring of surface currents and sea state conditions. These systems are characterised as radiating low-power signals, typically use compact low-gain antenna arrays and operate in the 5 to 27 MHz range. For coastal monitoring of surface currents and sea state conditions, system availability is not a critical factor. These systems can be configured to track larger vessels in relatively benign environments [1–3].

• Systems designed specifically for surveillance of ships and aircraft operating in maritime regions. These systems are characterised as radiating high-power signals and typically operate within the 3 to 15 MHz portion of the HF band. To detect smaller vessels, the radar employs large directive receive arrays and are required to maintain 24/7/365 availability. Consequently, these radars are significantly more complex than their ocean-remote sensing counterparts and incorporate advanced signal processing techniques to enable reliable operation in complex HF environments [4–7]. These systems can be sub-divided into three operating bands.

  • Low-band systems operating in the 3 to 5 MHz frequency band designed to track commercial ocean-going vessels throughout the EEZ.
  • Mid-band systems operate in the 5 to 10 MHz frequency band designed to detect smaller targets but to a lesser range. These systems are typically designed to monitor the mid-shore fisher-

Low-band *(day/night)*	X	X	125 / 110 nm	200 / 115 nm	200 / 140 nm
Mid-band *(day/night)*	75 / 65 nm	85 / 75 nm	100 / 95 nm	100 / 100 nm	100 / 100 nm
Vessel Type	Nearshore Commercial Fishing Vessel (16m)	Fast Patrol Vessel (22m)	Mid-shore Trawler (22m)	Offshore Trawler (30m)	Local Coastal (60m)

Where Low-band radar refers to operation between 3 and 5 MHz range and Mid-band
refers to operation in the 5 to 8 MHz range

Figure 10.2   *Detection range as a function of the band of operation and vessel*
type.

ies as well as protection of critical infrastructure by monitoring
approaching traffic.
*   High-band systems operate in the 10 to 15 MHz range and have
    the capability to track small, high-speed targets at ranges beyond
    the visible horizon.

A comparison of the typical performance of the low- and mid-band systems for
tracking of ships is presented in Figure 10.2.

10.2   HFSWR system configurations

As shown in Figure 10.3, HFSWR can be configured to operate in either monostatic
or bistatic modes.

10.2.1   Bistatic configuration

In a bistatic configuration, transmit and receive sites are separated by a considerable
distance. The separation enables these radars to operate using a continuous wave
(CW) without causing the receiver saturation.

CW operation has the benefit of requiring less transmit power when compared
to non-CW waveforms, but these CW systems require additional design considerations to allow detection of zero Doppler targets in the presence of the direct radiated

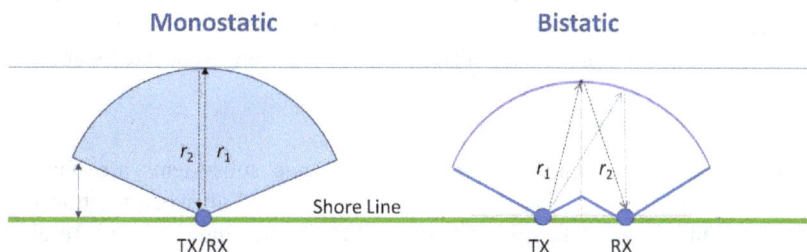

Figure 10.3   *Radar configurations: monostatic and bistatic*

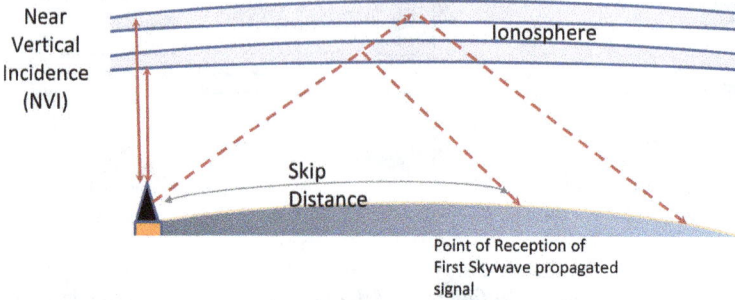

Figure 10.4   Geometry of skywave propagated signals

signal. One commonly used technique is to radiate an interrupted CW signal [8]. Another requirement is that the two sites must be synchronised using a highly stable external source.

The range of a bistatic radar is typically given in terms of distance from the receive site. The transmitter-to-target 'r_1' and target-to-receiver 'r_2' paths have, in general, different lengths. If the direct distance between transmit and receive sites is L and receive look direction is θ_R, then the target range from the receive site is given by

$$r_2 = L\sin\theta_R \pm \sqrt{r_1^2 - L^2\cos^2\theta_R} \tag{10.1}$$

It can be noted that the bistatic range has a greater number of variables than monostatic radar. Since each variable has an associated measurement error, it can be expected that the derived range will have a greater error relative to its monostatic equivalent. A similar argument holds true for Doppler. This hypothesis is supported by measurements comparing bistatic and monostatic radar configurations presented elsewhere, where they conclude that the range estimation accuracy and the resulting tracking accuracy of bistatic HFSWR is lower than those of an equivalent monostatic radar [8].

For long-range vessel detection, the bistatic configuration has a potential advantage related to ionospheric clutter. Figure 10.4 presents the geometry of skywave propagated signals. It can be observed that the separation of the transmit site from the receive site can minimise the impact of ionospheric clutter. This is a consequence of the receive array being located beyond the range of near-vertical reflections from the ionosphere and within the skip distance of the skywave propagated signal.

10.2.2   Monostatic

In a monostatic configuration, transmit and receive subsystems are physically located on the same sites. The receiver must be switched off when the transmitter is radiating, and likewise the transmitter must be off when the receiver is receiving. This requires that the radiated signal is interrupted resulting in either a pulse or interrupted CW transmissions. The ratio between the on and off times is known as the duty cycle, and the average radiated power is the peak radiated power multiplied by the duty cycle.

Range coverage of a monostatic radar is centred on the radar site with the range of a target being equal to half of the sum of the transmit path distance plus the receive path distance. For a linear array, azimuth coverage is typically limited to ±60 degrees centred on boresight.

Monostatic systems typically use an external source to provide a highly stable reference. However, the loss or disruption of the reference for a period of time will not significantly impact the radar performance as both transmit and receive subsystem remain locked.

Performance of bistatic and monostatic HFSWR are similar and selection of the mode of operation is often related to other practical factors such as real estate or the desire to maintain and provide security for a single site. The monostatic configuration is the assumed configuration in the remainder of this chapter.

10.3   HFSWR for target detection

HF radars are coherent devices that discriminate between echoes originating from targets and the usually more dominant sea echo, on the basis of their differing Doppler shifts. These echoes will experience a Doppler shift 'f_{dt}' that is proportional to the vessel's radial velocity 'V' with respect to the radar look direction:

$$f_{dt} = \frac{2Vf_c}{C} \tag{10.2}$$

A typical HFSWR site is illustrated in Figure 10.5. It can be observed that the site is located along a coastline and consists of a transmit antenna and a co-located linear array of receive elements. A typical radar site requires about 10 wavelengths of ocean

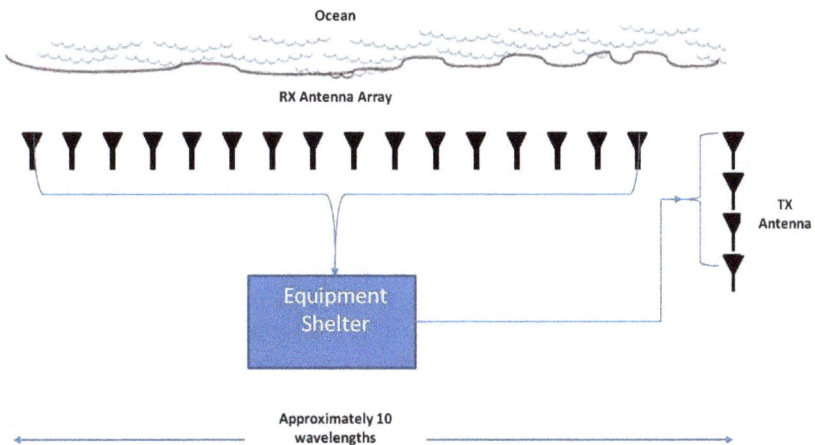

Figure 10.5   Typical HFSWR site layout for long-range surveillance of ships

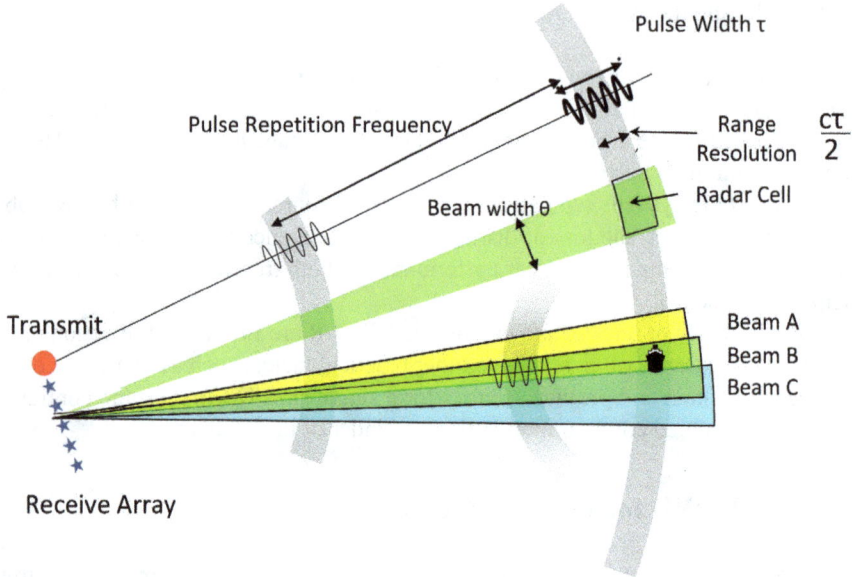

Figure 10.6   HFSWR general theory of operation

frontage, and for a low-band radar system this equates to approximately 700 m. The transmit array is designed to provide high gain and directional coverage over an arc of approximately 120 degrees. The receive array element typically consists of pairs of antennas phased to attenuate signals arriving from behind the array. Since the system is almost always externally noise-limited, low-gain antennas can be utilised and the key performance factor of the array is its directive gain. A detailed description of the physical implementation of a radar site is presented elsewhere [6].

Figure 10.6 presents the general theory of operation. The transmit antenna radiates a train of electromagnetic (EM) pulses that illuminate the desired surveillance area. Objects in the surveillance area reflect the EM pulses back towards the receiving array. The receive array is typically a linear array whose length is a compromise between the desire to maximise directive gain and azimuth resolution while minimising real estate requirements. The coverage area is symmetrical around the array boresight where the boresight is perpendicular to the axis of the array. The signals received on each antenna are digitally processed to enhance the signal-to-unwanted signal ratio, where the term unwanted signal includes signals such as sea clutter, external noise, ionospheric reflection as well as interference from other users of the band.

The basic signal processing architecture is illustrated in Figure 10.7. Each channel of time accumulated, range-gated data are processed using a Fast Fourier Transform (FFT) to convert from the time-domain to the Doppler-frequency domain. The frequency resolution of the Doppler spectrum is determined by the reciprocal of the coherent integration time (CIT). This is also referred to as the coherent integration period (CIP) or dwell time.

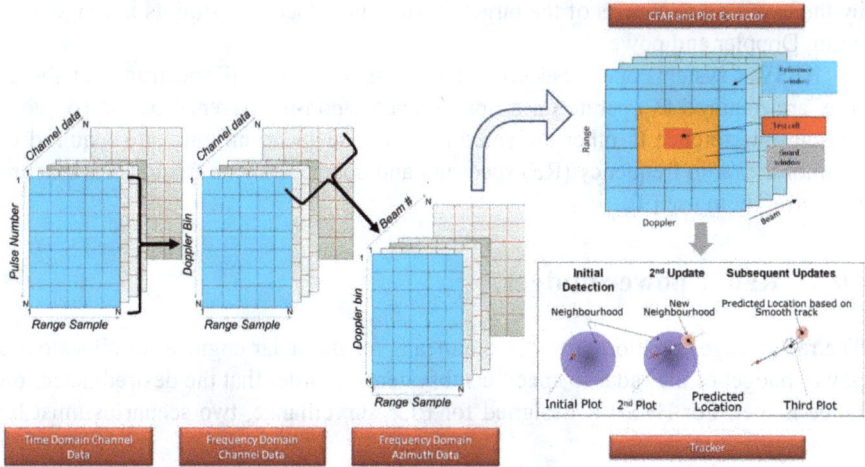

Figure 10.7   Basic data transformation process for HFSWR

The CIT determines the number of pulses that need to be collected for spectral analysis. Both the frequency resolution and coherent gain are proportional to the CIT: the longer the CIT, the better the Doppler resolution and the greater the gain. However, an upper limit is set by the requirement for the target to remain within the radar resolution cell during the processing period. Consequently, these radars generally operate with multiple, parallel, processing channels that are optimised for a range of target velocities.

The maximum velocity of the target and the radar carrier frequency determine the extent of the required Doppler bandwidth. For surface targets, this is signif-icantly less than the maximum unambiguous Doppler determined by the system pulse repetition frequency (PRF). Under-sampling or pulse rate decimation may be used to reduce the number of pulses to be processed while still maintaining the same CIT without loss of either Doppler resolution or processing gain.

Directional information is obtained by applying an FFT, or similar algorithm, across the range-gated, channel data at each Doppler bin. In addition to this basic processing, other algorithms are applied to remove unwanted signals.

The presence of a target is determined by using a constant false alarm rate (CFAR) algorithm where the presence of a vessel is determined by comparing the power in a given radar cell to that of its neighbours. The radar cell is bounded in azimuth, range and Doppler. A typical HFSWR may have 10 million cells per CIT. For optimum performance, a cell should contain a vessel return or white external noise. This background noise floor can also be raised randomly by the presence of unwanted signals such as impulsive noise, interference and clutter.

Resultant detections are then forwarded to the tracker that forms a track by associating consecutive detections that lie within the neighbourhood of the predicted future position of the previous detection or track. The neighbourhood is determined by the expected dynamics of the target motion that places constraints in range, azimuth, Doppler and power.

HFSWR systems operate as secondary users within the HF spectrum and, therefore, are required to operate on a non-allocated, non-interference basis. To avoid causing interference to other allocated users of the spectrum, they are required to monitor the radio frequency (RF) spectrum and operate only on frequencies that are not currently in use [9].

10.4   Radar power budget

The radar range equation provides the means for the radar engineer to allocate the power budget of the radar to specific subsystems in order that the desired detection range is met. For HFSWR designed for EEZ surveillance, two scenarios must be considered:

- Noise-limited environment: This is generally the case for vessel detection at long-range or high-speed targets at all ranges.
- Ocean clutter-limited environment: This is generally the case for vessel detection at near- and mid-ranges.

10.4.1   Radar range equation for a noise-limited environment

In a noise-limited scenario, the signal-to-noise ratio (SNR) of the return from the vessel after signal processing is given by [10]:

$$\frac{P_R}{N} = \frac{P_T G_T G_R \lambda^2 \tau t_c \sigma_t P_F}{2\left(4\pi\right)^3 d^4 \mathrm{pri}\, T_o K F_A L_{\mathrm{sys}}} W^4 \tag{10.3}$$

where
    P_R is the received power
    N is the noise
    P_T is the peak transmitted power
    G_T is the transmit antenna gain
    G_R is the receive antenna directive gain
    λ is the radar wavelength
    τ is the pulse duration t_c is the coherent integration interval
    σ_t is the aspect angle-averaged free space radar cross section (RCS) of the target
    P_F is the proximity factor and is the loss applied to the free space RCS to account for the image and enters twice since the vessel is both a receiver and a radiator.
    d is the vessel's range pri is the pulse repetition interval
    T_o is the absolute temperature
    K is Boltzmann's constant
    F_A is the external noise factor
    L_{sys} is the system loss
    W is the Norton surface wave attenuation factor at range "d".

The proximity factor compensates for the vessel not being in free space. The radar range equation already assumes that the target is located above perfect ground and includes the effect of the image. However, since a ship is located on the conducting plane, an additional factor must be added to accommodate for the coupling between the target and its image. The value is entered twice since the vessel acts as both a receiver and a radiator.

W^2 is the one-way transmission loss given by

$$W^2 = \frac{4\pi^2 d^2}{\lambda^2 L_{sw}}$$ (10.4)

where L_{sw} is the one-way surface wave propagation loss, and

$$P_A = P_T \frac{\tau}{\text{pri}}$$ (10.5)

where P_A is the average transmit power.

The radar range equation for noise-limited detection can, therefore, be written as

$$\frac{P_R}{N} = \frac{P_A G_T G_R t_c \sigma_t P_F \pi}{4\lambda^2 T_o K F_A L_{sys} L_{SW}}.$$ (10.6)

10.4.2   Radar range equation for an ocean clutter-limited environment

Detection of surface vessels at near- and mid-ranges is typically limited by the ocean clutter continuum. The clutter continuum at any given Doppler is dependent on the radar frequency as well as wind speed (sea state) and wind direction relative to the radar look direction.

For a clutter-limited detection scenario, the radar range equation simplifies to

$$\frac{P_{RT}}{P_{RC}}(\omega) = \frac{2\sigma_T}{dc\tau\theta\sigma_o}(\omega)$$ (10.7)

where
   ω is the Doppler frequency (radians/s)
   σ_o is the RCS of the ocean (free space) corresponding to the Doppler frequency
   c is the speed of light.

Therefore, to improve the detection range in a clutter-limited case, it is necessary to reduce the size of the resolution cell by either decreasing the receive antenna beamwidth, increasing the array aperture, decreasing the range cell or increasing the radar bandwidth.

From (10.7), it can be observed that for a vessel to be detected in a clutter-limited situation its RCS must be greater than that of the ocean patch at that range and Doppler.

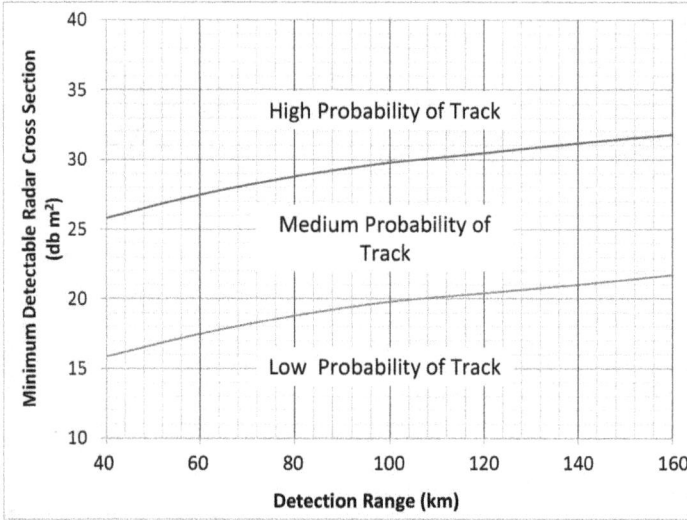

Figure 10.8   *Minimum detectable RCS as a function of range for a 16-element receive array. Vessels with an RCS above the upper curve have a high probability of detection, while vessels with an RCS below the lower curve have a low probability of detection.*

10.5   Ocean clutter

The ocean clutter spectrum for an HFSWR system is discussed in detail elsewhere [11]. In summary, in absolute terms, the magnitude of clutter continuum at any given Doppler will depend on the nature of the surface roughness generally dictated by wind velocity, the so-called modified surface impedance associated with the surface roughness and various operating parameters including radar frequency. Detection of smaller, low-speed targets will be heavily influenced by the continuum level, as observed between the Bragg lines, and hence wind speed and direction. Therefore, in the specification of the radar performance for low-speed vessels, a Doppler-averaged value of the clutter continuum, as observed between the Bragg lines, is generally quoted. This value assumes that the radar is operating above the critical frequency such that the sea is fully developed and in equilibrium. Under such conditions, an increase in wind speed results in an increase in the extent of the continuum 3-db bandwidth; however, the average magnitude of this continuum (as measured at the 3-dB bandwidth) remains approximately constant [6].

Since the average RCS of the continuum is known, the minimum detectable RCS of a vessel can be determined. Figure 10.8 plots the Doppler-averaged RCS of the ocean clutter (lower curve) for a fully developed sea as a function of range for a standard radar installation. Since the radar patch area increases with range so does its RCS.

For a vessel to be detected, it must have an RCS that is greater than that of the ocean. The figure includes a second curve that is 10 dB greater than the Doppler-averaged RCS of the ocean clutter. Vessels that have an RCS greater than this curve have a high probability-of-track (P_T), vessels whose RCS falls between the two curves have a medium P_T and those with RCS less than the lower curve have a low P_T.

10.6   Surface wave propagation

Surface wave propagation, also referred to as ground wave propagation, has been discussed in detail elsewhere [12, 13]. Surface wave signals propagate by diffraction to ranges beyond the horizon. They propagate most effectively at the lower end of the band (3–5 MHz) and over regions of the earth with high conductivity, such as saline oceans. Ranges can be in the order of 300 to 500 km over the ocean or as low as a few tens of kilometres over non-conductive terrain.

- Surface waves require a conducting surface to propagate efficiently
- They attenuate directly as functions of distance (range)
- They propagate efficiently in vertical polarisation only
- Their attenuation losses increase with increasing frequency as does the rate of increase
- Their additional losses due to sea state increase with frequency as does the rate of increase.

10.7   Maximum detection range

The maximum detection range of HFSWR depends on many factors. These include the transmit power, radar frequency, RCS of the target, target range, background noise, interference and clutter levels and ionospheric conditions. Of these, only the transmit power and transmit frequency are under the control of the radar. Both the noise and interference levels are dependent on the geographic location of the radar site, time of day, season and also the level of sunspot activity. Clutter levels, on the other hand, are determined by either ionospheric conditions or sea state.

In summary,

- The radar signal will experience a greater rate of attenuation as the radar frequency is increased or, for a given frequency, the sea state increases
- Detection of smaller vessels (typically less than 1 000 tons) is typically limited by the presence of ocean clutter. In this case, the detection range can only be increased by reducing the area of the radar clutter cell
- Detection of larger vessels (typically greater than 1 000 tons) is typically limited by external noise. In this case, the detection range can also be extended by increasing the effective radiated power.

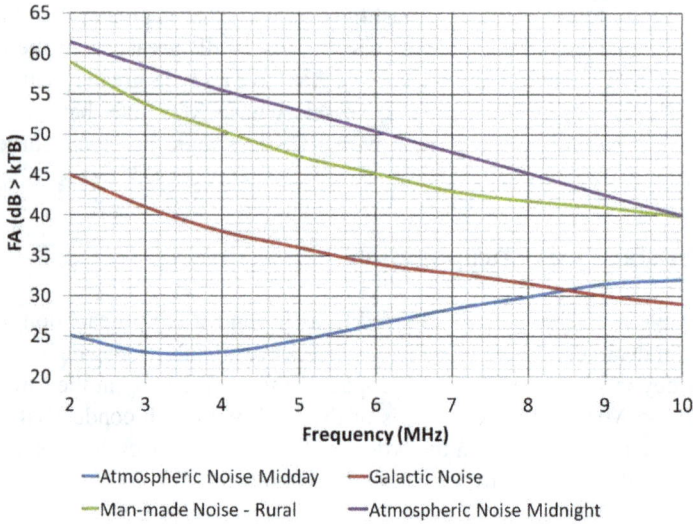

*Figure 10.9   Relationships between noise sources as a function of frequency at a
European location. Derived from [14].*

10.8   External noise

HFSWR sensitivity is limited by external noise, which ultimately limits the maximum achievable detection range. External noise consists of both the irreducible natural noise and the incidental man-made noise. The natural noise is the sum of the atmospheric noise (which is frequency-, location-, season- and time of day-dependent) and galactic noise (which is frequency-dependent). Typical relationships between the various noise sources, as a function of frequency, for an European location are presented in Figure 10.9 [14].

From Figure 10.9, it can be observed that man-made noise typically dominates during daylight hours and atmospheric noise generally dominates during darkness. Both atmospheric and man-made noises decrease as function of frequencies, and above approximately 10 MHz, there is little day/night dependency. It can also be observed that at frequencies below 10 MHz, the contribution of galactic noise can be ignored.

10.8.1   Man-made noise

The level of the man-made noise is determined by frequency and proximity to local noise-generating sources, such as:

- electrical machinery
- spark ignition systems
- leakage across insulators on high-voltage power lines
- switching transients on AC power lines

Table 10.1   CCIR derived values for constants c *and* d

Environmental category	c	d
Business stores, offices, industrial parks, etc.	76.8	27.7
Residential > 5 dwellings/Hectare	72.5	27.7
Rural < 0.5 dwellings/Hectare	67.2	27.7
Quite rural	53.6	28.6

- spurious radiation from local broadcast transmitters
- local HF radio emissions.

Man-made noise is dependent on the level of human activity and can be approximated from [15]:

$$N_m = c - d \log 10 f - 2 \qquad (10.8)$$

where N_m = man-made noise power in decibels below 1 W in a 1 Hz bandwidth, f = frequency in MHz, and c, d are constants derived from measurements

The constants c and d have been derived by the Comité Consultatif International des Radio Communications (CCIR) and are presented in Table 10.1.

It is preferable to select an HFSWR site where the level of man-made noise is low, and it is general to initially specify HF performance relative to a rural site.

10.8.2   Atmospheric noise level

Atmospheric noise is radio noise caused by natural atmospheric processes, primarily lightning discharges in thunderstorms. Figure 10.10 presents the predicted atmospheric noise level at 3.5 MHz for a location in Europe. It can be observed that the night-time atmospheric noise is considerably greater than the daytime atmospheric noise (in this example, ~10 dB), where the increase is due to impulsive noise arising as the result of lightning and is greatest at the equator and minimal at the poles. It can also be noted that there is a fluctuation in noise between the seasons.

10.8.3   Galactic noise

Galactic noise originates outside the earth's atmosphere and has similar characteristics as those of thermal noise. Galactic noise can be a contributing noise source at frequencies above 10 MHz.

10.9   Interference and clutter

Operation within the congested HF band requires that the radar signal processor must adequately suppress or otherwise mitigate against a wide variety of unwanted signals that include

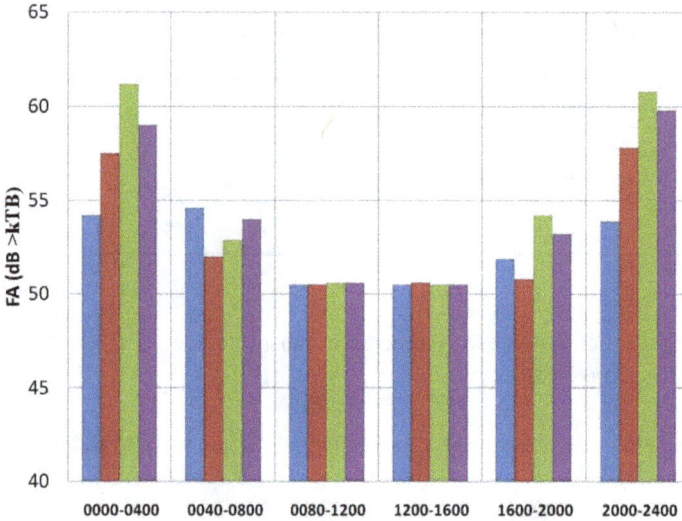

Figure 10.10   Predicted atmospheric noise level at 3.5 MHz for a location in Europe as a function of time of day and season

- External interference
- nwanted signals: interference from other users
- mpulsive noise
- elf interference (clutter)
- mbiguities: unwanted replicas of the wanted signal
- backscatter from the ocean: ocean clutter
- backscatter from the ionosphere: ionospheric clutter.

10.9.1   External interference

External interference is the result of unwanted, in-band signals that are independent of the radar operation, for example, co-channel interference and impulsive noise. External interference is also referred to as additive noise. Co-channel interference stems from local or faraway HF broadcasts with a frequency falling within the radar passband.

Most radio traffic is in the form of shortwave radio voice communications. A voice channel has a bandwidth of typically 3 kHz. From the viewpoint of the HFSWR, with a bandwidth of between 20 and 100 kHz, a 3 kHz voice channel can be considered as narrowband interference.

Typically, the HFSWR external interference environment changes from day to night. During the day, ionospheric propagation between 3 and 7 MHz is extremely lossy due to D-layer absorption. At night, the D-layer disappears, leaving the way open for shortwave broadcasts to be received from around the world.

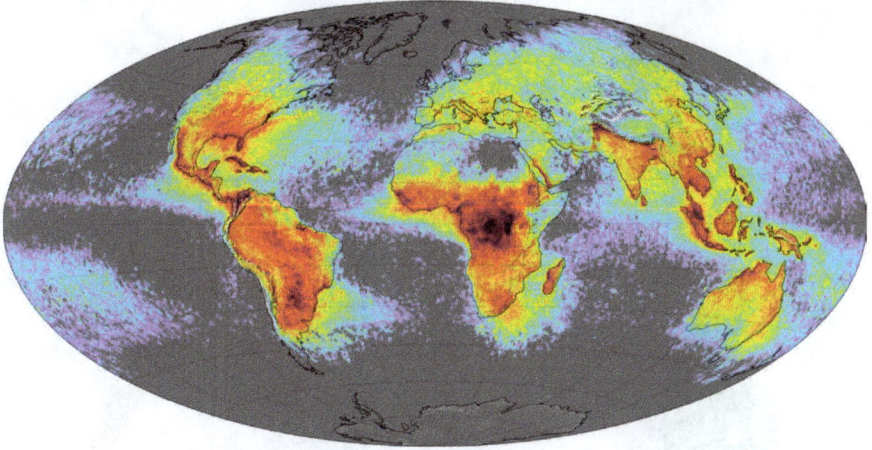

Figure 10.11   *Map showing the average yearly counts of lightning flashes per square kilometre based on data collected by NASA satellites between 1995 and 2002. Places where less than one flash occurred (on average) each year are grey or light purple. The places with the largest number of lightning strikes are deep red (credit NASA MSFC).*

One of the common contributors to external interference is impulsive noise caused by regional lightning. As illustrated in Figure 10.11, lightning occurs more frequently over land than over oceans because daily sunshine heats up the land surface faster than the ocean. The heated surface heats the air, and more hot air leads to stronger convection, thunderstorms and lightning. Lightning also occurs more frequently near the equator than near the poles. This pattern is also due to differences in heating. The equator is warmer than the poles, and convection, thunderstorms and lightning are widespread across the tropics every day [16].

10.9.2   Self interference (clutter)

Self interference or clutter is a collective term for unwanted signals that are a consequence of the radar's operation. Clutter is characterised as unwanted echoes that originate from a collection of spatially distributed scatterers and as such do not have the characteristic thumb tack ambiguity function of point targets. Since clutter is dependent on radar operation, it behaves in a multiplicative manner. Dominant forms of clutter include

- ionospheric clutter
- range wrap clutter
- meteor clutter
- ocean clutter.

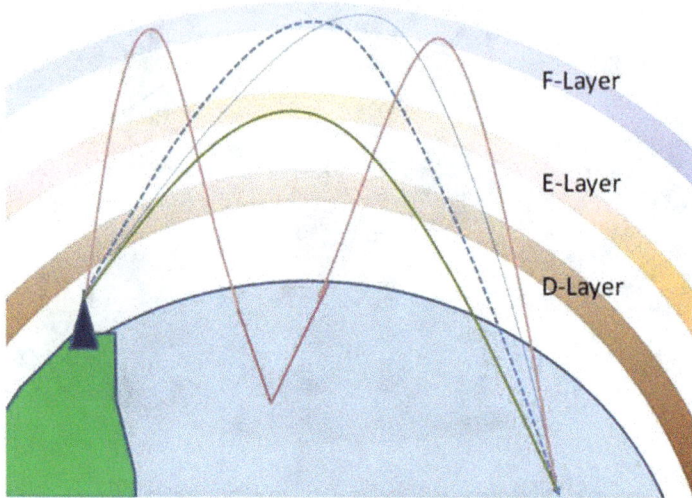

Figure 10.12   Potential skywave propagation paths

Detection that is clutter-limited can be improved by reducing the radar resolution cell.

10.9.3   Ionospheric clutter

Not all the energy emitted by the HFSWR propagates along the surface as a surface wave. Some of this energy is directed upward and may reflect from the ionosphere resulting in skywave propagation.

As illustrated in Figure 10.12, multiple skywave propagation modes may exist simultaneously, and the received signal is a vector sum of EM waves arriving from different angles, with different time delays, amplitudes and polarisation. These may cause severe distortion, deep and rapid fading (amplitude variation) as well as Doppler spread that result in unwanted signals being returned to the radar [17].

At low frequencies, ionospheric propagation is primarily a night-time phenomenon and is responsible for degradation in range performance during the hours of darkness. Night is defined as the period during which the sun is below the horizon and its light is not refracted over the horizon. The period between day and night is defined as the twilight or grey zone.

The day/night transition (and vice versa) happens quite abruptly at the equatorial zones, and much more slowly at the polar zones. The grey zone (twilight zone) is the area around the earth that separates day from night and is characterised by producing very unique propagation conditions at HF. In specifying the performance of HFSWR systems, it is also important to know how long these special propagation conditions exist and what is the effect on radar performance. The rate of change from darkness to daylight and vice versa depends upon the rate of sunrise or sunset. Two factors determine this rate: the season (the sun rises faster in summer than in winter) and the latitude of the location (the sun rises very high near the equator and much lower towards the poles).

At dawn, the E-layer that exists at approximately 100–130 km above the earth is beginning to form. At this time, the D-layer, which absorbs RF energy around 3 to 10 MHz range, has not yet been sufficiently established to prevent the near-vertical propagation of the radar pulse from penetrating the layer and being partially reflected from the evolving E-layer. Since the E-layer is forming, reflections will occur over a spread of ranges, resulting in a general increase in the 'background noise-ionospheric clutter' level of the radar, starting at a range of approximately 100 km and spreading outward. This phenomenon will continue until the D-layer establishes itself and absorbs the radar pulse as it enters the ionosphere.

10.9.4 Ionospheric clutter scattering modes

Ionospheric clutter is the result of the radar signal interacting with the ionosphere. Ionospheric clutter occurs in three dominant scattering modes that are defined here by the number of scattering points.

- First-order ionospheric clutter is a single scattering mechanism and is the result of direct overhead reflections from the ionosphere of the transmitted signal back to the receiver.
- Second-order ionospheric clutter is a double scatter that is the result of skywave propagation to the surface and propagation back to the radar via the surface wave (or vice versa).
- Third-order ionospheric clutter is a triple scatter that is the result of skywave propagation to the ocean surface and the return of that signal back to the receiver by skywave propagation.

First-order ionospheric clutter appears as narrow annuli bands of very strong clutter at ranges equalling that of the ionospheric layer heights. The ionosphere is very dynamic, and typically, the heights of the layers are both time- and frequency-dependent. In addition, plasma density irregularities in the ionosphere lead to the spreading of the clutter signal in the Doppler and angle-of-arrival domains [18]. The primary option for combating first-order ionospheric clutter is frequency agility. By increasing the radar carrier frequency, the layer critical frequency may be exceeded, and consequently the radar signal will penetrate the layer instead of reflecting from it. Similarly, the radar can be operated during the daytime at a lower frequency that does not support skywave propagation.

Second-order ionospheric clutter is a consequence of the skywave signal being returned to earth at an oblique angle and the signal returned to the radar via the surface wave or vice versa (surface wave out and skywave back). In general, for this clutter to dominate, it is required that Bragg resonant scatter occurs. For oblique skywave paths, this requires that ocean surface waves have a wavelength approximately equal to the radar carrier wavelength. Therefore, second-order ionospheric clutter is observed in higher sea conditions and at ranges greater than the height of the reflecting ionospheric layer. This is typically between 250 and 400 km.

The total path length of third-order ionospheric clutter typically places it at a range outside the system unambiguous range set by the PRF. In effect, the radar will receive returns from previous pulses while collecting data on the current transmit pulse. This type of interference is also referred to as 'Range Wrap Clutter' or 'Second-Time-Around Echoes'.

10.9.5   Range wrap clutter mitigation

In HFSWR, third-order ionospheric clutter can be treated as a second-time-around echo that results in range wrap. Range wrap clutter can be mitigated by changing the phase codes of successive pulses of the coherent pulse chain, and the second-time-around echo will be heavily attenuated by the filter that is matched to the current pulse and not the previous [19].

10.9.6   Meteor clutter

Meteoroids are small material particles, or assemblages of particles, in orbit around the sun. Those intercepted by the earth's atmosphere are heated by collision with atmospheric molecules and ions. The heating produces evaporation and the resulting expanding gas cloud produces, by further collisions, a trail of ionised gas behind the meteoroid. This ionisation trail can give rise to a transient echo. The phenomenon occurs at heights between 80 and 140 km, with a typical trail being 24-km long. More meteor echoes occur near dawn than near sunset, since the earth overtakes and meets more meteors than meteors overtake the earth [20].

Meteor clutter, as observed on surface wave radar, is usually the result of line-of-sight propagation at a slant range of 180 to 460 km. The echo signature usually appears as a large peak at a specific range.

Meteors' trails are typically visible for 5–30 s and appear as low Doppler (<1 Hz). They typically appear at ranges of at least a 100 km or more and, therefore, do not affect near range targets.

10.10   Radar cross section at HF

As illustrated in Figure 10.13, the RCS of targets falls into one of three regions. In the Raleigh region, the radar wavelength is similar or greater to the physical size of the target. In the resonance or Mie region, the wavelength is comparable in size of the target and in the optical region, the target size is significantly greater than the radar wavelength.

10.10.1   Definition of RCS at HF

The radar range equation, as presented in (10.3) specified the RCS of the target in terms of its free space value. For maritime surface targets, the RCS is measured in the presence of a ground plane and, therefore, as illustrated in Figure 10.14, includes

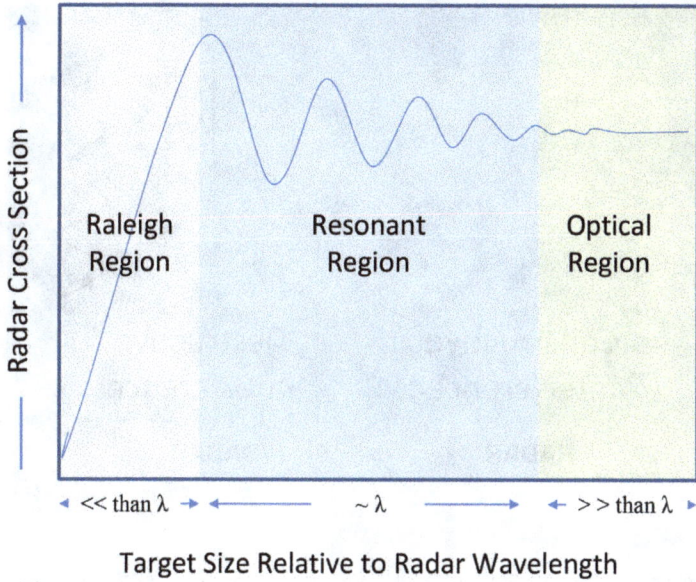

Target Size Relative to Radar Wavelength

Figure 10.13   RCS characteristics

the image of the vessel. The free space RCS is equivalent to the measured RCS minus the image. Since the vessel is both a receiver and a transmitter, the effect is included twice resulting in a measured RCS, which is 12 dB greater than the free space RCS.

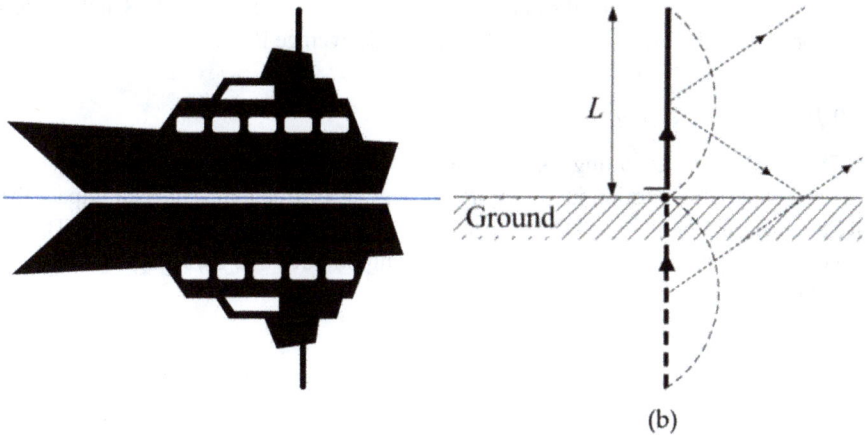

(b)

Figure 10.14   Measured RCS of surface vessels includes the image effect

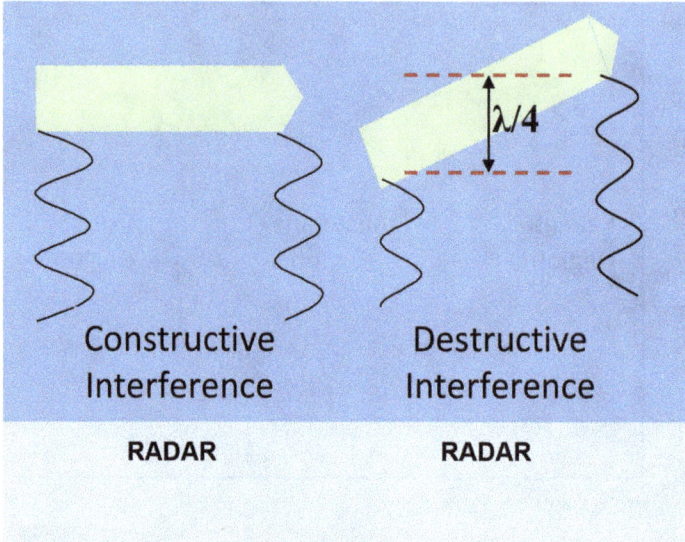

*Figure 10.15   Aspect angle dependence of RCS: constructive and destructive
interference*

10.10.2   RCS aspect angle dependency

The RCS of targets observed by HFSWR is typically determined by the sum of reflections of the radar signal from vertical structures. As illustrated in Figure 10.15, if the apparent distance between these vertical structures as seen by the radar is equal to an even number of quarter wavelengths, then constructive interference takes place. If the apparent distance is equal to an odd number of quarter wavelengths, then destructive interference occurs. This results in an aspect angle dependency of the RCS and by convention it is usual to specify the vessel aspect angle average RCS.

10.10.3   RCS sea state dependency

RCS is derived by assuming a stationary target. As illustrated in Figure 10.16, a target can experience motion in any of its six degrees of freedom. This can impact the RCS depending on the rate of motion compared to the pulse duration or CIT. This is accounted for in the radar range equation using one of three Swerling models [21] as defined below:

- **Swerling 1:** backscattered signal is relatively constant during the processing interval
- **Swerling 2:** backscattered signal fluctuates during the processing interval
- **Swerling 3:** backscattered signal fluctuates from one processing interval to another.

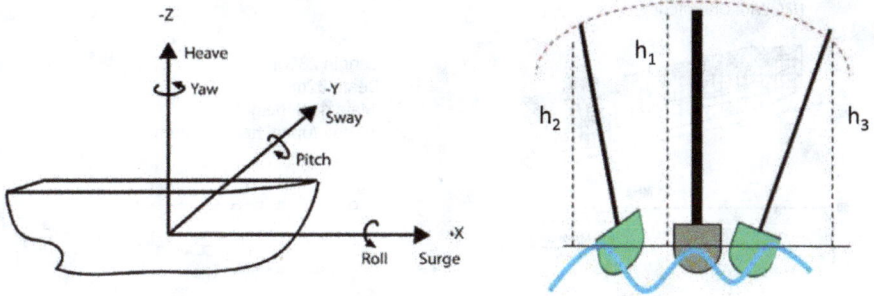

Figure 10.16   Vessel motion and impact on RCS

10.10.4   RCS and stealth

The RCS of a target as seen by an HFSWR is typically determined by metallic structures within the target. Traditional stealth technologies lower detectability by a combination of the external shape and the use of radar-absorbent materials (RAM). In general, as illustrated in Figure 10.17, a stealth target is designed to eliminate corner reflectors and maximise the amount of energy that is reflected away from the illuminating radar.

At microwave, the RCS is primarily determined by the skin of the target, while at HF, the RCS is primarily determined by the structure.

10.10.5   Modelling radar cross section of vessels

The RCS of vessels at HF is complex and has been treated extensively by others [22–24] using Numerical Electromagnetic Code (NEC) antenna modelling system. However, it has been shown [25] that for HFSWR, the RCS of vessels is dominated by their vertical metallic superstructure. If this superstructure is grounded to the ocean surface, then the RCS of the vessel can be approximated to that of a grounded monopole antenna [26]. When the metallic vertical superstructure is isolated from the ground, then the RCS can be approximated to that of a dipole antenna. Multiple structures will produce additive effects that can result in higher peaks in the RCS value as well as deeper nulls. An RCS model

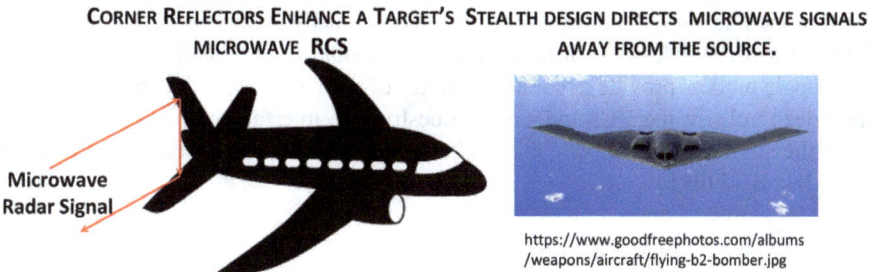

CORNER REFLECTORS ENHANCE A TARGET'S
MICROWAVE RCS

STEALTH DESIGN DIRECTS MICROWAVE SIGNALS
AWAY FROM THE SOURCE.

Microwave
Radar Signal

https://www.goodfreephotos.com/albums
/weapons/aircraft/flying-b2-bomber.jpg

Figure 10.17   Comparison between standard target and stealth target

NEC Wire Grid Model

Length 236m
Beam 32m
Main deck height ~20m
Bridge/funnel height ~ 30m

f = 4.1 MHz

NEC Wire Grid Model Results with
measured data overlay

Model based on Resonant Monopole Model
predictions

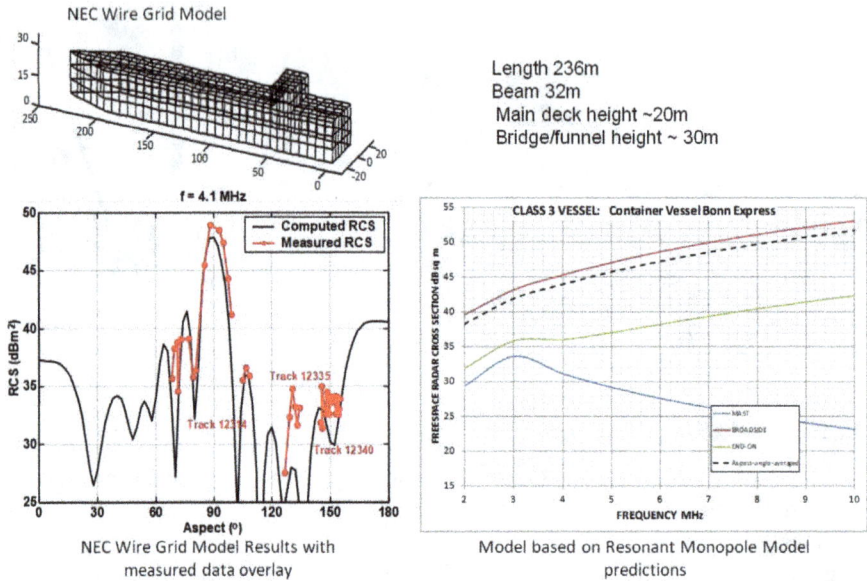

*Figure 10.18   Comparison of RCS estimation and measurement of a large bulk
carrier*

based on resonance monopoles developed by the author has been shown to produce accurate estimates of vessels RCS [27].

10.10.6   RCS of large vessel

Figure 10.18 presents the RCS of a large bulk carrier at HF. The vessel was modelled using NEC as well as the resonant monopoles method. Measured RCS results have been overlaid on the NEC predictions. It can be observed that both models produce very similar results that closely match measured values previously presented in [28]. The aspect angle-averaged free space RCS at 4.1 MHz is 43 dB m^2.

10.10.7   RCS of medium vessel

A comparison between the RCS of a medium size vessel using both NEC and resonant monopole model and measured results, previously presented in [28], is shown in Figure 10.19. It can again be observed that both the NEC and simple resonant monopole models provide accurate estimates of the vessel RCS. It can be observed that deep nulls in the RCS exist due to destructive interference from the A-frame near the stern and the vessels' mast. The aspect angle-averaged free space RCS at 4.1 MHz is 35 dB m^2.

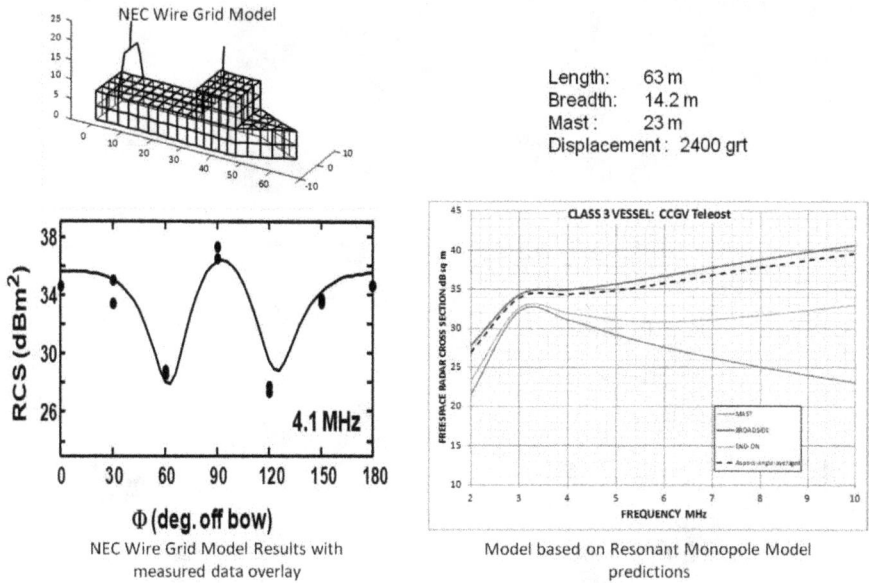

NEC Wire Grid Model

Length: 63 m
Breadth: 14.2 m
Mast: 23 m
Displacement: 2400 grt

NEC Wire Grid Model Results with
measured data overlay

Model based on Resonant Monopole Model
predictions

*Figure 10.19   Comparison of RCS estimation and measurement for a medium
                  size vessel*

10.10.8   RCS of small vessel

The RCS of a small 20-m fishing vessel is presented in Figure 10.20. Predictions
were made using the resonant monopole method. Measurements were presented by
RW Bogle and DB Trizna [29]. Again, it can be observed that there is close agreement between the measured and predicted aspect angle-averaged free space RCS.

10.10.9   RCS of very small vessels

The RCS of a fibreglass Go-Fast Boat (GFB) has been derived using an incline
wire model that modelled the reflectivity of the outboard motors and control cables.
Results are presented in Figure 10.21, from which it can be observed that the RCS
increases with frequency and with inclination (speed) [30].

The derived RCS was validated using a transportable HFSWR system [30].
A comparison between the measured signal strength to that predicted using a free
space RCS of 4 dB is presented in Figure 10.22.antenna measured in terms of the
Measurements were made with the GFB travelling inbound and outbound at maximum speed resulting in an inclination of approximately 10 degrees.

10.11   Resolution

Resolution is the radar's ability to resolve two objects in close proximity of each
other. For HFSWR, a target is an object in 3D space.

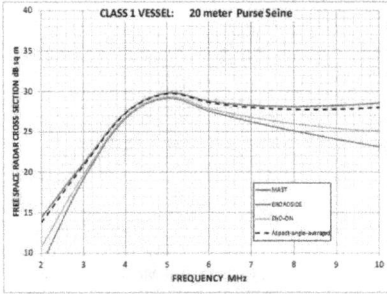

Model based on Resonant Monopole Model predictions (Ponsford)

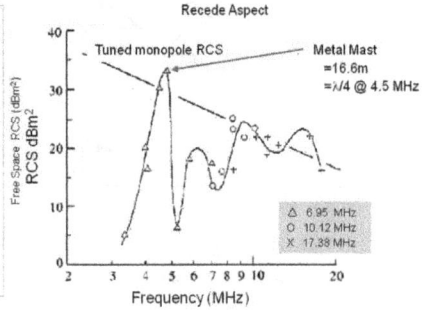

Model based on Resonant Monopole with measurements (RW Bogle & DB Trizna)

Figure 10.20   Comparison of RCS estimation and measurement for a small size vessel

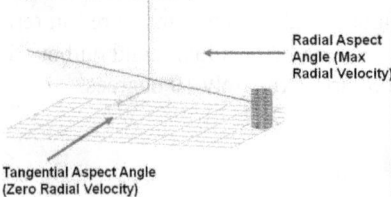

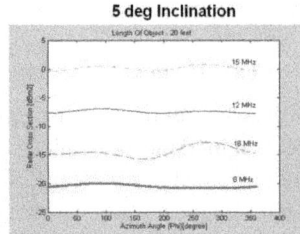

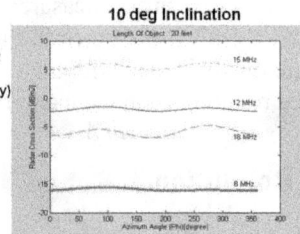

Figure 10.21   RCS model of GFB based on an inclined wire

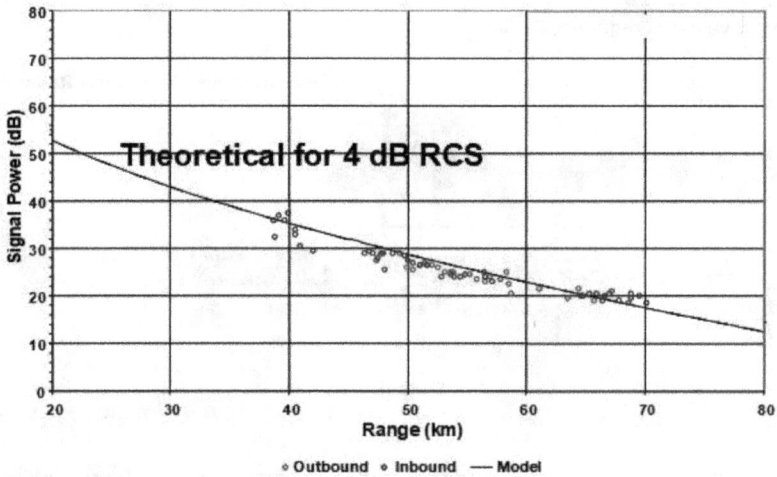

Figure 10.22   *Measured signal response as a function of range for radar frequency of approximately 15 MHz*

- The range resolution is directly proportional to the bandwidth of its transmitted waveform
- The velocity resolution is directly proportional to the CIT
- The azimuthal resolution is directly proportional to the aperture size of the antenna measured in terms of the radar's wavelength.

Azimuthal resolution may be further enhanced through the use of a virtual aperture array based on Multiple Input Multiple Output (MIMO) theory [31].

HFSWR radar does not have the inherent ability to provide altitude information (the fourth dimension) of a target. For two targets to be resolved, they must be separated into at least one dimension.

10.12   Accuracy of estimates

Another performance indicator of radar is the accuracy of the estimate of the target range, bearing and velocity. Although the resolution capabilities of the HFSWR are moderate, under a single-target scenario (i.e., only one target is in a range bin, at a given bearing), an estimated accuracy equal to one-tenth of the basic resolution can be achieved with a moderate SNR.

10.13   HFSWR and cognitive sensing

Cognitive sense-and-adapt technology and dynamic spectrum management ensures robust and resilient operation in the highly congested HF band. Dynamic spectrum access (DSA) enables the system to operate on a non-interference and non-protected

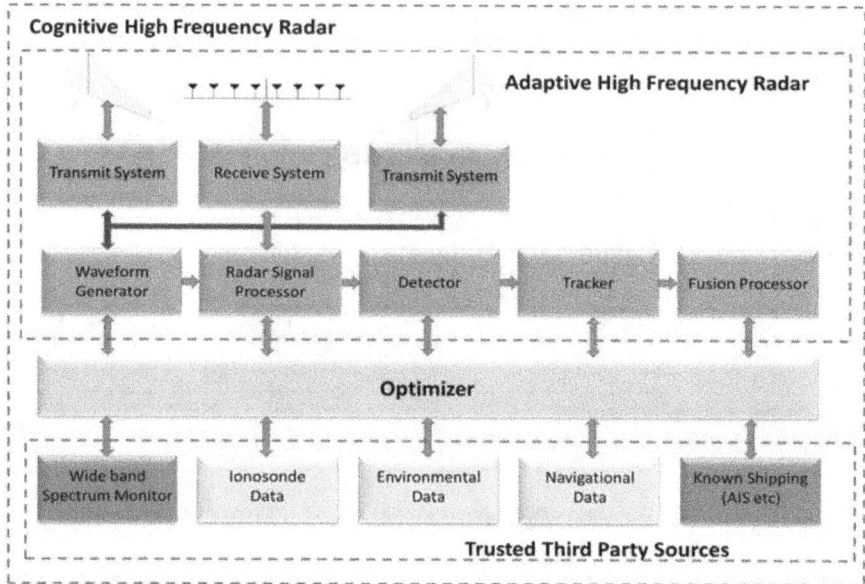

Figure 10.23   Key components of the cognitive third-generation HFSWR system

basis without impacting other spectrum users. DSA is also the primary line of defence against performance degradation by unwanted signals that are attempting to jam the radar.

Sense-and-adapt technologies ensure that the system instantaneously switches to a new vacant channel on detection of an unwanted jamming signal. Adaptive signal processing techniques mitigate against electrical noise, interference and clutter. Sense-and-adapt techniques applied at the detector and tracker stages maximise the probability of track while minimising the probability of false or otherwise erroneous track data.

The objective of cognitive radar is to utilise knowledge, obtained either by sensing the local environment or from trusted third-party sources, to maximise the probability of track initiation while minimising the probability of false or otherwise erroneous tracks. As illustrated in Figure 10.23, the Canadian third-generation HFSWR [32] is based on a cognitive architecture and consists of three major subsystems: an adaptive HFSWR, an optimiser and data from trusted third-party sources. For the demonstration system, these auxiliary data sources consisted of a wide-band spectrum monitoring system and a data fusion processor that associated HF radar-derived surface tracks with known shipping in the area obtained primarily from space-based interception of automatic identification system (AIS) reports. The data were combined to provide a comprehensive real-time overview of surface vessel activity [33].

The performance of the Canadian third-generation system is presented in Figure 10.24. [34].

(Instrumented Range 40 – 400 km)

Vessel Type			RCS FS dB m²	Sea State 3 (day/night)	Sea State 5 (day/night)	Sea State 7 (day/night)
Class D: Mid-Shore mechanical Fishing Vessels			15-20	230/210	–	–
Class C: Regional Vessels (Tugs, Offshore Fishing vessels, Barges, etc)			20-25	300/220	180/120	140/120
Class B: Local Coastal /Cabotage vessels			25-35	370/260	300/220	180/150
Class A: Large coastal and ocean going vessels,			>35	370/260	370/260	370/260

All ranges in kilometers, RCS are Free Space and Aspect Angle Averaged

Figure 10.24   *Performance specifications for the Canadian third-generation HFSWR system*

10.14   Challenges and ongoing research

As described previously, the key challenge to improving the performance of HFSWR is the removal or cancellation of unwanted signals such as noise and interference.

Developing advanced signal processing techniques for HF radar requires understanding of the expected clutter returns. In the Canadian context, the clutter includes returns from the ionosphere and the aurora. Researchers at the Univrsity of Toronto and Defence Research and Development Canada developed a propagation model for ionospheric clutter [35]. The model accounts for the spatial and temporal correlation of the ionospheric reflections using a geometric optics approach to determine the power spectral density of the clutter. Using a bulk Doppler shift, superimposed with a random component, the work was shown to produce realistic simulated data cubes for the clutter. This work was later extended to auroral clutter [36].

This modelling work allowed for the development of advanced signal processing techniques to suppress auroral and ionospheric clutter. Given the extreme heterogeneity of the clutter, traditional space-time adaptive processing techniques are not applicable. Researchers at the University of Toronto developed fast fully adaptive (FFA) processing to deal with non-homogeneous clutter. This multistage approach was shown to reduce errors introduced due to clutter non-homogeneities [37]. The clutter model also allowed for the development of waveform optimisation in the context of HF radar dealing with ionospheric and auroral clutter [38], including a training-based scheme to obtain the second-order transmit statistics using receive data [39].

Progress has also been made in the area of mitigation against low-level impulsive man-made and night-time (HF) noise [40]. Typically, impulse detectors detect the event by envelope power levels (in time domain) within the narrowband target-processing channels. This is effective against high-power impulses resulting from local lightning but insufficiently sensitive to identify distant lightning activity that dominates night-time noise and power lines and electrical machinery that limits installation of HF radar systems to rural locations. However, if a wide-band spectrum monitor channel is used to identify corrupted pulses via noise floor elevation then the identified pulses can be excised within the narrow band target processing channel, thus improving the detection process [41].

References

[1]   Wyatt L., IEEE. 'HF radar for coastal monitoring: a comparison of methods and measurements'. *Oceans 2005–Europe.* **1**; 2005. pp. 314–18.

[2]   Liu Y., Weisberg R.H., Merz C.R. 'Assessment of CODAR SeaSonde and WERA HF radars in mapping surface currents on the West Florida Shelf'. *Journal of Atmospheric and Oceanic Technology.* 2014;**31**(6):1363–82.

[3]   Barrick D., Everly J. Some aspects of detection of targets by radar near the surface of the sea. *Battelle Memorial Institute*; 1971.

[4]   Persistent Active Surveillance of the Exclusive Economic Zone. 2021Maerospace Corporation. Available from http://maerospace.com/pase/ [Accessed Accessed 17 Sept 2021].

[5]   Sevgi L., Ponsford A., Chan H.C. 'An integrated maritime surveillance system based on high-frequency surface-wave radars, in . *1. Theoretical Background and Numerical Simulations'. IEEE Antennas and Propagation Magazine.* 2001;**43**(4):28–43.

[6]   Ponsford A., Wang J. 'A review of high frequency surface wave radar for detection and tracking of ships'. *Special Issue on Sky- and Ground-Wave High Frequency (HF) Radars: Challenges in Modelling, Simulation and Application Turkish Journal of Electrical Engineering & Computer Sciences.* 2010;**18**(3):409–28.

[7]   Ponsford A., Dizaji R., McKerracher R. 'HF surface wave radar operation in adverse conditions'. *IEEE Proceedings of the International Conference on Radar*; Adelaide, Australia; 2003. pp. 3–5.

[8]   Sun W., Ji M., Huang W., Ji Y., Dai Y. 'Vessel tracking using bistatic compact HFSWR'. *Remote Sensing.* 2020;**12**(8):1266.

[9]   Riddolls R., Ponsford A. 'Review of the canadian east coast high frequency surface wave radar program and compatibility of HF radar operations with communication users'. 8th International Conference on Remote Sensing for Marine and Coastal Environments. Halifax(Canada), 17–19 May; 2005.

[10] Shearman E.D.R. 'Propagation and scattering in MF/HF groundwave radar'. *IEE Proceedings F Communications, Radar and Signal Processing.* 1983;**130**(7):579.

[11]   Huang W., Gill E.W. 'Ocean Remote Sensing Technologies: High Frequency'. *Marine and GNSS-Based Radar, IET*; 2020.

[12]   CCIR. 'GRWAVE model: user's manual personal computer version'. Geneva; 1990.

[13]   Rotheram S. 'Groundwave propagation, Part II: theory for long and medium distances and reference propagation curves'. *Proceedings of the Institution of Electrical Engineers*. 1981;**128**(5):275–95.

[14]   Spaulding A.D., Washburn J.S. Atmospheric radio noise: worldwide levels and other characteristics, National telecommunications and information administration. 85–173; 1985. p. 192.

[15]   CCIR. *Man-made radio noise, Report 258-4, CCIR*. Geneva: ITU; 1986.

[16]   King H.M. World Lightning Map. 2005. Available from https://geology.com/articles/lightning-map.shtml [Accessed Accessed 20 Feb 2020].

[17]   Ravan M., Riddolls R.J., Adve R.S. 'Ionospheric and auroral clutter models for HF surface wave and over-the-horizon radar systems'. *Radio Science*. 2012;**47**(3).

[18]   Vallières X., Villain J.-P., Hanuise C., André R. 'Ionospheric propagation effects on spectral widths measured by SuperDARN HF radars'. *Annales Geophysicae*. 2004;**22**(6):2023–31.

[19]   Ponsford A., Dizaji R., McKerracher R. 'HF surface wave radar operation in adverse conditions'. IEEE Proceedings of the International Conference on Radar, Adelaide, Australia, 3–5 September; 2003.

[20]   Thayaparan T. 'Strengths and limitations of HF radar for meteor backscatter detection'. IEEE 2000 International Radar Conference; USA; 2000. pp. 578–83.

[21]   Wolff C. Radar Tutorial. Available from http://www.radartutorial.eu/01.basics/Fluctuation%20Loss.en.html [Accessed 16 Aug 2019].

[22]   Sevgi L. 'Target reflectivity and RCS interactions in integrated maritime surveillance systems based on surface-wave high-frequency radars'. *IEEE Antennas and Propagation Magazine*. 2001;**43**(1):36–51.

[23]   Trueman C., Kubina S. 'RCS of fundamental scatterers in the HF band'. 7th Annual Review of Progress in Applied Computational Electromagnetics of the Applied Electromagnetics Computational Society, March 18–22. Monterey, California; 1991.

[24]   Leong H., Wilson H. 'An estimation and verification of vessel radar cross sections for HFSWR'. *IEEE Antennas and Propagation Magazine*. 2006;**48**(2):11–16.

[25]   Ponsford A.M. 'Surveillance of the 200 nautical mile exclusive economic zone (EEZ) using high frequency surface wave radar (HFSWR)'. *Canadian Journal of Remote Sensing*. 2001;**27**(4):354–60.

[26]   Bogle R., Trizna D. Small boat HF radar cross sections. 'NRL Memorandum Report 3322'; 1976.

[27]   Ponsford A. *Empirical formula for deriving the radar cross section of vesselsat HF. Formulation includes aspect angle dependency and impact ofMast Height [online]*. Available from https://www.researchgate.net/publication/

322962878_Empirical_Formula_for_Deriving_the_Radar_Cross_Section_ of_Vessels_at_HF_Formulation_includes_aspect_angle_dependency_and_ impact_of_Mast_Height [Accessed 5 Jan 2020].

[28]   Leong H., Wilson H. 'An estimation and verification of vessel radar cross sections for HFSWR'. *IEEE Antennas and Propagation Magazine.* 2006;**48**:11–16.

[29]   Bogle R., Trizna D. 'NRL Report 3322, July 1976: reproduced by trizna oceans'. BiStatic multi-frequency HF radar; Quebec City; 2008.

[30]   Ponsford A. 'Detection and tracking of go-fast boats using HFSWR'. SPIE Defence and Security Symposium in Orlando, April 17; 2004.

[31]   Wang J., Ponsford A., Wang E., McKerracher R. 'Virtual antenna extension for sampled aperture arrays'. Patent No. US 9638793; 2017.

[32]   Ponsford A., McKerracher R., Ding Z., Moo P., Yee D. 'Towards a Cognitive Radar: Canada's Third-Generation High Frequency Surface Wave Radar (HFSWR) for Surveillance of the 200 Nautical Mile Exclusive Economic Zone'. *Sensors.* 2017;**17**(7):1588.

[33]   Ponsford A.M., Sevgi L., Chan H.C, Chan A. 'An integrated maritime surveillance system based on high-frequency surface-wave radars. 2. operational status and system performance'. *IEEE Antennas and Propagation Magazine.* 2001;**43**(5):52–63.

[34]   Ponsford A., Moo P., McKerracher R., Ding Z., Yee D., Ramsden M. 'Canada's 3rd generation high frequency surface wave radar for persistent surveillance of the 200 nautical mile EEZ'. NATO SET 241; Quebec City, Canada; 2017.

[35]   Ravan M., Riddolls R.J., Adve R.S. 'Ionospheric and auroral clutter models for HF surface wave and over-the-horizon radar systems'. *Radio Science.* 2012;**47**(6).

[36]   Ravan M., Adve R. 'Modeling the received signal for the Canadian over-the-horizon radar'. Proceedings of the 2013 IEEE Radar Conference, April; 2013.

[37]   Saleh O., Ravan M., Riddolls R., Adve R. 'Fast fully adaptive processing: a multistage STAP approach'. *IEEE Transactions on Aerospace and Electronic Systems.* 2016;**52**(5):2168–83.

[38]   Gorji A.A., Riddolls R.J., Ravan M., Adve R.S. 'Joint waveform optimization and adaptive processing for random phase radar signals'. *IEEE Transactions on Aerospace and Electronic Systems.* 2015;**51**(4):2627–40.

[39]   Shaghaghi M., Adve R., Shehat G. 'Training-based adaptive transmit–receive beamforming for mimo ra-Based adaptive Transmit–Receive beamforming for MIMO Radars'. *IEEE transactions on aerospace and electronic systems.* 2017;**53**(6).

[40]   McKerracher R., Ponsford A. 'Impulse noise detection and removal for radar and communication systems'. US Patent 10288726; 2019.

[41]   McKerracher R., Moo P., Ponsford A. 'Spectrum utilization: sense and adapt: operation on a non-interference and non-protected basis'. *IEEE Aerospace and Electronic Systems Magazine*; 2017. pp. 30–4.

Chapter 11

Introduction to ocean remote sensing with marine radars

Merrick C. Haller[1]

Marine radars were (and still are) mainly designed for navigation and tracking of vessels at sea. However, as early as the 1940s, it was recognized that the intensity of the radar image background (i.e., the "sea clutter") was related to the sea surface roughness, which appeared to be accentuated along the crests of the visible surface waves [1]. Since the ocean surface roughness originates with the wind and is dependent on wind speed and direction, these early observations implied that both wind and wave information was encoded in the maps of radar backscatter derived from these systems. Also, spatial variability in the surface currents will locally enhance/ dampen the backscatter. Furthermore, sequences of images, and the time variability they capture, could provide even additional information through space-time analyses. Thus, they can provide a continuous synoptic picture of the ocean surface over scales of several kilometers, which can greatly enhance the interpretation of in situ observations, especially for event-based analyses (e.g., the passing of a front, eddy, or internal wave).

Ocean wind and wave parameters of environmental interest are typically measured by buoys moored offshore in intermediate depths (50–500 m). However, a buoy can only sample at a single location and much of the coastal oceans are relatively sparsely sampled with such buoys. Sparse point measurements cannot adequately describe the short-scale variability evident in the winds nor the local variations in the wave field induced as waves propagate over variable bathymetry, refract around headlands, interact with rivers and estuaries, or reflect from breakwaters and diffract through harbor entrances.

Remote sensing systems allow for the synoptic and non-intrusive sampling of large areas. Satellite sensing provides global ocean coverage, but the ability to observe regional-scale phenomena is limited by the relatively long revisit times (order of days/weeks). Land or ocean platform-based remote imaging systems offer the added capability of long dwell (i.e., collection of image sequences), while also

---

[1]School of Civil, Construction Engineering, Oregon State University, USA

avoiding the problems of deploying in situ instruments in difficult and sometimes hazardous sampling environments.

For coastal ocean processes, marine radars fill the remote sensing sampling gap that lies between the kilometer-scale sampling of high-frequency radar and the shorter-range, high-resolution sampling that is provided by beach cameras [2, 3]. In fact, there is a tremendous amount of environmental information encoded within the spatial patterns of short-scale ocean roughness and the modulation of that roughness by the larger-scale ocean waves. There is also a considerable body of literature regarding the extraction of surface wave and near-surface wind information via marine radars (see [4] for a recent review). Besides wind speeds and directions, wave imaging remote sensors like marine radars offer the potential of providing wave directions, periods, the space-time distribution of wave breaking, and wave heights. Furthermore, through the observed modifications of surface wave dispersion, radar data products can be combined with appropriate models in order to infer both surface currents and bathymetry (e.g., [5–7]).

Nonetheless, given all of this potential, there are still considerable challenges and limitations in our ability to extract this information from the marine radar observations. There still remain technological challenges with respect to the sources of low grazing angle (LGA) scattering, balancing the advantages and disadvantages of coherent versus non-coherent systems, and better specifying the relationship between water surface slope and modulations of the radar backscatter.

Chapters 12–17 review the state of the art of a subset of marine radar ocean remote sensing applications. Each of these chapters delivers a snapshot of the current science and additional details of recent advancements. They are not intended as a comprehensive review of all ocean-related marine radar work and applications but instead a summary of selected areas of emerging wave-estimation techniques, ocean current and bathymetry estimation, and wind field extraction.

In the rest of this first chapter, we first review background on marine radar hardware and software technology, and then provide a brief review of the technical challenges regarding the radar applications that are the subject of the subsequent chapters. Finally, each of the chapters is briefly introduced.

11.1   Marine radar ocean observing instrumentation

The standard commercial marine radar is a non-coherent imaging system using a rotating polarized antenna. The fundamental radar observational data are image sequences of relative (uncalibrated) backscatter intensity mapped to a geographical coordinate system and synchronized to a standard Global Positioning System (GPS) time signal. Coherent radars have the added capability of directly measuring the line-of-site velocity of the surface scatterers through their observed Doppler shift. However, they come at the price of requiring longer dwell (multiple pulses) in order to reduce noise in the Doppler spectra. Hence, terrestrial coherent radars operate with non- or slowly rotating antennas and are not able to produce large footprint two-dimensional images of the ocean surface [8–10]. There is, however, a newly

Figure 11.1   Tower-mounted marine radar with 2.74-m antenna

emerging "coherent-on-receive" hardware approach to marine radars that shows some promise [11]. An additional complication of the coherent information is that the measured Doppler shifts of the surface scatterers are due to some mixture of the orbital wave velocities (including Stokes drift), wind-driven flow, and tidal currents. Separating those components is often an under-specified problem.

A comprehensive description of the basic hardware and software requirements for ocean remote sensing using non-coherent, commercially available, marine radar systems was given in [12] and [13]. These are summarized in Sections 11.1.1–11.1.2.

11.1.1   Hardware

Marine radars operate either at X- or S-band microwave frequencies (~10 GHz and ~3 GHz, respectively). However, X-band radars are what are most commonly used for oceanographic sensing due to their higher frequency and shorter pulse lengths, which leads to finer spatial resolution. Though there have been some S-band research applications [14], S-band systems are mostly used for ship detection and navigation purposes, and are especially useful during rain and fog conditions.

The basic hardware components of the marine radar are the magnetron that produces the radar pulse and a linear, open-array, antenna mounted on a rotating pedestal (see Figure 11.1). In order to achieve the desired space and time resolution for the imaging of ocean surface gravity waves in the sea and swell bands, the basic requirements are an antenna length of at least 1.8 m and a rotation speed $\geq$ 24 rotations per minute (rpm). This antenna length translates to a horizontal beamwidth of approximately 1° producing 10-m azimuthal resolution at a range distance of 500 m. The 24 rpm rotation rate allows at least 0.4 Hz image sampling rate (revisit time). Of course, longer antennas and faster rotation rates will improve resolution and/or allow for pulse-to-pulse averaging to reduce noise.

A typical high-performance system will transmit at 9–10 GHz with a 2.4-m horizontally polarized antenna for both transmitting and receiving (HH). Standard marine radars have an adjustable pulse length with the shortest pulse length around ~50 nanoseconds translating to an intrinsic target discrimination resolution of 7.5 m in range. A significant polarization difference appears at LGAs whereby HH pol is more sensitive to breaking waves, which are most important close to shore.

11.1.2   Software

Early oceanographic efforts needed to use photographic image capture of the standard navigational radar displays (e.g., [15–17]). A long-standing, fully digital, commercial radar image data acquisition system is the WaMoS II software, initially described and tested in [18, 19] and [2]. The commercial software will often include some proprietary algorithms for estimating wave and current parameters as well. Individual groups also have developed more customized acquisition systems [20, 21], including research-level estimation algorithms for currents and bathymetry.

Marine radars sample in a polar coordinate system; individual pulses propagate in the range direction from the rotating antenna and each received echo is sampled over time with time-of-flight corresponding to twice the range distance. A single rotation of the antenna provides a circular image footprint with a radius of 2–12 km, depending on the duration of sampling (see Figure 11.2). This includes a short blanking distance close to the radar where the intensities often saturate the receiver. Although the resolution for target discrimination is governed by the pulse width, data acquisition systems often oversample to 3.0–7.5 m range resolutions. The image range limit, or total footprint size, is a user-specified parameter. For a pulse repetition frequency of 2 000 Hz, the theoretical range limit is in the neighborhood of 100 km. In practice, however, the limit of useful imaging is governed by the available signal-to-noise (SNR) ratio. The SNR, in turn, is governed by wind speed, antenna height, the occurrence of rain, the amount of pulse averaging, and hardware quality.

11.2   Applications

Marine radars occupy a niche in the overall set of microwave instruments in that they operate at very LGAs when compared to airborne and satellite systems. Also, they are rarely calibrated. Hence, the environmental signals are carried in relative measures of backscatter intensity, as opposed to absolute measures of radar cross-section. However, marine radars are highly effective at feature tracking, which means they are excellent sensors for quantifying feature orientation, speed and direction of propagation, length scales, and frequencies. These systems have been

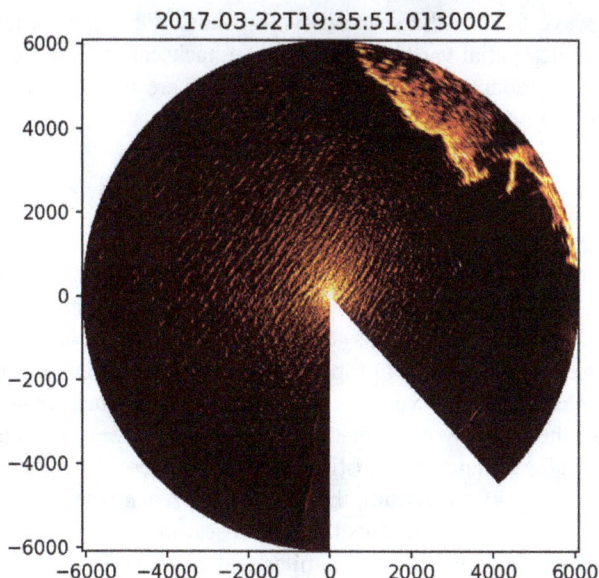

Figure 11.2   Marine radar image collected aboard ship, Monterey Bay, CA. Axes are shown in meters. Coastline can be seen in the upper right portion.

used for oceanographic applications since the 1960s [15]. The basic understanding of microwave backscatter from the ocean surface was laid out in the following works [22–25]; a nice historical summary of the development of radar-based wave measurements was given by Story *et al.* [26]. Since marine radars are installed on ocean- or shore-based platforms, they operate primarily at LGA or near grazing incidence. The primary source of LGA microwave returns for vertical polarization is Bragg scattering from capillary and capillary-gravity waves (Bragg waves) of approximately half the radar wavelength [24]. The magnitude of the radar returns depends primarily on the amplitude of the Bragg waves and the grazing angle, with wind being the initiator of the Bragg waves and the longer gravity waves then modulating the radar returns through the tilting of the LGA. Additional contributions to the modulation induced by the longer gravity waves come from the hydrodynamic modulation of the Bragg waves by the orbital velocities of the long waves, as well as optical shadowing in the lee of the long waves.

However, it should also be noted that many, perhaps most, marine radars operate at horizontal polarization (HH) and that, for HH polarizations, a primary source of LGA returns is surface roughness due to breaking waves [27–29], as HH Bragg scattering is weaker at these angles. The majority of phase-resolved wave reconstruction methods consider the dominant modulation mechanisms to be wave shadowing and surface tilt.

Below are outlined the marine radar applications that are considered in the subsequent chapters. Most of these applications are fundamentally based on the imaging of the ocean wave field in the sea and swell bands. Wind applications are based on the larger scale spatial variability of relative backscatter across a radar image. Additional marine radar applications not discussed here include oil spill detection [30] and sea ice monitoring [31].

11.2.1 Waves

The utility of X-band imaging systems for measuring open ocean wave spectra and surface currents was first established by Atanassov *et al.* [32] and Young *et al.* [5]. The original successful methodology relies on the three-dimensional fast Fourier transform (3D-FFT), which maps the observed wave field into frequency-wavenumber pairs. The signals of propagating ocean waves then lie primarily along the linear wave dispersion curve in the frequency-wavenumber domain and can be analyzed. Since the earlier works, notable improvements have been made in computing wave spectra for retrieval of statistical wave parameters [19, 33–36]. In addition to the radar-derived wave spectrum, the 3D-FFT approach was shown to allow for the determination of surface currents through observed current-induced distortions of the wavenumber distribution [5]. Similarly, the observed shifting of the wave dispersion curve due to the influence of the sea bottom in finite water depths also allows for bathymetric estimation [37, 38]. However, an ongoing challenge exists for applications where currents and bathymetry *simultaneously* influence dispersion [39–41]. This challenge is covered in Chapters 14–16.

In addition to the bulk wave parameters, there also exist methodologies to reconstruct the phase-resolved wave field using marine radar images [42–45]. These have mostly been driven by the need for short-term wave forecasting to support wave height-limited maritime operation tasks, such as cargo transfer, helicopter landing, or high-speed navigation [44, 46]. Recently, there have also been some applications for the control of wave energy converters [47]. These phase-resolved reconstruction methods often bootstrap the bulk parameter estimations made first from the 3D-FFT approach. The 3D spectra are subsequently filtered to only retain components satisfying the linear wave dispersion relationship and may be scaled using an external calibration factor to match in magnitude.

Another consistent challenge with wave-related marine radar applications is the occurrence of wave breaking and the distinctly different scattering mechanism that entails. This is especially important when imaging from shore-based platforms. For example, Haller and Lyzenga [48] compared simultaneous radar and optical observations of breaking waves in the surf zone and found that the breaking events accounted for 40–50% of the average backscattered power, which is much larger than for deepwater breaking events [49]. The scattering characteristics of nearshore breaking waves include larger backscattered intensities, especially at horizontal polarization, and a broad Doppler spectrum peaking at frequencies corresponding to the long-wave phase speed [50, 51]. However, the scattering physics underlying these characteristics is not understood well, despite a number of proposed non-Bragg models. To date, these models can only explain some and not all of the scattering characteristics. Nonetheless, the increased relative importance of wave breaking in marine radar imaging in the nearshore can be exploited. Wave breaking observations can be used to quantify wave dissipation and linked to bathymetry [52] and the transfer of momentum to nearshore currents and wave-induced setup [53, 54].

11.2.2 Currents

A fundamental distinction between noncoherent versus coherent radar systems is that the noncoherent systems can only estimate ocean currents through the indirect estimation of their effect on dispersion. Specifically, the imaged wave signals need to be first mapped into frequency and wavenumber space whereby the frequency-wavenumber pairs that lie along the curve predicted by linear wave theory (i.e., the surface gravity wave signal) are first isolated from everything else (i.e., the noise). Next, the effect of currents appears as a frequency-dependent shift in the dispersion curve, as determined from where the filtered image data is concentrated.

Recent advances on using marine radar for ocean currents were demonstrated by [55] and [56]. The most recent challenge for these methods is the determination of the depth variability of the observed currents, which manifest as a variation of the Doppler shift with wavenumber [57].

Currents also directly induce backscatter features that can be identified and tracked [12]. The imaging capability arises mainly due to patterns of divergence and strain in the surface current field, which amplify or attenuate the Bragg waves that are a dominant source of backscatter, especially when multiple images are averaged together over time. Some of the recent examples include the imaging of buoyant coastal fronts [58], estuarine ebb plumes [59], internal hydraulic jumps [60], internal waves [61, 62], and rip currents [63]. However, though these backscatter features are inherently related to the current field, their analysis does not lead directly to surface current estimates. Nonetheless, the features can be tracked to extract the velocities of the surface scatterers; yet, their relationship to the in situ current velocities can be complex.

11.2.3 Bathymetry

Estimating bathymetry via remote measurements has been a recognized application since at least World War II [64]. The basic approach is very similar to that previously described for ocean currents since wave dispersion is also directly influenced by water depth (for waves longer than 1/2 the local depth). Hence, the depth can be estimated via a curve fit to the frequency-wavenumber pairs derived from the image spectrum. The ongoing challenges with bathymetry estimation include the inherent coarse resolution of 3D-FFT approaches, which runs into challenges when nearshore waves are spatially inhomogeneous [65], as well as the separation of the dispersive effects of bathymetry versus currents when both are present [40, 41].

Similar to the increased backscatter due to persistent currents, time-averaging radar images to remove the wave-by-wave signals can isolate locations of persistent wave breaking. In the nearshore, these longer time-scale features in the radar backscatter can be associated with bathymetric features [52, 66], in a manner analogous to the video method first used by Lippmann and Holman [67]. There have also been marine radar applications for estimating shoreline position and foreshore slopes [68] and morphologic change [69].

11.2.4   Winds

The potential for retrieving winds from marine radar images offers the advantage of spatial mapping versus the point measurements of in situ sensors. For example, in situ wind sensors can be susceptible to interference from nearby structures and measurements from ships must necessarily account for the influence of the platform motion. Early on in the radar probing of the ocean surface investigators reported an upwind-downwind anisotropy in backscatter [70], which has been attributed to the slope anisotropy between the forward and backward faces of wind waves. For HH polarization specifically, there is an upwind peak that can be exploited to estimate the dominant wind direction.

On the other hand, wind speed cannot be measured directly from uncalibrated radars. However, backscatter from the ocean surface using horizontally polarized marine radars is strongly dependent on the wind speed, wind direction, range distance, and air-sea temperature differences. This provides the opportunity to develop empirical methods to extract the wind speed through calibration or the analysis of wind streaks using neural networks [71, 72], for example.

11.3   Recent developments included in this book

11.3.1   Chapter 12: Observation of sea surface waves by non-coherent X-band marine radar

The traditional 3D-FFT approach requires spatial homogeneity in the wave field (similar to stationarity in time for time series analysis) over long enough scales to achieve the necessary spectral wavenumber resolution. In this chapter, Chen *et al.* discuss the challenge of using the 3D-FFT approach in areas where the wave field is spatially inhomogeneous in nearshore regions or due to rain. They present an empirical orthogonal function method that decomposes the wave field variability at different scales and allows for the estimation of bulk wave parameters, including wave heights, using a self-calibration method without the need for in situ data. Finally, they also identify some continued challenges, such as the need to better understand the modulation transfer function, especially the role of shadowing.

11.3.2   Chapter 13: Wavelet-based methods to invert sea surfaces and bathymetries from X-band radar images

In this chapter, Chernyshov *et al.* directly address the limitations of the 3D-FFT approach in spatially inhomogeneous wave conditions by summarizing their previously presented work in [73] and [65]. Their approach utilizes the continuous wavelet transform, which is a modern variant of the traditional FFTs that allows a more local (i.e., higher resolution in space/time) spectral analysis. Furthermore, the method is applied to wave field reconstruction (synthetic data) through an alternative modulation transfer function and bathymetry estimation through a phase-based approach.

11.3.3 Chapter 14: Wave field reconstruction using orthogonal decomposition of Doppler velocities

In this chapter, Hackett provides a nice description of how many of the alternative approaches to the traditional FFT-based method can be properly classified under the umbrella of *proper orthogonal decomposition* (POD). A characteristic of these methods is that they do not make an a priori assumption of the form of the basis functions (sines and cosines for Fourier methods). This minimizes the number of modes required to describe the multi-scale variance of the image signals. In their application, they use the POD method to reconstruct phase-resolved ocean wave fields. Additionally, their methodology is innovative in that it utilizes a coherent-on-receive Doppler radar system. The POD methodology shows improved results for wave reconstruction when the wave field is more nonlinear, as demonstrated by energy lying off the dispersion curve but still associated with wave field through nonlinearity (i.e., the "group-line"). This work builds on the findings of Plant and Farquharson [74], which previously demonstrated that some relevant wave energy lies off the linear wave dispersion curve in frequency-wavenumber space.

11.3.4 Chapter 15: Current mapping from the wave spectrum

In this chapter, Smeltzer and Ellingsen provide an elegant review of the physics and methodologies behind estimating ocean currents through the effect they have on wave dispersion. This includes practical considerations of spectral resolution, the presence of higher harmonics, and spectral aliasing. Critically important for all of these current estimation methods is the recognition that currents are most often depth-dependent. Therefore this depth-dependence needs to be accounted for in the wave-current interaction model used to invert for currents. Often this has been avoided by simply assuming the currents to be depth-uniform but can be accounted for through the concept of "effective depth."

11.3.5 Chapter 16: Bathymetry (and current) retrieval: phase-based method

In this chapter, Honegger and Haller summarize the work of [38] and [41], which built on the earlier works of Plant *et al.* [75] and Holman *et al.* [76]. The approach they describe is a response to the challenge of overcoming the inherent limitation of the 3D-FFT methods. This challenge is the requirement of spatial homogeneity in the wave field over length scales 10 times the wavelength of the imaged waves. These authors use a wave phase-based approach (i.e., local time-domain only FFTs) that provides a way to bring the resolution of bathymetry and current estimates closer to the inherent spatial sampling resolution of the radar. However, this increased spatial resolution does come at the cost of some directional resolution in the directional wave spectrum. They also summarize how their approach has been applied in the challenging environments of coastal inlets.

11.3.6   Chapter 17: Wind parameter measurement using X-band marine radar images

In this chapter, Chen *et al.* summarize and intercompare the methods for extracting wind speed and direction parameters from marine radar images. These methods include the older technique of curve-fitting to the intensity levels in averaged backscattter images, as well as an update to the method for intensity-level selection. A texture-based method for removing the distorting effect of rain on the imaged ocean surface is described as well as ensemble-empirical-mode-decomposition methods for again dealing with image inhomogeneity. Finally, nonparametric regression-based methods are also summarized.

References

[1]   Goldstein H. 'Frequency dependence of the properties of sea echo'. *Physical Review*. 1946;**70**(11–12):938–46.

[2]   Wyatt L.R., Green J.J., Gurgel K.-W. 'Validation and intercomparisons of wave measurements and models during the EuroROSE experiments'. *Coastal Engineering*. 2003;**48**(1):1–28.

[3]   Holman R., Haller M.C. 'Remote sensing of the nearshore'. *Annual Review of Marine Science*. 2013;**5**:95–113.

[4]   Huang W., Liu X., Gill E.W. 'Ocean wind and wave measurements using X-band marine radar: a comprehensive review'. *Remote Sensing*. 2017;**9**(12):1261.

[5]   Young I.R., Rosenthal W., Ziemer F. 'A three-dimensional analysis of marine radar images for the determination of ocean wave directionality and surface currents'. *Journal of Geophysical Research*. 1985;**90**(C1):1049–59.

[6]   Bell P.S., Osler J.C. 'Mapping bathymetry using X-band marine radar data recorded from a moving vessel'. *Ocean Dynamics*. 2011;**61**(12):2141–56.

[7]   Wilson G.W., Özkan-Haller H.T., Holman R.A., Haller M.C., Honegger D.A., Chickadel C.C. 'Surf zone bathymetry and circulation predictions via data assimilation of remote sensing observations'. *Journal of Geophysical Research: Oceans*. 2014;**119**(3):1993–2016.

[8]   Hwang P.A., Sletten M.A., Toporkov J.V. 'A note on Doppler processing of coherent radar backscatter from the water surface: with application to ocean surface wave measurements'. *Journal of Geophysical Research Oceans*. 2010;**115**(C3).

[9]   Carrasco R., Streßer M., Horstmann J. 'A simple method for retrieving significant wave height from Dopplerized X-band radar'. *Ocean Science*. 2017;**13**(1):95–103.

[10]   Støle-Hentschel S., Seemann J., Nieto Borge J.C., Trulsen K. 'Consistency between sea surface reconstructions from nautical X-band radar Doppler and

amplitude measurements'. *Journal of Atmospheric and Oceanic Technology.* 2018;**35**(6):1201–20.

[11] Kammerer A.J., Hackett E.E., Erin Hackett. 'Use of proper orthogonal decomposition for extraction of ocean surface wave fields from X-band radar measurements of the sea surface'. *Remote Sensing.* 2017;**9**(9):881.

[12] Robinson I.S., Ward N.P., Gommenginger C.P., Tenorio-Gonzales M.A. 'Coastal oceanography applications of digital image data from marine radar'. *Journal of Atmospheric and Oceanic Technology.* 2000;**17**(5):721–35.

[13] Gommenginger C.P., Ward N.P., Fisher G.J., Robinson I.S., Boxall S.R. 'Quantitative microwave backscatter measurements from the ocean surface using digital marine radar images'. *Journal of Atmospheric and Oceanic Technology.* 2000;**17**(5):665–78.

[14] McGregor J.A., Poulter E.M., Smith M.J. '*S* band Doppler radar measurements of bathymetry, wave energy fluxes, and dissipation across an offshore bar'. *Journal of Geophysical Research: Oceans.* 1998;**103**(C9):18779–89.

[15] Ijima T., Takahashi T., Sasaki H. 'Application of radars to wave observations'. *Coastal Engineering Proceeding.* 1964;**30**:10–22.

[16] Williams P. 'Applications of Remote Sensing by Conventional Radars'. *Surveillance of Environmental Pollution and Resources by Electromagnetic Waves.* Springer; 1978. pp. 299–308.

[17] Watson G., Robinson I.S. 'A study of internal wave propagation in the Strait of Gibraltar using shore-based marine radar images'. *Journal of Physical Oceanography.* 1990;**20**(3):374–95.

[18] Reichert K., Hessner K., Nieto Borge J.C., Dittmer J. 'WaMoS II: a radar based wave and current monitoring system'. The Ninth International Offshore and Polar Engineering Conference; Brest, France; 1999.

[19] Borge J.C.N., Reichert K., Dittmer J. 'Use of nautical radar as a wave monitoring instrument'. *Coastal Engineering.* 1999;**37**(3–4):331–42.

[20] Trizna D.B. 'Errors in bathymetric retrievals using linear dispersion in 3-D FFT analysis of marine radar ocean wave imagery'. *IEEE Transactions on Geoscience and Remote Sensing.* 2001;**39**(11):2465–9.

[21] Ivonin D.V., Telegin V.A., Chernyshov P.V., Myslenkov S.A., Kuklev S.B. 'Possibilities of X-band nautical radars for monitoring of wind waves near the coast'. *Oceanology.* 2016;**56**(4):591–600.

[22] Valenzuela G.R. 'Theories for the interaction of electromagnetic and oceanic waves ? A review'. *Boundary-Layer Meteorology.* 1978;**13**(1):61–85.

[23] Alpers W.R., Ross D.B., Rufenach C.L. 'On the detectability of ocean surface waves by real and synthetic aperture radar'. *Journal of Geophysical Research.* 1981;**86**(C7):6481–98.

[24] Plant W.J. 'Bragg Scattering of Electromagnetic Waves from the Air/Sea Interface'. *Surface Waves and Fluxes.* Springer; 1990. pp. 41–108.

[25] Wetzel L.B. 'Electromagnetic Scattering from the Sea at Low Grazing Angles'. *Surface Waves and Fluxes.* Springer; 1990. pp. 109–71.

[26]   Story W.R., TC F., Hackett E.E. 'Radar measurement of ocean waves'. *International Conference on Offshore Mechanics and Arctic Engineering*; 2011. pp. 707–17.

[27]   Trizna D.B. 'Statistics of low grazing angle radar sea scatter for moderate and fully developed ocean waves'. *IEEE Transactions on Antennas and Propagation*. 1991;**39**(12):1681–90.

[28]   Smith M.J., Poulter E.M., McGregor J.A. 'Doppler radar measurements of wave groups and breaking waves'. *Journal of Geophysical Research: Oceans*. 1996;**101**(C6):14269–82.

[29]   Chen Z., He Y., Yang W. 'Study of ocean waves measured by collocated HH and VV polarized X-band marine radars'. *International Journal of Antennas and Propagation*. 2016;**2016**(1):1–12.

[30]   Zhu X., Li Y., Feng H., Liu B., Xu J. 'Oil spill detection method using X-band marine radar imagery'. *Journal of Applied Remote Sensing*. 2015;**9**(1):095985.

[31]   Lund B., Graber H.C., Persson P.O.G. 'Arctic sea ice drift measured by shipboard marine radar'. *Journal of Geophysical Research: Oceans*. 2018;**123**(6):4298–321.

[32]   Atanassov V., Rosenthal W., Ziemer F. 'Removal of ambiguity of two-dimensional power spectra obtained by processing SHIP radar images of ocean waves'. *Journal of Geophysical Research*. 1985;**90**(C1):1061–7.

[33]   Buckley J.R., Allingham M., Michaud R. 'On the use of marine radar imagery for estimation of properties of the directional spectrum of the sea surface'. *Atmosphere-Ocean*. 1994;**32**(1):195–213.

[34]   Wolf J., Bell P.S. 'Waves at Holderness from X-band radar'. *Coastal Engineering*. 2001;**43**(3-4):247–63.

[35]   Izquierdo P., Nieto Borge J.C., Guedes Soares C., Sanz González R., Rodríguez G.R. 'Comparison of wave spectra from nautical radar images and scalar buoy data'. *Journal of Waterway, Port, Coastal, and Ocean Engineering*. 2005;**131**(3):123–31.

[36]   Lund B., Collins C.O., Graber H.C., Terrill E., Herbers T.H.C. 'Marine radar ocean wave retrieval's dependency on range and azimuth'. *Ocean Dynamics*. 2014;**64**(7):999–1018.

[37]   Bell P.S. 'Shallow water bathymetry derived from an analysis of X-band marine radar images of waves'. *Coastal Engineering*. 1999;**37**(3–4):513–27.

[38]   Honegger D.A., Haller M.C., Holman R.A. 'High-Resolution bathymetry estimates via X-band marine radar: 1. beaches'. *Coastal Engineering*. 2019;**149**(4):39–48.

[39]   Senet C.M., Seemann J., Flampouris S., Ziemer F. 'Determination of Bathymetric and current maps by the method disc based on the analysis of Nautical X-band radar image sequences of the sea surface'. November 2007. *IEEE Transactions on Geoscience and Remote Sensing*; 2008. pp. 2267–79.

[40]   Hessner K., Reichert K., Borge J.C.N., Stevens C.L., Smith M.J. 'High-resolution X-band radar measurements of currents, bathymetry and sea state in highly inhomogeneous coastal areas'. *Ocean Dynamics*. 2014;**64**(7):989–98.

[41] Honegger D.A., Haller M.C., Holman R.A. 'High-resolution bathymetry estimates via X-band marine radar: 2. Effects of currents at tidal inlets'. *Coastal
Engineering*. 2020;**156**(7):103626.

[42] Nieto Borge J., RodrÍguez G.R., Hessner K., González P.I. 'Inversion of marine radar images for surface wave analysis'. *Journal of Atmospheric and
Oceanic Technology*. 2004;**21**(8):1291–300.

[43] Dankert H., Rosenthal W. 'Ocean surface determination from X-band radar-
image sequences'. *Journal of Geophysical Research. Oceans*. 2004;**109**(C4).

[44] Wijaya A.P., Naaijen P., Andonowati., Groesen Evan, van Groesen E.
'Reconstruction and future prediction of the sea surface from radar observations'. *Ocean Engineering*. 2015;**106**(C7):261–70.

[45] Qi Y., Xiao W., Yue D.K.P. 'Phase-resolved wave field simulation calibration of sea surface reconstruction using noncoherent marine radar'. *Journal
of Atmospheric and Oceanic Technology*. 2016;**33**(6):1135–49.

[46] Belmont M.R., Christmas J., Dannenberg J. 'An examination of the feasibility
of linear deterministic sea wave prediction in multidirectional seas using wave
profiling radar: theory, simulation, and sea trials'. *Journal of Atmospheric and
Oceanic Technology*. 2014;**31**(7):1601–14.

[47] Simpson A., Haller M., Walker D., Lynett P., Honegger D. 'Wave-by-Wave
forecasting via assimilation of marine radar data'. *Journal of Atmospheric
and Oceanic Technology*. 2020;**37**(7):1269–88.

[48] Haller M.C., Lyzenga D.R. 'Comparison of radar and video observations
of shallow water breaking waves'. *IEEE Transactions on Geoscience and
Remote Sensing*. 2003;**41**(4):832–44.

[49] Liu Y., Frasier S.J., McIntosh R.E. 'Measurement and classification of
low-grazing-angle radar sea spikes'. *IEEE Transactions on Antennas and
Propagation*. 1998;**46**(1):27–40.

[50] Lewis B.L., Olin I.D. 'Experimental study and theoretical model of high-resolution radar backscatter from the sea'. *Radio Science*.
1980;**15**(04):815–28.

[51] Catalán P.A., Haller M.C., Plant W.J. 'Microwave backscattering
from surf zone waves'. *Journal of Geophysical Research: Oceans*.
2014;**119**(5):3098–120.

[52] Ruessink B.G., Bell P.S., van Enckevort I.M.J., Aarninkhof S.G.J. 'Nearshore
bar crest location quantified from time-averaged X-band radar images'.
Coastal Engineering. 2002;**45**(1):19–32.

[53] Díaz Méndez G.M., Haller M.C., Raubenheimer B., Elgar S., Honegger D.A.
'Radar remote sensing estimates of waves and wave forcing at a tidal inlet'.
Journal of Atmospheric and Oceanic Technology. 2015;**32**(4):842–54.

[54] Díaz H., Catalán P., Wilson G. 'Quantification of two-dimensional wave
breaking dissipation in the surf zone from remote sensing data'. *Remote
Sensing*. 2018;**10**(1):38.

[55] Lund B., Haus B.K., Horstmann J. 'Near-surface current mapping by shipboard marine X-band radar: a validation'. *Journal of Atmospheric and
Oceanic Technology*. 2018;**35**(5):1077–90.

[56] Smeltzer B.K., Æsøy E., Ådnøy A., Ellingsen S.Å. 'An improved method for determining near-surface currents from wave dispersion measurements'. *Journal of Geophysical Research: Oceans.* 2019;**124**(12):8832–51.

[57] Campana J., Terrill E.J., de Paolo T. 'A new inversion method to obtain upper-ocean current-depth profiles using X-band observations of deep-water waves'. *Journal of Atmospheric and Oceanic Technology.* 2017;**34**(5):957–70.

[58] Horner-Devine A.R., Pietrzak J.D., Souza A.J. 'Cross-shore transport of nearshore sediment by river plume frontal pumping'. *Geophysical Research Letters.* 2017;**44**(12):6343–51.

[59] Pan J., Jay D.A. 'Dynamic characteristics and horizontal transports of internal solitons generated at the Columbia river plume front'. *Continental Shelf Research.* 2009;**29**(1):252–62.

[60] Honegger D.A., Haller M.C., Geyer W.R., Farquharson G. 'Oblique internal hydraulic jumps at a stratified estuary mouth'. *Journal of Physical Oceanography.* 2017;**47**(1):85–100.

[61] Ramos R.J., Lund B., Graber H.C. 'Determination of internal wave properties from X-band radar observations'. *Ocean Engineering.* 2009;**36**(14):1039–47.

[62] Lund B., Graber H.C., Xue J., Romeiser R. 'Analysis of internal wave signatures in marine radar data'. *IEEE Transactions on Geoscience and Remote Sensing.* 2013;**51**(9):4840–52.

[63] Haller M.C., Honegger D., Catalan P.A. 'Rip current observations via marine radar'. *Journal of Waterway, Port, Coastal, and Ocean Engineering.* 2014;**140**(2):115–24.

[64] Hart C., Miskin E. 'Developments in the method of determination of beach gradients by wave velocities'. Air Survey Research Paper No. 15; Directorate of Military Survey, UK War Office; 1945.

[65] Chernyshov P., Vrecica T., Streßer M., Carrasco R., Toledo Y. 'Rapid wavelet-based bathymetry inversion method for nearshore X-band radars'. *Remote Sensing of Environment.* 2020;**240**(3):111688.

[66] McNinch J.E. 'Bar and swash imaging radar (BASIR): a mobile X-band radar designed for mapping nearshore sand bars and swash-defined shorelines over large distances'. *Journal of Coastal Research.* 2007;**2007**(1):59–74.

[67] Lippmann T.C., Holman R.A. 'Quantification of sand bar morphology: a video technique based on wave dissipation'. *Journal of Geophysical Research.* 1989;**94**(C1):995–1011.

[68] Takewaka S. 'Measurements of shoreline positions and intertidal foreshore slopes with X-band marine radar system'. *Coastal Engineering Journal.* 2005;**47**(2–3):91–107.

[69] Galal E.M., Takewaka S. 'Longshore migration of shoreline mega-cusps observed with X-band radar'. *Coastal Engineering Journal.* 2008;**50**(03):247–76.

[70] Katzin M. 'On the mechanisms of radar sea clutter'. *Proceedings of the IRE.* 1957;**45**(1):44–54.

[71] Dankert H., Horstmann J., Rosenthal W. 'Ocean wind fields retrieved from radar-image sequences'. *Journal of geophysical research. Oceans.* 2003;**108**(C11):3352.

[72] Lund B., Graber H.C., Romeiser R. 'Wind retrieval from shipborne nautical X-band radar data'. *IEEE Transactions on Geoscience and Remote Sensing.* 2012;**50**(10):3800–11.

[73] Chernyshov P., Vrecica T., Toledo Y. 'Inversion of nearshore X-band radar images to sea surface elevation maps'. *Remote Sensing.* 2018;**10**(12):1919.

[74] Plant W.J., Farquharson G. 'Origins of features in wave number-frequency spectra of space-time images of the ocean'. *Journal of Geophysical Research. Oceans.* 2012;**117**(C6).

[75] Plant N.G., Holland K.T., Haller M.C. 'Ocean wavenumber estimation from wave-resolving time series imagery'. *IEEE Transactions on Geoscience and Remote Sensing.* 2008;**46**(9):2644–58.

[76] Holman R., Plant N., Holland T. 'cBathy: a robust algorithm for estimating nearshore bathymetry'. *Journal of Geophysical Research: Oceans.* 2013;**118**(5):2595–609.

Chapter 12

Observation of sea surface waves by noncoherent X-band marine radar

Zhongbiao Chen[1], Yijun He[1], and Weimin Huang[2]

12.1 Introduction

Ordinary X-band marine radar can image the sea surface with high spatial and temporal resolutions (typically less than 10 m and 1–3 s), and it has been used to observe sea surface wave [1–3], current [4], and wind [5, 6]. At low grazing angles, the incident electromagnetic waves interact with small capillary waves through Bragg scattering, and long gravity waves are imaged by modulating the short Bragg waves. Multiple modulation mechanisms contribute to the imaging of X-band marine radar, such as hydrodynamic modulation, tilt modulation, shadowing modulation, orbital modulation as well as wind modulation [7]. The modulations result in alternating bright and dark stripe radar images. Four gray-level images under typical conditions are shown in Figure 12.1.

Various algorithms have been developed to estimate wave parameters from X-band marine radar images. The widely used algorithms are based on three-dimensional (3D) fast Fourier transform (FFT) [1, 3], and they work well for homogeneous regions in the open sea (e.g., Figure 12.1(a)). To analyze inhomogeneous wave fields that are common in nature (e.g., Figure 12.1(b) and (c)), novel data analysis techniques are utilized to decompose the wave field into different scales or components, such as the two-dimensional (2D) continuous wavelet transform [8], the empirical orthogonal function (EOF) [9], the rotated EOF [10], the empirical mode decomposition [11], etc. However, those methods involve a calibration process using external instruments. To eliminate the need of calibration with buoy data, the algorithms based on theories of shadowing [12] and wavelet analysis [13] have been presented. Comprehensive review and comparison of the algorithms are provided in [14]. In addition, the performance of X-band marine radar can be affected by rain, which attenuates the radar backscatter from the sea surface and thus blurs the radar images (see Figure 12.1(d)), so the estimated wave parameters may be less accurate under rainy conditions [15].

[1]School of Marine Sciences, Nanjing University of Information Science and Technology, China
[2]Department of Electrical and Computer Engineering, Memorial University of Newfoundland, Canada

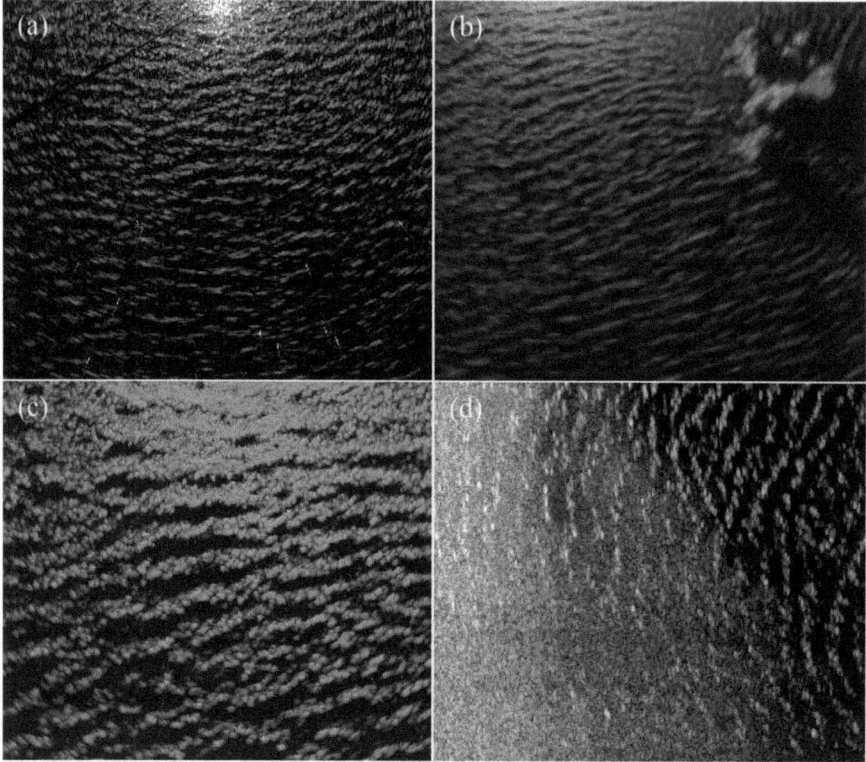

Figure 12.1   Four X-band marine radar images recorded under different conditions: (a) at open sea, (b) in coastal region, (c) under high sea state, and (d) under rain

In this chapter, two typical algorithms for retrieving wave information from X-band marine radar images are described, i.e., the traditional 3D FFT-based algorithm and the EOF-based algorithm, along with corresponding observation results.

12.2   FFT-based algorithms

The traditional algorithms for estimating wave parameters from X-band marine radar images are based on 3D FFT. By applying 3D FFT to a radar image sequence, the 3D image spectrum is obtained. Due to the nonlinearity of the modulation processes, the image spectrum is not a direct representation of the wave spectrum. The image spectrum is then filtered according to the dispersion relationship and multiplied by a modulation transfer function (MTF) to produce the 2D wave spectrum [16], from which the wave parameters (e.g., the wave period, wave direction, and wavelength) can be deduced. However, since the gray levels of X-band marine radar images are not calibrated, the significant wave height (SWH) cannot be obtained directly from

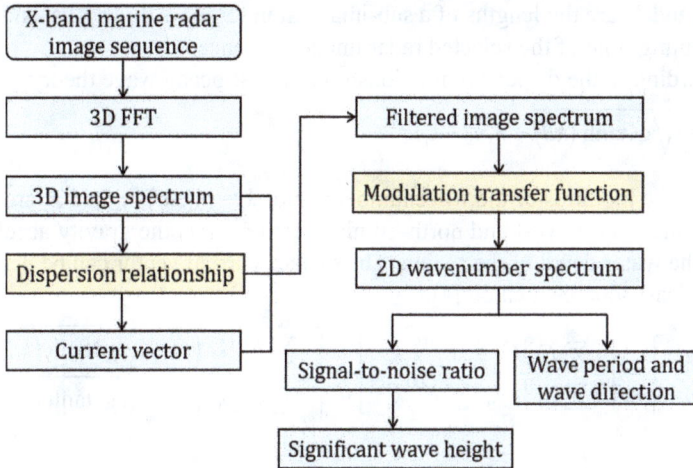

Figure 12.2   Schematic of the algorithm to retrieve wave parameters from X-band marine radar images based on 3D FFT

the spectrum, but using a linear relationship with the signal-to-noise ratio (SNR) [3, 17], which needs to be determined by using auxiliary data. The flow chart of the algorithms is shown in Figure 12.2, and their details are explained as follows.

12.2.1   Retrieval of wave spectrum

A rectangular sub-image is selected first from each gray-level radar image. The imaging of X-band marine radar is affected by azimuthal modulation [18], i.e., the wave patterns in the gray-level image are clear at the azimuths that are close to the peak wave direction, while they are vague at the perpendicular azimuths, so the sub-image is usually selected from the azimuths that are close to the peak wave direction. In order to mitigate the effect of noises and radial decay, the sub-images may be processed by digital filters, e.g., the Butterworth filters or Gaussian filters [19].

Then 3D FFT is applied to transform a sequence of sub-images into the spectral domain,

$$F_I^{(3)}\left(k_x, k_y, \omega\right) = \int_{-\infty}^{\infty} \int_{-\infty}^{\infty} \int_{-\infty}^{\infty} I\left(x, y, t\right) e^{-i\left(k_x x + k_y y - \omega t\right)} dx\, dy\, dt \qquad (12.1)$$

where the subscript "*I*" refers to the image spectrum, the superscript "(3)" refers to the 3D spectrum, *x* and *y* are the coordinates of the images, *t* is the time label of the image, $I(x,y,t)$ is the gray level of the pixel in the image, and k_x, k_y, and ω are the wavenumbers and angular frequency. The wavenumber and angular frequency resolutions of the spectrum are determined as

$$\Delta k_x = \frac{2\pi}{L_x}, \quad \Delta k_y = \frac{2\pi}{L_y}, \quad \Delta \omega = \frac{2\pi}{T}, \qquad (12.2)$$

where L_x and L_y are the lengths of a sub-image along x and y directions, and T is the total sampling time of the selected radar image sequence.

According to the dispersion relationship of linear ocean wave theory,

$$\omega = \sqrt{gk\tanh(kd)} + k_x u_x + k_y u_y \tag{12.3}$$

where k is the modulus of the wavenumber vector $\vec{k} = (k_x, k_y)$, u_x and u_y are the surface currents in east–west and north–south directions, g is the gravity acceleration, and d is the water depth of the region. The surface current vector can be determined using the least squares method [4],

$$\begin{bmatrix} u_x \\ u_y \end{bmatrix} = \begin{bmatrix} \sum F_I^{(3)} k_x^2 & \sum F_I^{(3)} k_x k_y \\ \sum F_I^{(3)} k_x k_y & \sum F_I^{(3)} k_y^2 \end{bmatrix}^{-1} \begin{bmatrix} \sum F_I^{(3)} k_x \left(\omega - \sqrt{gk\tanh(kd)}\right) \\ \sum F_I^{(3)} k_y \left(\omega - \sqrt{gk\tanh(kd)}\right) \end{bmatrix} \tag{12.4}$$

To estimate the wavenumber spectrum, the image spectrum (12.1) is filtered based on the dispersion relationship of linear waves (12.3),

$$F_F^{(3)}(k_x, k_y, \omega) = F_I^{(3)}(k_x, k_y, \omega) \cdot \delta\left(\omega - \sqrt{gk\tanh\left(|\vec{k}|d\right)} - k_x u_x - k_y u_y\right) \tag{12.5}$$

where the subscript "F" refers to the filtered image spectrum and $\delta(\cdot)$ is the Dirac's delta function. Next, the MTF is multiplied to produce the wave spectrum, which is denoted with a subscript "w,"

$$F_w^{(3)}(k_x, k_y, \omega) = F_F^{(3)}(k_x, k_y, \omega) \cdot |M(k)|^2 \tag{12.6}$$

where $M(k)$ is the MTF, which is often selected to be an empirical function of wavenumber k (see Section 12.2.3).

The 2D and one-dimensional (1D) wave spectra are derived by integrating the 3D wavenumber spectrum, respectively, as

$$F_w^{(2)}(\vec{k}) = \int_{\omega_{th}}^{\omega_c} F_w^{(3)}(k_x, k_y, \omega)\, d\omega \tag{12.7}$$

$$F_w^{(1)}(k) = \int_{-\pi}^{\pi} F_w^{(2)}(\vec{k}) k\, d\theta \tag{12.8}$$

where θ is the direction of the wavenumber vector $\vec{k}$, ω_{th} is the threshold of the angular frequency of gravity waves (e.g., 0.31 rad/s), and ω_c is the cutoff angular frequency that is determined as $\omega_c = \pi/\Delta t$, where Δt is the sampling period of radar.

12.2.2   Estimation of wave parameters

Wave parameters such as wave direction, wave period, and wavelength can be calculated from the derived 2D wavenumber spectrum (12.7). The peak wave direction and wavelength are obtained by

$$\theta = \arctan\left(\frac{k_{py}}{k_{px}}\right) \tag{12.9}$$

$$L = \frac{2\pi}{\sqrt{k_{px}^2 + k_{py}^2}} \tag{12.10}$$

where k_{px} and k_{py} are the peak wavenumbers of the wave spectrum along x and y directions.

The frequency spectrum $F(f)$ can be derived from the wavenumber spectrum by [20],

$$\int_{-k_c}^{k_c} F_w^{(1)}(k)\, k\, dk = \int_{f_{th}}^{f_c} F(f)\, df \tag{12.11}$$

where f is the wave frequency, the lower and upper bounds of wave frequency are $f_c = \omega_c/2\pi$ and $f_{th} = \omega_{th}/2\pi$, respectively, and k_c is the cutoff wavenumber and equals $k_c = \pi/\Delta r$, where Δr is the radar range resolution. The peak wave period and mean wave period are, respectively, calculated as

$$T_p = \frac{1}{f_p} \tag{12.12}$$

$$\bar{T} = \frac{\int_{f_{th}}^{f_c} F(f)\, df}{\int_{f_{th}}^{f_c} F(f)\, f\, df} \tag{12.13}$$

where f_p is the peak wave frequency.

Since the gray levels of X-band marine radar images are not calibrated, SWH H_s is estimated by assuming an empirical relationship with the square root of SNR [3],

$$H_s = A + B\sqrt{SNR} \tag{12.14}$$

where A and B are two undetermined coefficients. To determine the SNR, the signal contribution is assumed as the total energy of the wave spectrum estimated from sea clutter analysis, and that of noise is computed as the energy due to the speckle caused by sea surface roughness [3], i.e.,

$$SNR = \frac{\int_{\Omega_{\bar{k}}^{\alpha}} F_w^{(2)}(\bar{k})\, k\, dk\, d\theta}{\int_{\Omega_{BGN}} F_{BGN}^{(3)}(\bar{k}, \omega)\, k\, dk\, d\theta\, d\omega} \tag{12.15}$$

where

$$\Omega_{\bar{k}}^{\alpha} = \left\{ \bar{k} \in \left[-k_{xc}, k_{xc}\right) \times \left[-k_{yc}, k_{yc}\right) \mid F_w^{(2)}(\bar{k}) \geq \alpha \max\left[F_w^{(2)}(\bar{k}) \right], 0 \leq \alpha \leq 1 \right\},$$

$$\tag{12.16}$$

$$\Omega_{BGN} = \left[-k_{xc}, k_{xc}\right) \times \left[-k_{yc}, k_{yc}\right) \times \left[\omega_{th}, \omega_c\right) \tag{12.17}$$

$$F_{BGN}^{(3)}(\bar{k}, \omega) \approx F_I^{(3)}(\bar{k}, \omega) - F_F^{(3)}(\bar{k}, \omega) - F_{HH}^{(3)}(\bar{k}, \omega) \tag{12.18}$$

where k_{xc} and k_{yc} are the limits of the wavenumber of gravity waves and $F_{HH}^{(3)}(\bar{k}, \omega)$ is the high-harmonic spectrum [3].

The unknown coefficients in (12.14) are usually determined using in situ buoy data. Based on the SNR estimated from radar images and the SWH measured by buoy, the coefficients A and B can be determined through least square fitting.

Besides, the coefficients may change for different radar systems and observation regions, so they need to be determined for each application.

12.2.3   Modulation transfer function

As discussed earlier, to convert the image spectrum estimated from radar images to the ocean wave spectrum, an MTF is utilized in (12.6). The MTF depends on many factors, such as radar configurations, polarization, incidence angle, sea state, and atmospheric conditions, so it is often determined empirically as

$$|M(k)|^2 = \frac{F_I(k)}{F_b(k)} \tag{12.19}$$

where $F_b(k)$ is the wave spectrum measured by buoy and $F_I(k)$ is the 1D image spectrum obtained by integrating the 3D image spectrum in (12.1).

Various MTFs have been developed for different radar systems. For example, one commonly used MTF is a power function of wavenumber,

$$|M(k)|^2 \propto k^\beta \tag{12.20}$$

where β is a constant. The constant of $\beta = 1.2$ is used in [16], i.e.,

$$|M(k)|^2 \propto k^{1.2} \tag{12.21}$$

A piecewise function of wavenumber was used in [21],

$$|M(k)|^2 \propto \begin{cases} k^2, & \text{for } k < \pi/30 \\ k^{-1.8}, & \text{for } k \geq \pi/30 \end{cases} \tag{12.22}$$

Conventional X-band marine radar employs HH polarization, and most MTFs developed are only suitable for HH-polarized radar images in deep water [16]. To investigate the MTF for VV polarization in the coastal region, the spectra derived from VV-polarized radar images under different sea states are analyzed in Figure 12.3(a), 12.3(c), 12.3(e), and 12.3(g). Under low sea state (Figure 12.3(a)), the radar-derived spectrum agrees well with that of buoy. When the sea state increases (Figure 12.3(c), 12.3(e), and 12.3(g)), the differences between them become larger. Under high sea state (Figure 12.3(g)), the peak wavenumber from radar data deviates significantly from that of buoy because the shadowing modulation becomes more important when the SWH gets larger (Figure 12.1(c)).

The MTFs derived based on (12.19) are shown in Figure 12.3, and they are different from the linear MTF (12.20) in logarithm form. For the wavenumber of 0.03–0.31 rad/m, the MTF decreases first and then increases as the wavenumber increases, so a quadratic polynomial function is more appropriate to fit them [22],

$$\ln\left(|M(k)|^{-2}\right) = p_1 \ln^2 k + p_2 \ln k + p_3 \tag{12.23}$$

where $\ln(\cdot)$ is the natural logarithm and p_1, p_2, and p_3 are three coefficients that can be determined by the least square fitting method. For the wavenumber of 0.03–0.14 rad/m, the MTF decreases with wavenumber, so the power decay law (12.20) is used to determine the MTF for this range.

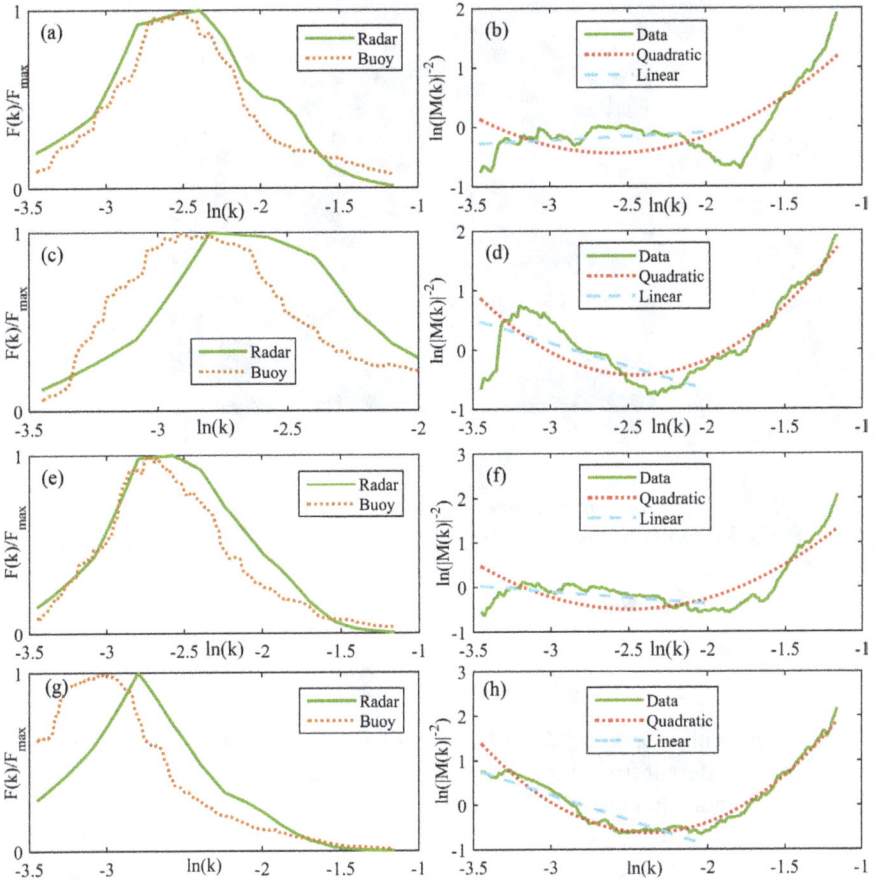

Figure 12.3   Left panels: normalized 1D wavenumber spectra derived from X-band marine radar image sequences and buoy data; right panels: MTF estimated from experimental data, fitted using a quadratic function and a linear function. They are measured under four typical sea state conditions when the SWHs are: (a) and (b): 0.40 m; (c) and (d): 1.51 m; (e) and (f): 2.25 m; and (g) and (h): 3.38 m, respectively.

By using the VV-polarized X-band radar images and simultaneous measurements of buoy, the coefficients of (12.23) are determined, and the MTFs for VV polarization are obtained as

$$\ln\left(|M(k)|^{-2}\right) = 1.16 \ln^2 k + 5.6 \ln k + 1.23 \tag{12.24}$$

for $0.03 \leq k \leq 0.31$ rad/m.

For comparison, the MTF in the form of power decay law (12.20) is also fitted and found to be

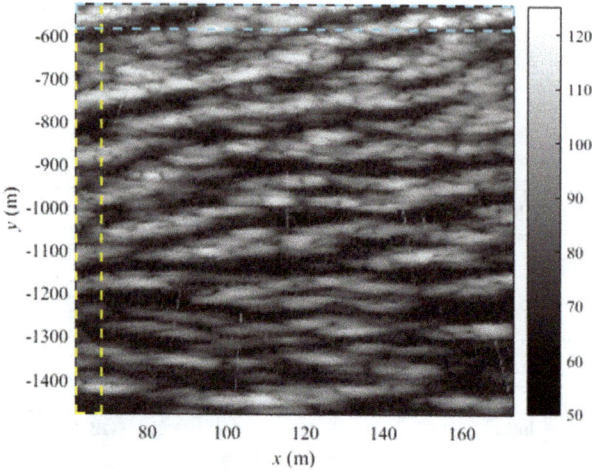

Figure 12.4　*A sub-image is selected from the gray-level X-band marine radar image in Figure 12.1(b). The sub-image is used for the 3D FFT-based algorithm, and the two dashed bins are used for the EOF-based algorithm.*

$$|M(k)|^2 \propto k^{0.55} \qquad\qquad (12.25)$$

which is only valid for $0.03 \leq k \leq 0.14$ rad/m. Thus, the MTF for VV polarization is significantly different from that for HH polarization. In addition, the quadratic MTF in (12.24) is more suitable for VV polarization in the coastal region. In the coastal region, the wave fields were inhomogeneous because of the shoaling of waves (Figure 12.1), so the modulations of waves on radar images may be more complex than those in deep water.

12.2.4　Example

The 3D FFT-based algorithm has been validated in previous studies [14]. To illustrate the implementation process of the algorithm, the X-band marine radar image in Figure 12.1(b) is used as an example. The radial resolution of the gray-level image was 3.75 m and the antenna rotation speed was 24 rotations per minute (rpm), so the cutoff wavenumber and cutoff angular frequency are 0.84 rad/m and 1.26 rad/s, respectively. Besides, 32 images were recorded in each image sequence. A sub-image is selected first from Figure 12.1(b) and shown in Figure 12.4. The wavelengths of the upper part of the sub-image (i.e., $y > -1200$ m) are larger than those of the bottom part (i.e., $y < -1200$ m), which indicates that the wave field was inhomogeneous.

Then 3D FFT is applied to the image sequence, and some of the wavenumber spectra are shown in Figure 12.5. The theoretical dispersion relation (12.3) in still water has also been added (the dashed circles) in the figures for comparison. As seen in Figure 12.5(a) and 12.5(b), for the angular frequencies of 0.08 rad/s and 0.31

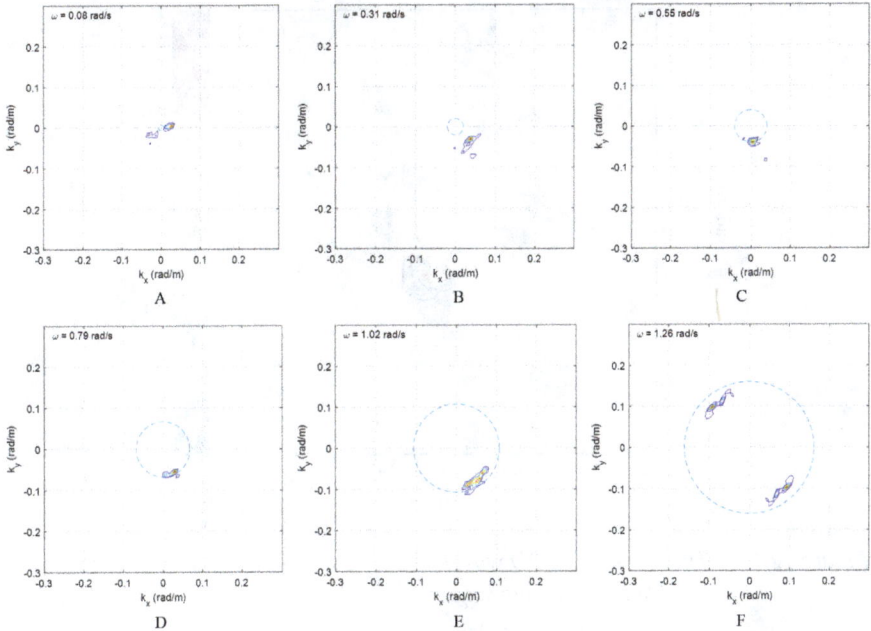

Figure 12.5   The normalized wavenumber spectra that are derived from an X-band marine radar image sequence for different angular frequencies. (a) ω = 0.08 rad/s, (b) ω = 0.31 rad/s, (c) ω = 0.55 rad/s, (d) ω = 0.79 rad/s, (e) ω = 1.02 rad/s, and (f) ω = 1.26 rad/s. The dashed circles are the theoretical dispersion relation with the current speed of 0.

rad/s, the wavenumber spectra deviate largely from the dispersion relation, so the spectra may not be associated with gravity waves. In this study, the threshold frequency of gravity waves is selected to be 0.31 rad/s. For the angular frequencies of 0.55–0.79 rad/s (see Figure 12.5(c) and 12.5(d)), the wavenumber spectra are consistent with the dispersion relation. For the angular frequencies of 1.02–1.26 rad/s (Figure 12.5(e) and 12.5(f)), the wavenumber spectra deviate slightly from the dispersion relation, and this deviation is caused by the surface current. By using (12.4), the surface currents of the region are derived to be u_x= 0.11 m/s and u_y= 0.07 m/s.

The 2D wavenumber spectrum that is derived using (12.5–12.7) is shown in Figure 12.6. Although the MTF is different for different radar systems and circumstances, the MTF is taken to be a constant of 1 here to illustrate the basic idea of the algorithm. By using (12.9–12.10), the peak wave direction and wavelength are determined to be 160.2° and 98.9 m, which agree with the wave patterns in Figure 12.4.

After that, the 1D frequency spectrum can be obtained according to (12.8) and (12.11), as shown in Figure 12.7. For comparison, the wave frequency spectrum

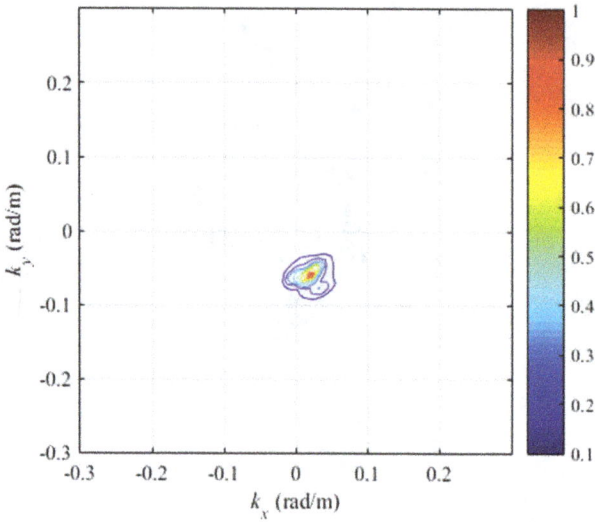

Figure 12.6   The normalized 2D wavenumber spectrum derived from X-band marine radar image sequence

measured by a simultaneous buoy is also shown. There are two peaks in the spectrum measured by buoy (i.e., 0.115 and 0.14 Hz) in the frequency range of 0.1–0.15 Hz, and there are three smaller peaks in the range of 0.15–0.25 Hz, which indicate that

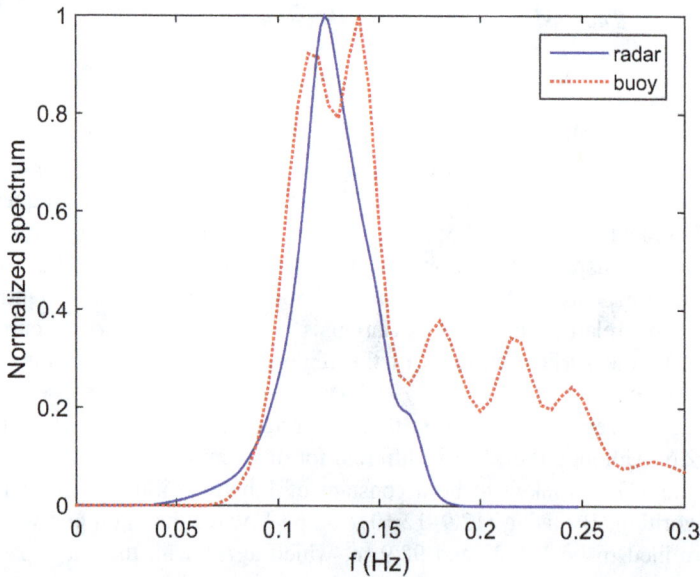

Figure 12.7   The frequency spectra measured by buoy and that derived from X-band marine radar image sequence

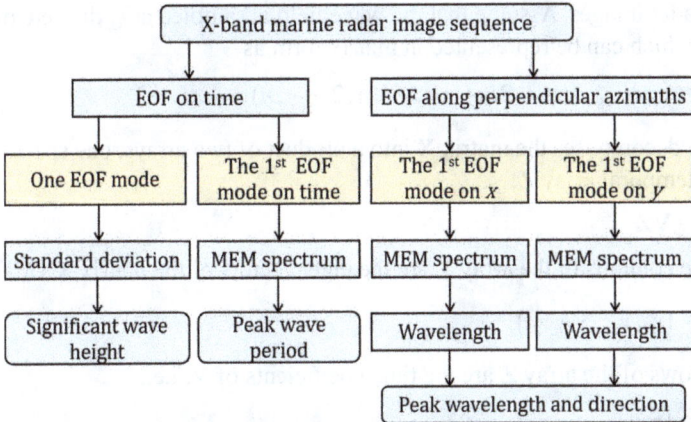

Figure 12.8   Flow chart of the EOF-based method to retrieve wave parameters from X-band marine radar images

the wave field was a mixed sea. However, there is only one peak in the spectrum derived from X-band marine radar images (i.e., 0.123 Hz), so the algorithm based on 3D FFT may lose information for inhomogeneous wave field. Moreover, the difference between the spectrum from buoy and that derived from radar images cannot be reduced by using different MTFs in (12.20–12.25), because the MTF only changes the locations of peaks rather than the number of peaks.

Finally, the SWH can be estimated by (12.14), and the coefficients need to be determined using buoy data.

12.3   The algorithm based on EOF analysis

Although the 3D FFT-based algorithms are widely used, several disadvantages should be noted. First, as mentioned earlier, the MTF can be affected by many factors, so it is difficult to be determined and different empirical MTFs need to be chosen for different radar systems and regions [16, 21, 22]. Second, the FFT analysis is suitable for the signal with spatial homogeneity and temporal stationarity, which are often violated in reality [8]. For example, Figure 12.1(b) shows the shoaling of waves in nearshore region. Figure 12.1(c) shows that the shadows of waves are significant in the middle of the region, while they are not so large on the right part of the image due to the inhomogeneous wave field under high sea state. To overcome such difficulties, the algorithms based on EOF analysis are introduced. The flow chart of the algorithm is shown in Figure 12.8.

12.3.1   EOF decomposition

The EOF is a powerful tool for data compression and dimensionality reduction, so it can be used to extract different modes of inhomogeneous wave field from X-band marine radar images. Assume that the wave field is sampled at m discrete points for n times, which can be represented in matrix form as

$$\mathbf{X} = \left(x_{ij}\right)_{m \times n}, (i = 1, 2, \cdots, m; \ j = 1, 2, \cdots, n) \tag{12.26}$$

The EOF decomposes the matrix $\mathbf{X}$ into a product of two arrays, one spatial array $\mathbf{V}$ and one temporal array $\mathbf{Z}$,

$$\mathbf{X} = \mathbf{VZ} \tag{12.27}$$

where the columns of the array $\mathbf{V}$ are the eigenvectors of the matrix $\mathbf{XX}^T$, i.e.,

$$v_j = \left(v_{j1}, v_{j2}, \cdots, v_{jm}\right)^T \tag{12.28}$$

and the rows of the array $\mathbf{Z}$ are the time coefficients of $\mathbf{V}$, i.e.,

$$z_i = \left(z_{i1}, z_{i2}, \cdots, z_{in}\right) \tag{12.29}$$

Thus the matrix $\mathbf{Z}$ can be obtained as

$$\mathbf{Z} = \mathbf{V}^T\mathbf{X} \tag{12.30}$$

In general, the EOF analysis expresses the discretely sampled field $\mathbf{X}$ as the superposition of m mutually orthogonal eigenvectors v_j modulated by m mutually uncorrelated time coefficients z_i. The eigenvectors provide the spatial patterns of the wave field, and the associated time coefficients (i.e., principal components) represent the significance of the spatial patterns at different times. The spatial patterns and principal components are matched pairs, which are generally referred to as EOF modes [23]. In the following, the principal components of EOF modes are used to estimate the wave parameters.

To evaluate the significance of one EOF mode, the fractional variance of the lth EOF mode can be written as

$$R_l = \lambda_l \left(\sum_{i=1}^{m} \lambda_i\right)^{-1}, \ (l = 1, 2, \cdots, m) \tag{12.31}$$

where λ_l is the lth eigenvalue of the matrix $\mathbf{XX}^T$, and the total variance of $\mathbf{X}$ is the sum of all the eigenvalues.

12.3.2   Estimation of wave parameters

12.3.2.1   Significant wave height

The basic idea to derive SWH from X-band marine radar images is that SWH is proportional to the energy of the wave field. Due to the mechanisms of Bragg scattering and the modulations of long waves [7], wave energy is related to the variance of the gray levels of X-band marine radar images. By decomposing one radar image sequence with EOF, the principal components can show the variations of the wave field. Thus, the wave energy is related to the standard deviation of the principal components, and a linear relationship between them is assumed as

$$H_s = A + B \cdot std\left(z_i\right) \tag{12.32}$$

where $std(\cdot)$ is the standard deviation, and A and B are the coefficients that need to be determined by using in situ buoy data. According to (12.32), SWH can be estimated from the principal components of different EOF modes, but the calibration coefficients A and B may be different for different modes.

12.3.2.2 Peak wave period

To estimate the peak wave period, the maximum entropy method (MEM) is used to derive the power spectral density (PSD) of the principal component of the first EOF mode. The MEM uses an autoregressive (AR) model for the PSD $S(f)$ estimation. For the time series of the first EOF mode z_1 (12.29),

$$S(f) = \frac{\sigma_{l_0}^2}{\left| 1 - \sum_{l=1}^{l_0} a_l^{(l_0)} e^{-i2\pi fl} \right|^2} \tag{12.33}$$

where l_0 is the order of the AR prediction model, $a_l^{(l_0)}$ is the coefficient of AR, and $\sigma_{l_0}^2$ is the error power estimate for order l_0. One criterion to determine the optimal order is based on the final prediction error (FPE)

$$FPE(l) = \frac{n+l}{n-l}\sigma_l^2 \quad (l = 1, 2, \cdots, n-1) \tag{12.34}$$

The optimal order is the value of l that minimizes (12.34). The Burg algorithm is chosen to estimate the parameters of the spectral model because it is suitable for resolving sharp peaks for short sequences. Then the PSD of the first EOF mode is obtained, and the peak wave period can be derived from the peak frequency.

12.3.2.3 Peak wavelength and wave direction

The wavelength is the spatial periodicity of a wave field, so it can be estimated in the same way as for wave period. Because of the directionality of the propagating waves, the true wavelength must be derived from the data in the peak wave direction. To be specific, the wave field is sampled at n discrete points for m times (different from the sampling way of (12.26)); then the principal components of the wave field can be obtained according to (12.27)–(12.30). The extracted principal components denote the spatial variation of the waves, and then the wavenumber spectrum can be estimated using (12.33)–(12.34). The spectrum signifies the wave energy with different spatial scales, and the peak wavelength can be estimated from the peak wavenumber.

To estimate the peak wave direction, the wave field is sampled along two perpendicular directions, as shown in Figure 12.9 (i.e., Bin-x and Bin-y), and the wavelengths of the two bins are estimated by using the above method (i.e., L_x and L_y), respectively. Then the peak wavelength L and wave direction θ can be determined according to the geometric relationship

$$L = \frac{L_x L_y}{\sqrt{L_x^2 + L_y^2}} \tag{12.35}$$

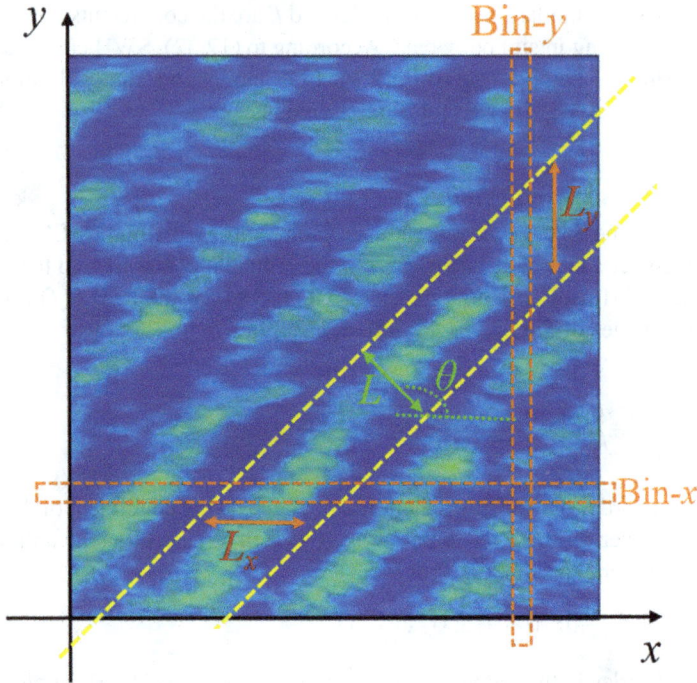

Figure 12.9   Schematic of the estimation of wavelength and wave direction from the first EOF mode of a radar image sequence. The yellow lines denote the wave troughs, and the green arrow marks the wavelength. The angle θ is the wave direction, and Bin-x and Bin-y are two bins in perpendicular directions.

and

$$\theta = \arctan \left(\frac{L_y}{L_x} \right).$$
(12.36)

There is an ambiguity of 180° in the estimated wave direction using (12.36), i.e., the waves may come from the direction θ or propagate toward the direction θ. This can be solved by using the cross spectrum of two or more adjacent images in a radar image sequence [24].

12.3.3   Physical interpretation of modes

To understand the physical meanings of the EOF modes, an X-band marine radar image sequence is decomposed into different EOF modes, and then they are used to reconstruct the image with individual EOF modes, as shown in Figure 12.10.

The sub-image reconstructed from the first EOF mode (Figure 12.10(b)) contains the main spatial pattern of the wave field (Figure 12.10(a)), which is in accordance with the fact that the first EOF mode makes the greatest contribution to the

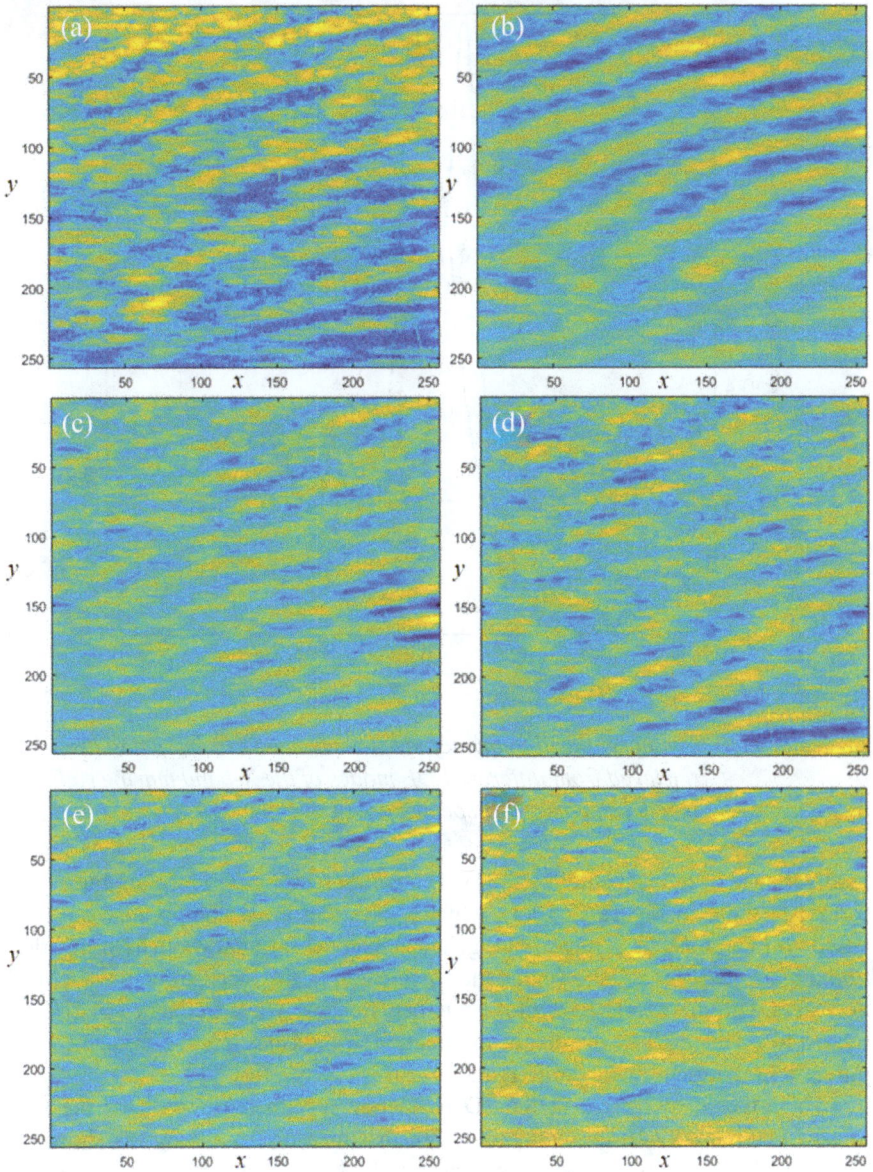

Figure 12.10   *(a) The original sub-image of an X-band marine radar image and the sub-images reconstructed from different EOF modes: (b) the 1st EOF mode, (c) the 5th EOF mode, (d) the 7th EOF mode, (e) the 11th EOF mode, and (f) the 17th EOF mode*

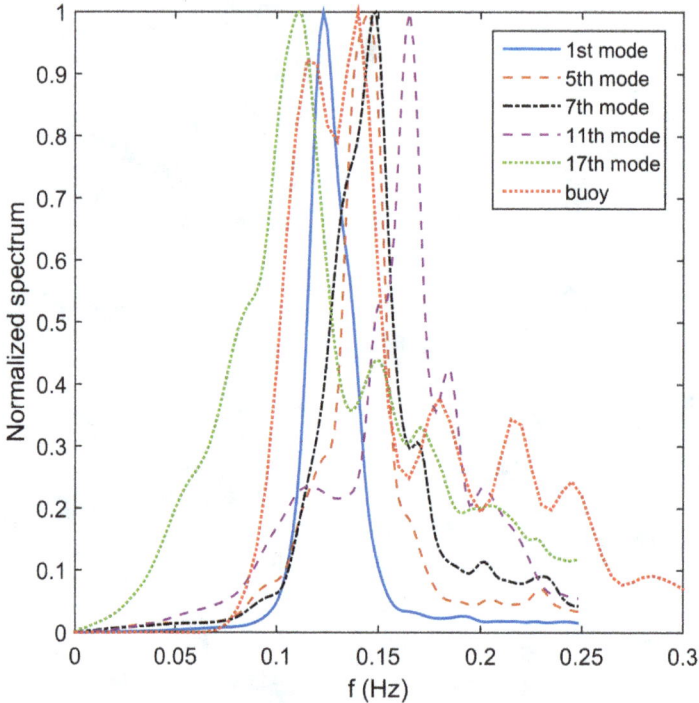

Figure 12.11   *The normalized frequency spectra measured by buoy and those retrieved from different EOF modes of the X-band marine radar image sequence. The spectra are derived by FFT.*

total variance of the wave field (12.31). The patterns in Figure 12.10 are more uniform than those in Figure 12.10(a), which shows that the decomposed wave field based on EOF analysis is more homogeneous than the original wave field. In addition, Figure 12.10(b)–(f) shows that the wavelength of the reconstructed wave field decreases with increasing EOF mode number. Because wave frequency is directly related to wavelength by the dispersion relation, it indicates that the frequencies of different EOF modes may be different.

To further demonstrate it, the 1D frequency spectra retrieved from the subimages in Figure 12.10 are compared with that measured by buoy, as shown in Figure 12.11. In the wave spectrum measured by buoy, there are two peaks in the range of 0.1–0.15 Hz (i.e., 0.115 and 0.14 Hz). The peak frequency of the first EOF mode (i.e., 0.123 Hz) is close to the first peak frequency of buoy, while the peak frequency of the fifth EOF mode (i.e., 0.145 Hz) is close to the second peak frequency of buoy. The peak frequency of the seventh EOF mode (i.e., 0.148 Hz) is close to that of the fifth EOF mode, but the spectral width of the seventh EOF mode is larger than that of the fifth EOF mode. Thus, the first several EOF modes mainly contribute to different frequency components of the wave field. There are two peaks in the spectrum of the 11th EOF mode (i.e., 0.165 and 0.184 Hz), which are close to the smaller peak frequency of buoy (i.e., 0.18 Hz), so high-order harmonics of the wave field contribute to the mode. The peak frequency of the 17th EOF mode is lower than that of buoy, and the spectral width of the 17th mode is much larger than those of other modes; it may be associated with noise of lower frequency.

Moreover, Figure 12.11 also shows that the spectral width of the first EOF mode is much less than that of the buoy. This is because the first EOF mode is mainly induced by the harmonics of peak wave frequency, while the spectrum measured by buoy is composed of wave harmonics of different frequencies. However, as demonstrated in [23], each individual EOF mode will not generally correspond to one individual dynamic mode and will not be statistically independent of other EOF modes. Therefore, the EOF modes that make less contribution to the total variance also contain wave information.

12.3.4 Discussion of mode choice for SWH estimation

According to (12.32), SWH can be retrieved from different EOF modes of X-band marine radar images. To analyze the performances of different modes, the root-mean-square error (RMSE) and bias between the SWH retrieved from different modes and that measured by buoy are shown in Figure 12.12(a). For the first 30 EOF modes, the RMSE of SWH is less than 0.3 m, and the bias is less than 0.05 m, indicating these modes are appropriate for the estimation of SWH.

Moreover, the SWH result that was retrieved from the third to the tenth EOF modes is better than that derived from the first EOF mode. Figure 12.12(c) shows that the contribution from each EOF mode is less than 8%, and the differences among the 11th–30th modes are small, so the wave energy is related to multiple modes rather than only the first EOF mode of the radar image sequence. This is reasonable because the nearshore wave field is complex and the imaging mechanisms of X-band marine radar are nonlinear. Therefore, each of the 30 EOF modes can be used to derive SWH from the X-band marine radar image sequence.

12.3.5 Example and validation

An example is presented to illustrate the main implementation process of the EOF-based algorithm, and then the algorithm is validated using experiment data.

12.3.5.1 Example

The gray-level radar image in Figure 12.4 is used as an example. Two small bins that are from two perpendicular directions are selected first. The sub-image in the yellow bin is used to derive SWH and peak wave period, while the sub-images in the yellow and cyan bins are used to estimate the peak wave direction and wavelength.

Then the sub-image sequence is decomposed by EOF analysis (12.27)–(12.30), and the principal component and eigenvector of the first EOF mode are shown in Figure 12.13. The principal component (Figure 12.13(a)) clearly shows the wave patterns in Figure 12.4. The eigenvector (Figure 12.13(b)) indicates the variations of the principal component at different times, e.g., the principal component in

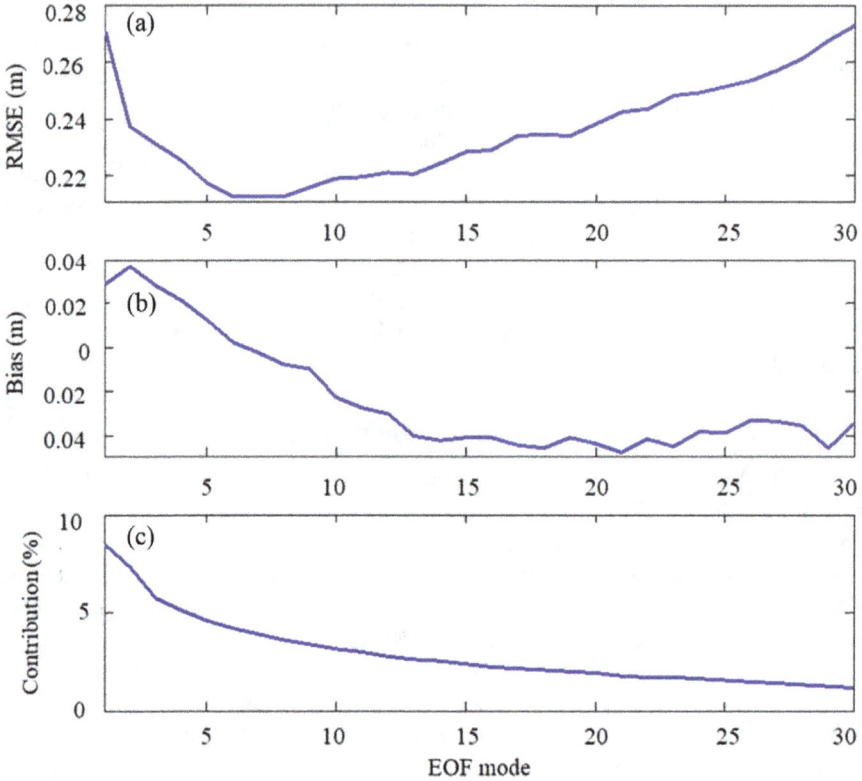

Figure 12.12   *The (a) RMSE and (b) bias between the SWH measured by buoy and those retrieved from the first 30 EOF modes of radar image sequences. (c) The average contributions of the first 30 EOF modes to the total variance.*

Figure 12.13(a) is significant at the time of 1 s, while the negative of the principal component is significant at the time of 6 s. Thus principal component and eigenvector of the first EOF mode depict the spatial and temporal variations of the wave field, and the wave parameters can be estimated from them.

The PSDs of the first and fourth principal components derived from the sub-image sequence by MEM are shown in Figure 12.14. Similar to the discussion in Section 12.3.3, the peak frequencies of the first and fourth principal components (i.e., 0.12 and 0.14 Hz) are close to the peak frequencies of the wave spectrum measured by buoy (i.e., 0.115 and 0.14 Hz). However, the peak frequency of the first principal component (i.e., 0.12 Hz) is smaller than the maximum frequency of the wave spectrum measured by buoy (i.e., 0.14 Hz), and there may be two reasons. First, the spectrum measured by buoy was the temporal variation of waves at a fixed position, while that derived from the radar image was the spatial variation of waves in a large area. Second, as stated in Section 12.3.3, each individual EOF mode will

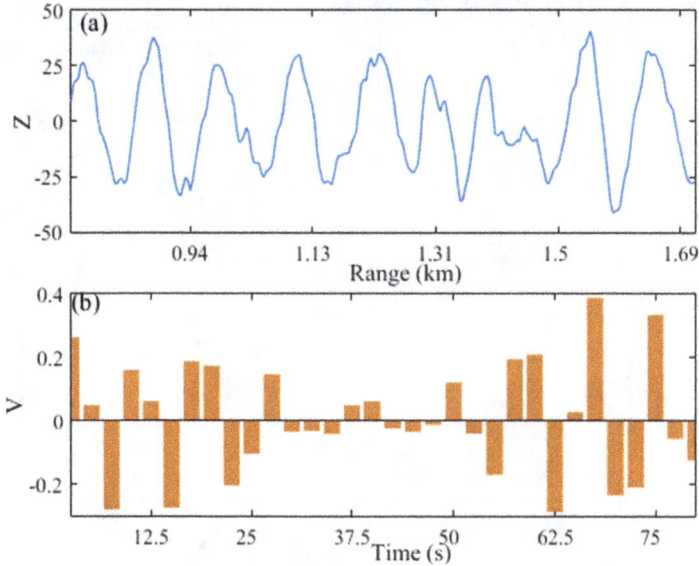

Figure 12.13   *The first EOF mode decomposed from the sub-images of an X-band marine radar image sequence: (a) principal component and (b) eigenvector*

not generally correspond to one individual dynamic mode, so the first EOF mode may not be the most significant dynamic wave harmonic. Besides, Figures 12.11 and 12.14 also show that the spectra derived by MEM are sharper than those derived by FFT, so the MEM algorithm is suitable for resolving sharp peaks for short radar image sequences.

12.3.5.2   Validation

The method based on EOF analysis has been validated in a few experiments with different X-band marine radar systems (see Table 12.1) [9, 25, 26]. In the experiments, the antennas of radars were mounted at 15–40 m above the sea level, and 10–100 m from the coastline. The analog-to-digital converter cards recorded the radar back-scatter from the sea surface and converted them into 8-bit gray-level images, with the resolutions of 3.75–7.5 m in range, and about 0.05°–0.1° in azimuth. The antenna rotation speeds were 24–48 rpm, i.e., the systems recorded one gray-level image every 1.25–2.5 s, and 32 or 64 images were recorded in each sequence.

As an example, two gray-level images that were recorded by two X-band marine radars with different polarizations are shown in Figure 12.15, and the wave patterns are distinct in the images although both propagated toward the shore. A pitch-and-roll wave buoy was placed about 1 km from the radar stations (the green asterisks in Figure 12.15). The buoy collected data with a sample rate of 2 Hz for 17 minutes every hour.

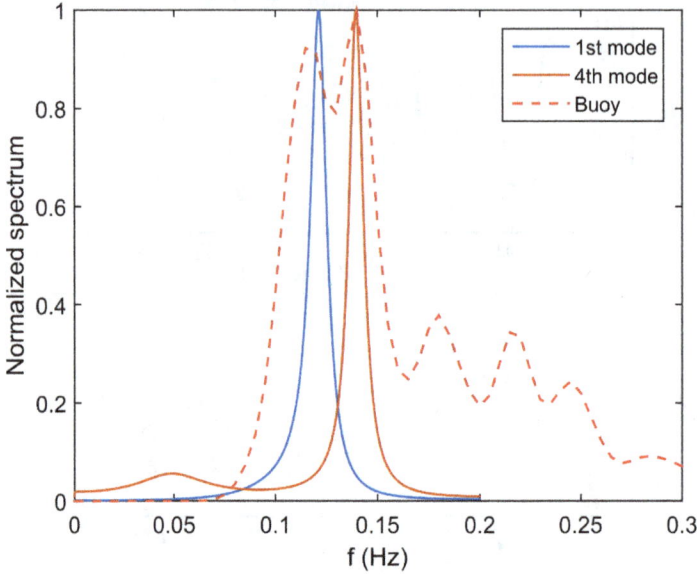

Figure 12.14   The wave spectrum measured by buoy and the spectra of the principal components of X-band marine radar image sequence derived by MEM

To evaluate the method, SWHs are estimated from three different regions at the ranges of 450.0–510.0 m (Region A1), 1 012.5–1 072.5 m (Region A2), and 1 575.0–1 635.0 m (Region A3), respectively. To determine the coefficients in (12.32), one-third of the SWHs measured by the buoy are randomly selected and utilized in the calibration, and then SWH is retrieved from the images of the two X-band marine radars with HH and VV polarizations (see Figure 12.16). For both the HH- and VV-polarized X-band marine radars, the SWHs estimated from different regions agree well with those measured by buoy. The RMSE between the SWH measured by buoy

Table 12.1   Configurations of the X-band marine radar systems

Property	Value
Polarization	HH/VV
Frequency	9 410 ± 30 MHz
Pulse width	50/70 ns
Pulse repetition frequency	1 800/3 000 Hz
Transmit power	25 kW
Antenna beam width at 3 dB	~1° horizontal, ~20° vertical
Antenna type	2.4-m slotted wave guide
Antenna rotation speed	24/30/42/48 rpm
Sample frequency	40/60 MHz

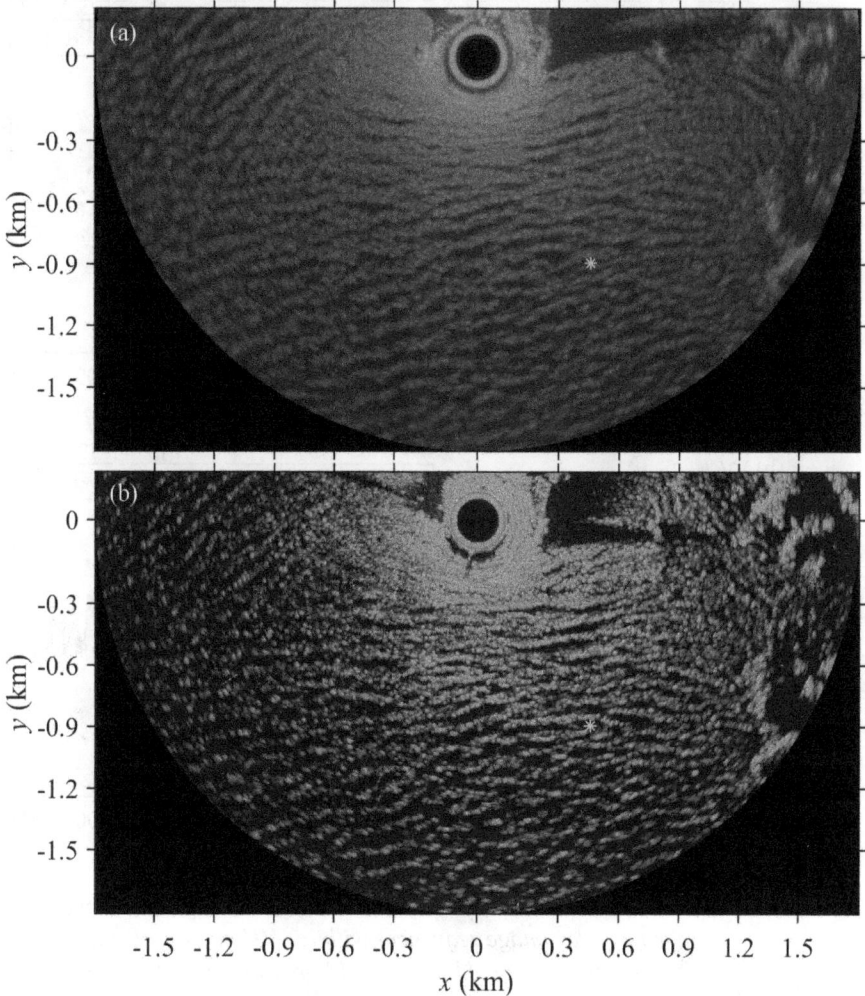

Figure 12.15   *The gray-level images recorded by two X-band marine radars on December 27, 2014, with (a) VV polarization and (b) HH polarization. The green asterisks mark the positions of the buoy.*

and that retrieved from radar images is 0.2–0.3 m, and the correlation coefficient (Corr) between them is higher than 0.9 for the regions investigated. Therefore, the estimated SWH correlates well with the in situ measurements, and (12.32) is valid for estimating SWH from X-band marine radar data.

Here the peak wave period is derived from the MEM spectrum of the first EOF mode of a radar image sequence, as shown in Figure 12.17. The RMSE, bias, and Corr between the estimated wave period and that measured by buoy are 0.84 s, 0.33 s, and 0.69, respectively. Using the SWH measured by buoy for reference, it shows

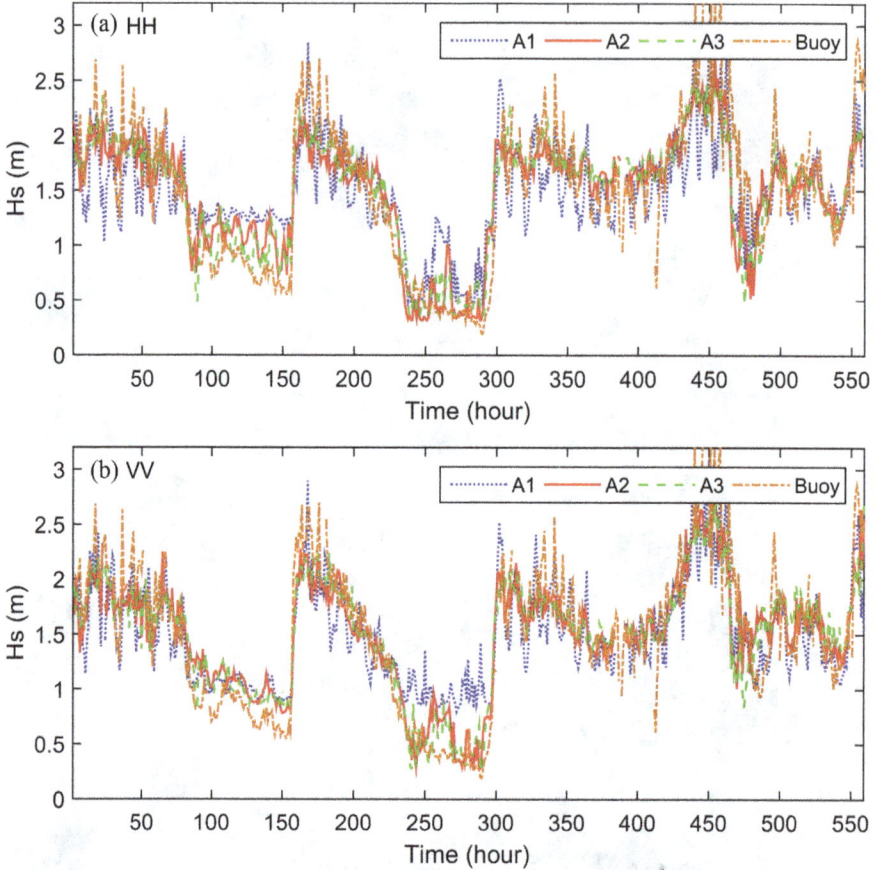

Figure 12.16   The SWH measured by buoy and those retrieved from X-band marine radar image sequences with (a) HH polarization and (b) VV polarization. A1, A2, and A3 are three different regions.

that large estimation error of peak wave period mostly occurs when SWH is lower than 1 m. The radar backscatter from the sea surface is weak when SWH is small, so the peak wave period estimation method is affected under low sea state.

To evaluate the wavelength and wave direction derived from the X-band marine radar image sequence, the observation area in Figure 12.18(a) is divided into different subareas, and the wavelength and wave direction in each subarea are estimated and shown in Figure 12.18(b). The waves came from northeast first, and then they propagated toward west and northwest. The wavelengths of the sub-areas that are far away from the land are larger than those near the shore, which are in accordance with the wave patterns in Figure 12.18(a). Besides, there are many abnormal wavelengths at the azimuths of 160°–190°, which is caused by the azimuthal modulation

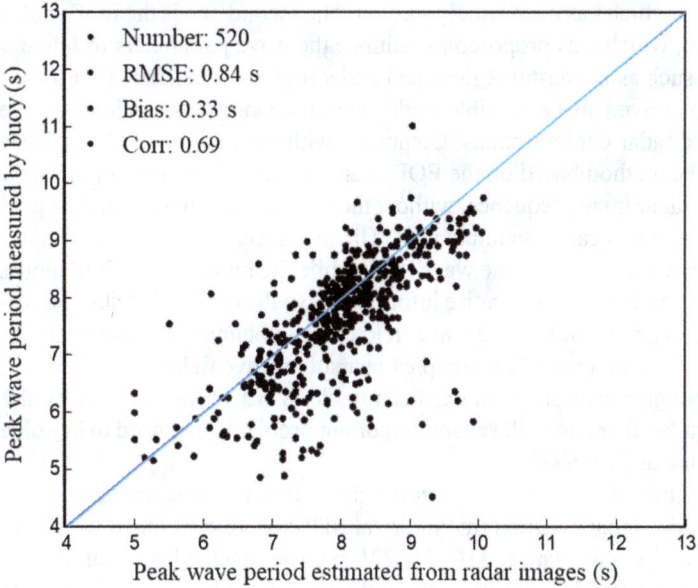

Figure 12.17   Comparison of the peak wave period measured by the buoy with that retrieved from X-band marine radar image sequence

of radar images [18], i.e., the wave patterns are almost absent when the antenna viewing angle is aligned with the wave crests.

12.4   Summary

In this chapter, two types of algorithms for estimating wave information from X-band marine radar images are illustrated. The first type is the traditional 3D FFT-based

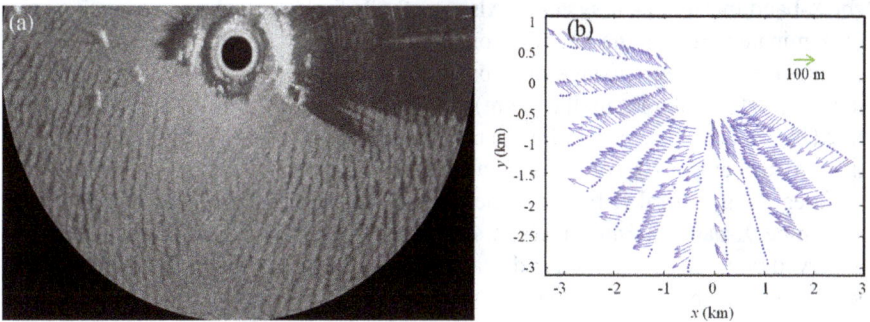

Figure 12.18   (a) One X-band marine radar image and (b) the peak wavelengths and wave directions retrieved from the radar image sequence. The dots denote the sub-areas with abnormal wavelengths.

algorithm, which has been widely studied. The second one is the EOF analysis-based algorithm, which was proposed to estimate the wave parameters of inhomogeneous regions, such as in coastal regions and under high sea state. The EOF-based method has been proved to be feasible under various conditions, including different sea states and radar configurations. Compared with the traditional 3D FFT-based algorithms, the method based on the EOF analysis derives the wave parameters directly from the radar image sequence, without the requirement of MTF and dispersion relation filter, so it is easier to implement. The first several EOF modes denote different frequency components of the wave field, while the higher order EOF modes contain both wave and background noise information, so the energy distributions of different frequency components of the wave field can be obtained, and the method may also be helpful for understanding complex nearshore wave fields.

Although much progress has been made in wave observations by the X-band marine radar, there are still several important problems that need to be solved. Some of them are as follows.

First, the MTF is important in the derivation of wave spectrum from X-band marine radar images. Different empirical MTFs were developed for different radar systems and circumstances [16, 21, 22], but the mechanisms that affect the MTF should be studied. For example, the influence of shadowing is different for different polarizations and sea states, rather than a purely geometrical effect [16, 25, 27]. The quadratic MTF (12.2) was derived using the VV-polarized radar images in the nearshore region, but the influences of polarization and water depth on the MTF are unknown.

Second, the algorithms for calibrating SWH adaptively should be further studied. Either the shadowing theory [12] or the theory of random ocean waves [26] is based on the assumption of Gaussian random sea surface [28], which may be unreasonable for the coastal regions with complex bathymetry or under high sea states.

Third, the influence of rain on X-band marine radar images should be studied quantitatively. Different indicative parameters of rain have been used to detect rain on radar images, but the thresholds were dependent on radar systems and regions [5, 6, 15], which limits the automatic recognition and removal of rain for different X-band marine radar systems. Also, methods for detecting and correcting rain-contaminated areas in the images should be developed [29, 30].

Fourth, one disadvantage of X-band marine radar in ocean engineering is that the observation area is small (~3 km), and parts of the field of view are often sheltered by land or ships, except for the regions that are affected by the azimuthal modulation mechanism. To solve the problem, observation networks may be constructed by integrating several X-band marine radars. For example, an observation network with nine X-band marine radars is shown in Figure 12.19; the wave parameters observed by them are merged and analyzed to get the sea state of the whole region, which extends the observation range greatly.

In addition, coherent marine radars are also used to observe wave parameters. Compared with the noncoherent radars, they directly measure the orbital velocity of waves, and then the SWH can be estimated without MTF and calibration, which will benefit the observation of waves [31–33].

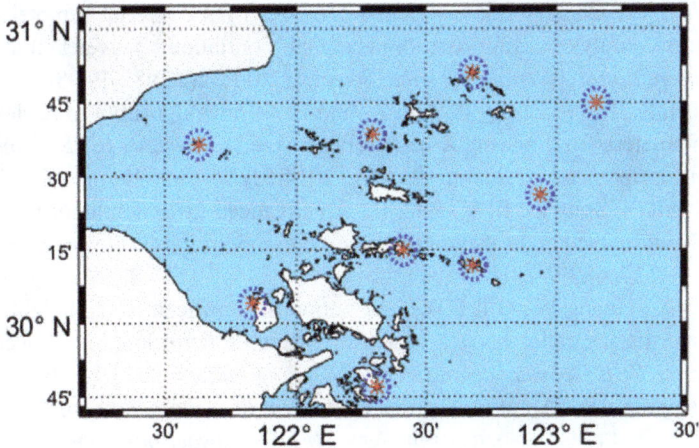

Figure 12.19   *An observation network constructed by different X-band marine radars. The dots and circles denote the radar stations and their observation regions.*

References

[1] Young I.R., Rosenthal W., Ziemer F. 'A three-dimensional analysis of marine radar images for the determination of ocean wave directionality and surface currents'. *Journal of Geophysical Research*. 1985;**90**(C1):1049–59.

[2] Seemann J., Ziemer F., Senet C.M. 'A method for computing calibrated ocean wave spectra from measurements with a nautical X-band radar'. Proceedings of MTS/IEEE Oceans Conference; Halifax, NS, Canada; 1997.

[3] Nieto-Borge J.C., Hessner K., Jarabo-Amores P., de la Mata-Moya D. 'Signal-To-Noise ratio analysis to estimate ocean wave heights from X-band marine radar image time series'. *IET Radar, Sonar & Navigation*. 2008;**2**(1):35–41.

[4] Gangeskar R. 'Verifying high-accuracy ocean surface current measurements by X-band radar for fixed and moving installations'. *IEEE Transactions on Geoscience and Remote Sensing*. 2018;**56**(8):4845–55.

[5] Lund B., Graber H.C., Romeiser R. 'Wind retrieval from shipborne nautical X-band radar data'. *IEEE Transactions on Geoscience and Remote Sensing*. 2012;**50**(10):3800–11.

[6] Liu Y., Huang W., Gill E.W., Peters D.K., Vicen-Bueno R. 'Comparison of algorithms for wind parameters extraction from shipborne X-band marine radar images'. *IEEE Journal of Selected Topics in Applied Earth Observations and Remote Sensing*. 2015;**8**(2):896–906.

[7] Alpers W.R., Ross D.B., Rufenach C.L. 'On the detectability of ocean surface waves by real and synthetic aperture radar'. *Journal of Geophysical Research*. 1981;**86**(C7):6481–98.

[8]   Wu L.C., Chuang L.Z.H., Doong D.J., Kao C.C. 'Ocean remotely sensed image analysis using two-dimensional continuous wavelet transforms'. *International Journal of Remote Sensing*. 2011;**32**(23):8779–98.

[9]   Chen Z., He Y., Zhang B., Qiu Z., Yin B. 'A new algorithm to retrieve wave parameters from marine X-band radar image sequences'. *IEEE Transactions on Geoscience and Remote Sensing*. 2014;**52**(7):4083–91.

[10]   Zhang S., Song Z., Li Y., Ying L. 'An advanced inversion algorithm for significant wave height estimation based on random field'. *Ocean Engineering*. 2016;**127**(4):298–304.

[11]   Liu X., Huang W., Gill E.W. 'Estimation of significant wave height from X-band marine radar images based on ensemble empirical mode decomposition'. *IEEE Geoscience and Remote Sensing Letters*. 2017;**14**(10):1740–4.

[12]   Gangeskar R. 'An algorithm for estimation of wave height from shadowing in X-band radar sea surface images'. *IEEE Transactions on Geoscience and Remote Sensing*. 2014;**52**(6):3373–81.

[13]   An J., Huang W., Gill E.W. 'A self-adaptive wavelet-based algorithm for wave measurement using nautical radar'. *IEEE Transactions on Geoscience and Remote Sensing*. 2015;**53**(1):567–77.

[14]   Huang W., Liu X., Gill E.W. 'Ocean wind and wave measurements using X-band marine radar: A comprehensive review'. *Remote Sensing*. 2017;**9**(1261):1261–39.

[15]   Chen Z., He Y., Zhang B., Ma Y. 'A method to correct the influence of rain on X-band marine radar image'. *IEEE Access*. 2017;**5**:25576–83.

[16]   Nieto Borge J., Rodríguez G.R., Hessner K., González P.I. 'Inversion of marine radar images for surface wave analysis'. *Journal of Atmospheric and Oceanic Technology*. 2004;**21**(8):1291–300.

[17]   Hasselmann K., Hasselmann S. 'On the nonlinear mapping of an ocean wave spectrum into a synthetic aperture radar image spectrum and its inversion'. *Journal of Geophysical Research*. 1991;**96**(C6):10713–29.

[18]   Lund B., Collins C.O., Graber H.C., Terrill E., Herbers T.H.C. 'Marine radar ocean wave retrieval's dependency on range and azimuth'. *Ocean Dynamics*. 2014;**64**(7):999–1018.

[19]   Antoniou A. *Digital Signal Processing: Signals, Systems and Filters*. New York: McGraw-Hill; 2006.

[20]   Plant W.J. 'The ocean wave height variance spectrum: wavenumber peak versus frequency peak'. *Journal of Physical Oceanography*. 2009;**39**(9):2382–3.

[21]   Ludeno G., Brandini C., Lugni C., *et al.* 'Remocean system for the detection of the reflected waves from the Costa Concordia ship wreck'. *IEEE Journal of Selected Topics in Applied Earth Observations and Remote Sensing*. 2014;**7**(7):3011–18.

[22]   Chen Z., Zhang B., He Y., Qiu Z., Perrie W. 'A new modulation transfer function for ocean wave spectra retrieval from X-band marine radar imagery'. *Chinese Journal of Oceanology and Limnology*. 2015;**33**(5):1132–41.

[23] Monahan A.H., Fyfe J.C., Ambaum M.H.P., Stephenson D.B., North G.R. 'Empirical orthogonal functions: the medium is the message'. *Journal of Climate*. 2009;**22**(24):6501–14.

[24] Chen Z., Zhang B., Kudryavtsev V., He Y., Chu X. 'Estimation of sea surface current from X-band marine radar images by cross-spectrum analysis'. *Remote Sensing*. 2019;**11**(9):1031.

[25] Chen Z., He Y., Yang W. 'Study of ocean waves measured by collocated HH and VV polarized X-band marine radars'. *International Journal of Antennas and Propagation*. 2016;**2016**:1–12.

[26] Chen Z., He Y., Zhang B. 'An automatic algorithm to retrieve wave height from X-band marine radar image sequence'. *IEEE Transactions on Geoscience and Remote Sensing*. 2017;**55**(9):5084–92.

[27] Plant W.J., Farquharson G. 'Wave shadowing and modulation of microwave backscatter from the ocean'. *Journal of Geophysical Research: Oceans*. 2012;**117**

[28] Smith B. 'Geometrical shadowing of a random rough surface'. *IEEE Transactions on Antennas and Propagation*. 1967;**15**(5):668–71.

[29] Chen X., Huang W., Zhao C., Tian Y. 'Rain detection from X-band marine radar images: a support vector machine-based approach'. *IEEE Transactions on Geoscience and Remote Sensing*. 2020;**58**(3):2115–23.

[30] Chen X., Huang W. 'Identification of rain and low-backscatter regions in X-band marine radar images: an unsupervised approach'. *IEEE Transactions on Geoscience and Remote Sensing*. 2020;**58**(6):4225–36.

[31] Hackett E.E., Fullerton A.M., Merrill C.F., Fu T.C .Comparison of incoherent and coherent wave field measurements using dual-polarized pulse-Doppler X-band radar'. *IEEE Transactions on Geoscience and Remote Sensing*. 2015;**53**(11):5926–42.

[32] Carrasco R., Streßer M., Horstmann J. 'A simple method for retrieving significant wave height from Dopplerized X-band radar'. *Ocean Science*. 2017;**13**(1):95–103.

[33] Kammerer A.J., Hackett E.E. 'Use of proper orthogonal decomposition for extraction of ocean surface wave fields from X-band radar measurements of the sea surface'. *Remote sensing*. 2017;**2017**(9):1–15.

Chapter 13

Wavelet-based methods to invert sea surfaces and bathymetries from X-band radar images

Pavel Chernyshov[1], Teodor Vrecica[2], and Yaron Toledo[1]

X-band radar-based sea state monitoring systems have become popular in the past decades. This is generally due to the following reasons: such systems cover a significant area, which allows spatio-temporal analysis with a quite good resolution. Unlike cameras, they do not depend on daylight. Another reason is that they are cost-effective with respect to a grid of in situ buoys or Acoustic Doppler Current Profiler measurements, especially taking into account their deployment and maintenance costs. Today X-band radar-based wave, current, and bathymetry monitoring systems are either self-designed or commercial products, such as WaMoS®. Incoherent radar systems provide the intensity information of the backscattered electromagnetic radiation (for an example of self-designed incoherent system, see e.g. [1]). Coherent systems augment this information by Doppler shifts, allowing the measurement of water surface velocities (e.g. [2]).

Conventional analysis techniques of radar intensity maps commonly apply a 3D Fast Fourier Transform (FFT) to the radar image sequences. This enables the separation of the main wave signal from noise and other effects (e.g. higher harmonics) by using a linear wave theory dispersion relation shell filter and high-pass filter in the wavevector–frequency domain [3–6] and others. This group of methods is quite efficient, since it allows the estimation of wave spectra and spectral mean sea state parameters, local depths, mean current vector, sea surface reconstruction, etc. It is also historically the most tested, since a vast body of work developed these conventional techniques, testing improved methodologies on real datasets and synthetically simulated maps with various sea state, geometrical, or environmental conditions ([7–10]). Nevertheless, this family of methods has their own limitations. Fourier transform requires signal's spatio-temporal periodicity and homogeneity – a property which is hardly fulfilled in a spatially changing nearshore environment. This obstacle is partially addressed by applying windowed Fourier transform with overlapping and/or adaptive window size. In the time domain, a shoaling wave signal

[1]School of Mechanical Engineering, Tel Aviv University, Israel
[2]Air-Sea Interaction Laboratory, Scripps Institute of Oceanography, USA

behaves as periodic; hence long sequences can be taken to attain better resolution in the frequency domain (for example, is used for the retrieval of shearing currents, when the high accuracy in the phase velocity shift is crucial). Unfortunately, long temporal sequences, acquisition is many times impossible. For instance, in the case of satellite probing, just a few successive images of the same location are available. The same is typical in the case of the moving ship installation. Even when such acquisition is possible, it may produce inaccuracies due to changing conditions.

In such cases, a different analysis approach utilizing Continuous Wavelet Transform (CWT) can provide many advantages. It enables resolution of the signal's local properties by means of translation, constriction, and rotation of the mother wavelet. Translation allows allocation of the center of the mother wavelet; constrictions–expansions and rotations enable estimation of local spectra of the signal. Directional properties of the signal are resolvable if the corresponding mother wavelet is anisotropic, i.e. its dilational and lateral sizes are not the same. CWT starts getting more attention for development of the improved radar signal analysis techniques (see e.g. [11, 12]).

This chapter discusses two problems of the radar intensity map sequence analysis using CWT instead of the commonly used FFT approach. The first is a CWT methodology for the inversion of these map sequences to sea surface elevation maps, and the second a CWT methodology for the reconstruction of the bathymetry map.

The sea surface reconstruction will be demonstrated on a single radar image. This is one of the advantages of the CWT, which can process each image by itself. This methodology includes a peak-dependent band-pass filter in the pseudo-wavevector domain, an alternative Modulation Transfer Function (MTF), and a phase correction of the tilt modulation effect. A statistical comparison between the original and the reconstructed surface elevations shows mean absolute error on the level of 6–14% of the significant wave height, which is quite an improvement, when compared to conventional 2D Fourier-based techniques and the 1D version of the proposed method (see [13]).

The bathymetry map reconstruction also requires the peak wave phase speed, since the dispersion relation is used for computing the depth. It is estimated by calculating the phase shifts of the corresponding peak components between successive radar images and then relating them to the time lag. The CWT method has several clear advantages. It requires image sequences; however, comparatively high accuracy can be achieved by much shorter measurement periods than those required for the 3D FFT approach. As well, the time steps between the images do not necessarily need to be uniform (in a conventional method, this would need an additional interpolation step). The method's ability to reconstruct the original bathymetry is shown to be robust in intermediate to shallow water depth conditions ($k_p h < 1$).

The CWT methodology is beneficial for analysis of nonperiodic and nonhomogeneous signals. It also shows great potential of its development in application to different fields dealing with compounded textures' analysis.

This chapter aims to provide a clear entry point to modeling of radar intensity images from sea surface elevation stochastic realizations for the purpose of development and testing of the new analysis approaches. It also aims to clarify the advantages of wavelet transform-based methods for the ocean remote sensing applications.

The chapter is structured as follows. Section 13.1 introduces sea surface and radar image realization simulation for shoaling wavefield. Section 13.2 describes the concept and application of the 2D direct and inverse CWT. In Section 13.3, 2D Wavelet-based Sea Surface Reconstruction (WSSR) method is introduced and applied to a realization of a shoaling wavefield. Section 13.4 describes and utilizes the new 2D Rapid Continuous Wavelet-based Bathymetry Inversion (RCWBI) method. The chapter is finalized by discussion and brief conclusions in Section 13.5.

13.1 Simulation of the sea surface elevation and radar images over a laterally uniform bottom profile

In this section, the simulation of the rough sea surface and the corresponding radar images will be briefly described. For the simulation of the rough sea surface elevations $\eta(\mathbf{r}, t)$ ($\mathbf{r} = (x, y)$), the discrete solution of the corresponding mild-slope equation (see e.g. [12, 14, 15]) will be used.

$$\eta(\mathbf{r}, t) = \sum_{i=1}^{N_\omega} \sum_{j=1}^{N_\theta} a_{ij}(\mathbf{r}) \cos\left(\omega_i t - k_{y_{ij}} - \int_0^x k_{x_{ij}}\, d\xi + \phi_{ij}\right), \tag{13.1}$$

where $k_{y_{ij}}(\mathbf{r}) = k_i(0)\sin(\theta_j(0))$ is the lateral wavenumber, $k_{x_{ij}}(\mathbf{r}) = k_i(\mathbf{r})\cos(\theta_j(\mathbf{r}))$ is the onshore wavenumber, $a_{ij}(\mathbf{r})$ is the wave amplitude (see 13.6), ϕ_{ij} is the initial uniformly distributed over $[0, 2\pi)$ random phase, and $\omega_i = 2\pi f_i$ is the angular wave frequency. N_ω and N_θ are the number of samples in frequency and direction domains, respectively (for values of all the parameters, refer to Table 13.1). Here and further, symbol 0 corresponds to the origin of coordinates in deeper water for the sea surface simulation stage. The angular frequency and the wave number satisfy the linear dispersion relationship without taking into account any ambient current influence:

$$\omega_i^2 = g k_i(\mathbf{r}) \tanh\big(k_i(\mathbf{r})h(\mathbf{r})\big), \tag{13.2}$$

which is used to define the wave number $k_i(\mathbf{r})$ as a solution of (13.2) for each point in space. Here g is acceleration of gravity and $h(\mathbf{r})$ is the local water depth at point $\mathbf{r}$.

In order to describe the linear shoaling and refraction of the wavefield, the following corollary of the energy flux conservation law is used:

$$a_{ij}(\mathbf{r}) = a_{ij}(0)\sqrt{\frac{C_{g_{ij}}^x(0)}{C_{g_{ij}}^x(\mathbf{r})}}, \tag{13.3}$$

where $C_{g_{ij}}^x(\mathbf{r})$ is the onshore group velocity expressed as

$$C_{g_{ij}}^x(\mathbf{r}) = \frac{\omega_i}{2k_i(\mathbf{r})}\left(1 + \frac{2k_i(\mathbf{r})h(\mathbf{r})}{\sinh(2k_i(\mathbf{r})h(\mathbf{r}))}\right)\cos(\theta_{ij}(\mathbf{r})), \tag{13.4}$$

with

$$\theta_{ij}(\mathbf{r}) = \arcsin\big(\sin(\theta_j(0))k_i(0)/k_i(\mathbf{r})\big) \tag{13.5}$$

Table 13.1   JONSWAP spectrum parameters used for surface modeling

Parameter	Value
Fetch (F)	700 (km)
Peak enhancement factor	3.3
Wind speed (U_{10})	6.64 (m/s)
Peak period (T_p)	10 (s)
Initial significant wave height ($H_s(0)$)	4.06 (m)
Incident wave peak in deep water	$\theta_p(0) = -40°$
Time step (dt)	2 (s)
Spatial step (dx, dy)	10 (m)
Number of frequencies (N_ω)	100
Number of directions (N_θ)	50
Spectral directions' range (θ)	$[-\pi, \pi]$
Number of time samples (N_t)	100
Number of range samples (N_x)	201
Number of azimuth samples (N_y)	101

obtained from Snell's law [16].

The original formulation of the JOint North Sea WAve Project (JONSWAP) spectra $S(\omega)$ [17] is used to define the energy distribution across all frequencies. The input parameters of the JONSWAP spectra for all simulations and the resulting wave parameters are given in Table 13.1. In order to better resolve the spectral peak, an elimination method, sometimes referred to as Neumann's method (see e.g. [18, 19]), is used.

The directional wave properties are described using a Mitsuyasu frequency-dependent spread function [20]. The directional wave bins are also determined using the elimination method around the mean wavefield direction. This process results in a much better frequency–direction resolution of the spectral peak (see Figure 13.1(c)). The initial individual wave amplitudes are defined as follows:

$$a_{ij}(0) = \sqrt{2 \int_{(\omega_i+\omega_{i-1})/2}^{(\omega_i+\omega_{i+1})/2} \int_{(\theta_j+\theta_{j-1})/2}^{(\theta_j+\theta_{j+1})/2} S\left(\omega, \theta\right) d\omega \, d\theta}. \tag{13.6}$$

The bathymetry is defined as a laterally uniform mildly changing beach with constant slope of 1.5% with the underwater mountain situated 800 m off the shore, which is modeled with a Gaussian bell shape:

$$h(\mathbf{r}) = 0.015x + 2 - 7\exp(-0.00001(x - 900)^2) \tag{13.7}$$

with depth value expressed in meters, where $\mathbf{r} = (x, y)$ (x corresponds to the onshore direction). Sea surface image sequence is constructed using (13.1) as hundreds of 2-second time steps, which correspond to a typical antenna revolution period and is enough for a statistical analysis of the obtained results. The incident wave peak in deep water propagates at $\theta_p(0) = -40°$ angle. As waves approach the coastline,

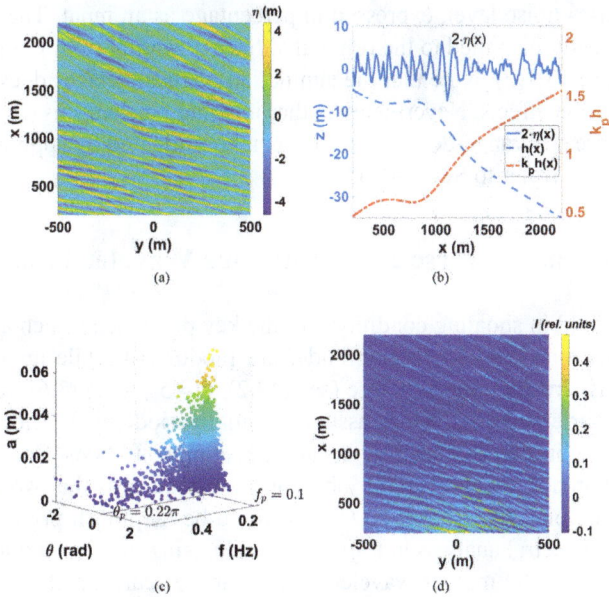

Figure 13.1   *(a) Example of a realization of the JONSWAP spectrum-based 2D shoaling wavefield; (b) its corresponding 1D cross section (two times magnified in solid blue), bathymetry profile defined in (13.7) (dashed blue line), relative water depth ($k_p h$) (dash-dotted red line). The shortening of the wavelength together with a mild amplitude modulation is clearly visible while entering shallower waters; (c) realization of the amplitude–frequency–direction spectrum obtained by applying the elimination method; (d) resulting simulated radar image for the original sea elevation from (a) (radar installation height is $H_r=60$ m and noise level is 5%).*

the wave fronts start to bend towards it, propagating almost perpendicularly to it in shallower waters. Wave amplitudes also change due to shoaling effects (13.3). For the following, the spatial coordinate origin is redefined as the radar installation point introducing the new coordinate system in order to simplify the radar image modeling. To ensure the image is in the far zone of the radar, an offset R_0 is added (here $R_0 = 200$ m). The simulation domain covers significant range in order to estimate the effect of shadowing. In Figure 13.1, the corresponding sea surface elevation is given as a function of the new coordinates.

As for the radar image model, one of the simplest, proposed originally in [3], is employed. This model takes into account geometrical shadowing, tilt modulations, and a random speckle noise. This model works well both for 1D and 2D spatial image simulations [13]. As an input, it uses radar installation height (H_r), a constant offset R_0 from the radar to the closest point of the image. The speckle noise is modeled as a multiplicative Gaussian noise using a simple "salt and pepper" model

[19], which uses noise level, expressed in percentage as an input. The result of the described procedure applied to the original surface elevation from Figure 13.1(a) is given in Figure 13.1(d). Note that the simulation of radar images does not include the radar equation effects. Nevertheless, other imaging mechanisms (shadowing, tilt modulation) are still range-dependent. This can be mitigated by applying the range trend subtraction (refer to Section 13.3).

13.2   Direct and inverse 2D Continuous Wavelet Transform

For the wavefield in shoaling conditions all the key parameters, such as wavenumber, wave direction, and wave amplitude, are modulated while approaching the coast as it was already noticed before (see (13.2), (13.5), and (13.6), respectively). Traditional Fourier analysis, which assumes spatial periodicity and homogeneity of the signal, has a limited applicability in such changing conditions. Consequently, an alternative approach accounting for inhomogeneities such as the two-dimensional Continuous Wavelet Transform (2D CWT) would be helpful. It principally allows to localize the spectral analysis in a spatial domain using the correlation of the analyzed signal and a 2D mother wavelet, which can be scaled, shifted, and rotated. The rotation allows for resolving the directional properties of the signal as long as the mother wavelet itself is anisotropic [21]. Here the analyzed signal (in our case, a spatial radar intensity map $I(\mathbf{r})$) is assumed to be locally integrable with its square. For the mother wavelet $\psi \in L_2(\mathbb{R}^2)$, the admissibility condition

$$C_\psi = (2\pi)^2 \int_{\mathbb{R}^2} |\boldsymbol{\xi}|^{-2} |\hat{\psi}(\boldsymbol{\xi})|^2 \, d\boldsymbol{\xi} < \infty \tag{13.8}$$

is assumed to be satisfied [22].

In order to employ the mother wavelet ψ in 2D space, triply indexed families of wavelets are formed:

$$\psi_{a,b,\theta}(\mathbf{r}) = a^{-1} \psi \left[\mathbf{R}_\theta^{-1} \left(\tfrac{\mathbf{r}-\mathbf{b}}{a} \right) \right], \tag{13.9}$$

where $a \in \mathbb{R}_+$ serves for the dilation of the mother wavelet (scaling factor), $\mathbf{b} \in \mathbb{R}^2$ is a positioning vector, and $\mathbf{R}_\theta \in SO(2)$ is a standard rotation matrix of a rotating angle θ. For the appropriate mother wavelet function $\psi(\mathbf{r})$, the 2D CWT of a signal $I(\mathbf{r})$ is given as follows:

$$W[I(\mathbf{r}), \psi(\mathbf{r})](a, \mathbf{b}, \theta) = C_\psi^{-1/2} a^{-1} \int_{\mathbb{R}^2} I(\mathbf{r}) \psi_{a,b,\theta}^*(\mathbf{r}) \, d\mathbf{r}, \tag{13.10}$$

where the asterisk denotes the complex conjugate. For a continuous 2D wavelet transform, an analog of Parseval's equality is [21]:

$$\int_0^{2\pi} \int_0^\infty \int_{\mathbb{R}^2} |W(a, \mathbf{b}, \theta)|^2 / a^3 \, d\mathbf{b} \, da \, d\theta = \|I(\mathbf{r})\|_{L_2(\mathbb{R}^2)}^2. \tag{13.11}$$

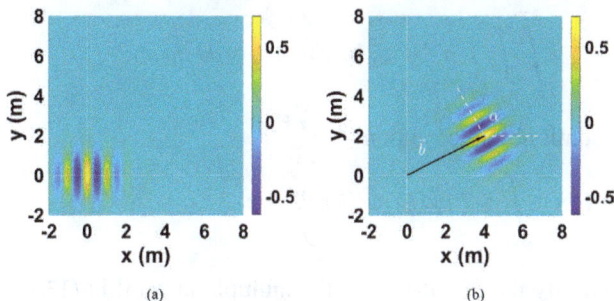

Figure 13.2   *(a) Original Morlet wavelet in the physical (x, y) space (see 13.11); (b) the same Morlet wavelet rotated ($\alpha = 2\pi/3$) and translated $\mathbf{b} = (4, 2)$. Triple $(k_0, \sigma, \varepsilon) = (6, 1, 2)$ was taken to demonstrate influence of the anisotropy parameter on the wavelet's shape.*

In linear water wave theory applications, the Morlet mother wavelet is usually used. The 2D Morlet wavelet in the spatial case can be represented in the following form:

$$\psi(x, y) = \frac{1}{\sqrt{2\varepsilon\sigma^2}} e^{-ik_0 x} e^{-1/2\sigma^2 \left(x^2/2 + (y/\varepsilon)^2 \right)}, \tag{13.12}$$

where k_0 is the central wavenumber, ε is the anisotropy parameter, and σ is the shape parameter (see original and transformed Morlet mother wavelets in Figure 13.2). In the Fourier domain, (13.12) is

$$\hat{\psi}(k_x, k_y) = e^{-\sigma^2 \left((k_x - k_0)^2 + (\varepsilon k_y)^2/2 \right)}. \tag{13.13}$$

The hat symbol (*cdot*) denotes the Fourier transform. Typically, the triple $(k_0, \sigma, \varepsilon) = (6 \text{ rad/m}, 1, 1)$ is taken. To fit the directional properties of the analyzed signal, the anisotropy parameter ε might be varied. The central wavenumber k_0 is taken to be 6 rad/m because of the additional requirements on the mother wavelet symmetries. During the analysis, it will be scaled by a to fit the real physical wavenumbers of the processed signal.

To optimize the numerical realization of the 2D CWT, the following identical representation of (13.10) is commonly used [23]:

$$W[I(\mathbf{r}), \psi(\mathbf{r})](a, \mathbf{b}, \theta) = C_\psi^{-1/2} a \int_{\mathbb{R}^2} \hat{I}(\mathbf{k}) \hat{\psi}^* \left(a \mathbf{R}_\theta^{-1}(\mathbf{k}) \right) e^{i\mathbf{b}\mathbf{k}} \, d\mathbf{k}, \tag{13.14}$$

which allows to replace the heavy calculation of the double integral in (13.10) with the calculation of two direct and one inverse 2D FFTs. In some technical software, this method may be referred to as 2D Continuous Wavelet Transform with Fourier Transform (2D CWTFT).

    The reconstruction process requires applying the corresponding inverse CWT, which can pose quite a complexity. Nevertheless, the mother wavelet function utilized for this operation (ψ') does not have to be the same one used before in the direct CWT (ψ):

$$I(\mathbf{r}) = C_{\psi,\psi'}^{-1} \int_0^{2\pi} \int_{\mathbb{R}_+} \int_{\mathbb{R}^2} a^{-3} W(a, \mathbf{b}, \theta) \psi'_{a,b,\theta}(\mathbf{r}) \, d\theta \, da \, d\mathbf{b}, \qquad (13.15)$$

with the new admissibility condition

$$C_{\psi,\psi'} = (2\pi)^2 \int_{\mathbb{R}^2} |\xi|^{-2} |\hat{\psi}(\xi)| \|\hat{\psi}'(\xi)| \, d\xi < \infty. \qquad (13.16)$$

In order to simplify the calculation of the multiple integral in (13.15), the mother wavelet function ψ' is chosen to be Dirac's delta (δ):

$$I(\mathbf{r}) = C_{\psi,\delta}^{-1} \int_0^{2\pi} \int_{\mathbb{R}_+} a^{-2} W(a, \mathbf{r}, \theta) \, d\theta \, da. \qquad (13.17)$$

Equations (13.10) and (13.17) describe direct and inverse 2D CWT, which will compose the core part of the wavelet and analysis and reconstruction procedure of the WSSR method.

Additional useful relationships for this procedure include the transformation from the pair (a, θ) to the pair (k_x, k_y). The relationship between the pseudo-wavenumber and the scaling factor is defined as

$$|\mathbf{k}| = \frac{k_0}{dx \cdot a}, \qquad (13.18)$$

where dx is the corresponding spatial resolution (refer to Table 13.1). In this chapter, the same resolution is used in onshore and longshore directions $dx = dy = 10$ m allowing to calculate the wavevector

$$\mathbf{k} = (k_x, k_y) = (|\mathbf{k}| \cos(\theta), |\mathbf{k}| \sin(\theta)) \qquad (13.19)$$

with θ taken from the set of angles defined for the rotations of the mother wavelet (spectral directions' set) given in Table 13.2.

13.3 The 2D wavelet-based sea surface reconstruction method

In this subsection, the 2D WSSR method for the reconstruction of sea surface elevation maps from a sequence of radar images will be described. A 2D CWTFT is used as a basis for the radar image inversion procedure that enables working with spatially inhomogeneous data. The CWTFT is applied to each 2D radar image for each time instant independently allowing to perform the inversion without using the dispersion relation fitting and windowing.

Since higher harmonics still appear due to the nonlinearity of the imaging mechanisms and the wavefield, a filtration of higher harmonics is required. This is performed in the wavevector domain using a specific band-pass peak-dependent filter.

The 2D WSSR method procedure is composed of the following steps (see also Figure 13.3):

Table 13.2   *Wavelet analysis parameters used for analysis of the sea surface and bathymetry reconstruction cases*

Parameter	Value
Simulated case 2D sea surface reconstruction	
$(k_0, \sigma, \varepsilon)$	$(6, 1, 2)$
$[a_{min}, a_{max}]$	$[4, 20]$
$[k_{min}, k_{max}]$	$[0.03, 0.15]$rad/m
$[\lambda_{min}, \lambda_{max}]$	$[42, 209]$m
$[\theta_{min}, \theta_{max}]$	$[-\pi/2, \pi/2]$
Number of scales (N_a)	250
Number of spectral directions (N_θ)	160
Simulated case of the bathymetry reconstruction	
$[\theta_{min}, \theta_{max}]$	$[-\pi/2, 0]$
Number of scales (N_a)	200
Number of spectral directions (N_θ)	100

1. Empirical scaling factors and spectral directions' definition. The empirical scaling factors are introduced as a vector in the interval

$$a \in [a_{min}, a_{max}] = k_0 T_p^2/dx[1/15, 1/3], \qquad (13.20)$$

which can be adjusted to fit local sea state and environmental conditions if needed. Estimation of the peak period T_p is performed using the radar image sequence or other independent measurements. Wavelet spectral directions are defined depending on the available memory resources and some reference information on the possible directional peak position distributions (for the cases regarded in this work, refer to Table 13.2).

2. Ramp (range trend) subtraction. In order to work with intensity maps without mean effects, the mean trend is subtracted as follows:

$$\tilde{I}(\mathbf{r}, t_n) = I(\mathbf{r}, t_n) - \frac{1}{N_t}\sum_{i=1}^{N_t} I(\mathbf{r}, t_i), \quad n = 1, \ldots, N_t, \qquad (13.21)$$

where N_t is the number of radar images.

3. 2D CWTFT. Calculation of the 2D CWTFT coefficients (13.14) transforming a set of corresponding successive radar images $\{\tilde{I}(\mathbf{r}, t_n)\}_{n=1}^{N_t}$ to a set of local directional complex wavelet spectra $\{W(\mathbf{r}, t_n|\mathbf{k}|, \theta)\}_{n=1}^{N_t}$ and calculation of the corresponding pseudo-wavevectors (13.18 and 13.19) with a taken from step 1.

4. Determination of the peak wavevector. In order to get a more accurate peak wavevector, the spectrum is averaged over time as follows:

$$W_{av}(\mathbf{r}, |\mathbf{k}|, \theta) = 1/N_t \sum_{n=1}^{N_t} \left| W(\mathbf{r}, t_n|\mathbf{k}|, \theta) \right|. \qquad (13.22)$$

The peak wavevector is then estimated as $\mathbf{k}_p(\mathbf{r}) = \arg \max_{(|\mathbf{k}|,\theta)} |W_{av}(r, |\mathbf{k}|, \theta)|$.

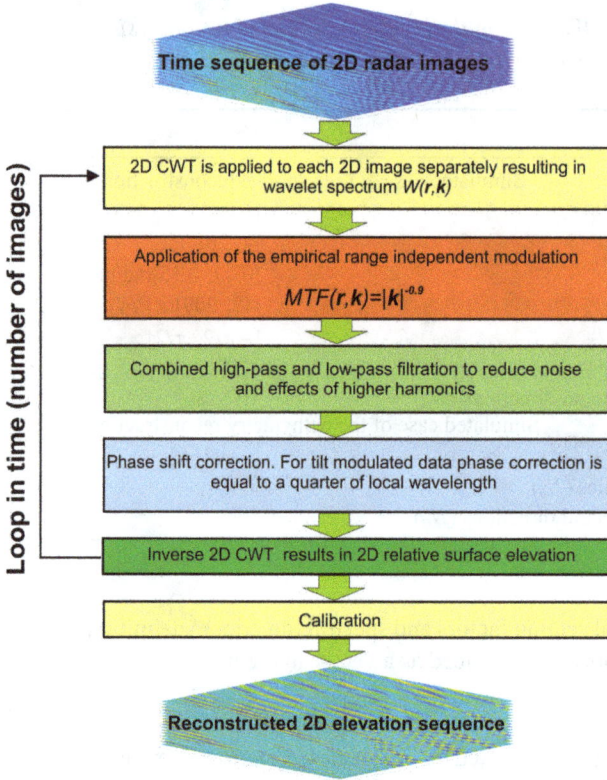

Time sequence of 2D radar images

Loop in time (number of images)

2D CWT is applied to each 2D image separately resulting in wavelet spectrum $W(r,k)$

Application of the empirical range independent modulation

$$MTF(r,k)=|k|^{-0.9}$$

Combined high-pass and low-pass filtration to reduce noise and effects of higher harmonics

Phase shift correction. For tilt modulated data phase correction is equal to a quarter of local wavelength

Inverse 2D CWT results in 2D relative surface elevation

Calibration

Reconstructed 2D elevation sequence

Figure 13.3   Flowchart of the 2D WSSR procedure

5. Band-pass filtering. In order to reduce noise and influence of higher harmonics, the corresponding filter F is applied in the wavevector–coordinate domain as follows:

$$\Phi(\mathbf{r}, \mathbf{k}) = \chi\left(\{(\mathbf{r}, \mathbf{k})|k_{th} < |\mathbf{k}| < lk_p(\mathbf{r}), \; \mathbf{r} \in [x_0, x_{max}] \times [y_0, y_{max}]\}\right), \qquad (13.23)$$

where χ denotes a characteristic function of a set of points and $k_{th} = 3T_p^{-2}$ is constant in space (already introduced in the first step). This high-pass component is exactly the same as regularly used in the conventional techniques (see e.g. [24]). The application of the low-pass filter is new and is defined to be peak-dependent. If $k_p(\mathbf{r})$ is the peak wavenumber calculated in the previous step and l is an empirical constant (here, the value $l = 3$ is used), the filtered wavelet spectrum is further defined as $W_F(\mathbf{r}, |\mathbf{k}|, \theta) = W(\mathbf{r}, |\mathbf{k}|, \theta)\Phi(\mathbf{r}, \mathbf{k})$.

6. Relative elevation reconstruction. The uncalibrated elevation $\eta_{uncalib}$ is calculated using (13.17) as follows:

$$\eta_{uncalib}(\mathbf{r}) = \text{const} \cdot \int_{\theta_{min}}^{\theta_{max}} \int_{k_{min}}^{k_{max}} \Re e(iW_F(\mathbf{r}, |\mathbf{k}|, \theta))|\mathbf{k}|^{1-0.9} dk \, d\theta. \qquad (13.24)$$

The imaginary unit i multiplying the wavelet spectrum relates to an application of a $\pi/2$ phase shift. This shift is applied according to the geometric tilt modulation relating the radar intensity to the surface elevation. The power 0.9 relates to the application of the empirical MTF in a wavevector–coordinate domain $MTF(\mathbf{r}, \mathbf{k}) = |\mathbf{k}|^{-\beta}$. For the reasoning of the alternative MTF power, refer to the discussion in the original paper [13]. Another power of $|\mathbf{k}|^1$ is coming from the Jacobian of the polar to Cartesian coordinates' transformation multiplied by the additional powers of $|\mathbf{k}|$ coming from (13.18).

7. Calibrated surface elevation reconstruction. The calibration correction is calculated as follows:

$$\eta_{rec}(\mathbf{r}) = \eta_{uncalib}(\mathbf{r})H_s^2/(16\sigma_{all}[\eta_{uncalib}]), \tag{13.25}$$

where H_s is the significant wave height (estimated from radar images in the deep sea or independent in situ measurements), and σ_{all} is the lateral coordinate and time-averaged standard deviation of a 3D array A. It is defined as follows:

$$\sigma_{all}[A] = \frac{1}{\sqrt{N_x-1}N_yN_t} \sum_{i=1}^{N_t} \sum_{j=1}^{N_y} \left(\sum_{k=1}^{N_x} \left| A(t_i x_k, y_j) - \frac{1}{N_x}\sum_{k=1}^{N_x} A(t_i x_k, y_j) \right|^2 \right)^{1/2} \tag{13.26}$$

Steps 5–7 can be repeated for each radar image to provide a time sequence of reconstructed surface elevation maps.

The calibration strategy presented in step 7 is applicable to real radar data. The corresponding calibration in simulated cases, where the whole original sea elevation is known, can be performed as follows:

$$\eta_{rec}(\mathbf{r}) = \eta_{uncalib}(\mathbf{r})\sigma_{all}[\eta(\mathbf{r})]/\sigma_{all}[\eta_{uncalib}(\mathbf{r})]. \tag{13.27}$$

The results of the corresponding sea surface reconstructions from simulated radar data are given in Figures 13.4 and 13.5.

13.4   Bathymetry reconstruction technique

Bathymetry reconstruction method was proposed and tested for both simulated and real data cases in [19]. The method known as a 2D Rapid Continuous Wavelet-based Bathymetry Inversion (RCWBI) method for bathymetry reconstruction is also described in this section. Unlike the traditional FFT analysis approach, the 2D CWTFT allows estimation of the peak wavenumber as a function of space $\mathbf{k}_p(\mathbf{r})$. In [25] and [26], it was proposed to track every wavenumber $\mathbf{k}(\mathbf{r}, \omega_n)$ in space. For this method, it is suggested to track only the peak spectral component $\mathbf{k}_p(\mathbf{r}) = \mathbf{k}(\mathbf{r}, \omega_p)$. So the bathymetry itself is estimated based on the peak wavevector's phase speed, which is calculated by relating spatial phase shift between the successive images and the known time lag. The dominant wave period $T_p(\mathbf{r})$ is assumed to be approximately

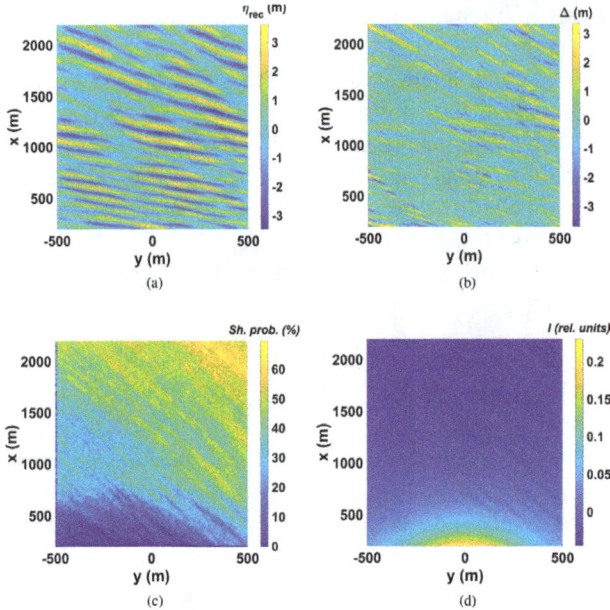

Figure 13.4   *(a) Reconstructed with 2D WSSR sea elevation $\eta_{rec}(\mathbf{r})$; (b) difference map $\eta_{rec}(\mathbf{r}) - \eta_{orig}(\mathbf{r})$; (c) shadowing probability obtained by the time averaging of the 100 shadowing maps for the image sequence; (d) range trend obtained by the time averaging of the 100 successive simulated radar images.*

constant and will be used to empirically estimate the scaling factor vector, corresponding to the primary peak component.

The bathymetry for this section is defined as a mildly changing beach in order to verify the reconstruction for a more natural bottom profile:

$$h(\mathbf{r}) = \begin{cases} 1/2(2\,000 - x)^{2/3}, & x \le 1\,800 \text{ (m)}, \\ 17.1 \times 10^{-(x-1\,800)0.00113}, & x > 1\,800 \text{ (m)}, \end{cases} \tag{13.28}$$

with depth value expressed in meters. The bathymetry is based on an equilibrium beach profile (see [27]) joined by a mild slope leading into shallow water. It is assumed there are no lateral changes to the beach profile.

The reconstruction procedure consists first of four preparatory steps, which is identical to the previously described sea surface reconstruction WSSR procedure. In addition, the following three steps are applied (see the complete flowchart in Figure 13.6):

1. Determination of the peak wavevector $\mathbf{k}_p(\mathbf{r}) = \arg\max_{(|k|,\theta)} |W_{av}(\tilde{r}, |k|, \theta)|$.
2. Calculation of the corresponding waves' phase velocities based on the phase shifts (dl):

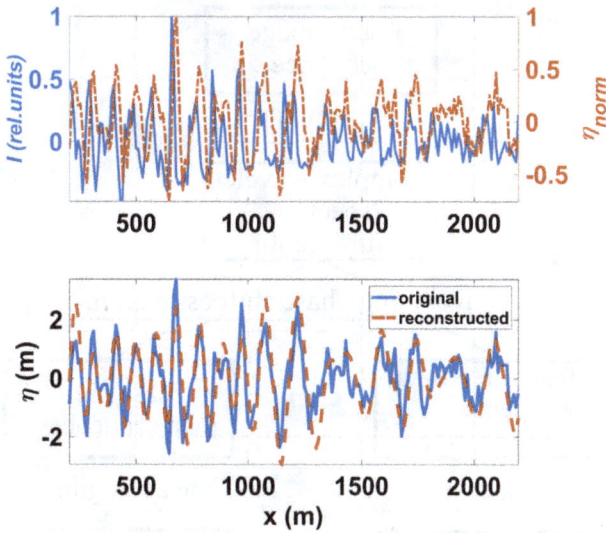

Figure 13.5   (upper panel) Relative fluctuations of the cross sections of the
initial surface elevation $\eta_{norm} = \eta / \max(\eta)$ and of the detrended
radar image intensity $\tilde{I} = (I - \bar{I})/ \max(I - \bar{I})$ for the simulation case.
A $\pi/2$ phase shift, resulting in a one-quarter local wavelength shift
in physical space is evident. (lower panel) Comparison between
η and η_{rec} for the original surface elevation and the 2D WSSR
method reconstructed sea surface elevations on the same cross
section.

$$dl(\mathbf{r}, t_n) = \mathrm{argc}\big(\widetilde{W}(\mathbf{r}, t_n\,|k_p|, \theta_p)\widetilde{W}^*(\mathbf{r}, t_{n+1}\,|k_p|, \theta_p)\big)/|k_p|, \tag{13.29}$$

where argc is the complex number argument. The time average of the waves'
phase speed is obtained as $c_p(\mathbf{r}) = \frac{1}{N_t dt} \sum\limits_{n=1}^{N_t-1} dl(\mathbf{r}, t_n)$.

3.   The solution of the corresponding dispersion relation to resolve depths

$$h(\mathbf{r}) = \left| \frac{\tanh^{-1}(c_p^2(\mathbf{r})|k_p(\mathbf{r})|/g)}{|k_p(\mathbf{r})|} \right| \tag{13.30}$$

results in a bathymetry map.

As one can notice, the wave–current interaction effect is neglected here as it
is commonly done in the nearshore area [28], since the magnitude of the corresponding onshore currents is assumed to be small. The longshore component can
be significant and regularly does not affect the wavefield, which is already refracted
and propagating almost shorewards. This is a common convention when talking
about radar-based bathymetry inversion. Introduction of the 3D CWT analysis could
resolve this limitation, enabling simultaneous estimation of the local depth and current vector in the pseudo-wavevector–frequency domain, but currently this has both

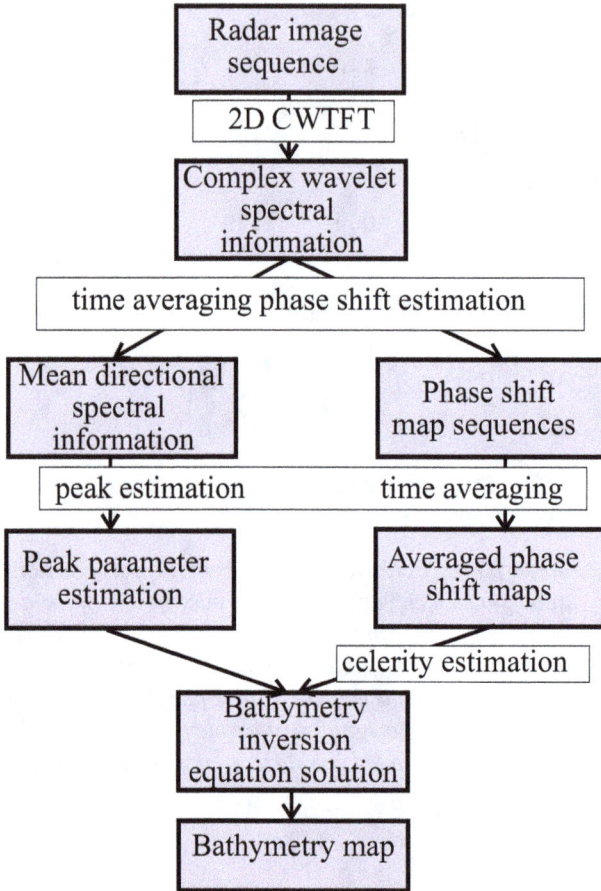

Figure 13.6   Block diagram of the 2D RCWBI bathymetry inversion process

mathematical and computational obstacles. Nevertheless, even in the current version of the method, it is possible to incorporate ambient currents estimated by radar-based methods, or independent in situ measurements.

In fact, the 2D RCWBI employs the most advanced techniques known as the FFT-based method, e.g. [29, 30]), since it realizes the "windowing" with an adaptive window size automatically due to the wave shoaling and nature of the CWT analysis. Results of the reconstruction procedure, including the difference map, the scatterplot of the results, and a comparison on a cross section are given in Figure 13.7. The obtained results demonstrate the ability of the proposed method to precisely retrieve bathymetry in the transition from intermediate to shallow water depths $k_p h < 1$. The corresponding bias on a lateral (y) coordinate slightly grows with depth as expected (see scatterplot in Figure 13.7(c)). The deviation from the original at shallow depths might be addressed as an edge effect of the FFT used for calculation inside the

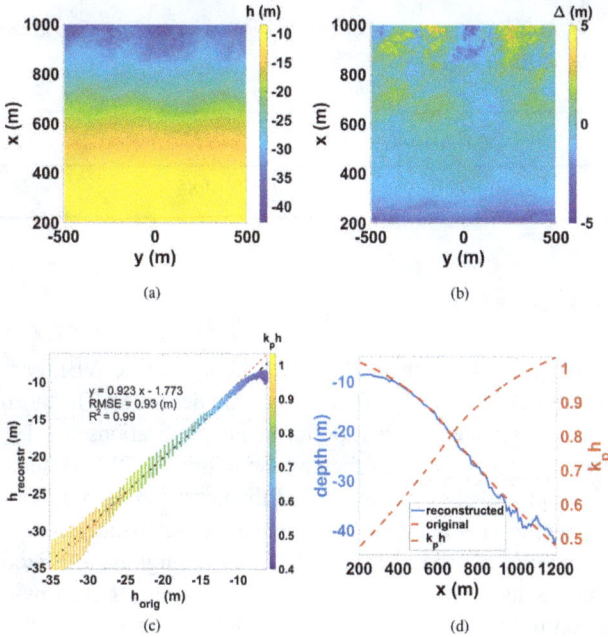

(a) (b)

(c) (d)

Figure 13.7 *(a) Reconstructed water depth h(**r**); (b) difference map between original and reconstructed water depths; (c) scatterplot of the original and reconstructed bathymetries showing the lateral distribution of the retrieved results, as well as the growing error dispersion in a deeper area. The dashed red line is a bisecting line; dashed black is the linear regression model. The well-pronounced bias of the resulting bathymetry in a shallow area (h < 10 m) is generally due to the edge effect of the 2D CWTFT; (d) comparison between original and reconstructed bathymetries on a cross section. The figure is taken from paper [19] by authors.*

2D RCWBT. This kind of deviation appears at both edges, but the deeper edge was cut out due to the dimensionless wavenumber limit mentioned above.

Mean value and standard deviation of the difference array is defined as

$$M_{all}(\Delta) = \frac{1}{N_x N_y} \sum_{i=1}^{N_x} \sum_{j=1}^{N_y} \Delta(x_i, y_j), \quad \sigma_{all}[\Delta] = \frac{1}{\sqrt{N_x - 1}N_y} \sum_{j=1}^{N_y} \left(\sum_{i=1}^{N_x} \left| \Delta(x_i, y_j) - \frac{1}{N_x} \sum_{i=1}^{N_x} \Delta(x_i, y_j) \right|^2 \right)^{1/2}.$$

(13.31)

The corresponding statistical estimates of the comparison are summarized in Table 13.3.

Table 13.3   Mean characteristics of the 2D bathymetry reconstruction with RCWBI method

Case(s)	$\sigma_{all}(|\Delta|)(m)$	$M_{all}(|\Delta|)(m)$	R^2
$T_p = 10(s)$	0.69	0.85	0.99

13.5   Conclusions

In this chapter, two CWT-based methods 2D WSSR and 2D RCWBI for reconstruction of sea elevations and bathymetries, respectively, were described. The thorough verification of the corresponding methods was performed in publications [13, 19, 31].

One of the main advantages of the corresponding 2D WSSR technique is that it can process comparatively large images of the nearshore region at once. There is also no need for windowing due to the fact that the wavelet analysis resolves energy-carrying shoaling wave components automatically through its localized nature. One of the downsides is its relative sensitivity to the amount of shadowing. Since the low-pass filtration in the wavevector domain is applied, as a part of the method, the corresponding reconstructed maps are much smoother than the original ones. For certain applications, high-frequency tails can be added aposteriori, as doing so (and after additional recalibration) would make reconstructed maps look more realistic. One possible approach is adding a cut-out high-frequency part from the results of the conventional method inversion, since this part of the spectrum is not so strongly affected by shoaling effects. Alternatively, high-frequency parts might be also added artificially by means of stochastic simulation of the nonlinear wave shoaling.

As for the bathymetry reconstruction method (2D RCWBI), it showed robust results. However, it did require longer image time sequences for the processing, in order to get the robust estimation of the corresponding peak waves' phase velocities. Anyway, these sequences are much shorter in comparison to the conventional (FFT-based) bathymetry reconstruction requirements. It was also reported in [19] that small-scale bathymetric features (sandbars, underwater rocks, etc.) cannot be always well-fitted by the long peak wave components. This feature can then be resolved by using smaller peak wave components, generated through nonlinear shoaling. Ambient currents are not taken into account in the current version of the method. Nevertheless, some external, independent estimations of the currents might be assimilated in the bathymetry estimation. Extension of the current method to 3D CWT or 2D CWT analysis but in the spatio-temporal domain may also contribute to the performance of the method.

It is also important to point out that the high capabilities of the 2D WSSR and RCWBI methods presented here are not limited to the use of X-band radars. This method can fit the analysis of other remote sensing equipment data, such as other types of radars and optical cameras mounted on various stationary and moving platforms.

Acknowledgment

The authors would like to acknowledge Ms Hazel Orpen for her kind help in the proofreading of this chapter.

References

[1] Ivonin D.V., Telegin V.A., Chernyshov P.V., Myslenkov S.A., Kuklev S.B. 'Possibilities of X-band nautical radars for monitoring of wind waves near the coast'. *Oceanology*. 2016;**56**(4):591–600.

[2] Carrasco R., Streßer M., Horstmann J. 'A simple method for retrieving significant wave height from Dopplerized X-band radar'. *Ocean Science*. 2017;**13**(1):95–103.

[3] Nieto Borge J., Rodríguez G.R., Hessner K., González P.I. 'Inversion of marine radar images for surface wave analysis'. *Journal of Atmospheric and Oceanic Technology*. 2004;**21**(8):1291–300.

[4] Hoogeboom F., Kleijweg J., Dv H. 'Seawave measurements using a ships radar'. *ESA*. 1986:819–23.

[5] Piotrowski C.C., Dugan J.P. 'Accuracy of bathymetry and current retrievals from airborne optical time-series imaging of shoaling waves'. *IEEE Transactions on Geoscience and Remote Sensing*. 2002;**40**(12):2606–18.

[6] Serafino F., Lugni C., Borge J.C.N., Zamparelli V., Soldovieri F, Nieto Borge J.C. 'Bathymetry determination via X-band radar data: a new strategy and numerical results'. *Sensors*. 2010;**10**(7):6522–34.

[7] Ludeno G., Brandini C., Lugni C. 'Remocean system for the detection of the reflected waves from the Costa Concordia ship wreck'. *IEEE Journal of Selected Topics in Applied Earth Observations and Remote Sensing*. 2014;**7**(7):3011–18.

[8] Wei Y., Zhang J.-K., Lu Z. 'A novel successive cancellation method to retrieve sea wave components from spatio-temporal remote sensing image sequences'. *Remote Sensing*. 2016;**8**(7):607.

[9] Holman R., Plant N., Holland T. 'cBathy: a robust algorithm for estimating nearshore bathymetry'. *Journal of Geophysical Research: Oceans*. 2013;**118**(5):2595–609.

[10] Rutten J., de Jong S.M., Ruessink G. 'Accuracy of nearshore bathymetry inverted from X-band radar and optical video data'. *IEEE Transactions on Geoscience and Remote Sensing*. 2017;**55**(2):1106–16.

[11] An J., Huang W., Gill E.W. 'A self-adaptive wavelet-based algorithm for wave measurement using nautical radar'. *IEEE Transactions on Geoscience and Remote Sensing*. 2015;**53**(1):567–77.

[12] Chuang L.Z.-H., Wu L.-C., Doong D.-J., Kao C.C. 'Two-dimensional continuous wavelet transform of simulated spatial images of waves on a slowly varying topography'. *Ocean Engineering*. 2008;**35**(10):1039–51.

[13] Chernyshov P., Vrecica T., Toledo Y. 'Inversion of nearshore X-band radar images to sea surface elevation maps'. *Remote Sensing*. 2018;**10**(12):1919.

[14] LC W., Chuang L.Z.H., Doong D.J. 'Quantification of non-homogeneity from ocean remote-sensing images using two-dimensional continuous wavelet transforms'. *International Journal of Remote Sensing*. 2011;**32**(5):1303–18.

[15] Bredmose H., Agnon Y., Madsen P.A., Schäffer H.A. 'Wave transformation models with exact second-order transfer'. *European Journal of Mechanics – B/Fluids*. 2005;**24**(6):659–82.

[16] Dean R.G., Dalrymple R.A. *Water Wave Mechanics for Engineers and Scientists*. **2**. World Scientific Publishing Co Inc; 1991.

[17] Hasselmann K., Barnett T., Bouws E. 'Measurements of wind-wave growth and swell decay during the joint north sea wave project (JONSWAP)'. *Deutches Hydrographisches Institut*. 1973.

[18] Litvenko K.V., Prigarin S.M. 'The error analysis for spectral models of the sea surface undulation'. *Russian Journal of Numerical Analysis and Mathematical Modelling*. 2014;**29**(4):239–50.

[19] Chernyshov P., Vrecica T., Streßer M., Carrasco R., Toledo Y. 'Rapid wavelet-based bathymetry inversion method for nearshore X-band radars'. *Remote Sensing of Environment*. 2020;**240**(3):111688.

[20] Mitsuyasu H., Tasai F., Suhara T., *et al.* 'Observations of the directional spectrum of ocean waves using a cloverleaf buoy'. *Journal of Physical Oceanography*. 1975;**5**(4):750–60.

[21] Antoine J.-P., Murenzi R. 'Two-dimensional directional wavelets and the scale-angle representation'. *Signal Processing*. 1996;**52**(3):259–81.

[22] Daubechies I. *Ten lectures on wavelets*. SiaM; 1992.

[23] Wang N., Lu C. 'Two-dimensional continuous wavelet analysis and its application to meteorological data'. *Journal of Atmospheric and Oceanic Technology*. 2010;**27**(4):652–66.

[24] Nieto Borge J.C., Guedes Soares C., Borge J.C.N., Soares C.G. 'Analysis of directional wave fields using X-band navigation radar'. *Coastal Engineering*. 2000;**40**(4):375–91.

[25] Hessner K., Reichert K., Rosenthal W. 'Mapping of sea bottom topography in shallow seas by using a nautical radar'. 2nd Symposium on Operationalization of Remote Sensing; 1999.

[26] Poupardin A., Idier D., de Michele M., Raucoules D. 'Water depth inversion from a single SPOT-5 dataset'. *IEEE Transactions on Geoscience and Remote Sensing*. 2016;**54**(4):2329–42.

[27] Bruun P. *Coast erosion and the development of beach profiles*; 1954.

[28] Bell P. 'Bathymetry derived from an analysis of X-band marine radar images of waves'. *Proceedings of Oceanology '98, vol. 3*; Brighton, U.K; 1998. pp. 535–43.

[29] Ludeno G., Reale F., Dentale F. 'An X-band radar system for bathymetry and wave field analysis in a harbour area'. *Sensors*. 2015;**15**(1):1691–707.

[30] Ludeno G., Postacchini M., Natale A. 'Normalized scalar product approach for nearshore bathymetric estimation from X-band radar images: an

assessment based on simulated and measured data'. *IEEE Journal of Oceanic Engineering.* 2018;**43**(1):221–37.

[31] Chernyshov P., Vrecica T., Nauri S., Toledo Y. 'Wavelet-based 2D sea surface reconstruction method from nearshore X-band radar image sequences'. IEEE Transactions on Geoscience and Remote Sensing; 2021.

Chapter 14

Wave field reconstruction using orthogonal decomposition of Doppler velocities

Erin E Hackett[1]

Methods for estimating wave field statistics from either radar backscatter or Doppler velocities of the sea surface have developed extensively over the last several decades as described in Chapters 12 and 13. As previously described, these methods typically involve fast Fourier transform (FFT)-based techniques that filter 2D or 3D wave spectra using the linear dispersion relationship for ocean surface waves in order to isolate the wave field from other contributions to the backscatter or Doppler/ surface velocities. Implicit in this filtering is removal of nonlinear aspects of the wave field such as wave breaking and wave grouping effects [1, 2]. When examining wave statistics, presumably, the removal of such features generally has a small effect on the wave statistics [3]; however, there is ever increasing interest in reconstructing phase-resolved wave fields from radar measurements [4–6]. The removal of some effects of wave grouping, for example, could have a significant influence on instantaneous wave heights at certain locations at particular times. Thus, there is a need for exploring alternative techniques that could enable more of these effects to be included in phase-resolved wave field reconstructions. For naval and shipping applications, it is this "anomalous" wave that can be highly relevant in predictions when, for example, activities involve moving items between two ships or between an offshore platform and a ship.

In this chapter, the use of orthogonal decomposition is examined to isolate the wave field and generate 2D phase-resolved time series of orbital velocity, which could be converted to sea surface elevation using the transformations discussed in Chapter 12. In this chapter, this orthogonal decomposition will be referred to as proper orthogonal decomposition, or POD, but orthogonal decompositions go by many names across a variety of fields or subdisciplines, such as empirical orthogonal functions (EOF), principal component analysis (PCA), Karhunen–Loève transform (KLT), or singular value decomposition (SVD) [7]. All these techniques are more similar than they are different and can be considered the same for the purposes of this chapter. POD is a technique used to provide low-dimensional representations

[1]Department of Marine Science, Coastal Carolina University, United States

of complex, high-dimensional systems. The POD method decomposes a signal into a set of modes (or basis functions) and time or spatial series coefficients. The measured signal can be reconstructed by the summation of the modes multiplied by the corresponding coefficients. A reduced order representation of the signal can be created using a subset of the modes; this is referred to as a reconstruction. This technique makes no assumption of the basis function's form *a priori*; instead, the basis functions are determined by the data. This feature differentiates POD from FFT methods in that FFT methods assume sinusoidal basis functions. The determination of the basis functions *a posteriori* minimizes the number of POD modes required to account for the majority of the variance of the signal, which is the premise upon which the basis functions are determined. While there is no inherent physical interpretation to the mode basis functions, there is an increasing body of work associating POD basis functions with physical significance [3, 8–13].

The POD method can be used to reconstruct phase-resolved ocean wave fields from Doppler radar measurements using the leading mode functions as a filter to separate wave contributions to the radar signal from non-wave contributions, assuming some of the basis functions can be associated with the physics of the ocean surface waves while others can be associated with unwanted artifacts of the radar measurement. Also implied is the assumption that the majority of the variance of the measured Doppler velocities are associated with the wave field (see Chapter 12). If feasible, then the need to filter data based on the linear dispersion relationship is removed, potentially allowing nonlinear features, such as nonlinear aspects of wave grouping and other wave–wave interactions, to be included in the retrieved wave fields.

14.1   Potential limitations of the FFT-based wave field processing

Regardless of whether using a coherent or incoherent system, marine radar measurements of the sea surface contain contributions to the measurement from "non-wave-" related factors. For Doppler measurements, where the ideal measurement is wave orbital velocity, contaminating sources include phase speeds of capillary waves, phase speeds of breaking waves, ocean currents, and even hard targets like boats or buoys. FFT methods exploit the relationship between wavelength and period for linear ocean surface waves to identify and extract the wave field information. However, there are potential wave contributions to the radar signal that are not associated with the linear dispersion relationship. Thus, FFT-based dispersion filtering would eliminate such wave contributions to the radar measurement and more generally any nonlinear features of the wave field. These eliminated contributions may be important in obtaining accurate measurements of the ocean surface under some conditions.

Most wavenumber–frequency (k–ω) spectra of radar images of the sea surface exhibit a "group line" feature: a low–frequency feature below the first-order dispersion relationship that passes through the origin. Figure 14.1 shows an example of

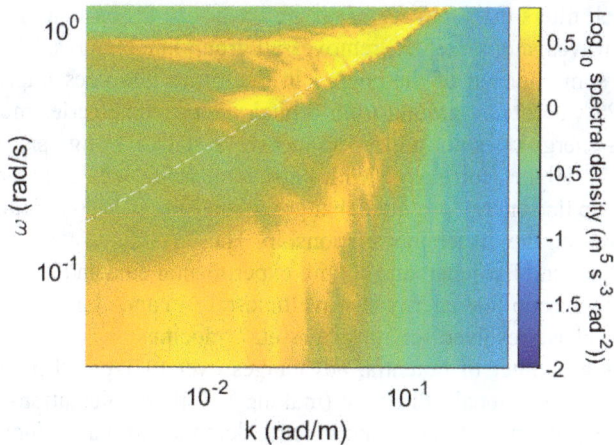

Figure 14.1   An example wavenumber–frequency spectrum of Doppler velocities with the dispersion relationship shown as a white dashed line. Apparent at frequencies lower than the dispersion relationship is the "group line" [12].

this group line feature. Numerous explanations exist for the origin of this feature that are wave- and non-wave-related, such as shadowing, nonlinear wave–wave interactions, contamination (e.g., hard targets in the images), turbulence advected by winds, interference-induced wave breaking, and nonlinear scattering effects [2, 14–19].

Plant and Farquharson [2] provide numerical evidence that interference-induced wave breaking from the interaction of linear wave fields are a primary source of the group line feature. They demonstrate that the superposition of wind waves and swell can generate steep short gravity waves that break near the local maxima of surface slope resulting in Doppler measurements of the phase speed of these steep short gravity waves. Such effects as noted by Plant and Farquharson [2] as well as nonlinear second order wave-wave interactions may be features that should be accounted for in the generation of instantaneous (phase-resolved) sea surface elevation maps produced by radar (e.g., [16, 20]). However, it is difficult to separate these wave contributions to the group line feature to validate whether they are important to the reconstruction of sea surface elevation maps because other non-wave-related effects may also occupy the same frequency–wavenumber space (e.g., shadowing).

Kammerer and Hackett [11] use POD to reconstruct ocean surface orbital velocities from X-band Doppler measurements of the sea surface, which permits the inclusion of some of the spectral energy in the group line feature in the reconstruction of instantaneous orbital velocities. The inclusion of the group line energy is based on how much it contributes to the overall variance of the measured spatial series [3]. They evaluate the importance of group line associated energy to accurate phase-resolved reconstructions of ocean surface wave orbital velocities by comparing POD-reconstructed orbital velocity maps to the more conventional FFT-based method, which filters the k–ω spectra on the linear dispersion relationship for surface gravity waves and, therefore, removes all group line features. They show that inclusion of some portion of the energy in the group line does improve correlation with GPS wave buoy ground truth orbital velocity time series measurements, although this energy does not greatly impact wave statistics (e.g., significant wave height) [3, 12]. Higher correlation with buoy time series was demonstrated when including group line energy provided that group line energy is comparable or higher than the energy on the dispersion relationship. These results support the numerical findings of Plant and Farquharson [2] with experimental data and show that at least a portion of the group line energy is wave field-related and contributes to accurate instantaneous phase-resolved sea surface orbital velocities.

POD has a number of potential advantages over FFT-based processing such as increased computational efficiency (making real-time calculations more feasible), eliminating the need to enter the spectral domain and thus reducing sampling requirements (e.g., spectral resolution is set by the length of the time series), reduction of large dataset storage size (because of the reduction in dimensions of the data), elimination of spectral artifacts from the FFT/IFFT and filtering process, and the potential retention of features off the dispersion curve. Although POD is a linear technique, it is the optimal linear technique for representation of nonlinear processes [7]. In contrast, due to the empirical nature of POD, there is no innate relationship between the POD modes and the physics of the wave field; thus, the content of the POD modes must be investigated to establish a connection to the physics of the measured wave field. Hackett *et al.* [10] showed numerically that when POD is applied to idealized radar measurements of synthetic sea surfaces, the leading mode functions are associated with the physics of the wave field. In addition, Zhang *et al.* [21], Zhongbiao *et al.* [13], and Chen *et al.* [22] discuss similar methods for obtaining a transfer function for wave height estimation using radar backscatter. In the next section, the POD method is explained along with a discussion of the physical significance of the POD modes when applied to rotating Doppler radar measurements. That section is followed by an evaluation of the POD method for both wave field statistics and phase-resolved wave field retrievals. The chapter concludes with a summary and discussion of limitations of the POD method.

14.2 Proper orthogonal decomposition for wave field reconstruction

14.2.1 Data

The POD decomposition technique will be demonstrated using data collected from two rotating Doppler radar systems. An overview of these data is provided in this section, but a more thorough review of the data can be found in Kammerer and Hackett [3] and Kammerer [12]. The first radar system was a rotating coherent-on-receive Doppler system, which will be referred to as UM. The center frequency of the radar was 9.41 GHz (X-band) [23] and rotated at 24 RPM. Pulse-pair processing

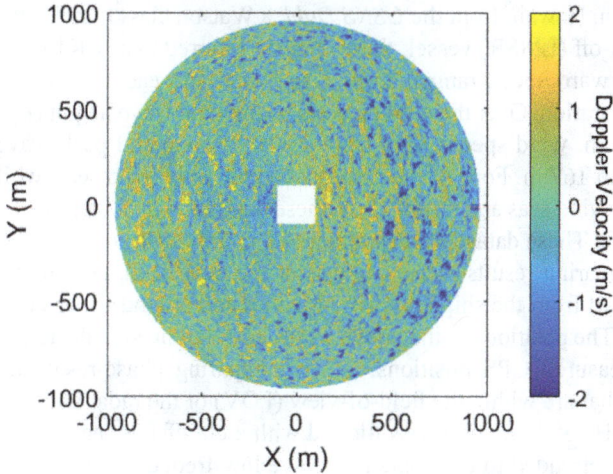

Figure 14.2   Sample Doppler velocity distribution for a single frame [3]

is used to estimate the Doppler velocity [24, 25]. The Doppler estimates are averaged over 12 pulse-pairs for noise reduction leading to a Doppler range distribution approximately every 0.86 degrees of rotation of the antenna. The resulting Doppler data distribution is a function of range (r), time (t), and azimuth (ϕ). The range extent of the data is out to 960 m with a 100 m radial blanking region surrounding the radar to eliminate high power returns. The range resolution is 3.75 m. One complete frame (one revolution) of Doppler data is generated every 2.5 s. An example of a single frame of radar data is shown in Figure 14.2.

The second radar system, which will be referred to as APS, involves four coherent antennas mounted at 90 degrees to each other, each rotating at 5 RPM. Each of the four antennas is identical. Each has a center frequency of 9.2 GHz and a pulse repetition frequency of 25 kHz. Data are a function of range (r), time (t), azimuth (ϕ), and antenna (A). FFT processing is used to produce Doppler estimates over 64 pulses [26], yielding a Doppler range distribution every 1.23° for a rotation rate of 12 s. One quarter rotation of the system yields a complete frame of data every 3 s because the data from each of the four antennas are combined to generate one frame. The potential advantage of this configuration is that a slower rotation rate permits more pulses to go into each Doppler estimate (referred to as the dwell time), which should make the Doppler estimate more accurate. The tradeoff is the slow rotation rate can result in aliasing because the revisit time to the same patch of ocean surface is longer than many of the ocean surface wave periods. The four antennas mitigate this tradeoff. The resulting Doppler distributions have a range resolution of 4.8 m and cover a range of 998 m. The radar had a blanking range of 100 m around the vessel to eliminate high power return.

Ten datasets were collected from the R/V Melville from September 14–17, 2013, south of the Channel Islands offshore of Los Angelos, CA; these data are referred to as Melville. Four other datasets were taken from May 17–21, 2015, off the east coast of Oahu, Hawaii, from the *USNS Dahl*, a Watson class Large, Medium-Speed, Roll-on/Roll-off (LMSR) vessel; these data are referred to as CK15.

Ship forward speed ranged from 1 to 3 m/s and each dataset collected was 2–3 min in duration. Over these 14 datasets, significant wave height (H_s) varied from 0.10 to 2.15 m, wind speeds (U_w) from 2.3 to 15 m/s, and peak wavelengths (λ_p) spanned 77 to 167 m. Four of the datasets contain only wind seas, while the others contain both wind seas and swell. Thus, these datasets span a variety of environmental conditions. These datasets are summarized in Table 14.1.

For comparing results of the analysis of the radar data, GPS mini wave buoys were deployed from the ship and drifted within and around the radar measurement region [27]. The positions of the buoys were geo-referenced to the radar using buoy- and ship-measured GPS positions. When comparing phase-resolved time series, only buoys that are within the field-of-view (FOV) of the radar are used. The buoys sample at 1 Hz and are high-pass filtered with a cutoff frequency of 0.05 Hz (20 s period) and detrended to eliminate non-wave low-frequency signals mostly related to the transitioning of GPS satellites. The buoy time series are also subsequently down-sampled to match the temporal resolution of the radar data. More information about the wave buoys can be found in Drazen *et al.* [27].

14.2.2   *Proper orthogonal decomposition*

14.2.2.1   **Preparing data**
First, frames are reoriented such that the zero azimuth is aligned with north rather than the ship's heading and are linearly detrended along each azimuth to remove currents and ship forward speed. Kammerer and Hackett [3] and Kammerer [12] show the most accurate POD wave field reconstructions occur when the peak wave propagation direction is aligned with the direction of mode functions. Aligning the dominant wave propagation direction along the direction of the mode functions (see 14.2.2.2) enables them to encapsulate the peak wavelength of the wave system more accurately. The polar Doppler velocity data ($\widetilde{D}(r,\phi,t,)$) are rotated such that the zero azimuth is aligned with the dominant wave propagation direction. Then, the data are converted to a Cartesian coordinate system ($\widetilde{D}(x,y,t)$) such that the dominant wave direction is aligned with x direction of the Cartesian coordinate system. The dominant wave direction is determined based on the directional Doppler velocity spectrum (2D FFT of an individual radar frame) as well as the time series of radar images due to the 180-degree ambiguity innate to the wavenumber spectrum of an individual radar frame [28]. An example of directional wave spectrum from a single radar frame is shown in Figure 14.3.

14.2.2.2   **Decomposition**
The data from a single Doppler velocity frame, $D(x,y)$, are organized into a matrix **D**. A singular value decomposition is used to decompose the matrix **D**:

$$\mathbf{D} = \boldsymbol{B\Sigma P}^T \tag{14.1}$$

Table 14.1 Datasets and the associated range of environmental conditions covered. H_s, λ_p and RMS of wave orbital velocities (V_{rms}) were measured using GPS mini wave buoys. U_w was measured using a ship-mounted anemometer. $\Delta\Theta$, the difference in angle between swell and wind wave propagation directions, was calculated from the radar measured directional wave spectrum. The dataset number, associated test, date and time of the measurements, number of wave systems, and which radar-systems collected data at this time are also provided [3].

Dataset	Test	Date	Time	APS	UM	H_s (m)	U_w (m/s)	λ_p (m)	V_{rms} (m/s)	Wave systems	$\Delta\Theta$ (°)
1*	CK15	May 17, 2015	20:08:00	✓	✓	0.1	6	167	0.03	2	53
2§	CK15	May 17, 2015	21:20:00	✗	✓	0.19	7.5	111	0.09	1	0
3	Melville	September 15, 2013	17:10:40	✓	✓	1.62	12	105	0.39	1	0
4§	Melville	September 16, 2013	01:36:00	✓	✓	1.42	15	95	0.46	1	0
5*	Melville	September 16, 2013	15:43:00	✓	✓	1.29	11	82	0.43	2	23
6*	Melville	September 16, 2013	18:29:32	✓	✓	1.29	12	98	0.44	2	50
7	Melville	September 17, 2013	00:08:45	✓	✓	1.65	11	77	0.54	2	41
8	Melville	September 17, 2013	00:33:34	✓	✓	1.48	11	78	0.49	2	46
9	Melville	September 17, 2013	00:58:10	✓	✓	1.68	11	78	0.52	2	44
10	Melville	September 17, 2013	02:00:00	✓	✓	1.59	12	92	0.51	2	30
11†	CK15	May 17, 2015	22:20:00	✗	✓	1.64	7	111	0.53	2	61
12‡	CK15	May 21, 2015	20:32:00	✓	✓	1.07	2.3	97	0.28	1	0
13‡	Melville	September 17, 2013	14:28:53	✓	✓	2.1	8.9	108	0.63	2	59
14†	Melville	September 17, 2013	19:27:33	✓	✓	2.15	9.9	105	0.632	2	0

*Lowest H_s for respective test.
†Highest H_s for respective test.
‡Lowest U_w for respective test.
§Highest U_w for respective test.

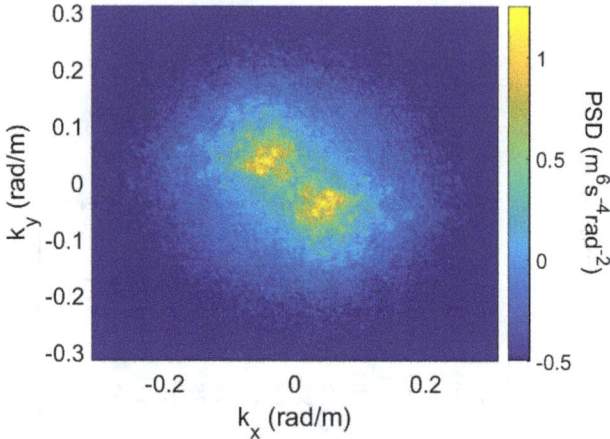

Figure 14.3   Sample radar-based directional wave spectrum for a single frame
(PSD, power spectral density). Note the 180-degree ambiguity
innate to a directional spectrum based on a single radar
frame [12].

where $\mathbf{B}$, $\mathbf{\Sigma}$, and $\mathbf{P}$ are matrices, and ^{T} indicates matrix transpose. The mode functions are encompassed in $\mathbf{P}$, and the diagonal elements of $\mathbf{\Sigma}$ are the singular values of $\mathbf{D}$. If we let $\mathbf{Q} = \mathbf{B}\mathbf{\Sigma}$, then,

$$\mathbf{D} = \mathbf{Q}\mathbf{P}^{T} = \sum_{k=1}^{M} q_k p_k^T \qquad (14.2)$$

where q_k are the spatial coefficients (columns of $\mathbf{Q}$), p_k^T are the basis functions of the Doppler velocity or the proper orthogonal modes (transposed columns of $\mathbf{P}$), and M is the number of samples in the x direction. The singular values occur in ranked order along the diagonal of $\mathbf{\Sigma}$ and signify the relative importance of each mode. The modes or basis functions are determined from the data using this decomposition and ranked such that the first mode accounts for the most variance of the signal, the second mode the second order contributor to the variance, and so on. Thus, if the variance of the Doppler measurements is dominated by ocean surface orbital velocity variations, then one would expect the leading modes to be associated with the wave field. This assumption is reasonable given that most of the other contributors to the Doppler velocity that are not wave orbital velocities are relatively constant (or contribute a smaller amount of variance due to being intermittent). A summation of all the modes multiplied by the corresponding coefficients results in the reconstruction of the original matrix, $\mathbf{D}$. An example mode function is shown in Figure 14.4.

A reduced-order representation or reconstruction of the data is produced by summing over a subset of the modes in (14.2), i.e., taking the summation from $k = 1$ to n where $n < M$. The amount of total variance (or energy) captured by a reduced order representation can be examined by computing the cumulative energy with an increasing number of modes included. As an example, Figure 14.5 shows

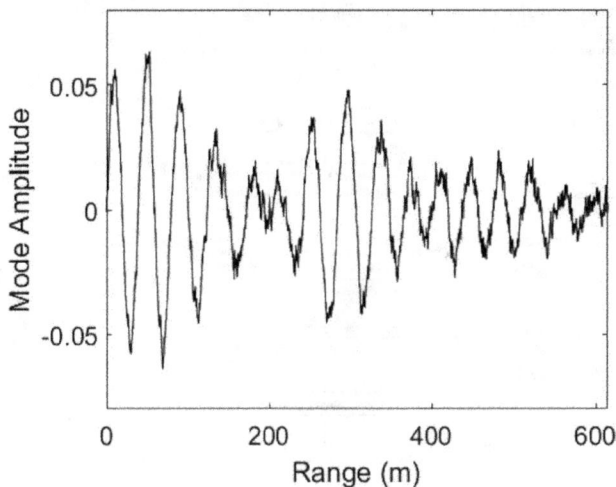

Figure 14.4   Example mode function [12]

the cumulative energy versus the number of modes included in the reconstruction, where the reconstructed energy is normalized by the total energy (or variance) in the data at the peak wavelength. By the 20th mode, approximately 50% of the overall variance is accounted for and by 40 modes 90% of the energy is included in the reconstruction. While it is not known exactly how much of the variance of the original

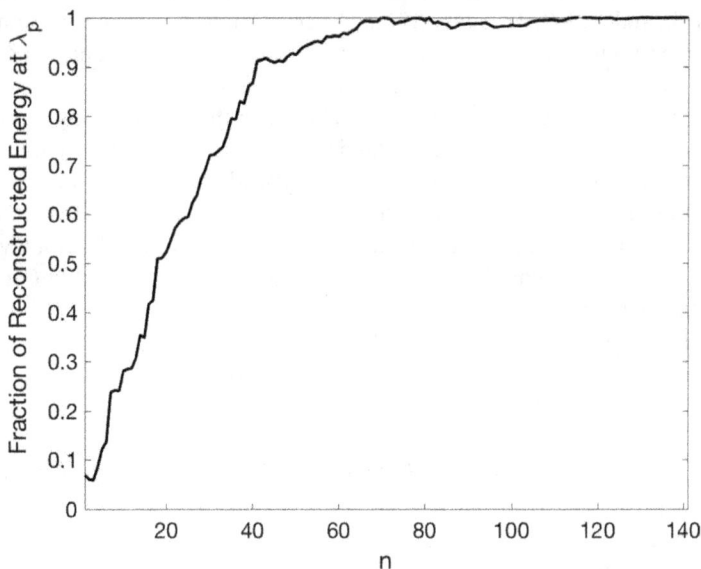

Figure 14.5   Example cumulative normalized energy distribution for 1 to n mode reconstructions at the peak wavelength [3]

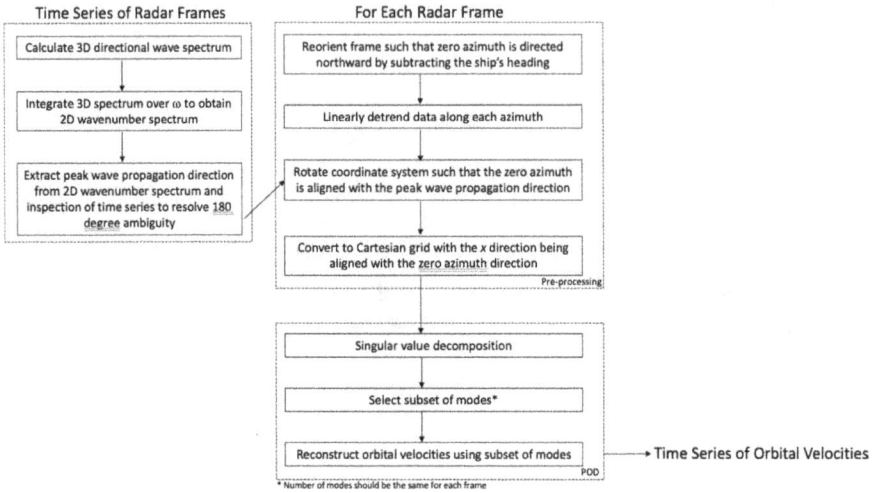

Figure 14.6   Flow chart of steps for the POD method

Doppler measurements can be attributed to the wave field, it is reasonable to assume that the majority of it should be related to the wave field as most other contributors to the Doppler velocity presumably contribute much less to the variance. The choice of the number of modes to include to accurately reconstruct the wave field could vary with dataset and environmental conditions and is a primary weakness of this method; mode selection will be discussed in Section 14.2.3. Nevertheless, with proper mode selection, the reduced representation could filter or extract the wave field signal from the radar measurements and discard unwanted artifacts. This procedure is performed for each frame in the radar time series. This method is presented in Kammerer and Hackett [3] and Kammerer [12], which was based on the methodology presented in Hackett *et al.* [10]. The result of the method is a time series of orbital velocities from which wave statistics can be computed using typical methods of estimating wave statistics from orbital velocity time series. It is important to note that velocities that are not along wave propagation directions are projections of the full orbital velocity, which needs to be accounted for using methods like those described in Carrasco *et al.* [29]. In this chapter, wave statistics are derived along the peak wave propagation direction (±5 degrees and averaged) and thus such directional effects are assumed to minimally impact results. A summary of the steps to obtain the time series of orbital velocities is presented in Figure 14.6.

14.2.3   *Mode selection and physical significance of the POD modes*

In this section, evidence of the leading mode functions being associated with the wave field is demonstrated along with discussion of mode selection. As shown in Figure 14.4, the leading mode functions are oscillatory as would be expected for orbital velocities. Hackett *et al.* [10] showed similar leading order basis functions are generated from idealized radar measurements of synthetic sea surfaces.

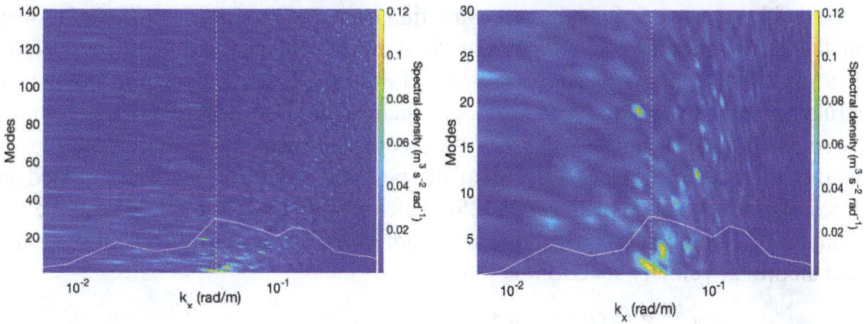

Figure 14.7   *1D wavenumber spectra of individual mode velocity*
reconstructions versus mode number. The left panel shows the
full range of POD modes, and the right panel shows a zoom-in of
the leading 30 modes. The solid white line is the corresponding
average spectrum of all the buoys amplified by 2 000 for (left)
and by 500 for (right) for visibility. The vertical white dashed line
denotes the peak wavelength [3].

To further demonstrate that leading order basis functions can be associated with the
wave field, one can perform reconstructions using each individual basis function
(i.e., performing the summation in 14.2 over one mode), compute the 2D wavenumber spectrum of this reconstructed velocity distribution, and then integrate the 2D
wavenumber spectrum over k_y (wavenumber in y direction). This procedure yields
a 1D wavenumber spectrum for each mode reconstruction. These 1D spectra can be
compiled into a spectrogram as that shown in Figure 14.7. The white solid line in
this figure is the average spectrum from all the wave buoys and the dashed white
vertical line is the peak wavelength. The spectrogram shows that energy in the lowest modes ($<\sim 20$) is concentrated around the wavelengths with largest energy from
the buoy spectra while at higher modes (>20) energy is spread across many wavelengths and is diffuse. In other words, aside from these leading modes, the variance
in the modes is not concentrated at specific length scales. This energy is presumably
part of non-wave contributions to the radar Doppler velocity measurement because
it is not directly associated with any known dominant spatial scale of the ocean surface waves. Such results strongly suggest that the leading modes are capturing variability associated with the wave field and are consistent with the findings presented
in Hackett *et al.* [10] based on synthetic data. These results are also consistent with
results from EOF analysis of gray scale images of radar backscatter, which showed
that the main EOF modes are related to different frequency components of the wave
field and higher EOF modes contain information about smaller-scale waves and
noise [13]. Kammerer [12] also showed that 1D spectra of leading order modes are
similar to those measured by a buoy.

With verification of the leading modes capturing wave field physics, the question turns to how many modes are optimal. Kammerer and Hackett [3] compute the significant wave height for reconstructions using all possible number of modes in the reconstruction and defines the "best mode" reconstruction to be the one that matches the buoy-measured H_s most closely. H_s is computed from the reconstructions based on radial transects of velocity along the peak wave direction (x direction) plus or minus ~5 degrees and averaged. Each 1D velocity spectrum is converted to a sea surface displacement spectrum using the method outlined in Hackett et al. [30] and Hwang et al. [25], which is similar to the method described in Chapter 12. Specifically, the full linear relationship between a wave height spectrum and an orbital velocity spectrum is [30]:

$$F_{vv}(k) = \int_0^{2\pi} \frac{g^2 k^2}{\omega^2} \frac{\cosh^2[k(h+z)]}{\cosh^2(kh)} F_{nn}(k)D(\theta)\cos^2(\theta - \theta_r)d\theta \qquad (14.3)$$

where h is water depth, g is gravitational acceleration, $D(\theta)$ is a directional distribution, θ is the wave propagation direction, θ_r is the radar look direction, and z is the vertical coordinate, which is defined as zero at the mean free surface. F_{vv} indicates a (1D) velocity spectrum, and F_{nn} indicates a (1D) sea surface elevation spectrum. By assuming the measurements are performed at the surface ($z = 0$), neglecting directional effects, including the assumption ($\theta - \theta_r = 0$), and use of the still water dispersion relationship, this expression simplifies to:

$$F_{vv}(k) = \frac{\omega^2}{\tanh^2(kh)} F_{nn}(k) \qquad (14.4)$$

(14.4) is used to compute $F_m(k)$, which is subsequently converted to a frequency spectrum [31], $F_m(f)$, and integrated to estimate:

$$H_s = 4\sqrt{\int F_{nn}(f)df} \qquad (14.5)$$

which assumes deep water, and where f is linear frequency. The resulting distribution for the best mode for the various datasets is shown in Figure 14.8. The figure shows that the "best mode" is not consistent across datasets, environmental conditions, or across radar systems; however, in nearly all cases it is below the 20th mode. The best mode shows a weak trend with increasing significant wave height, particularly for the coherent-on-receive (UM) radar, but there is a large spread. As wave height increases, the complexity of the wave field increases, which might explain this slight increase in the number of required modes to accurately reconstruct the wave field. Clearly, if one is to determine the mode selection *a priori* then analysis into which modes contain considerable energy near the peak wavelength is a good starting point. The spectrogram shown in Figure 14.7 can provide this type of insight. Zhongbiao et al. [13] applied EOF analysis to radar backscatter and found that H_s is most accurately estimated using individual middle modes. This result may differ relative to those shown in Figure 14.8 because the decomposition is being applied to Doppler velocities rather than backscatter, which inherently has different modulation factors and may also be related to the fact that Zhongbiao et al. [13] do not do any pre-processing, which might cause the first several modes to show large-scale trends (e.g., backscatter intensity drop-off with range) rather than the wave patterns in their analysis. A calibration for a particular system over a range

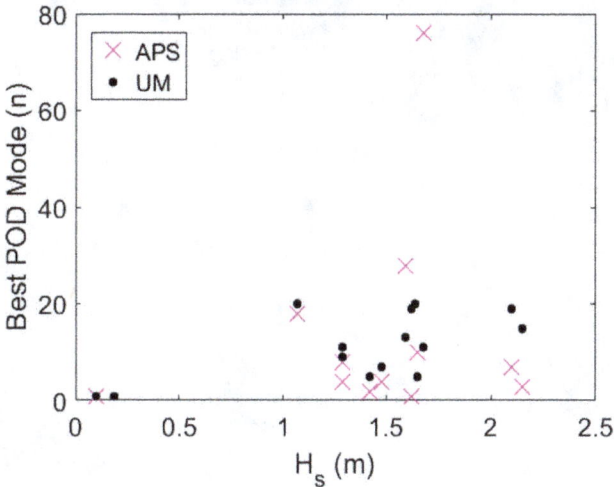

Figure 14.8   Most accurate mode (reconstruction of modes 1 through n) with respect to buoy-measured H_s for two different radar systems [12]

of conditions could be another approach for determining the number of modes to retain. Clearly, this aspect of this method warrants further investigation and is still an area of open research.

14.3   Evaluation of POD-based wave field reconstructions

For the evaluation of POD-based wave field reconstructions, the mode selection will be based on the "best mode" in comparison to the buoy-based measurements. Thus, the comparisons shown indicate the best possible representation of the data using the POD method.

14.3.1   Wave field statistics

Figure 14.9 shows comparison of H_s estimated using the POD method to that based on the buoy versus wind speed for two different radar systems. The error bars show 95% confidence intervals. Clearly, the POD method is able to replicate the buoy-based measurements for differing radar systems except in low wind conditions, which is a shortcoming of any radar measurement as discussed in prior chapters.

Figure 14.10 shows a comparison between H_s estimates based on the classic FFT approach of determining wave field statistics described in prior chapters along with the POD-based estimates and buoy-based estimates. For the FFT-based approach, similar to the POD, the width of the dispersion curve filter is varied and the width that produces the best match to buoy-based H_s is used in these results. The results of the POD approach yield similar statistical results as the FFT approach with the exception of one dataset where the POD-based approach was able to match the buoy-measured H_s within statistical uncertainty while the FFT approach did not. The

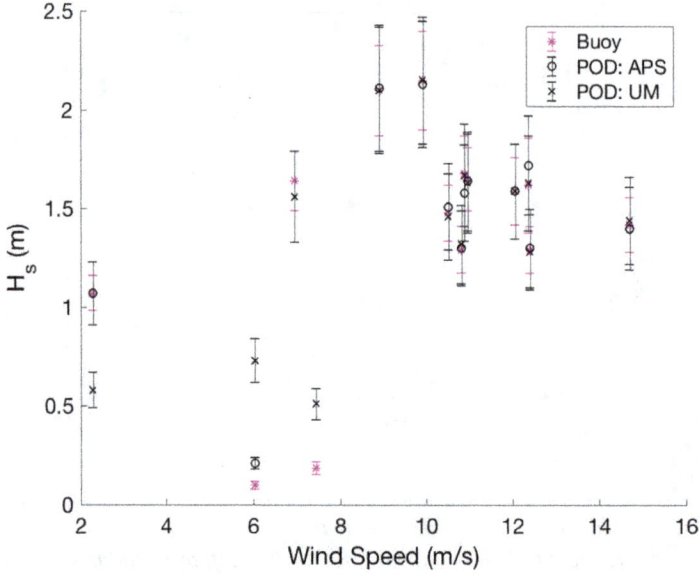

Figure 14.9 H_s estimates with 95% confidence intervals for the radar-based POD method (black) along with buoy-measured estimates (magenta asterisks) for two different radar systems versus wind speed [3]

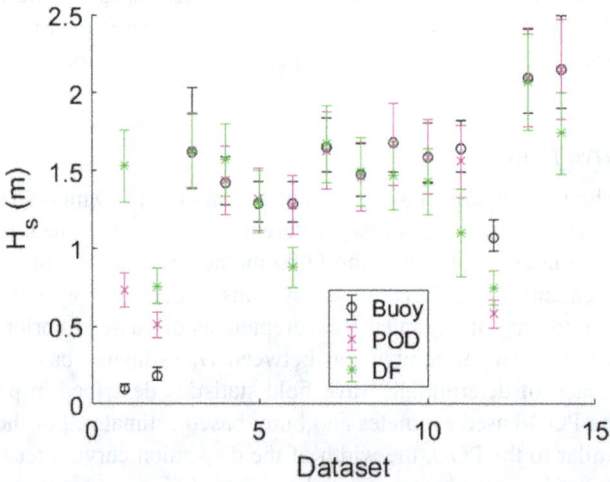

Figure 14.10 H_s estimates with 95% confidence intervals for radar-based POD and FFT-based dispersion filtered (DF) methods along with buoy-measured values for UM radar [12]

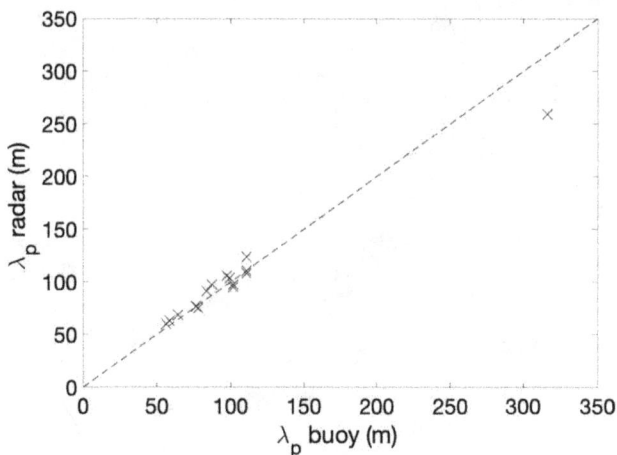

Figure 14.11   *Scatterplot of peak wavelength (λ_p) based on POD reconstruction (vertical axis) and that based on the buoy (horizontal axis) for UM radar. The solid line is a one-to-one line.*

three lowest H_s cases occur in low-wind conditions, which explains the shortcomings of both the FFT and POD methods because this discrepancy is associated with the raw radar data and not the methodology.

Determination of other wave field statistics from the reconstructions, such as the peak wavenumber, is also accurate [3]. For example, Figure 14.11 shows a comparison of buoy-based and POD reconstruction-based peak wavelengths. Note that the peak frequency from the buoy spectra is converted to wavenumber using the method outlined by Plant [31]. As can be seen in this figure, the peak wavelengths determined from the POD reconstructions are consistent with those based on the buoy data.

14.3.2   Phase-resolved wave field comparisons

The prior section demonstrated that with proper mode selection, the POD method can reproduce wave field statistics as accurately as FFT-based approaches. Because the performances of the two methods are similar, there appears to be little advantage to the POD approach, but this situation can differ when considering phase-resolved wave field reconstruction, as will be discussed in this section. Presumably, the difference stems from the fact that nonlinear aspects of the wave field have relatively small impacts on wave statistics (except in extreme conditions, e.g., sea state 5, where nearly all the waves are strongly nonlinear) but can have significant effects on an instantaneous wave height at a particular location in space and time.

Wave field reconstructions are carried out using the aforementioned methods to generate a time series of reconstructed orbital velocity maps. The instantaneous velocity fields can be compared to velocity measurements by collocated synchronous

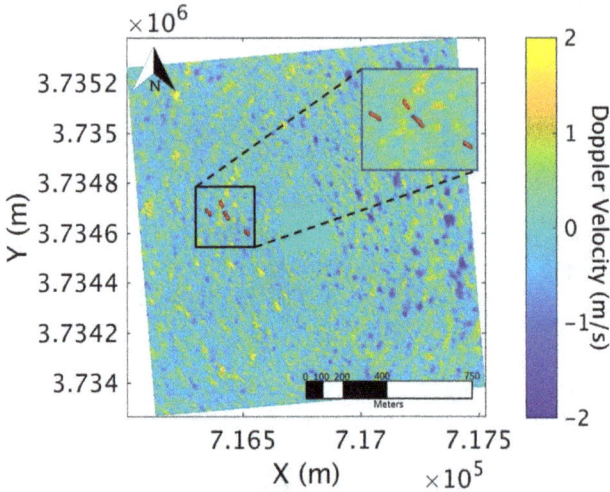

Figure 14.12   Example of tracks from four GPS wave buoys in the radar field of view during radar data collection, shown in red, overlaid on an example frame of Doppler velocity data. The black box shows a zoom-in of that area [11].

buoy measurements. Buoys that are drifting through the field of view of the radar, as exemplified in Figure 14.12, are used to generate time series of orbital wave velocity. Importantly, the buoy time series is down-sampled to match the temporal resolution of the radar to enable more accurate comparisons (see Section 14.2.1). Time series generated from IFFT of dispersion curve filtered (radar) 3D spectra is also compared. As previously mentioned, one of the main differences between the POD-based approach and the dispersion filtering approach is that energy off the dispersion relationship is retained [11].

The energy off the dispersion curve can be significant even for relatively similar wave conditions. This point is exemplified with two of the datasets, specifically datasets 7 and 8 from Table 14.1. Datasets were selected based on the availability of wave buoys in the field of view of the radar and for their unique group line features. Dataset 8 was taken approximately 25 min after dataset 7 and during this time winds were decreasing in magnitude (decaying seas). Both datasets were collected under bimodal seas with wind waves and swell present, which are consistent with conditions discussed in Plant and Farquharson [2] that can create "group line" energy. Dataset 7 has a strong group line feature with a high magnitude of group line energy relative to dispersion curve associated energy, while dataset 8 shows a weaker group line feature relative to dispersion curve associated energy. The primary difference between datasets 7 and 8 is the amount of group line energy relative to dispersion curve associated energy because they were obtained in such close proximity in time to each other. During the data collection of dataset 7, four GPS buoys were in the field of view of the radar, while only 3 buoys were present in the radar field of view for dataset 8.

Figure 14.13 shows the measured 2D k–ω spectrum of Doppler velocities for these two radar datasets, and Figure 14.14 shows the 2D k–ω spectrum of the POD reconstruction of the orbital velocity field for each dataset along with that based on dispersion filtering. In both figures, the 2D k–ω spectrum is based on integration of the 3D k–ω spectrum over k_y (recall the data were rotated such that the peak wave direction is along the x-direction). Note that energy that appears above and to the left of the dispersion curve in Figure 14.14c and d is an artifact of the 3D dispersion cone filtering integrated into 2D for presentation of the figure. As previously stated, in one case (Figure 14.13a), the energy off the dispersion relationship is similar or larger than the energy on the dispersion curve, while in the other case (Figure 14.13b) the energy on the dispersion curve is larger than that off the dispersion curve. In the POD reconstructions (Figure 14.14a and b), it is evident that energy on the dispersion curve is retained, which further supports the concept that the leading order modes are dominated by the wave orbital velocities; but, evident in these subfigures is that a portion of energy off the dispersion curve is also retained, which, of course, is not retained in the dispersion filtered approach (Figure 14.14c and d) [11].

One must consider whether the energy retained off the dispersion curve improves or deteriorates comparisons with buoy-measured orbital velocities at specific times and locations. Figure 14.15 shows time series comparisons for dataset 7, and Figure 14.16 shows comparisons for dataset 8. Recall dataset 7 has 4 buoys in the radar field of view, and dataset 8 has 3 buoys in the radar FOV, with 1 of the 3 buoys being in the field of view for only part of the dataset (thus the time series comparison is only possible when the buoy is in the radar FOV). The number of leading POD modes used for the example reconstructions shown in Figures 14.14–14.16 was selected to be representative of peak correlation between the buoy and POD method. Dataset 7 is shown using an $n = 32$ mode reconstruction, and dataset 8 is shown using an $n = 6$ mode reconstruction [11]. With more energy off the dispersion curve in dataset 7, it seems logical that more modes are needed to reconstruct the time series if it is assumed that a portion of the energy off the dispersion curve is

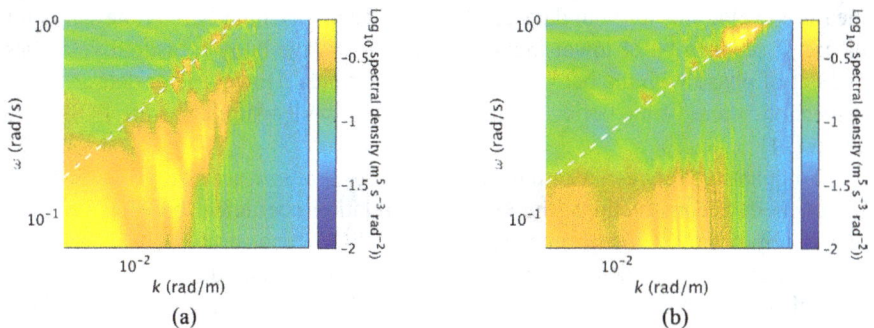

Figure 14.13   *Measured 2D k–ω spectrum of Doppler velocities for datasets 7 and 8 described in Table 14.1 [11]*

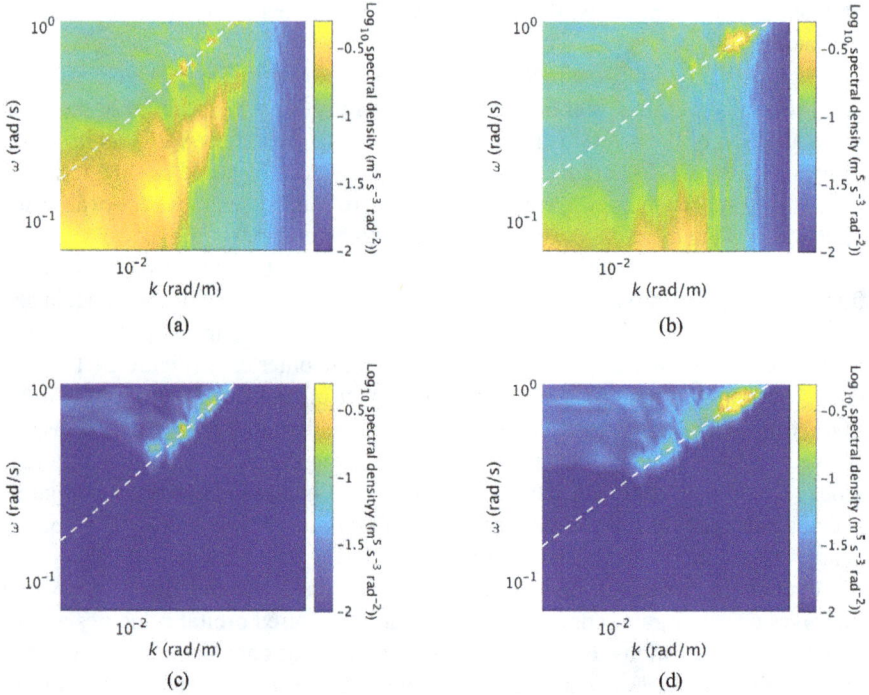

Figure 14.14   (a) k–ω spectrum of the POD reconstruction for dataset 7 using
32 modes (n = 32). (b) k–ω spectrum of the POD reconstruction
for dataset 8 using 6 modes (n = 6). (c) k–ω spectrum for the
dispersion curve filtered result for dataset 7, and (d) k–ω spectrum
for the dispersion curve filtered result for dataset 8 [11]

associated with the wave field; thus, the wave field in that case is more complex. It
can be seen in Figures 14.15 and 14.16 that the POD reconstructed orbital velocity
time series are generally in-phase and of comparable magnitude to the wave buoy-
measured velocities for both datasets 7 and 8. The dispersion filtered time series for
dataset 7 seem to be of lower magnitude than the wave buoy-measured velocities
and out of phase with the buoy measurements at times. However, for dataset 8, the
dispersion filtered time series seem to perform visually similar to the POD time
series [11].

    To quantitatively evaluate the comparisons of the time series for various num-
ber of mode reconstructions, the average correlation coefficient between the GPS
wave buoy orbital velocity time series and the POD reconstructed time series for
various number of modes used in the reconstruction is computed. The correlation
coefficients are averaged over all the available buoys. Figure 14.17 shows these
average correlation coefficients versus the number of modes used in the reconstruction along with the average correlation between the buoy time series and the time
series produced from inverse fast Fourier transform (IFFT) of the dispersion curve

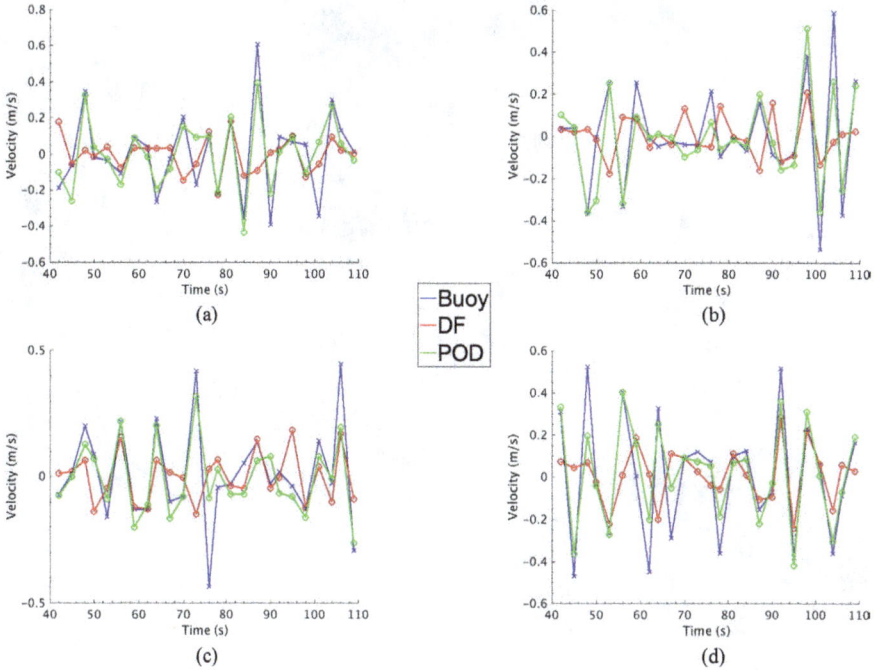

Figure 14.15   *Time series comparisons between the GPS wave buoys (blue lines), POD reconstructions using the leading 32 modes (n = 32) (green lines), and dispersion filtered (DF) reconstructions (red line) for dataset 7 (Table 14.1). Panels (a)–(d) show comparisons for each wave buoy in the FOV [11].*

filtering methodology. For dataset 7 (Figure 14.17a), an average correlation coefficient of 0.8 is attained within the leading 10 modes and the correlation peaks at ~0.9. For this dataset, the dispersion filtering method generated substantially lower correlation with the buoy time series than any of the POD-based time series. In contrast, for dataset 8 (Figure 14.17b), the POD-based method and the dispersion filtering method yield similar results. Recall for dataset 7, the group line energy is stronger relative to dataset 8, and POD reconstructions achieve significantly higher correlation with ground truth wave buoy measurements than the FFT method regardless of the number of modes used in the POD reconstruction. In contrast, for dataset 8, in which dispersion curve energy is significantly higher relative to group line energy, the FFT and POD methods result in similar correlation coefficients regardless of the number of modes used for the POD reconstruction. Further, when spectral energy is limited to primarily dispersion curve associated energy (as in dataset 8), the POD method reaches a correlation "plateau" in fewer modes than when the energy spectrum is more complex (i.e., containing group line energy in addition to dispersion curve energy). In summary, POD reconstructed orbital velocity time series correlate highly with buoy-measured wave orbital velocity time series for both datasets,

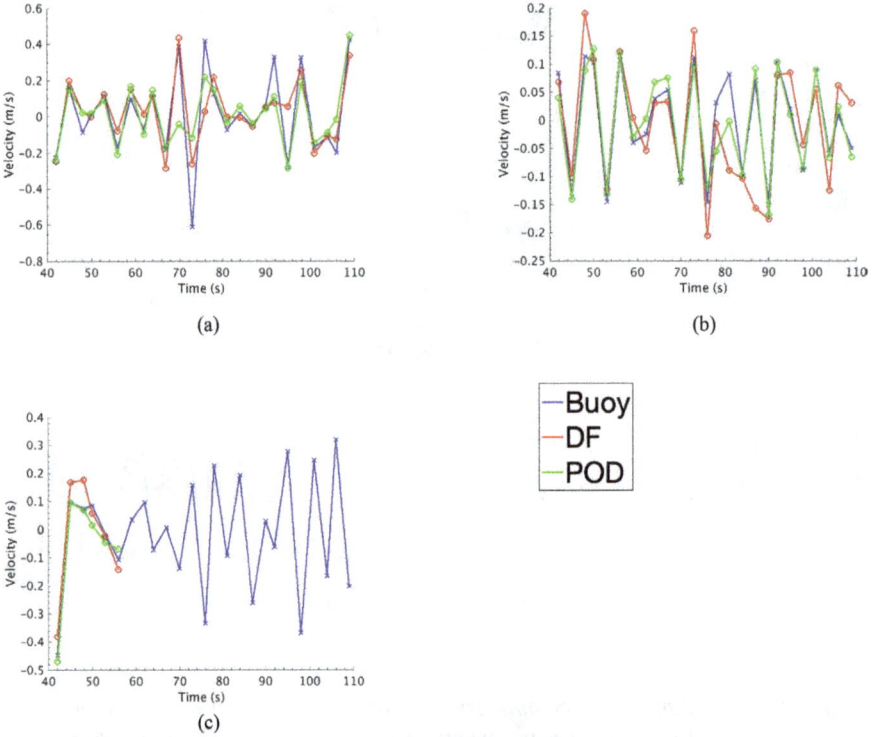

Figure 14.16   Time series comparisons between the GPS wave buoys (blue lines), POD reconstructions using the leading 6 modes (n = 6) (green lines) and dispersion filtered (DF) reconstructions (red line) for dataset 8 (Table 14.1). Panels (a)–(c) show comparisons for each wave buoy in the FOV [11].

regardless of group line to dispersion curve energy ratio; when the ratio of group line energy relative to dispersion curve energy is high, the POD method attains significantly higher correlations with buoy measurements than the conventional dispersion curve filtering method [11]. This difference in the correlation coefficients is attributed to the inclusion of group line energy in the POD reconstructions. Nevertheless, the selection of the optimal number of leading modes is non-trivial, but any mode selection achieved higher correlations than dispersion-filtered time series.

14.4   Summary and limitations of POD-based wave field reconstructions

Use of orthogonal decomposition methods to extract wave field information from Doppler measurements of the sea surface was explained and demonstrated using examples from two ship-based rotating Doppler radar systems. The wave

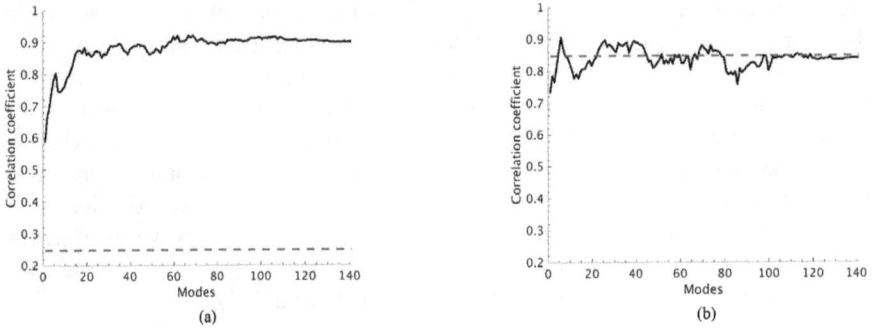

Figure 14.17   *Average correlation coefficient between POD reconstructed orbital velocity time series and wave buoy orbital velocity time series for each (1 to n) mode reconstruction (black line) and average correlation coefficient between FFT-based dispersion curve orbital velocity time series and wave buoy time series (red dashed line). Correlation coefficients are averaged over all available buoys. Panel (a) shows dataset 7 and panel (b) shows dataset 8 (from Table 14.1) [11].*

statistics extracted from the POD-based method was consistent with buoy-based measurements as well as with traditional FFT-based approaches. In contrast, the reconstructed time series of orbital velocity distributions from the POD-based approach was more highly correlated with co-located buoy orbital velocity time series when the radar measurements and, presumably, the wave field were more complex, i.e., showing significant energy off the dispersion curve. In the case where the radar measurement was dominated by energy on the dispersion curve, the methods performed similarly, as they did with the wave field statistics. Based on these results, it is conjectured that the statistics of the wave field is accurately captured by either method, but in order to obtain more accurate phase-resolved wave fields, a portion of the energy off the dispersion curve is needed in some cases. Without this energy, the phasing of the waves may be impacted as well as the wave height at isolated times and locations [11]. These results support the numerical findings of Plant and Farquharson [2], which demonstrated that at least a portion of the group line energy is related to the ocean surface wave field.

Although the main advantage of the POD method is inclusion of energy off the dispersion curve, another significant advantage is the reduced data size. Storage of all the data retrieved from a scan of the sea surface by a Doppler radar can be quite large. Using the POD method, one can reduce the size of the dataset to just the leading order modes and their coefficients for long term storage of the data. This reduction in data size enables longer time series to be obtained and stored locally.

The primary weakness of the POD method is the need to select the number of modes for the reconstruction *a priori*, whose optimization appears to be dependent on the conditions and the radar system used (e.g., Figure 14.8). However, there appears to be relatively little sensitivity to the selection in some cases as shown in the examples in Figure 14.17. Nevertheless, examining the frequency/wavenumber content of individual modes, as that shown in Figure 14.7, is a good starting point for determining the number of modes to include *a priori*. This aspect of this method still needs further research; however, the results presented in this chapter clearly demonstrate the viability of the POD approach for extraction of wave field information and may be the best option for recovery of phase-resolved wave fields in mixed seas (wind waves and swell) because of the influence wave interference has on the k–ω spectrum, specifically the group line feature [2]. These interference effects are speculated to impact the phasing of the wave field [11], and, therefore, in order to obtain accurate phase-resolved wave field information, such features need to be included. Because these features are intermittent and related to phasing of the surface, they do not impact wave statistics dramatically [3, 11].

The application of this method to radar data has also been limited to a couple studies; thus, application of this method to a wider range of conditions using a wider range of radar systems is needed to further validate the approach.

References

[1]   Smith M.J., Poulter E.M., McGregor J.A. 'Doppler radar measurements of wave groups and breaking waves'. *Journal of Geophysical Research: Oceans*. 1996;**101**(C6):14269–82.

[2]   Plant W.J., Farquharson G. 'Origins of features in wave number-frequency spectra of space-time images of the ocean'. *Journal of Geophysical Research: Oceans*. 2012;**117**(C6):C06015.

[3]   Kammerer A.J., Hackett E.E. 'Use of proper orthogonal decomposition for extraction of ocean surface wave fields from X-band radar measurements of the sea surface'. *Remote Sensing*. 2017;**881**.

[4]   Holman R., Haller M.C. 'Remote sensing of the nearshore'. *Annual Review of Marine Science*. 2013;**5**(1):95–113.

[5]   Carrasco R., Streßer M., Horstmann J. 'A simple method for retrieving significant wave height from Dopplerized X-band radar'. *Ocean Science*. 2017;**13**(1):95–103.

[6]   Lyzenga D.R. 'Polar Fourier transform processing of marine radar signals'. *Journal of Atmospheric and Oceanic Technology*. 2017;**34**(2):347–54.

[7]   Liang Y.C., Lee H.P., Lim S.P., Lin W.Z., Lee K.H., WU C.G. 'Proper orthogonal decomposition and its APPLICATIONS—PART I: theory'. *Journal of Sound and Vibration*. 2002;**252**(3):527–44.

[8]   Kerschen G., Golinval J.C. 'Physical interpretation of the proper orthogonal modes using the singular value decomposition'. *Journal of Sound and Vibration*. 2002;**249**(5):849–65.

[9]   Diamessis P.J., Gurka R., Liberzon A. 'Spatial characterization of vortical structures and internal waves in a stratified turbulent wake using proper orthogonal decomposition'. *Physics of Fluids*. 2010;**22**(8):086601.

[10] Hackett E.E., Merrill C.F., Geiser J. 'The application of proper orthogonal decomposition to complex wave fields'. Proceedings of the 30th Symposium on Naval Hydrodynamics; Hobart, Australia; November 2–7 2014. p. 9.

[11] Kammerer A.J., Hackett E.E. 'Group line energy in phase-resolved ocean surface wave orbital velocity reconstructions from X-band Doppler radar measurements of the sea surface'. *Remote Sensing*. 2019;**11**(1):71.

[12] Kammerer A.J. 'The application of proper orthogonal decomposition to numerically modeled and measured ocean surface wave fields remotely sensed by radar. Master of Science Thesis'. USA: Coastal Carolina University, Conway, SC; 2017.

[13] Zhongbiao C., Yijun H., Biao Z., Zhongfeng Q., Baoshu Y. 'A new method to retrieve significant wave height from X-band marine radar image sequences'. *International journal of remote sensing*. 2014;**35**:11-12–4559.

[14] Dugan J.P., Fetzer G.J., Bowden J., *et al.* 'Airborne optical system for remote sensing of ocean waves'. *Journal of Atmospheric and Oceanic Technology*. 2001;**18**(7):1267–76.

[15] Dugan J.P., Piotrowski C.C. 'Surface current measurements using airborne visible image time series'. *Remote Sensing of Environment*. 2003;**84**(2):309–19.

[16] Dugan J.P., Piotrowski C.C. 'Measuring currents in a coastal inlet by advection of turbulent eddies in airborn optical imagery'. *Journal of Geophysical Research*. 2012;**117**:1–15.

[17] Frasier S.J., McIntosh R.E. 'Observed wavenumber-frequency properties of microwave Backscatter from the ocean surface at near-grazing angles'. *Journal of Geophysical Research: Oceans*. 1996;**101**(C8):18391–407.

[18] Stevens C.L., Poulter E.M., Smith M.J., McGregor J.A. 'Nonlinear features in wave-resolving microwave radar observations of ocean waves'. *IEEE Journal of Oceanic Engineering*. 1999;**24**(4):470–80.

[19] Rino C.L., Eckert E., Siegel A., *et al.* 'X-band low-grazing-angle ocean backscatter obtained during Logan 1993'. *IEEE Journal of Oceanic Engineering*. 1997;**22**(1):18–26.

[20] Nwogu O.G., Lyzenga D.R. 'Surface-wavefield estimation from coherent marine radars'. *IEEE Geoscience and Remote Sensing Letters*. 2010;**7**(4):631–5.

[21] Zhang S., Song Z., Li Y, Ying L. 'An advanced inversion algorithm for significant wave height estimation based on random field'. *Ocean Engineering*. 2016;**127**(4):298–304.

[22] Chen Z., He Y., Zhang B. 'An automatic algorithm to retrieve wave height from X-band marine radar image sequence'. *IEEE transactions on geoscience and remote sensing : a publication of the IEEE Geoscience and Remote Sensing Society*. 2017.

[23] Smith G.E., O'Briend A., Pozderac J., *et al.* 'High power coherent-on-receive radar for marine surveillance'. Proceedings of the 2013 International Conference on Radar; Adelaide, SA, Australia; September 2013. pp. 434–9.

[24] Miller K., Rochwarger M. 'A covariance approach to spectral moment estimation'. *IEEE Transactions on Information Theory*. 1972;**18**(5):588–96.

[25] Hwang P.A., Sletten M.A., Toporkov J.V. 'A note on Doppler process-
     ing of coherent radar backscatter from the water surface: with application
     to ocean surface wave measurements'. *Journal of Geophysical Research.*
     2010;**115**(C3):1–8.

[26] Thompson D.R., Jensen J.R. 'Synthetic aperature radar interferometry ap-
     plied to ship-generated internal waves in the 1989 Loch Linnhe experiment'.
     Journal of geophysical research. Oceans. 1993;**98**(C6):10259–69.

[27] Drazen D., Merrill C., Gregory S., Fullerton A. 'Interpretation of in-situ ocean
     environmental measurements'. Proceedings of the 31st Symposium on Naval
     Hydrodynamics; Monterey, CA, USA; September 11-16 2016. p. 8.

[28] Young I.R., Rosenthal W., Ziemer F. 'A three-dimensional analysis of marine
     radar images for the determination of ocean wave directionality and surface
     currents'. *Journal of Geophysical Research.* 1985;**90**(C1):1049–59.

[29] Carrasco R., Horstmann J., Seemann J. 'Significant wave height measured
     by coherent X-band radar'. *IEEE Transactions on Geoscience and Remote
     Sensing.* 2017;**55**(9):5355–65.

[30] Hackett E.E., Fullerton A.M., Merrill C.F., Fu T.C. 'Comparison of incoherent
     and coherent wave field measurements using dual-polarized pulse-Doppler
     X-band radar'. *IEEE Transactions on Geoscience and Remote Sensing.*
     2015;**53**(11):5926–42.

[31] Plant W.J. 'The ocean wave height variance spectrum: wavenumber peak ver-
     sus frequency peak'. *Journal of Physical Oceanography.* 2009;**39**(9):2382–3.

Chapter 15

Current mapping from the wave spectrum

Benjamin K Smeltzer[1] and Simen Å Ellingsen[1]

In this chapter, we review methods by which near-surface ocean currents can be measured remotely using images of the water surface, as obtained by X-band radar in particular. The presence of a current changes the dispersive behavior of surface waves, so our challenge is to solve the inverse problem: to infer the spatially varying current from measurements of the wavy surface. Measuring near-surface currents in the ocean is important for a large variety of applications. Examples include oil spill tracking, understanding the transport of microplastics, predicting loads on marine structures, fuel optimization for ships, and providing data to inform oceanographic and climate models, among many others [1–8]. Many in situ methods for measuring currents experience a broad range of challenges such as time-consuming and expensive deployment and maintenance, noise in the measurements from wave or platform motions, and fouling [6]. Remote sensing of currents is an attractive alternative to in situ measurements, as currents can be mapped over a finite areal extent simultaneously, and the deployment and retrieval of sensors are not needed.

We here examine how remote sensing of currents is achieved in practice by analyzing the wave spectrum, as may be measured by X-band radar, e.g. [9–11]. A set of consecutive backscatter images recorded as a function of time is Fourier-transformed to produce the spectrum, which gives information concerning the propagation of waves whose dispersion is altered by currents. X-band radar images measure the wave field over multiple square kilometers, and analyzing various spatial subsets of the images allows a map of the spatial variation of the currents to be reconstructed, e.g. [5, 6, 12]. An example map where arrows show surface currents calculated from wave spectra is shown in Figure 15.1, demonstrating the areal coverage and spatial resolution that may be achieved. Thus, with a single sequence of radar images, the currents can be mapped over a large area simultaneously, where achieving the same degree of spatial coverage using in situ point sensors would require a large-scale deployment effort.

We describe herein methods for reconstructing currents from a measured wave spectrum. It is worth mentioning that although the methods presented in this chapter

[1]Department of Energy and Process Engineering, Norwegian University of Science and Technology, Norway

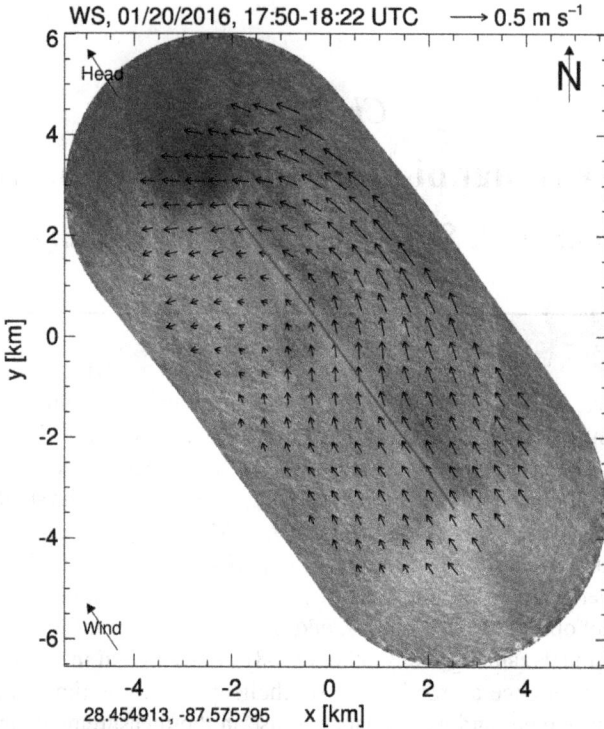

Figure 15.1   An example of a current vector field obtained from X-band radar.
Figure taken from [13].

are described in the context of marine radar, they may be readily applied to wave spectra measured by other means as well, e.g. [3, 14, 15]. The starting point for this chapter is thus the directional frequency-wavenumber spectrum, which may, in theory, be obtained by a variety of means.

15.1   Wave propagation atop background currents

We briefly introduce the dispersion relation for waves traveling atop background currents, which governs wave propagation and is the basis for determining currents in measurements of the wave spectrum. The dispersion relation describes the relationship between the wave frequency and wavenumber, which for quiescent waters reads

$$\omega_0(\mathbf{k}) = \sqrt{gk \tanh kh}, \tag{15.1}$$

where ω_0 is the wave angular frequency, $k = |\mathbf{k}|$, g is the acceleration due to gravity, and h is the water depth. Surface tension has been neglected, which is a valid assumption for waves measured by marine radar. In cases where the water depth is greater than roughly half the relevant wavelength, $\tanh kh \approx 1$ and the deepwater

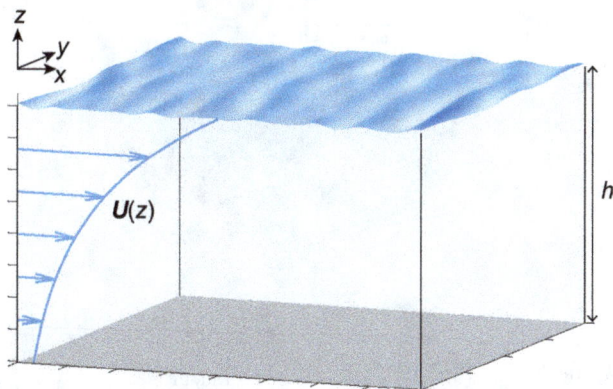

Figure 15.2   *The basic geometry and coordinate system used in this chapter. A general functional form of the current profile U(z) is shown. In part of the chapter a depth-uniform profile is assumed.*

limit may be used ($\omega_0 = \sqrt{gk}$). Throughout the chapter, we use ω_0 to denote the wave frequency in quiescent waters, with the implicit understanding that finite depth must be taken into account when relevant. We consider a background current moving in the horizontal plane: $\mathbf{U}(\mathbf{r}) = [U_x(\mathbf{r}), U_y(\mathbf{r})]$, where $\mathbf{r} = [x, y, z]$. The geometry and coordinate system are shown in Figure 15.2. It is noted that some qualitative information concerning currents moving the vertical direction may be obtained by considering the spatial variation of the horizontal current components together with the continuity equation (e.g. to identify regions of upwelling or downwelling), but herein we neglect any discussion of vertical currents. The entire spatial domain of a marine radar image spans multiple kilometers, within which the strength and direction of currents may vary significantly (see Figure 15.1). To proceed, we consider a small subset spatial window of the domain in which it will be assumed that the variation of $\mathbf{U}$ in the x and y directions is negligible, such that $\mathbf{U} = \mathbf{U}(z)$, as sketched in Figure 15.2. This assumption drastically simplifies the analysis of the wave spectrum as refraction effects may be neglected. The subset spatial window we consider corresponds in Figure 15.1 to a local spatial extent centered on the location of one of the current vectors. By reconstructing the currents within each individual subset spatial window, a full map of the horizontal variation of the currents can be achieved at the resolution of the window size. Smaller window sizes thus give higher resolution in the reconstructed current field but at the cost of lower spectral resolution, which may decrease the accuracy of the sensed currents (discussed in Section 15.2.1.1).

In addition to varying in the horizontal plane, $\mathbf{U}$ may also vary with depth, particularly in the vicinity of the surface. This variation is often neglected as it complicates the analysis, and the current is, therefore, often assumed to be depth-uniform, equal to its surface value at all z. We use this assumption for the first part of the chapter in Section 15.3. However, the depth dependence may also be extracted from the wave spectrum, and we describe methods to achieve this in Section 15.4.

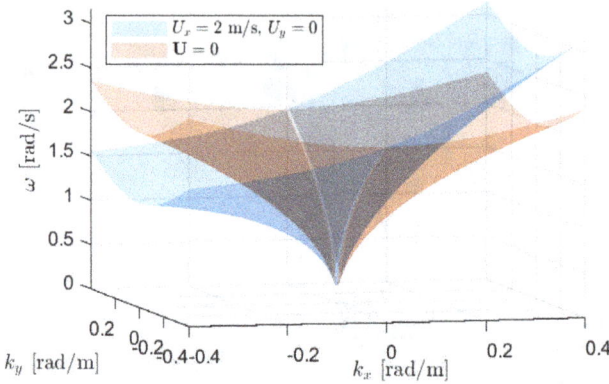

Figure 15.3   *The dispersion surface ω_{DR} for the case of a current of 2 m/s in the x-direction and in quiescent waters*

For now, we assume the currents are constant in all spatial dimensions, in which case, the dispersion relation may be written as

$$\omega_{DR}(\mathbf{k}) = \omega_0 + \mathbf{k} \cdot \mathbf{U}. \tag{15.2}$$

Examples of the dispersion surface $\omega_{DR}(\mathbf{k})$ for the case of $\mathbf{U} = (2 \text{ m/s}, 0)$ and $\mathbf{U} = 0$ are shown in Figure 15.3.

It is noted that ω_{DR} and the velocity $\mathbf{U}$ are defined in the reference frame of the radar system, a distinction relevant when mounted on a moving platform such as a ship, and $\mathbf{U}$ in (15.2) is often termed the "velocity of encounter." In such cases, accurate geographic registration of the radar images is important [16, 17]. "Calibration" methods are presented in [16] and [17] for handling errors in ship heading measurements as well as other error sources.

15.2   Appearance of the linear dispersion relation in the spectrum

When taking a fast Fourier transform (FFT) of a set of radar backscatter images of the sea surface, the resulting spectrum includes components spanning frequencies and wavenumbers up to the Nyquist frequency in both time and space. The spectrum is thus a function of three variables: wavenumber components k_x and k_y, and frequency ω. From the wave dispersion relation (15.2), we see that only certain combinations of (k_x, k_y, ω) or "triplets" are allowed by the physics: for a particular wavevector, the linear dispersion relation gives a unique frequency component. Thus, in examining the spectrum obtained by FFT, the greatest signal is expected to lie in frequency-wavenumber triplets that satisfy the dispersion relation.

The spectrum is typically defined as:

$$P(k_x, k_y, \omega) = |\text{FFT}\{I(x, y, t)\}|^2, \tag{15.3}$$

where I is the radar backscatter image intensity. Given that I is a real quantity and P the square magnitude of the Fourier transform, a point symmetry about the origin $P(0,0,0)$ applies:

$$P(k_x, k_y, \omega) = P(-k_x, -k_y, -\omega). \tag{15.4}$$

Due to the symmetry expressed in (15.4), there are two frequencies $\omega_\pm$ in practice associated with the dispersion relation for a particular wavevector: $\omega_+(\mathbf{k}) = \omega_{DR}(\mathbf{k})$, and $\omega_{-(\mathbf{k})} = -\omega_{DR}(-\mathbf{k})$. Given appropriate handling of this symmetry in algorithms, in practice, only half the spectrum is necessary for analysis, which reduces computational demands.

It is noted that the spectrum P may be normalized by a function $N(\omega, k)$ defining the background noise to produce a signal-to-noise (SNR) frequency-wavenumber spectrum, e.g. [10]. The noise spectrum is typically greater at lower frequencies and wavenumbers, and the normalization results in an SNR spectrum where the values of the peaks are more uniform over frequencies and directions. We use the spectrum P in this chapter with the understanding that it may correspond to that precisely defined by (15.3) or the SNR spectrum.

15.2.1 Practical considerations

Though the maximum signals in the measured wave spectrum are expected to lie on the linear dispersion relation surface, practical analysis of the spectrum is complicated by several factors. At root, some of these factors arise from the finite sampling in space and time of the radar images, as well as their finite temporal and spatial extent. Other factors are due to imperfections of the radar imaging of waves as well as the underlying physics of the waves themselves, which result in additional signatures in the spectrum than simply the linear dispersion relation. We outline several of these practical issues in this section.

15.2.1.1 Spectral resolution

The finite extent of the recorded radar images of the waves in space and time determines the resolution of the wave spectrum in wavenumber and frequency. For spatial extent L_x, L_y, and T in the spatial and temporal dimensions, the extent of one pixel in the spectral domain is $\frac{2\pi}{L_x} \times \frac{2\pi}{L_y} \times \frac{2\pi}{T}$. The finite spectral resolution directly affects the precision with which the location of the energy peaks corresponding to the dispersion relation may be determined. It is thus desirable to maximize the spectral resolution especially when considering lower wavenumbers whose frequencies are less sensitive to currents due to the $\mathbf{k}$-proportionality in (15.2). In the spatial domain, this entails larger spatial windows, which then decreases the spatial resolution of the reconstructed current field map. In the temporal domain, T may be increased by recording more images (provided the currents do not vary appreciably during this duration), which increases the size of the dataset and slows the processing. In practice, T may be limited in cases where the radar system is mounted on a moving platform.

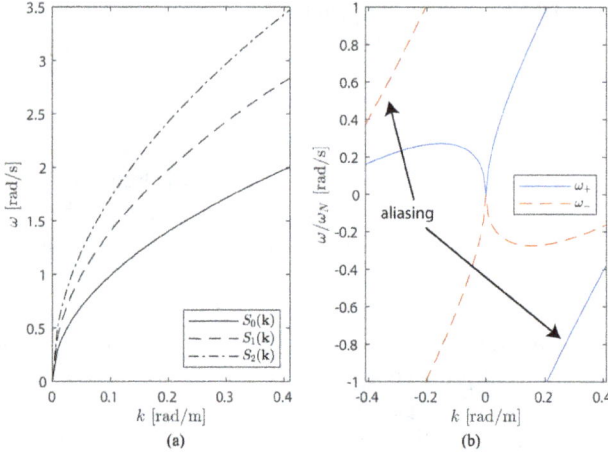

Figure 15.4   (a) Equation (15.5) for the fundamental harmonic (p = 0) and
first two harmonics. (b) Frequencies $\omega_\pm + 2n\omega_N$ are plotted for the
case of a strong current as a function of wavenumber component k
aligned with the current. The $2\omega_N$ periodicity has been included to
illustrate the effects of aliasing in the spectrum.

15.2.1.2   Harmonics

Harmonics of the linear dispersion relation arise in the spectrum from multiple sources, e.g. [18]. First, there is the nonlinearity of the radar imaging system:
the mapping between wave height and radar signal intensity is not entirely linear.
This results in components in the spectrum at integer multiples of the frequency-
wavenumber combinations lying along the linear dispersion relation

$$S_p(\mathbf{k}) = \pm(\sqrt{p+1})\omega_0(\mathbf{k}) + \mathbf{k} \cdot \mathbf{U}, \tag{15.5}$$

where p is a positive integer ≥ 0. Examples of S_p are shown in Figure 15.4a for
$p = 0, 1, 2$.

There is additional nonlinearity from the surface waves themselves, which do
not form perfect sinusoidal surface profiles when finite in amplitude. Their harmonic
signatures in the spectrum may also be described by (15.5), making them in practice difficult to distinguish from imaging nonlinearities. For most realistic cases, the
wave nonlinearity spectral structure is likely smaller than that due to imaging non-
linearities, except perhaps under extreme sea states.

15.2.1.3   Aliasing

Aliasing is an artifact of under-sampling either in time or in space. For spectra
obtained by marine radar, temporal under-sampling is often more relevant given the
relatively slow rotation rate of a radar antenna, as well as the possibility for large
velocities of encounter when mounted on a moving platform. Given a sampling fre-
quency f_S, all frequencies $|\omega| > \pi f_S$ will be under-sampled. The frequency $\pi f_S \equiv \omega_N$

is known as the Nyquist frequency, representing the largest frequency (in magnitude) that will be adequately sampled.

To understand the signatures in the wave spectrum from aliasing, we consider two spectra P_1 and P_2, consisting only of monochromatic waves at a frequency ω_1 in the case of P_1 and $\omega_2 = \omega_1 + 2n\omega_N$ for P_2, where n is an integer. Due to the finite sampling frequency, it can be shown that $P_1 = P_2$, thus wave components with frequencies differing by $2n\omega_N$ are indistinguishable in the spectrum. An under-sampled wave frequency ω_{aliased} will appear in the spectrum at a frequency

$$\omega = \omega_{\text{aliased}} + 2n\omega_N, \tag{15.6}$$

with values of n such that ω lies within the interval $[-\omega_N, \omega_N]$.

When examining the wave spectrum, aliasing manifests itself as extra artifacts of energy located away from the linear dispersion relation. An illustrative example is shown in Figure 15.4b. Frequencies $\omega_\pm + 2n\omega_N$ are plotted for the case of a strong current aligned with the waves (k, in this case, is the component of $\mathbf{k}$ along the direction of the current). Considering ω_+, the wave frequency is greater than the Nyquist frequency for wavenumbers above ~ 0.2 rad/m resulting in aliased components appearing at negative frequencies corresponding to the $n = -1$ branch in (15.6). The ω_- curve displays an analogous characteristic. Equation (15.6) and the symmetry relation (15.4) may be used to perform de-aliasing to determine the true under-sampled wave frequency ω_{aliased} to which the spectral signal corresponds [18]. It is noted that the harmonics in (15.5) also will be aliased resulting in a more complicated spectrum to interpret, especially in the presence of an unknown current velocity.

15.3 Extracting currents from the spectrum

We have seen thus far how the wave dispersion relation, indicative of wave propagation, manifests itself as peaks in the measured wave spectrum. The location of the peaks thus allows a measurement of the dispersion relation. Assuming a depth uniform flow, the wave dispersion relation is described by (15.2), and the goal is to determine the unknown current velocity $\mathbf{U}$. We now examine algorithms to extract the current by analyzing the peaks in the spectrum.

15.3.1 Least squares method

A least squares (LS) method was proposed for extracting currents from the wave spectrum by Young and Rosenthal [9]. Wavenumber-frequency triplets $(k_{x,i}, k_{y,i}, \omega_i)$ corresponding to the linear dispersion relation are identified as values of the spectrum above a certain threshold C_1 of the maximum value satisfying

$$P(k_{x,i}, k_{y,i}, \omega_i) \geq C_1 \max\{P(k_x, k_y, \omega)\}. \tag{15.7}$$

An example of a set of triplets is shown in Figure 15.5 as black circles. Then, an error parameter is defined as

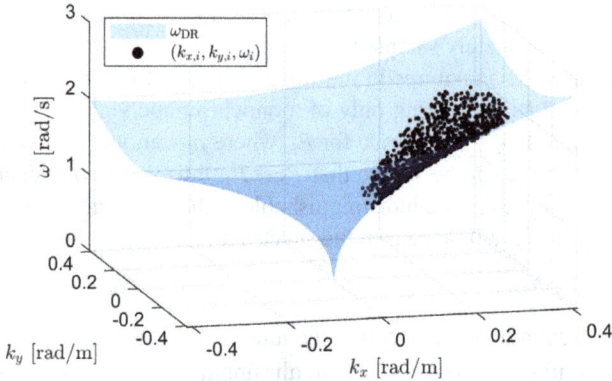

Figure 15.5   *Illustration of wavenumber-frequency triplets (black circles)*
extracted from peaks in the spectrum corresponding to the linear
dispersion relation (blue surface), using mock data

$$Q(\mathbf{U}) = \sum_{i=1}^{N_1}(\omega_i - \omega_0(\mathbf{k}_i) - \mathbf{k}_i \cdot \mathbf{U})^2, \tag{15.8}$$

where the summation is over all N_1 triplets selected by the threshold criteria (15.7).
Q is then minimized with respect to $\mathbf{U}$ to obtain the unknown current components.

The choice of the threshold value is a tradeoff between two factors. On the one
hand, the precision of the LS fit generally improves with an increasing number of
selected triplets N_1, achieved with a lower threshold value. On the other hand, as the
threshold decreases, other signatures in the spectrum such as harmonics, aliasing,
and noise will begin to be selected, which corrupt the fit. Ideally, the optimal threshold value would thus be as small as possible while avoiding artifacts and noise. In
practice, a value of 0.2 is common [19]. Several related LS algorithms have been
proposed, such as in [20].

Referring to Section 15.2.1, the signatures in the spectrum from harmonics and
aliasing also contain information about the currents that may be used in the fit if we
modify the LS fitting algorithm to include them. The advantage is a greater number
of triplets and increased precision. Senet *et al.* [18] proposed such an algorithm that
is iterative in nature and termed it the iterative least squares (ILS) method. First,
the conventional LS method (15.8) is used to obtain a coarse guess of the current.
Second, the value of the current is used to select triplets in the spectrum corresponding to both the fundamental and the higher harmonics using a much smaller threshold parameter C_2 performing de-aliasing as outlined in Section 15.2.1.3 using (15.6),
(15.5), and (15.4). Further practical details concerning an approach to de-aliasing
are given in [18]. De-aliasing approaches arrive at a function $S_p(\mathbf{k})$ expressing the
de-aliased frequencies at harmonic p using the current velocity of iteration j. A corresponding error parameter is then defined analogous to (15.8):

$$Q(\mathbf{U}) = \sum_{i=1}^{N_1}(\omega_i - S_p(\mathbf{k}_i) - \mathbf{k}_i \cdot \mathbf{U})^2, \tag{15.9}$$

which is then minimized to find an updated current velocity. The process is then repeated, with the de-aliasing being performed with the updated current velocity. The algorithm may be set to run for a fixed number of iterations or until satisfactory convergence is reached.

15.3.2   Normalized scalar product method

A conceptually related algorithm to the LS involves the maximization of the normalized scalar product (NSP) between the spectrum and a characteristic function G, defining the dispersion relation shell [19, 21, 22]:

$$G(\mathbf{k},\omega, \mathbf{U}) = \begin{cases} 1, & \text{if } |\omega_0(k) + \mathbf{k} \cdot \mathbf{U} - \omega| \leq \Delta\omega/2 \\ 0, & \text{otherwise} \end{cases} \tag{15.10}$$

where $\Delta\omega$ is the frequency resolution of the spectrum (see Section 15.2.1.1). The characteristic function defines the dispersion shell having a full width of $\Delta\omega$ in frequency for each wavevector. In theory, with the correct current velocity $\mathbf{U}$, the characteristic function should overlap the peaks in the measured spectrum. The NSP expresses this overlap and it is typically defined as

$$V(\mathbf{U}) = \frac{\langle |F_I(\mathbf{k},\omega)|, G(\mathbf{k},\omega,\mathbf{U}) \rangle}{\sqrt{P_F P_G}}, \tag{15.11}$$

where $F_I = \sqrt{P}$, and P_F and P_G are the power of F_I and G, respectively. The NSP method maximizes V by searching over appropriate ranges of current velocities $\mathbf{U}$. Compared with the LS and ILS methods, the NSP method does not involve the choice of a threshold parameter, reducing the user-input parameters that may affect the results. A potential disadvantage with the NSP method is that the maximized metric V tends to be weighted most by lower wavenumbers, which typically have the largest energy in the spectrum and are less sensitive to currents than higher wavenumbers.

Another potential drawback of the NSP method is the computational cost, as (15.11) must be evaluated for many values of the current velocity in the search process. To reduce computational demands, a variable search range has been used [19]. Typically, an initial wide search range with coarse velocity resolution is used to produce a rough estimate of the current velocity. Then, a narrower range with finer resolution is used centered on the initial estimate to find a higher precision current velocity. The latter step may be repeated to give successively higher precision if desired.

15.3.3   Polar current shell method

An algorithm that transforms the wavenumber plane into polar coordinates to determine the current is known as the polar current shell (PCS) method [22, 23]. The first step is analogous to the LS algorithm: wavenumber triplets corresponding to the linear dispersion relation are identified in the spectrum by locating peaks in the frequency for a given wavevector. The wavenumber coordinates of the triplets are then converted into polar coordinates (k, θ), where $k_x = k \cos \theta$ and $k_y = k \sin \theta$, with

θ being the angle in the x, y plane from the positive x-axis. It is noted that (15.2) expressed in polar coordinates may take the form

$$\omega_{\text{DR}}(k, \theta) = \omega_0(k) + kU\cos(\theta - \theta_U), \tag{15.12}$$

where $U = |\mathbf{U}|$, and θ_U is the angle reflecting the direction of the current. We see from the inspection of (15.12) that when ω_{DR} is evaluated at a particular wavenumber as a function of θ, i.e. along the azimuthal direction, the frequency is a sum of a θ-independent component ω_0 and an oscillating component $kU\cos(\theta - \theta_U)$. Considering the latter, the oscillation amplitude is proportional to the strength of the current, while the phase of the oscillation determines the current direction.

The PCS algorithm analyzes the θ-dependence of the frequency at each wavenumber to find the current from the set of triplets. The LS fit is performed to extract the current magnitude and direction from a subset of triplets, where the wavenumber is a constant. The result is a set of current velocities over a range of wavenumbers, which are then averaged to give a single vector. Further practical details on the implementation of the PCS are given in [23] and [22].

A related algorithm which is essentially a modified implementation of the PCS was developed by Smeltzer *et al.* [24]. Instead of extracting triplets from the spectrum, an NSP was defined using a characteristic function expressing the azimuthal dependence of (15.12) for constant wavenumber. The characteristic function may be expressed (slightly modified from [24]) as

$$G_i(\omega, \theta, U_i, \theta_{U,i}) = \exp\left[\frac{(\omega - \omega_{\text{DR}(k_i, \theta)})^2}{4a}\right] \tag{15.13}$$

where a is the frequency width parameter (typically set to a value on the order of the frequency resolution) and the subscript i denotes a discrete wavenumber in the spectrum at which the NSP is evaluated. The dependence of the right-hand side on U_i and $\theta_{U,i}$ is implicitly included in ω_{DR}. Using (15.13), the NSP was maximized analogously to (15.11) for each separate wavenumber.

15.3.4   Algorithm comparison

The authors of [19, 22] have compared the performance of some of the algorithms described above and largely found similar accuracy. The ILS and NSP methods show comparable accuracy, with ILS being an improvement over the LS. Comparison to the PCS algorithm also showed comparable accuracy in another study [22]. The authors, however, note that the comparison was performed for low velocities of encounter with minimal aliasing. One motivation for developing the NSP [21] was to overcome the problem of increasing errors of the LS and ILS methods for large velocities of encounter, so it is possible the NSP is a better choice in such situations. Referring to the similar performance of the various algorithms, the authors in [22] conclude "This implies that the technology of current measurement using X-band marine radar has become sufficiently mature and the emphasis perhaps may be legitimately shifted from research methodology toward applications."

15.4   Reconstructing depth-dependent flows

So far in this chapter, we have made the assumption that the currents extracted from the wave spectrum are uniform in depth. This assumption drastically simplifies the analysis of the wave spectrum. However, in some realistic situations, currents have significant variation with depth, such as created by wind forcing or a river plume. The wave dispersion may be approximated by a different relation:

$$\omega_{DR}(\mathbf{k}) = \omega_0(\mathbf{k}) + \mathbf{k} \cdot \tilde{\mathbf{c}}(k), \tag{15.14}$$

where $\tilde{\mathbf{c}}$ is a wavenumber-dependent Doppler shift velocity from the background current. The wavenumber-dependence is the key difference to (15.2). The Doppler shift velocity is a weighted average of the current as a function of depth, approximated as [25]

$$\tilde{\mathbf{c}}(k) = 2k \int_{-\infty}^{0} \mathbf{U}(z) e^{2kz} dz \tag{15.15}$$

in deep water, and in finite water depth as [26, 27]

$$\tilde{\mathbf{c}}(k) = \frac{2k}{\sinh(2kh)} \int_{-h}^{0} \mathbf{U}(z) \cosh[2k(h+z)] dz. \tag{15.16}$$

As may be noticed from inspecting both expressions, shorter wavelengths are thus influenced by currents in close vicinity to the surface, while longer wavelengths are influenced by currents at greater depths.

Equations (15.15) and (15.16) reveal the nature of the error one makes by making the common assumption of a depth-uniform current as in Section 15.3 when in fact $\mathbf{U}(z)$ has a significant variation with depth. What is measured is then not the surface current as commonly reported, but rather a weighted average of the current in the topmost part of the water column. The weighting factor $\exp(2kz)$ means that the influence nearest the surface is the strongest and it rapidly decreases at a rate that, importantly, depends strongly on the wavelength. The current at depths down to a quarter of the wavelength or so is significant. In effect, the current is thus measured at some depth beneath the surface, where $\mathbf{U}(z)$ equals the measured velocity. It is not straightforward to surmise the exact depth because the surface current as found in Section 15.3 is determined from a spectrum of different wavelengths and the form of $\mathbf{U}(z)$ is *a priori* unknown. This point becomes particularly important when comparing radar-derived currents to in situ measurements, which are typically point measurements at a given depth. When $\mathbf{U}(z)$ is approximately constant, the two may be directly compared but, in general, the comparison is more complicated.

Starting again from the measured wave spectrum, reconstructing depth-dependent currents involves two general steps. First, Doppler shift velocities are extracted from the spectrum at a range of wavenumber values. Second, the set of Doppler shifts are used to estimate the unknown profile $\mathbf{U}(z)$. The first step is similar to the methods described in Section 15.3. The difference is that while a single current velocity was derived from the spectrum spanning all wavenumbers in Section 15.3, multiple velocities representing the second term in (15.14) are found, each corresponding to a unique wavenumber. In practice, this is accomplished by only considering a narrow range of wavenumbers at a time (a bin) and then using one of the methods described in Section 15.3 to find a velocity that is assigned to the center-wavenumber value of the particular bin. The result is a set of velocities, each corresponding to a discrete wavenumber value. Example results are shown in figure 6 of [10] and 9 and 10 of [28]. Small wavenumbers to a larger extent represent the current at greater depths compared to high wavenumbers, something which can be seen by inspecting the integrand in (15.15).

The second step uses an inversion method to find the depth profile from the Doppler shift velocities. As described above, the Doppler shift velocities reflect a weighted average of the current profile over different depth ranges depending on the wavenumber. Thus, the best performance is obtained when there are Doppler shifts for a wide range of wavenumbers. In addition, the depth range over which the currents can be reconstructed is also dependent on the wavenumber range: the smallest wavenumbers influence the greatest depth at which the waves "see" the flow. In addition, the process of determining the unknown current profile from the Doppler shift velocities is an ill-posed problem mathematically. The resulting current profile is not necessarily mathematically unique, and furthermore, errors in the Doppler shifts tend to be amplified in the inversion process. Because of these challenges, many inversion methods use a priori assumptions and constraints on the functional form of the current profile to produce realistic estimates. In this section, we describe several inversion methods that have been used to reconstruct a depth profile estimate from a set of Doppler shift velocities measured at discrete wavenumbers.

15.4.1   Effective depth method

Assuming a profile where the current strength varies linearly with depth, $U(z) = U'z + U_0$, with U' being the shear-strength (vorticity) and U_0 the surface current. Assuming deep water and using (15.15), the Doppler shifts can be expressed as [25].

$$\tilde{c}(k) = -\frac{U'}{2k} + U_0 = U(z = -(2k)^{-1}). \tag{15.17}$$

We see from inspecting (15.17) that the Doppler shifts are equal to the current profile at a depth $Z_{eff} = -(2k)^{-1}$. This effective depth is roughly 8% of the wavelength.

Similarly for a logarithmic profile, $U(z) = U_0 - \frac{u^*}{\kappa} \log \frac{z}{z_0}$, where u^* is the friction velocity, κ the von Kármán constant, and z_0 the roughness length. The above parameters characterize a turbulent boundary layer that has been hypothesized to be a reasonable model of a wind-driven shear flow near the water surface [29]. Again using (15.15), the Doppler shifts are evaluated as [30]

$$\tilde{c}(k) \approx U_0 - \frac{u^*}{\kappa} \log \left(\frac{1}{2kr}\frac{1}{z_0} \right) = U(z = -(3.56k)^{-1}), \tag{15.18}$$

with $r = 1.78$. We see, as with (15.17), that the Doppler shifts are equal to the current profile at a particular depth $Z_{eff} = -(3.56k)^{-1}$, or 4.5 % of the wavelength. The only difference from a linear profile is the proportionality of the inverse

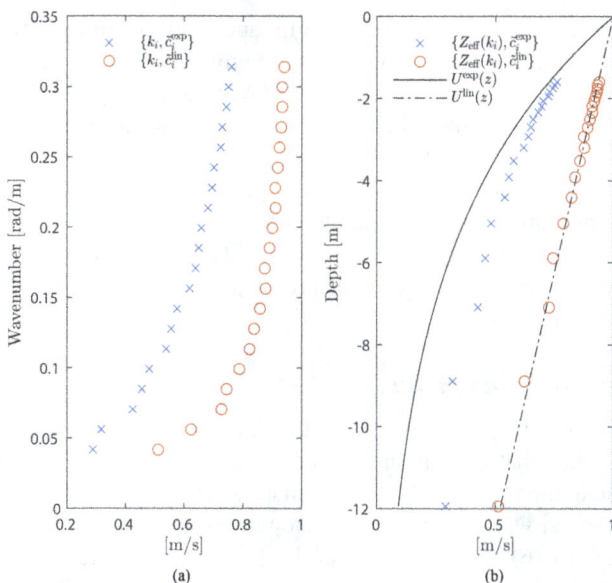

Figure 15.6   *(a) Doppler shift velocities in the direction of the current as a function of wavenumber for an exponential profile and linear profile (superscripts "exp" and "lin," respectively). (b) Mapped velocities assuming a linear current profile, with the true current profiles shown as the solid (exponential) and dashed-dotted (linear) curves for comparison. See text for the parameters defining the current profiles.*

wavenumber-dependence, where Doppler shifts are mapped to shallower depths relative to a linear current profile.

We term this mapping process the effective depth method (EDM). The EDM has been used extensively in the past with both linear and logarithmic profile assumptions to produce estimates of the depth profile [3, 4, 10, 25, 31, 32]. The main advantage is the simplicity of implementation: once the Doppler shift velocities have been extracted from the spectrum, all that remains is the mapping step using $Z_{eff}(k)$ to produce current velocities at various depths. If a smooth functional profile is desired, the set of velocity-depth pairs may be fit to some function to result in a continuous depth profile. The fitting process may introduce additional parameters associated with the fit. The main drawback of the EDM is the necessary a priori assumption of the functional form of the depth profile. In cases where the true profile does not resemble a linear or logarithmic function, the mapping results in errors in the velocities at the mapped depths.

The EDM is illustrated in Figure 15.6 using mock data with a small degree of noise artificially added, considering the horizontal velocity component of the Doppler shifts aligned with the current. Figure 15.6a shows the Doppler shift velocities as a function of wavenumber (vertical axis) for the case of an exponential profile

$U^{\text{exp}}(z) = U_0 e^{z/d}$ with $U_0 = 1$ m/s and $d = 5$ m, and a linear profile $U^{\text{lin}}(z) = U_0 + Sz$, with $S = 0.04$ s^{-1}. The assumption of a linear profile, $Z_{\text{eff}}(k) = -(2k)^{-1}$, is used to map the wavenumbers to depths, shown in Figure 15.6b. For comparison, the true current profiles are also shown. For the linear profile, the EDM maps the velocities to the correct depths as expected, given a correct assumption concerning the functional form of the profile. In the case of the exponential profile, there is some deviation between the mapped currents and the true profile, since the mapping function is not appropriate in this case. Figure 15.6b highlights the main drawback of the EDM: the method leads to errors when the true profile differs from the assumed functional form used in the mapping.

15.4.2   Ha-Campana method

To avoid a priori assumptions of the functional form of the current profile, Ha [33] proposed a method that directly inverts the integral of (15.15) to find the unknown current profile in the integrand by approximating the integral using Gaussian quadrature. The method was later further developed and extended to finite depth using the integral of (15.16) by Campana *et al.* [34]. We here outline the method for the case of infinite depth (the finite depth version is described in [34]).

An integral of a function $f(x)$ (assumed to be smooth) can be approximated as

$$\int_{-1}^{1} f(x)dx \approx \sum_{j=1}^{n} f(x_j)w_j, \tag{15.19}$$

where x_j are quadrature points, w_j are weights, and n is the order of the Legendre polynomial. To match the integral limits of (15.15) to (15.19), the coordinate transformation $x = 2e^{-2k_0 z} - 1$ is made, where k_0 is a reference wavenumber chosen to minimize quadrature error [11, 33].

For a set of Doppler shifts measured at discrete wavenumbers, (15.15) may be approximated using a matrix equation:

$$\mathbf{f} = \mathbf{A} \cdot \mathbf{u}, \tag{15.20}$$

where column vector $\mathbf{f}$ contains the measured Doppler shifts, $\mathbf{A}$ is a matrix of coefficients derived from (15.19), and $\mathbf{u}$ is a column vector with the unknown current velocities at discrete depths defined by the coordinate transformation (details given in [33] and [11]). The form of (15.20) is an over-determined linear system of equations and may be solved in the LS sense for the unknown values of the current $\mathbf{u}$ given a set of Doppler shift-wavenumber pairs.

Up to this point, no assumptions concerning the functional form of the current profile have been made, a clear advantage relative to the EDM. However, the biggest challenge and drawback of the method concern how errors in the Doppler shift velocities propagate through the inversion process to corresponding errors in the current depth-profile solution. Ideally, the resulting errors in the depth profile would be similar in magnitude to those of the input Doppler shifts. However, the direct inversion of the integrals (15.15) and (15.16) results in a severe amplification of the error: small errors in the Doppler shift velocities can result in large errors in the resulting depth profile. The error amplification means that (15.20) is impractical to be used in reality without some additional constraints on the velocity solutions. The constraints that one typically imposes limit the second derivative of the velocity profile with respect to depth and limit the distance from an initial guess. A cost function is typically defined and minimized, with increasing cost as the curvature or distance from the initial guess increases. The introduction of the constraints results in smoother profiles that suppress the amplification of the errors, yet the current profile solutions may depend on the values of the parameters weighting the constraints. The challenge then is how to choose optimal values of the constraint parameters. Campana *et al.* [11] offer an empirical method for choosing the curvature constraint, though the universality of the method remains unclear. The Ha-Campana method demonstrates comparable accuracy relative to the EDM when compared to acoustic Doppler current profiler (ADCP) truth measurements, while reconstructing the current profile at a deeper range of depths.

15.4.3   Polynomial effective depth method

Another method for reconstructing the depth profile without assumptions to the functional form was proposed by Smeltzer *et al.* [24]. The method starts from the conventional EDM, fits the profile to a polynomial form, and then scales the coefficients based on a simply derived relation to produce an improved estimate of the true current profile. We outline the method below, considering one horizontal component of the Doppler shift velocity vector and current profile, expressed here as $\tilde{c}$ and $U(z)$, respectively. The method is, in practice, applied to each velocity component separately.

If we assume a polynomial form to the current profile, i.e. $U(z) = \sum_{n=0}^{\infty} u_n z^n$, evaluation of the resulting Doppler shifts using (15.15) yields

$$\tilde{c}(k) = \sum_{n=0}^{\infty} n! u_n \left(-\frac{1}{2k}\right)^n. \tag{15.21}$$

By inspecting (15.21), we notice that the $(-2k)^{-1}$ term is equal to the mapping function $Z_{\text{eff}}(k)$ of the EDM for a linear profile (15.17). If we substitute the EDM mapping function, we then obtain the current profile

$$U_{\text{EDM}}(z) = \sum_{n=0}^{\infty} n! u_n z^n. \tag{15.22}$$

We see that $U_{\text{EDM}}(z)$ differs from the true profile $U(z)$ only by a factor $n!$ for the n-th order term. The $n!$ characterizes the error of the EDM profile in cases where the linear assumption is invalid. For the term $n < 2$, (15.22) matches the true profile as expected, while differing for higher-order terms that represent a profile with nonlinear functional form. The similarity of (15.22) to the form of the true profile motivates the method described here, the polynomial effective depth method (PEDM), which attempts to correct for the discrepancy in terms of $n \geq 2$ by simply scaling the higher-order polynomial coefficients by a factor $n!$. Quoted from [24], the PEDM procedure consists of three steps:

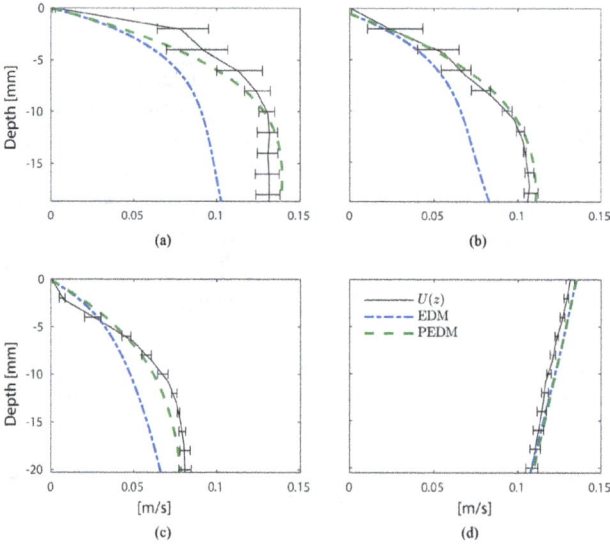

Figure 15.7   A comparison of the PEDM and EDM methods applied to experimentally measured Doppler shifts along the direction of the flow for four different current profiles in panels (a–d). In situ truth measurements are denoted as U(z). Data from [35] (presented in [24]).

1. For each of the measured values $\tilde{c}_i$, assign effective depths $z_i = -(2k_i)^{-1}$ according to the EDM procedure of (15.17) using $Z_{\text{eff}}(k)$.

2. Obtain $U_{\text{EDM}}(z)$ by fitting the set of points $\{z_i, \tilde{c}_i\}$ to a polynomial of degree n_{max}:

$$U_{\text{EDM}}(z) \approx \sum_{n=0}^{n_{max}} u_{\text{EDM},n} z^n,$$ (15.23)

where $u_{\text{EDM},n}$ are the coefficients obtained in the polynomial fit.

3. Then the improved PEDM estimate is

$$U_{\text{PEDM}}(z) = \sum_{n=0}^{n_{max}} \frac{1}{n!} u_{\text{EDM},n} z^n.$$ (15.24)

Equation (15.24) follows immediately by comparing (15.22) and (15.23), where $u_{\text{EDM},n} = n! u_n$. Additional practical details concerning the implementation of the method are given in [24].

The PEDM has been tested on Doppler shifts measured in a laboratory with currents of variable depth dependence, with in situ particle image velocimetry serving as "truth" measurements. The laboratory setup may be considered as a scale model of the oceanographic currents, including wind-drift profiles. The wave spectrum was measured by optical means, yet representative of what may be obtained by X-band radar for scaled-down length dimensions. Example results are shown in Figure 15.7 for four different current profiles. Using the PEDM resulted in a >3 times accuracy

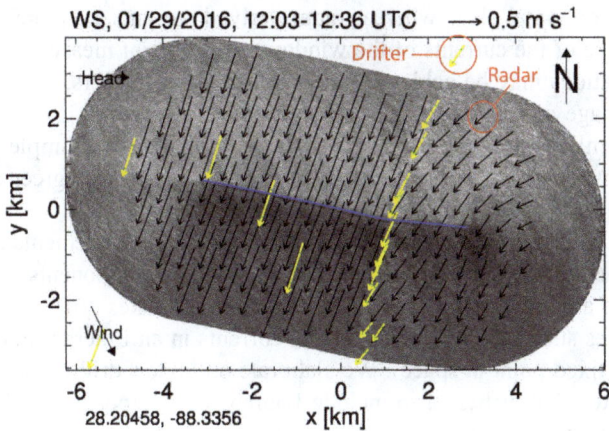

Figure 15.8 A comparison of radar and drifter currents. Reproduced from [13].

improvement relative to the EDM for current profiles with significant near-surface curvature (profiles a–c). For one profile that varied approximately linearly with depth (profile d), the PEDM and EDM resulted in essentially indistinguishable profiles as expected, since in that case, the assumptions inherent to the EDM were fulfilled.

15.5 Challenges and further work

So far in this chapter, we have examined different methods for extracting currents from the wave spectrum, both assuming a depth-uniform as well as depth-varying profile. We now outline some challenges associated with evaluating and interpreting the results, which may be the focus of future efforts within the field.

15.5.1 Validation

A key area of work within X-band current mapping concerns the validation of the methods described previously in this chapter and the evaluation of their accuracy. Of particular interest is the absolute accuracy, as well as identifying what factors affect the accuracy and reliability of reconstructed current maps. Comparison is typically performed relative to in situ measurements such as ADCP or drifters. Field studies, such as [5] with selected data in Figure 15.8 have demonstrated agreement down to nearly cm/s-scale between radar-derived and in situ measurements assuming a depth-uniform flow.

One challenge when compared to in situ measurements is due to different sampling of the current field between the two approaches: the currents obtained from the radar are representative of the current over a finite areal extent in the horizontal plane whereas in situ measurements are point measurements. In areas with strong horizontal current shear or local variations within the radar spatial window, in situ results may give different results, simply because they sample only a single discrete point in the horizontal plane whereas radar-derived currents are representative of a spatial average of the currents of the window extent. Point measurements at many different locations may be achieved conveniently using drifters, though they have the disadvantage of tending to congregate at convergent zones [5] and thus potentially not evenly sample the horizontal area. An illustrative example is shown in Figure 15.8, where the drifters (marked as the yellow arrows with green dots) clearly congregate near a convergent zone of the currents.

Another challenge when compared to some in situ measurements concerns the fact that radar-derived currents may include Lagrangian components (following the movement of a particular fluid parcel) such as waves Stokes drift whereas some in situ techniques such as ADCP measure the currents in an Eulerian framework (the velocity at a fixed point in space). We elaborate on Stokes drift in the next subsection. It is noted that drifters also include Lagrangian components and may offer a more direct comparison in cases where such Lagrangian current components are relevant.

In addition, ADCP may provide measurements of the depth profile, but many in situ techniques measure the current at a single depth. When using a depth-uniform assumption, the comparison is thus more complicated since the radar-derived currents represent a weighted average over a depth range determined by the wavenumbers as discussed in Section 15.4. Discrepancies between radar and in situ measurements may, in many cases, be due to the different depths to which the currents correspond. A few field studies have reconstructed the depth profile using one of the methods described in Section 15.4 and compared it to in situ measurements. In general, more validation is required for testing the accuracy of the methods in Section 15.4. A particular challenge concerns reliable truth measurements of the depth profile in the upper meters of the water column, a relative "blind spot" of current sensing technology. Currents reconstructed from waves is an attractive method in the near-surface regime, which introduces a paradox of sorts: radar-derived currents are attractive since they can provide measurements in this blind spot where few other reliable methods exist, yet this, in turn, makes validation a challenge because truth measurements are difficult to obtain.

15.5.2   *Interpretation of the currents: Stokes drift*

As mentioned above, radar-derived currents may include Lagrangian components, namely the wave-induced Stokes drift. A fluid parcel considered in a Lagrangian framework perturbed by waves follows an elliptical trajectory. In addition, there is a net translation of the parcel in the direction of the wave (except in the limit of infinitesimal wave amplitude), which is known as Stokes drift. It has been proposed that the currents measured by radar systems are a sum of background Eulerian currents (such as a wind drift of tidal current) and a Stokes drift component [10]:

$$\mathbf{U}_R = \mathbf{U}_E + \mathbf{U}_{SS}, \tag{15.25}$$

where $\mathbf{U}_E$ and $\mathbf{U}_{SS}$ are the Eulerian and dominant Stokes drift components, respectively, and $\mathbf{U}_R$ is the total current measured from the wave spectrum. The dominant

Stokes drift component is a function of the wave energy spectrum reflecting the heights of the waves and has been suggested to be expressed as [36]

$$\mathbf{U}_{SS} = 4\pi \int_0^{f_B} \int_0^{2\pi} f\mathbf{k}(f)E(f,\theta)\mathrm{d}f\mathrm{d}\theta, \tag{15.26}$$

where frequency $f = \omega/2\pi$, $E(f, \theta)$ is the wave energy density spectrum as a function of frequency and direction θ, $\mathbf{k}(f)$ is an alternate form of the dispersion relation now with frequency as the independent variable, and f_B is the frequency of the Bragg-resonant wave determined by the radar wavelength. The Stokes drift component decays rapidly with depth, complicating the interpretation of $\mathbf{U}_R$ at greater depths. Given the measurements of $E(f, \theta)$, the Stokes drift component may be estimated and subtracted using (15.26), allowing separate analysis of the Eulerian current components [10]). It has been found that the Stokes drift under relevant conditions may be on the order of 5–10 cm/s as discussed in which is [28], non-negligible relative to the current strengths typically derived from radar measurements.

However, there remains ongoing debate about the extent to which Stokes drift is a component of radar-derived currents [37, 38]. Different theoretical formulations have been proposed, as summarized in [38]. Further research is required on the matter, which is clearly important for an increased understanding of the physical interpretation of the currents measured from radar images.

15.6 Summary

In this chapter, we have examined how spatially varying ocean currents can be extracted from remote measurements of the wave spectrum. For marine radar images, the spatial variation of currents may be mapped within the full radar field of regard, an attractive means of current remote sensing that has multiple advantages to in situ point sensors. Currents are extracted by analyzing their effect on wave propagation, appearing as frequency-dependent shifts in the linear dispersion relation curve along which the wave spectrum is strongly peaked. Several algorithms were described for obtaining empirical dispersion relations from the measured spectrum and extracting the currents: the so-called LS and ILS method, the NSP method, and the PCS method, all of which are reviewed and compared.

We go on to describe how the same methods and algorithms can be extended to also allowing the depth-dependence of the current to be determined. Multiple velocities must now be extracted over different wavenumber bins, whereupon the set of velocities at varying wavenumber is further analyzed using an inversion method to find the depth dependence.

Reasonable agreement between radar-derived currents and in situ measurements has been demonstrated in multiple field measurements. However, more validation is necessary especially in the context of depth-varying flows, where comparison is more difficult and in situ data are scarce. Understanding the extent to which the Lagrangian current from the waves, Stokes drift, is measured as part of the radar-derived current is not well-understood, yet important for the interpretation of the mapped currents and comparison with in situ measurements.

References

[1] Kudryavtsev V., Shrira V., Dulov V., Malinovsky V. 'On the vertical structure of wind-driven sea currents'. *Journal of Physical Oceanography.* 2008;**38**(10):2121–44.

[2] Belcher S.E., Grant A.L.M., Hanley K.E. 'A global perspective on Langmuir turbulence in the ocean surface boundary layer'. *Geophysical Research Letters.* 2012;**39**(18):L18605.

[3] Laxague N.J.M., Haus B.K., Ortiz-Suslow D.G. 'Passive optical sensing of the near-surface wind-driven current profile'. *Journal of Atmospheric and Oceanic Technology.* 2017;**34**(5):1097–111.

[4] Laxague N.J.M., Özgökmen T.M., Haus B.K. 'Observations of near-surface current shear help describe oceanic oil and plastic transport'. *Geophysical Research Letters.* 2018;**45**(1):245–9.

[5] Lund B., Haus B.K., Horstmann J. 'Near-surface current mapping by shipboard marine X-band radar: a validation'. *Journal of Atmospheric and Oceanic Technology.* 2018;**35**(5):1077–90.

[6] Gangeskar R. 'Verifying high-accuracy ocean surface current measurements by X-band radar for fixed and moving installations'. *IEEE Transactions on Geoscience and Remote Sensing.* 2018;**56**(8):4845–55.

[7] Zippel S., Thomson J. 'Surface wave breaking over sheared currents: observations from the mouth of the Columbia river'. *Journal of Geophysical Research. Oceans.* 2017;**122**(4):3311–28.

[8] Dalrymple R.A. 'Water wave models and wave forces with shear currents'. Coastal and Oceanographic Engineering Laboratory, University Florida, Gainesville (FL), Technical Report No; 2002. p. 20.

[9] Young I.R., Rosenthal W., Ziemer F. 'A three-dimensional analysis of marine radar images for the determination of ocean wave directionality and surface currents'. *Journal of Geophysical Research.* 1985;**90**(C1):1049–59.

[10] Lund B., Graber H.C., Tamura H. 'A new technique for the retrieval of near-surface vertical current shear from marine X -band radar images'. *Journal of Geophysical Research. Oceans.* 2015;**120**(12):8466–86.

[11] Campana J., Terrill E.J., de Paolo T. 'A new inversion method to obtain upper-ocean current-depth profiles using X-band observations of deep-water waves'. *Journal of Atmospheric and Oceanic Technology.* 2017;**34**(5):957–70.

[12] Hessner K., Reichert K., Borge J.C.N., Stevens C.L., Smith M.J. 'High-resolution X-band radar measurements of currents, bathymetry and sea state in highly inhomogeneous coastal areas'. *Ocean Dynamics.* 2014;**64**(7):989–98.

[13] Lund B., Haus B. Radar measurements collected during the LAgrangian Submesoscale ExpeRiment (LASER) experiment aboard R/V Walton Smith cruise WS16015 in the gulf of mexico from 2016-01-20 to 2016-02-12.

distributed by: Gulf of Mexico Research Initiative Information and Data Cooperative (GRIIDC), Harte Research Institute, Texas A&M UIniversity-Corpus Christi. [online]. 2018. Available from http://doi.org/10.7266/N7N01550.

[14] Horstmann J., Stresser M., Carrasco R. 'Surface currents retrieved from airborne video'. Proceedings OCEANS 2017-Aberdeen; 2017.

[15] Streßer M., Carrasco R., Horstmann J. 'Video-based estimation of surface currents using a low-cost quadcopter'. *IEEE Geoscience and Remote Sensing Letters*. 2017;**14**(11):2027–31.

[16] Lund B., Graber H.C., Hessner K., Williams N.J. 'On shipboard marine x-band radar near-surface current "calibration"'. *Journal of Atmospheric and Oceanic Technology*. 2015;**32**(10):1928–44.

[17] Mccann D.L., Bell P.S. 'A simple offset "calibration" method for the accurate geographic registration of ship-borne X-band radar intensity imagery'. *IEEE Access*. 2018;**6**(13):13939–48.

[18] Senet C.M., Seemann J., Ziemer F. 'The near-surface current velocity determined from image sequences of the sea surface'. *IEEE Transactions on Geoscience and Remote Sensing*. 2001;**39**(3):492–505.

[19] Huang W., Gill E.W. 'Surface current measurement under low sea state using dual polarized X-band nautical radar'. *IEEE Journal of Selected Topics in Applied Earth Observations and Remote Sensing*. 2012;**5**(6):1868–73.

[20] Gangeskar R. 'Ocean current estimated from X-band radar sea surface, images'. *IEEE Transactions on Geoscience and Remote Sensing*. 2002;**40**(4):783–92.

[21] Serafino F., Lugni C., Soldovieri F. 'A novel strategy for the surface current determination from marine X-band radar data'. *IEEE Geoscience and Remote Sensing Letters*. 2010;**7**(2):231–5.

[22] Huang W., Carrasco R., Shen C., Gill E.W., Horstmann J. 'Surface current measurements using X-band marine radar with vertical polarization'. *IEEE Transactions on Geoscience and Remote Sensing*. 2016;**54**(5):2988–97.

[23] Shen C., Huang W., Gill E.W., Carrasco R., Horstmann J. 'An algorithm for surface current retrieval from X-band marine radar images'. *Remote Sensing*. 2015;**7**(6):7753–67.

[24] Smeltzer B.K., Æsøy E., Ådnøy A., Ellingsen S. Å. 'An improved method for determining near-surface currents from wave dispersion measurements'. *Journal of Geophysical Research: Oceans*. 2019;**124**(12):8832–51.

[25] Stewart R.H., Joy J.W. 'HF radio measurements of surface currents'. *Deep Sea Research and Oceanographic Abstracts*. 1974;**21**(12):1039–49.

[26] Skop R.A. 'Approximate dispersion relation for wave-current interactions'. *Journal of Waterway, Port, Coastal, and Ocean Engineering*. 1987;**113**(2):187–95.

[27] Kirby J.T., Chen T.-M. 'Surface waves on vertically sheared flows: approximate dispersion relations'. *Journal of Geophysical Research*. 1989;**94**(C1):1013–27.

[28] Lund B., Haus B.K., Graber H.C. 'Marine x-band radar currents and bathym-
      etry: an argument for a wave number-dependent retrieval method'. *Journal of
      Geophysical Research. Oceans*. 2020;**125**(2).

[29] Wu J. 'Wind-induced drift currents'. *Journal of Fluid Mechanics*.
      1975;**68**(01):49–70.

[30] Plant W.J., Wright J.W. 'Phase speeds of upwind and downwind traveling short
      gravity waves'. *Journal of Geophysical Research*. 1980;**85**(C6):3304–10.

[31] Fernandez D., Vesecky J., Teague C. 'Measurememts of upper ocean surface
      current shear with high-frequency radar'. *Journal of Geophysical Research*.
      1996;**101**(28):615–25.

[32] Teague C.C., Vesecky J.F., Hallock Z.R. 'A comparison of multifrequency
      HF radar and ADCP measurements of near-surface currents during COPE-3'.
      IEEE Journal of Oceanic Engineering. 2001;**26**(3):399–405.

[33] Ha E.-C. 'Remote sensing of ocean surface current and current shear by
      HF backscatter radar'. *Stanford University, Stanford, CA, Technical Report
      Number D415-1*. 1979

[34] Campana J., Terrill E.J., de Paolo T. 'The development of an inversion tech-
      nique to extract vertical current profiles from X-band radar observations'.
      Journal of Atmospheric and Oceanic Technology. 2016;**33**(9):2015–28.

[35] Smeltzer B.K. *Replication data for: New method to determine near-surface
      currents from measurements of the wave spectrum, dataverseno, v1 [online]*.
      2019. Available from https://doi.org/10.18710/8JBWCJ [Accessed 2 Sept
      2020].

[36] Ardhuin F., Marié L., Rascle N., Forget P., Roland A. 'Observation and estima-
      tion of Lagrangian, Stokes, and Eulerian currents induced by wind and waves
      at the sea surface'. *Journal of Physical Oceanography*. 2009;**39**(11):2820–38.

[37] Röhrs J., Sperrevik A.K., Christensen K.H., Broström G., Breivik Øyvind.
      'Comparison of HF radar measurements with Eulerian and Lagrangian sur-
      face currents'. *Ocean Dynamics*. 2015;**65**(5):679–90.

[38] Chavanne C. 'Do high-frequency radars measure the wave-induced Stokes
      drift?' *Journal of Atmospheric and Oceanic Technology*. 2018;**35**(5):1023–31.

Chapter 16

Bathymetry (and current) retrieval: phase-based method

David A. Honegger[1] and Merrick C. Haller[1]

16.1 Introduction

A fundamental goal of bathymetry retrieval is the reproduction of bedforms that affect navigation and recreational safety, and that have dominant roles in wave and circulation dynamics. Some bedforms span only a few ocean wavelengths, such as sand bars, rip channels, and complex tidal shoal/channel networks. The constituent local depths may be estimated by exploiting the dispersive relationship between water column thickness and the celerity (i.e., phase speed) of surface gravity waves, provided that the corresponding spatial variations in celerity can be spatially resolved. Extensively developed estimation techniques that can resolve the full directional spectrum of the surface gravity wave field (such as 3D-Fourier techniques) assume spatial homogeneity over analysis tiles that must span ten or more wavelengths [1]. However, direct inspection of variations in wave phase can achieve spatial estimate resolution on the order of one ocean wavelength and bathymetric structures with spatial scales of about two wavelengths [2]. With increased spatial resolution comes decreased directional resolution, which can be as stringent as the assumption of a single representative wave train at each frequency. This tradeoff of directional resolution for spatial resolution can be enabling, particularly when the observation record length supports incorporating multiple estimates in time. In addition, the method can operate on spatially scattered time series. It can therefore honor the native polar resolution of marine radars, with higher cross-range (azimuthal) resolution near the radar and lower azimuthal resolution at far ranges. Herein, we describe this method of exploiting the spatial structure of surface gravity wave phase in X-band radar imagery for the primary goal of estimating bathymetry. Two depth retrieval examples using a shore-based radar are shown in Figures 16.1 and 16.2. In locations where ambient flow modulates the wave field, information about the currents may be extracted as well.

We proceed directly to a detailed description of the method, foregoing a review of the history of estimating depth from spatial variations in wave phase (or

[1]School of Civil & Construction Engineering, Oregon State University, USA

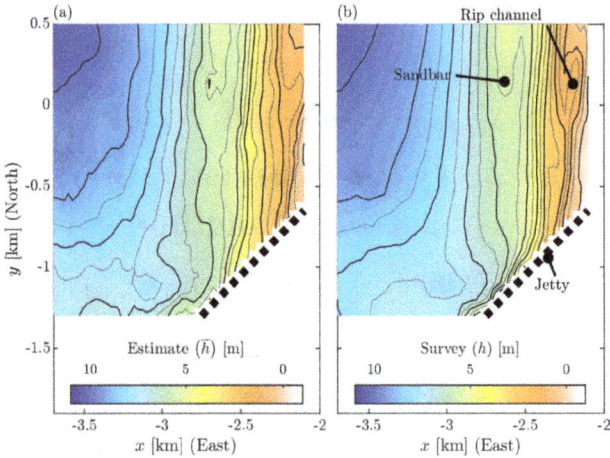

Figure 16.1　*Comparison of (a) a bathymetry estimate at a beach environment (Benson Beach, adjacent to the Columbia River Mouth, Washington, USA) using a radar-adaptation of the phase-based cBathy [3] algorithm, and (b) a concurrent gridded soundings survey (c/o Peter Ruggiero, Oregon State University). The X-band marine radar location is located at the coordinate system origin. [Modified from [4], c.f. figure 6].*

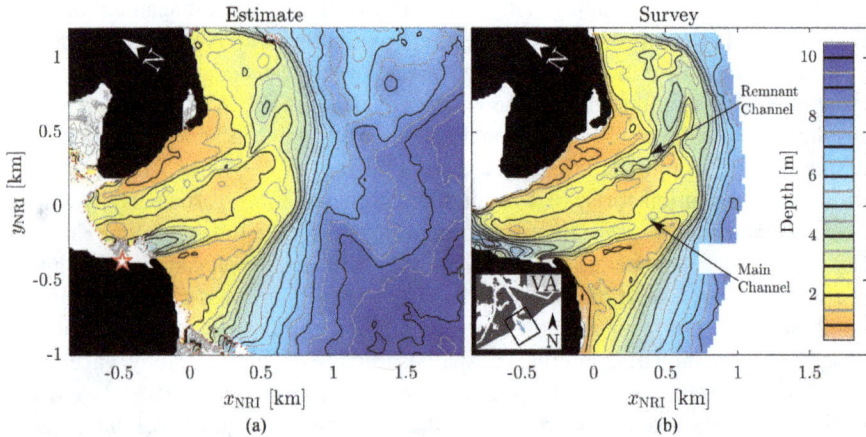

Figure 16.2　*Comparison of (a) a bathymetry estimate at a shallow tidal inlet (New River Inlet, North Carolina, USA) using a radar-adaptation of the phase-based cBathy [3] algorithm, and (b) a concurrent gridded soundings survey (c/o the United States Army Corps Field Research Facility). The X-band marine radar location is marked by a red star in (a). [Reprinted from [5], c.f. figure 1].*

wavelength), which includes developments using aerial reconnaissance platforms during World War II [6–8], shore-based optical video [2, 3, 9–11], as well as marine radar [4, 5, 12–16]. This description most closely follows recent applications [4, 5] and adaptations [16] of cBathy [3], an algorithm initially developed for optical video that does not prefilter the radar signal information in wavenumber space [14, 15].

16.2 Brief overview

The foundation of phase-based bathymetry and current inversion is similar to the methods described elsewhere in this book. Namely, depth and currents are retrieved through their known hydrodynamic relationships with quantities that can be directly estimated from observations of ocean wave frequency and wavenumber. Some of these relationships are complex, involving feedback between bathymetry, currents, and waves, which can be estimated via empirical covariance matrices built using an ensemble of forward-coupled model runs [17–19]. A simpler but nonetheless useful analytical relationship is the linear water wave dispersion relation

$$\left(\omega - \vec{k} \cdot \vec{u}\right)^2 = gk \tanh(k\tilde{h}),$$

(16.1)

where the aim is to estimate depth, $\tilde{h}$, and potentially the depth uniform current vector, $\vec{u}$, given radar-derived estimates of frequency, ω, and wavenumber, $\vec{k}$ (with magnitude $|\vec{k}| = k$), as well as known gravitational acceleration, g. The estimation procedure described below makes use of the common Fourier series description of the imaged ocean wave field:

$$I(\vec{x}, t) = \sum_{\omega, \vec{k}} I_0(\omega, \vec{k}) \cos [\phi(\omega, \vec{k})],$$

(16.2)

where I is the microwave backscatter intensity, I_0 is the wave signal amplitude, and ϕ is the planar wave phase:

$$\phi(\omega, \vec{k}) = \vec{k} \cdot \vec{x} \mp \omega t + \phi_0.$$

(16.3)

Here, $\vec{x}$ is the spatial position vector, t is time, and ϕ_0 is a random phase offset. Given a time series of pixel intensity that adequately resolves an energetic (and depth-dispersive) band of the surface gravity wave spectrum, frequency content can readily be identified via the Fourier transform and wavenumber can be estimated at high spatial resolution as the spatial gradient of wave phase, i.e.,

$$\vec{k} = \nabla \phi(\omega, \vec{x}) = \left(\frac{\partial \phi}{\partial x}, \frac{\partial \phi}{\partial y}\right).$$

(16.4)

Specifically, the image time series is first decomposed into its constituent temporal Fourier frequencies, each of which is examined to assess spatial coherence and cross-spectral signal complexity. The phase of each frequency with coherent and simple cross-spectral phase structure is then modeled as a single wave train (one wavenumber vector), which is estimated via an iterative nonlinear least squares method. A second least squares fit to a dispersion relationship is performed to estimate instantaneous depth from the set of retrieved frequency-wavenumber pairs. A final step of filtering in time allows the bathymetry product to be updated using objective reliability metrics. The remainder of this chapter will describe these steps in greater detail.

16.3   Frequency and wavenumber estimates

16.3.1   *Fourier series representation of the imaged wave field*

Let the discretely sampled radar backscatter time series be denoted $I_p(t_j) = I(\vec{x}_p, t_j)$, where $\vec{x}_p$ is the location of the pth pixel. The spatial coordinate system may be defined on either a geographic Cartesian grid, (x_p, y_p), or a polar grid with the antenna located at the origin, (r_p, θ_p); t_j is the jth time step in a record of N_t samples (with temporal resolution δt) and $I_p(t_j)$ may be represented by a sum of propagating waves:

$$I_p^s(t_j) = \Re \left\{ \sum_{\vec{k}, \omega} I_0(\vec{k}, \omega) e^{i[\vec{k}\cdot\vec{x}_p \mp \omega t_j + \phi_0(\vec{k},\omega) + \phi_p^s(\omega)]} \right\}, \tag{16.5}$$

where each wave component, defined by its $\{\vec{k}, \omega\}$ pair, has a unique amplitude, I_0, and phase offset, ϕ_0. We retain the complex, polar form of the wave signal to explicitly keep track of the wave phase, noting that only the real (cosine) component of the wave field is physically relevant. A spatially varying sampling phase offset, ϕ_p^s, is present due to the finite rotation rate of the radar antenna, i.e., each radar "image" is not generated instantaneously but rather constructed one radial pulse at a time. Assuming that the antenna rotates at a constant rate, the sampling offset is

$$\phi_p^s = \omega \frac{\theta_p - \theta_0}{\omega_a}, \tag{16.6}$$

where ω_a is the angular antenna rotation rate and θ_0 is the reference azimuth (e.g., North).

16.3.2   *Compute the temporal discrete Fourier transform*

Frequency content of the observed signal is extracted using the temporal discrete Fourier transform,

$$\tilde{I}_p^s(\omega_n) = \sum_{j=0}^{N_t-1} I_p(t_j) e^{-i[\omega_n t_j]} \delta t; \quad \omega_n = \frac{2\pi n}{N_t}; \quad n = -\frac{N_t}{2}, \dots, \frac{N_t}{2}, \tag{16.7}$$

which produces an estimate of the Fourier coefficient at each discrete frequency, ω_n. The complex coefficient corresponding to the nth frequency is

$$\tilde{I}_p^s(\omega_n) = I_0(\vec{k}, \omega_n) e^{i[\vec{k}\cdot\vec{x}_p + \phi_0 + \phi_p^s]} N_t \delta t. \tag{16.8}$$

The sampling phase offset, ϕ_p^s, contributes a spurious azimuthal component to the resulting wavenumber vector that becomes significant (10 percent of the true wavenumber) typically only near the antenna for fast-moving long-period waves. Nevertheless ϕ_p^s may be easily subtracted from the total Fourier coefficient phase at each calculated ω_n using known ω_a, θ_p, and θ_0. Wavenumber content may then be extracted from these corrected coefficients:

$$\tilde{I}_p = \tilde{I}_p^s e^{-i\phi_p^s}, \tag{16.9}$$

via Fourier transform-derived estimates of the wavenumber spectrum, or by directly examining local variations of the phase (16.4). Here, the latter approach is taken.

Direct inspection of the frequency-dependent phase can clearly identify the planar structure of a simple, unidirectional wave train phase [13]. However, not all frequencies contain meaningful information and some wave fields may be directionally complex. As an example, a series of phase maps, $\phi(\vec{x}, \omega_n)$, are shown in Figure 16.3 alongside a single-rotation X-band marine radar image. Local planar phase structure is apparent at the peak frequency of $2\pi/10$ rad/s and at a secondary peak at $2\pi/15$ rad/s, and not clearly apparent toward the lower and higher ends of the spectrum. It is therefore useful to focus the computational cost of wavenumber estimation on frequencies at which the phase structure is spatially coherent. This may be accomplished via the estimation of the cross-spectral coherence spectrum between each pair of pixels.

16.3.3 Compute the cross-spectral coherence spectrum

The cross-spectral coherence spectrum between each pair of pixels (p and q) is estimated as

$$\Gamma_{pq}(\omega_b) = \frac{\langle \tilde{I}_p \tilde{I}_q^* \rangle}{\langle |\tilde{I}_p|^2 \rangle^{\frac{1}{2}} \langle |\tilde{I}_q|^2 \rangle^{\frac{1}{2}}}, \tag{16.10}$$

where the conjugate transpose is denoted by an asterisk ($*$), and ensemble- or band-averaging (denoted by angle brackets, $\langle \cdot \rangle$) is required for the coherence estimate to be meaningful. In the case of band averaging, Γ_{pq} is defined for a reduced set of frequencies, ω_b. Γ_{pq} may be written in polar complex form as

$$\Gamma_{pq}(\omega_b) = \gamma_{pq}(\omega_b) e^{i[\Phi_{pq}(\omega_b)]}, \tag{16.11}$$

where the coherence magnitude, $\gamma_{pq} = |\Gamma_{pq}|$, is a normalized cross-spectral power, and the cross-spectral phase, Φ_{pq}, is the frequency-dependent difference between the complex Fourier coefficient phases at the pth and qth pixels,

$$\Phi_{pq}(\omega_b) = \vec{k} \cdot \vec{x}_p - \vec{k} \cdot \vec{x}_q = \vec{k} \cdot (\vec{x}_p - \vec{x}_q). \tag{16.12}$$

Wavenumber information may be derived from this cross-spectrum, since Γ_{pq} is directly estimated from the observed image time series and the matrix of pixel component distances, $\vec{x}_p - \vec{x}_q$ is known [2]. As few as two pixels are required to estimate each directional component of the wavenumber, given a sufficiently high signal-to-noise ratio (SNR) and provided that the pixels are not so far apart that there exists a 2π-phase ambiguity. Real cases often do not have sufficiently high SNR, and the wavelength is not known *a priori*, so a localized analysis tile comprising N_p pixels is typically used to increase the robustness of the estimate. However, modeling the full $N_p \times N_p$ matrix is computationally inefficient when the first eigenvector often represents the vast majority of the signal variance.

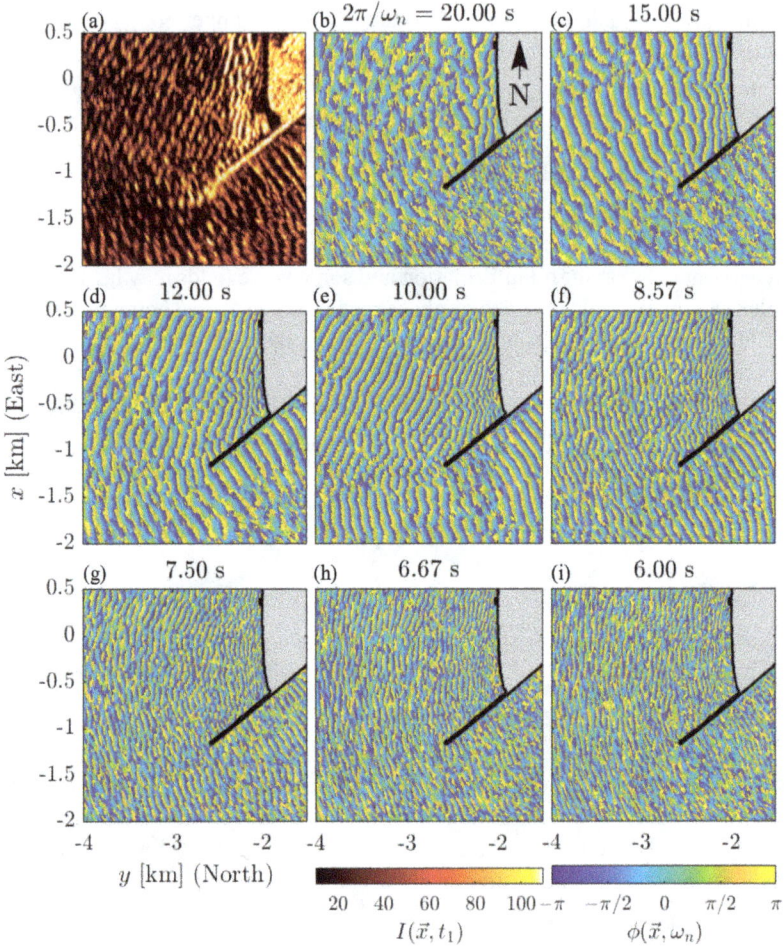

Figure 16.3   (a) Example of single-rotation image and (b–i) example of Fourier frequency phase maps of a wave field. Waves incident from the west (−y) approach a beach and an estuary mouth, separated by an angled jetty (Columbia River Mouth, Washington, USA). Note the coherent structure of the wave field at the peak period of 10 s, the coherent structure at a secondary peak with period 15 s, and the decreasing coherence at the shortest and longest periods. Most readily observed in (c), also note the change in cross-shore (x) distance over which the phase wraps from −π to π, which corresponds directly to wavenumber variations as the waves shoal. [More about these observations can be found in [4] and [5]].

16.3.4 Extract the dominant cross-spectral eigenvector

The cross-spectral matrix, $\Gamma_{pq}(\omega_b)$, at the bth coherent frequency may be decomposed as

$$\Gamma_{pq}(\omega_b) = \sum_{n=1}^{N_p} \lambda_n v_{pn} v_{qn}^*, \tag{16.13}$$

where λ_n is the set of magnitude-sorted eigenvalues, and both v_{pn} and v_{qn} are the corresponding magnitude-sorted complex eigenvectors. Asterisk ($*$) superscripts denote the complex conjugate. The dominant eigenvector, $v_{n=1} = \check{v}$, which corresponds to the dominant eigenvalue, $\lambda_{n=1} = \check{\lambda}$, has the same number of elements and the same complex structure as each column of Γ_{pq}:

$$\check{v}_p = \check{\gamma}_p e^{i[\check{\Phi}_p]} \tag{16.14}$$

Here, $\check{\gamma}_p$ is the dominant eigenvector amplitude and $\check{\Phi}_p$ is the dominant eigenvector phase, which is unique to a constant phase offset, $\check{\Phi}_0$. This constant phase offset may be interpreted as an arbitrary reference location for the pixel component distance vector, i.e., $\check{\Phi} = \vec{k} \cdot (\vec{x}_p - \vec{x}_0) = \vec{k} \cdot \vec{x}_p + \check{\Phi}_0$, where $\vec{k} \cdot \vec{x}_0 = \check{\Phi}_0$.

16.3.5 Minimize a cost function to estimate wavenumber

To estimate the bth frequency-dependent wavenumber vector, the dominant complex eigenvector is then modeled as

$$\check{v}_{bp} = \check{\gamma}_{bp} e^{i[\vec{k}_b \cdot \vec{x}_p + \Phi_{0,b}]}, \quad p = 1, \dots, N_p, \tag{16.15}$$

which can be expanded into the individual wavenumber vector components in either Cartesian form

$$\check{v}_{bp} = \check{\gamma}_{bp} e^{i[k_{x,b} x_p + k_{y,b} y_p + \Phi_{0,b}]}, \tag{16.16}$$

or in polar form

$$\check{v}_{bp} = \check{\gamma}_{bp} e^{i[k_b(x_p \cos \alpha_b + y_p \sin \alpha_b) + \Phi_{0,b}]}, \tag{16.17}$$

where α_b is the wave direction.

For each frequency ($b = 1, \dots, N_b$) and at each location on a predefined grid (x_m, y_m), a least squares problem can then be constructed:

$$\operatorname*{argmin}_{\beta_{mb}} \left[\sum_{p=1}^{N_p} w_{mp} R_{mbp}^2 (\beta_{mb}, \vec{x}_p) \right], \tag{16.18}$$

where w_{mp} is the vector of pixelwise weights and R_{mbp} is the vector of residuals between a calculated quantity from the remote sensing observations and a corresponding model, $f(\beta_{mb}, \vec{x}_p)$, with fit parameters, β_{mb}. The residual function may be defined using either the complex eigenvector

$$R_{mbp} = \check{v}_{bp} - \check{\gamma}_{bp} e^{i\left[k_{x,mb} x_p + k_{y,mb} y_p + \Phi_{0,mb}\right]} \tag{16.19}$$

or the eigenvector phase

$$R_{mbp} = \left[\tan^{-1} \frac{\Im(\check{v}_{bp})}{\Re(\check{v}_{bp})} \right] - \left[k_{x,mb} x_p + k_{y,mb} y_p + \Phi_{0,mb} \right], \tag{16.20}$$

where in each case the fit parameters are

$$\beta_{mb} = (k_{x,mb}, k_{y,mb}, \Phi_{0,mb}). \tag{16.21}$$

Cost function weights are used to spatially constrain each fit with a localization function, $w_{mb} = \ell_{mb}$. One such function is the Hann kernel:

$$\ell_{mp} = \cos^2\left[\frac{\pi}{2}\frac{(x_p - x_m)}{L_x}\right]\cos^2\left[\frac{\pi}{2}\frac{(y_p - y_m)}{L_y}\right], \tag{16.22}$$

which smoothly tapers from 1 at the estimate location (x_m, y_m) to 0 at $(x_m \pm L_x, y_m \pm L_y)$. Choice of the localization length scales L_x and L_y should take into consideration the highest frequency (shortest wavelength) waves that are temporally and spatially resolved by the pixel spacing, the SNR of the imaged wave signal, and the expected length scales of wavenumber (bathymetric) variation. The scales must be large enough to resolve the phase slope of the shortest wavelength (i.e., produce a statistically robust estimate), and small enough to minimize both the influence of cross-spectral phase ambiguities and the warping of the phase plane due to spatial changes in wavenumber. Although wavelength is not known *a priori*, it may be roughly estimated to be a reasonable fraction of the deepwater wavelength of the highest frequency coherent wave. L_x and L_y may also be chosen to remain constant, producing well-resolved wavenumber estimates at the highest frequencies and an array of noisier wavenumber estimates at the lower frequencies, or to be frequency-dependent, producing fewer but wavelength-tuned estimates. Horizontal spacing between the analysis tile centers is then typically $(\Delta x_m, \Delta y_m) = (L_x/2, L_y/2)$.

Solution of the nonlinear least squares problem (16.18) can be carried out using the Levenberg-Marquardt algorithm, which is popular due to its rapid rate of convergence. Dropping the *m*- and *b*- subscripts for simplicity and starting with an initial guess, β_a, the vector of fit parameters is updated for each iteration as

$$\beta_{a+1} = \beta_a + \left[\mathbf{J}_a^T\mathbf{W}\mathbf{J}_a + \lambda_D\text{diag}\left(\mathbf{J}_a^T\mathbf{W}\mathbf{J}_a\right)\right]^{-1}\mathbf{J}_a^T\mathbf{W}R_a, \tag{16.23}$$

eventually reaching an estimate of β, here denoted $\hat{\beta}$. The model Jacobian matrix evaluated at the current step:

$$\mathbf{J}_a = \left.\frac{\partial f(\beta, \vec{x})}{\partial \beta}\right|_{\beta=\beta_a}, \tag{16.24}$$

describes the local sensitivity of the model function, $f(\beta, \vec{x})$ to the estimate parameters. The weights are $\mathbf{W} = \text{diag}(w_p)$ and λ_D is the damping parameter. In the Levenberg-Marquardt algorithm, λ_D is initially small ($\lambda_D = 0.1$) but is increased by a factor of 10 if the convergence criteria (e.g., sum of squared errors) are not improved. An equivalent but more stable and complex version of (16.23) is computed in statistical analysis software packages (e.g., MATLAB®nlinfit and Python scipy.optimize.leastsquares).

Figure 16.4 shows the results of one such fit, located at the analysis tile marked by a red box in Figure 16.3e. The spatial structure of the observed phase (filled points) appears to be adequately modeled by a plane (gray surface), but this may not be the case at every location and at each time step. Wavenumber fit quality affects

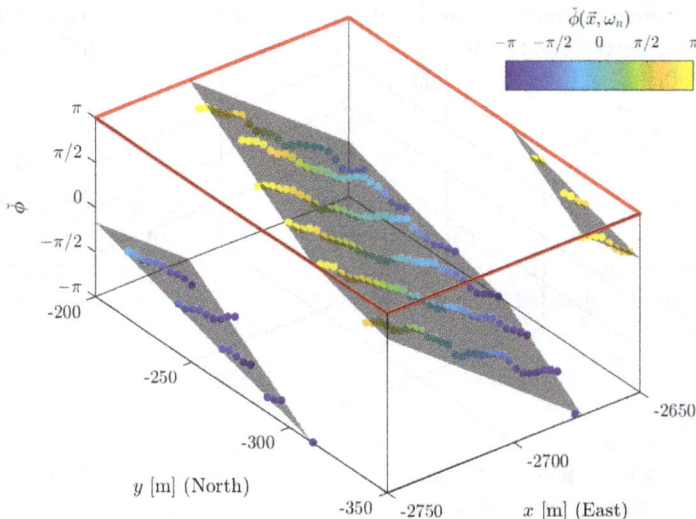

Figure 16.4   *First eigenvector phase, $\check{\phi}$, corresponding to the coherent frequency, $\omega_n = 2\pi/10$ rad/s, and localized analysis tile corresponding to the red rectangle in Figure 16.3e. Observed phase is marked by filled circles (colored in accordance with the colormap in Figure 16.3), and the nonlinear fit phase plane is shown as a gray surface. The fit parameters are $\hat{k}_x = -0.070$ rad/m, $\hat{k}_y = 0.023$ rad/m, and $\hat{\phi}_0 = 0.75$ rad/m, corresponding to a wavelength of about 85 m. [More about these observations can be found in [4] and [5].]*

the accuracy and precision of the depth (and current) estimates, and is addressed next in Section 16.3.6.

16.3.6   *Wavenumber estimate quality metrics*

It is important to calculate objective quality control metrics to cull cases in which fundamental assumptions are violated and to supply uncertainty estimates to depth and/or current inversion routines. First, to reduce the computational cost associated with performing an eigen-decomposition and iterative search for a wavenumber vector at every band- or ensemble-averaged frequency, the set of ω_b can be sorted by the mean squared coherence over the full analysis tile:

$$\overline{\gamma^2}(\omega_b) = \frac{1}{N_p(N_p-1)} \sum_{p,q} \gamma_{pq}^2, \tag{16.25}$$

and only a reasonable, predefined number of the most-coherent frequencies is chosen for wavenumber estimation. Alternatively, a threshold may be placed on the statistical significance (e.g., 95 percent confidence level) of the mean squared coherence. The presence of multiple wave trains, i.e., a directionally multimodal sea state, may be present at a single frequency. The effects of these wavenumber vectors on the estimated cross-spectral phase must either be modeled directly or be identified and discarded or downweighted. Normalizing the first eigenvalue of the cross-spectral matrix by the mean of the eigenvalues

$$\overline{\lambda} = \frac{\check{\lambda}}{\frac{1}{N_p} \sum_{n=1}^{N_p} \lambda_n}, \tag{16.26}$$

provides a metric for the dominance of the first eigenvector. Large values of $\overline{\lambda}$, typically greater than 10, are indicative of the presence of a simple, high-SNR cross-spectral signal. This value can be used to weigh the cost function for the depth estimate fit detailed in Section 16.4. The skill of the wavenumber fit may be defined as

$$s = 1 - \frac{\|R\|^2}{\sigma_{\text{Obs}}^2}, \tag{16.27}$$

where $\|R\|$ is the norm of the residuals (16.19) and σ_{Obs}^2 is the variance of the observed dominant eigenvector, $\check{v}$, or dominant eigenvector phase, $\check{\Phi}$, depending on the construction of the cost function. (Subscripts have been dropped here for simplicity.) Values near 1 indicate excellent agreement between observations and the final model fit, and typically $s \geq 0.5$ is sufficient. Another metric is an estimate of the wavenumber component uncertainties, $(\hat{\sigma}_{k_x}, \hat{\sigma}_{k_y})$ or $(\hat{\sigma}_k, \hat{\sigma}_\alpha)$, depending on whether (16.19) or (16.20) is used in the fit. These uncertainties are estimated from the residuals and model Jacobian, under the assumption that the errors are random and normally distributed. A conservative uncertainty estimate is the wider of the two bounds of the 95 percent confidence interval:

$$\hat{\sigma}_\beta = \max \left| \pm t_{(0.975, N_p - N_\beta)} \text{diag} \left[\left(\frac{\|R\|^2}{(N_p - N_\beta)} \left[\mathbf{J}^T \mathbf{J} \right]^{-1} \right)^{1/2} \right] \right|, \tag{16.28}$$

where $t_{(0.975, N_p - N_\beta)}$ is the $(100 - 5/2)$-th percentile of Student's t-distribution with $N_p - N_\beta = N_p - 3$ degrees of freedom. An equivalent but more stable and more complex version of (16.28) is computed in MATLAB's `nlparci` script. In environments where an ambient current is not expected to substantially alter the wave celerity, a direct estimate of depth using each estimated $\omega_{mb}, \vec{\hat{k}}_{mb}$ pair may be calculated:

$$\hat{h}_b = \frac{1}{\hat{k}_b} \tanh^{-1} \left(\frac{\omega_b^2}{g \hat{k}_b} \right). \tag{16.29}$$

An unreasonably shallow or deep $\hat{h}_b$ indicates that the wavenumber estimate is not useful. The subscript m is omitted here (i.e., $\hat{h}_b = \hat{h}_{mb}$) to emphasize the utility of $\hat{h}_b$ as a pointwise quality control metric rather than a direct estimate of depth. More robust depth estimate procedures are described in Section 16.4.

16.4　Depth inversion

The unique basis of phase-based depth (and current) estimation lies in the wavenumber estimate procedure detailed in Section 16.3, which can resolve spatial wavenumber variations of roughly one ocean wavelength. Various methods of retrieving depth and currents from these high-resolution wavenumber estimates are possible, from simple inversions of the linear water wave dispersion relation to complex inversions using data assimilation frameworks and empirically constructed covariance matrices. Here we focus on a relatively simple method that uses the linear water wave dispersion relation (16.1) to analytically link depth to the set of wavenumber–frequency vector pairs estimated in Section 16.3:

$$\{\omega_{mb}, \hat{\vec{k}}_{mb}\} = \{\omega_b(x_m, y_m), \hat{\vec{k}}_b(x_m, y_m)\}; \quad \begin{array}{l} b = 1, \ldots, N_b \\ m = 1, \ldots, N_m \end{array}, \tag{16.30}$$

where N_b wavenumbers have been estimated at each of the N_m locations.

16.4.1   Problem formulation

Initial estimates of depth are produced at each location on a predefined grid, (x_d, y_d), by iteratively solving (16.23), a nonlinear least squares problem, which incorporates the frequency- and spatially dependent wavenumber estimates:

$$\operatorname*{argmin}_{\beta_d} \left[\sum_{m,b} w_{dmb} R_{dmb}^2 (\beta_d, \omega_{mb}, \hat{\vec{k}}_{mb}) \right], \tag{16.31}$$

where the set of fit parameters, β_d, includes depth, $\tilde{h}_d = \tilde{h}(x_d, y_d)$. Weights, w_{dmb}, are used to localize the fit to the region immediately surrounding the depth estimate, and to incorporate wavenumber quality control metrics. Again, localization may be carried out with a Hann kernel, here written explicitly in terms of the depth estimate grid:

$$\ell_{dm} = \cos^2 \left[\frac{\pi}{2} \frac{(x_m - x_d)}{L_x} \right] \cos^2 \left[\frac{\pi}{2} \frac{(y_m - y_d)}{L_y} \right]. \tag{16.32}$$

The wavenumber estimate quality control metrics described in Section 16.3.6, including the eigenvector dominance $\overline{\lambda}_{mb}$, fit skill, s_{mb}, and direct uncertainty estimates, $\sigma_{\hat{k},mb}$, may be included in w_{dmb} in order to weigh more heavily those $\{\omega_{mb}, \hat{\vec{k}}_{mb}\}$ in which there is greater confidence. The product $w_{dmb} = \ell_{dm}\overline{\lambda}_{mb}s_{mb}$ was used in [3–5], and [16] opted for $w_{dmb} = \ell_{dm}\overline{\lambda}_{mb}s_{mb}/\sigma_{\hat{k},mb}$. Prior knowledge of the environment and reasonable assumptions can aid in the choice of model derived from the dispersion relation (16.1). Below are the two approaches that have previously been taken.

16.4.1.1   Depth only

The simplest and most common method of estimating depth from this information is to model only depth-dispersive effects on the wave field, neglecting modulations due to currents. The effect of currents on the observed wave field is small in environments where either the ambient flow is slow relative to the phase speed of the incident wave ($u \ll \omega/k$) or the propagation direction of a narrowly spread wave field is oriented roughly perpendicular to the ambient flow direction ($\vec{u} \perp \vec{k}$). Assuming then that $\vec{u} \cdot \vec{k} = 0$ in (16.1), this model has the familiar form

$$\omega^2 = gk \tanh\left(k\tilde{h}\right). \tag{16.33}$$

In defining the cost function, it is natural to directly compare the set of estimated or "observed" wavenumbers $\hat{k}_{mb}$ to those modeled. Although (16.33) cannot be solved explicitly for k, iterative (e.g., Newton-Raphson) or power series approximations are alternatives. Hunt's power series approximation of $(k\tilde{h})^2$ is adequately accurate and is significantly faster to compute than an iterative solver [6]:

$$k_H = k_H(\omega, \tilde{h}) = \frac{1}{\tilde{h}}\left[\left(k_0\tilde{h}\right)^2 + \frac{(k_0\tilde{h})}{1+\sum_{n=1}^{6}\xi_n(k_0\tilde{h})^n}\right]^{1/2}, \tag{16.34}$$

where $(k_0\tilde{h}) = \omega^2\tilde{h}/g$, and

$$\begin{aligned} \xi_1 &= 0.6666666666 \quad \xi_4 = 0.0632098765 \\ \xi_2 &= 0.3555555555 \quad \xi_5 = 0.0217540484 \\ \xi_3 &= 0.1608465608 \quad \xi_6 = 0.0065407983. \end{aligned} \tag{16.35}$$

The residual function can then be written as

$$R_{dmb} = \hat{k}_{mb} - k_H(\tilde{h}_d, \omega_{mb}) \tag{16.36}$$

where $\beta = \tilde{h}_d$ is the sole fit parameter.

16.4.1.2   Depth and currents

In some environments, such as coastal regions with strong tidal currents, ambient flows can significantly affect the radar-observed wave field and the Doppler term, $\vec{u} \cdot \vec{k}$, cannot be neglected. Equation (16.1) represents the lowest-order effect of currents on the wave field, which, after moving the Doppler term to the right-hand side is

$$\omega = \left[gk \tanh\left(k\tilde{h}\right)\right]^{1/2} + \vec{u} \cdot \vec{k}. \tag{16.37}$$

Note that the ambient current vector, $\vec{u}$, is assumed depth-uniform. Although the Doppler term consists of two-directional components

$$\vec{u} \cdot \vec{k} = u_x k_x + u_y k_y, \tag{16.38}$$

the wavenumber estimates are often narrowly spread about a dominant direction that may not be aligned with the x- or the y-axis. This narrow spread is primarily due to the single wavenumber vector estimated at each frequency, which, again, is the cost associated with directly modeling the wave phase with a single plane wave. This narrow spread is demonstrated clearly in Figure 16.5, in which a set of phase-based wavenumber estimates (white circles) is overlain on the dispersion surface (16.37). It is useful to define a rotated coordinate system local to each depth estimate location:

$$(\vec{e}_{\|d}, \vec{e}_{\perp d}) = \left[\vec{e}_{\|}(x_d, y_d), \vec{e}_{\perp}(x_d, y_d)\right] \tag{16.39}$$

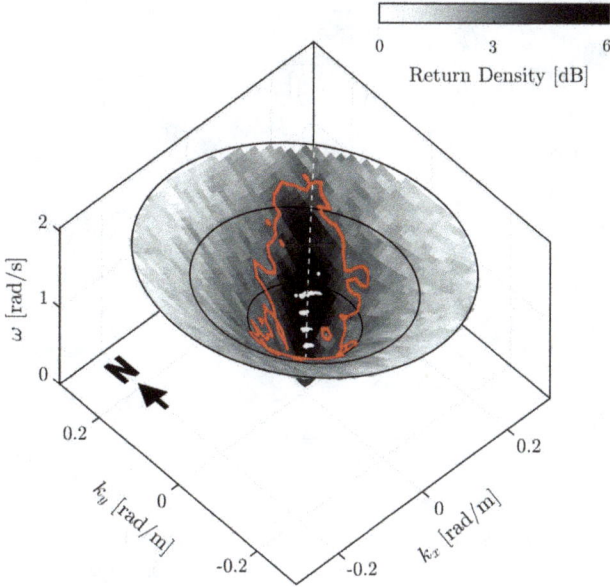

Figure 16.5 Phase-based $\{\vec{k}_{bm}, \omega_{bm}\}$ estimates, and 3D-FFT variance density mapped onto the theoretical dispersion surface (16.37) with known depth and depth-average current velocity. Phase-based estimates (white circles) span a narrower range of wavenumber directions and frequencies than the set of variance density spectral cells with SNR>2 (area encompassed by red contour) but are derived from a region 1/25th the area (and weighted most heavily over a region 1/100th the area) of the 3D-FFT tile. [Reprinted from [5], c.f. figure 8].

in which $\vec{e}_\parallel$ is oriented in the direction of the skill-weighted mean wavenumber,

$$\vec{e}_{\parallel d} = \frac{\sum_{m,b} \ell_{dm} s_{mb} \left(\hat{\vec{k}}_{mb} / \hat{k}_{mb} \right)}{\sum_{m,b} \ell_{dm} s_{mb}} \tag{16.40}$$

and $\vec{e}_\perp$ is orthogonal to $\vec{e}_\parallel$. The ℓ_{dm} weights (16.32) localize the fit to the dth analysis location and s_{mb} weights (16.27) drive the mean toward high-skill wavenumber estimates. The Doppler term can then be restated in the local, rotated coordinate system:

$$\vec{u} \cdot \vec{k} = u_\parallel k_\parallel + u_\perp k_\perp \tag{16.41}$$

where

$$\begin{aligned}
k_\parallel &= \left(\vec{e}_\parallel \cdot \vec{e}_x \right) k_x + \left(\vec{e}_\parallel \cdot \vec{e}_y \right) k_y, & u_\parallel &= \left(\vec{e}_\parallel \cdot \vec{e}_x \right) u_x + \left(\vec{e}_\parallel \cdot \vec{e}_y \right) u_y \\
k_\perp &= \left(\vec{e}_\perp \cdot \vec{e}_x \right) k_x + \left(\vec{e}_\perp \cdot \vec{e}_y \right) k_y, & u_\perp &= \left(\vec{e}_\perp \cdot \vec{e}_x \right) u_x + \left(\vec{e}_\perp \cdot \vec{e}_y \right) u_y.
\end{aligned} \tag{16.42}$$

The least squares residual function may then be formulated as

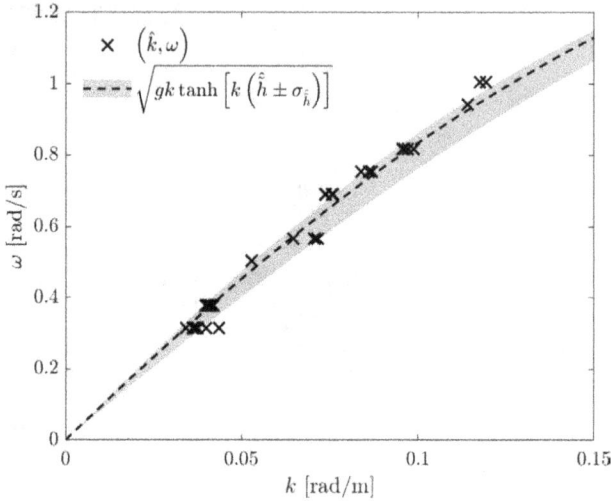

Figure 16.6   Least squares fit (dashed line and shading, $\hat{\tilde{h}} \pm \sigma_{\hat{\tilde{h}}} = 8.8 \pm 0.5$ m) of the depth-only dispersion curve (16.33) from a set of wavenumber–frequency pairs, $(\hat{k}, \omega)$

$$R_{dmb} = \omega_{mb} - \left[g\hat{k}_{mb} \tanh\left(\hat{k}_{mb}\tilde{h}_d\right) \right]^{1/2} - u_{\|d}\,\hat{k}_{\|mb}u_{\perp d}\,\hat{k}_{\perp mb}, \tag{16.43}$$

where the set of fit parameters is

$$\beta_d = \left(\tilde{h}_d, u_{\|d}, u_{\perp d}\right). \tag{16.44}$$

By casting the problem in a local coordinate system, the resulting current estimate, $\hat{\vec{u}}_d$, is naturally expressed in terms of a better-constrained component, $\hat{u}_{\|d}$, and a more loosely constrained component, $\hat{u}_{\perp d}$.

Following the same Levenberg-Marquardt algorithm (16.23) and confidence interval estimate procedure (16.28) as in each wavenumber fit, initial depth (and current vector) estimates and associated uncertainties are produced:

$$\hat{\tilde{h}}(x_d, y_d) \pm \sigma_{\hat{\tilde{h}}}(x_d, y_d) \tag{16.45}$$

neglecting the Doppler term (16.33), or

$$\begin{aligned}
\hat{\tilde{h}}(x_d, y_d) &\pm \sigma_{\hat{\tilde{h}}}(x_d, y_d) \\
\hat{u}_{\|}(x_d, y_d) &\pm \sigma_{\hat{u}_{\|}}(x_d, y_d) \\
\hat{u}_{\perp}(x_d, y_d) &\pm \sigma_{\hat{u}_{\perp}}(x_d, y_d)
\end{aligned} \tag{16.46}$$

including the Doppler term (16.37).

Figure 16.6 shows an example fit of the dispersion relation, modeling depth only (16.33) using the residual function (16.36). Some spread in the wavenumber estimates contributes to depth estimate uncertainty about the estimated depth. An analogous fit of (16.37) using the residual function (16.43) is a 3-D surface similar to that shown in Figure 16.5.

16.4.2 Remove temporal water level trends

The instantaneous water column thickness at each estimate location, $\tilde{h}(x_d, y_d)$, is composed of a water surface elevation, $\eta(t)$, a bathymetric component, $h(x_d, y_d)$, as

$$\tilde{h}(t) = h + \eta(t), \tag{16.47}$$

where $\eta(t)$ is associated with hydrodynamic variability, including astronomical tides, storm surge, wave-driven setup, or even large changes in river discharge. It is important to separate these two components of total depth because responsible public dissemination of bathymetry must be referenced to a common vertical datum and because multiple estimates in time may be combined to improve estimate precision.

In locations where tide gauge information is available, either the nearest or nearest two gauges can typically provide sufficient information about both the more easily predicted tidal component as well as the more challenging episodic components of η. Elsewhere, a tidal model such as the Oregon State University TPXO9 [20] may be a viable alternative. If none of these options are available or sufficiently accurate, deviation from an estimate of the local mean sea level may be calculated using a multi-tidal record of radar-derived individual depth estimates, $\tilde{h}_d(t)$ [16].

An example of a real depth estimate time series at a single location is provided in Figure 16.7. The initial depth estimates, derived from the least squares fit ($\hat{\tilde{h}}$; 16.31), exhibit substantial variability in time, including an apparent semidiurnal periodicity. Subtraction of local tidal harmonics (referenced to NAVD88) from this time series significantly decreases the temporal variability. And, since bottom morphology, $\hat{h}$, typically varies more slowly than the hourly separation between depth estimates, temporal smoothing may be applied. This procedure is described in Section 16.5.

16.5 Temporal updates

Temporally and/or spatially local processes can drastically alter the quality of individual depth estimates, including natural factors (e.g., wind too weak or too strong, waves too small or highly nonlinear, rain cells) and anthropogenic factors (e.g., mechanical failures; vessel traffic). Due to the relatively small number of pixels entrained in each depth estimate, phase-based bathymetry retrieval can be particularly susceptible to some of these processes. Since morphology varies rapidly in time only in rare cases such as large storm events, estimate precision can be increased substantially by combining multiple independent estimates at each location. Previous demonstrations of phase-based bathymetry retrieval have utilized a few different approaches to filtering the estimates in time. Here, we let h_{di} represent the ith datum-referenced estimate at time t_i, but since each filter below is applied independently at each estimate location, $\vec{x}_d$, we again drop the subscript d for notational simplicity.

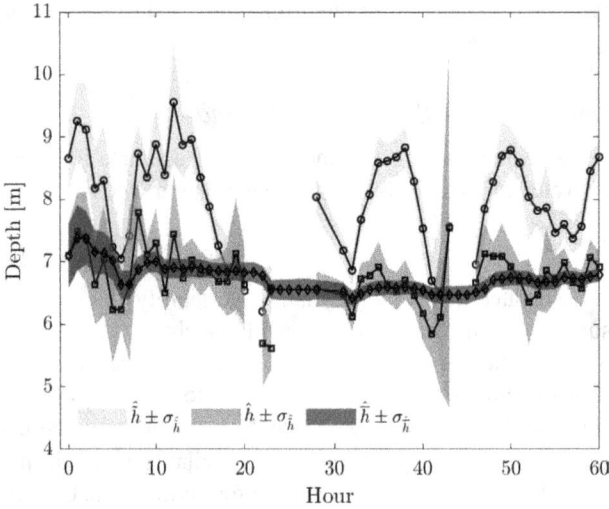

Figure 16.7   *The 60-day time series of estimated depth and estimated uncertainty at a single location. Results of the nonlinear least squares fit (light gray shading, circles) are corrected for tidal variations (medium gray shading, squares), and then smoothed in time with a Kalman filter (dark gray shading, diamonds).*

16.5.1 Kalman filter

A more complex but powerful filter is the Kalman filter, which is similar to an exponentially weighted moving average but also allows for morphological change. Depth estimates are updated in time as

$$\overline{h}_i = \overline{h}_{(i-1)} + K_{(i-1)}\left[h_{(i)} - \overline{h}_{(i-1)}\right], \tag{16.48}$$

where the Kalman gain at the ith update step, K_i,

$$K_{(i)} = \frac{(Q\Delta t)_{(i)} + \hat{\overline{\sigma}}^2_{h(i-1)}}{(Q\Delta t)_{(i)} + \hat{\overline{\sigma}}^2_{h(i-1)} + \hat{\sigma}^2_{h(i)}}. \tag{16.49}$$

The gain, which controls how much the ith estimate causes changes to the previous smoothed estimate, is updated using the estimated uncertainty variances of the current estimate, $\hat{\sigma}^2_{h(i)}$, the estimated uncertainty variance of the prior filtered estimate, $\hat{\overline{\sigma}}^2_{h(i-1)}$, and the time-integrated process error, $(Q\Delta t)_{(i)}$. The process error rate, Q, is included to account for natural morphological change, which varies in space and in time. A form for Q at an open, barred beach (Duck, North Carolina, USA) was proposed in [3] to be a Gaussian curve as a function of cross-shore distance x, with a maximum located at the bar crest, x_c,

$$Q(x,t) = C_Q H^2_{m0}(t)e^{-[(x-x_c)/\sigma_x]^2}, \tag{16.50}$$

where the magnitude varies with wave height, H_{m0}, and a constant, $C_Q = 0.067$ m²/day, that was calibrated using a 39-day record of almost daily bathymetry surveys. The spread of the Gaussian curve about the bar crest, σ_x, was chosen to be 100 m based on previously observed interannual variability of bar position. Although the extensive prior bathymetry observations and ongoing measurements of bulk wave height that informed this Q are rarely enjoyed at other locations, it is reasonable to assume that morphological activity is greater during periods of more energetic sea states and in shallower regions more prone to wave breaking. Rough estimates of wave breaking location [21] and incident bulk wave height may be estimated from just the marine radar image time series. The calibrated C_Q may be used as a first guess at locations with similar morphology and a similar wave climate as the calibration case. A warning is given in [11], however, that no statistically significant relationships were detected between the [3] formulation of Q and yearlong, continuous observations of bathymetric change at three altimeter (in situ point sensor) locations spanning the sand bar at the same site. Nevertheless, the Kalman filter is a temporal updating scheme option that, like the simple moving average, has an adjustable filter timescale.

Figure 16.7 additionally shows the real depth estimate time series with a Kalman filter applied. The filtered time series, $\hat{\bar{h}}$, initially tracks the datum-referenced (tide-corrected) time series, $\hat{h}$, and then converges to a more slowly evolving estimate. Over a similar timescale, the estimated uncertainty, $\sigma_{\hat{\bar{h}}}$ decreases as more estimates are entrained in the filter, eventually converging to a dynamic equilibrium between update uncertainty and the process error. The filter also elegantly handles missing data (e.g., hours 24–27), slowly increasing uncertainty about the previous estimate until a new update is available.

16.5.2 Moving average

A simpler but easily understood filter is the uncertainty-weighted moving average:

$$\bar{h}_i = \frac{\sum_{i-N_\tau}^i \sigma_{h_i}^{-2} h_i}{\sum_{i-N_\tau}^i \sigma_{h_i}^{-2}}, \tag{16.51}$$

where N_τ is the number of averaged estimates, spanning an appropriate temporal window. This window size may vary, depending on the location and presence of periodic processes (e.g., unmodeled tidal effects) and episodic events. While a few hours may be sufficient in environments with high SNR and randomly occurring adverse events, more challenging locations may call for a week or more. For instance, concurrent inversion of both components of velocity along with depth using a set number of $\{\vec{k}, \omega\}$ pairs triples the number of free parameters, necessarily reducing precision on depth (while typically gaining accuracy).

16.6   Revisit current estimation

A significant hindrance to the quality of current vector estimates, $\hat{\vec{u}}$, is the lack of constraint on depth, a co-estimated parameter in (16.37). Temporal averaging of consecutive current estimates is often not appropriate because current fields can vary rapidly in coastal environments. However, one option is to revisit the current estimation procedure after a tighter constraint on depth is achieved (e.g., via a Kalman filter). In other words, a more precise set of current vector estimates may be generated by treating each filtered depth estimate as a "known" parameter in the dispersion relation, (16.37). The residual function to be fit via least squares (16.31) is then constructed as

$$R_{dmb} = \omega_{mb} - \left[\sqrt{g\hat{k}_{mb} \tanh\left[\hat{k}_{mb} \left(\hat{\bar{h}}_d + \eta \right) \right]} + \hat{u}_{\|d}\, \hat{k}_{\|mb} + \hat{u}_{\perp d}\, \hat{k}_{\perp mb} \right], \qquad (16.52)$$

where the set of fit parameters is

$$\beta_d = \{u_{\|d}, u_{\perp d}\}. \qquad (16.53)$$

The (tidal) water surface elevation time series, η, is explicitly included here, as it was subtracted before $\hat{\bar{h}}$ was calculated from consecutive datum-referenced estimates. It is important to note that accuracy of these updated current vector estimates hinges on the accuracy of the corresponding depth estimate.

16.7   Performance

The most direct evaluations of radar-derived, phase-based estimates of depth and currents were carried out in [4, 5], wherein depth estimates were compared against bathymetry surveys conducted during the period of radar observations and current estimates were compared against concurrent flow velocity measurements. Application of the cBathy algorithm at two open beaches resulted in a bulk, domain-wide root mean squared error (RMSE) of about 0.4 to 0.5 m with a shallow bias of 0.2 to 0.4 m. In the example case at Benson Beach, Washington, shown in Figure 16.1, the bulk RMSE is 0.35 m and the (shallow) bias is 0.11 m. These bulk metrics are similar to those derived from equivalent phase-based estimates using optical video instead of radar. Comparison of concurrent radar- and optical video-derived estimates indicated better performance from the optical video in the shallow surf zone, where pixel resolution of the X-band radar marginally resolved dominant wavelengths.

   Both [16] and [5] took phase-based depth estimation to inlet environments with significant tidal current effects. The authors of [16] employed frequency-dependent localization length scales, estimated tidal elevation using only the radar observations, produced final depth estimates using a running one-week temporal average, and achieved bulk RMSE of about 1.1 m and essentially zero bias. In the example case shown in Figure 16.2, [5] used constant localization length scales of about the same size as the shortest used in [16], nearby water surface elevation observations,

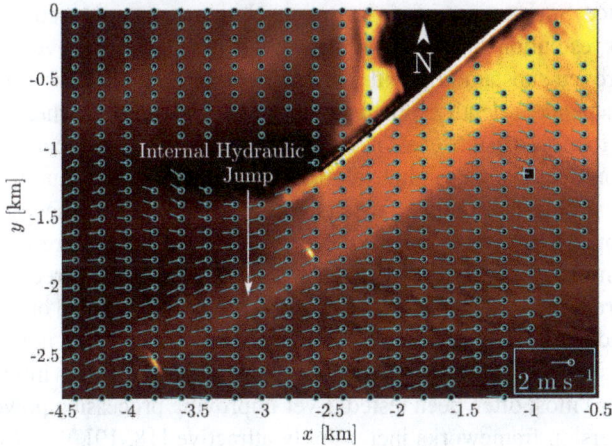

Figure 16.8   Estimated (circle markers) and measured (square marker at (x, y) = (−1 km, −1.2 km)) current vectors overlain on a time-averaged radar backscatter image at the Columbia River Mouth, Washington, USA, during westward ebb flow. An internal hydraulic jump, associated with a sharp deflection of the ebbing flow, occurs along the linear region of increased backscatter running parallel to the North Jetty [22]. [Reproduced from [5], c.f. figure 11].

and a Kalman filter over a tidal running average, and achieved bulk RMSE of 0.35 m and a (shallow) bias of 0.02 m. Honegger *et al.* [5] took the estimate procedure to a deep estuary with particularly strong currents, and after concurrently estimating currents with depth (instead of tidal averaging) and applying a Kalman filter, produced a bathymetric estimate with about 0.9 m RMSE and 0.4 m bias. The performance of retrieved currents at the shallow and deep inlets by comparing time series at point sensor locations was also assessed in [5]. As expected, only the current velocity component in the mean wavenumber direction was found to have reasonable accuracy, with an RMSE of 0.3 m/s in the presence of about 2 m/s amplitude tidal currents. Additional spatial smoothing was required to bring the off-mean velocity component to a level that could produce a reasonable current vector map (e.g., Figure 16.8).

16.8 Summary and future work

In many coastal environments, bathymetric structures (and even current fields) can cause rapid wave field transformations at length scales that cannot be adequately resolved by finding the wavenumbers associated with 3D-FFT variance density maxima (which is about 10 ocean wavelengths). For simple wave fields with one dominant direction at each frequency, spatial wavenumber resolution can be brought down to the order of 1 wavelength by calculating the rate at which wave phase moves in space. Such an increase in spatial resolution comes at the cost of directional wavenumber resolution and, to a degree, estimate precision. However, this precision may be improved substantially without sacrificing spatial resolution by combining multiple, independent depth estimates over time. Application of these methods has accordingly focused on retrieval of slowly varying depth, with currents estimated only as a byproduct. Depth estimate algorithms that extract wavenumber from wave phase structure have been applied extensively to optical video observations, including a free, open-source package (CIRN: www.coastal-imaging-research-network. github.io). Importantly, optical video applications have provided predictive value for rip current predictions [23]. Although the method has been applied only to a handful of marine radar datasets, reported performance evaluations have nonetheless been promising, especially given the dynamically challenging inlet environments where the method has most often been tested. Ever improving processing power is making complex inversion frameworks increasingly attractive [18, 19], thereby suggesting that optimization efforts might best be directed toward the wavenumber estimation step. For instance, it is argued in [14] that more directional information from the wave field may be accessible by estimating local wavenumbers on a series of wave phase subgrids that are each generated by directionally bandpassing the wave signal on a grid with coarser, traditional 3D-FFT spatial resolution.

References

[1]   Piotrowski C.C., Dugan J.P. 'Accuracy of bathymetry and current retrievals from airborne optical time-series imaging of shoaling waves'. *IEEE Transactions on Geoscience and Remote Sensing.* 2002;**40**(12):2606–18.

[2]   Plant N.G., Holland K.T., Haller M.C. 'Ocean wavenumber estimation from wave-resolving time series imagery'. *IEEE Transactions on Geoscience and Remote Sensing.* 2008;**46**(9):2644–58.

[3]   Holman R., Plant N., Holland T. 'cBathy: a robust algorithm for estimating nearshore bathymetry'. *Journal of Geophysical Research: Oceans.* 2013;**118**(5):2595–609.

[4]   Honegger D.A., Haller M.C., Holman R.A. 'High-resolution bathymetry estimates via X-band marine radar: 1. beaches'. *Coastal Engineering.* 2019;**149**(4):39–48.

[5]   Honegger D.A., Haller M.C., Holman R.A. 'High-resolution bathymetry estimates via X-band marine radar: 2. Effects of currents at tidal inlets'. *Coastal Engineering.* 2020;**156**(7):103626.

[6]   Hunt J.N. 'Direct solution of wave dispersion equation'. *Journal of the Waterway, Port, Coastal and Ocean Division.* 1979;**105**(4):457–9.

[7]   Fuchs R. 'Depth estimation on beaches by wave velocity methods'. *Institute of Engineering Research Wave Research Laboratory.* 1953.

[8]   Williams W.W. 'The determination of gradients on enemy-held beaches'. *The Geographical Journal.* 1947;**109**(1/3):76–90.

[9]   Stockdon H.F., Holman R.A. 'Estimation of wave phase speed and nearshore bathymetry from video imagery'. *Journal of Geophysical Research: Oceans (1978–2012)*. 2000;**105**(C9):22015–33.

[10]   Bergsma E.W.J., Conley D.C., Davidson M.A., *et al.* Video-based nearshore bathymetry estimation in macro-tidal environments'. *Marine Geology*. 2016;**374**:31–41.

[11]   Brodie K.L., Palmsten M.L., Hesser T.J., *et al.* Evaluation of video-based linear depth inversion performance and applications using altimeters and hydrographic surveys in a wide range of environmental conditions'. *Coastal Engineering*. 2018;**136**:147–60.

[12]   Bell P.S. 'Shallow water bathymetry derived from an analysis of X-band marine radar images of waves'. *Coastal Engineering*. 1999;**37**(3):513–27.

[13]   Hessner K., Reichert K., Rosenthal W. 'Mapping of sea bottom topography in shallow seas by using a nautical radar'. 2nd Symposium on Operationalization of Remote Sensing; 1999.

[14]   Senet C.M., Seemann J., Flampouris S., *et al.* Determination of bathymetric and current maps by the method disc based on the analysis of nautical X-band radar image sequences of the sea surface'. *IEEE Transactions on Geoscience and Remote Sensing*. 2008;**46**(8):2267–79.

[15]   Flampouris S., Ziemer F., Seemann J. 'Accuracy of bathymetric assessment by locally analyzing radar ocean wave imagery (February 2008)'. *IEEE Transactions on Geoscience and Remote Sensing*. 2008;**46**(10):2906–13.

[16]   Zuckerman S., Anderson S. 'Bathymetry and water-level estimation using X-band radar at a tidal inlet'. *Journal of Coastal Research*. 2018;**345**(5):1227–35.

[17]   Wilson G.W., Özkan-Haller H.T., Holman R.A. 'Data assimilation and bathymetric inversion in a two-dimensional horizontal surf zone model'. *Journal of geophysical research*. 2010;**115**:C12–12057.

[18]   Wilson G.W., Özkan-Haller H.T., Holman R.A., Haller M.C., Honegger D.A., Chickadel C.C. 'Surf zone bathymetry and circulation predictions via data assimilation of remote sensing observations'. *Journal of Geophysical Research: Oceans*. 2014;**119**(3):1993–2016.

[19]   Moghimi S., Özkan-Haller H.T., Wilson G.W., *et al.* Data assimilation for bathymetry estimation at a tidal inlet'. *Journal of Atmospheric and Oceanic Technology*. 2016;**33**(10):2145–63.

[20]   Egbert G.D., Erofeeva S.Y. 'Efficient inverse modeling of barotropic ocean tides'. *Journal of Atmospheric and Oceanic Technology*. 2002;**19**(2):183–204.

[21]   Díaz Méndez G.M., Haller M.C., Raubenheimer B., *et al.* Radar remote sensing estimates of waves and wave forcing at a tidal inlet'. *Journal of Atmospheric and Oceanic Technology*. 2015;**32**(4):842–54.

[22]   Honegger D.A., Haller M.C., Geyer W.R., *et al.* Oblique internal hydraulic jumps at a stratified estuary mouth'. *Journal of Physical Oceanography*. 2017;**47**(1):85–100.

[23]   Radermacher M., de Schipper M.A., Reniers A.J.H.M. 'Sensitivity of RIP current forecasts to errors in remotely-sensed bathymetry'. *Coastal Engineering*. 2018;**135**:66–76.

Chapter 17

Wind parameter measurement using X-band marine radar images

Xinwei Chen[1], Weimin Huang[1], and Björn Lund[2]

Ocean wind parameter measurement is important for the safety and efficiency of various on- and off-shore activities, such as coastal construction, ship voyages, and marine resource development [1]. It also facilitates the study of many ocean and atmospheric processes. Traditionally, wind information can be obtained through in situ sensors (e.g., anemometers) and spaceborne/airborne scatterometers. However, in situ wind measurements from towers, ships, and buoys are often susceptible to blockage effects and turbulence caused by the sensor platform, while spaceborne/ airborne wind measurements have coarse spatial resolution. In the past two decades, X-band marine radars have been exploited to retrieve wind parameters due to their high spatial and temporal resolution. Also, in contrast to in situ sensors, the retrieval of wind information from the radar backscatter intensity of the ocean surface is not affected by the sensor's installation height and motion [2].

It is well known that the radar backscatter from the ocean surface is mainly caused by the small-scale roughness ($\sim$ 3 cm) on the sea surface, which is mostly dependent on the local wind speed and wind direction [3, 4]. The roughness is in turn modulated by long surface gravity waves, and then surface waves can be imaged [5]. For an X-band marine radar operating at grazing incidence with horizontal-transmit-horizontal-receive (HH) polarization, it has been found that its backscatter only exhibits one peak, which lies in the upwind direction. Thus, wind direction can be derived from the dependence of radar image pixel intensities on the relative azimuth between antenna look direction and wind direction. As for the wind speed, it can be retrieved by establishing an empirical model relating wind speed to infor-mation extracted from image pixels and other parameters (e.g., air-sea temperature differences). In this chapter, multiple types of methods concerning wind direction and wind speed estimation will be introduced and compared with each other.

[1]Faculty of Engineering and Applied Science, Memorial University of Newfoundland, St. John's, NL, Canada

[2]Rosenstiel School of Marine and Atmospheric Science, University of Miami, Miami, FL, USA

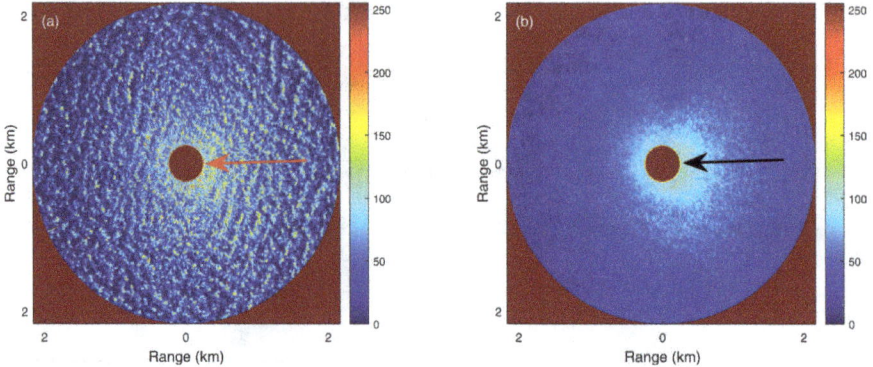

Figure 17.1   *(a) An X-band marine radar image with surface wave signatures.*
The arrow indicates the direction from which the wind blows,
measured by an anemometer. (b) The temporally integrated
image of 32 consecutive radar images including (a). The arrow
indicates the direction from which the wind blows, measured by an
anemometer. It can be observed that wave signatures are
filtered out.

17.1   Wind streaks/wind gusts-based methods

17.1.1   Local-gradient-based method

The presence of wind-induced streaks can be observed in temporally integrated
X-band marine radar images [6]. Based on the observation that those streaks are
aligned in local wind direction with a typical spacing of the order of 100 m, Dankert
et al. [6–8] proposed a local-gradient-based method (LGM) for wind direction estimation. The procedures of the proposed method are illustrated below.

- **Temporal integration**: Due to the presence of signatures with higher variability in time, e.g., surface waves, wind streaks are mostly concealed and hard
  to observe in raw marine radar images such as in Figure 17.1(a). As shown in
  Figure 17.1(b), those signatures can be filtered out by integrating a radar image
  sequence over time.
- **Resolution reduction and gradient calculation**: After temporal integration,
  the radar image is subjected to a resolution reduction process, in which the
  image is smoothed using a Gaussian average and downsampled for multiple
  times. The goal of resolution reduction is to obtain the local gradients, which
  are computed with the optimized Sobel operators proposed in the paper by
  Koch [9].
- **Ambiguity removal**: While the orientation of wind-induced streaks is normal
  to local gradients, a 180° ambiguity still exists. Thus, the last step is to remove
  the 180° ambiguity by employing one of the following techniques. The first
  technique is based on the cross-correlation function (CCF) of two temporally

integrated radar image sub-sequences (typically 24 images). The respective propagation direction of wind gusts can be indicated by the location of the CCF peak, thus removing the 180° ambiguity. The second technique is based on the Fourier spectrum of the CCF. The movements of different harmonic waves can be derived from the phases of the complex-valued spectrum. Since the resulting motion directions of the harmonic waves are always within 90° of the down-wind direction, the 180° ambiguity can be resolved.

- **Validation**: In the studies by Dankert *et al.* [6–8], the proposed method was tested using two sets of data. The first dataset was collected from an X-band marine radar installed on Platform "2/4k" of the Ekofisk oil field in the central North Sea from February to June 2001, with wind speeds up to 18 m/s. The second dataset was obtained from August 2003 until November 2004 using a Furuno radar mounted aboard the research platform FINO-I, which is located at German Bight of the southern North Sea, with wind speeds up to 16 m/s. The comparison between in situ and radar-derived wind directions showed a correlation coefficient of 0.99 and a root mean square difference (RMSD) of 14.2° (the first dataset) and 12.8° (the second dataset).

17.1.2 Optical flow-based method for wind vector retrieval

Dankert *et al.* [10] observed that the movement of wind gusts could be extracted from X-band marine radar sequences and utilized for wind vector retrieval. Similar to the LGM, the radar image sequence is first integrated and then downsampled using a Gaussian pyramid. Given an integrated and downsampled radar image $G(\vec{\gamma}, t)$, with $\vec{\gamma} = (x, y)$ and t being space and time coordinates, it was assumed that the values of $G(\vec{\gamma}, t)$ only changed due to an optical flow, i.e., the movement of wind gusts. Thus, the total derivative of $G(\vec{\gamma}, t)$ should be zero, i.e.,

$$(\nabla G)^T f + G_t = 0, \tag{17.1}$$

where ∇G and G_t are the spatial gradient and time derivatives of $G(\vec{\gamma}, t)$, respectively; $f = [f_x, f_y]^T$ denotes the optical flow, i.e., the local wind vector. Compared with the LGM, both wind direction and wind speed can be obtained using the optical-flow-based method (OFM). Nevertheless, testing results obtained from radar data collected from the same platform in the study by Dankert *et al.* [6] indicate a significant deviation between in-situ measurements and radar-derived wind parameters (32.1° and 2.68 m/s of RMSDs in wind direction and wind speed, respectively).

17.2 Intensity-information and curve-fitting-based methods

Among various wind parameter estimation algorithms, a large portion of them follow two major steps. As shown in Figure 17.2, after image preprocessing, pixel intensity-based parameters (features) are extracted from X-band marine radar images. In the second step, those parameters are used to build empirical functions

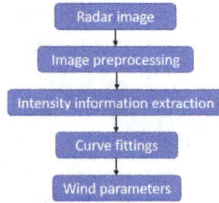

Figure 17.2　*General scheme of the methods using intensity information extraction algorithms and curve-fitting techniques*

for wind parameter retrieval using curve-fitting techniques. In this section, several typical algorithms following those two steps will be introduced.

17.2.1　*Single curve-fitting-based algorithm*

In order to accurately estimate wind parameters from a shipborne X-band marine radar, Lund *et al.* [2] developed a novel method using intensity information extraction and curve-fitting techniques.

- Wind direction estimation: Given a radar image in range-azimuth coordinates, the average intensity over range for each azimuthal direction is calculated and denoted as σ_θ. A cosine-squared function is used to relate azimuth θ to σ_θ, which can be expressed as

$$\sigma_\theta = a_0 + a_1 \cos^2(0.5(\theta - a_2)), \tag{17.2}$$

where a_0, a_1, and a_2 are parameters determined through least squares fitting. It should also be noted that σ_θs obtained from azimuths shadowed by ship structures were not used for curve fitting. An example of the curve-fitting result is presented in Figure 17.3. The estimated wind direction corresponds to the azimuth located at the peak of the fitted function.

- Wind speed estimation: Since the wind speed is dependent on the average backscatter intensity, the cosine-squared function is first integrated over the azimuth, which can be expressed as

$$\bar{\sigma} = \frac{1}{2\pi} \int_0^{2\pi} \left(a_0 + a_1 \cos^2(0.5(\theta - a_2))\right) d\Theta. \tag{17.3}$$

Then, as shown in Figure 17.4, a third-degree polynomial model between $\bar{\sigma}$ and wind speed (denoted as w_{spd}) is trained using images obtained under different wind speeds with the least squares method. The function can be expressed as

$$\bar{\sigma} = b_3 w_{spd}^3 + b_2 w_{spd}^2 + b_1 w_{spd} + b_0. \tag{17.4}$$

- **Validation:** Data used to train the polynomial function and validate the proposed method were collected from a standard Furuno FAR2117BB X-band marine radar installed on a ship travelling in the Philippine Sea near Taiwan

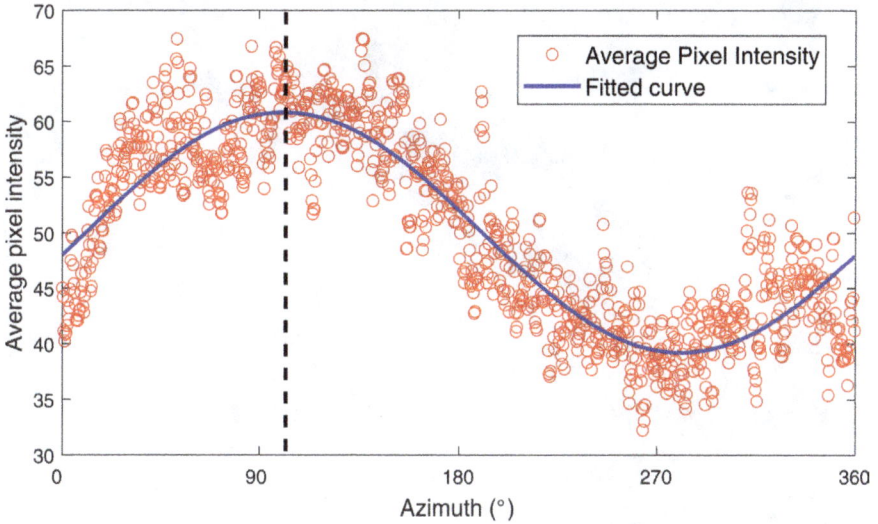

Figure 17.3   *Average pixel intensity as a function of azimuth for a horizontal-transmit-horizontal-receive (HH) polarized X-band marine radar image. The corresponding best-fit curve is shown in blue. The estimated wind direction corresponds to the azimuth located at the peak of the fitted function, which is indicated by the black-dashed line.*

island during two storms from August 6 to August 9 and from September 12 to September 17, 2010, with simultaneous wind speeds ranging from just above 0 m/s to almost 15 m/s. A parameter named zero pixel percentage (ZPP) was first extracted from the radar image and used as a threshold to determine the presence of rain in the radar image. Images with ZPPs lower than the threshold value were regarded as rain-contaminated and excluded from the study. The reference wind data were obtained from two anemometers mounted above the ship. The RMSDs found for the test dataset were 0.78 m/s for the wind speed and 17.4° for the wind direction.

17.2.2   Two-model curve fitting for rain mitigation

The presence of rain alters the relationship between wind speed and average backscatter intensity. A comparison between rain-free and rain-contaminated radar images is presented in Figure 17.5. In consequence, the third-degree polynomial function trained from rain-free data cannot be applied to rain-contaminated data. Thus, in order to mitigate the effect of rain on wind speed estimation, Huang and Gill [11] developed a two-model-based approach to estimate wind speed from rain-contaminated X-band marine radar images. To be more specific, after classifying images into rain-free and rain-contaminated types using ZPP, two third-degree polynomial functions were trained for two types of data, respectively. Shipborne marine Decca radar data collected during a sea trial off the east coast of Canada

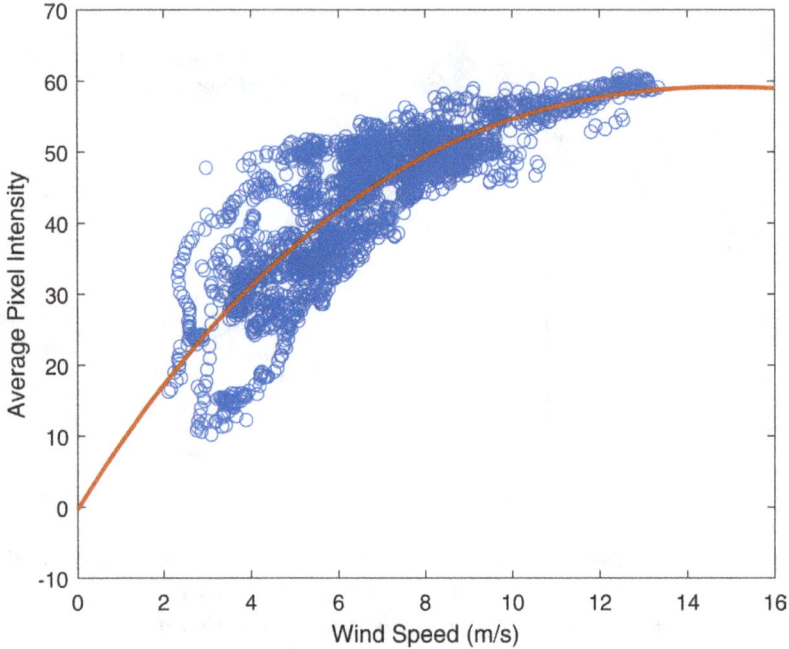

Figure 17.4   Scatter plot showing the wind speed from an anemometer, corresponding radar image average pixel intensity, and the best-fit curve based on a third-degree polynomial function

in late November 2008 were used to validate the proposed method. The experimental results showed that compared with the single model, the RMSD between radar-derived wind speeds and anemometer measurements can be reduced by about

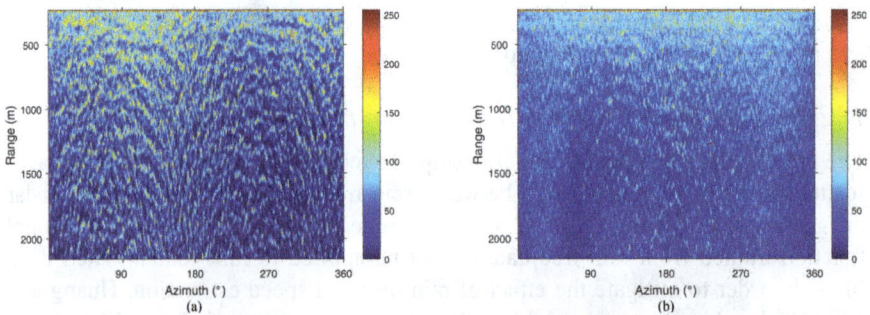

Figure 17.5   (a) A rain-free image in range-azimuth coordinates with a zero pixel percentage (ZPP) value of 14.2%. Wave signatures can be clearly observed. (b) A rain-contaminated image with a ZPP value of 0.5% and blurred wave signatures.

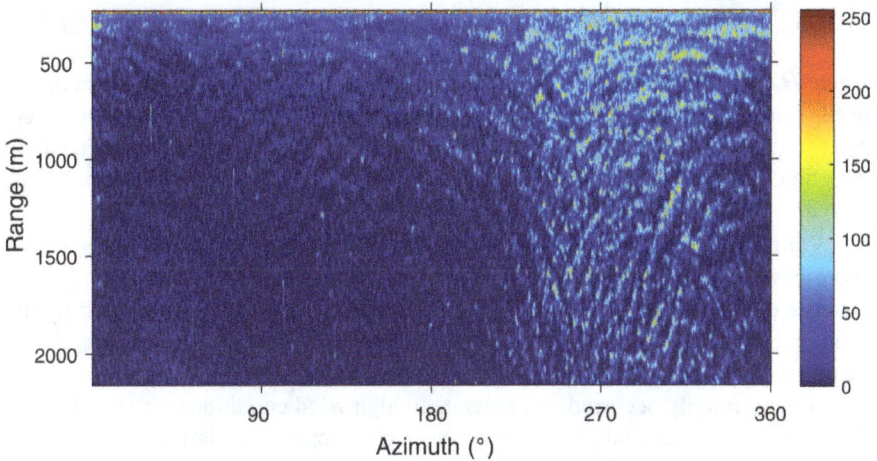

Figure 17.6   *A radar image obtained in low sea states with very low intensities in most azimuth directions*

1.2 m/s using two separate models for rain-free and rain-contaminated data, respectively.

17.2.3   Dual curve fitting for low sea state cases

Liu *et al.* [12] found that many radar images obtained in low sea states consisted of a large portion of pixels with very low intensities; an example of this is shown in Figure 17.6. As a result, wind directions estimated using the curve-fitting algorithm in the study by Lund *et al.* [2] were not accurate. Thus, a dual curve-fitting scheme was proposed. After obtaining the first estimated wind direction result (denoted as θ_0) using the cosine-squared function, the σ_θs that are located within 60° of θ_0 were selected for another cosine-squared curve fitting. The second- and final-estimated wind direction corresponds to the azimuth located at the peak of the second-fitted function.

The proposed algorithm was implemented using the radar image data described in Section 17.2.2. Results showed that compared with the single curve fitting, wind direction estimation using the dual-curve-fitting scheme reduced the RMSD by 8°.

17.2.4   Significant wave height incorporated curve fitting

The pixel intensities of the marine radar image are not only dependent on wind condition but also on sea states, especially in nearshore regions where swell appears frequently [5]. Therefore, Chen *et al.* [5] first used a linear function to fit between wind speed and the average pixel intensity of the radar image, and then considered the influence of sea states by incorporating significant wave height (SWH) into the wind speed estimation function, which can be expressed as

$$w_{spd} = (p_1\sigma_{avg} + p_2)\left[1 + \exp(-\frac{p_3}{H_s})\right], (p_3 > 0), \tag{17.5}$$

where H_s is the SWH retrieved from radar image sequences, p_1 and p_2 are parameters in the linear function determined by the least squares method, and p_3 is an under-determined coefficient. As for the wind direction, a sine function and a term called direction deviation were used to fit between σ_θ (the average intensity over range for each azimuthal direction) and wind direction.

Validation: Datasets used to test the proposed function were collected by a shore-based Furuno X-band marine radar located on Haitan Island, China, from October to December 2010. During the experiment, wind speeds measured by the anemometer ranged from 0 m/s to 20 m/s. An RMSD of 1.37 m/s in wind speed esti-mation was obtained using the proposed function. It was also found that the influence of sea states mainly occurred on moderate to high wind conditions. For wind direc-tion, since the directional resolution of the anemometer was relatively low (22.5°), the radar-derived wind directions were not very accurate, with an RMSD of 26°.

17.2.5 Intensity level selection algorithms

Vicen-Bueno *et al.* [13] proposed an intensity-level-selection (ILS)-based method for wind parameter estimation using X-band marine radar image sequences. The procedures of the method are introduced below.

- **Range vector generation**: First, a set of intensity levels L_is are determined based on the range of radar image pixel intensities. Each radar image sequence in range-azimuth coordinates is integrated over time. Then, in order to reduce the noise in the range direction, the temporally integrated image is smoothed in the range direction using an averaging technique. The range distance vector $r_i(\theta)$ for each L_i is computed, which consists of the maximum range distance where the intensity is equal to or greater than L_i in every azimuth direction. $r_i(\theta)$ is then smoothed in the azimuthal direction by conducting a 5° moving average. Figure 17.7 shows the smoothed $r_i(\theta)$ (indicated by the red curve) for $L_i = 40$ in a temporally integrated radar image.
- **Wind direction estimation**: For the kth-integrated and smoothed image, its lowest L_i (denoted as $L_i(k)$) with all its associated smoothed maximum ranges being greater than an inner distance boundary (determined based on the range coverage of the radar) is designated as the selected intensity level. As shown in Figure 17.7, the estimated wind direction corresponds to the azimuth angle with the maximum value in the selected $r_i(\theta)$. After applying the above procedures to the first 16 integrated and smoothed images, rather than all the L_is, only the selected intensity level from the previous image (i.e., $L_i(k - 1)$) and its upper- and lower-defined intensity levels (i.e., $L_{i-1}(k - 1)$ and $L_{i+1}(k - 1)$) are used for generating the range distance vectors, which makes the algorithm more efficient.
- **Wind speed estimation**: Wind speed (denoted as w_{spd}) can be determined by relating to the maximum range of the smoothed $r_i(\theta)$ (i.e., $\max\{r_i(\theta)\}$), which can be expressed as

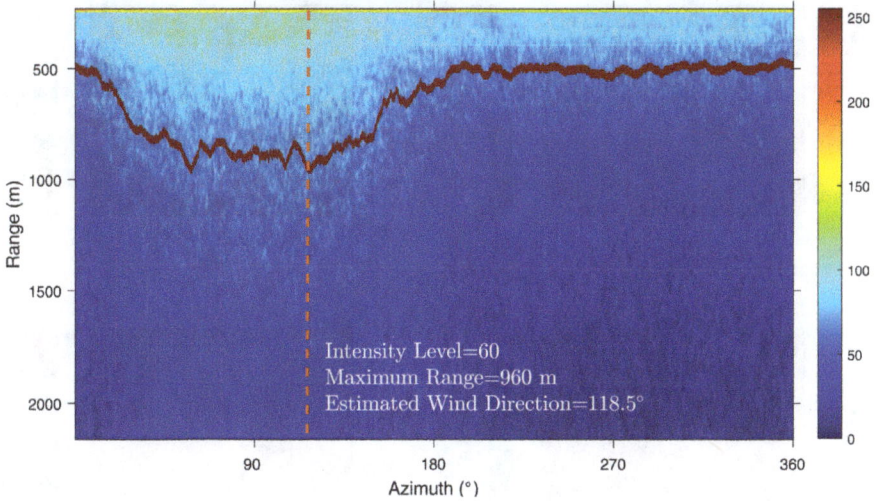

Figure 17.7   An example of a temporally integrated radar image. The selected intensity level is 40. The red line corresponds to the smoothed maximum range distance in each azimuth. The dashed line corresponds to the azimuth with the greatest distance, which is regarded as the radar-derived wind direction.

$$w_{spd} = \alpha_i \times \max\{r_i(\theta)\}, \tag{17.6}$$

where α_i refers to a conversion rate and its value can be obtained from the selected intensity level L_i by fitting a third-order polynomial continuous function. The function can be expressed as

$$\alpha = \beta_3 L^3 + \beta_2 L^2 + \beta_1 L + \beta_0, \tag{17.7}$$

where $\beta_0, ..., \beta_3$ are parameters that can be found using the total least squares algorithm.

- **Validation**: The proposed method was tested using image sequences collected from a Furuno 2117BB radar mounted 30 m above a research floating platform located approximately 30 km northwest of Bodega Bay, California. The experiment was conducted during four time periods in June 2010, while the wind speeds varied from 4 m/s to 22 m/s. The database was divided into the design dataset and the validation dataset. Image sequences in the design dataset were used to obtain the parameters in (17.7). The comparison between simultaneous anemometer data and radar-derived wind directions and wind speeds showed an RMSD of 14° and 0.6 m/s,

17.2.6   Modified ILS

Liu *et al.* [12] proposed a modified ILS-based method with higher robustness. In order to avoid the interference caused by an island in the far range, an outer distance

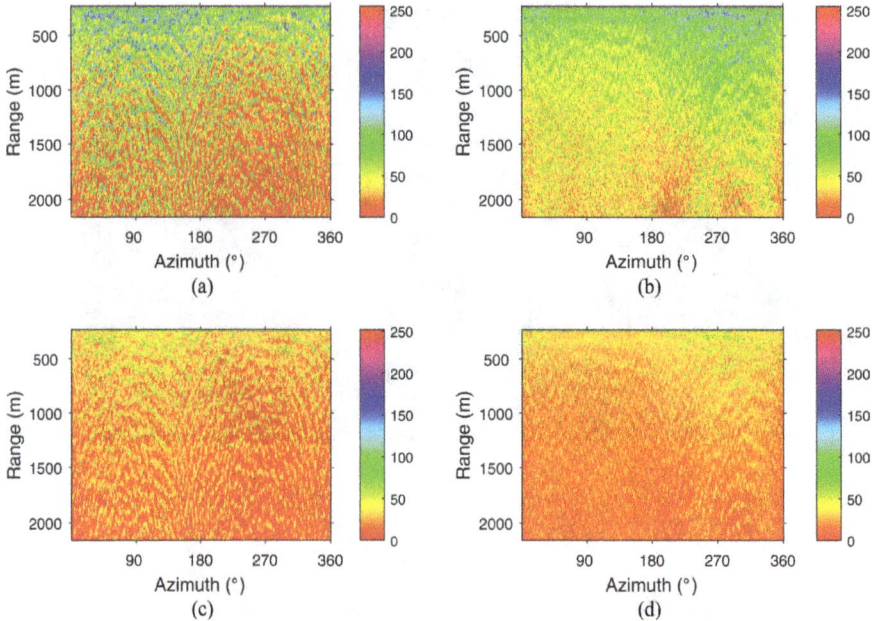

Figure 17.8   Examples of marine radar images and their texture maps (adapted from Huang et al. [14]): (a) radar image without rain contamination; (b) rain-contaminated radar image; (c) texture map of (a); (d) texture map of (b).

boundary is added to limit the selection range distance vectors. Also, a threshold-based method was introduced to detect and exclude the blocked areas in radar images. Since the range distance vectors do not contain elements in the azimuthal directions of the obstructed areas, the cosine-squared function in (17.2) was introduced to fill the gap by fitting the selected range distance vector as a function of azimuth. The azimuthal direction of the function's peak corresponds to the estimated upwind direction.

Shipborne radar data collected during a sea trial off the east coast of Canada in late October, late November, and early December 2008 were used to test the proposed method. It was found that compared to the original ILS algorithms, the wind direction estimation results were further improved using the modified ILS, with a reduction of RMSD by around 4.9°.

17.2.7 Texture analysis incorporated ILS

- **The influence of rain on image texture**: Huang *et al.* [14] observed that since rain may blur wave signatures, the texture of rain-contaminated echoes was different from that of less-contaminated or rain-free echoes. A comparison between a rain-free and rain-contaminated radar image and their texture maps is shown in Figure 17.8. Therefore, texture analysis was incorporated into the ILS algorithm in order to detect the presence of rain and measure wind parameters from rain-contaminated radar images with higher accuracy.

- **Texture-incorporated ILS**: The algorithm proceeds as follows. First, the texture map of each radar image in range-azimuth coordinates (denoted as T_{xy}) is calculated as

$$T_{xy} = \sqrt{\frac{1}{9} \sum_{i=-1}^{i=1} \sum_{j=-1}^{j=1} (I_{xy} - I_{x+i,y+j})^2}, \tag{17.8}$$

where I_{xy} represents the radar image intensity of pixel (x, y) in azimuth (x) and in range (y). Then, if the number of pixels with intensity higher than a threshold is less than a specified value (20 in the study by Huang *et al.* [14]) in a certain azimuth, such a direction will be regarded as rain-contaminated. The value of the threshold value is determined adaptively for each texture image of the rain-contaminated cases. The data in the rain-contaminated directions of the radar image will be removed from the wind-retrieval process. The ILS algorithm is then applied to the texture map to obtain wind direction. Since the original range vector $r_i(\theta)$ of the selected intensity level only contains elements from rain-free azimuths, the cosine-squared function in Equation (17.2) is used to fit the smoothed $r_i(\theta)$ to every azimuth. The azimuth with the greatest range distance value will be the estimated upwind direction. Then, Equation (17.6) and Equation (17.7) are utilized to estimate the wind speed.

- **Validation**: The shipborne marine radar data described in Section 17.2.2 were utilized to test the accuracy of the proposed method. Compared with the original ILS-based algorithm, the proposed method with texture analysis significantly improved the measurement accuracy, with a reduction of 14.5° and 1.3 m/s in RMSDs for wind direction and wind speed, respectively.

17.3 Transform-domain and curve fitting-based methods

Instead of directly extracting information from radar images in spatial and time domains, some methods exploit the intrinsic characteristics of radar images by conducting some sorts of transformation of radar images. After preprocessing, information extracted from transform domain analysis is fitted to empirical functions for wind parameter estimation. Here, three transform-domain-based methods will be introduced.

17.3.1 Spectral noise-based algorithm

Ocean wave-directional spectrum extracted from X-band marine radar sequences using three-dimensional Fast Fourier Transform (3D-FFT) has long been used for surface wave and current parameter measurement. In the study by Izquierdo *et al.* [15], after conducting 3D-FFT, spectral noise (denoted as $F_n(\vec{k}, \omega)$) was separated from the ocean wave signal and was used to infer wind speed through a linear equation expressed as

$$\sum_{\vec{k},\omega} F_n(\vec{k}, \omega) = q_1 + q_2 w_{spd}, \tag{17.9}$$

where $\sum_{\vec{k},\omega} F_n(\vec{k}, \omega)$ represents the summation of spectral noise in both wave number and frequency. The coefficients q_1 and q_2 can be obtained by a linear regression. Image data collected from an X-band radar system close to the Port of Sines in Portugal, from September 2000 to December 2001 were used to test the proposed scheme. However, as the number of data samples was relatively small and RMSD was relatively large (2.86 m/s), the method needs further improvement and validation.

17.3.2   Spectral integration-based algorithm

Huang and Wang [16, 17] proposed a novel spectral-integration-based algorithm, which is able to measure wind parameters from both rain-free and rain-contaminated images.

- **Rain-contaminated image detection**: It has been observed by Huang and Wang [16] that under high wind speeds, the presence of rain may not blur wave signatures because wind force dominates the generation of sea surface roughness. Therefore, two parameters, that is, ZPP proposed by Lund *et al.* [2] and high pixel percentage (HPP), were extracted from radar images and used as thresholds to classify images with rain into two cases: low wind speed rain and high wind speed rain. Cases of high wind speed rain can be treated as rain-free images in wind parameter estimation.
- **Wind direction estimation**: The spectral value of wave number k in the θth direction $E_\theta(k)$ is calculated as

$$E_\theta(k) = \sum_{n=0}^{N-1} I_\theta(n) e^{-j\frac{2\pi}{N} kn}, \tag{17.10}$$

where N is the total number of pixels in one azimuth, $I_\theta(n)$ is the nth pixel intensity in direction θ. For low wind speed rain cases, an integral of the normalized spectra over wave number range [0.01, 0.2] (denoted as S_θ) is calculated for each azimuth direction,

$$S_\theta = \int_{0.01}^{0.2} |E_\theta(k)| dk, \tag{17.11}$$

where S_θ is the curve fitted using the cosine-squared function in Equation (17.2). For cases of high wind speed rain and rain-free cases, the spectral values at the zero wave number, that is, $|E_\theta(0)|$, are calculated and used for curve fitting. Wind direction is determined as the peak of the fitted curve.

- **Wind speed estimation**: A spectral summation is first conducted on the spectral values of discrete wave number k_m, that is, $|E_\theta(k_m)|$, of all the azimuths, which is expressed as

$$S = \frac{1}{\Theta} \sum_{\theta=0}^{\Theta-1} \sum_{m=0}^{M} |E_\theta(k_m)|, \tag{17.12}$$

where Θ is the total number of azimuth directions in a radar image, $M = N/2$, and

$$k_m = \frac{2m\pi}{N\Delta r}, \tag{17.13}$$

with Δr being the radar range resolution. The relationship between S and wind speed w_{spd} can be expressed as

$$w_{spd} = e^{\frac{s-t_0}{t_1}} - t_2. \tag{17.14}$$

Therefore, a logarithmic function can be derived from Equation (17.14) to fit S and w_{spd} using least squares fitting techniques, expressed as

$$S = t_0 + t_1 \ln(w_{spd} + t_2), \tag{17.15}$$

where t_0, t_1, and t_2 can be determined through least squares fitting.
- **Validation**: Radar data introduced in Section 17.2.2 were used to train and validate the proposed function for wind parameter estimation. The RMSDs for wind direction and wind speed estimation using the testing data were 15.8° and 1.6 m/s, respectively. In addition, compared with the original curve-fitting-based method in Section 17.2.1, the proposed method significantly reduced the RMSDs of rain-contaminated radar-image-derived wind direction and wind speed estimation results by 25.1° and 5.9 m/s, respectively.

17.3.3   Ensemble empirical mode decomposition-based methods

Empirical mode decomposition (EMD), also known as Hilbert-Huang transform, was first proposed by Huang *et al.* [18, 19] as an effective method for adaptive time-frequency analysis of non-linear and non-stationary data. Liu *et al.* [20, 21] proposed three ensemble empirical mode decomposition (EEMD)-based methods (i.e., 1D-EEMD, 2D-EEMD, normalization-incorporated EEMD) for wind parameter estimation using both rain-free and rain-contaminated X-band marine radar images. The procedures of those proposed methods are introduced below.

- **1D-EEMD algorithm**: Suppose a polar radar image I has M rows (ranges) and N columns (azimuths); after image preprocessing, the intrinsic mode function (IMF) components and a residual of pixels from each azimuth are extracted using EMD. Details of the algorithms can be found in the study by Liu *et al.* [20]. Therefore, if I_n denotes the intensity sequence of the nth azimuth, it can be decomposed as follows after applying 1D-EEMD:

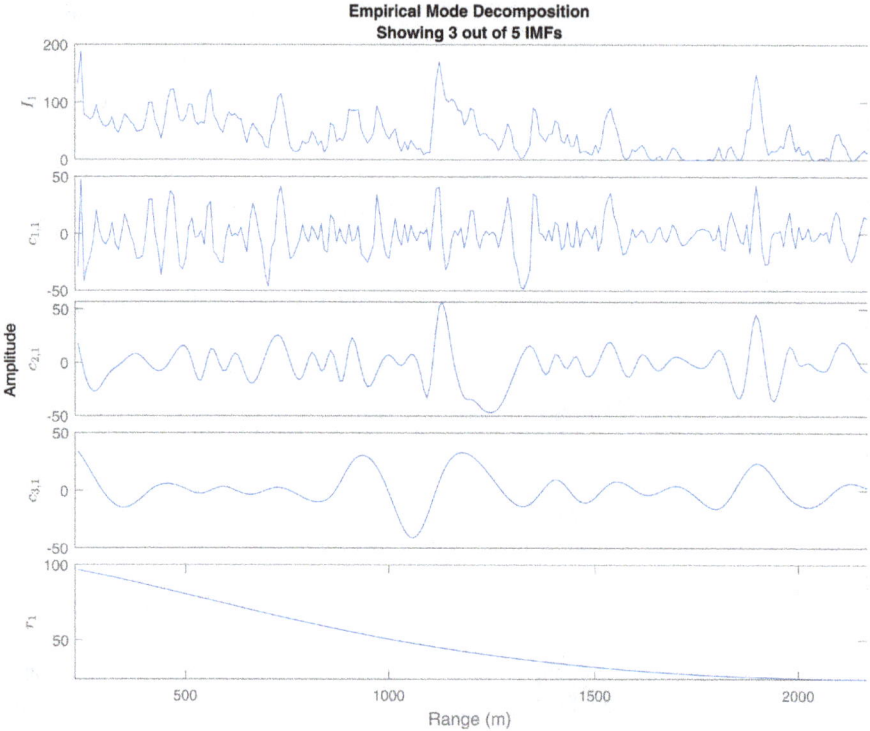

Figure 17.9   1D ensemble empirical mode decomposition (EEMD) result for I_l in a radar image (adapted from Huang et al. [21]). IMF, intrinsic mode function.

$$I_n = \sum_{j=1}^{J} c_{j,n} + r_n, \tag{17.16}$$

where $c_{j,n}$ is the jth IMF in the nth azimuth direction, J is the number of IMFs obtained, and r_n is the residual in the nth azimuth. An example of the EMD results using the first azimuth direction of a radar image is shown in Figure 17.9. The standard deviation of the third IMF components is curve fitted as a function of the azimuth using the cosine-squared function in Equation (17.2).

- **2D-EEMD algorithm**: In addition to 1D-EEMD, Liu *et al.* [20] also applied 2D-EEMD for wind direction estimation, which is composed of two times the 1D-EEMD operation. If C_j denotes the matrix consisting of the jth IMF components in all azimuths, the EEMD is applied to each row of C_j in order to obtain $H_{j,k}$, that is, the kth IMF components of C_j with k ranging from 1 to K, K being the number of IMFs. Thus, the radar image can be decomposed as

$$I = \sum_{j=1}^{J} \sum_{k=1}^{K} H_{j,k}. \tag{17.17}$$

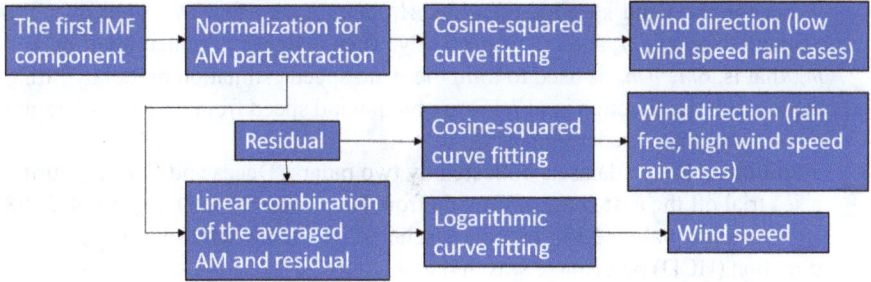

Figure 17.10   Scheme of the normalization-incorporated ensemble empirical mode decomposition (EEMD)-based algorithms for wind parameter estimation. AM, amplitude modulation; IMF, intrinsic mode function.

Then, based on a comparable minimal scale combination principle [22], the lth IMF component of the 2D-EEMD of I, that is, C_l, is expressed as

$$C_l = \sum_{j=1}^{J} H_{j,l} + \sum_{k=l+1}^{K} H_{l,k}. \tag{17.18}$$

The RMSD of the first IMF components of the 2D-EEMD (C_1) is curve fitted as a function of the azimuth using the cosine-squared function in Equation (17.2).

Rain-contaminated radar images introduced in Section 17.2.2 were used to test both 1D-EEMD- and 2D-EEMD-based methods. The RMSDs between the radar-derived wind directions and anemometer measurements for the 1D-EEMD and 2D-EEMD algorithms were 12.7° and 11.4°, respectively.

- **Normalization-incorporated EEMD**: Later, Huang *et al.* [21] incorporated the normalization scheme into the 1D-EEMD-based method and proposed an algorithm that could extract wind direction and wind speed from all radar data. The flowchart in Figure 17.10 shows the procedures of the algorithm after conducting EEMD on radar images. After obtaining the IMF components and the residual of each azimuth using 1D-EEMD, the first IMF component, that is, c_1, is selected and separated into amplitude modulation (AM) and frequency modulation (FM) parts using the normalization scheme. Hence, each column of c_1 can be decomposed as

$$c_{1,n} = a_{1,n} \cdot {*} f_{1,n}, \tag{17.19}$$

where $a_{1,n}$ and $b_{1,n}$ are the AM and FM parts of the first IMF in the nth azimuth, respectively, ".$*$" denotes the multiplication with corresponding elements. Then, for rain-contaminated images obtained during relatively low wind speeds, the average of $a_{1,n}$ in each azimuth (denoted as $A_{1,n}$) is used to fit the cosine-squared function in (17.2) for the estimation of wind direction, while for rain-free and high wind speed rain cases, the average of residual values in each azimuth (denoted as R_n) is selected for curve fitting.

As for the wind speed estimation, $A_{1,n}$ and R_n are first averaged over the azimuth to obtain m_a and m_r, respectively. Then, a linear combination of m_a and m_r, that is, $6m_a + m_r$, is used to train the wind speed estimation model by fitting the logarithmic function in (17.15) to obtain wind speed from both rain-free and rain-contaminated radar images.

- **Validation**: Image datasets collected by two radars (Decca and Furuno) during a sea trial off the east coast of Canada from November 26 to December 4, 2008 were used to validate the proposed scheme. A parameter called high clutter direction (HCD) percentage was first extracted as a threshold to detect the presence of rain in radar images. Then, low wind speed and high wind speed rain cases were separated using the HPP threshold proposed by Wang and Huang [16]. The RMSD for wind direction and wind speed measurement are 13.1° and 1.48 m/s for rain-free images, and 13.5° and 1.54 m/s for rain-contaminated images, respectively.

17.4   Nonparametric regression-based methods

Previous works introduced above applied various types of empirical functions for wind speed estimation, which are straightforward to implement. However, the performance of those functions has only been validated using one certain radar dataset, and it is hard to determine which function will produce the best result given a new set of radar images. Compared with traditional curve-fitting techniques, nonparametric regression-based methods do not require an assumption to be made about parameters and thus can be applied to different datasets directly. A general scheme of nonparametric regression-based methods for estimation of wind speed is shown in Figure 17.11. In this section, three nonparametric regression-based methods for estimation of wind speed will be illustrated.

17.4.1   Neural network-based method

Dankert *et al.* [6] used the mean radar cross sections (RCSs) in the cross-wind direction at four different ranges, together with the radar-derived wind direction and air-sea temperature difference as input to train the neural network (NN)-based wind speed estimation model, after obtaining temporally integrated radar images, The proposed NN consisted of two hidden layers with five and two neurons, and an output layer with one parameter. Results obtained using radar data introduced in

Figure 17.11   Scheme of the nonparametric regression-based algorithms for wind speed estimation

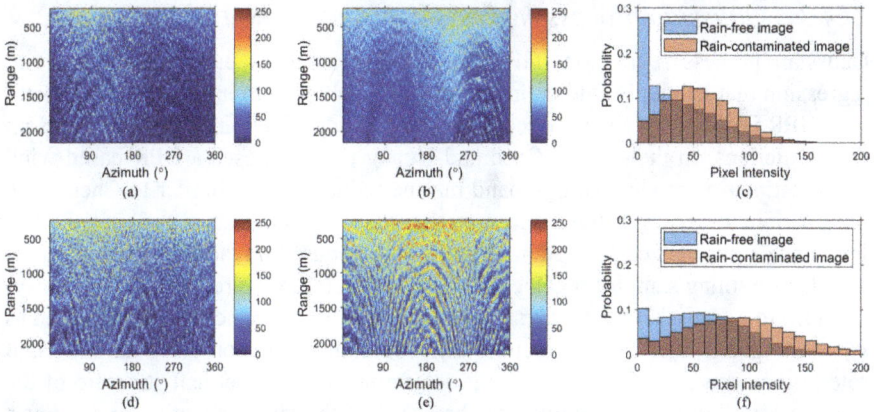

Figure 17.12   *Rain-free and rain-contaminated X-band marine radar images*
with different wind speeds and their normalized histograms,
adapted from Chen et al. [23]. (a) An example of a rain-free
image, with the wind speed measured by anemometers as 3.9 m/s.
(b) An example of a rain-contaminated image, with the wind speed
measured by anemometers as 3.1 m/s. (c) The normalized 20-bin
histogram of (a) and (b). (d) An example of a rain-free image
with the wind speed measured by anemometers as 13.6 m/s. (e)
An example of a rain-contaminated image with the wind speed
measured by anemometers as 12.1 m/s. (f) The normalized 20-bin
histogram of (d) and (e).

the LGM section show an RMSD of 0.85 m/s compared with in situ measurements. Although the accuracy is relatively high, parameters such as air-sea temperature difference cannot be derived from radar images and need to be obtained from other instruments.

17.4.2   Support vector regression-based method

Recently, Chen *et al.* [23] observed that both wind speed and the presence of rain affect the pixel intensity distribution of radar images. Examples of rain-free and rain-contaminated images obtained under different wind speeds and their normalized histograms are shown in Figure 17.12. For each radar image, the bin values of its normalized histogram were first extracted and combined as the feature vector. The feature vectors extracted from training sets were used to train a support vector regression (ε-SVR)-based wind speed estimation model with a Gaussian kernel using a sequential minimal optimization (SMO) algorithm. Data described in Section 17.3.3 were used to train the SVR-based estimation model and validate its accuracy. The RMSDs of wind speed estimation for rain-free and rain-contaminated data were 1.27 m/s and 1.42 m/s, respectively.

17.4.3   Gaussian process regression-based method

Gaussian process regression (GPR) is a nonparametric, Bayesian approach to regression that is widely studied in the area of machine learning. Given small datasets, GPR is still able to work effectively and provide uncertainty measurements on the predictions. In their study, Chen and Huang [24] proposed a GPR-based wind speed estimation model using X-band marine radar images. Similar to Chen *et al.* [23], the feature vector extracted from each radar image consists of the bin values of the image's normalized histogram. Then, it is assumed that the simultaneous wind speeds of training samples obtained from anemometer measurement (denoted as y) conform to a Gaussian process with zero mean and a covariance matrix (denoted as k_y) is computed from feature vectors. The covariance function plays an important role in the accuracy of the model as it determines the geometrical structure of the training samples, which is similar to the effect of the kernel function in a support vector machine. As the joint distribution of y and the estimated wind speeds of testing samples (denoted as μ_*) was assumed to be Gaussian as well, μ_* can be calculated as

$$\mu_* = k_*^T k_y^{-1} y, \qquad (17.20)$$

where k_* denotes the covariance matrix computed from feature vectors extracted from testing images.

Data described in Section 17.3.3 were used to train two GPR-based wind speed estimation models using rain-free and rain-contaminated images, respectively. The HCD percentage proposed by Huang *et al.* [21] was used as a threshold to classify between rain-free and rain-contaminated images. The RMSDs of wind speed estimation for rain-free and rain-contaminated data were 1.29 m/s and 1.36 m/s, respectively.

17.5   Error mitigation

- **Ship motion**: Wind streaks and wind gusts-based methods (i.e., LGM and OFM) cannot be directly applied to shipborne marine radar data because the radar's field of view changes with the ship's motion. Since the spacing between wind streaks is of the order of 100 m, integrating the image sequence without correcting the ship's motion may not obtain the desired wind streak information [2]. Instead of temporally integrated images, many methods only require single radar images for wind parameter estimation, which can be applied to shipborne marine radar data directly.
- **Blockage**: The blockage portion in radar images should be excluded from wind parameter estimation. For the original ILS-based algorithm, it cannot be applied to radar images with blockage because the elements of the range distance vector $r_i(\theta)$ obtained from the blocked azimuths are missing. In order to fill the gap caused by blocked azimuths, curve-fitting techniques can be introduced to fit the extracted parameter (e.g., the range distance vector extracted by Vicen-Bueno *et al.* [13], the average backscatter intensity of each azimuth extracted

by Lund *et al.* [2]) as a function of azimuth. In the study by Liu *et al.* [12], compared to original ILS, the curve-fitting-incorporated ILS reduces the mean differences and standard deviation of wind direction between the radar and the anemometer results by about 3° and 4.9°, respectively.

- **Rain**: The existence of rain may blur sea surface wave signatures and negatively affect wind measurement accuracy. For example, it has been found by Liu *et al.* [25] that under rainy conditions, the standard deviations of wind speed and wind direction using the modified ILS-based method increase by about 1.6 m/s and 10.7°, respectively, compared with the results obtained from rain-contamination-free data. That is because rain alters radar backscatter in two ways: (1) rain in the intervening atmosphere scatters and attenuates X-band radiation and (2) rain impinging upon the ocean surface alters its scattering and changes the relationship between the surface wind and the centimeter scale wave field [26].

As it has been observed that rain generally enhances the average pixel intensity of the radar image, some sorts of parameters (e.g., ZPP, HPP, and HCD) can be extracted from the image and used as thresholds to determine whether an image or a certain azimuth is contaminated by rain. Instead of excluding those images with rain, recent studies [11, 14, 16, 17, 20, 21, 23, 24] applied novel methods and models to rain-contaminated data, which mitigated the influence of rain and improved wind parameter measurement accuracy. For instance, in the study by Huang *et al.* [14], by incorporating texture analysis-based rain mitigation techniques into the ILS algorithm, the standard deviations of wind direction and speed can be reduced by about 14.5° and 1.3 m/s, respectively.

17.6   Conclusions and outlook

This chapter reviews the wind parameter estimation using X-band marine radar images. A summary and inter-comparison of the reviewed methods are provided in Table 17.1. It should be noted that since different datasets were used in different studies, it may not be appropriate to compare the performance of different methods based only on measurement accuracy (RMSD). The effectiveness of the different methods can also be evaluated and compared by referring to their applicable scenarios (e.g., ship motion, blockage, low sea states, rain). Although many recently proposed methods are still based on previously proposed techniques (e.g., curve fitting, ILS algorithm), the novel incorporated algorithms (e.g., texture analysis, EEMD) further extend the applicable conditions and improve measurement accuracy.

Future works can focus on the following points.

- **Theoretical challenges**: It should be acknowledged that marine radars are not radiometrically calibrated instruments. Hence, marine radar wind sensors require independent wind measurements for calibration purposes. Complicating matters further, we are still lacking a complete physical understanding of the

Table 17.1 A summary and comparison of the reviewed algorithms for wind parameter estimation

Method	Measurement accuracy (RMSD)		Images required	Applicable under		Techniques involved		Others
	Wind direction (○)	Wind speed (m/s)		Ship motion	Blockage	Rain detection	Rain mitigation	
LGM	12.8–14.2	–	Multiple	×	✓	×	×	–
OFM	32.1	2.68	Multiple	×	✓	×	×	–
Curve fitting original	14.2–17.4	0.79–0.88	Single	✓	✓	✓	×	–
Two-model curve fitting	–	1.7	Single	✓	✓	✓	✓	–
Dual curve fitting	6.6–16.6	0.8–1.9	Single	✓	✓	×	×	Low sea state mitigation
SWH-incorporated curve fitting	26.0–26.5	1.37–1.40	Single	✓	✓	×	×	Ship detection and removal, sea state mitigation
Original ILS	14.3	0.8	Multiple	✓	×	×	×	–
Texture analysis incorporated ILS	15.9–19.9	1.4–2.0	Multiple	✓	✓	✓	✓	Island interference mitigation
Spectral noise	–	2.86	Multiple	✓	×	×	×	–
Spectral integration	15.8	1.6	Single	✓	✓	✓	✓	–
1D- / 2D-EEMD	11.4–12.7	–	Single	✓	✓	✓	✓	–

(Continues)

Table 17.1 *Continued*

Method	Measurement accuracy (RMSD)		Images required	Applicable under		Techniques involved		
	Wind direction (∘)	Wind speed (m/s)		Ship motion	Blockage	Rain detection	Rain mitigation	Others
Normalization-incorporated EEMD	13.1–13.5,† 8.1–11.3*	1.48–1.54,† 0.95–1.05*	Single	✓	✓	✓	✓	–
NN-based	–	0.42–0.85	Multiple	×	✓	×	×	–
SVR-based	–	1.27–1.42	Single	✓	✓	✓	✓	–
GPR-based	–	1.29–1.36	Single	✓	✓	✓	✓	–

*Furuno radar data.
†Decca radar data.

grazing incidence X-band radar backscatter intensity's dependency on antenna polarization, antenna look direction, and range. These challenges are likely the main reasons why marine radar wind sensors have not yet been embraced by the broader oceanographic/atmospheric science community as an alternative to standard anemometers. However, in contrast to standard anemometers, marine radar backscatter intensity measurements hold valuable and unique information on the wind's spatial variability and, more specifically, wind gusts. The retrieval of wind gusts from radar measurements is therefore a highly promising (albeit difficult) field for future research.

- **Rain mitigation**: Although the threshold-based methods for rain-contaminated image detection are relatively straightforward to implement, the threshold values may vary significantly between different radar systems and it is hard to determine without a large amount of radar data [27]. In addition, it has also been found that the influence of rain on different regions of the radar image may differ. Specifically, the wave signatures in some regions may be blurred by rain while others remain less affected or unaffected. Also, under high wind speeds, rain may not blur the wave signatures in radar images because wind force dominates the generation of surface roughness. In consequence, rain only causes additional radar backscatter and does not affect the accuracy of wind parameter measurement significantly. Recently, an unsupervised-learning based method has been proposed concerning rain-contaminated region identification of X-band marine radar images [28]. Also, a novel method has been proposed to correct the influence of rain on radar images [29]. Therefore, those novel techniques should be incorporated into future works in order to further improve the measurement accuracy using rain-contaminated radar images.
- **Swell influence mitigation**: In addition to wind conditions, the presence of swell also affects the pixel intensities of the radar image. For example, when swell appears with wind simultaneously, the radar backscatter generated by local winds may be offset by swell because they usually propagate in different directions, which reduces the average pixel intensities of the radar image [5]. In consequence, wind speed may be underestimated. Therefore, future work should incorporate techniques that can detect and mitigate the influence of swell in X-band marine radar images.
- **Nonparametric methods**: Nonparametric methods have been applied for wind speed estimation and have shown satisfactory results. Future works should further improve the robustness of those methods and apply them to wind direction estimation.

References

[1] Huang W., Liu X., Gill E.W. 'Ocean wind and wave measurements using X-band marine radar: a comprehensive review'. *Remote Sensing*. 2017;**9**(12):1261.

[2] Lund B., Graber H.C., Romeiser R. 'Wind retrieval from shipborne nautical X-band radar data'. *IEEE Transactions on Geoscience and Remote Sensing.* 2012;**50**(10):3800–11.

[3] Lee P.H.Y., Barter J.D., Caponi E., *et al.* 'Wind-speed dependence of small-grazing-angle microwave backscatter from sea surfaces'. *IEEE Transactions on Antennas and Propagation.* 1996;**44**(3):333–40.

[4] Trizna D.B., Carlson D.J. 'Studies of dual polarized low grazing angle radar sea scatter in nearshore regions'. *IEEE Transactions on Geoscience and Remote Sensing.* 1996;**34**(3):747–57.

[5] Chen Z., He Y., Zhang B., Qiu Z. 'Determination of nearshore sea surface wind vector from marine X-band radar images'. *Ocean Engineering.* 2015;**96**(7):79–85.

[6] Dankert H., Horstmann J., Rosenthal W. 'Ocean wind fields retrieved from radar-image sequences'. *Journal of Geophysical Research.* 2003;**108**(C11).

[7] Dankert H., Horstmann J., Rosenthal W. 'Wind- and wave-field measurements using marine X-band radar-image sequences'. *IEEE Journal of Oceanic Engineering.* 2005;**30**(3):534–42.

[8] Dankert H., Horstmann J. 'A marine radar wind sensor'. *Journal of Atmospheric and Oceanic Technology.* 2007;**24**(9):1629–42.

[9] Koch W. 'Directional analysis of SAR images aiming at wind direction'. *IEEE Transactions on Geoscience and Remote Sensing.* 2004;**42**(4):702–10.

[10] Dankert H., Horstmann J., Rosenthal W. 'Ocean surface winds retrieved from marine radar-image sequences'. Proceedings of IEEE International Geoscience and Remote Sensing Symposium, vol. 3.; 2004. pp. 1903–6.

[11] Huang W., Gill E.W. 'Ocean remote sensing using X-band shipborne nautical radar'applications in eastern canada'. Coastal Ocean Observing Systems. Elsevier; 2015. pp. 248–64.

[12] Liu Y., Huang W., Gill E.W., Peters D.K., Vicen-Bueno R. 'Comparison of algorithms for wind parameters extraction from shipborne X-band marine radar images'. *IEEE Journal of Selected Topics in Applied Earth Observations and Remote Sensing.* 2015;**8**(2):896–906.

[13] Vicen-Bueno R., Horstmann J., Terril E., de Paolo T., Dannenberg J. 'Real-time ocean wind vector retrieval from marine radar image sequences acquired at grazing angle'. *Journal of Atmospheric and Oceanic Technology.* 2013;**30**(1):127–39.

[14] Huang W., Liu Y., Gill E.W. 'Texture-analysis-incorporated wind parameters extraction from rain-contaminated X-band nautical radar images'. *Remote Sensing.* 2017;**9**(2):166.

[15] Izquierdo P., Guedes Soares C., Soares C.G. 'Analysis of sea waves and wind from X-band radar'. *Ocean Engineering.* 2005;**32**(11-12):1404–19.

[16] Wang Y., Huang W. 'An algorithm for wind direction retrieval from X–band marine radar images'. *IEEE Geoscience and Remote Sensing Letters.* 2016;**13**(2):252–6.

[17] Huang W., Wang Y. 'A spectra-analysis-based algorithm for wind speed estimation from X-band nautical radar images'. *IEEE Geoscience and Remote Sensing Letters*. 2016;**13**(5):701–5.

[18] Huang N.E., Shen Z., Long S.R., *et al.* 'The empirical mode decomposition and the Hilbert spectrum for nonlinear and non-stationary time series analysis'. *Proceedings of the Royal Society of London. Series A: Mathematical, Physical and Engineering Sciences*. 1998;**454**(1971):903–95.

[19] Huang N.E., Shen Z., Long S.R. 'A new view of nonlinear water waves: the Hilbert spectrum'. *Annual Review of Fluid Mechanics*. 1999;**31**(1):417–57.

[20] Liu X., Huang W., Gill E.W. 'Wind direction estimation from rain-contaminated marine radar data using the ensemble empirical mode decomposition method'. *IEEE Transactions on Geoscience and Remote Sensing*. 2017;**55**(3):1833–41.

[21] Huang W., Liu X., Gill E.W. 'An empirical mode decomposition method for sea surface wind measurements from X-band nautical radar data'. *IEEE Transactions on Geoscience and Remote Sensing*. 2017;**55**(11):6218–27.

[22] Wu Z., Huang N.E., Chen X. 'The multi-dimensional ensemble empirical mode decomposition method'. *Advances in Adaptive Data Analysis*. 2009;**01**(03):339–72.

[23] Chen X., Huang W., Yao G. 'Wind speed estimation from X-band marine radar images using support vector regression method'. *IEEE Geoscience and Remote Sensing Letters*. 2018;**15**(9):1312–16.

[24] Chen X., Huang W. 'Gaussian process regression for estimating wind speed from X-band marine radar images'. Proceedings of MTS/IEEE Oceans; Charleston, USA; 2018. pp. 1–4.

[25] Liu Y., Huang W., Gill E.W. Analysis of the effects of rain on surface wind retrieval from X-band marine radar images. Proceedings of MTS/IEEE Oceans; St. John's, Canada; 2014. pp. 1–4.

[26] Contreras R.F., Plant W.J. 'Surface effect of rain on microwave backscatter from the ocean: measurements and modeling'. *Journal of Geophysical Research*. 2006;**111**(C8).

[27] Chen X., Huang W., Zhao C., Tian Y. 'Rain detection from X-band marine radar images: a support vector machine-based approach'. *IEEE Transactions on Geoscience and Remote Sensing*. 2020;**58**(3):2115–23.

[28] Chen X., Huang W. 'Identification of rain and low-backscatter regions in X-band marine radar images: an unsupervised approach'. *IEEE Transactions on Geoscience and Remote Sensing*. 2020;**58**(6):4225–36.

[29] Chen Z., He Y., Zhang B., Ma Y. 'A method to correct the influence of rain on X-band marine radar image'. *IEEE Access*. 2017;**5**(576–25):25576–83.

Chapter 18

Introduction to remote sensing using GNSS signals of opportunity

Adriano Camps[1]

This chapter provides an introduction to remote sensing by means of Reflectometry using Global Navigation Satellite Systems (GNSS) signals of opportunity, in short GNSS-R. This technique was originally proposed in the late 1980s and proven in the mid 1990s. However, it took two more decades until satellite data were widely available, and more affordable and reliable ground-based and airborne instruments were designed, before the number of researchers and applications began to significantly increase.

The chapter is structured as follows:

- First, a brief historical review is presented. This is very much needed to provide the context and perspective of the evolution of the different GNSS-R techniques that exist today, with emphasis on + remote sensing.
- Then, the basics of GNSS will be presented. This is important as these signals are the ones that will be used. Since most of the GNSS systems operate at L-band, where ionospheric effects are important, and satellites can be observed over a wide solid angle, this leads to the transmission of circularly polarized signals that are immune to Faraday rotation and do not require antenna pointing to mitigate polarization matching losses. It is very important to know the properties of the GNSS signals (type, structure, frequency bands and bandwidths used, modulation, received power, polarization, etc.) to understand what can and what cannot be done with GNSS-R.
- In the next section, the basics of GNSS-R are introduced. GNSS-R is a passive system because it does not transmit a signal. As such, some calibration techniques of GNSS-R receivers are inherited from microwave radiometry. GNSS-R can also be understood as a multi-static radar with many transmitters (i.e., navigation satellites) that are in the view of the receiver. As such, GNSS-R can be analyzed as a particular case of a multi-static radar. The different GNSS-R

[1]CommSensLab Research Center, Dept. of Signal Theory and Communications, Universitat Politècnica de Catalunya-BarcelonaTech, Spain

techniques will be presented with their pros and cons. The discussion will focus on the use of GNSS signals of opportunity, but actually the concepts are applicable to any other signal of opportunity, provided its structure is known. Some GNSS-R techniques are applicable even if the signals are not known and vary over time. These signals typically use code-division multiple access (CDMA) modulations, or spread spectrum techniques such as the Orthogonal Frequency-Division Multiple Access (OFDMA) used in Long-Term Evolution (LTE) or the Coded Orthogonal Frequency-Division Multiplexing (COFDM) used in Digital Video Broadcasting - Terrestrial (DVB-T) that enlarge and flatten the signal's spectrum making it look like white noise.

• Then, the different GNSS-R techniques will be described.
• Finally, different ocean applications will be briefly introduced, grouped as scatterometric, altimetric, and imaging, which will be described in more detail in the following chapters.

18.1   A quick historical review of GNSS-R

The origin of GNSS-R dates back to 1988, when Hall and Cordey proposed the concept of multi-static scatterometry using Global Positioning System (GPS) signals [1]. In July 1991, an incident with a French Alpha jet aircraft testing a GPS receiver showed that GPS navigation signals scattered on the sea surface could be collected and tracked. Even though these results were known by part of the navigation community, the results were not made public until the investigation was finished in 1994 [2]. In 1993, the concept of reflectometry using GNSS-R was proposed for mesoscale altimetry as a way to reduce the revisit time [3], since reflections from multiple transmitters could be measured simultaneously (Figure 18.1).

In the 1990s, the first GPS-Reflectometry (GPS-R) observations from an aircraft were collected using an ad hoc receiver [4], and the first GPS-R data were "found" in segments of NASA/JPL Spaceborne Imaging Radar-C (SIR-C) mission data acquired in 1994 when radar returns were not present [5] (Figure 18.2).

In 2005, the first GPS-R data were acquired from the UK Disaster Monitoring Constellation (UK-DMC) satellite using a dedicated spaceborne instrument [6]. This experiment allowed to record few tens of data segments that showed different features of the GPS reflected signals when scattered over the ocean, land, or ice.

The UK TechDemoSat-1 (TDS-1) was launched on July 8, 2014, carrying onboard the SGR-ReSI from Surrey Satellite Technology Ltd (SSTL) as a secondary payload, an instrument designed to demonstrate accurate measurement of ocean wind speeds around the globe using GNSS-R [7]. As GNSS-R data become more widely available, the number of researchers has increased, expanding to other applications such as ice detection, soil moisture, coarse resolution altimetry, or phase altimetry of the ice sheets.

After this success, on December 15, 2016, NASA launched the Cyclone Global Navigation Satellite System (CYGNSS) mission consisting of 8 micro-satellites in a

Passive Reflectometry and Interferometry System

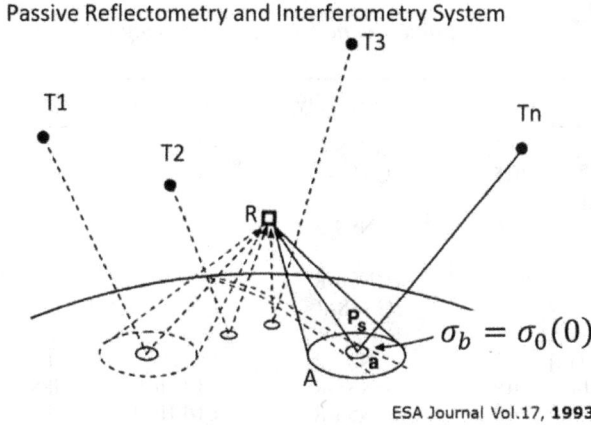

$$\sigma_b = \sigma_0(0)$$

ESA Journal Vol.17, **1993**

Figure 18.1   The Passive Reflectometry and Interferometry System (PARIS) concept proposal (European Space Agency ESA Patent 321) (adapted from Martin-Neira [3])

low-inclination orbit of 35° to measure wind speeds over the Earth's tropical oceans, increasing the ability of scientists to understand and predict hurricanes [8]. Based on the same GNSS-R instrument as the UK TDS-1, each satellite acquires information from four GPS satellites. The large amount of data generated and the very short revisit time have opened new applications such as soil moisture [9] and map flooding [10], vegetation biomass estimation [11], and so on.

Apart from these dedicated missions, after the NASA Soil Moisture Active Passive (SMAP) radar failed, the receiver was tuned to the L2 (1227.60 MHz) band to collect GPS reflections at horizontal and vertical polarizations, with a very high signal-to-noise ratio (*SNR*), thanks to the large antenna reflector [12].

Table 18.1 summarizes the past, present, and known planned missions carrying on board GNSS-R payloads. Apart from those, several other missions were planned, but unfortunately not selected such as the PAssive Reflectometry and Interferometry System In-orbit Demonstrator (PARIS-IoD) [20], the GNSS rEflectometry, Radio

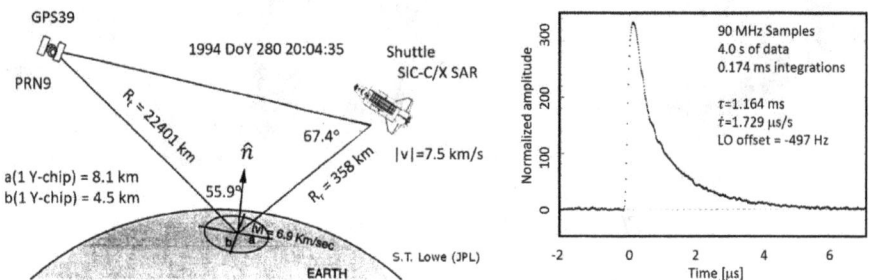

Figure 18.2   First GPS-Reflectometry (GPS-R) data collected from space in segments of SIR-C data (adapted from [5]).

Table 18.1   Past, current, and planned missions carrying GNSS-R instruments

Mission	Date	GNSS-R type	Band/Pol used	GNSS system used
UK-DMC [6]	2003	cGNSS-R	L1/LHCP	GPS
UK-TDS-1 [7] (Figure 18.3a)	2015	cGNSS-R	L1/LHCP	GPS
CYGNSS [8] (Figure 18.3b)	2016	cGNSS-R	L1/LHCP	GPS
³Cat-2 [13]	2016	cGNSS-R rGNSS-R iGNSS-R	L1 + L2/ LHCP + RHCP	GPS/GLONASS Galileo/Beidou
SMAP GNSS-R [12]	2017	cGNSS-R	L2/H+V	GPS
BuFeng-1 A/B [14]	2019	cGNSS-R	L1/LHCP	GPS/Beidou
Spire [15]	2019	cGNSS-R	L1/LHCP	GPS/Galileo
FSSCat (³Cat-5 A) [16]	2020	cGNSS-R	L1/ LHCP	GPS/Galileo
Fengyun-3 series [17]	2020	cGNSS-R	L1/ LHCP	GPS/Gailleo/Beidou
³Cat-4 [18]	2021	cGNSS-R	L1+ L2/ LHCP	GPS/Galileo
PRETTY [19]	2021	iGNSS-R	L1/RHCP	GPS/Galileo

Detailed description of the acronyms will be provided in the next sections.

Occultation and Scatterometry (GEROS) mission onboard the International Space Station [21] or the Global navigation satellite system (GNSS) Transpolar Earth Reflectometry exploriNg system (G-TERN) proposal as an Earth Explorer 9 [22].

As of June 2020, a total of 135 navigation satellites are actively transmitting navigation signals: 31 GPS, 26 GLONASS, 22 Galileo, 44 Beidou, 4 QZSS (Japan's Quasi-Zenith Satellite System) [23], and 8 IRNSS (India's Regional Navigation Satellite Systems). Therefore, the unique feature of remote sensing using

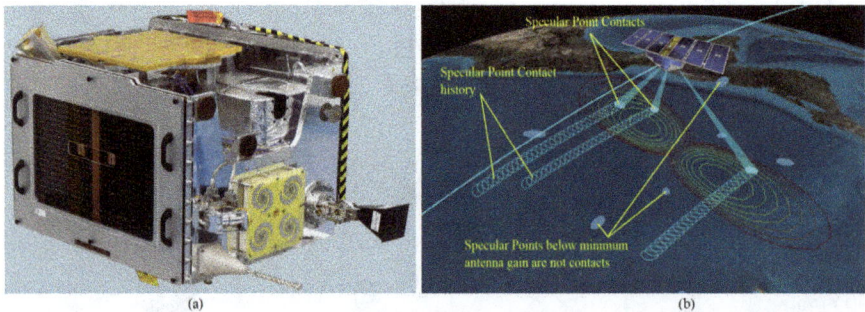

Figure 18.3   a) UK TDS-1 with the 2×2 array of spiral patch antennas for Global Navigation Satellite System-Reflectometry (GNSS-R) [7], b) artist's view composition of one of the eight spacecrafts of NASA Cyclone Global Navigation Satellite System (CYGNSS) constellation

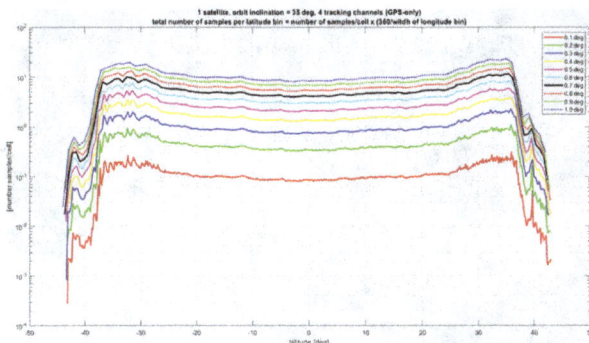

Figure 18.4   *Average number of samples per day and per cell, for different cell sizes (from 0.1° × 0.1° to 1° ×1°) for a Cyclone Global Navigation Satellite System (CYGNSS)-like payload capable of tracking four different global positioning system (GPS) satellites simultaneously from a 500 km orbit height inclined at 35° (adapted from [24]).*

these highly precise and continuous signals is an unbeatable inherent wide swath (≥1 000 km), and short revisit time. As an example, Figure 18.4 shows the average number of samples per day and per cell, for different cell sizes (from 0.1°×0.1°) for a CYGNSS-like payload capable of tracking 4 different GPS satellites simultaneously from a 500 km orbit height inclined at 35°.

Before entering into the details of GNSS-R systems, it is important to review the main properties of the GNSS signals to understand their advantages and "limitations" as compared to conventional radars.

18.2   Basic concepts on GNSS

18.2.1   Measurement principle

All GNSS systems are based on the measurement of the Time of Arrival (ToA). GNSS satellites have atomic clocks synchronized among them, and with the control. A GNSS receiver receives the signal from the different satellites in view and determines their position and time of emission of the signal from the ephemeris and almanac data, which are also transmitted by each satellite. The transit time is computed by measuring the time from transmission to reception, and the so-called "pseudo-range" (PR) or raw distance from the satellite to the receiver is estimated by multiplying the speed of light by the transit time. For each satellite, the geometric figure of all points satisfying the measured PR is a sphere centered at the GNSS satellite position with radius PR. A point (x, y, z) in space is determined by the intersection of three spheres. However, the receiver's clock accuracy is typically much poorer than that of the transmitting satellites, and its error has to be determined as well. This introduces a fourth unknown, that requires at least a fourth equation, which is

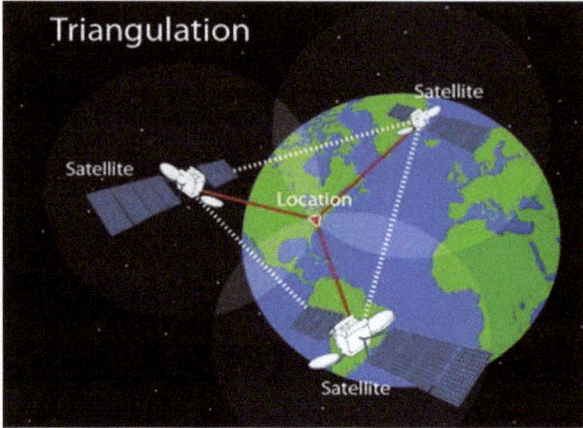

Figure 18.5   *Determination of the three-dimensional position by triangulation of the times of arrival of at least Global Navigation Satellite System (GNSS) signals [25]*

created from the observation of the fourth satellite. This process is schematically shown in Figure 18.5.

The way the receiver distinguishes the signal coming from one satellite from that of another one was different in the first two GNSS systems. In the United States, GPS transmitted signals were separated using a technique called CDMA. Each signal was modulated using a finite sequence of pseudo-random numbers or PRN that when convolved with itself has a very narrow auto-correlation function (ACF) and when convolved with a different code leads to a very small cross-correlation (ideally zero). In the Russian GLONASS, transmitted signals were first separated by Frequency-Division Multiple Access (FDMA). The upcoming GLONASS-K2 series satellites will include CDMA signals as well, and the following series GLONASS-V will include only CDMA signals. Other GNSS constellations such as the European Galileo or the Chinese Beidou use CDMA as well. From now on, we will focus on CDMA GNSS systems, but the basic principles are the same for FDMA ones although the receiver becomes more complex.

In addition to the fact that multiple transmitters can use the same frequency band, other advantages of CDMA modulation are the increased tolerance to multipath and jamming as the spectrum enlarges during the modulation process, but then compresses again in reception, while jammers' spectra enlarge, and their spectral density decreases. On top of the PRN signal, the navigation messages are modulated at a much lower speed (50 MHz in GPS L1 Coarse Acquisition (C/A) code), which are ultimately decoded by the navigation receiver. In principle, in GNSS-R, the navigation bit is not of interest and does not require to be demodulated unless long coherent integration times are required.

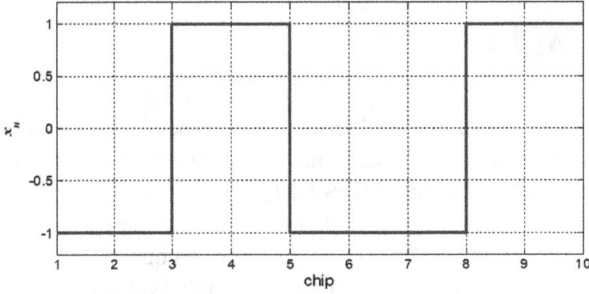

Figure 18.6   *Sample random sequence of pulses $x_n = \{-1,-1,+1,+1,-1,-1,-1, +1,+1...\}$*

A few more considerations of these systems are summarized below, and they will be dealt with in more detail in subsequent sections. Ionospheric effects depend on the solar activity, which modulates the Vertical Total Electron Content (VTEC).

1. Since the ionospheric delay varies as the squared inverse of frequency, this dependence can be removed up to more than 99.9% using two frequency measurements.
2. The ionosphere is also responsible for the Faraday rotation, which alters the polarization plane.

This effect can be neglected if circularly polarized signals are transmitted, which also helps in reception as there is no need to point the antenna towards each transmitting satellite to minimize polarization losses, so all visible satellites can be received simultaneously. The interested reader can refer to numerous books and publications on this topic [26–28].

18.2.2   Structure of the GNSS signals

Understanding the structure of the GNSS signals is critical as, nowadays, most GNSS-R instruments implement a technique called "conventional GNSS-R", or cGNSS-R in short, which requires knowledge of the transmitted signals.

For the sake of simplicity, let us consider first the case of a pure random train of pulses of width τ_c (Figure 18.6 and (18.1)):

$$P(t) = \sum_{n=-\infty}^{+\infty} x_n \cdot \Pi\left(\frac{t-n\cdot\tau_c}{\tau_c}\right), \tag{18.1}$$

where x_n takes the values ± 1 with equal probability.

Each individual pulse that composes the sequence is known as a "chip" as opposed to a "bit", since it does not carry any information. The auto-correlation of $P(t)$ is approximately a triangle function given by:

$$R_p\left(\tau\right) \approx \bigwedge\nolimits_{\tau_c}\left(\tau\right) = \begin{cases} 1 - \frac{|\tau|}{\tau_c}, & |\tau| < \tau_c \\ 0, & \text{elsewhere} \end{cases}, \tag{18.2}$$

where τ is the time lag. For practical reasons and to speed up the acquisition process, the codes used in navigation systems have a finite length and then repeat indefinitely. In the case of the so-called GPS L1 ($f_{L1} = 1\,575.42$ MHz) C/A code used for the open-access civil service, it has 1 023 chips per period and repeats every 1 ms. This leads to a chip duration of $\tau_c = 0.977$ μs (clock frequency = 1.023 MHz), which corresponds to a distance of $c \cdot \tau_c = 293$ m. Note that this value is much larger than the resolution of conventional radar altimeters, which operate with larger bandwidths and higher SNRs.

C/A codes are generated from the product of two $2^n - 1$ long "Maximal Length Sequences" (MLS) G1 and G2 generated from an *n*-stage Linear Feedback Shift Register (LFSR), so that the cross-correlation properties of the single MLS are improved. Both G1 and G2 are generated by LFSR of 10 stages driven by a 1.023 MHz clock. The actual satellite identifier (ID) is determined by the relative delay between G1 and G2. This delay is determined by the position of the two connectors of the cells that compose the G2 LFSR. For example, PRN 1 is generated when taps 2 and 6 are selected, and PRN 31 with taps 3 and 8. There are only 37 delay combinations: 32 of them are reserved for the satellites and 5 are used for other applications, such as ground transmissions. Figure 18.7 summarizes the code generation.

The resulting C/A codes have a high auto-correlation peak that clearly indicates that a given satellite is in view and has low cross-correlation peaks so that the satellites do not interfere among themselves (Figure 18.8). In order to distinguish a weak signal surrounded by strong ones, it is necessary for the auto-correlation peak of the weak

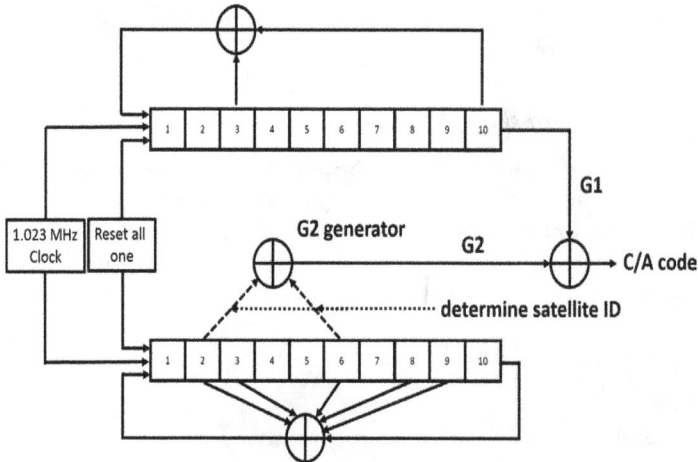

Figure 18.7   *Generation of the coarse acquisition (C/A) code as the product of two Maximal Length Sequences (MLS) (adapted from [26])*

signal to be higher than the cross-correlation peaks of the stronger signals. In the ideal case of using totally random sequences, the codes would be orthogonal and the cross-correlations become zero. The used PRN codes are almost orthogonal, and the cross-correlation values are as low as $-65/1023$ during 12.5% of the time, $-1/1023$ during 75% of the time, or $63/1023$ during the remaining 12.5% of the time.

To overcome the accuracy limitations of the GPS system, larger bandwidth signals (i.e., shorter "chips") are transmitted as well. The GPS system also uses the "Precise" code (or P code) for the restricted military signal. It has a chipping rate ten times faster than the C/A code (10.23 MHz) that results in a ten-fold increase of the PR accuracy. The code period is much longer (one week) so that the direct acquisition of the code (i.e., the estimation of the code offset) is nearly impossible, so the acquisition is done first with the C/A code. To acquire the P code, special data fields of the navigation frames are used (Z-count and Time of Week (ToW)). To increase the code robustness even more, it is possible to switch the system operation to use an encrypted version of the P code, noted as P(Y) [26]. The C/A ($CA(t)$) and P ($P(t)$) codes are modulated in-phase and quadrature on the L1 carrier (ω_1) as shown in (18.3):

$$
\begin{aligned}
S_1(t) &= \sqrt{2 \cdot P_{C/A_1}} D(t) \cdot CA(t) \cdot \cos(\omega_1 \cdot t + \phi_1) \\
&+ \sqrt{2 \cdot P_{P_1}} D(t) \cdot P(t) \cdot \sin(\omega_1 \cdot t + \phi_1),
\end{aligned}
\tag{18.3}
$$

where $S_1(t)$ is the signal transmitted by a given GPS satellite, P_{C/A_1} is the transmitted power for the civil signal at L1, P_{P_1} is the transmitted power for the restricted signal at L1, and ϕ_1 is the phase of the carrier.

To compensate for the ionospheric delay, the P code is also broadcast in the L2 band ($f_{L2} = 1227.60$ MHz) (18.4):

$$
S_2(t) = \sqrt{2 \cdot P_{P_2}} \cdot P(t) \cdot \cos(\omega_2 \cdot t + \phi_1),
\tag{18.4}
$$

and using this property, advanced navigation and GNSS-R receivers are able to infer the P(Y) code needed to demodulate the signal received. This technique is called the "reconstructed" code technique and provides "access" to a large bandwidth signal

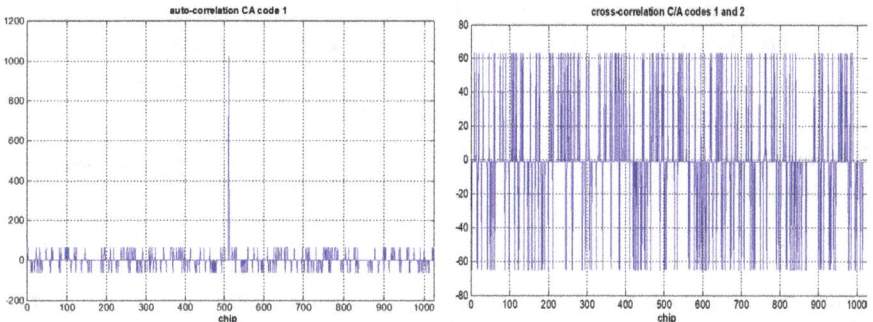

Figure 18.8 *Auto-correlation of the coarse acquisition (C/A) code 1 (left) and cross-correlation of C/A codes 1 and 2 (right)*

without knowing it. The advantages and disadvantages of the different types of GNSS-R instruments will be described later on.

GNSS systems use many other types of signals and modulations, for example, the LC2 sequence has the same chip rate of the C/A signal, but it is composed of two PRN codes of different lengths generated using the same 27-state LFSR, which is restarted at every one moderate length code (CM) or long code (CL) period. The CM is 10230 chips long and repeats every 20 ms, and it is modulated with navigation data, while the CL has 767250 chips and repeats every 1.5 s and has no data modulation [29].

Other satellite navigation systems such as Galileo use other PRN sequences, which are not necessarily generated using shift registers, but using look-up tables instead, and instead of a Binary Phase Shift Keying (BPSK) modulations in the in-phase and/or quadrature components, advanced modulation techniques such as Binary Offset Carrier (BOC) signals are used to increase the ranging accuracy with the same signal bandwidth [30]. This modulation is the result of the multiplication of the PRN code with a sub-carrier which is equal to the sign of a sine or a cosine waveform, yielding the so-called sine-phased or cosine-phased BOC signals. The BOC signal is commonly referred to as BOC(m,n), where $f_s = m \cdot 1.023$ and $f_c = n \cdot 1.023$, and unless indicated in a different way, when talking about BOC signals it will always be understood as the sine-phased variant.

For the sine-phased BOC signals (i.e., L1M, E1B, and E1C), or the cosine-phased BOC signals (i.e. E1A), the ACF can be expressed as an addition of triangles [31]. Finally, the ACF for the Galileo E5 signal, by far the most sophisticated signal among all the signals used for GNSS, can be closely approximated using the general expression of a Complex Double-Binary-Offset-Carrier (CDBOC) modulation [32].

As an example, Figure 18.9 shows the absolute value of the ACFs for infinite bandwidth signals (i.e., with sharp transitions). The time axis is expressed in C/A chips units for easier comparison with the ACF given in (18.2), with a width of two chips. Figure 18.9a is the ACF of the GPS L1 signal which is the composition of three functions: a triangle of base $[-1, +1]$ C/A code chips (corresponding to the L1 C/A signal), another triangle of base $[-0.1, +0.1]$ C/A code chips (corresponding to the P code), and two side peaks (corresponding to the M code, if available, depending on the satellite). Figure 18.9b shows the ACF of the Galileo E1 signal, and Figure 18.9c,d the ACFs of the GPS L5 and Galileo E5 signals.

18.2.3   Received power of the GNSS signals

The minimum received power (P_R) is computed using the standard propagation equation:

$$P_R = P_T \cdot G_T \cdot \frac{1}{4\pi R^2} \cdot A_{eff} \cdot L, \qquad (18.5)$$

where P_T is the transmitted power, G_T is the gain of the transmitting antenna, $1/4\pi R^2$ is the propagation losses, and A_{eff} is the effective area of the receiving antenna, which is related to the receiving antenna gain (G_R) and the electromagnetic wavelength (λ) by:

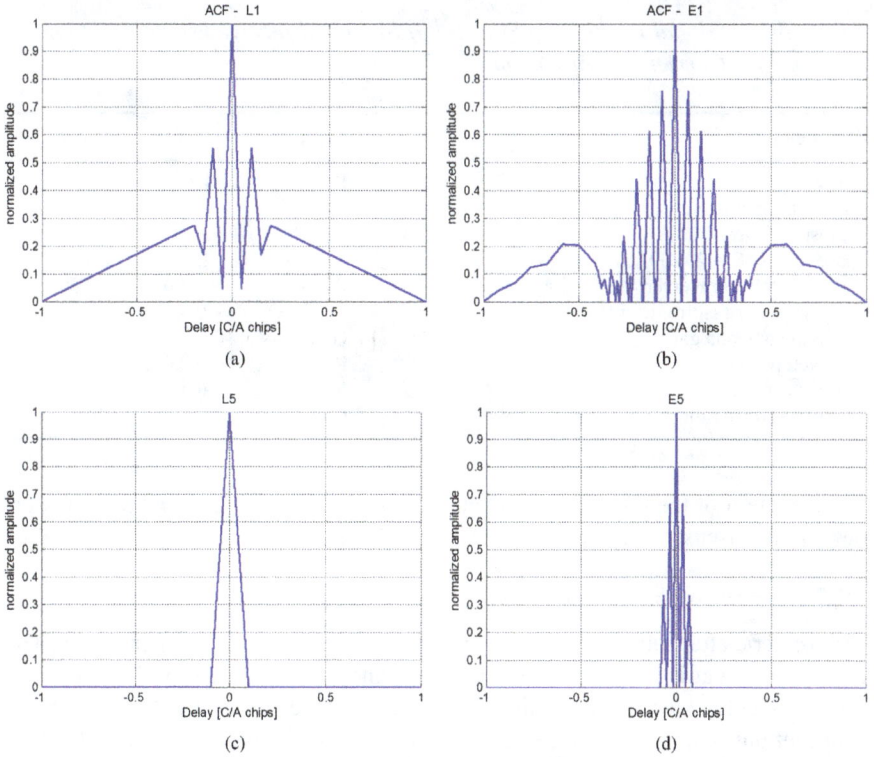

Figure 18.9   *Absolute value of the auto-correlation functions (ACFs) of a)*
Global Positioning System (GPS) L1, b) Galileo E1, c) GPS L5, and
d) Galileo E5 signals. C/A, coarse acquisition.

$$A_{eff} = \frac{\lambda^2}{4\pi} \cdot G_R,$$ (18.6)

and L is the loss in the atmosphere and other media between transmitter and receiver. Assuming an isotropic receiving antenna ($G_R = 0$ dBi), Table 18.2 summarizes the received power and Carrier-to-Noise ratio (C/N0) for the GPS L1 C/A code signal.

The C/N0 for the direct signal ranges between 39 dBHz and 52 dBHz, depending on the geometry, the receiving antenna gain, and the actual transmitted power, which varies from satellite to satellite, and over time due to aging or it is selectively adjusted over certain regions. Taking into account the 2.46 MHz bandwidth, and the −130 dBm received power at L1 C/A, the *SNR* is approximately −20 dB which is well under the noise level. The key is the correlation gain introduced during the demodulation, which is equal to the sequence length ($10 \cdot \log_{10}(1\,023) = 30.1$ dB), which makes the post-correlation *SNR* approximately equal to 10 dB. For the L1 P signal, the received power is −133 dBm, 3 dB less than for the C/A signal. Within the GPS satellite antenna Field of View (FoV), the different signal attenuations due to different propagation losses and atmospheric absorption are nearly compensated

Table 18.2   Global Positioning System (GPS) typical received signal level and Carrier-to-Noise ratio (C/N0)

Parameter	Symbol	Value	Units
Transmitter power (Coarse Acquisition (C/A) code)	P_T	14.3	dBW
Transmitting antenna gain	G_T	10.2	dB
Atmospheric loss	L	−2	dB
Space loss @ R = 20 200 km		−157.1	dB/m^2
Power density at receiver	P_D	−134.6	dBW/m^2
Effective area of isotropic antenna		−25.4	dBm^2
Receiving antenna gain	G_R	0	dB
Received power	P_R	−160.0	dBW
Noise spectrum	N_0	−204	dBW/Hz
Carrier-to-Noise ratio	C/N_0	44	dB-Hz

by the pattern of the transmitting antenna itself. The main signals and their properties will be discussed in Section 18.2.5.

18.2.4   Atmospheric and ionospheric effects

Atmospheric attenuation at L-band is nearly negligible, even under heavy rain. The bottom part, that is, the troposphere is denser and more variable and introduces an additional delay with respect to the free space. It can be modeled as the sum of two contributions: the hydrostatic zenith delay (*HZD*), and the wet zenith delay (*WZD*):

$$\rho_{tropo} = m_{hz} \cdot HZD + m_{wz} \cdot WZD, \tag{18.7}$$

where m_{hz} and m_{wz} are the mapping functions that project *HZD* and *WZD*, from the zenith to slant directions of observation and depend on the elevation angle, the geographical coordinates of the receiver, and the day of the year. The "dry component" (*HZD*) is larger than the "wet component" (*WZD*), and it can be accurately computed from atmospheric pressure data at the receiver site/level, while *WZD* cannot.

In ground-based GNSS-R instruments, both the direct and the reflected signals are affected in a similar way by tropospheric effects. In airborne instruments, it depends on the flight height, and in a Low Earth Orbit (LEO) instrument, only the reflected signal is affected twice (down-welling and up-welling paths). While multi-frequency radiometers (23.8 GHz and 36.5 GHz, at least) are used in nadir-looking radar altimeters to infer the "wet delay", due to the wide range of GNSS-R instruments, corrections should rely on proper modeling [33, 34].

The ionosphere extends from ~60 km until more than 2 000 km in height and contains a partially ionized medium, as a result of the solar radiation and the incidence of charged particles. The propagation of GNSS electromagnetic waves through the ionosphere depends on the electron density, its daily and seasonal variations, and its local irregularities. The main ionospheric effects are summarized in Table 18.3, modified from ITU-R P.531-12 recommendation [35]. Faraday rotation effects are mitigated by using circular polarized signals. Refraction is used in GNSS-Radio Occultations (GNSS-RO) to infer atmospheric water vapor and temperature

Table 18.3   Estimated maximum ionospheric effects at 1 GHz for elevation angles of about 30° one-way traversal (from [35]).

Effect	Magnitude	Frequency dependence
Faraday rotation	108°	$1/f^2$
Propagation delay	0.25 µs	$1/f^2$
Refraction	<0.17 mrad	$1/f^2$
Variation in the direction of arrival	0.2 min of arc	$1/f^2$
Absorption (polar cap absorption)	0.04 dB	$\sim 1/f^2$
Absorption (auroral + polar cap absorption)	0.05 dB	$\sim 1/f^2$
Absorption (mid-latitude)	<0.01 dB	$1/f^2$
Dispersion	0–4 ns/MHz	$1/f^3$
Scintillation intensity/phase	Varies	Varies

profiles, but can be neglected for our purposes. Absorption and dispersion – for the signal bandwidths involved – can be neglected as well. The two most important effects are the propagation delay, as it translates into a PR error, the phase delay, and the scintillations. To the first order, the propagation (group) delay (in units of (m)) can be computed as

$$I_g = \frac{40.3 \cdot 10^{16}}{f^2} \cdot STEC, \tag{18.8}$$

where f is the frequency in Hz, and *STEC* is the Slant Total Electron Content, or the integral of the electron density along the propagation path in TECU or TEC units (1 TECU = 10^{16} electrons/m²). The phase delay (in m) is equal to the group delay, but with the opposite sign. That is, phase measurements advance when crossing the ionosphere, that is, a negative delay, and code measurements get delayed.

If more than one frequency band is used (f_1 and f_2), the ionospheric delay can be estimated:

$$\Delta t_1 = \frac{f_2^2}{f_1^2 - f_2^2} \cdot \delta(\Delta t), \tag{18.9}$$

where Δt_1 is the time delay at the f_1 due to the ionosphere, and $\delta(\Delta t)$ is the measured time difference between the two frequencies f_1 and f_2. If the receiver has only one band, ionospheric models such as Klobuchar's [36] or NeQuick [37] are used.

Ionospheric scintillation occurs mostly in equatorial and high-latitude regions, and their behavior is different. Equatorial scintillations occur around ±20° of latitude of the magnetic equator, after sunset and before midnight, and are produced by convective plasma processes. High-latitude scintillations occur in the polar region, mostly during the dark months at all local times, while auroral zones are observed during the night-time. Mid-latitude scintillations occur as an extension of phenomena occurring at equatorial and auroral latitudes or due to an intense sporadic E layer during the daytime. While in Equatorial regions amplitude and phase scintillations occur, in high-latitude regions, phase scintillation is dominant. Amplitude ionospheric scintillation is usually characterized by the S_4 parameter [38], phase scintillation by its standard deviation or σ_φ [39], and the slope of the power spectrum of signal phase (p).

Ionospheric scintillation does not affect the code-based range measurements of the navigation signals like the ionospheric delay, since it does not affect the propagation time. However, strong scintillation may lead to sudden lowering of the received power such that it results in loss of lock.

18.2.5   Satellite navigation systems

In previous sections, the discussion has focused mainly on the GPS system because it was the first one being fully deployed and operational and because most of the concepts that apply to the simple L1 C/A code apply to other frequency bands and signal modulations. In this section, other GNSSs are presented for the sake of completeness. In addition to the GPS, other systems are the Russian GLONASS, the European Galileo, the Chinese Beidou, and the regional ones from Japan, the QZSS, and India, the IRNSS. As it is evident from Figure 18.10, several of these systems share the same frequency bands to increase the interoperability, notably at L1 (1 575.42 MHz) and L5 (1 176.45 MHz), including the modernized GLONASS system.

Table 18.4 shows the main signal properties for the GPS L1 and L5, and Galileo E1 and E5 bands. The carrier frequency and bandwidths are denoted by F_c and BW, and the chipping rate and the BOC sub-carrier frequency are named as f_c and f_s, respectively. Bandwidths and received powers are the ones defined in the GPS Interface Specification (IS) documents [41] and the Galileo Interface Control Document [30], except for the L1M power and for the E1A power, which is assumed to be equally distributed within the E1 band. These documents describe the minimum received power and, therefore, may lead to pessimistic performance estimates. It is known that, actually, typical power values for the GPS signals are about 3 dB higher than the specified minimum.

18.3   GNSS-R

GNSS-R can be understood as a "multi-static radar" using signals from transmitters of opportunity because transmitters and receivers are separated by a considerable distance. The GNSS-R concept is sketched in Figure 18.11, showing a receiver collecting the direct and reflected (actually scattered) navigation signals emitted by a satellite, but in practice, they can collect as many reflections as satellites are in view.

GNSS-R instruments can be operated either as a scatterometer when the power of the forward-scattered signal is measured or as an altimeter when the time delay between the direct and the reflected signals is measured or as a (unfocused) synthetic aperture radar (SAR) when the observable is the whole delay-Doppler map (DDM) that is deconvolved to derive a radar cross-section density (σ^0) image. Depending on the application and the receiver's platform (from static ground-based to spaceborne), instruments will have different requirements. For scatterometric purposes, bandwidth is not an issue and a single frequency receiver suffices. For altimetric

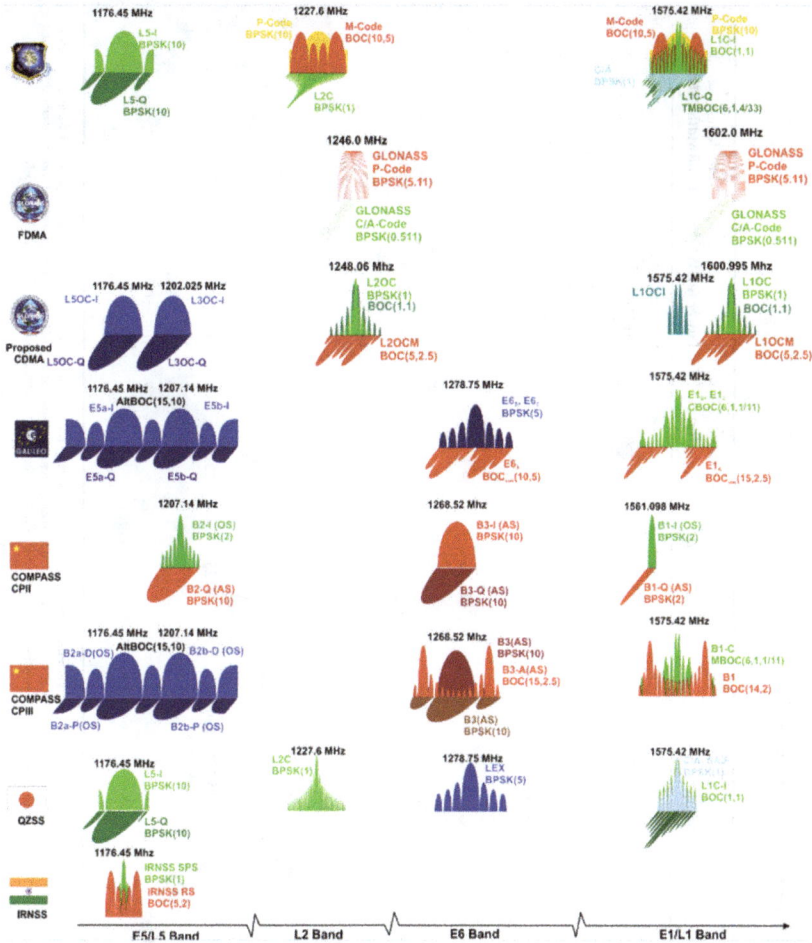

Figure 18.10   Frequency bands used by the different Global Navigation Satellite Systems (GNSS) [40]

applications, having a large bandwidth is a must, since it determines the achievable altimetric accuracy, and dual frequency instruments are required to compensate for the ionospheric delay. Instruments measuring DDMs will be much more complex, since they will require a large number of complex correlators for each delay-Doppler bin, while for scatterometry and altimetry applications, just a few correlators are needed to measure a delay cut along the central Doppler frequency.

Before entering in the particularities of the different GNSS-R techniques and applications, it is convenient to briefly review some elementary concepts of a bistatic radar.

Table 18.4   GNSS signals main parameters (adapted from Pascual et al. [31])

GNSS	Band and bandwidth (MHz)	Service	Component	Modulation		f_c (MHz)	f_s (MHz)	Power (dBW)[1,3,4,6]	Main lobe bandwidth (MHz)[7]
GPS	L1[1] Fc = 1575.42 BW$_1$ = 20.46 BW$_2$ = 30.69	P(Y)[2]	DATA	BPSK-R10		10.23	-	Min: −161.5 Typ: −158.5 Max: −155.5	20.46
		C/A	DATA	BPSK-R1		1.023	-	Min: −158.5 Typ: −155.5 Max: −153	2.046
		M[2]	N/A	BOCs(10,5)		5.115	10.23	Min: −157 Typ: −154 Max: −150	30.69
	L5[2] Fc = 1176.45 BW = 24	SoL	DATA (L5I)	BPSK-R10		10.23	-	Min: −157.9 Typ: −154.9 Max: −150	20.46
			DATA (L5Q)	BPSK-R10				Min: −157.9 Typ: −154.9 Max: −150	20.46
GALILEO	E1 Fc = 1575.42 BW = 24.552 BW$_{assumed}$ = 32	PRS[2]	DATA (E1A)	BOCc(15,2.5)		25.575	15.345	Min: −157[5] Typ: −154 Max: −150	35.805
		OS,SoL,CS	DATA (E1B) PILOT (E1C)	CBOC (6,1,1/11)	BOCs(1,1) BOCs(6,1)	1.023	1.023 6.138	Min: −157 Typ: −154 Max: −150	4.092 14.322
	E5 Fc = 1191.795 BW = 51.15 Fca = 1176.45 Fcb = 1207.14	OS	DATA (E5aI) PILOT (E5aQ)	AltBOC(15,10) + constant envelope		10.23	15.345	Min: −155 Typ: −152 Max: −148	51.15
		OS,SoL,CS	DATA (E5bI) PILOT (E5bQ)					Min: −155 Typ: −152 Max: −148	

[1]IS defined RF bandwidths; [2]Restricted codes; [3]GPS signals: Received minimum RF signal strength on Earth's surface when the Space Vehicle (SV) is above 5° user elevation angle with a 3 dBi linearly polarized antenna, [4]Galileo signals: Received minimum RF signal strength on Earth's surface when the SV is above 10° user elevation angle with an ideally matched and isotropic 0 dBi antenna and lossless atmosphere; [5]Equal power distribution assumed; [6]For the GPS signals, the typical value is obtained by increasing 3 dB the specified minimum; [7]for the BOC, signals the bandwidth is defined between the outer nulls of the largest spectral lobes. Basic Concepts of GNSS-R.

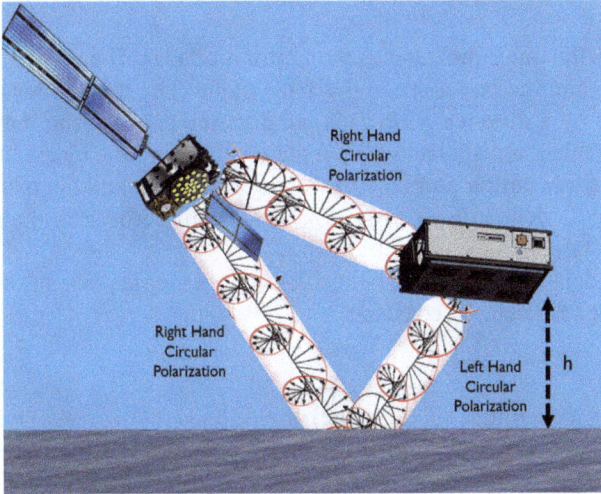

Figure 18.11   Graphical representation of a small satellite collecting the direct right-hand circular polarization signal, and the reflected left-hand circular polarization signal

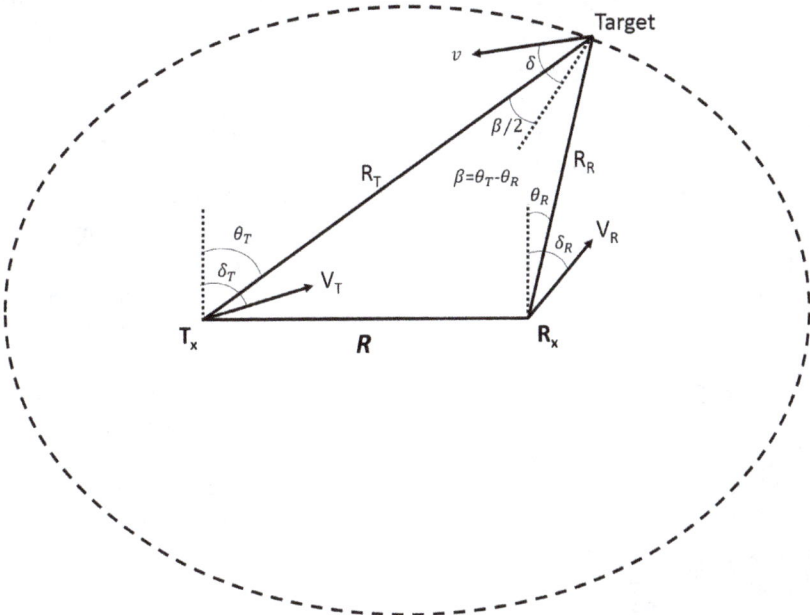

Figure 18.12   Geometry of a bistatic Doppler (adapted from Klobuchar [36])

18.3.1   Spatial resolution

Figure 18.12 illustrates the geometry of a bistatic radar. The transmitter (T_x), the receiver (R_x), and the target are in the same plane. The transmitter and receiver are separated by a distance R, $R_T(t)$ is the distance between the transmitter and the target $R_R(t)$ is the distance between the receiver and the target, the angles θ_T and θ_R are the transmitter and receiver angles of arrival, and the bistatic or scattering angle β is defined as $\beta = \theta_T - \theta_R$ and it will be used in the calculations related to the target.

The "iso-delay" contours are defined by the geometric place of the points that satisfy that the total delay is constant:

$$\tau = \frac{R_T(t) + R_R(t)}{c} = \text{const.} \tag{18.10}$$

These points make ellipses (in the plane) or ellipsoids (in the space) with the transmitter and receiver in the two foci (Figure 18.12). Taking the receiver's position as a reference and defining $R_{tot}(t) = R_T(t) + R_R(t)$, the equations of the ellipses are given by

$$R_R(t) = \frac{R_{tot}^2(t) - L^2}{2(R_{tot}(t) + L \cdot \sin\theta_R)}. \tag{18.11}$$

The "iso-Doppler" contours are given by

$$f_D = -\frac{1}{\lambda}\frac{d}{dt}\{R_T(t) + R_R(t)\} = \text{const.,} \tag{18.12}$$

where λ is the electromagnetic wavelength. In the particular case of a stationary (or quasi-stationary) target (e.g., the Earth's or the ocean's surface), (18.12) becomes

$$f_D = \frac{\vec{V}_T \cdot \hat{n}_{Tx-target}}{\lambda} + \frac{\vec{V}_R \cdot \hat{n}_{Rx-target}}{\lambda}. \tag{18.13}$$

where $\hat{n}_{Tx-target}$ and $\hat{n}_{Rx-target}$ are the unit vectors from the transmitter and receiver to the target. For a flat surface, the iso-Doppler contours become hyperbolae. Figure 18.13 shows the iso-delay (or iso-range) contours together with the iso-Doppler contours. As it can be noticed, each pixel in the (x,y) plane is associated with a delay-Doppler pair (τ, f_D), but this correspondence is not unique since two pairs of (x, y) points correspond to the same (τ, f_D) values (gray areas in Figure 18.13). In a synthetic aperture radar, this ambiguity is solved by steering the antenna towards an unambiguous zone away from the y-axis, but in GNSS-R, in principle, this is not often the case since the antennas point towards the specular direction point (sp) at the origin of the coordinates in Figure 18.13. This issue will be addressed in a subsequent section when dealing with GNSS-R as imagers.

In the case where the surface is perfectly flat, the scattering is coherent and it comes from a region around the specular point (sp in Figure 18.13a) determined by the size of the first Fresnel zone, an ellipse with semi-axes:

$$a = \sqrt{\lambda \cdot \frac{|\vec{R}_T| \cdot |\vec{R}_R|}{|\vec{R}_T| + |\vec{R}_R|}}, \tag{18.14a}$$

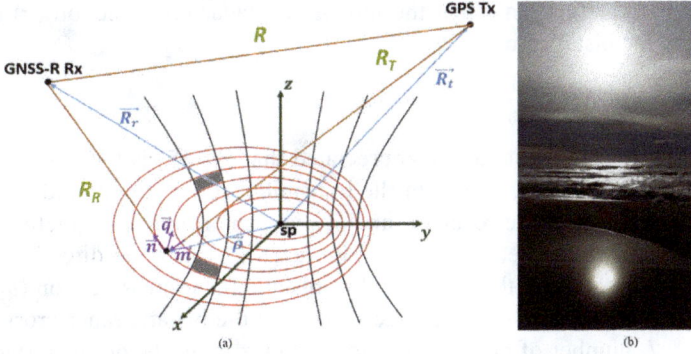

Figure 18.13   a) Geometry of the iso-delay (ellipses) and iso-Doppler (hyperbolae) contours overlaid. $\vec{R}_t$, $\vec{R}_r$ and $\vec{\rho}$ are the position vectors of the transmitter, receiver, and scattering point. $(\vec{m}, \vec{n}, \vec{q})$ are the local scattering vectors. b) Scattered waves over a rough surface come from a larger area because the increased slopes (delay-Doppler bins farther away from the specular reflection point), and scattered waves over a flat surface come from a small disk (author: A. Camps).

$$b = \frac{a}{\cos(\theta_i)},$$ (18.14b)

and θ_i is the local incidence angle. From a LEO satellite, the spatial resolution is then determined by the size of the first Fresnel zone, around 300–500 m regardless of the antenna footprint (see e.g. [42, 43] for a more detailed discussion on the achievable spatial resolution).

On the other hand, the size of the first iso-delay disk is $\tau = (R_T(t) + R_R(t))/c = \tau_c$, ~10 km for a transmitter in the zenith, and the separation of the iso-Dopplers (1 kHz, corresponding to 1 ms coherent integration) is >14 km (for a rate of change of >70 Hz/km). It should be noted that in the case of a rough surface, the scattering is incoherent, and it comes from a region around the specular reflection point, whose size increases with the surface roughness; therefore, the spatial resolution is much worse than in the coherent scattering case.

18.3.2   Received power: coherent and incoherent components

Later on, when explaining the GNSS-R observables, the most widely used expression for the "Delay Doppler Map" will be presented. Here, the discussion will focus on the received power, making the link to the radar theory. In a bistatic radar, the total power is computed as the sum of the scattered power (dP_R) from every dS area in the surface, and it can be computed in a similar way as for a monostatic radar but taking into account that R_T and R_R are different, and so the antenna gains of the transmitting (G_T) and receiving antennas (G_R) must be evaluated in the direction of the target, and then the monostatic radar cross-section (σ) must be replaced by the bistatic one (σ_b):

$$dP_R = \frac{P_T \cdot G_T}{4\pi R_T^2} \cdot \sigma_b \cdot \frac{1}{4\pi R_R^2} \cdot \frac{G_R \cdot \lambda^2}{4\pi} \cdot L. \tag{18.15}$$

As compared to (18.5), it can be appreciated that (18.15) is the cascade of two equations such as (18.5), one from the transmitter to the target, and the second one from the target to the receiver, and σ_b plays the role of an effective area (τ, f_D) cell of area dS marked in gray in Figure 18.13 times the directivity of the target in the direction of the receiver. The bistatic radar cross-section (σ_b) has to be computed as the area of the surface dS times the bistatic radar cross-section density σ^0. A number of models exist to predict σ^0 over the ocean surface more accurately than the simple qualitative expressions given above. One of the most widely used models in GNSS-R is the Kirchhoff model under the stationary phase approximation,[a] or tangent plane, which assumes that scattering occurs only for points on the surface for which there are specular reflections, and local diffraction effects are excluded.

$$\sigma^0\left(\vec{r}\right) = \pi \cdot k^2 \left|\Re\right|^2 \cdot \frac{q^2}{q_z^4} \cdot P\left(Z_x, Z_y\right), \tag{18.16}$$

where $k = 2\pi/\lambda$ is the electromagnetic wavenumber, $\Re$ is the Fresnel reflection coefficient (actually it is $\Re_{rs}$ to indicate the incident polarization "r" and the scattered polarization "s"), $\vec{q} = q_x \cdot \hat{x} + q_y \cdot \hat{y} + q_z \cdot \hat{z} \triangleq k \cdot \left(\hat{k}_s - \hat{k}_i\right)$, $q = \left|\vec{q}\right|$, and $P\left(Z_x, Z_y\right)$ is the probability density function of the surface slopes along the x- and y-directions (Z_x and Z_y) and can be obtained from the sea surface wave spectrum or from the slopes statistics (e.g., [44] or [45]).

Since navigation systems use Right-Hand Circular Polarization (RHCP or R) to avoid ionospheric effects and antenna pointing requirements at receiver level, the Fresnel reflection coefficients at incidence angle θ and r, s polarizations (r, $s = L$, R, H or V, i.e., Left/Right-hand circularly polarized, Horizontal or Vertical) to be used in (18.16) are given by [46]:

$$\Re_{RR} = \Re_{LL} = \frac{\Re_{VV} + \Re_{HH}}{2}, \tag{18.17a}$$

$$\Re_{RL} = \Re_{LR} = \frac{\Re_{VV} - \Re_{HH}}{2}, \tag{18.17b}$$

where

---

[a]There are other techniques to compute the bistatic scattering coefficient such as the Small Perturbation Method (SPM), the two-scale model (TSM), the Integral Equation Model (IEM), and the Small Slope Approximation. However, because of its simplicity, and when polarimetric studies do not have to be performed, the Kirchhoff model in the Geometric Optics limits is the one most often used in GNSS-R.

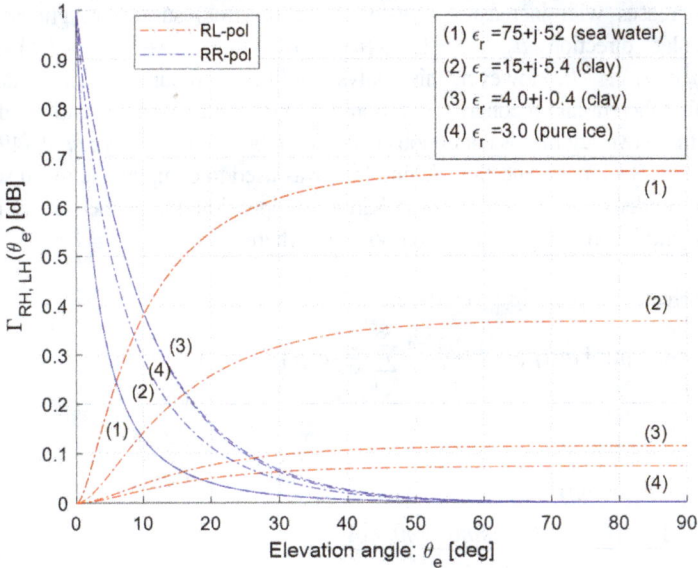

Figure 18.14   *The reflection coefficients as a function of the elevation angle for an incidence wave at Right-Hand Circular Polarization (RHCP), and a reflected one at RHCP (blue), or Left-Hand Circularly Polarized (LHCP)*

$$\mathfrak{R}_{VV} = \frac{\varepsilon_r \cdot \sin(\theta) - \sqrt{\varepsilon_r - \cos^2\theta}}{\varepsilon_r \cdot \sin(\theta) + \sqrt{\varepsilon_r - \cos^2\theta}}, \tag{18.18a}$$

$$\mathfrak{R}_{HH} = \frac{\sin(\theta) - \sqrt{\varepsilon_r - \cos^2\theta}}{\sin(\theta) + \sqrt{\varepsilon_r - \cos^2\theta}}, \tag{18.18b}$$

with ε_r being the dielectric constant of the medium where the reflection is taking place, assuming that the dielectric constant of the first medium is equal to 1. Figure 18.14 shows the reflection coefficients as a function of the elevation angle for an incidence wave at RHCP, and a reflected one at RHCP (blue), or left-hand circularly polarized (LHCP) for different values of the dielectric constant. Note that at low-elevation angles, the polarization of the reflected signal is RHCP regardless of the material, but for elevation angles larger than 7°–30,° the reflected signal is mostly LHCP.

The Kirchhoff model under the Physical Optics (PO) approximation involves the integration of the scattered fields over the entire rough surface, not just the facets contributing to the specular reflection. Unlike the Geometric Optics (GO) approximation, the PO approximation also predicts a coherent component, which is given by Long and Ulaby [47]:

$$\sigma_{pq}^{0,\,coh} = \pi k^2 \left| a_0 \right|^2 \delta \left(q_x \right) \delta \left(q_y \right) e^{-q_z^2 \cdot \sigma^2}, \tag{18.19}$$

which decreases with increasing root-mean-square (rms) surface roughness (σ). In the specular direction $a_{0,\,VV} = +2\Re_{VV}\left(\theta_i\right) \cdot cos\left(\theta_i\right)$, $a_{0,\,HH} = -2\Re_{HH}\left(\theta_i\right) \cdot cos\left(\theta_i\right)$, and $a_{0,\,VH} = a_{0,\,HV} = 0$. However, this analysis is limited to surfaces with small slopes [48]. Both coherent and incoherent components exist simultaneously, and the transition between these two regimes is continuous was shown by Alonso-Arroyo *et al.* [49], where the model by Beckmann and Spizzichino [50] was used to compute the mean scattering coefficient under the assumption of the Kirchoff approximation as the sum of coherent ($\langle\rho\rangle\,\langle\rho^*\rangle$) and incoherent ($Var\,\{\rho\}$) components, where

$$\text{coherent: } \langle\rho\rangle\,\langle\rho^*\rangle = |\rho_0|^2 \cdot \exp\left(-g\right), \tag{18.20}$$

$$\text{incoherent: } Var\,\{\rho\} = \frac{\pi T^2 F^2 e^{-g}}{A} \sum_{m=1}^{\infty} \frac{g^m}{m!\cdot m} exp\left(-\frac{u_{xy}^2 T^2}{4m}\right), \tag{18.21}$$

where

$$g = \left(2k\sigma_h cos\left(\theta_i\right)\right)^2, \tag{18.22}$$

$$F = \frac{1 + cos\theta_i\,cos\theta_s - sin\theta_i\,sin\theta_s\,cos\phi_s}{cos\theta_i\,\left(cos\theta_i + cos\theta_s\right)}, \tag{18.23}$$

and

$$u_{xy}^2 = k^2\left(sin^2\theta_i - 2sin\theta_i\,sin\theta_s\,cos\phi_s + sin^2\theta_s\right). \tag{18.24}$$

In (18.21), T is the correlation length of the surface and A is the scattering area, and in (18.22), σ_h is the surface rms height, θ_i and θ_s are the incidence and scattering angles, and ϕ_s is the azimuthal scattering angle. For $g \ll 1$, the surface is considered smooth, and for $g \gg 1$, the surface is considered rough.

As it can be seen, the power of the coherent component does not decrease as $1/(R_T^2 \cdot R_R^2)$ in (18.14), but as in (18.5) replacing R by $\left(R_T + R_R\right)$, and multiplying it by (18.20). A similar result was derived by Voronovich and Zavorotny [51], where the reflection coefficient is replaced by $\Re_{pp}\left(\theta_i\right) \cdot \exp\left(-4R_a^2\right)$, where the Rayleigh parameter R_a is given by $R_a = k\sigma_h cos\left(\theta_i\right)$.

Finally, the calculation of *SNR* requires dividing the received power ((18.5) with $\left(R_T + R_R\right)$ for the coherent case or integrating (18.15) for the incoherent case) by the thermal noise introduced by the receiver:

$$N = k_B \cdot T_{sys} \cdot B, \tag{18.25}$$

where k_B is the Boltzmann's constant ($k_B = 1.38 \cdot 10^{-23}$ J/K), $T_{sys} = T_A + T_R$ is the system temperature, which includes the noise collected by the antenna (T_A) and the noise generated internally by the receiver (T_R), and B is the noise bandwidth. However, the *SNR* is not only affected by thermal noise but also increased by the power within the band of other interfering signals, including other GNSS signals from the same system, from a different constellation, from signals with a different structure, belonging or not to the same system

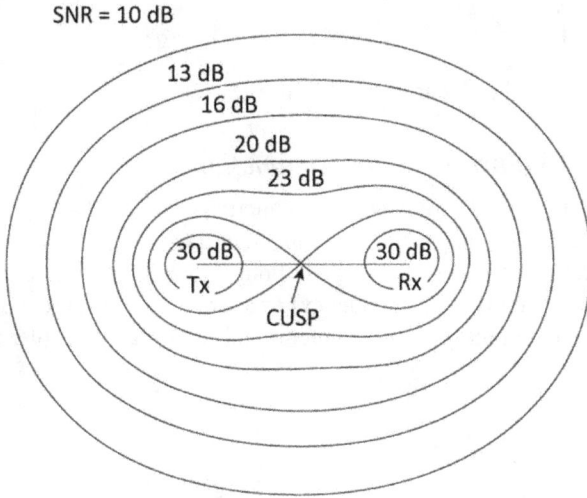

Figure 18.15 *Contours of constant signal-to-noise ratio (SNR), or ovals of Cassini, where the baseline = L and k = 30 L (adapted from Skolnik [52]).*

[46, 47], and by speckle noise,[b] which is often the dominant term. A rough approximation to assess the effect of speckle noise is by adding it to the thermal noise as

$$NSR'_{R/cr} = NSR_{D/cr} + NSR_{speckle},$$
(18.26)

where the speckle noise-to-signal ratio $NSR_{speckle} = 1/3.63$, or -5.6 dB in logarithmic units, and it is independent of the bandwidth. Speckle noise can only be reduced by incoherent averaging, that is, by averaging the absolute values of N_i continuous samples (ideal reduction factor equal to $\sqrt{N_i}$).

The regions of constant SNR can be derived by isolating $R_T \cdot R_R$ from the ratio of (18.15) and (18.25), leading to:

$$\left(R_T \cdot R_R\right)\big|_{max} = \sqrt{\frac{P_T \cdot G_T \cdot G_R \cdot \lambda^2 \cdot \sigma_b \cdot L}{(4\pi)^3 \cdot k_B \cdot T_{sys} \cdot B \cdot SNR_{min}}} = \kappa,$$
(18.27)

or

$$SNR = \frac{k}{R_T^2 \cdot R_R^2},$$
(18.28)

with $k \triangleq \kappa^2 \cdot SNR_{min}$.

Finally, the contours $r(\theta)$ of constant SNR on any bistatic plane can be derived as

[b]Speckle noise is a "multiplicative noise", also found in other imaging instruments with a coherent illumination source, such as synthetic aperture radars or laser. It is caused by the coherent sum of the scattered waves in many different "facets" of the surface with different phases. Speckle noise is especially significant in the ocean and it can only be reduced by incoherent averaging.

$$R_T^2 \cdot R_R^2 = \left(r^2 + \tfrac{L^2}{4}\right)^2 - r^2 \cdot L^2 \cdot \cos^2\theta, \tag{18.29}$$

which are known as the Cassini ovals (Figure 18.15).

18.3.3   The Woodward ambiguity function

In radar theory, the Woodward Ambiguity Function (WAF) [53] is used to characterize the radar performance, as there is an inherent trade-off in the ability of a signal $u(t)$ to accurately measure simultaneously both the range (linked to the delay) and velocity (linked to the Doppler frequency) of a target. In the context of GNSS-R, it is the response to a point target as a function of delay τ and Doppler frequency f_D, and it is defined as

$$\left|\chi\left(\tau, f_D\right)\right| = \left|\int_{-\infty}^{+\infty} u\left(t\right) \cdot u^*\left(t - \tau\right) \cdot e^{j2\pi f_D t} \cdot dt\right| \tag{18.30}$$

Because of the non-linear relationship between the delay and the range (18.10), and the Doppler frequency and the speed (18.12) [54], in the bistatic case, the shape of the ambiguity function gets distorted.

Let us focus on the relevant case of a rectangular pulse of duration T as given by (18.31):

$$s\left(t\right) = \tfrac{1}{\sqrt{T}}\Pi\left(\tfrac{t}{T}\right). \tag{18.31}$$

The associated WAF is given by (18.32):

$$\left|\chi\left(\tau, f_D\right)\right| = \left|\left(1 - \tfrac{|\tau|}{T}\right) \cdot sinc\left(T \cdot f_D \cdot \left(1 - \tfrac{|\tau|}{T}\right)\right)\right|, \tag{18.32}$$

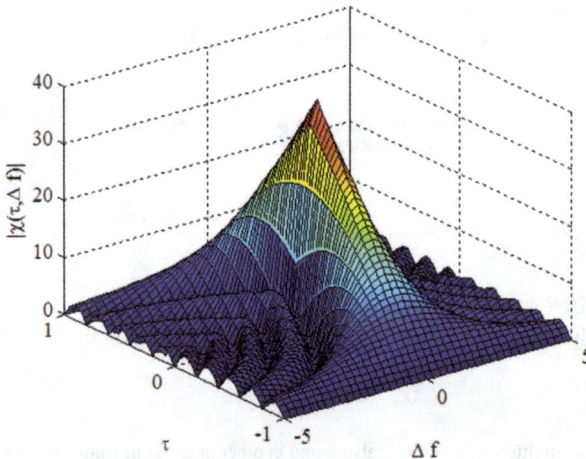

Figure 18.16   *Woodward ambiguity function for rectangular pulse (normalized duration = 1)*

which is plotted in Figure 18.16 for $T = 1$. The width along the τ direction for $\Delta f = f_D = 0$ is 1 unit ($1/T$), while in frequency, along $\tau = 0$, it is a sinc function. Therefore, a rectangular pulse can achieve high temporal resolution (i.e., range) if the pulse is short ($T \to 0$), but then, the frequency resolution degrades ($1/T \to \infty$). In the case of the GPS L1 C/A code, pulses are ~1 μs long, so the range resolution is ~300 m, and since the coherent integration is performed during 1 ms (1 023 samples), the Doppler resolution is ~1 kHz. Away from the maximum, $|\chi|$ vanishes and it is almost zero although there are residual contributions away from them. The resolution cells or "delay-Doppler" bins are then the regions where WAF is larger than half its peak value.

18.3.4   GNSS-R observables and techniques

GNSS-Reflectometers correlate the received reflected signal $s_R(t)$ with either:

1. A locally generated replica of the transmitted signal $a(t)$, only if the codes are publicly available. This technique is called conventional GNSS-R or cGNSS-R:

$$Y^c(t, \tau, f_d) = \int_t^{t+T_c} s_R(t') \, a^*(t' - \tau) \, e^{-j2\pi(F_c + f_d)t'} \, dt'. \tag{18.33}$$

2. The direct signal itself $s_D(t)$ collected by an up-looking antenna. This technique is called interferometric GNSS-R or iGNSS-R:

$$Y^i(t, \tau, f_d) = \int_t^{t+T_c} s_R(t') \, s_D^*(t' - \tau) \, e^{-j2\pi(F_c + f_d)t'} \, dt'. \tag{18.34}$$

3. A reconstructed version of the direct signal $\hat{s}_D(t)$. Although some chips in this "reconstructed" or regenerated direct signal may be wrong, overall, the *SNR* is improved. This technique is called reconstructed GNSS-R or rGNSS-R:

$$Y^r(t, \tau, f_d) = \int_t^{t+T_c} s_R(t') \, \hat{s}_D^*(t' - \tau) \, e^{-j2\pi(F_c + f_d)t'} \, dt'. \tag{18.35}$$

In (18.33)–(18.35), t is the time when the coherent integration starts, F_c is the carrier frequency, and T_c is the coherent integration time. The coherent integration time has to be a multiple of the code duration including secondary codes – if any – that is 1 ms for GPS L1 C/A, or 4 ms for Galilo E1B and E5b-I, 20 ms for E5a-I, or 100 ms for Galileo E1C and E5a/b-Q [55]. The *SNR* can be increased by increasing the coherent integration time over several code duration periods provided the navigation signal does not introduce a sign change. However, the coherent integration time must also be smaller than the coherence time, which depends on the receiver dynamics (e.g., ground-based, airborne, or spaceborne) and the dynamics of the surface where the scattering is taking place, for example, even for a ground-based receiver, the behavior of GNSS-R signals over land or the ocean will be different.

Since the reflected signal is even weaker than the direct signal, which is typically under the noise level, the signal-to-(thermal) noise ratio is even poorer. In addition, since the scattered signal comes from multiple scatterers over the surface with different phases within the resolution cell, it also suffers from speckle noise, and a large number of incoherent averages (N_i) are required in order to improve the SNR of $Y^{c,i,r}(t, \tau, f_d)$:

$$\left\langle \left| Y^{c,i,r}(\tau, f_d) \right|^2 \right\rangle \approx \frac{1}{N_i} \sum_{n=1}^{N_i} \left| Y^{c,i,r}(t_n, \tau, f_d) \right|^2, \tag{18.36}$$

However, the rate of change of the delay and Doppler frequencies must be tracked properly in both delay and Doppler so that the GNSS-R observables do not get blurred [56, 57]. These dynamics depend on the receiver type and are faster in cGNSS-R than in iGNSS-R.

A detailed analysis of $\left\langle \left| Y(\tau, f_d) \right|^2 \right\rangle$ (superscripts not shown unless required from now on) was performed by Zavorotny and Voronovich [46]:

$$\left\langle \left| Y(\tau, f_d) \right|^2 \right\rangle = \frac{P_T \cdot \lambda^2}{(4\pi)^3} T_c^2 \iint \frac{G_T\left(\overrightarrow{r}\right) G_R\left(\overrightarrow{r}\right)}{4 \cdot R_T^2\left(\overrightarrow{r}\right) \cdot R_R^2\left(\overrightarrow{r}\right)}$$
$$\left| \chi \left(\tau - \frac{R_T\left(\overrightarrow{r}\right) + R_R\left(\overrightarrow{r}\right)}{c}, f_D\left(\overrightarrow{r}\right) - f_c \right) \right|^2 \sigma^0\left(\overrightarrow{r}\right) \cdot d^2r, \tag{18.37}$$

The physical interpretation of $\left\langle \left| Y(\tau, f_d) \right|^2 \right\rangle$ in (18.37) is the incoherent sum of the received power[c] by a bistatic radar from all the points in the surface where the incident wave is scattered. The following considerations must be made.

In (18.37), the WAF $\left| \chi \right|^2$ is maximum around surface points that satisfy that the delay and Doppler are:

$$\tau \approx \frac{R_T\left(\overrightarrow{r}\right) + R_R\left(\overrightarrow{r}\right)}{c}, \tag{18.38}$$

and

$$f_D \approx f_c. \tag{18.39}$$

The WAF shape can be computed using (18.30), and a negligible error is committed if the time and frequency dependences are separated so that it can be approximated by the product of the ACF of the codes used and the sinc($f \cdot T_c$). In the case of the GPS C/A code, (18.32) is approximated by

$$\left| \chi(\tau, f_D) \right| \approx \left| S\left(\frac{|\tau|}{T_c} \right) \text{sinc}\left(f \cdot T_c \right) \right|. \tag{18.40}$$

where S ($\cdot$) is an ACF, for example, a triangular function for GPS L1 C/A code.

---

[c]Equation (18.36) is to be compared with (18.15).

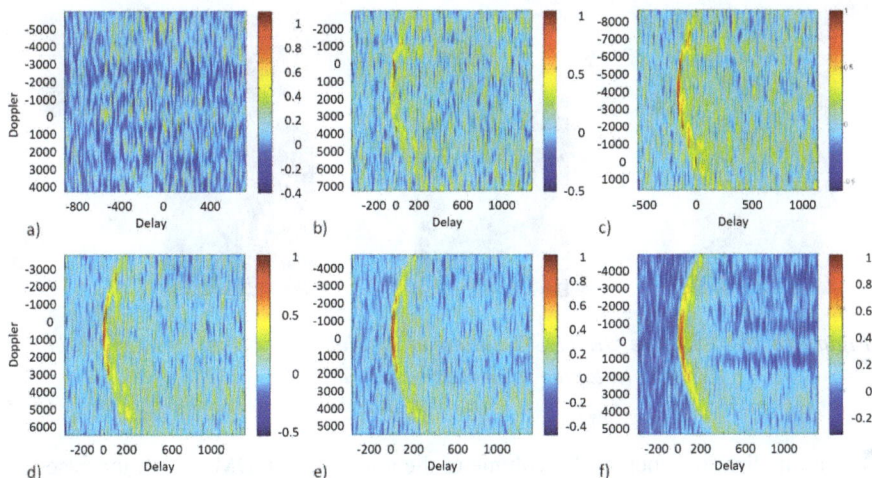

Figure 18.17   Sample UK Disaster Monitoring Constellation (UK-DMC) Global Positioning System-Reflectometry (GPS-R) data processed with a software receiver for a coherent integration time of 1 ms, and different incoherent integration times: a) 10 ms, b) 40 ms, c) 80 ms, d) 120 ms, e) 200 ms, and f) 800 ms

Using the above approximation, the WAFs for other GPS, Galileo, and so on, navigation signals can be readily computed.

Figure 18.17 shows sample UK-DMC GPS C/A code – reflected data processed with a software receiver for a coherent integration time of 1 ms and different incoherent integration times from 10 ms to 800 ms over the ocean. As it can be appreciated, the incoherent integration time must be increased significantly to reduce the noise, and still the *SNR* is moderate. It is also important to note the spread over a large Doppler frequency range ~10 kHz, and delays up to ~10–15 C/A chips. In the case of the ocean, this spread increases with increasing wind speeds as waves exhibit larger slopes, and therefore the scattered signal can arrive from points that are farther away from the specular reflection point (18.16).

In the case of reflections over land (Figure 18.18 right) or ice (Figure 18.18 left), the delay and Doppler spreads are much smaller than over the ocean (Figure 18.18 center) as surface roughness is also much smaller. Note also the differences in the *SNR*s as the reflection coefficients over land are much smaller.

At this point and in view of Figures 18.17 and 18.18, different GNSS-R observables can be defined:

- DDM is the most complete GNSS-R observable, and it is simply (18.33) computed for a number of delay and Doppler frequency bins around the cross-correlation peak. DDM is depicted in Figures 18.17 and 18.18, and it contains all the information to derive, for example, wind speed and direction over the ocean. A number of features of the DDM can be computed, typically after

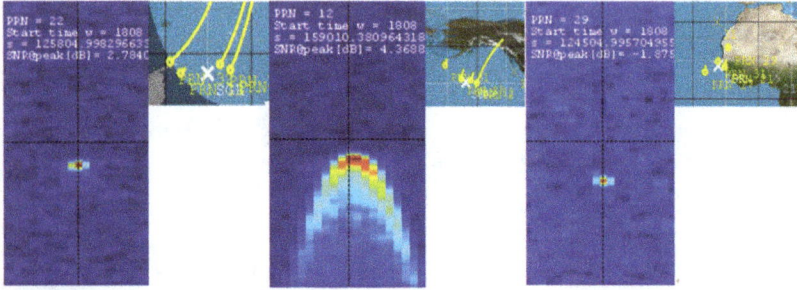

Figure 18.18   Sample UK TDS-1 data over ice (left), the ocean (center), and land (right) (adapted from merrbys.co.uk/)

normalization, such as the volume of the normalized DDM, which increases as DDM spreads, the Delay-Doppler Map Average (DDMA), the Delay-Doppler Map Variance (DDMV), the Allan Delay-Doppler Map Variance (ADDMV), the Leading Edge Slope (LES), the Trailing Edge Slope (TES), and the Minimum Variance (MV). These observables can be combined appropriately in order to obtain a generalized observable that maximizes the *SNR*.

*   The waveform (WF) is the cut of the DDM in the delay direction through the peak. This observable contains the minimum information required to derive the altimetric information from the position of the peak of its derivative in the leading edge (LE), or biomass information from the peak of its derivative in the trailing edge (TE), or sea ice detection from the Doppler Integrated Waveform (DIW), or to perform the sea state correction in sea surface salinity retrievals using microwave radiometry at L-band from the normalized waveform area.
*   The Waveform Peak Amplitude (WPA) and the Waveform Peak Phase (WPP). The WPA observable contains the minimum scattering information required, for example, to derive wind speed, soil moisture, or vegetation biomass. This observable can be either absolute or differential (relative amplitude between the peaks measured by an up-looking and a down-looking antenna). It can also be measured in reflection or in transmission. The WPP observable contains information of the phase of the reflection coefficient and the distance (i.e., satellite height and topography).
*   The ratio of the WPAs at two polarizations, for example, LHCP and RHCP. This observable contains the scattering information required, for example, to derive wind speed, although it depends on the polarization purity of the transmitted signals and the cross-polarization level of the receiving antennas.

Figure 18.19 shows the main types of GNSS-R instruments: a) the cGNSS-R, (18.29), b) the iGNSS-R, (18.30), and c–d) the reconstructed code GNSS-R (rGNSS-R, (18.31)) [58, 59]. In rGNSS-R, the locked C/A code model is used to form an L1P model, which is then applied to the direct signal (center left), and after integration, over ~0.5 MHz W-chips to estimate their signs, it is combined with the P-code model

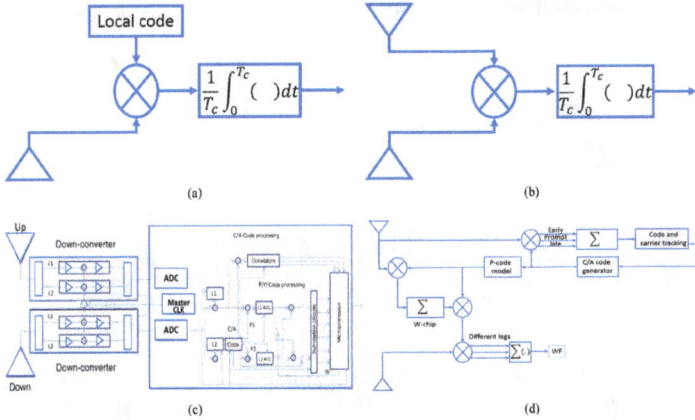

Figure 18.19 *Block diagrams of different types of Global Navigation Satellite System-Reflectometry (GNSS-R) instruments: a) conventional GNSS-R (cGNSS-R), b) interferometric GNSS-R (iGNSS-R), c) rGNSS-R [58], d) rGNSS-R [59]*

to form an L1 Y-code model, which is used to correlate with the down-looking channel. The advantages of this technique rely mainly on the larger bandwidth of the P(Y) codes as compared with the C/A ones, and the large *SNR*, despite the losses of the semi-codeless approach. In addition, there is a fourth one that has been proposed but, to the authors' knowledge, has never been implemented: the partial iGNSS-R or piGNSS-R [60]. In the piGNSS-R, the signal from the uplooking antenna is used to estimate the C/A code, which is then multiplied by the direct signal. The product signal then only contains the P and M codes, and it is ultimately cross-correlated against the signal from the down-looking antenna as in an iGNSS-R. The advantage of this technique is an even better range resolution as compared to the iGNSS-R one as the rms bandwidth (defined in (18.37)) is larger, but at the expense of a 3 dB signal loss (C/A code has been removed and also its associated power), which needs to be compensated by a 3 dB larger antenna directivity .

The selection of the particular GNSS-R technique to be used and the dimensions of the τ-f_D space (i.e., number and width of the delay and Doppler frequency bins) to be measured will depend on the platform height (the higher the platform, the larger the spread of the DDM), speed (larger Doppler), and application. For example, in a ground-based platform, the delay and Doppler frequency spread is negligible, so measuring the peak of the DDM is enough.

Regarding the application, cGNSS-R is enough for scatterometry, since the peak of the waveform does not depend much on the bandwidth of the signals. Therefore, measuring a few delay-Doppler bins around the DDM peak is enough, or even a few delays around the WF peak. For altimetry, due to the larger signal's bandwidth, the slope of the leading edge of the iGNSS-R or rGNSS-R waveforms is steeper than that of the cGNSS-R waveform (see Figure 18.20), which should lead to a better delay (range) resolution (i.e., smaller σ_τ). Under the assumption of uncorrelated

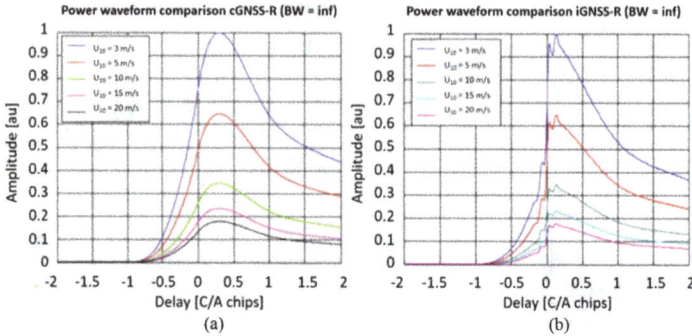

Figure 18.20   Normalized power waveforms for different wind speeds normalized to U_{10} = 3 m/s, for h = 700 km and θ_i = 0°, for a) interferometric GNSS-R (iGNSS-R), and b) conventional GNSS-R (cGNSS-R) (adapted from Martín et al. [61]). GNSS-R, Global Navigation Satellite System-Reflectometry.

additive white Gaussian noise (AWGN), the delay estimation error is given by the Cramer-Rao bound[d] (CRB):

$$\sigma_\tau^2 \geq \frac{1}{SNR \cdot \beta^2},$$ (18.41)

where β is the so-called rms or Gabor bandwidth, defined as:

$$\beta^2 \triangleq \frac{\int_0^{min\{B,B_{IS}\}} f^2 |S(f)|^2 \, df}{\int_0^B |S(f)|^2 \, df},$$ (18.42)

where B and B_{IS} are the baseband bandwidths of the receiver's filter and the transmitted signal according to the IS documents [30, 41], and $|S(f)|^2$ is its spectrum. Other effects not discussed in this chapter but that may have a non-negligible impact are: 1) the error introduced by inter-satellite cross-talk due to the residual cross-correlation between different signals in cGNSS-R [63] or because two or more satellites fall within the antenna beam in iGNSS-R [64] and 2) error correlation between waveform samples [61, 65].

At this point, it is convenient to introduce three other GNSS-R techniques that are applicable only for ground-based receivers at a height that is a fraction of the chip length ($\tau_c \cdot c$). This way, the direct and reflected signals arrive at the receiving antenna within the same chip. Somehow, the first two inherit the pioneering work by Kavak *et al.* [66],

---

[d]For low SNRs (<5 dB) the Ziv-Zakai bound [62] provides a better indication of the magnitude of the estimation errors, which can be actually much larger than the CRB ones.

where they observed that reflected GPS signals exhibit fadings that depend – among other parameters – on the soil moisture through the dielectric constant ε_r.

The first of these techniques is the *SNR* technique [67] that relies on the observation of multi-path effects due to ground reflections measured by geodetic GPS receivers with an RHCP Zenith-looking antenna. While the period of the *SNR* oscillations depends on the receiver height, the amplitude and phase are related to the soil moisture (ε_r). This technique is limited to very low-elevation angles (below the Brewster angle) because the reflected signal must be kept as the RHCP so that it can be collected by the antenna. Data gathered by networks of geodetic receivers are used to routinely produce soil moisture and snow maps [68]. An extension of this technique to also include phase measurements was presented by Löfgren and Haas [69] to measure sea tides, showing that SNR analysis was possible even for high winds, while the phase delay analysis became difficult above moderate winds.

The second of these techniques is the Interference Pattern Technique (*IPT*) [70] that overcomes the limitation of the elevation angle by using linearly polarized antennas (vertical and horizontal) with symmetric antenna patterns pointing horizontally. In this configuration, the direct (E_i) and reflected (E_r) signals are affected by the antenna radiation pattern ($F_n(\theta)$) in the same way, and the received power at *p*-polarization can be expressed as

$$P \propto \left|E_i + E_r\right|^2 = \left|E_{0i}\right|^2 \cdot F_n(\theta) \cdot \left|1 + \Gamma_p(\theta, \varepsilon_r) \cdot e^{j\Delta\phi}\right|^2 \tag{18.43}$$

$$\Delta\phi = \tfrac{4\pi}{\lambda} \cdot h \cdot \sin(\theta), \tag{18.44}$$

where Γ_p is the reflection coefficient, $\Delta\phi$ is the phase difference between the direct and reflected signals, and *h* is the antenna phase center height with respect to the specular reflection point. Since the frequency of the oscillations is related to the height where the reflections take place, this technique was also used to determine the water level height [71], surface topography [72], and even the significant wave height (SWH) [49] because when the surface is rough, the coherence of the reflected signal is destroyed and the interference pattern no longer exists.

Other properties that can be exploited by the IPT are that – at vertical polarization – the oscillatory pattern of the received power vanishes at the Brewster angle, which is a very precise indirect measurement of the dielectric constant (related to soil moisture over land) [70, 72]. Additionally, the presence of vegetation over land or snow over sea ice (or soil) produces multiple reflections in the air-canopy/air-snow and canopy-soil/snow-ice interfaces which produce several notches in the oscillatory pattern. The angular position of these notches is related to the vegetation height [72] or snow thickness [73]. The combined use of vertical and horizontal polarizations can be ultimately used to compensate for surface roughness effects [74].

Finally, the third technique is called the Interferometric Complex Field (ICF). For a static receiver, it can be demonstrated that the measured waveforms ((18.33)–(18.35)) are random processes that exhibit the following functional relationship [75]:

$$\left\langle Y(t, \tau, f_C) \cdot Y^*\left(t + \tilde{t}, \tau, f_C\right)\right\rangle \propto \exp\left\{-4k^2 \cdot \sigma_z^2 \cdot \tfrac{\tilde{t}}{2 \cdot \tau_z^2} \cdot \sin^2\gamma\right\}, \tag{18.45}$$

where $k = 2\pi/\lambda$ is the electromagnetic wavelength, σ_z is the standard deviation of the surface elevation, τ_z is the correlation time, and γ is the elevation angle. Therefore, measuring the correlation time of $Y(\tau_F, \tau, f_C)$ provides τ_F:

$$\frac{\tau_F}{\lambda} \cdot \pi \cdot sin\gamma = \frac{\tau_z}{SWH},$$
(18.46)

which is related to the *SWH*. Numerical simulations with Elfouhaily's spectrum show good results with the following semi-empirical fit:

$$\tau_F' = \frac{\lambda}{\pi} \cdot \left(\frac{a+b\cdot SWH}{SWH}\right),$$
(18.47)

where $\tau_F' = \tau_F \cdot sin\gamma$ is the effective correlation time.

Table 18.5 summarizes the main advantages and disadvantages of the above GNSS-R techniques. In all the previous discussions, it has been assumed that the GNSS signals are properly tracked in delay and Doppler frequency, but if they are not, the DDMs or WFs get blurred, the peak decreases, and the slope of the leading edge is reduced.[e]

18.3.5   SNR computation

The *SNR* (thermal noise only) at the input of the correlator is given by

$$SNR = \frac{\left[P_R(\tau, f_d) * |\chi(\tau, f_d)|^2\right]}{k_B \cdot T_{sys} \cdot min\{B, B_{IS}\}},$$
(18.48)

where P_R is the total received reflected power and χ is the WAF of the transmitted signal. For a cGNSS-R instrument, in which the scattered signal is cross-correlated with a locally generated replica of the transmitted signal, the SNR at the correlator's output is also given by (18.40) but replacing $min\{B, B_{IS}\}$ by the inverse of the coherent correlation time $(1/T_c)$. For an iGNSS-R instrument, the *SNR* at the correlator's output is given by Martin-Neira *et al.* [20]:

$$SNR = \frac{SNR_{cr}}{1 + \frac{1+SNR_R}{SNR_D}},$$
(18.49)

where SNR_{cr}, SNR_R, and SNR_D are the *SNR*s of the clean-replica cross-correlation (as in cGNSS-R), reflected and direct signals. If $SNR_D \gg 1$, then $SNR \to SNR_{cr}$, but this requires very large and directive antennas.

In addition, the *SNR* is affected by not only thermal noise but speckle noise can be even dominant, as discussed in (18.26).

18.4   GNSS-R ocean applications

In what follows, the geophysical information will be extracted from different GNSS-R observables. The SNR and IPT techniques can only be applied from ground, and they have been successfully used as sea tide gauges to measure water level height

---

[e]The effect of tracking or non-tracking in the DDM formation is illustrated in https://www.youtube.com/watch?v=fII6_bPxsEE (non-tracking case:) and https://www.youtube.com/watch?v=gQJdsEBXoSo (tracking case).

Table 18.5   Advantages and disadvantages of the different GNSS-R techniques

GNSS-R type	Advantages	Disadvantages
cGNSS-R	Code replica generated locally ⇒ high Signal-to-Noise Ratio (SNR), smaller antennas Transmitters separable by code	Only public codes can be used Limited bandwidth ⇒ limited Resolution Large dynamics of τ and f_d ⇒ more complex tracking
iGNSS-R	Any opportunity signal can be used (e.g. TV): larger bandwidths and transmitted power ⇒ better SNR Full bandwidth can be used ⇒ improved resolution Differential processing ⇒ reduced delay (τ) and Doppler frequency (f_d) dynamics ⇒ simpler tracking	Direct signal has poor SNR ⇒ larger up-looking antennas required Transmitters have to be separated by different τ, f_d in delay-Doppler domain Prone to inter-satellite cross-talk
piGNSS-R	Improved resolution as compared to iGNSS-R	Poorer SNR than iGNSS-R ⇒ +3 dB larger directivity required.
rGNSS-R	Large bandwidth code replica generated locally using pseudo-correlation techniques ⇒ high SNR (<cGNSS-R) ⇒ smaller antennas Transmitters separable by code	Large dynamics of τ and f_d
SNR	Uses existing geodetic GNSS receivers Precise	Low elevation angles only (<Brewster angle) Static receivers Maximum height limited by chip length
IPT	Use commercial off-the-shelf (COTS) GNSS receivers Precise	Requires dual-polarization linear antenna Static receivers Maximum height limited by chip length
ICF	Use COTS GNSS receiver front-ends Precise	More sophisticated processing than SNR or IPT Static receivers Maximum height limited by chip length

in lakes and ponds, or SWH in low-to-moderate sea state conditions. All other techniques can also be applied from airborne and spaceborne platforms. In the coming sections, the following ocean applications will be briefly presented: wind speed, wind direction, swell, sea state correction for sea surface salinity, delay and phase altimetry, and σ^0 mapping for oil spills – as an introduction to the subsequent chapters: Chapter 19 on the modeling and simulation of GNSS-R DDMs over the ocean, a first step towards numerical retrieval of other geophysical variables; Chapters 20 and 21 on wind speed retrieval and ocean altimetry applications; Chapter 22 on the emerging topic of sea ice mapping with GNSS-R, which is receiving more attention due to the reduction of the sea ice extent and thickness due to global warming; and finally, Chapter 23 on Taiwan's Triton – GNSS-R mission.

18.4.1   Ocean scatterometry
18.4.1.1   Wind speed

Ocean wind retrieval was the first application of GNSS-R and it was demonstrated with NASA's Delay-Mapping Receiver (DMR) instrument. Several techniques have been developed to infer wind speed from GNSS-R observables:

- Parameter *estimation* is based on finding the best fit of a model to the measured waveform $|Y(\tau,f_D)|^2$. Recalling from (18.37) that the wind speed information lies in σ^0 and that this parameter depends on the slopes PDF, a simple expression of the slopes pdf must be used in the inversion, for example, the up/cross-wind variances or the wind speed dependence as in the Cox and Munk model [44] properly scaled at L-band [45].

- Other techniques [76] adjust the waveforms using a series of exponentials parameterized with just a few variables, such as the slopes' mean squared slope (mss), the peak delay, and a scaling factor.

- Other techniques [77] derived the wind speed and direction by minimizing the total rms residual of waveforms from two different satellites at different azimuths.

- Full DDM inversion using data obtained with a GNSS-R software receiver has also been performed [78].

- Other approaches found in the literature [79] are derived from the DDM image (in decibels) truncated $10 \cdot log_{10}(e^{-1}) = -4.34$ dB, either using the weighted area $S_{DDM} = \sum_{\tau}\sum_{f_D} I(\tau,f_D) \cdot d\tau \cdot df_D$, the two-norm Euclidean distance from the center of mass $(CM_\tau, CM_{f_D}) = (\sum\sum f_D \cdot I(\tau,f_D) \cdot d\tau \cdot df_D, \sum\sum \tau \cdot I(\tau,f_D) \cdot d\tau \cdot df_D)$
  $/ \sum\sum I(\tau,f_D) \cdot d\tau \cdot df_D$ to the maximum value of the DDM, the distance from the geometric center to the maximum value of the DDM, or the one-norm Euclidean distance ("taxicab" distance) defined as $d = |x_1 - x_2| + |y_1 - y_2|$ from the center of mass to the maximum value of the DDM. These methods have been analyzed using airborne data acquired with the NOAA Gulfstream-IV jet aircraft on January 24, 2010, finding that, after selecting the most suitable coherent integration time ($T_c = 1$ ms) and the threshold ($10 \cdot log_{10}(e^{-1}) = -4.34$ dB), the wind retrieval error was minimum for the taxicab method with a 0.76 m/s bias and standard deviation 0.79 m/s. However, these results may not be extrapolated to a spaceborne case.

- Schiavulli *et al.* [80] proposed to use the polarimetric ratio computed as the ratio between the maximum of the measured LHCP DDM and the RHCP one as a metrics to infer wind speed, showing a dynamic range of ~15 dB for wind speeds from 5 m/s to 20 m/s at 45° incidence angle.

- Finally, it is worth describing briefly one of the approaches used in NASA CYGNSS mission [81]. Two DDM "observables", DDMA and LES, are derived first, then wind speeds are estimated from both observables using a Geophysical Model Function (GMF) empirically constructed for each observable that maps each observable value into a wind speed value, and finally wind estimates from DDMA and LES are linearly combined using a best-weighted estimator with weights computed using an adaptive covariance method.

18.4.1.2   Wind direction

A wind direction signature exists in the asymmetry of the DDM shape. Several approaches are found in the literature:

- As said in Section 18.4.1, the first attempt to retrieve wind speed and also the wind direction minimized the total rms residual of waveforms from two different satellites at different azimuths [78].
- Ten years later, using airborne data acquired with the NOAA Gulfstream-IV jet aircraft on January 24, the DDM asymmetry (skewness) was computed and used to infer the wind direction [82]. To measure the DDM skewness, the skewness angle ($\phi_{1, skew}$) was defined as the angle formed in the τ-f_D from the DDM peak to the "center of mass" (CM_{skirt}) of the truncated DDM from -8 dB to -5 dB under the peak (DDM skirt). For a single DDM measurement, there are four possible solutions due to the double ambiguity of the problem, although one of the solutions was always close to the ground-truth wind direction. To solve for one of the two ambiguities, simultaneous DDM measurements for different space vehicles had to be used achieving a final rms error of 22.°
- Finally, Guan *et al.* [83] analyzed the feasibility of wind direction retrieval in the spaceborne case. Two metrics were defined: the azimuth of the vector pointing from the specular point to the center of mass of the DDM and the azimuth of vector pointing from the center of mass of the DDM to the center of mass of the DDM skirt. It concluded that the first metric was more robust with respect to noise and that $SNR > 11$ dB was required to retrieve the wind direction.

18.4.1.3   Swell

Recently, the first experimental evidence of wind and swell wave signatures in L5/E5a GNSS-R waveforms obtained from an aircraft experiment over the Bass Strait (Australia) was presented [84]. Thanks to the higher SNR and narrower ACF, WFs with two and three peaks were systematically collected. It was found that the second and third peaks correspond to swell waves, which could not be resolved using cGNSS-R with L1 C/A codes.

18.4.1.4   Sea state corrections for sea surface salinity retrievals

Valencia *et al.* [85] showed that GNSS-R data could be directly linked to the L-band brightness temperature variations caused by the sea state effect without using emission/scattering models or sea spectra models. The measured L-band brightness temperatures on a field experiment in the Baltic sea were corrected for sea state effects using collocated GNSS-R observables and, in particular, the area under the normalized waveforms. It was shown that salinity retrieval error was reduced from 2.8 psu (practical salinity units, approximately g/l) for the uncorrected brightness temperatures down to 0.51 psu for the GNSS-R corrected brightness temperatures. More recently, this concept has been proven from space using FSSCat mission data [86].

18.4.1.5   Sea ice

Other applications of the above techniques to sea ice monitoring include sea ice classification [87] and sea ice altimetry [88]. More recently, the combination of GNSS-R and L-band microwave radiometer data has also been used for improving sea ice extent and concentration determination [89] and to support sea ice thickness estimations [90]. However, recent results from the Multidisciplinary drifting Observatory for the Study of Arctic Climate (MOSAiC) experiment have proven that this last geophysical variable is a very challenging problem, and multi-frequency and multi-angular observations are required to isolate the effect of the snow layer on top of the ice layer [91].

18.4.2   Ocean altimetry

The first proposal to use GNSS-R for altimetry applications dates back to 1993 [3] with the PARIS concept. Much has been advanced since then, including two competitive ESA Phase A studies for a PARIS In-Orbit Demonstration mission. GNSS-R altimetry differs from conventional nadir-looking radar altimetry in the multi-static geometry, the frequency band (L-band, instead of C to Ka bands, for which atmospheric effects – mainly hydrometeors – are less important, but ionospheric effects are far more important), the available signal bandwidths (~2 MHz to ~20 MHz for the GPS C/A and P(Y) codes as compared with the ~300 MHz for some radar altimeters), the identification of the specular reflection point in the waveform shape, and the scattering area which is neither beam-limited (when highly directive antennas are used) nor pulse-limited (when short pulses are used) but more "roughness limited" as the sun glint over the ocean (Figure 18.13b).

Altimetry observations can be performed using group delay or phase delay. While group delay provides (coarse) absolute range measurements, phase delays provide precise, although not accurate, differential range measurements. Moreover, GNSS-R from Earth's surfaces tends to be incoherent except for ice and some smooth land surfaces, so only group delay between the transmission and reception is usually feasible.

18.4.2.1 Delay altimetry

In delay altimetry, the observable (ρ_i) is the delay between transmission and reception at a given frequency band (f_i). The delay associated with the geometry (ρ_{geo}), which is the altimetry observable, is given by

$$\rho_{geo} = \left| \overrightarrow{R}_{spec} - \overrightarrow{R}_T \right| + \left| \overrightarrow{R}_R - \overrightarrow{R}_{spec} \right|, \tag{18.50}$$

where $\overrightarrow{R}_T$, $\overrightarrow{R}_{spec}$, and $\overrightarrow{R}_R$ are the vectors indicating the position of the transmitter, specular reflection point, and receiver, respectively. However, in addition to ρ_{geo}, there are other error contributions:

$$\rho_i = \rho_{geo} + \rho_{iono,i} + \rho_{tropo} + \rho_{EM\ bias} + \epsilon_{multipath} + \epsilon_{instr} + \epsilon_n, \tag{18.51}$$

the ionosphere ($\rho_{iono,i}$), which – at L-band – can introduce error delays of up to ~40 m; the troposphere (ρ_{tropo}), which is largely driven by the "dry component" (about 2.3 m), and the "wet component", which is smaller but more variable geographically; and the electromagnetic bias ($\rho_{EM\ bias}$), which is due to the asymmetry of the surface waves, which scatter the electromagnetic radiation differently. It is a small, although not negligible, contribution on the order of 1% of the SWH and other errors induced by multi-path ($\epsilon_{multipath}$), instrumental errors (ϵ_{instr}), or noise errors (ϵ_n).

In many ground-based and airborne experiments, however, a differential measurement (reflected minus direct ranges: $\rho_{i,\ r} - \rho_{i,\ d}$) is used, because some errors cancel out ($\rho_{iono,i}$ and ρ_{tropo}, for low altitude flights, ϵ_{instr}) and – assuming a "flat Earth" – the expression becomes very simple:

$$\rho_{geo} = \left| \overrightarrow{R}_{spec} - \overrightarrow{R}_T \right| + \left| \overrightarrow{R}_R - \overrightarrow{R}_{spec} \right| - \left| \overrightarrow{R}_R - \overrightarrow{R}_T \right| + \rho_{ant\ offset} \approx 2 \cdot h \cdot sin\left(\theta_e\right) + \rho_{ant\ offset}, \tag{18.52}$$

where h is the receiver height above the "flat Earth", θ_e is the elevation angle, and $\rho_{ant\ offset}$ is an offset term associated with the fact that the phase centers of the up- and down-looking antennas are not physically in the same point. $\rho_{ant\ offset}$ is computed as the projection of the offset vector between the up- and down-looking antennas in the direction of the reflected signal.

The altimetry observables from a bistatic waveform can then be obtained either by fitting a model of the waveform, from the slope of the geometric term corrected by antenna offsets:

$$H = \frac{1}{2} \cdot \frac{\partial \rho_{geo}}{\partial (sin(\theta_e))} \tag{18.53}$$

or from the delay from the peak of the waveform of the direct signal to the maximum peak of the derivative of the waveform of the reflected signal. This approach was proposed by Hajj and Zuffada [92] and demonstrated later by Rius *et al.* [93].

18.4.2.2 Phase altimetry

Phase altimetry is estimated from the phase of the complex waveform (I and Q components). If the GNSS-R receiver includes a phase-locked loop, then phase variations are dynamically compensated and the delay can be readily estimated from the PRs and a model of the scattering geometry. If the GNSS-R receiver is an open-loop one, then the local oscillator phase appears at the phases of the output waveforms (direct and reflected fields) and has to be stopped

$$\varphi\left(t\right) = arg\left\{E_r \cdot E_d^*\right\}, \tag{18.54}$$

before the phase unwrapping; E_r and E_d are the electric fields of the reflected and direct signals, respectively.

As said, phase altimetry is only feasible for flat surfaces such as ice or at near grazing angles because the Rayleigh parameter tends to be zero.

18.4.2.3   The electromagnetic bias in GNSS-R altimetry

The electromagnetic (EM) bias is the correction for the measurement bias introduced by the varying reflectivity of wave crests and troughs. It is defined as the normalized correlation between the radar cross-section and the sea surface elevation [94]:

$$\rho_{EM\ bias} = \frac{\langle \xi \cdot \sigma^0 \rangle}{\langle \sigma^0 \rangle}. \tag{18.55}$$

A few studies have been performed on the EM bias at L-band and in a bistatic configuration [95–97]. At nadir incidence, the predicted EM bias is less than 2 cm for wind speeds up to 18 m/s, but it rapidly increases with increasing incidence angles (up to 10 cm at 18 m/s, for $\theta_i = 25°$) and shows a non-negligible azimuthal dependence. As GNSS-R altimeters (e.g. PARIS IoD, GEROS ISS) are targeting for decimetric accuracy, EM bias is an important error source that will have to be corrected taking into account the observation geometry and the wind speed conditions.

18.4.3   Ocean imaging

One of the most recent approaches of GNSS-R is the imaging of the bistatic radar cross-section by deconvolving the DDM. Actually, (18.32) can be interpreted as the two-dimensional convolution of WAF $\left|\chi\left(\tau, f_D\right)\right|^2$ and the function $\Sigma\left(\tau, f_D\right)$ [98]:

$$\left\langle \left|Y\left(\tau, f_d\right)\right|^2 \right\rangle \equiv \left|\chi\left(\tau, f_D\right)\right|^2 ** \Sigma\left(\tau, f_D\right), \tag{18.56}$$

with

$$\Sigma\left(\tau, f_D\right) = \frac{P_T \cdot \lambda^2}{(4\pi)^3} T_c^2 \iint \frac{G_T\left(\vec{r}\right)G_R\left(\vec{r}\right)}{4 \cdot R_T^2\left(\vec{r}\right) \cdot R_R^2\left(\vec{r}\right)} \sigma^0\left(\vec{r}\right) \cdot \delta\left(\tau\right) \cdot \delta\left(f_D\right) \cdot d^2r, \tag{18.57}$$

which can be written as

$$\Sigma\left(\tau, f_D\right) = \frac{P_T \cdot \lambda^2}{(4\pi)^3} T_c^2 \frac{G_T\left(\vec{r}\right)G_R\left(\vec{r}\right)\sigma^0\left(\vec{r}\right)}{4 \cdot R_T^2\left(\vec{r}\right) \cdot R_R^2\left(\vec{r}\right)} \cdot \left|J\left(\tau, f_D\right)\right|, \tag{18.58}$$

where the term $\left|J\left(\tau, f_D\right)\right|$ is the Jacobian function resulting from the change of variables (x, y) to (τ, f_D) as defined in [98]. The term $\Sigma\left(\tau, f_D\right)$ can be considered as an imaged scene including the scattering coefficients σ^0 of the glistening zone.

However, this technique suffers from the ambiguity of spatial location, since two (x, y) points map into the same (τ, f_D) bin. As in conventional SAR, the only way to eliminate this is by tilting (electronically or mechanically) the antenna away from the specular reflection point so that the antenna pattern attenuates the alias image or by using a more sophisticated antenna that can introduce nulls in the direction of the alias [99, 100].

This technique has been tested for oil slick detection [101], hurricane wind mapping, and target detection over the ocean [102] or even above the Earth's surface [103]. The estimate on ground resolutions compares well with the expected range of values (Δr) and azimuth (Δa) resolutions found in bistatic SAR theory [104]:

$$\Delta r = \frac{c}{B \cdot sin(\theta_i)}, \tag{18.59a}$$

$$\Delta a = \frac{1}{T_c} \cdot \frac{\lambda \cdot h_R}{V_R \cdot cos(\theta_i)}. \tag{18.59b}$$

18.5   Conclusions

In this chapter, a brief historical review of GNSS-R has been presented to provide the context and perspective of the evolution of the different GNSS-R techniques that exist today. The basic concepts of GNSS have been introduced including the signal characteristics that ultimately determine the properties of GNSS-R systems: wave polarization, range and Doppler resolutions, SNR, required integration time, and so on.

Then, the different GNSS-R techniques have been described indicating their pros and cons. The chapter provides a non-exhaustive list of ocean applications, grouped as scatterometric, altimetric, and imaging, and indicating which particular GNSS-R observable is used in each case. An extensive, but not a comprehensive, list of references is also provided for the interested readers, as the literature in the field is growing very rapidly [105] because of the increased interest that this field has triggered in the Earth Observation scientific community.

Acknowledgment

The effort devoted to prepare this material is part of projects "SPOT: Sensing with Pioneering Opportunistic Techniques grant RTI2018-099008-B-C21/ AEI/10.13039/501100011033", and "Unidad de Excelencia Maria de Maeztu MDM-2016-0600". The author would like to thank the editor and reviewers for their support to improve the quality and clarity of the text.

References

[1]   Hall C.D., Cordey R.A. 'Multistatic scatterometry'. 1988 IEEE International Geoscience and Remote Sensing Symposium, IGARSS '88; 1988. pp. 561–2.

[2]   Auber J.C. Characterization of multipath on land and sea at GPS frequencies. Proceedings of 7th International Technical Meeting of the Satellite Division of the Institute of Navigation, Part 2, ION GPS 94, Sept; 1994. pp. 1155–71.

[3]   Martin-Neira M. 'A passive reflectometry and interferometry system (PARIS) – application to ocean altimetry'. *ESA Journal*. 1993;**17**(4):331–55.

[4]   Garrison J.L., Katzberg S.J. 'The application of reflected GPS signals to ocean remote sensing'. *Remote Sensing of Environment*. 2000;**73**(2):175–87.

[5]   Lowe S.T. 'First spaceborne observation of an earth-reflected GPS Signal'. *Radio science*. 2020;**37**:0048–6604.

[6]   Gleason S., Hodgart S., Yiping Sun. 'Detection and processing of bistatically reflected GPS signals from low earth orbit for the purpose of ocean remote sensing'. *IEEE Transactions on Geoscience and Remote Sensing*. 2005;**43**(6):1229–41.

[7]   TechDemoSat-1 (Technology Demonstration Satellite-1)/TDS-1 [online]. Available from https://earth.esa.int/web/eoportal/satellite-missions/t/techdemosat-1 [Accessed 23 Jun 2020].

[8]   CYGNSS (Cyclone Global Navigation Satellite System) [online]. Available from https://earth.esa.int/web/eoportal/satellite-missions/c-missions/cygnss [Accessed 23 Jun 2020].

[9]   Chew C.C., Small E.E. 'Soil moisture sensing using spaceborne GNSS reflections: comparison of CYGNSS reflectivity to SMAP soil moisture'. *Geophysical Research Letters*. 2018;**45**(9):4049–57.

[10]   Chew C., Reager J.T., Small E. 'CYGNSS data map flood inundation during the 2017 Atlantic Hurricane season'. *Scientific Reports*. 2018;**8**(1):9336.

[11]   Carreno-Luengo H., Luzi G., Crosetto M., *et al.* 'Above-ground biomass retrieval over tropical forests: a novel GNSS-R approach with CyGNSS'. *Remote Sensing*. 2020;**12**(9):1368.

[12]   Rodriguez-Alvarez N., Misra S., Podest E., Morris M., Bosch-Lluis X. 'The use of SMAP-Reflectometry in science applications: calibration and capabilities'. *Remote Sensing*. 2019;**11**(20):2442.

[13]   Carreno-Luengo H., Camps A., Via P., *et al.* '³Cat-2—an experimental nanosatellite for GNSS-R earth observation: mission concept and analysis'. *IEEE Journal of Selected Topics in Applied Earth Observations and Remote Sensing*. 2016;**9**(10):4540–51.

[14]   Jing C., Niu X., Duan C., Lu F., Di G., Yang X. 'Sea surface wind speed retrieval from the first Chinese GNSS-R mission: technique and preliminary results'. *Remote Sensing*. 2019;**11**(24):3013.

[15]   Spire for Third Party Mission Programme [online]. Available from https://earth.esa.int/eogateway/documents/20142/37627/Spire-Product-Guide.pdf [Accessed 28 Apr 2021].

[16]   Munoz-Martin J.F., Capon L.F., Ruiz-de-Azua J.A., Camps A. 'The flexible microwave payload-2: a SDR-based GNSS-Reflectometer and *L*-band radiometer for CubeSats'. *IEEE Journal of Selected Topics in Applied Earth Observations and Remote Sensing*. 2020;**13**:1298–311.

[17] Bai W., Wang G., Sun Y., *et al.* 'Application of the Fengyun 3 C GNSS occultation sounder for assessing the global ionospheric response to a magnetic storm event'. *Atmospheric Measurement Techniques.* 2019;**12**(3):1483–93.

[18] Ruiz-de-Azua J.A. ³Cat-4 Mission: A 1-Unit CubeSat for earth observation with a L-band radiometer and a GNSS-reflectometer using software defined radio. 2019 IEEE International Geoscience and Remote Sensing Symposium, IGARSS '19; 2019. pp. 8867–70.

[19] Technology CubeSats [online]. Available from https://www.esa.int/ Enabling_Support/Space_Engineering_Technology/Technology_CubeSats [Accessed 28 Apr 2021].

[20] Martin-Neira M., D'Addio S., Buck C., Floury N., Prieto-Cerdeira R. 'The PARIS ocean Altimeter in-orbit demonstrator'. *IEEE Transactions on Geoscience and Remote Sensing.* 2011;**49**(6):2209–37.

[21] Wickert J., Cardellach E., Martin-Neira M., *et al.* 'GEROS-ISS: GNSS reflectometry, radio occultation, and scatterometry onboard the International Space Station'. *IEEE Journal of Selected Topics in Applied Earth Observations and Remote Sensing.* 2016;**9**(10):4552–81.

[22] Cardellach E., Flato G., Fragner H., *et al.* 'GNSS transpolar earth reflectometry exploring system (G-TERN): mission concept'. *IEEE Access.* 2018;**6**:13980–4018.

[23] Quasi-Zenith Satellite System (QZSS). *List of positioning satellites* [online]. 2020. Available from https://qzss.go.jp/en/technical/satellites/index.html [Accessed 23 Jun 2020].

[24] Molina C., Camps A. 'First evidences of ionospheric plasma depletions observations using GNSS-R data from CYGNSS'. *Remote Sensing.* 2020;**12**(22):3782.

[25] How does a GPS work. Available from http://www.howtechnologywork. com/how-does-a-gps-work/ [Accessed 23 Jun 2020].

[26] Tsui J.B.-Y. *Fundamentals of Global Positioning System Receivers: A Software Approach.* Ed. John Wiley & Sons; 2004.

[27] Borre K. *A Software-Defined GPS and Galileo Receiver: A Single-Frequency Approach (Applied and Numerical Harmonic Analysis.* Ed. Springer; 2006.

[28] Kaplan E., Hegarty C. 'Understanding GPS/GNSS: Principles and Applications (GNSS Technology and Applications Series)'. *Ed Artech-House, ISBN: 9781630810580.* Third Edition; 2017.

[29] Fontana R.D. 'The new L2 civil signal'. Proceedings of ION GPS; Salt Lake City, UT; 2001. pp. 617–31.

[30] European GNSS (Galileo) open service signal-in-space interface control document, issue 1.3 [online]. December 2016. Available from https://www. gsc-europa.eu/sites/default/files/sites/all/files/Galileo-OS-SIS-ICD.pdf [Accessed 23 Jun 2020].

[31] Pascual D., Camps A., Martin F., Park H., Arroyo A.A., Onrubia R. 'Precision bounds in GNSS-R ocean altimetry'. *IEEE Journal of Selected Topics in Applied Earth Observations and Remote Sensing.* 2014;**7**(5):1416–23.

[32]   Shivaramaiah N.C. 'The Galileo E5 AltBOC: understanding the signal structure'. *International Global Navigation Satellite Systems Society IGNSS Symposium 2009, Holiday Inn, Surfers Paradise; Qld, Australia, 1–3 Dec; 2009*.

[33]   Tropospheric Delay [online]. Available from https://gssc.esa.int/navipedia/index.php/Tropospheric_Delay [Accessed 23 Jun 2020].

[34]   Galileo Tropospheric Correction Model [online]. Available from https://gssc.esa.int/navipedia/index.php/Galileo_Tropospheric_Correction_Model [Accessed 23 Jun 2020].

[35]   Recommendation ITU-R P.531-12. *Ionospheric propagation data and prediction methods required for the design of satellite services and systems (Question ITU-R 218/3) [online]*. Available from https://www.itu.int/dms_pubrec/itu-r/rec/p/R-REC-P.531-12-201309-S!!PDF-E.pdf [Accessed 23 Jun 2020].

[36]   Klobuchar J. 'Ionospheric time-delay algorithm for single-frequency GPS users'. *IEEE Transactions on Aerospace and Electronic Systems*. 1987;**AES-23**(3):325–31.

[37]   Nava B., Coïsson P., Radicella S.M., *et al.* 'A new version of the NeQuick ionosphere electron density model'. *Journal of Atmospheric and Solar-Terrestrial Physics*. 2008;**70**(15):1856–62.

[38]   Briggs B.H., Parkin I.A. 'On the variation of radio star and satellite scintillations with zenith angle'. *Journal of Atmospheric and Terrestrial Physics*. 1963;**25**(6):339–66.

[39]   Yeh C., Chao-Han L. 'Radio wave scintillations in the ionosphere'. *Proceedings of the IEEE*. 1982;**70**(4):324–60.

[40]   Navipedia. 2020. Available from https://gssc.esa.int/navipedia/images/c/cf/GNSS_All_Signals.png [Accessed 23 Jun 2020].

[41]   GPS Interface Control Documents. Available from https://www.gps.gov/technical/icwg/ [Accessed 23 June 2020].

[42]   Camps A. 'Spatial resolution in GNSS-R under coherent scattering'. *IEEE Geoscience and Remote Sensing Letters*. **17**; 2020. pp. 32–6.

[43]   Camps A., Munoz-Martin J.F. 'Analytical computation of the spatial resolution in GNSS-R and experimental validation at L1 and L5'. *Remote Sensing*. 2020;**12**(23):3910.

[44]   Cox C., Munk W. 'Measurement of the roughness of the sea surface from photographs of the Sun's glitter'. *Journal of the Optical Society of America*. 1954;**44**(11):838–50.

[45]   Katzberg S.J., Torres O., Ganoe G., *et al.* 'Calibration of reflected GPS for tropical storm wind speed retrievals'. *Geophysical Research Letters*. 2006;**33**(18):L18602.

[46]   Zavorotny V.U., Voronovich A.G. 'Scattering of GPS signals from the ocean with wind remote sensing application'. *IEEE Transactions on Geoscience and Remote Sensing*. 2000;**38**(2):951–64.

[47]   Long D., Ulaby F.T. 'Microwave radar and radiometric remote sensing'. Artech House; 2015.

[48]   Ticconi F. *Models for scattering from rough surfaces [online]*. Available from http://cdn.intechopen.com/pdfs-wm/16082.pdf [Accessed 23 Jun 2020].

[49]   Alonso-Arroyo A., Camps A., Park H., Pascual D., Onrubia R., Martin F. 'Retrieval of significant wave height and mean sea surface level using the GNSS-R interference pattern technique: Results from a three-month field campaign'. *IEEE Transactions on Geoscience and Remote Sensing*. **53**; 2015. pp. 3198–209.

[50]   Beckmann P., Spizzichino A. *The Scattering of Electromagnetic Waves from Rough Surfaces*. Norwood, MA, USA: Artech House; 1987.

[51]   Voronovich A.G., Zavorotny V.U. 'Bistatic radar equation for signals of opportunity revisited'. *IEEE Transactions on Geoscience and Remote Sensing*. 2018;**56**(4):1959–68.

[52]   Skolnik M. *Introduction to Radar Systems*. McGraw-Hill; 2002.

[53]   Woodward P.M. *Probability and Information Theory, with Applications to Radar*. Pergamon Press, 1953; reprinted by Artech House; 1980.

[54]   Tsao T., Slamani M., Varshney P., Weiner D., Schwarzlander H., Borek S. 'Ambiguity function for a bistatic radar'. *IEEE Transactions on Aerospace and Electronic Systems*. 1997;**33**(3):1041–51.

[55]   Shivaramaiah N.C. 'Exploiting the secondary codes to improve signal acquisition performance in galileo receivers'. Proceedings of the 21st International Technical Meeting of the Satellite Division of the Institute of Navigation (ION GNSS 2008); Savannah, GA; 2008. pp. 1497–506.

[56]   Park H., Valencia E., Camps A., Rius A., Ribo S., Martin-Neira M. 'Delay tracking in spaceborne GNSS-R ocean altimetry'. *IEEE Geoscience and Remote Sensing Letters*. 2013;**10**(1):57–61.

[57]   Park H., Pascual D., Camps A., Martin F., Alonso-Arroyo A., Carreno-Luengo H. 'Analysis of spaceborne GNSS-R delay-Doppler tracking'. *IEEE Journal of Selected Topics in Applied Earth Observations and Remote Sensing*. 2014;**7**(5):1481–92.

[58]   Carreno-Luengo H., Camps A., Ramos-Perez I., Rius A. 'Experimental evaluation of GNSS-reflectometry altimetric precision using the P(Y) and C/A signals'. *IEEE Journal of Selected Topics in Applied Earth Observations and Remote Sensing*. 2014;**7**(5):1493–500.

[59]   Lowe S.T., Meehan T., Young L., *et al.* 'Direct signal enhanced semi-codeless processing of GNSS surface-reflected signals'. *IEEE Journal of Selected Topics in Applied Earth Observations and Remote Sensing*. 2014;**7**(5):1469–72.

[60]   Li W., Yang D., D'Addio S., *et al.* 'Partial interferometric processing of reflected GNSS signals for ocean altimetry'. *IEEE Geoscience and Remote Sensing Letters*. 2014;**11**(9):1509–13.

[61]   Martín F., Camps A., Park H., D'Addio S., Martin-Neira M., Pascual D. 'Cross-correlation waveform analysis for conventional and interferometric GNSS-R approaches'. *IEEE Journal of Selected Topics in Applied Earth Observations and Remote Sensing*. 2014;**7**(5):1560–72.

[62]   Kay S.M. *Fundamentals of Statistical Signal Processing: Estimation Theory, Prentice Hall, Upper Saddle River*; 1993.

[63]   GNSS Interference Model [online]. Available from https://gssc.esa.int/navipedia/index.php/GNSS_Interference_Model [Accessed 23 Jun 2020].

[64]   Onrubia R., Pascual D., Park H., *et al.* 'Satellite cross-talk impact analysis in airborne interferometric global navigation satellite system-reflectometry with the microwave interferometric reflectometer'. *Remote Sensing.* 2019;**11**(9):1120.

[65]   Martín F., D'Addio S., Camps A., Martin-Neira M. 'Modeling and analysis of GNSS-R waveforms sample-to-sample correlation'. *IEEE Journal of Selected Topics in Applied Earth Observations and Remote Sensing.* 2014;**7**(5):1545–59.

[66]   Kavak A., Vogel W.J., Xu G., *et al.* 'Using GPS to measure ground complex permittivity'. *Electronics Letters.* 1998;**34**(3):254–5.

[67]   Larson K.M., Small E.E., Gutmann E., Bilich A., Axelrad P., Braun J. 'Using GPS multipath to measure soil moisture fluctuations: initial results'. *GPS Solutions.* 2008;**12**(3):173–7.

[68]   PBO H2O Data Portal [online]. Available from http://cires1.colorado.edu/portal/index.php?product=soil_moisture [Accessed 23 Jun 2020].

[69]   Löfgren J.S., Haas R. 'Sea level measurements using multi-frequency GPS and GLONASS observations'. *EURASIP Journal on Advances in Signal Processing.* 2014;**50**(2014).

[70]   Rodriguez-Alvarez N., Bosch-Lluis X., Camps A., *et al.* 'Soil moisture retrieval using GNSS-R techniques: experimental results over a bare soil field'. *IEEE Transactions on Geoscience and Remote Sensing.* 2009;**47**(11):3616–24.

[71]   Rodriguez-Alvarez N. 'Water level monitoring using the interference pattern GNSS-R technique'. 2011 IEEE International Geoscience and Remote Sensing Symposium, IGARSS '11; Vancouver, BC; 2011. pp. 2334–7.

[72]   Rodriguez-Alvarez N., Camps A., Vall-llossera M., *et al.* 'Land geophysical parameters retrieval using the interference pattern GNSS-R technique'. *IEEE Transactions on Geoscience and Remote Sensing.* 2011;**49**(1):71–84.

[73]   Rodriguez-Alvarez N. 'Snow monitoring using GNSS-R techniques'. 2011 IEEE International Geoscience and Remote Sensing Symposium, IGARSS '11; Vancouver, BC; 2011. pp. 4375–8.

[74]   Alonso Arroyo A., Camps A., Aguasca A., *et al.* 'Dual-polarization GNSS-R interference pattern technique for soil moisture mapping'. *IEEE Journal of Selected Topics in Applied Earth Observations and Remote Sensing.* 2014;**7**(5):1533–44.

[75]   Ruffini G., Soulat F. 'On the GNSS-R interferometric complex field coherence time'. *arXiv:physics/0406084 [physics.ao-ph].*

[76]   Garrison J.L., Komjathy A., Zavorotny V.U., Katzberg S.J. 'Wind speed measurement using forward scattered GPS signals'. *IEEE Transactions on Geoscience and Remote Sensing.* 2002;**40**(1):50–65.

[77] Komjathy A., Armatys M., Masters D., Axelrad P., Zavorotny V., Katzberg S. 'Retrieval of ocean surface wind speed and wind direction using reflected GPS signals'. *Journal of Atmospheric and Oceanic Technology*. 2004;**21**(3):515–26.

[78] Germain O., Ruffini G., Soulat F., Caparrini M., Chapron B., Silvestrin P. 'The eddy experiment: GNSS-R speculometry for directional sea-roughness retrieval from low altitude aircraft'. *Geophysical Research Letters*. 2004;**31**(21):L21307.

[79] Rodriguez-Alvarez N., Akos D.M., Zavorotny V.U., Smith J.A., Camps A., Fairall C.W. 'Airborne GNSS-R wind retrievals using delay–Doppler maps'. *IEEE Transactions on Geoscience and Remote Sensing*. 2013;**51**(1):626–41.

[80] Schiavulli D., Ghavidel A., Camps A., Migliaccio M. 'GNSS-R wind-dependent polarimetric signature over the ocean'. *IEEE Geoscience and Remote Sensing Letters*. 2015;**12**(12):2374–8.

[81] Clarizia M.P., Ruf C.S. 'Wind speed retrieval algorithm for the cyclone global navigation satellite system (CYGNSS) mission'. *IEEE Transactions on Geoscience and Remote Sensing*. 2016;**54**(8):4419–32.

[82] Valencia E., Zavorotny V.U., Akos D.M., Camps A. 'Using DDM asymmetry metrics for wind direction retrieval from GPS ocean-scattered signals in airborne experiments'. *IEEE Transactions on Geoscience and Remote Sensing*. 2014;**52**(7):3924–36.

[83] Guan D., Park H., Camps A., *et al.* 'Wind direction signatures in GNSS-R observables from space'. *Remote Sensing*. 2018;**10**(2):198.

[84] Munoz-Martin J.F., Onrubia R., Pascual D., *et al.* 'Experimental evidence of swell signatures in airborne L5/E5a GNSS-reflectometry'. *Remote Sensing*. 2020;**12**(11):1759.

[85] Valencia E. 'Improving the accuracy of sea surface salinity retrieval using GNSS-R data to correct the sea state effect'. *Radio science*;46.

[86] Munoz-Martin J.F., Camps A. 'Sea surface salinity and wind speed retrievals using GNSS-R and L-band microwave radiometry data from FMPL-2 onboard the FSSCat mission'. *Remote Sensing*. 2021;**13**(16):3224.

[87] Rodriguez-Alvarez N., Holt B., Jaruwatanadilok S., Podest E., Cavanaugh K.C. 'An Arctic sea ice multi-step classification based on GNSS-R data from the TDS-1 mission'. *Remote Sensing of Environment*. 2019;**230**(4):111202.

[88] Cardellach E., Li W., Rius A., *et al.* 'First precise Spaceborne sea surface Altimetry with GNSS reflected signals'. *IEEE Journal of Selected Topics in Applied Earth Observations and Remote Sensing*. 2020;**13**:102–12.

[89] Llaveria D., Munoz-Martin J.F., Herbert C., Pablos M., Park H., Camps A. 'Sea ice concentration and sea ice extent mapping with L-Band microwave radiometry and GNSS-R data from the FFSCat mission using neural networks'. *Remote Sensing*. 2021;**13**(6):1139.

[90] Herbert C., Munoz-Martin J.F., Llaveria D., Pablos M., Camps A. 'Sea ice thickness estimation based on regression neural networks using L-Band microwave radiometry data from the FSSCat mission'. *Remote Sensing*. 2021;**13**(7):1366.

[91]   Munoz-Martin J.F., Perez A., Camps A., *et al.* 'Snow and ice thickness retrievals using GNSS-R: preliminary results of the mosaic experiment'. *Remote Sensing.* 2020;**12**(24):4038.

[92]   Hajj G.A., Zuffada C. 'Theoretical description of a bistatic system for ocean altimetry using the GPS signal'. *Radio Science.* 2003;**38**(5):1089.

[93]   Rius A., Cardellach E., Martin-Neira M. 'Altimetric analysis of the sea-surface GPS-reflected signals'. *IEEE Transactions on Geoscience and Remote Sensing.* 2010;**48**(4):2119–27.

[94]   Elfouhaily T., Thompson D., Vandemark D., Chapron B. 'Weakly nonlinear theory and sea state bias estimations'. *Journal of Geophysical Research: Oceans.* 1999;**104**(C4):7641–7.

[95]   Picardi G., Seu R., Sorge S.G., Neira M.M. 'Bistatic model of ocean scattering'. *IEEE Transactions on Antennas and Propagation.* 1998;**46**(10):1531–41.

[96]   Ghavidel A., Schiavulli D., Camps A. 'Numerical computation of the electromagnetic bias in GNSS-R Altimetry'. *IEEE Transactions on Geoscience and Remote Sensing.* 2016;**54**(1):489–98.

[97]   Park J., Johnson J.T., Lowe S.T., *et al.* 'A study of the electromagnetic bias for GNSS-R ocean altimetry using the choppy wave model'. *Waves in Random and Complex Media.* 2016;**26**(4):599–612.

[98]   Marchan-Hernandez J.F., Camps A., Rodriguez-Alvarez N., Valencia E., Bosch-Lluis X., Ramos-Perez I. 'An efficient algorithm to the simulation of delay–Doppler Maps of reflected global navigation satellite system signals'. *IEEE Transactions on Geoscience and Remote Sensing.* 2009;**47**(8):2733–40.

[99]   Park H. 'New approach to sea surface wind retrieval from GNSS-R measurements'. 2011 IEEE International Geoscience and Remote Sensing Symposium, IGARSS '11; Vancouver, BC; 2011. pp. 1469–72.

[100]   Camps, A., Park, H., Alonso-Arroyo A. 'Wind speed mapping from the ISS using GNSS-R? a simulation study'. 2013 IEEE International Geoscience and Remote Sensing Symposium, IGARSS '13; Melbourne, VIC; 2013. pp. 382–5.

[101]   Valencia E., Camps A., Rodriguez-Alvarez N., Park H., Ramos-Perez I. 'Using GNSS-R imaging of the ocean surface for oil Slick detection'. *IEEE Journal of Selected Topics in Applied Earth Observations and Remote Sensing.* 2013;**6**(1):217–23.

[102]   Di Simone A., Park H., Riccio D., Camps A. 'Sea target detection using Spaceborne GNSS-R Delay-Doppler maps: theory and experimental proof of concept using TDS-1 data'. *IEEE Journal of Selected Topics in Applied Earth Observations and Remote Sensing.* 2017;**10**(9):4237–55.

[103]   Hu C., Benson C., Park H., Camps A., Qiao L., Rizos C. 'Detecting Targets above the Earth's Surface Using GNSS-R Delay Doppler Maps: Results from TDS-1'. *Remote Sensing.* 2019;**11**(19):2327.

[104]   Moccia A., Renga A. 'Spatial resolution of Bistatic synthetic aperture radar: impact of acquisition geometry on imaging performance'. *IEEE Transactions on Geoscience and Remote Sensing.* 2011;**49**(10):3487–503.

[105] GNSS+R Bibliography. Available from https://www.ice.csic.es/personal/rius/gnss_r_bibliography/ [Accessed 23 Jun 2020].

Chapter 19

Modeling and simulation of GNSS-R delay-Doppler maps over the ocean

Maurizio di Bisceglie[1] and Carmela Galdi[1]

19.1 Introduction

This chapter develops the main concepts concerned with modeling and computer simulation of the Global Navigation Satellite System (GNSS) bistatic ocean-scattered signal and illustrates how this signal can be processed to obtain an observable that is generally referred to as delay-Doppler map (DDM). The variability of the sea surface geometry in space and time makes this modeling task especially difficult, also in consideration of the fact that a full understanding of how physical processes translate into scattering mechanisms has not been acquired at this time.

After the pioneering work by Martìn-Neira [1], where the idea of using signals reflected by GNSSs for ocean altimetry and, in general, for surface measurements was first conceived, a first attempt to study the structure of the global positioning system (GPS) scattered signal is found in the study by Clifford *et al.* [2], where the authors provide a clear evidence of wind signatures in the reflected waves. A few years later, a significant step was moved forward with Zavorotny and Voronovich's publication [3], where the authors derived a DDM of the average signal power received using Kirchhoff approximation in the geometric optics limit. The bistatic radar equation, in the form of an integral over the scattering surface, shows that the key roles are played by the ambiguity function of the GPS signal and by the probability density function (PDF) of the sea surface slopes. The comprehension of the received signal structure and of its dependence on geophysical and signal parameters encouraged the development of retrieval algorithms that, in turn, paved the way for satellite missions [4, 5]. During the next ten years, a first investigation was carried out with the UK Disaster Monitoring Constellation (UK-DMC) platform, launched in 2003, to demonstrate the potential for GPS reflectometry. Despite the moderate link budget margin, the experiment was successful and reflected signals were detected over the ocean and land. From 2005 to 2010, new research studies provided successful inversion of the DDM observable for retrieval of the mean square slopes and wind speed. The receiver and the real-time processor were optimized with the next generation

[1]Engineering Department, Università degli Studi del Sannio, Benevento, Italy

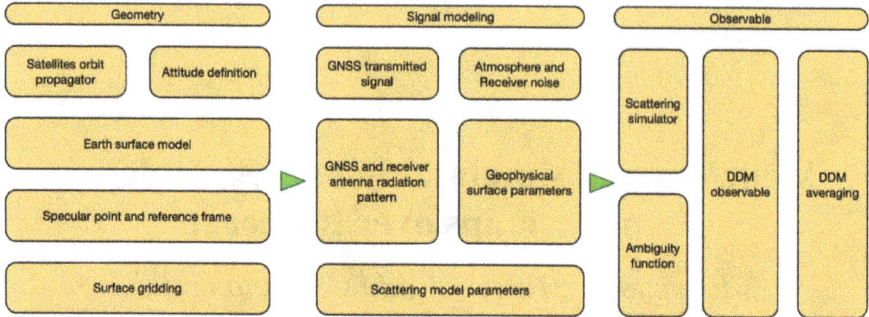

Figure 19.1   Schematic of the simulation process for ocean Global Navigation Satellite System-Reflectometry (GNSS-R). DDM, delay-Doppler map.

payload [6] onboard TechDemoSat-I, launched in 2014. The same payload was used, later, in the Cyclone GNSS (CYGNSS), a NASA venture project for a constellation of eight microsatellites launched in December 2016 with the main mission of studying the development and dynamics of tropical cyclones. The enduring activity of CYGNSS satellites (instruments are still healthy and acquiring data) pushed forward the research activity aimed to refining models and technologies. Electromagnetic models have been improved by introducing two modifications: on one hand, a generalized bistatic equation that incorporates coherent and incoherent reflected components, not necessarily requiring the geometric optics approximation [7]; on the other hand, new modeling approaches based on first- and second-order small slopes approximation [8]. Besides these theoretical works, a remarkable activity emerged for studying the retrieval of wind speed fields from the DDM observable on a global scale. The GNSS-Reflectometry (GNSS-R) operational wind inversion relies on the empirical geophysical model function (GMF) approach. The novel approach presented by Clarizia and Ruf [9] was improved by much additional processing for decoupling the effects of the geometry and the Woodward's ambiguity function (WAF), and for including the dependence on the significant wave height and wave fetch [10, 11].

The material covered in this chapter is aimed to discuss the main methods and models that were developed during the two decades of scientific research, highlighting the advantages and application limits. An overview of the organization is presented in Figure 19.1, with the understanding that many of the topics can be found at an "advanced design level" in more specific textbooks – such as, for instance, orbit propagation, which is exhaustively studied in books of astrodynamics, or signal attenuation and bending in the atmosphere, which is also a theme in telecommunication standards and radiation transfer – and will not be material of this work. The description of the forward scattering models used in the simulation process takes a relevant part of the chapter; it is without doubts the ground plan for building up the simulation structure. Around this core, another main component of the simulation process develops, that is, the definition of the various coordinate systems involved in the representation of signals and vectors.

From a very general viewpoint, an end-to-end simulator for applications to ocean GNSS-R is built around the following combinations of functional modules:

1. The geometry subsystem includes a space segment with satellite orbit propaga-
    tor and attitude determination. The satellite position and orientation, with the
    Earth surface model, allow determination of the nominal specular point posi-
    tion. For simulations over the ocean, a digital elevation model is not generally
    needed. Once the specular point position is known, a surface grid is computed
    with a regular or otherwise defined grid spacing.
2. The signal modeling subsystem provides models for the transmitted signal, its
    propagation in the atmosphere, and the forward scattering by the ocean surface.
3. The simulated product is generated using the scattered signal simulator and the
    ambiguity function of the transmitted waveform. Incoherent averaging provides
    the DDM observable. This process may include conventional processing, where
    the incoming signal is correlated with a replica of the transmitted code, and
    interferometric processing, where the code is acquired from the direct signal.

After an introduction to the observation geometry and signal propagation for spatial platforms, we will consider, in Section 19.3, the models and spectral proper-ties of the ocean surface. The purpose of these two sections is to present the basic material for deriving the bistatic scattering models considered in ocean GNSS-R. Section 19.4 illustrates the electromagnetic models for GNSS signals scattered by a rough surface. The last sections bridge the final gap between modeling and simula-tion and present the final remarks and conclusions.

19.2   Observation geometry and GNSS signal propagation

GNSS-R simulators use several reference systems to define the observation geom-etry. The satellite positions as well as the position of the specular point are given with respect to the Earth-centered Earth-fixed (ECEF) coordinates' frame having the Earth's equator as the fundamental plane and the origin located in the center of mass of the Earth. The x'-axis intersects the Greenwich meridian, the z'-axis is collinear with the Earth's rotation axis in the direction of the north, and the y'-axis completes the right-handed set. The ECEF frame rotates with the Earth and, therefore, coordi-nates of a point on the Earth's surface do not change with time.

Points on the Earth's surface are also located with terrestrial latitude, longitude, and height with respect to the WGS84 coordinate system that defines the standard ellipsoid. The geodetic latitude is the angle between the normal to the Earth surface at a given point and the plane of the equator; the longitude is the angle between the vector from the center of the Earth to the point and the plane of the Greenwich meridian.

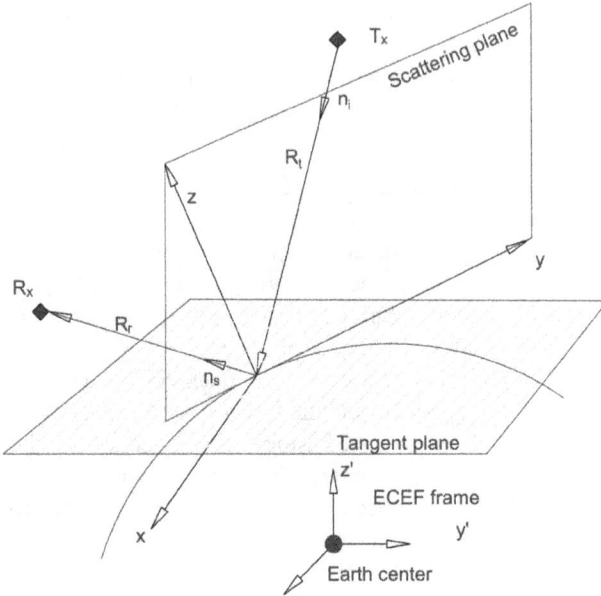

Figure 19.2   *Reference frames in Global Navigation Satellite System-Reflectometry (GNSS-R). The specular reference frame is centered in the specular point with tangent plane as fundamental reference. The Earth-centered Earth-fixed (ECEF) frame is centered in the Earth's center of mass with the plane of the equator as reference. The ECEF frame is body-fixed, that is, fixed to the rotating Earth.*

A more accurate definition of the Earth's height is based on the mean sea level which provides the geoid undulations with respect to the WGS84 ellipsoid. The mean sea level is given by the Earth Gravitational Model 2008 (EGM2008).

A prominent role in GNSS-R is played by the specular reference frame (SRF) that is centered in the specular point, with vertical axis z normal to the tangent plane and oriented towards the free space, axis y defined by the intersection of tangent plane and scattering plane (the plane defined by transmitter, receiver, and specular point positions), oriented in the direction of the transmitter, and axis x completing a right-handed set. The ECEF and the SRF frames are represented in Figure 19.2.

The transmitted signal used on all of the presently operating and planned GNSSs (United States' Global Positioning System (GPS), European Union's Galileo, Russia's GLObal NAvigation Satellite System (GLONASS), and China's BeiDou) can be represented in the general form

$$s(t) = d(t)p(t)\exp(-j2\pi f_0 t) \tag{19.1}$$

in which $p(t)$ is the baseband spreading signal, $d(t)$ is the data message, and f_0 is the carrier frequency (1575.42 MHz for the GPS L1 channel). The spreading signal is a pseudorandom noise (PRN) code which can be generated by a simple algorithm or a lookup table on both the transmitting satellite and in the receiver. GNSS uses various modulations for $p(t)$, including binary phase-shift keying (BPSK), binary offset carrier (BOC), and AltBOC [4].

The Earth's atmosphere affects the amplitude and phase of the GNSS reflected signal. Both effects can be neglected in ocean reflectometry applications where the attenuation, less that 0.1 dB, and phase are not of interest. Recent feasibility demonstrations of satellite phase altimetry at a low grazing angle may require modeling of the signal propagation path by the troposphere and ionosphere for simulation purposes. We recall shortly the simplest models used for the ionosphere and troposphere.

19.2.1 Ionospheric delay

The ionospheric delay can be modeled as

$$\delta_I = \frac{40.35}{f_0^2} \text{sTEC} \tag{19.2}$$

where f_0 is the carrier frequency in Hz and sTEC denotes the slant total electron content, related to the vertical total electron content (vTEC) as

$$\text{sTEC} = \text{vTEC} \left[1 - \left(\frac{R_E \cos(\gamma)}{R_E + h_{\text{pp}}} \right)^2 \right] \tag{19.3}$$

where R_E is the Earth's radius, γ is the GNSS satellite elevation angle, and h_{pp} is the height of the ionosphere shell.

19.2.2 Tropospheric delay

The total tropospheric delay can be divided into hydrostatic and wet components. The hydrostatic component is about 90% of the total delay; therefore the wet component can be neglected in the GNSS-R ocean simulators. The Zenith hydrostatic delay ΔL_h^z can be calculated as

$$\Delta L_h^z = \frac{0.0022768 \, P}{1 - 0.00266 \cos(2\vartheta) - 0.28 10^{-6} h} \tag{19.4}$$

where P is the ground pressure, ϑ is the geodetic latitude, and h is the height above the ellipsoid. The Zenith component can be mapped into the slant component using the mapping function [12].

19.3 Statistically rough surfaces

The ocean surface appears as a moving rough surface whose characteristics can be defined by geometric or statistical models. The statistical approach is the most widely used in models aimed at the description of the electromagnetic scattering and, in this context, the widespread approach is to model the two-dimensional PDF of the sea surface slope and the spatial autocorrelation function of the surface heights, in the hypothesis that the surface is locally stationary. If we indicate $\zeta(\mathbf{r})$ as the surface height at a point $\mathbf{r} \equiv (x, y, 0)$, and $\zeta'(\mathbf{r}) = [\zeta_x(x, y), \zeta_y(x, y)]$ as its slope, the characterization of a random surface can be given in terms of the two-dimensional slopes pdf $f_{\zeta_x \zeta_y}(u, v)$ and the height correlation function at lag $\mathbf{s}$

$$r_\zeta(\mathbf{s}) = \mathbb{E}[\zeta(\mathbf{r})\zeta(\mathbf{r} + \mathbf{s})]. \tag{19.5}$$

The statistical distribution of sea-surface elevations assumes a zero mean distribution around the reference height and is often modeled as Gaussian for linear waves. The model relies on the superposition of sinusoidal water waves with random phase and may adapt to an isotropic or non-isotropic model of the slopes [13]. In a more general framework, the energy balance of the ocean waves is controlled by several competing processes: the input from the wind, the energy transfer by nonlinear wave-to-wave interactions, and the energy lost by viscosity and wave breaking. Nonlinear coupling of the ocean waves generates a continuous redistribution of the energy and momentum of the waves. In this case, the statistics are no longer Gaussian. Non-Gaussian distributions can be considered using the Gram-Charlier series, as suggested by Cox and Munk [14] and Longuet-Higgins [15]. This method introduces deviations from a two-dimensional Gaussian distribution to account for skewness and peakedness in the statistics of the surface slopes. An estimation of the surface slope asymmetry in GNSS-R applications was recently proposed by Cardellach and Rius [16].

The basic method for studying the correlation properties of the sea surface is the two-dimensional wavenumber power spectrum $S(\mathbf{k})$, where $\mathbf{k} = (k_x, k_y)$ is the wavenumber vector, defined as the Fourier transform of the autocorrelation of the sea surface height field, namely

$$S(\mathbf{k}) = \frac{1}{2\pi} \iint r_\zeta(\mathbf{s})e^{-i\,\mathbf{k}\cdot\mathbf{s}}d\mathbf{s}. \tag{19.6}$$

The spectral properties of the sea surface have been studied in the pioneering experimental works of Phillips and Pierson-Moskowitz [17, 18]. Today the description of long waves has reached a mature state after work by Elfouhaily [19], which unifies the previous models. The description is based upon three key parameters:

a.   the wind speed at 10 m over the surface;
b.   the duration of wind blowing action;
c.   the fetch, which can be defined as the distance on the water over which the wind blows without appreciable change in direction.

The expression of the Elfouhaily spectrum in the wavenumber magnitude, k, and phase, ϕ, has the form

$$S(k, \phi) = \frac{1}{2\pi}k^{-4}[S_l(k) + S_h(k)][1 + \Delta(k)\cos(2(\phi - \phi_0))] \tag{19.7}$$

where $S_l(k) + S_h(k)$ represents the omnidirectional spectrum, divided into low- and high-frequency components, and $[1 + \Delta(k)\cos(2\phi)]$ is the spreading function which is symmetric with respect to the upwind and crosswind directions, with ϕ_0 being the direction of travel of the waves.

The omnidirectional spectrum, which is obtained by integrating the wavenumber spectrum in polar coordinates with respect to the angle variable, that is,

$$S(k) = \int_{-\pi}^{\pi} k S(k, \phi) d\phi \tag{19.8}$$

includes gravity and capillary wave scales with boundary between scales $k_m = 370$ rad/m, corresponding to wavelength $\lambda = 2\pi/370 = 0.017$ m. For such a value of spatial frequency, the gravity and surface tension forces (which tend to bring a wave surface back to an unperturbed level surface) are balanced. The function $\Delta(k)$ describes the anisotropy of the spreading function and can be interpreted as the amplitude of the second harmonic in the Fourier series expansion of the angular spreading function. For simplicity, the components of the omnidirectional spectrum are reported in Table 19.1.

The upwind and crosswind mean square slopes can be determined from the ocean surface elevation spectrum in the Cartesian wavenumber domain as

$$
\begin{aligned}
\sigma_u^2 &= \iint k_x^2 S(k_x, k_y) dk_x dk_y \\
\sigma_c^2 &= \iint k_y^2 S(k_x, k_y) dk_x dk_y
\end{aligned}
\tag{19.9}
$$

For scattering calculations with Kirchhoff approximation, only the strong specular-like reflections, that is, with sufficiently large-scale roughness, must be considered. This means that the mean square slopes in (19.9) are determined mostly by the low-frequency region of the spectrum (expressed for $k \leq k_c$). The upper integration limit is determined by a *scale dividing* constant that is experimentally found as $k_c = 2\pi/3\lambda \cos(\theta)$, where θ is the incidence angle. As to the lower integration limit, considering that the decay to zero of the wavenumber spectrum is very quick to the left of the maximum, a reasonable value for the minimum wavelength value would correspond to the peak of the Elfouhaily power spectrum, given by k_p. In practical calculations, it is reasonable and safe to consider slightly smaller values of the minimum for calculation of the mean square slope.

The Elfouhaily model is widely employed in GNSS-R, but more recent and detailed studies have focused on L-band roughness, especially in tropical cyclone conditions, to improve wave spectrum models in the transition zone between the short and long scales [20, 21]. Additionally, besides the empirical models whose main parameters are wind speed and wave age, spectral models based on the numerical solution of the energy balance equation are also of great interest. Among them WAVEWATCH3, a model run by the National Weather Service with a number of source inputs, including wind, nonlinear interaction, dissipation, and bottom friction [22], has been used to develop a revised forward model for GNSS-R scattering, with an interesting potential for both modeling and retrieval of mean square slopes [23].

Table 19.1   Definitions for Elfouhaily spectrum function and constants

$S_l(k)$	$0.5\alpha_p(c_p/c)F_p$	[19, eq. (31)]
$S_h(k)$	$0.5\alpha_m(c_m/c)F_m$	[19, eq. (40)]
$\Delta(k)$	$\tanh[a_o + a_p(c/c_p)^{2.5} + a_m(c_m/c)^{2.5}]$	[19, eq. (57)]
F_p	$L_{PM}J_p \exp[-0.3162\,\Omega(\sqrt{k/k_p} - 1)]$	[19, eq. (32)]
F_m	$L_{PM}J_p \exp[-0.25(k/k_m - 1)^2)]$	[19, eq. (41)]
L_{PM}	$\exp[-1.25(k_p/k)^2]$	[19, eq. (2)]
J_p	$\gamma^{\,\Gamma}$	[19, eq. (3)]
Γ	$\exp\left[-1/2\sigma^2\left(\sqrt{k/k_p} - 1)^2\right)\right]$	[19, eq. (3)]
U_{10}		Wind speed in ms^{-1}, 10 m above msl
g	9.81 ms^{-2}	Acceleration of gravity
u^*	$\sqrt{Cd_{10N}}\,U_{10}$	Friction velocity [19, eq. (61)]
a_m	$0.13\,u^*/c_m$	[19, eq. (59)]
c_m	0.23 m/s	Phase speed of wave for $k = k_m$
α_p	$0.006\,\Omega^{0.5}$	[19, eq. (34)]
α_m	$0.01[1 + \ln(u^*/c_m)]$	$u^* < c_m$; [19, eq. (44)]
	$0.01[1 + 3\ln(u^*/c_m)]$	$u^* > c_m$
k_m	370.0 rad/m	Boundary between scales
Ω	0.84	Wave age for "fully developed" sea
	1	Wave age for "mature" sea
	2 to 5	Wave age for "young" sea
Cd_{10N}	0.00144	Drag coefficient
a_o	0.1733	[19, eq. (59)]
a_p	4.0	
γ	1.7	$0.84 \leq \Omega \leq 1$; [19, eq. (3)]
	$1.7 + 6\log_{10}\Omega$	$1 < \Omega \leq 5$; [19, eq. (3)]
σ	$0.08\,(1 + 4\Omega^{-3})$	
k_o	g/U_{10}^2	
k_p	$k_o\Omega^2$	Spatial frequency at spectrum max
c_p	$\sqrt{g/k_p}$	Phase speed of wave for $k = k_p$
c	$\sqrt{(g/k)(1 + (k/k_m)^2)}$	Phase speed of wave

19.4   Scattering models for GNSS-R signal simulation

Models of electromagnetic scattering from rough surfaces are widely discussed and well established in the literature [24, 25]. In this section, we limit the study to three main models that are useful for simulation of the scattered signal: a) modeling based on physical optics (PO) scattering with a physical three-dimensional representation

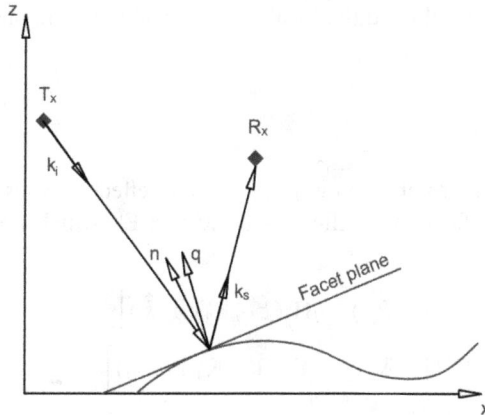

*Figure 19.3 Geometry of facet scattering. The scattering vector **q** is the bisector of the angle formed between the transmitter and the receiver with the facet center. The incident field and scattered field unit vector are indicated with $\hat{k}_i$ and $\hat{k}_s$, respectively.*

of the sea surface, referred to as the facet approch; b) modeling based on the Zavorotny-Voronovich bistatic equation; c) modeling based on ensemble scattering by specular points, referred to as the statistical scattering model.

19.4.1 Facet approach

In the facet-based model, the ocean surface is represented as an elevation map where each facet is a planar surface with the local normal tilted by the underlying long sea wave [26, 27]. With reference to Figure 19.3, the incident wave at a distance R_T from the transmitter is a spherical wave with amplitude E_0 directed along the unit vector $\hat{k}_i$ that may be written as

$$\mathbf{E}^i = \frac{E_0 e^{-ikR_T}}{4\pi R_T} \hat{e}^{\,i} \; e^{-ik\hat{k}_i \cdot \mathbf{r}} \tag{19.10}$$

where $\hat{e}^{\,i}$ is the incident unit polarization vector and k the wavenumber of the propagation medium. The Stratton-Chu equation provides an integral expression of the field scattered by a surface S with arbitrary orientation, in the form [28]

$$\mathbf{E}^s = \frac{-ik}{4\pi R_R} e^{-ikR_R} \; \hat{k}_s \times \iint_S \left[\hat{n} \times \mathbf{E} - \eta \hat{k}_s \times (\hat{n} \times \mathbf{H}) \right] e^{ik\hat{k}_s \cdot \mathbf{r}} d\mathbf{r} \tag{19.11}$$

where R_R and $\hat{k}_s$ are distance and unit vector, respectively, from the center of the facet to the receiver, η is impedance of the propagation medium, $\hat{n}$ is the local normal, and $\mathbf{E}$ and $\mathbf{H}$ are the total electric and magnetic fields on the interface. For a given polarization direction, the incident and reflected electric and magnetic fields can be expressed as the sum of the component perpendicular to the plane of incidence (the plane of the incident field and the local normal) and the component parallel to the plane of incidence,

$$\mathbf{E}^{i,r} = \mathbf{E}_\perp^{i,r} + \mathbf{E}_\parallel^{i,r}$$
$$\mathbf{H}^{i,r} = \mathbf{H}_\perp^{i,r} + \mathbf{H}_\parallel^{i,r}. \tag{19.12}$$

The tangent plane approximation implies that the reflected fields can be calculated from the Fresnel reflection coefficients as $\mathbf{E}^r = \mathbb{R}\, \mathbf{E}^i$, which provide, after some algebra

$$\hat{n} \times \mathbf{E} = \left[\hat{n} \times \mathbf{E}_\perp^i (1 + \mathbb{R}_\perp) + \eta \mathbf{H}_\parallel^i (1 - \mathbb{R}_\parallel)(\hat{n} \cdot \hat{k}_i) \right]$$
$$\hat{n} \times \mathbf{H} = \left[\hat{n} \times \mathbf{H}_\parallel^i (1 + \mathbb{R}_\parallel) - \frac{1}{\eta} \mathbf{E}_\perp^i (1 - \mathbb{R}_\perp)(\hat{n} \cdot \hat{k}_i) \right]. \tag{19.13}$$

The orthogonal component of the incident electric field and the parallel component of the incident magnetic field can be expressed as

$$\mathbf{E}_\perp^i = \frac{E_0 e^{-ikR_T}}{4\pi R_T} (\hat{e}^i \cdot \hat{t})\, \hat{t} e^{-ik\hat{k}_i}$$
$$\mathbf{H}_\parallel^i = \frac{E_0 e^{-ikR_T}}{4\pi R_T} (\hat{e}^i \cdot \hat{d})\, \hat{t} e^{-ik\hat{k}_i} \tag{19.14}$$

where the three unit vectors $\hat{k}_i$, $\hat{t} = \hat{k}_i \times \hat{n}/|\hat{k}_i \times \hat{n}|$, and $\hat{d} = \hat{k}_i \times \hat{t}$ define an orthogonal coordinate system. Rearranging the terms we get

$$\mathbf{E}^s = \frac{-ik}{(4\pi)^2 R_T R_R} e^{-ik(R_T + R_R)}\, \hat{k}_s \times \iint_S \mathbf{p}\, e^{i\,\mathbf{q}\cdot\mathbf{r}} d\mathbf{r} \tag{19.16}$$

where $\mathbf{p}$ is a polarization-dependent term that, for a planar facet, can be taken outside the integral and $\mathbf{q}$ is the scattering vector, defined as

$$\mathbf{q} = k \left(\frac{\mathbf{R}_R}{R_R} - \frac{\mathbf{R}_T}{R_T} \right) = (\hat{k}_s - \hat{k}_i) \equiv (q_\perp, q_z), \tag{19.17}$$

directed along the bisector of the angle formed by the incidence and scattering directions, which also splits into a component $q_\perp$ lying in the tangent plane and a component q_z that is orthogonal to such a plane.

Solving the integral in (19.16), after some algebra, the expression for pq polarization combination can be arranged in the form

$$E_{pq}^s (\mathbf{r}_k) = \frac{-ikE_0}{(4\pi)^2 R_R R_T} e^{-ik(R_R + R_T)} e^{-i\mathbf{q}\cdot\mathbf{r}_k} \left(1 + \zeta_x^2 + \zeta_y^2 \right)^{1/2} \mathbb{S}_{pq}$$
$$\times L_x L_y \operatorname{sinc}\left[L_x \frac{q_x + \zeta_x q_z}{2} \right] \operatorname{sinc}\left[L_y \frac{q_y + \zeta_y q_z}{2} \right] \tag{19.18}$$

In the above equation, $\mathbf{r}_k$ is the center of the facet; L_x and L_y are the facet projections along x and y; ζ_x and ζ_y are the slopes of the facet plane along x and y; and $\mathbb{S}_{pq}$ is a polarization-dependent coefficient defined as

$$\mathcal{S}_{HH} = \gamma \ \mathbb{R}_V(\hat{h}_s \cdot \hat{k}_i)(\hat{h}_i \cdot \hat{k}_s) + \mathbb{R}_H(\hat{h}_s \cdot \hat{k}_i)(\hat{v}_i \cdot \hat{k}_s)$$
$$\mathcal{S}_{VV} = \gamma \ \mathbb{R}_V(\hat{v}_s \cdot \hat{k}_i)(\hat{v}_i \cdot \hat{k}_s) + \mathbb{R}_H(\hat{h}_s \cdot \hat{k}_i)(\hat{h}_i \cdot \hat{k}_s)$$
$$\mathcal{S}_{HV} = \gamma \ \mathbb{R}_V(\hat{h}_s \cdot \hat{k}_i)(\hat{v}_i \cdot \hat{k}_s) - \mathbb{R}_H(\hat{v}_s \cdot \hat{k}_i)(\hat{h}_i \cdot \hat{k}_s)$$
$$\mathcal{S}_{VH} = \gamma \ \mathbb{R}_V(\hat{v}_s \cdot \hat{k}_i)(\hat{h}_i \cdot \hat{k}_s) - \mathbb{R}_H(\hat{h}_s \cdot \hat{k}_i)(\hat{v}_i \cdot \hat{k}_s)$$

$$(19.19)$$

that depends on the unit polarization vectors $\hat{h}_i$, $\hat{h}_s$, $\hat{v}_i$, and $\hat{v}_s$, such that the incident polarization vector is $\hat{e}^i = \hat{h}_i + \hat{v}_i$ and $\hat{e}^s = \hat{h}_s + \hat{v}_s$, and the constant

$$\gamma = \frac{|\mathbf{q}||q_z|}{kq_z[(\hat{h}_s \cdot \hat{k}_i)^2 + (\hat{v}_s \cdot \hat{k}_i)^2]}. \tag{19.20}$$

The size of the facet is related to the applicability of the Kirchhoff approximation, which translates into identifying a region on the tangent plane with linear dimensions which are large with respect to the electromagnetic wavelength, but does not deviate noticeably from the surface at the edges of the region. This condition can be satisfied if the local incidence angle θ, surface curvature r_c, and the size of the facet satisfy the conditions

$$\frac{1}{k\cos(\theta)} \ll L/2 \ll \sqrt{\left[\frac{\cos(\theta)}{k}\right]^2 + 2\frac{r_c \cos\theta}{k}} \tag{19.21}$$

For the GNSS center frequency corresponding to $\lambda = 19$ cm, incidence angle $\theta_i = 20°$ and local scattering angle that ranges from $-10°$ to $50°$ from the specular incidence angle point, we require the facet size to be greater than 0.06 m and less than 1.6 m.

The total field scattered by a delay-Doppler cell is obtained by coherent accumulation of fields scattered by individual facets F_k. For each delay and Doppler cell $C(\tau, f)$, centered in (τ, f), the accumulated field is

$$\mathbf{E}^s_{pq}(\tau, f) = \sum_n \mathbf{E}^s_{pq}(\mathbf{r}_n) \qquad n : (\tau(\mathbf{r}_n), f(\mathbf{r}_n)) \in C(\tau, f). \tag{19.22}$$

19.4.2   Zavorotny-Voronovich bistatic equation model

Rather than considering the scattering contribution from each facet, the idea here is to define a grid of points and calculate the total electric field at the receiver using the bistatic radar equation, assuming a uniformly distributed phase of the complex received signal. The electric field at each surface cell here is modeled as an ensemble variable with lower spatial resolution. The fundamental equation for modeling the received signal in the bistatic framework represented in Figure 19.2 was derived by Zavorotny and Voronovich [3] under the Kirchhoff approximation with geometric optics (GO) limit. First applications of a refined GO model to real GPS data, reflected by the ocean surface and received by an aircraft platform, can be found in [29]. Here, we consider as starting point the equation of the received signal after downconversion,

$$u_R(t) = \iint \mathcal{D}(\mathbf{r})a\left(t - \frac{R_T(\mathbf{r};t) + R_R(\mathbf{r};t)}{c}\right)g(\mathbf{r};t)d\mathbf{r} \tag{19.23}$$

reporting the scattered field under the Kirchhoff approximation, assuming a large distance from the transmitter to the scattering point and from the scattering point to the receiver. In (19.23), $\mathbf{r}$ is the vector of coordinates $(x, y, 0)$, $\mathcal{D}(\mathbf{r}) = \Theta_T(\mathbf{r})\Theta_R(\mathbf{r})$ is the footprint function that depends on the transmitting and receiving antenna radiation patterns projected onto the ground, $a(t)$ is the transmitted waveform, and

$$g(\mathbf{r};t) = -\frac{\mathbb{R}}{4\pi i} \frac{q(\mathbf{r};t)^2}{q_z(\mathbf{r};t)} \frac{e^{ik(R_T(\mathbf{r};t) + R_R(\mathbf{r};t))}}{R_T(\mathbf{r};t)R_R(\mathbf{r};t)} \tag{19.24}$$

is a function accounting for propagation and scattering from the ocean surface. At the receiver stage, the signal is cross-correlated with a frequency-shifted signal that can be a replica of the PRN code transmitted by a GNSS transmitter or the signal directly received by a different antenna pointed in the space hemisphere. The cross-correlation $Y(t_0, \tau, f)$ is calculated over the *coherent integration time T* that usually matches the 1 ms length of the transmitted PRN sequence. Denoting t_0 as the initial time, τ the delay, and f the frequency shift, we have

$$Y(t_0, \tau, f) = \int_{t_0}^{t_0+T} u_R(t)a(t - \tau)e^{j2\pi f t}dt. \tag{19.25}$$

All terms in (19.24) are time-dependent because the transmitter and receiver are moving, but in a coherent integration interval, which evolves on a millisecond scale, all range variations can be neglected except for the phase term, where a first-order approximation provides

$$R_T(\mathbf{r};t) + R_R(\mathbf{r};t) \simeq R_T(\mathbf{r};t_0) + R_R(\mathbf{r};t_0) + (t - t_0)[\mathbf{v}_T(\mathbf{r};t_0) \cdot \hat{k}_i - \mathbf{v}_R(\mathbf{r};t_0) \cdot \hat{k}_s] \tag{19.26}$$

where $\mathbf{v}_T$ and $\mathbf{v}_R$ are the transmitter and receiver velocities, respectively, at time t_0. Defining the Doppler frequency

$$\frac{1}{\lambda}(\mathbf{v}_T(\mathbf{r};t_0) \cdot \hat{k}_i - \mathbf{v}_R(\mathbf{r};t_0) \cdot \hat{k}_s) = f_{d0}(\mathbf{r}) \tag{19.27}$$

and the delay

$$\frac{R_T(\mathbf{r};t_0) + R_R(\mathbf{r};t_0)}{c} = \tau_0(\mathbf{r}), \tag{19.28}$$

after substitution in (19.25), we obtain

$$Y(t_0, \tau, f) = T\iint \mathcal{D}(\mathbf{r})g(\mathbf{r}, t_0)\chi[\tau - \tau_0(\mathbf{r}), f - f_{d0}(\mathbf{r})]d\mathbf{r} \tag{19.29}$$

The function

$$\chi(\tau, f) = \frac{1}{T}\int_T a(t)a(t + \tau)e^{-i2\pi f t}dt \tag{19.30}$$

is the WAF of the transmitted waveform that plays an important role in the analysis of the scattered signal. We observe that the correlated signal received in (19.29) is given as a superposition of scattered waves having amplitude and phase determined by the function $g(\mathbf{r}, t_0)$, by transmitter and receiver antenna footprints, and weighted by the ambiguity function of the transmitted signal centered at the delay and Doppler values of the wave. In the spatial domain, the iso-delay lines are the intersection with the Earth's surface of the locus of points with a constant sum of delays, from the transmitter and receiver, which is an ellipsoid. These curves are approximately ellipses. The iso-Doppler lines correspond to the points where (19.27) is constant over the Earth's surface and are approximately hyperbolae. A delay-Doppler cell is thus a surface patch determined by the intersection (possibly not connected) between two contiguous iso-Doppler and iso-delay lines.

Interestingly, the received signal can be also rewritten as a two-dimensional convolution in the delay-Doppler plane [30]. To see more clearly this correspondence, it is necessary to make a change of variables from the Earth's surface domain represented in the coordinates $(x, y, z(x, y))$ and the delay-Doppler domain $(\tau = \tau(x, y,), f_d = f_d(x, y))$. The surface differential for this transformation is calculated as

$$
d\mathbf{r} = \det\left[J(\tau, f_d)\right] d\tau df_d = \det\begin{bmatrix} \frac{\partial x}{\partial \tau} & \frac{\partial x}{\partial f_d} \\ \frac{\partial y}{\partial \tau} & \frac{\partial y}{\partial f_d} \end{bmatrix} d\tau df_d. \tag{19.31}
$$

The transformation from delay-Doppler to spatial coordinates is not one-to-one because a delay-Doppler point may turn into two points or one point on the Earth's surface, or none at all. We have the so-called ambiguity problem for all points except for the singular solutions corresponding to the points where iso-Doppler hyperbole and iso-delay ellipses are tangent, which are not affected by the ambiguity. There are no solutions when the iso-Doppler line lies outside the elliptical region corresponding to the given delay. The two solutions corresponding to different regions of the observed surface must be combined; substitution into (19.30) yields

$$
Y(t_0, \tau, f) = \gamma(\tau, f; t_0) * \chi[\tau - \tau_0(\mathbf{r}), f - f_{d0}(\mathbf{r})] \tag{19.32}
$$

$$
\gamma(\tau, f; t_0) = T\sum_{i=1,2} \mathcal{D}(\tau_i, f_i)g(\tau_i, f_i; t_0) \det[J_i(\tau, f)] \tag{19.33}
$$

The representation (19.33) of the received signal as a two-dimensional convolution emphasizes that the ambiguity function acts as a spatial filter for the surface-dependent function $\gamma(\tau, f_d; t_0)$ and, therefore, has a main impact in the spatial resolution of the sensing system.

The cross-correlated signal is not used directly as observable of the sea surface parameters because the scattered wave is a random process corrupted by thermal noise. The variance of the signal can be reduced by an incoherent temporal averaging that is usually carried over one second or half a second, that is over 500 or 1 000 received signal waveforms. The resulting observable, after the incoherent average of N signals, is the DDM

$$\langle |Y(\tau,f)|^2\rangle = \frac{1}{N}\sum_{n=1}^{N}|Y(\tau,f;t_n)|^2. \tag{19.34}$$

The expression of the DDM can be further elaborated when the temporal averaging can be replaced by an ensemble averaging. Here we do not report the details of the procedure that is discussed by Zavorotny and Voronovich [3] leading to the expression

$$\langle |Y(\tau,f)|^2\rangle = \frac{\lambda^2 T^2}{(4\pi)^3}P_t \iint \frac{|\mathcal{D}(\mathbf{r})|^2}{R_T^2(\mathbf{r})R_R^2(\mathbf{r})}|\chi(\tau,f)|^2\sigma_0\ d\mathbf{r} \tag{19.35}$$

where

$$\sigma_0 = \pi |\mathbb{R}|^2 \frac{|q|^4}{q_z^4}f_{\xi_x\xi_y}\left(-\frac{q_x}{q_z},-\frac{q_y}{q_z}\right) \tag{19.36}$$

is the normalized bistatic radar cross section (NBRCS) of the rough surface which is a function of the two-dimensional PDF $f_{\xi_x\xi_y}(\ \cdot\ )$ of the large-scale sea surface slopes and P_t is the transmitted power.

The present finding should be interpreted in light of several limitations. First of all, in the derivation of (19.35) it was assumed that the scattering regime is largely diffusive. If it is not true, a coherent component that is not accounted for by (19.35) emerges, with a peak in the specular direction. A second limitation arises because the transmitted signal is not purely monochromatic. Indeed, the linear part of the time-frequency excursion of the signal generates a Doppler frequency shift that is not accounted into (19.35) and may become significant at low grazing angles in satellite reflectometry. This component combines with the Doppler shift generated by the platforms' orbital speed. Finally, the Kirchhoff geometric optics approximation used in (19.35) can be applied when the surface roughness is smooth on a scale comparable to the wavelength of the incident radiation. In case the surface is not smooth but its slope remains small, the small slope approximations of the first order (SSA1) or second order (SSA2) remain applicable but the bistatic radar cross section (BRCS) is no longer expressed by (19.36) (see Voronovich and Zavorotny [31, 32] for further details).

A new form of the bistatic radar equation was derived by Voronovich and Zavorotny [7], which is not affected by such limitations. It is found that the expression of received power requires an additional term accounting for the coherently reflected power but the term which is responsible for non-coherent scattering maintains the same structure as in (19.35).

19.4.3   Statistical scattering model

This model develops around the idea that the main contribution to the scattered field is generated by a small region of the sea wave whose slope allows a specular reflection of the transmitted signal towards the receiver direction. The original idea dates back to 1966, in a paper by Kodis [33], recently reformulated for GNSS-R signal modeling in [34]. From (19.23) and (19.26), the received signal can be written as

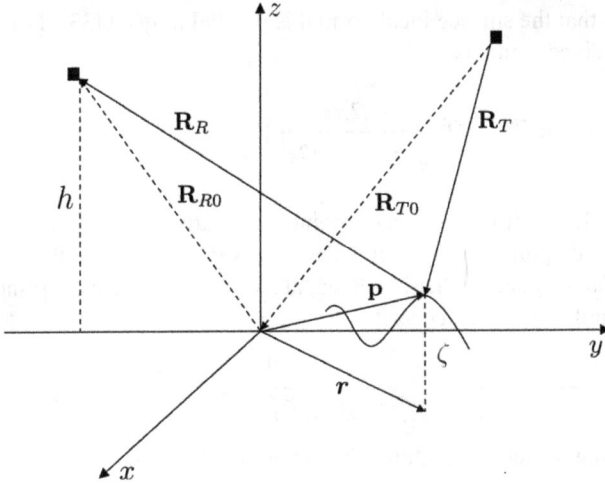

Figure 19.4   Scattering geometry for local specular point analysis

$$u_R(t) = \iint \mathcal{D}(\mathbf{r})a\left(t - \tau_0(\mathbf{r})\right) g(\mathbf{r}; t_0)\, e^{i2\pi f_{d0}(\mathbf{r})(t-t_0)} d\mathbf{r} \tag{19.37}$$

where

$$g(\mathbf{r}; t_0) = -\frac{\mathbb{R}}{4\pi i} \frac{q(\mathbf{r}; t_0)^2}{q_z(\mathbf{r}; t_0)} \frac{e^{ik(R_T(\mathbf{r}; t_0) + R_R(\mathbf{r}; t_0))}}{R_T(\mathbf{r}; t_0)R_R(\mathbf{r}; t_0)} \tag{19.38}$$

$$= G(\mathbf{r}; t_0)\, e^{ik(R_T(\mathbf{r}; t_0) + R_R(\mathbf{r}; t_0))}$$

Considering the geometry in Figure 19.4, for each point $\mathbf{r}$ at time t_0, we have

$$\begin{aligned}
\mathbf{R}_T &= R_T \hat{k}_i = \mathbf{R}_{T0} + \mathbf{p} \\
\mathbf{R}_R &= R_R \hat{k}_s = \mathbf{R}_{R0} - \mathbf{p}.
\end{aligned} \tag{19.39}$$

The transmitter and receiver can be assumed to be at a large distance from the scattering points such that the vectors from the platforms to the Earth's surface can be considered approximately parallel; therefore $\mathbf{R}_T \simeq R_T \hat{k}_i$ and $\mathbf{R}_R \simeq R_R \hat{k}_s$ and we have that

$$\begin{aligned}
k[R_T(\mathbf{r}) + R_R(\mathbf{r})] &= k[R_{T0} + R_{R0}] - k(\hat{k}_s - \hat{k}_i) \cdot \mathbf{p} \\
&= k[R_{T0} + R_{R0}] - \mathbf{q} \cdot \mathbf{p}.
\end{aligned} \tag{19.40}$$

It can be recognized that the function

$$f(\mathbf{r}; t) = \mathcal{D}(\mathbf{r})G(\mathbf{r}; t_0)a(t - \tau_0(\mathbf{r}))e^{ik(R_{T0}+R_{R0})}\, e^{i2\pi f_{d0}(\mathbf{r})(t-t_0)} \tag{19.41}$$

is a slowly varying function of $\mathbf{r}$ whereas the function $e^{-i\mathbf{q}(\mathbf{r}; t_0) \cdot \mathbf{p}}$ is rapidly oscillating. The integral (19.37) can be approximated using the stationary phase method resulting in a sum of signals scattered by the ensemble of specular points $\{\mathbf{r}_j\}$ satisfying the condition that the surface local normal is parallel to $\mathbf{q}(\mathbf{r}_j)$ [33]. For any specular point, the received signal is

$$
u_j(t) = f(\mathbf{r}; t)e^{-i\mathbf{q}(\mathbf{r}; t_0)\cdot\mathbf{p}} \left. \frac{-i2\pi\alpha}{q_z|\zeta_{xx}\zeta_{yy} - \zeta_{xy}^2|^{\frac{1}{2}}}\right|_{\mathbf{r}=\mathbf{r}_j} \tag{19.42}
$$

where $\alpha = \{-1, 1, i\}$ if the stationary point is a relative maximum, a relative minimum, or a saddle point, respectively. The quantity $\zeta_{xx}\zeta_{yy} - \zeta_{xy}^2$ is the determinant of the surface *shape operator* that is related, at any point, to the local principal radii of curvature r_x and r_y of a surface as

$$
|r_x r_y| = \frac{(1 + \zeta_x^2 + \zeta_y^2)^2}{|\zeta_{xx}\zeta_{yy} - \zeta_{xy}^2|} = \frac{|q(\mathbf{r}; t_0)|^4}{q_z^4(\mathbf{r}; t_0)} \frac{1}{|\zeta_{xx}\zeta_{yy} - \zeta_{xy}^2|}. \tag{19.43}
$$

In terms of the surface curvature, the scattering from the j-th specular point is expressed as

$$
u_j(t) = \mathcal{D}(\mathbf{r})\, \mathbb{R}(\mathbf{r})\, \frac{e^{ik[R_T(\mathbf{r}; t_0) + R_R(\mathbf{r}; t_0)]}}{2R_T(\mathbf{r}; t_0)R_R(\mathbf{r}; t_0)}
$$

$$
\times\, a[t - \tau_0(\mathbf{r})]e^{i2\pi f_{d0}(\mathbf{r})(t - t_0)}\alpha|r_x r_y|^{1/2}\Big|_{\mathbf{r}=\mathbf{r}_j}. \tag{19.44}
$$

The signals scattered by the ensemble of N_m specular points lying in the delay-Doppler cell, here denoted with the subscript m, add coherently at the receiver antenna such that the overall signal is given by

$$
u_m(t) = \sum_{j=1}^{N_m(t)} u_j(t). \tag{19.45}
$$

The physical picture of the observation process is concerned with random placement of specular points on a plane that are observed from different positions at different times and are subject to internal forces and the force of gravity. The evolution model can be formally defined through a birth, death, and immigration process [35, 36] where two parameters, usually denoted as λ_0 and μ_0, define the *birth* and *death* rates per unit time and per population element, and ν_0 is the *immigration rate*, independent of the population size. The mean number of specular points in a rough surface can be calculated with a zero-crossing approach as demonstrated by Barrick [37], where it is assumed that the surface height is an isotropic Gaussian process with Gaussian correlation. Later, the approach was extended by Beltramonte *et al.* [38] to a random surface with the Elfouhaily power spectrum. A first result is that the mean number of specular points, assuming cut-off wavelengths as discussed in Section 19.3, ranges in the order of unity per square meters, which translates into a number of the order of 10^8 for a delay-Doppler cell in a satellite configuration. A second result is that the number of specular points can be modeled as a stochastic birth-death-immigration process. This final consideration touches the interesting and debated question of temporal statistics of the evolving sea surface, as well as of the number of specular points that we do not approach here, leaving the interested reader to more specialized references on the theme.

To derive an usable expression of $u_m(t)$, we may observe that all quantities in (19.44) that are slowly varying functions of $\mathbf{r}$ can be approximated with the corresponding value in the center of the delay-Doppler cell $\mathbf{r}_m$ and, therefore, do not depend on the specular point. Precisely,

1. The antenna function $\mathcal{D}(\mathbf{r})$, the reflection coefficient $\mathbb{R}(\mathbf{r})$, the quantities $R_T(\mathbf{r}; t)$ and $R_R(\mathbf{r}; t)$, and the signal delay $\tau_0(\mathbf{r})$ can be approximated with their value at the cell center.
2. The phase shift corresponding to the geometric range is subject to variations that are much longer than the wavelength from one specular point to another and the point locations are not predictable in space and time. For this reason, the phase can be written as the sum of the value $k[R_T(\mathbf{r}_m; t) + R_R(\mathbf{r}_m; t)]$ at the cell center plus a random term ϕ_j with uniform phase in $(0, 2\pi)$, accounting for the unknown position of the specular point.
3. The curvatures r_x and r_y are also random variables, reasonably independent from one specular reflector to another.

Rearranging (19.45) and letting for simplicity $R_T(\mathbf{r}_m; t_0) = R_{Tm}$ and $R_R(\mathbf{r}_m; t_0) = R_{Rm}$, we get

$$u_m(t) = \mathcal{D}_m \mathbb{R}_m \frac{e^{-ik(R_{Tm}+R_{Rm})}}{2R_{Tm}R_{Rm}} a(t - \tau_0(\mathbf{r}_m)) e^{j2\pi f_{d0}(\mathbf{r}m)(t-t_0)} \sum_{j=1}^{N_m(t)} A_j e^{j\phi_j} \tag{19.46}$$

with the assumption that the phase and curvature of specular points do not change in the coherent integration time. The sum

$$Z_m(t) = X_m(t) + jY_m(t) = \sum_{j=1}^{N_m(t)} A_j e^{j\phi_j} \tag{19.47}$$

is a *random walk* in two dimensions with statistically independent steps. The probability distribution of Z for large values of the mean number of specular points $\bar{N}$ is defined by the following asymptotic theorem [35].

Let $\{Z_m\}$ be a sequence of N independent, identically distributed zero-mean complex random variables, with arbitrary distribution, and N a discrete random variable with probability distribution $P_N(n)$ and mean $\bar{N}$. If all cumulants of Z_m are finite, then

$$\lim_{\bar{N} \to \infty} \frac{Z}{\sqrt{\bar{N}}} = \sqrt{S}G \tag{19.48}$$

where the equality is in distribution, S is a random variable with cumulative distribution

$$F_s(x) = \lim_{\overline{N} \to \infty} P\left(\frac{N}{\overline{N}} \leq x\right) \tag{19.49}$$

and G is a complex Gaussian random variable with zero mean and covariance matrix **C**.

The multiplicative model (19.48) is known as the *Compound-Gaussian* model. The non-negative random variable S modulates the power of the Gaussian variable, which is often referred to as speckle component, accounting for fast signal fluctuations. The main considerations stemming from representation (19.48) can be summarized as follows:

1. For large, finite values of $\overline{N}$, the random sum in (19.46) can be approximated as the product of a Gaussian random variable, depending on the specular points' curvature and phase and a non-negative random variable, modeling the distribution of the number of the specular points.
2. There is no need to consider a statistical characterization of the elements of the random sum other than the covariance matrix **C**.
3. The phase is uniformly distributed in $(0, 2\pi)$, therefore the real and imaginary parts of the Gaussian component are statistically independent with the circularly symmetric joint PDF, and the covariance matrix reduces to

$$C = \begin{bmatrix} \sigma_c^2 & 0 \\ 0 & \sigma_c^2 \end{bmatrix}; \tag{19.50}$$

4. The statistical characterization of the random variable S can be found through the limit (19.49) once the discrete distribution for the number of specular points is known. In any case, the definition (19.49) implies $\mathbb{E}[S] = 1$.

The stationary solution of the birth-death-immigration process is the negative binomial distribution that in the limit (19.49) provides the gamma PDF

$$f_S(x; \beta) = \frac{\beta^\beta}{\Gamma(\beta)} x^{\beta-1} e^{-\beta x}; \quad x \geq 0, \ \beta \geq 0 \tag{19.51}$$

where $\Gamma(\cdot)$ is the Eulerian gamma function and $\beta = v_0/\lambda_0$ is a shape parameter. The product $Z = \sqrt{S}G$ is a K-distributed complex random variable whose real and imaginary parts have PDF

$$f_Z(z; \beta) = \frac{b}{\sqrt{\pi}\Gamma(\beta)} \left(\frac{b|z|}{2}\right)^{\beta-\frac{1}{2}} K_{\beta-1/2}(b|z|). \tag{19.52}$$

where $b^2 = 2\beta/\sigma_c^2$ and $K_\alpha(\cdot)$ is the K-Bessel function. The rate of decay of the tails of the gamma distribution increases as β increases and the distribution converges to the Dirac delta function $\delta(x - 1)$ in the limit $\beta \to \infty$. In this case, the compound-Gaussian variable is simply Gaussian.

After substitution of the limit (19.48) into (19.46) and assuming that the process is frozen during coherent integration time, that is, $\sqrt{S_m(t)}G_m(t) \simeq \sqrt{S_m}G_m$, the expression of the signal scattered by a delay-Doppler cell becomes

$$u_m(t) = \mathcal{D}_m \mathbb{R}_m \frac{e^{-ik(R_{Tm}+R_{Rm})}}{2R_{Tm}R_{Rm}} a(t-\tau_m)e^{-2\pi i f_{dm}t} \sqrt{S_m}G_m. \tag{19.53}$$

This shows that the stochastic term $\sqrt{S_m}G_m$ plays the role of the *complex scattering coefficient* for the ensemble of specular points in a delay-Doppler cell. The variance of the real and imaginary parts of the scattering coefficient is given by

$$\sigma_m^2 = \frac{\overline{N_m}}{2}\mathbb{E}[A_j^2]. \tag{19.54}$$

Assuming that the two-dimensional PDF of the sea surface slopes is Gaussian and using results given by Barrick [37] for the mean number of specular points per unit area, we have that

$$\sigma_m^2 = A_m f_{\zeta_x \zeta_y}\left(-\frac{q_\perp(\mathbf{r}_m)}{q_z(\mathbf{r}_m)}\right) \mathbb{E}\left[|\zeta_{xx}\zeta_{yy} - \zeta_{xy}^2|\right]\frac{\mathbb{E}[A_j^2]}{2} \tag{19.55}$$

with ζ_{xx}, ζ_{xy}, and ζ_{yy} being the random variables associated to the second derivatives of the sea surface heights, $f_{\zeta_x \zeta_y}$ the joint Gaussian probability distribution of the sea surface slopes and A_m the area of the delay-Doppler cell centered in $\mathbf{r}_m$. The mean square value of the curvature term A_j is calculated as

$$\mathbb{E}[A_j^2] = \mathbb{E}[|r_x r_y|] = (1 + \zeta_x^2 + \zeta_y^2)^2 \mathbb{E}\left[\frac{1}{|\zeta_{xx}\zeta_{yy} - \zeta_{xy}^2|}\Bigg|_{r=r_m}\right] \tag{19.56}$$

where ζ_x and ζ_y are the surface slopes in the stationary point, approximated as the surface slopes in the center of the cell. The approximation $\mathbb{E}[1/|J|] \approx 1/\mathbb{E}[|J|]$, with $J = \zeta_{xx}\zeta_{yy} - \zeta_{xy}^2$ is reasonable in the geometric optics approximation; therefore, the average $\mathbb{E}[A_j^2]$ can be recast in the form

$$\mathbb{E}[A_j^2] \simeq \frac{(1 + \zeta_x^2 + \zeta_y^2)^2|_{r=r_m}}{\mathbb{E}\left[|\zeta_{xx}\zeta_{yy} - \zeta_{xy}^2|\right]} = \frac{q^4(\mathbf{r}_m)}{q_z^4(\mathbf{r}_m)}\frac{1}{\mathbb{E}\left[|\zeta_{xx}\zeta_{yy} - \zeta_{xy}^2|\right]} \tag{19.57}$$

and after substitution of (19.57) into (19.54), the variance σ_m^2 takes the form

$$\sigma_m^2 = \frac{A_m}{2}\frac{q^4(\mathbf{r}_m)}{q_z^4(\mathbf{r}_m)}f_{\zeta_x\zeta_y}\left(-\frac{q_\perp(\mathbf{r}_m)}{q_z(\mathbf{r}_m)}\right). \tag{19.58}$$

After the cross-correlation in the GNSS-R receiver between the received signal and a frequency-shifted version of a local replica of the transmitted signal, the signal at the output of the correlator is given by

$$Y(\tau,f) = T\sum_m \frac{\mathcal{D}_m \mathbb{R}_m}{2R_{Tm}R_{Rm}}e^{-ik(R_{Tm}+R_{Rm})}\sqrt{S_m}G_m\chi\left(\tau-\tau_m, f-f_{dm}\right), \tag{19.59}$$

as the superposition of ambiguity functions $\chi\left(\tau - \tau_m, f - f_{dm}\right)$ centered in the delay and Doppler shift at the center of each resolution cell.

The average power of the DDM can be easily calculated by exploiting the property that the random variables $\sqrt{S_m} G_m$ in (19.60) are statistically independent. It follows that all mutual terms vanish and, after the substitution $\mathbb{E}[S_m] = 1$ and $\mathbb{E}[|G_m|^2] = 2\sigma_m^2$ in the remaining terms, it is found that

$$
\mathbb{E}\left[|Y(\tau, f)|^2\right] = \frac{T^2}{4} \sum_m \frac{A_m \mathcal{D}_m^2 |\mathbb{R}_m|^2}{R_{Tm}^2 R_{Rm}^2} \frac{q^4(\mathbf{r}_m)}{q_z^4(\mathbf{r}_m)} |\chi\left(\tau - \tau_m, f - f_{dm}\right)|^2 f_{\zeta_x \zeta_y}\left(-\frac{q_\perp(\mathbf{r}_m)}{q_z(\mathbf{r}_m)}\right).
$$

(19.60)

The result of this derivation can be interpreted as a discrete form of the average power reported in (19.35), with the integral operator replaced by the sum and the differential surface element replaced by the area A_m of the delay-Doppler cell. This is in agreement with the approximations considered in Section 19.4.2 for the derivation of the scattered field that, in both cases, are consistent with the geometric optics approximation in the regime of a large Rayleigh parameter.

19.4.4   *Comments to modeling equations*

The central theme of this section was the presentation of bistatic scattering models for rough surfaces. All models rely on the Kirchhoff approximation, with the assumption that the mean radius of curvature of the scattering surface is much greater than the wavelength of the incident radiation. The meaning of rough surface is, instead, related to the surface height variations that are supposedly large enough to generate uncorrelated phase values in the scattered wave. These conditions are generally applicable to well developed large sea waves.

In the facet model, a rough surface that complies with the Kirchhoff approximation is obtained by considering two different dimension scales. At a small scale, the surface is approximated by a tangent plane centered at a given point, with linear dimensions which are large with respect to wavelength. At a large scale, the facets interpolate a random sea surface with phase variations from facet to facet that are generated by the underlying spectral properties of the sea surface. The facet model may accommodate different surface properties and different electromagnetic scattering models at the expense of an increased numerical complexity of the simulation. On the other hand, the model fails to account for small-scale roughness, typically found over mild ocean surfaces or calm water ponds. Another consideration is that the facet model provides the scattered electric field at facet-size resolution whereas the other two models considered in the previous analysis give an expression of the DDM. Having in mind the final objective of simulation, the scattered signal from each facet can be accumulated to form the received signal that is processed to obtain the DDM. Differently, a single realization of the DDM can be generated via two-dimensional convolution of the scattered field with WAF.

The bistatic equation model, in its simplest derivation, is built around an incoherent scattering model. At a macroscopic level, the surface roughness is introduced through the variance of the sea surface slopes in the two-dimensional PDF. The refined formulation [7] may accommodate more complicated scattering models and surface statistics but the implementation of such models is difficult and heavy in terms of computing resources.

The statistical scattering model provides the PDF of the signal scattered by an ensemble of specular points in a limited region of space having homogeneous parameters. It is similar to the facet approach but without the need to generate high-resolution returns because the statistics provide the accumulated scattered field in one shot. On the other hand, the electromagnetic modeling is the same as in the geometric optics approximation, therefore, from this point of view, there is a similarity with the bistatic equation models and, on average, the two models provide the same results. Another consideration is strictly related to the statistics of the scattered signal, which turns out to be non-Gaussian, in general.

19.5   Simulation of the GNSS-R signal

A first definition of a GNSS-R end-to-end ocean simulator could be of an object that is able to reproduce as better as possible the output of an operational receiving platform. The idea is that a simulator should be able to provide valuable information for the design and development of onboard signal processors and for future applications based on observed data.

The goal is challenging in any respect and requires considerable effort for putting together all pieces of a setup that includes signal transmission, a forward scattering model, and signal processing to produce the observable. The main modules of a simulator were presented at the beginning of this chapter, in Figure 19.1. Often, there are many alternate ways of approaching a single operation, and this is clear if we look at the different approaches used in forward scattering models. Often, the computational complexity can be a main factor in choosing among different methods. We start to approach the problem with a description of the processes involved in GNSS-R simulations.

19.5.1   Observation geometry and specular point calculation

The definition of observation geometry requires the state vectors of the transmitting and receiving platforms in the form of position and velocity in ECEF coordinates. There are two different ways for determining such information: one makes use of orbit propagation, thus resorting to the orbital elements of the platforms and a software propagator; the other relies on real satellite positions that are extracted from platform metadata of a specific orbital segment. The second approach is mostly useful when the purpose is to compare real data with models. Positions are used for calculation of the specular point that defines the origin of the specular reference frame. The specular point search is generally formulated as an optimization problem in which the optical path length is minimized with the constraint the specular point is located on the Earth's geoid [39]. Denoting $\mathbf{T}$ and $\mathbf{R}$ as the transmitter and receiver coordinates and $\mathbf{S}$ as the coordinate of a reflecting point, the search can be implemented as an iterative steepest descent algorithm where the function $f(\mathbf{S}) = |\mathbf{T} - \mathbf{S}| + |\mathbf{R} - \mathbf{S}|$ is updated using a correction proportional to the projection of the gradient

$$\nabla f(\mathbf{S}) = \frac{\mathbf{S} - \mathbf{T}}{|\mathbf{T} - \mathbf{S}|} + \frac{\mathbf{S} - \mathbf{R}}{|\mathbf{R} - \mathbf{S}|} \tag{19.61}$$

onto the tangent plane of the surface. The initial location of the specular point is usually determined as the vector from the Earth's center to the receiver's position with magnitude equal to the Earth's radius of the WGS84 ellipsoid at the receiver's latitude. The iterative path-length optimization procedure can be summarized as follows

1. Initialization:

$$\mathbf{S}_0 = \frac{\mathbf{R}}{|\mathbf{R}|} R_T (\vartheta) \tag{19.62}$$

   with R_T the radius of the WGS84 ellipsoid at the latitude ϑ;
2. Iterations: from position $\mathbf{S}_n$, apply a correction in the direction of the steepest descent as

$$\widehat{\mathbf{S}}_n = \mathbf{S}_n - K\nabla f(\mathbf{S}_n) \tag{19.63}$$

   and calculate the vector projection onto the tangent plane to the surface point;
3. Constrain the estimate to the surface of the WGS84 geoid

$$\mathbf{S}_{n+1} = \frac{\widehat{\mathbf{S}}_n}{|\widehat{\mathbf{S}}_n|} R_T (\theta_n) \tag{19.64}$$

4. Check if displacement

$$\epsilon_n = |\mathbf{S}_{n+1} - \mathbf{S}_n| \tag{19.65}$$

   is less than a predefined tolerance.

Other methods have been investigated to improve the convergence and computational cost of the specular point search. Among them, we mention the unit difference method by Southwell and Dempster [40] and the geoid gridding search by Gleason *et al.* [41].

19.5.2   Surface gridding

Gridding is the process used for partitioning the glistening zone into small surface patches. In GNSS-R simulators, it is performed with two different approaches. We shortly discuss the two methods and their applications.

1. *Direct gridding*, based on making a regular grid of latitude and longitude values around the specular point. Simulators based on facet approach with synthetic sea surface simulation require regular spatial sampling that is naturally achieved by direct gridding. The sea surface is generated at a very high resolution and the scattered signal is calculated after coherent accumulation of the complex returns from each facet.

2. *Transformed domain gridding*, based on defining a grid in a transformed domain and by finding the corresponding latitude and longitude as the inverse image of these points. Simulators based on surface integration of the scattered field may adopt both direct and transformed domain gridding. We find two different transformations in literature. A first approach is based on defining a regular grid in the delay-Doppler domain and by finding the inverse image of this grid of points over the glistening zone. With respect to uniform gridding, this method defines a non-uniform mesh that is closer to what is really integrated into the delay-Doppler receiver architecture. Additionally, the received signal, both as the surface integral in (19.29) and as the sum in (19.59), can be efficiently calculated as a convolution between the scattering function and the ambiguity function. The main difficulties in the application of this method are due to the inversion of the mapping equation and the computation of the Jacobians, in particular, in the region where the iso-Doppler hyperbolae are tangent to the iso-delay ellipses.

Another approach has been derived by Garrison [42], where the spatial coordinates are transformed into an elliptical-polar pair where the range moves along an iso-delay ellipse and an angular azimuth coordinate φ is used instead of Doppler as the second variable. This implies that the transformation from delay-azimuth to spatial coordinates is one-to-one and that all azimuth domains in the range $(-\pi, \pi)$ produce meaningful points.

19.5.3 Simulation based on a facet scattering model

One main point to be considered in this approach is the simulation of a realistic three-dimensional model of the sea surface. As discussed in Section 19.3, a random surface with Gaussian elevation is a more than reasonable choice. Each element of the random surface is associated with one point of the Earth's surface grid and the required spectral properties are applied by filtering a two-dimensional white Gaussian process with the directional wave slope spectrum. The model for sea surface simulation is shown in Figure 19.5. A matrix of independent zero-mean, unit-variance Gaussian random variables is generated and transformed into the spatial frequency domain via two-dimensional Fast Fourier Transform. The wavenumber spectrum with selected wind and wave parameters is sampled at the same spatial frequencies and its square root is multiplied by the random Gaussian surface. Note that the Kirchhoff approximation does not model a small-scale surface feature and, therefore, a wavenumber cut-off needs to be chosen to identify the surface components for large-scale roughness. In Figure 19.6, two simulated maps, with different wind speed, are shown as an example.

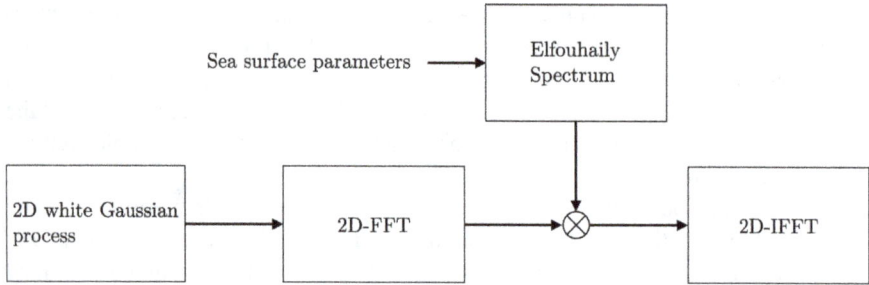

Figure 19.5   *Generation of correlated sea surface elevations via two-dimensional filtering of a white Gaussian random field. 2D-FFT, two-dimensional Fast Fourier Transform. 2D-IFFT, two-dimensional Inverse Fast Fourier Transform.*

Sea surface elevations are used to generate the slopes of plane surface facets that are supposed to have an equal projected area. The scattering coefficient for each facet is calculated by computing the angles between the facet normal and the scattering vector. The complex scattering coefficient, as modeled in (19.19), is coherently accumulated for each facet to generate the complex scattering at delay-Doppler cell resolution and the scattering map is convolved with the WAF of the pseudorandom GPS sequence. The DDM formation process requires averaging the square modulus of many complex realizations, each one computed by slightly different positions of the platforms and with some degree of correlation between one realization of the random sea surface elevation map and the following one.

The main advantage of this method is its ability to produce polarimetric simulations that are tightly connected with high resolution and accurate realizations of

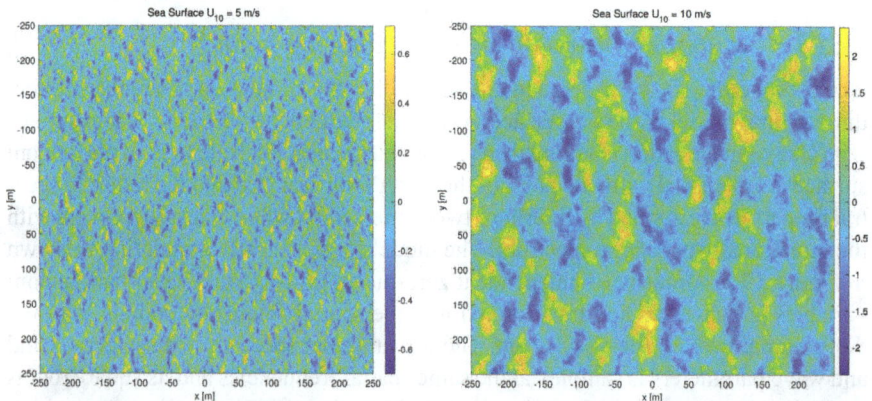

Figure 19.6   *Sea surface elevation maps simulated according to the Elfouhaily power spectrum with 0.2 m spatial resolution, for a wind speed of 5 m/s (left) and 10 m/s (right) and a wind direction parallel to x -axis (0°). The spatial resolution is 0.2 m.*

sea surface elevation maps. On the other hand, the computational requirements are very demanding and generation of satellite DDMs turns out practically unfeasible without resorting to further simplifying assumptions.

19.5.4 Simulation based on the Zavorotny-Voronovich equation

This simulation model is based on the average DDM power in (19.35), assuming that the observations of the complex received signal are originated by a two-dimensional Gaussian random process filtered by the ambiguity function [39, 43, 44]. The process is simulated by generating a matrix of independent identically distributed complex Gaussian random variables with variance

$$\sigma^2 = \frac{\lambda^2 T^2}{(4\pi)^3} P_t \frac{|\mathcal{D}(\mathbf{r})|^2}{R_T^2(\mathbf{r}) R_R^2(\mathbf{r})} \sigma_0 . \tag{19.66}$$

The variance (19.66) is also determined by the covariance matrix of the wind-dependent surface slopes' PDF in the expression of the NBRCS σ_0 reported in (19.36). The covariance matrix can be expressed as $\mathbf{C} = \mathbf{R}(\theta)\Lambda(\sigma_u^2, \sigma_c^2)\mathbf{R}^T(\theta)$, where $\mathbf{R}$ is a rotation matrix in two dimensions and Λ is a diagonal matrix, depending on the upwind and crosswind mean square surface slopes. The numerical evaluation of the surface slopes can be carried out using the Elfouhaily model for the sea-surface elevation spectrum [19], integrating between the low-frequency region of the spectrum, determined by the constant k_l and the scale dividing value k_c. For high wind speed a possible alternative for relating mean square slopes and wind speed is the semi-empirical model proposed by Katzberg *et al.* in [45]. In the simulation setup, a grid of points is defined by one of the following two methods:

1. By forming a fine grid of points over the Earth's surface that are mapped in the delay-Doppler plane and assigned to rectangular domains determined by the center of the cell and its boundaries. This method requires gridding at high resolution to fill the delay-Doppler cells with a reasonable number of points, especially in the boundaries of the horseshoe shape of the DDM.
2. By inverse mapping the center of the delay-Doppler cell into spatial coordinates and accounting for modification of the scattering area with the Jacobians of transformation. This method is more efficient but requires implementation of complicate equations for inverse mapping and expressions of the Jacobians. The surface area calculated this way is subject to quick variations in the DDM region close to the boundaries of the horseshoe shape, where interpolation or oversampling methods are necessary.

Each realization of the complex scattered random field is calculated as the two-dimensional convolution in the delay-Doppler domain between the two-dimensional Gaussian random field and the ambiguity function of the PRN sequence. The DDM observable is formed by incoherent averaging of the realizations over 500 ms or 1 000 ms.

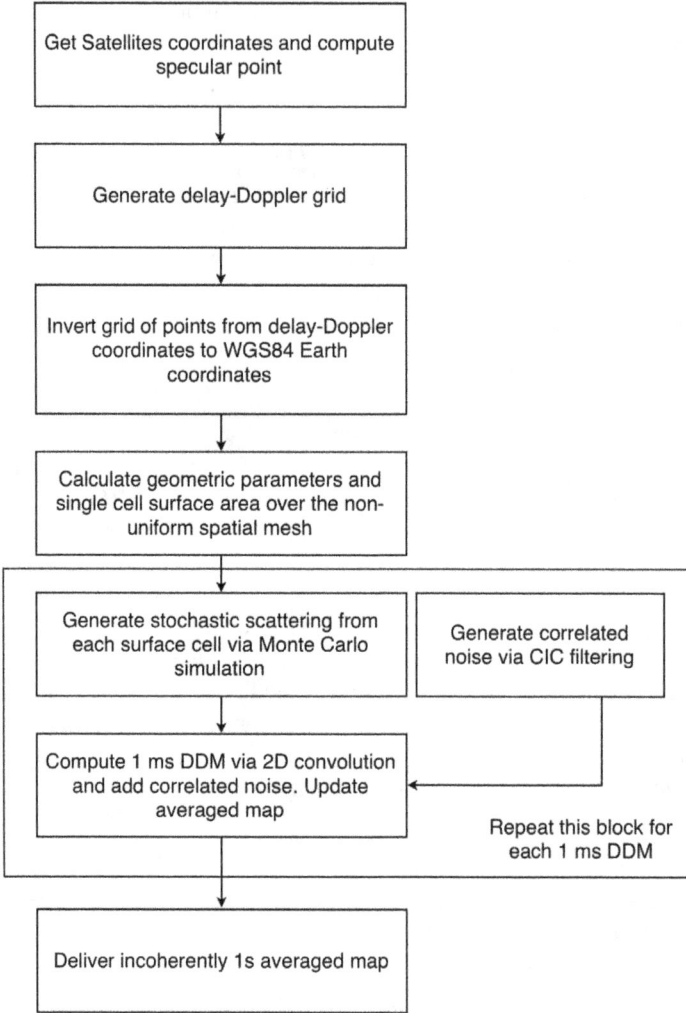

*Figure 19.7   Simulation process flow using a stochastic model. DDM, delay-
Doppler map. CIC, Cascaded Integrator-Comb filter.*

19.5.5   Simulation based on a stochastic model

The implementation of this method is similar, to a certain extent, to the description
given for the Zavorotny-Voronovich method. The main difference is in the generation of the stochastic term, that is non-Gaussian, here. The simulation procedure is
sketched in Figure 19.7. Further details can be found in [46].

The initial setup includes computation of platform and environmental variables
and definition of surface points via transformed-domain gridding. Then, the stochastic scattering, in the form of the product of a cell-dependent complex Gaussian
random variable and the square root of a Gamma random variable, is generated for

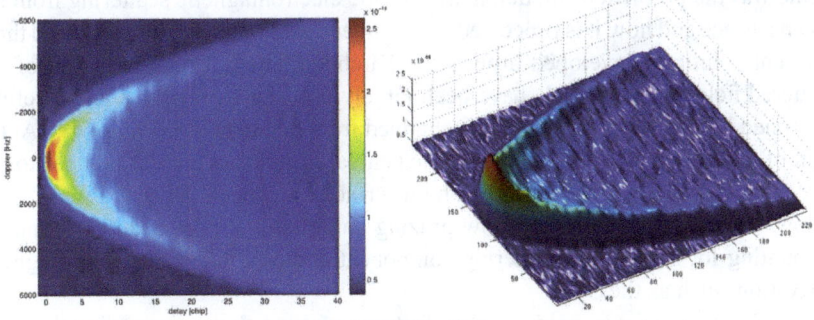

Figure 19.8   *Representation of received power as a delay-Doppler map after averaging 1 000 realizations obtained using a stochastic approach with 1 ms coherent integration time. The image on the right side shows a three-dimensional representation with evidence of speckle and thermal noise components. A thermal noise plateau can be observed in the region outside the horseshoe shape of the map. The wind speed is 5 m/s, oriented 40° off the x-axis. Thermal noise has been generated with 6.2 dB signal-to-noise ratio.*

each cell. The parameters of the transmitted waveform are applied through a two-dimensional convolution of the scattered field with the ambiguity function. Finally, few variables are updated for sequential generation of observations with a 1 ms acquisition step and the averaged map is formed after adding the correlated thermal noise.

A different approach is based on computation of the terms in (19.53). The generation consists of setting up three terms: the delayed and Doppler-shifted waveform $a(t - \tau_m)e^{-2\pi i f_{dm} t}$; the cell-dependent complex constant $\mathcal{D}_m \mathbb{R}_m \frac{e^{-ik(R_m+R_{0m})}}{2R_m R_{0m}}$, and the complex scattering. For each delay-Doppler point, the spatial parameters are calculated by inverse mapping. The received signal, prior to correlation, is the sum of all random waveforms (19.53), scattered from the delay-Doppler cells lying within the footprint of the receiving antenna. The receiver output (19.59) can be directly simulated as correlation between the received signal and delay and Doppler-shifted replicas of the PRN sequence, obtaining an instantaneous DDM. A simulated DDM, obtained averaging 1 000 realizations, is presented at the left-hand side of Figure 19.8. The image shows both thermal and speckle noise: the first, more clearly visible outside the horseshoe shape of the DDM and the second, inside. The different levels of the two components is more evident in the three-dimensional map at the right-hand side of the figure.

19.6 Conclusions

This chapter covered a wide range of topics, with the main objective of giving first insight into the architecture of the GNSS-R signal simulation process. The main theme was the problem of modeling the bistatic electromagnetic scattering from the ocean surface and how it is processed by a conventional receiver. We examined three different solutions developed around the Kirchhoff integral representation of the scattered field. These models have been largely used for simulations and carefully examined in comparison with results obtained from real data [26, 39, 46–49]. At the light of this broad experimental work, the general idea is that there is no need to use more refined models in the regime of high surface roughness. When the wind speed is low, ocean observations at a low grazing angle may require a different model, accounting for a coherent scattering component and/or a different electromagnetic derivation, such as the SSA.

Little attention was paid to other features such as ocean wave directionality, electromagnetic field polarization, and time-varying sea surface. At a first attempt, these issues can be discarded. Wave directionality has a minor impact on the DDM observable, limited to regions of the DDM far away from the specular point [50, 51]. Polarizations different from the left-hand circular polarized (LHCP) signal for the receiving antenna are also of little interest over the ocean surface where the right-hand circular polarized (RHCP) signal remains more than 20 dB under the LHCP component for incidence angles less than 70°. Finally, the sea surface correlation can be safely discarded because the velocity of the satellite platforms is large enough to decorrelate the signal after a few milliseconds. There are, however, some applications of reflectometry where the need to consider longer coherent integration time or the use of fixed or slowly moving platforms may require some additional modeling effort. The problem was considered by Garrison, You *et al.* and Principe *et al.* [42, 52, 53], which the interested reader may consider as a useful resource.

References

[1]   Martin-Neira M. 'A passive reflectometry and interferometry system (Paris): application to ocean altimetry'. *ESA Journal.* 1993;**17**(4):331–55.

[2]   Clifford S.F., Tatarskii V.I., Voronovich A.G., Zavorotny V.U. 'GPS sounding of ocean surface waves: theoretical assessment'. Proceedings of IEEE International Geoscience and Remote Sensing. Symposium; 1998. pp. 2005–7.

[3]   Zavorotny V.U., Voronovich A.G. 'Scattering of GPS signals from the ocean with wind remote sensing application'. *IEEE Transactions on Geoscience and Remote Sensing.* 2000;**38**(2):951–64.

[4]   Gleason S., Gebre-Egziabher D. 'GNSS applications and methods'. *Artech House.* 2009.

[5]   Zavorotny V.U., Gleason S., Cardellach E., Camps A. 'Tutorial on remote sensing using GNSS bistatic radar of opportunity'. *IEEE Geoscience and Remote Sensing Magazine.* 2014;**2**(4):8–45.

[6]   Jales P. 'Spaceborne receiver design for scatterometric GNSS reflectometry'. University of Surrey, UK: Technical Report; 2012.

[7]   Voronovich A.G., Zavorotny V.U. 'Bistatic radar equation for signals of opportunity revisited'. *IEEE Transactions on Geoscience and Remote Sensing*. 2018;**56**(4):1959–68.

[8]   Voronovich A.G., Zavorotny V.U. 'Full-polarization modeling of monostatic and bistatic radar scattering from a rough sea surface'. *IEEE Transactions on Antennas and Propagation*. 2014;**62**(3):1362–71.

[9]   Clarizia M.P., Ruf C.S. 'Wind speed retrieval algorithm for the cyclone global navigation satellite system (CYGNSS) mission'. *IEEE Transactions on Geoscience and Remote Sensing*. 2016;**54**(8):4419–32.

[10] Lin W., Portabella M., Foti G., Stoffelen A., Gommenginger C., He Y. 'Toward the generation of a wind geophysical model function for spaceborne GNSS-R'. *IEEE Transactions on Geoscience and Remote Sensing*. 2019;**57**(2):655–66.

[11] Wang T., Zavorotny V.U., Johnson J., Yi Y., Ruf C. 'Integration of CYGNSS wind and wave observations with the WAVEWATCH III numerical model'. in Proceedings of IEEE International Geoscience and Remote Sensing Symposium; 2019. pp. 8350–3.

[12] Davis J.L., Herring T.A., Shapiro I.I., Rogers A.E.E., Elgered G. 'Geodesy by radio interferometry: effects of atmospheric modeling errors on estimates of baseline length'. *Radio Science*. 1985;**20**(6):1593–607.

[13] Elfouhaily T., Thompson D.R., Linstrom L. 'Delay-Doppler analysis of bistatically reflected signals from the ocean surface: theory and application'. *IEEE Transactions on Geoscience and Remote Sensing*. 2002;**40**(3):560–73.

[14] Cox C., Munk W. 'Measurement of the roughness of the sea surface from photographs of the Sun's glitter'. *Journal of the Optical Society of America*. 1954;**44**(11):838–50.

[15] Longuet-Higgins M. 'The statistical analysis of a random, moving surface'. *Philosophical Transactions of the Royal Society of London. Series A, Mathematical and Physical Sciences*. **249**; 1957. pp. 321–87.

[16] Cardellach E., Rius A. 'A new technique to sense non-Gaussian features of the sea surface from L-band bi-static GNSS reflections'. *Remote Sensing of Environment*. 2008;**112**(6):2927–37.

[17] Phillips O.M. 'The equilibrium range in the spectrum of wind-generated waves'. *Journal of Fluid Mechanics*. 1958;**4**(4):426–34.

[18] Pierson W.J., Moskowitz L. 'A proposed spectral form for fully developed wind seas based on the similarity theory of S. A. Kitaigorodskii'. *Journal of Geophysical Research*. 1964;**69**(24):5181–90.

[19] Elfouhaily T., Chapron B., Katsaros K., Vandemark D. 'A unified directional spectrum for long and short wind-driven waves'. *Journal of Geophysical Research: Oceans*. 1997;**102**(C7):15781–96.

[20] Hwang P.A., Ainsworth T.L. 'L -Band ocean surface roughness'. *IEEE Transactions on Geoscience and Remote Sensing*. 2020;**58**(6):3988–99.

[21] Hwang P.A., Fan Y., Ocampo-Torres F.J., García-Nava H. 'Ocean surface wave spectra inside tropical cyclones'. *Journal of Physical Oceanography*. 2017;**47**(10):2393–417.

[22] The WAVEWATCH III Development Group (WW3DG) User manual and system documentation of wavewatch III version 5.16. Tech.Note 329, NOAA/NWS/NCEP/MMAB, College Park, MD, USA, Technical Report; 2016.

[23] Chen-Zhang D.D., Ruf C.S., Ardhuin F., Park J. 'GNSS-R nonlocal sea state dependencies: model and empirical verification'. *Journal of Geophysical Research: Oceans.* 2016;**121**(11):8379–94.

[24] Voronovich A. 'Wave Scattering from Rough Surfaces'. *Springer*; 1999.

[25] Elfouhaily T.M., Guérin C.-A. 'A critical survey of approximate scattering wave theories from random rough surfaces'. *Waves in Random Media.* 2004;**14**(4):R1–40.

[26] Clarizia M.P., Gommenginger C., di Bisceglie M., Galdi C., Srokosz M.A. 'Simulation of L-band bistatic returns from the ocean surface: a facet approach with application to ocean GNSS reflectometry'. *IEEE Transactions on Geoscience and Remote Sensing.* 2012;**50**(3):960–71.

[27] Clarizia M.P. 'Investigating the effect of ocean waves on GNSS-R microwave remote sensing measurements'. University of Southampton, UK: Technical Report; 2012.

[28] Ulaby F.T., Moore R.K., Fung A.K. *Microwave Remote Sensing, Active and Passive, Volume II: Radar Remote Sensing and Surface Scattering and Emission Theory.* Addison-Wesley; 1982.

[29] Thompson D.R., Elfouhaily T.M., Garrison J.L. 'An improved geometrical optics model for bistatic GPSs scattering from the ocean surface'. *IEEE Transactions on Geoscience and Remote Sensing.* 2005;**43**(12):2810–21.

[30] Marchan-Hernandez J.F., Camps A., Rodriguez-Alvarez N., Valencia E., Bosch-Lluis X., Ramos-Perez I. 'An efficient algorithm to the simulation of delay–Doppler maps of reflected global navigation satellite system signals'. *IEEE Transactions on Geoscience and Remote Sensing.* 2009;**47**(8):2733–40.

[31] Voronovich A.G., Zavorotny V.U. 'Full-polarization modeling of monostatic and bistatic radar scattering from a rough sea surface'. *IEEE Transactions on Antennas and Propagation.* 2014;**62**(3):1362–71.

[32] Voronovich A.G., Zavorotny V.U. 'The transition from weak to strong diffuse radar bistatic scattering from rough ocean surface'. *IEEE Transactions on Antennas and Propagation.* 2017;**65**(11):6029–34.

[33] Kodis R. 'A note on the theory of scattering from an irregular surface'. *IEEE Transactions on Antennas and Propagation.* 1966;**14**(1):77–82.

[34] Giangregorio G., di Bisceglie M., Addabbo P., Beltramonte T., D'Addio S., Galdi C. 'Stochastic modeling and simulation of delay–Doppler maps in GNSS-R over the ocean'. *IEEE Transactions on Geoscience and Remote Sensing.* 2016;**54**(4):2056–69.

[35] di Bisceglie M., Galdi C. 'Random walk based characterisation of radar backscatter from the sea surface'. *IEE Proceedings – Radar, Sonar and Navigation.* 1998;**145**(4):216–25.

[36] Jakeman E. 'On the statistics of K-distributed noise'. *Journal of Physics A: Mathematical and General.* 1980;**13**(1):31–48.

[37] Barrick D. 'Rough surface scattering based on the specular point theory'. *IEEE Transactions on Antennas and Propagation*. 1968;**16**(4):449–54.

[38] Beltramonte T., di Bisceglie M., Galdi C., Ullo S. 'Space-time statistics of the number of specular points in sea-surface GNSS reflectometry'. IEEE International Geoscience and Remote Sensing Symposium; 2014. pp. 3818–21.

[39] O'Brien A. 'End-to-end simulator technical memo'. Ohio State University, Columbus, OH, USA: Technical Report; 2014.

[40] Southwell B.J., Dempster A.G. 'A new approach to determine the specular point of forward reflected gnss signals'. *IEEE Journal of Selected Topics in Applied Earth Observations and Remote Sensing*. 2018;**11**(2):639–46.

[41] Gleason S., Ruf C.S., O'Brien A.J., McKague D.S. 'The CYGNSS level 1 calibration algorithm and error analysis based on on-orbit measurements'. *IEEE Journal of Selected Topics in Applied Earth Observations and Remote Sensing*. 2019;**12**(1):37–49.

[42] Garrison J.L. 'A statistical model and simulator for ocean-reflected gnss signals'. *IEEE Transactions on Geoscience and Remote Sensing*. 2016;**54**(10):6007–19.

[43] Park H., Marchan-Hernandez J.F., Rodriguez-Alvarez N., *et al.* 'End-to-end simulator for global navigation satellite system reflectometry space mission'. 2010 IEEE International Geoscience and Remote Sensing Symposium; 2010. pp. 4294–7.

[44] Park H., Camps A., Pascual D., Alonso A., Martin F., Carreno-Luengo H. 'Improvement of the PAU/PARIS end-to-end performance simulator (P2EPS in preparation for upcoming GNSS-R missions)'. 2013 IEEE International Geoscience and Remote Sensing Symposium – IGARSS; 2013. pp. 362–5.

[45] Katzberg S.J., Torres O., Ganoe G. 'Calibration of reflected GPS for tropical storm wind speed retrievals'. *Geophysical Research Letters*. 2006;**33**(18):n/a.

[46] Addabbo P., Giangregorio G., Galdi C., di Bisceglie M. 'Simulation of TechDemoSat-1 delay-Doppler maps for GPS ocean reflectometry'. *IEEE Journal of Selected Topics in Applied Earth Observations and Remote Sensing*. 2017;**10**(9):4256–68.

[47] Li B., Zhang B., Yu Y., *et al.* 'A random model and simulation for generating GNSS ocean reflected signals'. *IEEE Geoscience and Remote Sensing Letters*. 2019;**16**(7):1036–40.

[48] Fabra F., Cardellach E., Li W., Rius A. 'WAVPY: a GNSS-R open source software library for data analysis and simulation'. 2017 IEEE International Geoscience and Remote Sensing Symposium (IGARSS). Fort Worth, TX: IEEE; 2017. pp. 4125–8.

[49] Pascual D., Park H., Camps A., Arroyo A.A., Onrubia R. 'Simulation and analysis of GNSS-R composite waveforms using GPS and Galileo signals'. *IEEE Journal of Selected Topics in Applied Earth Observations and Remote Sensing*. 2014;**7**(5):1461–8.

[50]   Park J., Johnson J.T. 'A study of wind direction effects on sea surface specular scattering for GNSS-R applications'. *IEEE Journal of Selected Topics in Applied Earth Observations and Remote Sensing*. 2017;**10**(11):4677–85.

[51]   Giangregorio G., Galdi C., di Bisceglie M. 'Wind direction estimation by deconvolution of GNSS delay–Doppler maps: a simulation analysis'. *IEEE Journal of Selected Topics in Applied Earth Observations and Remote Sensing*. 2020;**13**:2409–18.

[52]   You H., Garrison J., Heckler G., Zavorotny V. 'Stochastic voltage model and experimental measurement of ocean-scattered GPS signal statistics'. *IEEE Transactions on Geoscience and Remote Sensing*. 2004;**42**(10):2160–9.

[53]   Principe S., Beltramonte T., di Bisceglie M., Galdi C. 'GNSS ocean bistatic statistical scattering in the time-varying regime: modeling and correlation properties'. *IEEE Transactions on Geoscience and Remote Sensing*. 2021:1–8.

Chapter 20

Wind estimation

*Benjamin J Southwell[1], Joon W Cheong[2], and
Andrew G Dempster[2]*

Due to the ubiquitous nature of Global Navigation Satellite System (GNSS) transmitters, GNSS-Reflectometry (GNSS-R) is particularly suited for remote sensing of
Earth's oceans which are vast and have difficult environments that prohibit exhaustive in-situ measurements. Furthermore, with a space-borne receiver operating in
low Earth orbit, the achievable spatio-temporal coverage is extraordinary and scales
well with multiple satellites. Therefore, in this chapter, we focus on ocean wind
estimation using space-borne platforms such as the retired TechDemoSat-1 (TDS-1)
satellite and the currently operational Cyclone GNSS (CyGNSS) constellation.

20.1   Modelling ocean-reflected GNSS signals

The bistatic configuration is defined by the location of the transmitter, T, the receiver,
R, and the specular point, S. A multi-static configuration exists whenever multiple
specular points are considered. This can be a result of the utilisation of multiple
transmitters and/or multiple receivers. Often, the terms bistatic and multi-static are
used interchangeably in the GNSS-R literature, typically, to describe single receiver
multi-transmitter configurations.

   Ocean wind speeds, typically defined at a height of 10 m above the sea surface
and denoted by U_{10}, can be estimated using reflected GNSS signals and scatterometric techniques. When the receiver is at a significant altitude, the reflected signal
arrives at the receiver sufficiently after the direct signal such that they do not coherently interfere. Furthermore, when reflected off a rough ocean surface, the reflected
signal is also spread in delay as non-specular reflections are supported in the direction of the receiver. Moreover, if the receiver's velocity is significant then a Doppler
spread is also present in the received reflected signal. A map of the reflected power
as a function of delay and Doppler is known as the Delay Doppler Map (DDM) and
is the fundamental observable produced by GNSS-R receivers.

[1]Dolby Laboratories, Australia
[2]Australian Centre for Space Engineering Research, University of New South Wales, Australia

20.1.1   *Woodward's ambiguity function*

The GNSS-R receiver produces DDMs with a matched filter by correlating the received signal with a replica of the GNSS signal. The impulse response of such a receiver is Woodward's Ambiguity Function (WAF) [1], χ, which describes the uncertainty in a scatterer's range and velocity from its radar return. It is widely used as a tool to evaluate the performance of a radar system [2–4] transmitting the signal $s(t)$. In asymmetric form, it can be written as [5]

$$\chi(\Delta\tau, \Delta f_D) = \int_{-\infty}^{\infty} s^*(t)\, s\,(t - \Delta\tau)\, e^{j2\pi\Delta f_D}\, dt \tag{20.1}$$

where $\Delta\tau$ and Δf_D are the relative time delay and Doppler shift of the scatterer. This allows χ to be defined for all $(\bar{\tau}, \bar{f}_D)$ cells in a DDM, where

$$\begin{aligned} \Delta\tau &= \bar{\tau} - \tau \\ \Delta f_D &= \bar{f}_D - f_D \end{aligned} \tag{20.2}$$

and (τ, f_D) is the actual delay-Doppler (DD) coordinate of the scatterer. In the context of remote sensing of extended surfaces, the DD coordinates are often represented as a function of the scattering facet's position, p, i.e., we use $\tau(p)$ and $f_D(p)$ in (20.2).

The velocity and acceleration of the radar configuration impact WAF [6]. However, assuming a coherent integration time of 1 ms, for the GPS L1 C/A signal, the time-bandwidth product is sufficiently small such that WAF can be approximated with [7]

$$\left\langle |\chi(\Delta\tau, \Delta f_D)|^2 \right\rangle \simeq \Lambda^2(\Delta\tau)\mathrm{Sinc}^2(\Delta f_D) \tag{20.3}$$

which is a separable function comprised of the triangle function, Λ, and the Sinc function, which are the idealised autocorrelation functions for delay and Doppler mismatch, respectively.

There are many various but similar definitions of WAF [8]. The definition in (20.3) is particularly convenient as it can be computed beforehand provided the receiver's configuration is known. Thus, it is the most commonly used definition in the GNSS-R community. It should be noted that the representation in (20.3) ignores the non-idealities associated with the autocorrelation [9] and cross-correlation functions [10, 11].

20.1.1.1   Delay Doppler mapping in the spatial domain

The area on the Earth's surface surrounding the specular point that supports reflected power towards the receiver, thus contributing to the DDM, is known as the Glistening Zone (GZ). The GZ is composed of a continuum of surface scattering facets, p, over an extended surface area such that there is a significant spread in the delay and Doppler of the received signal.

The delay associated with a point on the reflecting surface is

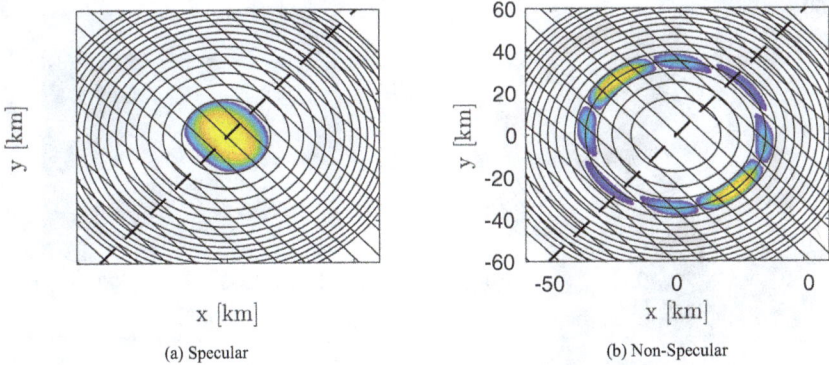

(a) Specular

(b) Non-Specular

Figure 20.1   *The iso-DD contours for a space-borne receiver configuration are shown as solid black lines at 1 chip and 500 Hz intervals. The receiver's heading is indicated by the dashed black line. The χ^2 responses to various DD coordinates are shown in colour on a dB scale down to −30 dB. (a) The specular response. (b) A non-specular response.*

$$\tau(\boldsymbol{p}) = \|\boldsymbol{T} - \boldsymbol{p}\| + \|\boldsymbol{R} - \boldsymbol{p}\| \tag{20.4}$$

Geometrically, the iso-delay lines can be described as the intersection of the scattering surface with an iso-delay ellipsoid defined by the $\boldsymbol{R}$ and $\boldsymbol{T}$ positions as foci. These iso-delay lines can be approximated as ellipses [12].

The Doppler frequency associated with a point on the reflecting surface is

$$f_D(\boldsymbol{p}) = \frac{-f_0}{c}\left(\dot{\boldsymbol{T}}^T\hat{\boldsymbol{m}} + \dot{\boldsymbol{R}}^T\hat{\boldsymbol{n}}\right) \tag{20.5}$$

A hat accent indicates a normalised vector; a dot accent, the time derivative; and the superscript T, the transpose operator.

The iso-delay Doppler (iso-DD) lines for a space-borne receiver are shown in Figure 20.1 along with the χ^2 response in colour. The bistatic configuration is typical for the space-borne receiver and was generated using metadata from the TDS-1 satellite [13]. Figure 20.1a shows the response to the specular point, i.e., $(\bar{\tau}, \bar{f}_D) = (0, 0)$, while Figure 20.1b shows a non-specular response: $(\bar{\tau}, \bar{f}_D) = (4 \text{ chips}, 2\,500\,\text{Hz})$. The iso-delay lines (ellipses) are plotted at 1-chip intervals and the iso-Doppler lines (hyperbole) at 500 Hz intervals. These plots are made assuming a coherent integration time of 1 ms so the Sinc^2 component of the χ^2 response has nulls spaced at 1 000 Hz. The Λ function has finite support, limiting the χ^2 responses in Figure 20.1 to within ±1 chip. The receiver's velocity is approximately 42 degrees measured counter-clockwise from the x-axis indicated by the dashed line, i.e. it is moving from the bottom-left to the top-right. The iso-Doppler lines are a function of the receiver's velocity and the incidence angle and are often approximated as uniformly spaced parallel lines [14]. Furthermore, the iso-Doppler frequencies increase in the

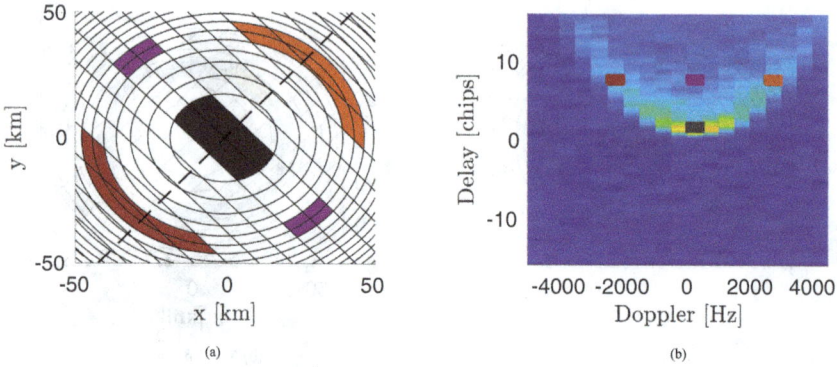

Figure 20.2   Mapping of power from the spatial domain to the DD domain

direction of the receiver's velocity, that is, the iso-Doppler lines towards the upper-right in Figure 20.1 are highest.

WAF acts as a spatial filter. This is evident in Figure 20.1, where the χ^2 response to various $(\bar{\tau}, \bar{f}_D)$ coordinates has been shown in colour on a dB scale. This response depicts the weighting of the surface facet's response when mapped to that particular coordinate in the DD domain. In these figures, it is also evident that the mappings of the iso-DD lines cause the response of WAF to be spatially variant.

20.1.2   The delay Doppler map

The DD mapping across the GZ causes the received signal to be spread in delay and Doppler. This gives the DDM its characteristic horseshoe shape (for a space-borne receiver) when it is composed of a response from an extended scattering surface such as a rough ocean surface. Figure 20.2 shows the mapping of select points in the spatial domain to a typical horseshoe DDM. There are some cells in the DDM that do not map to points on the Earth's surface and as a result, these are often referred to as forbidden cells. These cells cannot contain any reflected signal power and thus can be used to estimate the noise power in the reflected channel. The DDMs collected by SGR-ReSI [15, 16] onboard TDS-1 and CyGNSS present a Doppler spread of $\approx \pm 4$ kHz and have observable power to the edge of the DDM window at +15 chips when rough ocean surfaces are encountered.

20.1.2.1   The ambiguity free line

We can observe in Figure 20.2 that many cells in the DDM will be composed of power reflected from two distinct spatial zones on the Earth's surface. These two zones straddle are called the Ambiguity Free Line (AFL), which is where a folding occurs when mapping from the spatial domain to the DD domain. Figure 20.3 shows the χ^2 response to an ambiguous DD coordinate for the configuration that is used to generate Figures 20.1 and 20.2. Here, the two distinct spatial zones that map to the same DD cell are separated by significant distance and potentially consist of differing ocean responses to the incident GNSS signal. However, we are unable to resolve these individual responses because a

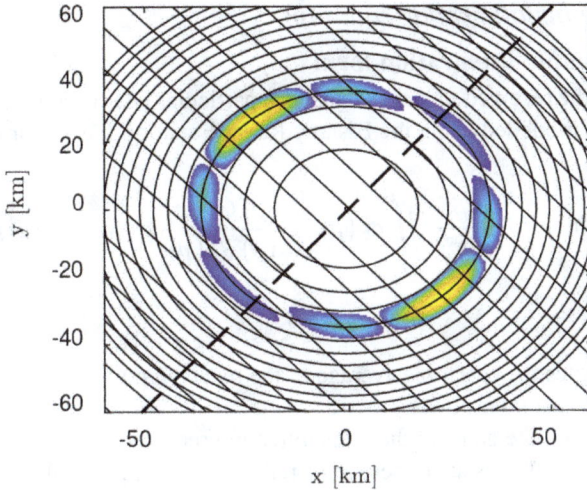

Figure 20.3 An ambiguous χ^2 response to 4 chips and 0 Hz for the same bistatic configuration used to generate the plots in Figures 20.1 and 20.2. The χ^2 response is shown in colour on a dB scale down to -30 dB.

single DDM contains just one observation of the composite response. This is the root of the infamous (within the GNSS-R community) ambiguity problem. As such, for the purposes of wind estimation, only points along the AFL have been utilised in state-of-the-art algorithms. Consequentially, as discussed later in this chapter, the state-of-the-art data products are extracted using only a small portion of the DDM which is not sensitive to wind direction.

The AFL is also the contour where the delay is minimum for a given Doppler frequency [14]. This is true in both the spatial and the DD domains. In the DD domain, the AFL defines the leading edge of the DDM, or equivalently the contour dividing the forbidden cells from those that can contain power reflected off the Earth's surface. In the spatial domain, the AFL is also approximated by the direction of the specular point's velocity [17] which itself is dominated by the receiver's velocity.

Monostatic radars do not typically suffer from the ambiguity problem when employing both delay and Doppler filtering as the iso-delay contours result from the intersection of the Earth's surface and an iso-delay sphere centred on the radar platform. This results in a large angular separation (opposite hemispheres) between the two ambiguous points which means that simple antenna-based techniques can be used to avoid the problem.

20.1.3   The bistatic radar equation

20.1.3.1   Double-integration form

The power of the received reflected signal at a particular DD coordinate in the DDM, $(\bar{\tau}, \bar{f}_D)$, can be modelled using the Bistatic Radar Equation (BRE) for extended targets [18].

$$\left\langle \left| Y(\bar{\tau}, \bar{f}_D) \right|^2 \right\rangle = T_i^2 \frac{P_T \lambda^2}{(4\pi)^3} \iint\limits_{GZ} \sigma_0\left(\boldsymbol{p}\right) \frac{G_T\left(\boldsymbol{p}\right) G_R\left(\boldsymbol{p}\right)}{R_R^2\left(\boldsymbol{p}\right) R_T^2\left(\boldsymbol{p}\right)} \chi^2(\Delta\tau, \Delta f_D)\, d^2\boldsymbol{p} \qquad (20.6)$$

where

- $\left\langle \left| Y \right|^2 \right\rangle$ denotes the ensemble average of the received power
- P_T is the transmitter power
- G_T and G_R are the gains of the transmitter and receiver antennas, respectively
- R_T and R_R are the distances between a reflecting facet, $\boldsymbol{p}$, and the transmitter and receiver, respectively
- λ is the electromagnetic wavelength of the signal, $\approx 19\,\text{cm}$ for L1
- T_i is the coherent integration time of the receiver
- σ_0 is the extended radar cross section of the scattering surface.

The double integral is performed over the entire GZ such that the power reflected of all surface scattering facets within the GZ is mapped to the DDM. The reflectivity of the surface σ_0 is in units of $\left(\frac{\text{m}^2}{\text{m}^2}\right)$, i.e. it is dimensionless and is referred to as the Normalised Bistatic Radar Cross Section (NBRCS). After integration over the entire area of the GZ, the total Bistatic Radar Cross Section (BRCS) is found so we can consider the size of the GZ to be analogous with an aperture gain. This is obviously more complicated due to the weighting of the numerous other parameters in the BRE.

20.1.3.2   Convolution form

The DDM can be described as a convolution of the illumination scene, Σ, with χ^2 [2, 19, 20]. Applying this, the convolution form of the BRE can be given as [20]

$$\left\langle \left| Y\left(\bar{\tau}, \bar{f}_D\right) \right|^2 \right\rangle = \chi^2\left(\Delta\tau, \Delta f_D\right) \circledast \Sigma\left(\bar{\tau}, \bar{f}_D\right) \qquad (20.7)$$

where $\circledast$ is the convolution operator and

$$\Sigma\left(\bar{\tau}, \bar{f}_D\right) = T_i^2 \left(\frac{\sigma_0\left(\boldsymbol{p}\left(\bar{\tau}, \bar{f}_D\right)\right) G_T\left(\boldsymbol{p}\left(\bar{\tau}, \bar{f}_D\right)\right) G_R\left(\boldsymbol{p}\left(\bar{\tau}, \bar{f}_D\right)\right)}{4\pi R_R^2\left(\boldsymbol{p}\left(\bar{\tau}, \bar{f}_D\right)\right) R_T^2\left(\boldsymbol{p}\left(\bar{\tau}, \bar{f}_D\right)\right)} \left| \mathcal{J}(\tau, f_D) \right| \right) \qquad (20.8)$$

is the illumination scene after a change of variables. Thus, this method requires $\mathcal{J}$, the Jacobian of $\boldsymbol{p}_\perp = \left(p_x, p_y\right)$ with respect to (τ, f_D). Furthermore, due to the ambiguous mapping of points from the spatial domain to the DD domain, there will be two

Σ responses away from the AFL, and (20.7) can be written as the sum of these two responses after their respective convolution with χ^2.

This form has led to the development of techniques that deconvolve the DDM in order to produce radar images in the spatial domain, e.g. [21], for a variety of applications including wind estimation [22]. However, these methods either ignore the ambiguity problem altogether or presume it to be solved with antenna-based methods where the antenna is steered (in a manner similar to [23, 24]) so that only one side of the AFL contributes significantly to the DDM. Much of the analysis of these solutions is only performed on the nadir scattering geometry, which is the bistatic configuration that occurs the least [25]. Additionally, away from nadir scattering geometries, the angular size of the GZ from Low Earth Orbit (LEO) is so small that the practical antenna beamwidth realisable may result in poor spatial filtering leading to an increasingly ill-posed problem.

20.1.3.3   The scattering frame

The Scattering Reference Frame (SRF) is a Cartesian coordinate system located at the specular point and provides a basis for modelling in the spatial domain. Its z-axis points in the direction of the surface normal at the specular point, N; R and T, lie in the xz plane; the x-axis is tangential to the surface at S and is in the direction of the receiver; the y-axis completes a right-handed system.

The scattering facets are located in the xy plane if a flat Earth is assumed which is typical for airborne receivers. For space-borne receivers, the GZ extends well over 100 km [26, 27] on the ocean surface and the curvature of the Earth becomes non-negligible. That is, when we model the reflection coming from a facet, p, that is in the SRF and on the sea surface, the vertical component p_z becomes significant due to the curvature of the Earth. Assuming a locally spherical Earth, we can compute the vertical component with

$$p_z = \sqrt{r_e^2\left(S\right) - p_\perp^2} - r_e\left(S\right) \tag{20.9}$$

where $r_e\left(X\right)$ returns the radius of the Earth along the radial path to the point X.

The specular point is computed from the bistatic geometry and is used to steer the correlators in the matched filter producing the DDM. In addition to this, the computed specular point is used to map the DDM and/or observables and estimates derived from it back to the surface of the Earth. Recent work [28] has demonstrated that the most popular method for determining the specular point, which relies on minimising the reflected path length, converges incorrectly on the WGS84 ellipsoid with errors up to 20 km occurring at intermediate latitudes. This work also proposed a new cost function that is based on Snell's law explicitly, which does not suffer from these errors.

20.1.4   *Electromagnetic scattering model*

In order to adapt the general BRE to a particular sensing configuration, e.g. passive bistatic sensing of Earth's oceans, a specific scattering model is used for σ_0.

By applying the Geometric Optics (GO) limit to Kirchhoff's Approximation (KA) [29–31], we can model the scattering off the ocean surface due to the large-scale surface structure with [18]

$$\sigma_0(p) = \frac{\pi |\Re|^2 \bar{q}^4}{\bar{q}_z^4} P\left(-\frac{\bar{q}_\perp}{\bar{q}_z}\right) \tag{20.10}$$

where $\Re$ is the Fresnel reflection coefficient; q is the bistatic vector which is the bisector of the anti-incident and scattered vector,

$$q(p) = \hat{m}(p) + \hat{n}(p) \tag{20.11}$$

and $q_\perp$ is the horizontal component of q and q_z the vertical. Fresnel's reflection coefficient is a complicated function depending on the incidence angle at the scattering facet, the polarisation of the transmitting and receiving antennae and the complex dielectric permittivity of the ocean which itself is a strong function of water temperature and salinity [32].

The anti-incident vector, $\hat{m}$, and scattered vector, $\hat{n}$, are given by

$$\hat{m}(p) = \frac{(T - p)}{\|T - p\|}, \quad \hat{n}(p) = \frac{(R - p)}{\|R - p\|} \tag{20.12}$$

The angle between $\hat{m}(S)$ and $N(S)$ is the incidence angle θ_i, which is complementary to the elevation angle, θ_e, of the transmitter at S. In (20.10), we scale q by the electromagnetic wave number, K.

$$\bar{q} \equiv Kq(p) \quad \text{where} \quad K = \frac{2\pi}{\lambda} \tag{20.13}$$

$P(\cdot)$ is the Probability Density Function (PDF) expressing the probability of surface slopes supporting scattering towards the receiver. In [18], a bivariate Gaussian distribution is utilised for $P(\cdot)$, which is parameterised by the Mean Square Slopes (MSS) of the ocean surface.

$$P(s) = \frac{1}{2\pi \sigma_{s_x} \sigma_{s_y} \sqrt{1 - b_{xy}^2}} \exp\left[-\frac{1}{2(1 - b_{xy}^2)}\left(\frac{s_x^2}{\sigma_{s_x}^2} + \frac{s_y^2}{\sigma_{s_y}^2} - 2b_{xy}\frac{s_x s_y}{\sigma_{s_x}\sigma_{s_y}}\right)\right] \tag{20.14}$$

where $\sigma_{s_x}^2$ and $\sigma_{s_y}^2$ are the MSSs in the x- and y-direction of the SRF, respectively. The correlation between the MSSs in the x- and y-direction, b_{xy}^2, will be zero when the wind is blowing in the direction of either axes.

We can consider

$$s = -\frac{\bar{q}_\perp}{\bar{q}_z} \tag{20.15}$$

as the domain that the surface slopes PDF is evaluated in while (20.11) and (20.13) provide the means for transformation from the spatial domain to the s domain. This allows us to interpret the bistatic vector as a sampling coordinate of the surface slopes PDF. At the specular point, $q = \hat{m}(S) + \hat{n}(S) \parallel N(S)$ and $q_\perp = 0$, so a maximum of both $P(\cdot)$ and σ_0 occurs.

The use of a bivariate Gaussian for the PDF results in there being a $\pm 180°$ ambiguity in wind direction [33]. However, the sea surface has demonstrated sensitivity to up- vs downwind directions [34], leading to others utilising various PDFs such as the Gram-Charlier [35], first proposed by [36], to capture the bias of the PDF towards the downwind direction.

Equation (20.10) is referred to as the ZV model, named after the authors of [18]. As this model utilises the GO limit to KA, it is referred to as a KAGO approach.

Since the ZV model was developed, millions of DDMs have been collected by the TDS-1 and CyGNSS missions using SGR-ReSI. In many of these, a strong coherent component over smooth waters has been detected [37–42]. This coherent component results in the model underestimating the power in the specular zone of the DDM and a composite model comprising the sum of both an incoherent and coherent component was proposed [7].

$$\left\langle \left| Y\left(\bar{\tau}, \bar{f}_D \right) \right|^2 \right\rangle = \left\langle \left| Y\left(\bar{\tau}, \bar{f}_D \right) \right|^2_{coh} \right\rangle + \left\langle \left| Y\left(\bar{\tau}, \bar{f}_D \right) \right|^2_{inc} \right\rangle \tag{20.16}$$

where the subscripts *coh* and *inc* denote the coherent and incoherent components, respectively. The incoherent component cannot be expressed with (20.6) and

$$\left\langle \left| Y\left(\bar{\tau}, \bar{f}_D \right) \right|^2_{coh} \right\rangle = \frac{P_T G_T G_R \lambda^2 \chi^2 \left(\Delta\tau, \Delta f_D \right) \left| \Re \right|^2 e^{-4R_a^2}}{\left(4\pi \right)^2 \left(R_T + R_R \right)^2} \tag{20.17}$$

noting that all of the variables are evaluated at the specular point and R_a is the Rayleigh number. This model is valid for coherent reflections off the ocean surface, and reflections off sea-ice, land, etc. require alternate models even if they are coherent.

In addition to the KAGO approach, there exist numerous alternative approaches and variations, e.g. [39, 43–47]. However, any reasonable model for σ_0 may be used in the BRE [48].

The range of $p \in GZ$, where $\sigma_0(p)$ is sufficient to support reception of the reflected signal, defines the extent of the GZ [27]. Thus, for rough ocean surfaces, (20.10) and (20.14) predict the GZ will be quite large, while a smooth surface will result in a small GZ. As $\sigma_{sx,sy} \to 0$, the coherent component becomes dominant until the diffuse incoherent component disappears completely for a perfectly flat surface. Unlike the diffuse component, the delay range comprising the coherent component is limited to the first Fresnel zone [48–50], not the first iso-delay chip. This results in an iso-delay resolution on the order of hundreds of metres compared to the ≈ 25 km figure often cited for the C/A code resolution [51, 52]. Support for this phenomenon has been observed experimentally [49, 53], suggesting that higher resolution GNSS-R data products can be developed using techniques such as subspace-based super-resolution, e.g. [54–56].

20.1.5 Sea surface models

While (20.10) and (20.14) provide a model for reflectivity as a function of ocean roughness explicitly through the MSS, for wind speed applications, a model is required to link the surface wind speed with the MSS. This can be done using models of the sea surface elevation spectrum.

Some spectrum models are based on the energy balance equation, e.g. [57, 58], while others parameterise the spectrum based on the wind speed and age, e.g. [59]. Despite not accounting for non-wind effects, the latter model is the most widely used in the GNSS-R community. Energy transfer from the wind to the ocean is a non-linear process and here we assume that Fully Developed Seas (FDS) are present, i.e., the inverse wave age, $\Omega = 0.84$, and large fetch conditions exist. A directional sea surface elevation spectrum, Ψ, can be estimated and the MSS derived from it [59].

$$\sigma^2_{sx,sy} = \iint k^2_{x,y} \Psi\left(k_x, k_y\right) \, dk_x \, dk_y \tag{20.18}$$

where the integration is performed over wave numbers up to the cut-off k^*. Various values for k^* have been presented in the literature. Originally, with (20.10), [18] proposed

$$k^* = \frac{2\pi}{3\lambda} \tag{20.19}$$

which, while in agreement with [60], does not capture the effect of the changing apparent roughness due to the scattering geometry; thus, [19] proposed a new cutoff, which is a function of the elevation angle

$$k^* = \frac{2\pi \sin\left(\theta_e\right)}{3\lambda} \tag{20.20}$$

The wave number computed in (20.20) ensures that at all incidence angles, the region of the elevation spectrum, which is considered rough, contributes to the MSS.

20.2   Processing delay Doppler maps

20.2.1   Feature extraction

There are numerous features that have been extracted from the DDM for a range of applications. In this section, we will briefly introduce only some of the most popular features that are relevant for wind speed estimation.

20.2.1.1   Noise power

In Figure 20.4a, multiple boxes bounding regions of interest are depicted. It is common for the power in these boxes of the DDM to be averaged in order to extract features. The noise-box is located at delays significantly less than specular, which corresponds to points that are above the Earth's surface and thus do not contain any reflected signal power. By averaging the DDM over the noise-box, we can obtain an estimate of the noise power in the DDM.

$$\tilde{P}_N[n] = \frac{1}{IJ}\sum_i \sum_j \left\langle |Y(i,j,n)|^2 \right\rangle \tag{20.21}$$

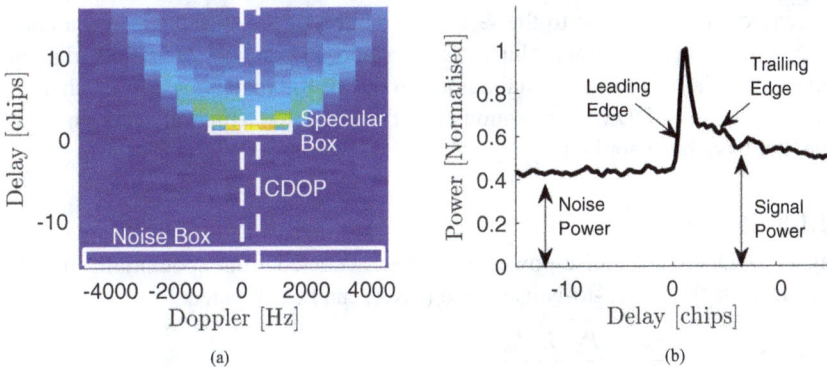

Figure 20.4   *(a) A DDM collected by TDS-1. The specular and noise-boxes are shown with solid lines. These boxes are used to estimate the signal and noise power. Additionally, the Central DOPpler (CDOP) column is shown with the dashed line. (b) The Delay Waveform (DW), which is the CDOP slice of the DDM in (a).*

where i and j define the range of delay and Doppler cells defining the noise-box, and I and J are the numbers thereof. It is common to use all Doppler cells for a few of the largest delay rows in the DDM which for the case of SGR-ReSI onboard TDS-1 yields $\bar{\tau} \in [-15, \ldots, -14]$ chips and $\bar{f}_D \in [-5\,000, \ldots, 4\,500]$ Hz.

20.2.1.2   DDM average

For the specular box, the mean is referred to as the DDM Average (DDMA) and is computed in a similar fashion to the noise power

$$\tilde{P}_R[n] = DDMA[n] = \frac{1}{IJ} \sum_i \sum_j \left\langle \left| Y(i,j,n) \right|^2 \right\rangle \tag{20.22}$$

where i and j now define the specular box of which a common definition yields $\bar{\tau} \in [-0.25, \ldots, 0.75]$ chips and $\bar{f}_D \in [-1\,000, \ldots, 1\,000]$ Hz. It is worthwhile to note that the DDMA can be extracted from DDMs of various variables such as raw receiver counts, power received in Watts and the σ_0 DDM, for example. In addition to the averaging performed across the DD domain at a particular epoch, temporal averaging can also be employed to further reduce variance in the extracted features. That is,

$$\left\langle \tilde{P}_R[n] \right\rangle = \left\langle DDMA[n] \right\rangle = \frac{1}{MIJ} \sum_m \sum_i \sum_j \left\langle \left| Y(i,j,n+m) \right|^2 \right\rangle \tag{20.23}$$

where M is the number of contiguous DDMs to average. This can be considered additional incoherent integration which will reduce the variance in the DDM sequence. A common misconception is that further incoherent integration will improve the SNR of the DDM. Incoherent integration can only reduce the variance in the estimated (received) DDM. To increase the SNR of the underlying DDM that the receiver produces a noisy estimate thereof, the coherent integration time must be increased. However, due to the ≈ 6 km/s motion of the specular point, a commensurate $6M$ km along-track blurring occurs and, in practice, M is limited so an acceptable spatial resolution is maintained [61]. For the remainder of this chapter, we will use the term DDMA ambiguously in the sense that temporal averaging may or may not have been applied.

20.2.1.3   Signal-to-noise ratio

Using the DDMA and noise-power estimates extracted from (potentially uncalibrated) DDMs, the Signal-to-Noise Ratio (SNR) can be estimated:

$$\text{SNR} = \frac{\tilde{P}_R - \tilde{P}_N}{\tilde{P}_N} \approx \frac{P_R}{P_N} \frac{G_{int} L_N}{L_S} \tag{20.24}$$

where P_R is the actual received signal power; P_N is the actual noise power; G_{int} is the integration gain associated with the matched filtering process; and L_S and L_N are the implementation losses associated with the signal and noise, respectively, in the GNSS-R receiver.

20.2.1.4   Signal-minus-noise

An alternate to the SNR is the Signal-Minus-Noise (SMN):

$$\text{SMN} = \tilde{P}_R - \tilde{P}_N \approx G \frac{G_{int} P_R}{L_S} \tag{20.25}$$

where G is the total RF gain. When the Automatic Gain Controller (AGC) is operational, the analogue gain of the receiver is varied in order to maintain a constant (optimal) magnitude bit ratio [62] and is, thus, driven by the noise power unless the antenna gain is extremely high [63]. When the AGC is off (fixed gain mode), the noise floor of the SMN feature varies; however, as G and L_S have a lower expected variation when compared to N, it is preferred over the SNR in a fixed gain regime [64].

20.2.1.5   Integrated delay waveform

The DW is the CDOP column of the DDM. It is the main observable for airborne or ground-based receivers where there is no significant Doppler spread over the GZ. The CDOP slice of the DDM in Figure 20.4a is shown in Figure 20.4b. For the space-borne case, incoherent integration (averaging) of the DDM over a range of Doppler columns produces the Integrated Delay Waveform (IDW), which is an intermediate data product from which features are extracted.

20.2.1.6   Leading edge slope

The leading edge is highlighted in Figure 20.4b. The Leading Edge Slope (LES) is typically extracted from the IDW [61] and is strongly correlated with the DDMA, which is unsurprising since they both originate from the specular zone. However, the LES and DDMA are sensitive to different components of the underlying sea state and also contain uncorrelated noise, thus both can be used for wind speed estimation, e.g. [65].

20.2.1.7 Trailing edge slope

The trailing edge is also highlighted in Figure 20.4b. Unlike the LES, which originates from near-specular scattering facets, the Trailing Edge Slope (TES) contains power from scattering facets that are far from the specular point. The power in the trailing edge is most sensitive to wind speed [18] but suffers from low SNR due to the lower signal power at high delay values. This can be explained by examining the surface slopes scattering PDF in the BRE, wherein the near-specular points are located in the s domain close to the peak of the PDF, while the scattering facets far from the specular point map to locations in the s domain corresponding to the tails of the PDF, which are most sensitive to variations in the MSS. This holds when even non-Gaussian – but still reasonable – surface slope models and PDFs are used. Moreover, we can also use this reasoning to see how near-specular geometry is also insensitive to the wind direction in the bivariate (and similar) Gaussian distribution.

20.2.2 *Calibration*

The DDMs produced by a basic processor will be images of raw arbitrary counts output by the matched filtering process from which the SNR can still be estimated using (20.22) and (20.21). In early operations of TDS-1, the dependency of the raw DDM SNR on wind speed was observed and an empirical model was produced using the first significantly sized space-borne dataset collected by TDS-1 [66]. However, many of the non-geophysical parameters in (20.6) and their effect, i.e. variance in the estimated wind speed, can be removed from the DDM through calibration. The Receiver Corrected Gain (RCG), is defined as

$$\bar{\Gamma} = \frac{G_R}{R_R^2 R_T^2} \tag{20.26}$$

captures a significant portion of these non-geophysical parameters. The SNR of the DDM is strongly dependent on the RCG and is also often used for quality control purposes. Various data product levels were defined during the TDS-1 mission [16], specifying the amount of processing that had been applied to the DDM and what the DDM is imaging, e.g. a DDM of power in Watts as opposed to a DDM of arbitrary counts. Afterwards, even though using the same receiver, the CyGNSS mission redefined these levels to align more closely with other remote sensing data products delivered by sensors such as SeaWinds and ASCAT [67–69] and are summarised in Table 20.1.

In addition to the data products shown in Table 20.1, there are higher-level data products such as the L3-gridded wind product. This product is the result of a fine re-estimation of the scattering point to produce the gridded product and a blending of this product can be done to deliver an L4 product. Both of these are topics of ocean processes/data assimilation and are beyond the scope of this chapter.

Table 20.1   CyGNSS data product-level definitions

Data product	Type	Estimate
L0a	Raw Intermediate Frequency (IF) samples	N/A
L0b	DDM	Raw (arbitrary units)
L1a	DDM	Power (Watts)
L1b	DDM	σ_0 (dimensionless)
L2	Data point	Wind speed

The raw IF samples are the output of the analogue-to-digital converter in the front end of the receiver. Due to the nature of GNSS signals, 2-bits of resolution is commonly employed by GNSS(-R) receivers. The raw IF data is the input to the matched filtering process which produces the raw L0B DDMs. The systematic calibration of raw DDMs to L1a DDMs of Power and L1b DDMs of σ_0 for the CyGNSS constellation is detailed in [70–73] and has been done in a similar fashion for TDS-1 [64, 74]. An estimate of the power DDM, $\hat{P}_R\left(\bar{\tau},\bar{f}_D\right)$, is made by normalising the raw DDM with the receiver processing gain, G_S. By periodically using DDMs formed over the open ocean and when the receiver's front end is switched to a black body, the noise power and gain of the receiver can be estimated.

$$\hat{P}_R\left(\bar{\tau},\bar{f}_D\right) = \frac{\left\langle \left| Y\left(\bar{\tau},\bar{f}_D\right)\right|^2\right\rangle}{G_S} \tag{20.27}$$

To estimate the σ_0 DDM, the bistatic system gain must be removed from the power DDM. Looking at (20.6), we can see that this gain can be written as

$$\Gamma'\left(\bar{\tau},\bar{f}_D\right) = \frac{P_T\lambda^2 G_T\left(\bar{\tau},\bar{f}_D\right) G_R\left(\bar{\tau},\bar{f}_D\right)}{\left(4\pi\right)^3 R_R^2\left(\bar{\tau},\bar{f}_D\right) R_T^2\left(\bar{\tau},\bar{f}_D\right)} \tag{20.28}$$

and the total BRCS can be estimated by normalising the power DDM by this. However, for wind speed estimation, the goal is to estimate the NBRCS. Thus, we are required to compute the area of the χ^2 footprint on the sea surface.

$$A_{\chi^2}\left(\bar{\tau},\bar{f}_D\right) = \iint\limits_{GZ} \chi^2\left(\Delta\tau,\Delta f_D\right) d^2\boldsymbol{p} \tag{20.29}$$

which we can consider to be a gain due to the aperture of the scattering surface and is also referred to as the area DDM. Finally, we can now express the total system gain as

$$\Gamma\left(\bar{\tau},\bar{f}_D\right) = \Gamma'\left(\bar{\tau},\bar{f}_D\right) A_{\chi^2}\left(\bar{\tau},\bar{f}_D\right) \tag{20.30}$$

which is used to estimate the σ_0 DDM

$$\hat{\sigma}_0\left(\bar{\tau},\bar{f}_D\right) = \frac{\hat{P}_R\left(\bar{\tau},\bar{f}_D\right)}{\Gamma\left(\bar{\tau},\bar{f}_D\right)} \tag{20.31}$$

This is a relatively simple model. Factors such as propagation and implementation losses as well as variable transmitter EIRP and receiver AGC values are not considered. However, this does convey the basics of the GNSS-R calibration concept. A more detailed description can be found in [71, 75] and [72] to produce radiometrically calibrated DDMs in units of Watts and NBRCS, respectively.

20.3 Retrieval techniques

The use of empirically derived Geophysical Model Functions (GMFs) for wind speed estimation is popular in remote sensing in general and for GNSS-R specifically. In this section, we will review the development and use of empirical GMFs for space-borne GNSS-R-based wind speed estimation. This is followed by the application of more general machine learning-based approaches to the wind speed prediction problem which have only recently been proposed. Finally, we will also review a lesser-known technique called stare processing, which is a promising multi-look scheme.

20.3.1 Empirical wind speed estimation

Since wind speed cannot be measured directly and inverse models are either intractable or do not exist, GMFs are commonly employed for wind speed estimation using remote sensing, e.g. [3, 68]. GMFs are empirical models built by fitting models relating DDM features (in the case of GNSS-R) with ground truth wind speeds. The sources of ground truth utilised by the GNSS-R community include in-situ measurements both from buoys, estimates made by other established remote sensing platforms such as ASCAT, and from numerical predictions such as the European Centre for Medium-Range Weather Forecasting (ECMWF) [76]. Simulated DDMs have even proven useful for developing GMFs [65]. In this section, we will review some of the fundamental steps involved in developing and utilising GMFs.

A GMF can be built by fitting a parametric model using features extracted from the DDM that are matched up with coincident ground truth observations. Alternatively, the data can be binned and averaged such that a Look-Up-Table (LUT) is built. Additional constraints, such as monotonicity, are often imposed on the resultant GMF [65, 77], whether it be in parametric or a LUT form. In Figure 20.5, a log-density plot of SNR against wind speed obtained at space-borne geometries is shown. These data were produced by simulating DWs using WavPy [78], which is an open-source set of GNSS-R tools for simulating and processing GNSS-R data. A GMF is produced by fitting a model of the form $\hat{U}_{10} = a\left(SNR + b\right)^c$ to the scatter data and is shown with a white line. The features of the GMF and the density in Figure 20.5 are representative of actual data. The most important of which is the decreasing sensitivity to wind speed as it increases and the distribution

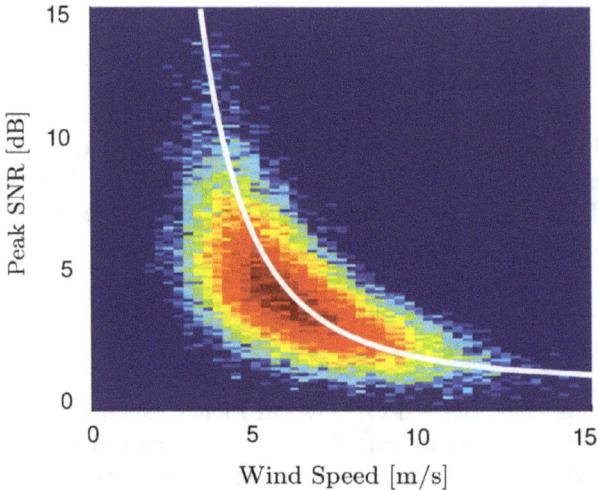

Figure 20.5   *A log-density plot of SNR values extracted from DWs simulated using WavPy against wind speed. Overlaid in white is a parametric GMF fitted to this data.*

of actual ocean wind speeds resulting in most observations being made at moderate wind speeds.

Using a limited dataset produced by the UK-DMC mission, [79] extracted the SNR from L0B DDMs and then produced a biased estimate of the BRCS using a quasi-calibration technique similar to that outlined in Section 20.2.2. Absolute calibration was not achievable due to the technical limitations of the UK-DMC configuration, notably the different gains in the zenith and nadir signal path along with an active AGC. Nevertheless, this work produced the first space-based GNSS-R scatterometric observations of Earth's oceans wherein MSS estimates were made from the BRCS and wind speed estimates thereon. Also using data collected by UK-DMC, in addition to in-situ measurements collected by buoys, a Minimum Variance Estimator (MVE) was built using five DDM features including the DDMA, DDM Variance, Allen DDM Variance, TES and LES [80]. While these features are correlated, they also contain uncorrelated noise which motivated the development of the MVE.

The TDS-1 mission saw the first significant amount of data collected by a space-borne platform. TDS-1 flew the SGR-ReSI GNSS-R receiver, which utilises the MAX2769 front end to receive the L1 GPS signal [13]. Thus, during this time, with the gain fluctuating according to the noise power and the actual gain value not available to the GNSS-R processor, the SNR was extracted from the L0B DDMs as it is not dependent on the gain and radiometrically calibrated DDMs could not be produced. Based on the SNR extracted from L0B DDMs, two wind-speed inversion algorithms termed Fast Delivery Inversion (FDI) and the BRE algorithm were developed [81]. The FDI algorithm requires just the L0B SNR and the receiver antenna gain towards the specular point using an empirical model of the form

$$\hat{U}_{10} = a \left(SNR - b_1 G_R \left(\mathbf{S} \right) + b_2 \right)^c \tag{20.32}$$

where a, b_1, b_2 and c are coefficients to be found when fitting the model which parameterises the GMF. As a result of ignoring the majority of variables in (20.6), the FDI algorithm produced wind speed estimates with a relatively large RMS error of approximately 4 m/s. The BRE algorithm involves numerically computing and removing the impact of the terms in (20.6) with the exception of P_T and G_T to provide $\breve{\sigma}_0$ which is an estimate of the NBRCS derived from the SNR which is, in turn, used as an input for the GMF

$$\hat{U}_{10} = a e^{b \breve{\sigma}_0 (\text{SNR})} + c \tag{20.33}$$

that produced unbiased estimates with an RMS error of 2.2 m/s for winds between 3 and 18 m/s [66]. This work also discovered that when the SNR was below 3 dB, the sensitivity of the estimated wind speed to parameters not modelled by the BRE was significant. The BRE algorithm can also be viewed as a process with implicit calibration where the calibrated DDMs are not exposed explicitly and/or delivered, and only the DDM cells from which features are extracted are calibrated. This is highlighted by the fact that the single input to the GMF, $\breve{\sigma}_0$, is estimated from the SNR of a DDM that was not radiometrically calibrated.

Looking at (20.24), we can see that GMFs based on the SNR extracted from L0B DDMs are sensitive to the noise power and additional considerations must be made when considering fixed vs automatic gain control regimes. With regards to calibrated DDMs, errors are introduced when the transmitter power and antenna gain terms are considered constant which is a simplifying assumption that has been somewhat popular in the past. However, over 4 dB variation in the transmitted power has been reported [75] and, furthermore, the block IIR-M and II-F GPS satellites have a flex power mode where over 10 dB variation in some signals is intentionally introduced on a regional basis [82], which results in both calibrated and uncalibrated observations not being valid for inversion. In addition to this, antenna gain patterns are not widely available for all blocks, e.g. block II-F satellites which are currently not used for CyGNSS-based wind speed estimation [65].

After commissioning, the CyGNSS mission has been able to utilise calibrated L1B DDMs. The wind speed GMF utilised by the CyGNSS mission uses both the $\hat{\sigma}_0$ DDMA and the LES observables [61, 83] which produced the most correlated retrievals of the five observables extracted from DDMs collected by UK-DMC [80]. The GMFs are built using the Global Data Assimilation System (GDAS) [84] and ECMWF-derived wind speeds [65]. Furthermore, the GMFs for each observable are composite: one for moderate and one for low wind speeds. For example,

$$\hat{\sigma}_0 = \begin{cases} a U_{10}^{-1} + b U_{10}^{-2} + c, & \text{if } U_{10} \leq 15 m/s, \\ a_2 U_{10}^{-1} + b_2 U_{10}^{-2} + c_2, & \text{otherwise.} \end{cases} \tag{20.34}$$

is the observation model for the $\hat{\sigma}_0$ DDMA and a similar pair of models exist for the LES observable (moderate speed cut-off is just 10 m/s). The two models in (20.34) are combined using a spline fit to produce a composite GMF for the $\hat{\sigma}_0$ DDMA observable and likewise for the two observation models of the LES resulting in two composite GMFs. These GMFs produce individual wind speed estimates which are combined using an MVE to produce the final wind speed estimate. At wind speeds ≤20 m/s, the RMS error between the CyGNSS and ECMWF wind speed estimates was found to be 1.96 m/s [85]; however, if the uncertainty in the ECMWF-based wind speed estimates is removed, this falls to just 1.44 m/s.

In wind speed retrievals from DDMs collected by TDS-1, an additional structured dependence on the incidence angle after compensating for the parameters in the BRE was observed [81]. This has since motivated the development of 2-dimensional (2D) GMFs that are dependent on the incidence angle in addition to the observable extracted from the DDM, e.g. [61], which is now a popular approach.

In addition to the incidence angle, the DDM is impacted by many factors – both non-geophysical and geophysical – of which most have a complex non-linear impact on the response. Many non-geophysical factors are removed during the calibration process imperfectly, e.g. incomplete knowledge of the receiver antenna gain pattern, noisy platform attitude estimates, etc. Additionally, the sources of ground truth are often noisy themselves and rely on close but not exactly coincident observations which may even be binned incorrectly according to errors in the computed specular point position [28]. Moreover, the sources of the ground truth observations may be sensitive to differing parts of the sea surface slope spectrum compared to the GNSS-R receiver. This is further compounded by interfering geophysical factors that are not modelled such as the non-wind-driven component of the sea surface roughness which contribute to σ_0. Comprehensive analyses of both geophysical and non-geophysical effects on the performance of GNSS-R-based wind speed estimation algorithms have been carried out, see [85–87].

Observing the GMF shown in Figure 20.5, we can see that with increasing wind speed the expected retrieval error grows as the GMF saturates and the SNR of the reflected signal decreases. At low wind speeds, the contribution of the non-wind-driven component of the sea surface slope spectrum, MSS and, ultimately, σ_0 becomes a significant source of error that results in an overestimation of wind speed [86]. The Significant Wave Height (SWH) number, which is more sensitive to the longer wavelengths of the sea surface slopes spectrum, has been observed to be strongly correlated with wind speed retrieval errors [88]. As a result, it has been proposed to compute the excess MSS due to non-wind-driven ocean roughness and use this to correct the $\hat{\sigma}_0$ estimates before retrieving wind speed [88]. This work demonstrated the capability of such a technique using CyGNSS data; however, only a limited amount of data was processed and further work is required. In other work, using a priori SWH statistics, a Bayesian approach, realising the minimum mean square error estimator, was applied to data collected by TDS-1 [89] to mitigate the impact of non-wind-driven ocean roughness on wind speed retrieval errors. However, this approach introduced significant biases at higher wind speeds and a composite strategy was proposed where the GMF was still utilised at higher wind speeds.

For environments such as hurricanes, where wind fields are not constant and speeds are extremely high, the FDS assumption does not hold and an underdeveloped Young Seas with Limited Fetch (YSLF), i.e. $\Omega > 0.84$, occurs. Extending upon the excess MSS approach using SWH [88] and including inverse wave age estimates from WaveWatch III [58] have been demonstrated to be able to further improve wind speed estimates [90]. In addition to the FDS GMFs based on ECMWF and GDAS data matchups, the CyGNSS mission also employs a set of YSLF GMFs based on matchups with the NOAA P3 Hurricane Hunter [91]. Again the GMFs for the $\hat{\sigma}_0$ DDMA and LES are composites of two models similar to (20.34) with the exception that the high wind speed model is only a first-order model [65]. The YSLF GMF-based wind speed retrievals unsurprising perform better than the FDS estimates in extreme wind speeds such as those encountered during hurricanes but are obviously drastically incorrect when sea conditions are moderate.

It has been shown that if observables and the wind speed are monotonically related, then wind speed estimates can be made using their respective empirically derived Cumulative Distribution Functions (CDFs) [92]. This has been recently applied to CyGNSS data, where estimates based on the $\hat{\sigma}_0$ DDMA and LES observables were then fed into the MVE where improvements over the traditional GMF-based observables were demonstrated [93]. A considerable advantage of this approach is that ground truth matchups and the associated data and model processing are not required.

20.3.2 Machine learning

Empirical GMFs are produced by regressing the ground truth winds (target variable) with the observables (feature vector) extracted from the DDM and receiver metadata. The goal of supervised Machine Learning (ML) is to learn a mapping of inputs to outputs given a labelled set of training data [94] and thus we can consider the empirical GMFs covered earlier in this chapter as rudimentary ML models. However, categorising a technique as ML typically implies the use of somewhat more general models that can be considerably more expressive than empirical GMFs even when we consider just shallow networks and thus, in this chapter, we differentiate between empirical GMF- and ML-based models.

We have already seen how the inclusion of additional inputs to the inversion scheme, such as $G_R(S)$ and θ_i [61, 65, 81, 89, 93], improved the performance of GMF-based predictions as previously unmodelled effects could be compensated for. In the case of θ_i as an additional input, this often resulted in the development of multiple GMFs for a multitude of θ_i ranges, e.g. [61, 65], effectively dropping the requirement to be able to fit a model describing the effect of θ_i and replacing it with the need to fit a multitude of models. Moreover, the reduction of the DDM into a few variables, such as the DDMA or LES, does throw some information away (albeit a significant portion remains) and including additional features, such as receiver metadata, that we have high confidence can explain variability in the DDM not due to wind speed will improve performance and is desirable. This has naturally led to the application of ML to GNSS-R wind speed estimation.

20.3.2.1   Wind speed prediction using neural networks

The Multi-Layer Perceptron (MLP) is the quintessential example of a (Deep) Neural Network (D)NN [95] and is the architecture recently employed by [96–99] for GNSS-R-based wind speed estimation. The MLP, also known as an artificial neural network, is a feed-forward network, comprised of multiple layers, of which the first is given by

$$h^{(1)} = g^{(1)}(W^{(1)T}x + b^{(1)}) \tag{20.35}$$

where the superscript 1 denotes the first layer, x is the input feature vector to the network, $h^{(1)}$ is the output of the first layer, $g^{(1)}(\cdot)$ is an activation function chosen by the designer, while $W^{(1)}$ and $b^{(1)}$ are the weights and biases of the first layer, respectively, that are found during the training process. Layers are cascaded in a feed-forward manner such that the output of one is the input of the next. Thus, the second layer is given by

$$h^{(2)} = g^{(2)}(W^{(2)T}h^{(1)} + b^{(2)}) \tag{20.36}$$

and so on for as many layers compose the network, at which point the output of the final layer is also the output of the network. The dimensionality of W and b are chosen by the designer with the only requirement being that the input and output dimensions match that of the feature vector and target variable, respectively. The activation function, $g(\cdot)$, is typically a non-linear function such as the sigmoid, hyperbolic tangent or rectified linear unit; however, there is a vast range of possible activation functions which is still an active area of research [95]. For regression applications, the activation function of the final layer J should be a fully linear function, i.e. an identity transform such that $g^{(J)}(x) = x$, enabling the output variable $h^{(J)} = \hat{U}_{10}$, the predicted wind speed in the context of GNSS-R-based wind speed estimation, to take on any real number. The architecture design of an MLP network involves the selection of the number of layers, the number of neurons (dimensionality of W and b) for each hidden layer (layers which are not directly connected to the input or output) and the activation function used by each layer. Note that a network is said to be deep if it has at least two hidden layers.

Using an MLP NN with one hidden layer with six neurons and a feature vector comprised of $\hat{\sigma}_0$, θ_i, θ_e, G (S), the latitude of the specular point, the z-coordinate of the receiver in the ECEF frame and the DDM numerical scaling, extracted from DDMs and metadata collected by TDS-1, unbiased wind speed estimates were produced with an RMSE of 2.2 m/s [98]. Also using an MLP NN, however, with an undisclosed number of hidden layers and neurons, [97] reported unbiased wind speed predictions with an RMSE of 1.79 m/s when using the DDMA, LES, θ_i and the DDM windowed to $\bar{\tau} \in [-0.25, \ldots, 0.75]$ chips and $\bar{f}_D \in [-1\,000, \ldots, 1\,000]$ Hz as inputs from data collected by the CyGNSS mission. Once again, using an MLP NN, this time with 2 hidden layers each with 16 neurons, and the sigmoid activation function, unbiased wind speeds were produced with an RMSE of 1.51 m/s after applying a CDF correction to the network's output [99]. This was achieved using CyGNSS data, specifically the DDMA, LES, θ_i, RCG and the specular point latitude and longitude, as inputs to the network.

One reason why DNNs can be so powerful comes from their ability to build increasingly more abstract and complex representations of features at deeper layers [100]. While this enables the development of networks that perform remarkably well, obtaining a physical interpretation of the network's behaviour, and thus validating it, becomes increasingly difficult as the complexity of the network grows. However, we still must be careful even with relatively small networks. For example, the network developed by [99] saw an RMSE improvement of 7% after including the specular points latitude and longitude in the input feature vector. This was attributed by the authors to the network learning some form of geographical variation in the wind speed range and distribution, which is a natural conclusion to be drawn towards. However, without analysing the trained network and its feature representations, we could equally surmise that the network had instead learnt how the pointing performance of the satellite degrades at a particular location in its orbit or how the antenna noise temperature changes over the surface of the Earth. Systematic methods for interpreting the behaviour of NNs, e.g. [101], should be utilised to ensure proper conclusions are drawn from the observant behaviour of the network and to ensure proper feature representations are present therein [101, 102] as part of a validation strategy that extends beyond comparing unseen ground truth data with predictions.

The increasingly more abstract representation of features at deeper layers also suggests that architectures such as that developed in [97], where both a DDM and metadata were fed into the network at the same point and propagated through the same layers, are likely to be outperformed by multi-modal approaches [103, 104]. Such an approach involves feeding the data corresponding to the different modalities through different paths in the network whereby individual representations are learnt of each modality which are then combined in a deeper layer to construct a shared representation. For GNSS-R, the DDM could comprise one modality while the remainder of inputs explored so far in this chapter such as metadata, DDMA, LES, etc. could comprise a second modality. More concretely, the representation of the DDM would contain information that is related to wind speed and the shared representation, when combined with the representation of the metadata, would form a joint distribution of the features extracted from the DDM and metadata modalities [104] at which point subsequent layers would be able to more efficiently learn how to utilise information in the metadata to compensate for variation in the DDM not due to wind speed. Moreover, at this point, it is useful to note that simply including all of the receiver metadata as input to the network will, in general, lead to suboptimal results, and efforts should be made to engineer the feature vectors; this is especially true for GNSS-R as we have a vast amount of domain knowledge at our disposal.

Experimenting with various activation functions is a common practice to optimise a network architecture. Using an MLP NN consisting of 1 hidden layer with 16 neurons, [96] reported, unsurprisingly, large variations in performance between networks that varied only in the choice of the activation function. The sigmoid, hyperbolic tangent and linear (identity transform) were utilised in all possible combinations for both the hidden and output layers. One configuration was comprised of a linear activation for both the hidden and output layer which meant that the network could not learn any non-linear functions and thus could never be expressive enough to model even the most basic GMFs such as those we introduced in earlier sections of this chapter, e.g. (20.32), (20.32), or that depicted in Figure 20.5. No number of additional neurons or layers could compensate for this. Other configurations saw the sigmoid and hyperbolic tangent activation functions being applied to the output layer which has limited output ranges of $(0, 1)$ and $(-1, 1)$, respectively, and, in general, should not be used on the output layer of networks for regression applications for this reason as the predicted variable, wind speed here, is then bound to these ranges. If alternates to the identity transform are desired for the output layer, functions such as the rectified linear unit and similar activation functions may be of interest as they have no upper bound and the acvaa wind speed is strictly positive.

20.3.2.2   Predicting extreme winds and handling rare events

The network developed by [99] was observed to over(under) predict wind speeds at low(high) wind speeds which has also been pronounced in empirical GMF-based predictions, e.g. [85]. This was attributed to a large discrepancy in the number of training data points at extreme wind speeds [99] and we can explain this by the fact that rare extreme wind events do not contribute significantly to the overall cost function when training the predictor resulting in the optimisation routine placing little emphasis on obtaining high accuracy at extreme winds which are underrepresented in the training data. This has been addressed somewhat for the empirical GMF approach by the development of composite GMFs where the segregation of the data into low and moderate winds results in a more uniform distribution of wind speeds within each range of wind speeds and thus the pair of resultant GMFs is more accurate at extreme wind ranges that are of interest. Much research has been done to address imbalanced data in ML in general, resulting in the development of three main approaches: (1) weighting the cost function being minimised to emphasised underrepresented regions of the dataset, (2) altering the form of the models and (3) data sampling techniques [105]. It should be noted that the vast majority of the literature addressing imbalanced data is concerned with classification applications, not regression, and there remains significant work to be done in this area generally.

One data sampling technique addressing the imbalanced data problem is known as the synthetic minority over-sampling technique (SMOTE) [106], which has been applied to regression tasks and is known as SMOTE for Regression (SMOTER) [107]. SMOTER involves oversampling the underrepresented data (e.g. extreme winds) by way of generating synthetic data points [107] and has been extended with an under-sampling of overrepresented data and the addition of noise (SMOGN) [108]. While being introduced in the traditional ML framework, all of these techniques also have the potential to improve the performance of not only the ML-based wind speed estimations but also the empirical GMF-based estimators described earlier in this chapter and, thus, is of significant interest to the entire GNSS-R community. Furthermore, with a number of forward geophysical models and simulators already in existence, the generation of synthetic data can be complemented with physics-based models rather than just data-driven techniques such as SMOTER and SMOGN.

Addressing the data imbalance problem may allow the development of a single wind speed prediction model that can be fitted with sufficient accuracy across the entire range of wind speeds that GNSS-R receivers are sensitive to, resulting in composite GMFs becoming obsolete. However, as we incorporate more diverse data into the feature vector of models, we must also consider the distribution of the data across these new dimensions. For example, when we consider the input feature vector used by [98], which contains the space vehicle number, we are now considering ground truth data that are not just distributed across a range of wind speeds but across all of the GPS blocks in addition to whatever dimensions correspond to the remaining inputs in the feature vector. The Block II-A GPS satellite is not only underrepresented in, but a mode of, the underlying input data distribution that contains a significant amount of information due to its antenna pattern being notably different than the other blocks [109] and thus it is desirable that the trained model is sensitive to it. Furthermore, as an example, we can see how the distribution, and joint distributions, of other interesting metadata such as the specular point position and θ_i (see [25]) are appreciably non-uniform due to the bistatic configuration and dynamics of a receiver in low Earth orbit. Much of the metadata that can explain variance in the DDM due to non-wind effects, and thus can improve prediction accuracy, will not be uniformly distributed and a significant amount of work remains to investigate the extent of and to find resolutions to the imbalanced data problem for both ML-based and GMF-based wind speed prediction.

20.3.3 Stare processing

We have seen how GMF-based wind speed estimates are made using features such as the $\hat{\sigma}_0$ DDMA and LES extracted from DDMs which are often temporally averaged. This temporal averaging reduces the variance of the DDM and its features which, in turn, reduces the variance of the wind speed estimates produced by an empirical model linking the two. This is traded off against a decrease in spatial resolution as the specular point moves along the ocean surface and a commensurate spatial blurring occurs. Since we have each individual DDM, we could potentially average the power in the DDMs after projecting the unambiguous points back to the spatial domain. That is, we could then average the power in spatial bins during the course of temporal averaging of the DDM sequence and avoid the blurring and loss of spatial resolution. However, we can see from (20.10) and (20.14) that σ_0 is a function of the incidence angle θ_i through q. Therefore, we are motivated to keep a series of the estimated $\hat{\sigma}_0$ and q tuples at the spatial bin p instead of averaging them. This is justified when we recall that the performance of GMF-based wind estimates was improved through the development of 2D GMFs that are also dependent on incidence angle. The information in q is more comprehensive than ofthat contained in the incidence angle: q provides us with a sampling coordinate of the surface slopes PDF which, in turn, allows us to directly estimate the underlying MSS that parameterises it.

The processing strategy just described is similar to that performed by traditional scatterometers which focus on a fixed point on the ocean surface. One such method is the SHARP processing method [110], which involves taking measurements along the specular point track to avoid the ambiguities associated with the DD mapping on the ocean surface. In the context of GNSS-R, an interpretation of this method known as stare processing was first analysed in [111] and later applied to data collected by UK-DMC [17] and TDS-1 [64] to estimate isotropic MSSs. The remainder of this section briefly outlines the stare processing technique.

20.3.3.1   Profile extraction

Stare processing involves tracking a point, P, that is fixed on the Earth's surface to obtain multiple looks in order to improve wind speed and MSS estimates by combining multiple observations. These looks provide a sampling of the surface slopes PDF, (20.14), at a multitude of s coordinates as the bistatic vector $q(P)$ changes during the course of the observation [112]. This point moves through the DD domain due to the dynamic scattering geometry and, for a space-borne receiver, a stare point may be observable for up to 20 contiguous DDMs. The stare point is selected as the specular point at the midpoint of the observation period so that it does not deviate far from the AFL and thus remains unambiguous throughout the entire observation period. This is important because if the stare point is ambiguous, the achievable resolution is limited not by the effective area of the χ^2 footprint, but by the box bounding the two distinct ambiguous χ^2 responses, e.g. Figure 20.3. Thus, we can obtain up to 20 independent measurements without the approximate 180 km ($\approx$ 6 km/s $\times$ 20 seconds) of blurring in the along-track direction that would occur if temporal averaging were applied.

In Figure 20.6, the stare points P_o and P_{o-1} corresponding to observations o and $o - 1$ are highlighted in blue and red, respectively. Note that only the two penultimate and specular-coincidence epochs are depicted for clarity; practically, there would be approximately 10 epochs before and after specular coincidence. Since the stare point is selected as the specular point at the midpoint of the observation, we can see that $P_o = S|_t$ and $P_{o-1} = S|_{t-1}$. As the receiver moves over time, t, the incidence angle of the point changes providing the multi-look sampling of the surface slopes PDF. In Figure 20.7a, a typical trajectory of a stare point in the DD domain is depicted at 1 Hz intervals on a single DDM to illustrate how a sequence of DDMs would be sampled to extract the stare profile. The set of observations extracted from the DDM sequence by sampling it at these points is referred to as a stare profile. An example profile, extracted from DDMs collected by TDS-1, is shown in Figure 20.7b. Illustrated in Figure 20.6 is the fact that multiple stare points are tracked at once. The number of stare points in the DDM at any given time is equal to the number of epochs the stare points are observable, i.e. approximately twenty.

A single stare profile is used to estimate the MSS at the stare point alone, so we can combine approximately twenty observations over time which do not result in a significant loss of spatial resolution in the direction of the receiver velocity. Figure 20.8 shows the WAF response at a few epochs including specular coincidence,

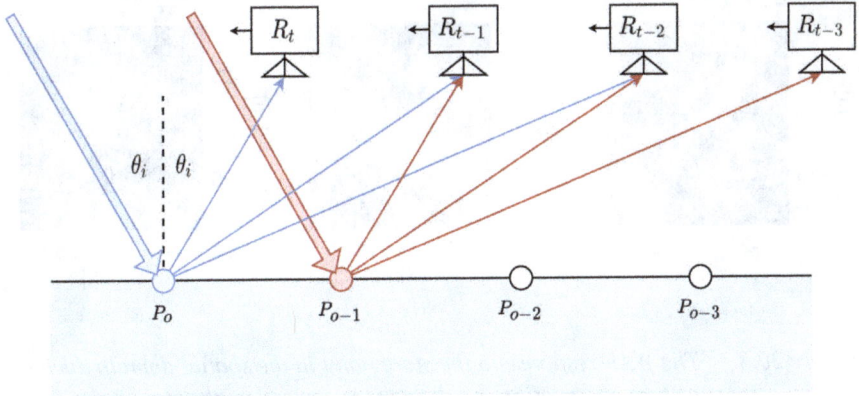

Figure 20.6   *The stare processing geometry. The stare points, **P**, are selected so they coincide with the specular point midway through the observations period. During the course of the observation, the geometry changes which provides diversity in the sampling of the surface slopes distribution.*

i.e. $t = 0$, where we can see how the χ^2 footprint varies over time. Away from specular coincidence, the WAF footprint is elongated in the cross-track dimension while compressed in the along-track. This has motivated the development of an adaptive window [14, 25] to exploit the spatially variant DD fields [113] to increase the number of DD cells included in the overall stare profile while maintaining an along-track

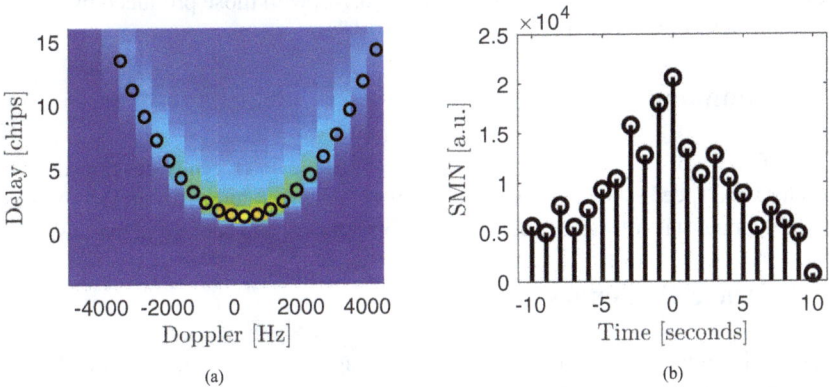

(a)

(b)

Figure 20.7   *(a) The stare point's trajectory in the DDM at 1 Hz intervals which moves from positive to negative Doppler frequencies along the AFL which is the leading edge of the DDM; (b) the SMN profile extracted from the DDMs using a simple nearest-neighbour sampling.*

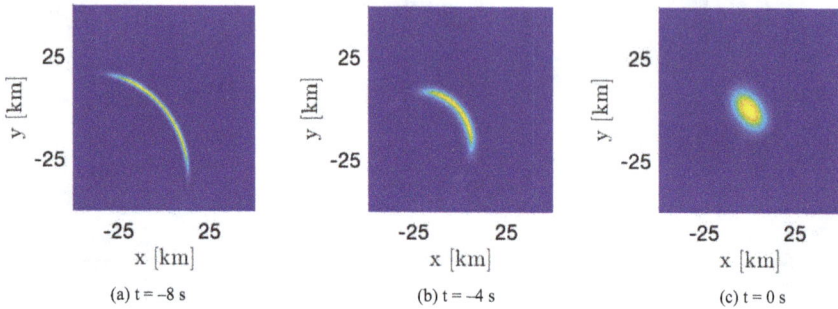

(a) t = –8 s   (b) t = –4 s   (c) t = 0 s

Figure 20.8   *The WAF response to the stare point in the spatial domain during*
a stare observation. Each frame is centred on the stare point and
the shape of the WAF response varies as the GZ passes over it. The
receiver's velocity is 42° measured anticlockwise from the x-axis.

resolution that is comparable to traditional techniques which temporally average the
DDMA over a shorter amount of time.

20.3.3.2   Estimating mean square slopes

Using the SMN profile extracted from sequences of DDMs produced by TDS-1,
the MSS of an isotropic surface slopes PDF can be estimated by fitting simulated
profiles using a least squares approach [64]. The Γ term was computed and used to
correct the SMN profile for all of the non-geophysical BRE effects in a similar way
which was done for the GMF-based inversion using TDS-1 data. This was the first
attempt at applying stare processing to a large dataset and the resultant MSS estimates had a correlation coefficient of 0.684 and 0.742 with those produced by WW3
at large and small scales, respectively.

20.4   Summary

In Table 20.2, we briefly tabulate some of the most notable and recently developed techniques for wind speed estimation that can be found in the literature. This includes works
that are no longer state-of-the-art but are useful for historical and contextual reasons.

20.5   Future challenges

Even though wind estimation has been a principle application of interest for GNSS-R
since its conception, there exist many unresolved issues and unexplored techniques.
For example, the CyGNSS mission has recently demonstrated the effectiveness of a
small constellation of satellites in low Earth orbit in delivering fast and comprehensive observations. Notably, the collection of a large corpus of data – from independent receivers – has presented an enormous opportunity to the GNSS-R community

Table 20.2 Techniques for wind speed estimation

Class	Input	Note	Reported performance	References
Two-dimensional (2D) GMF	L0B SNR & G_R (S)	FDI algorithm	RMSE : 4 m/s bias: 0.2 m/s	[81]
One-dimensional (1D) GMF	$\breve{\sigma}_0$, DDMA	BRE algorithm which encompasses both implicit quasi-calibration and inversion	RMSE : 2.2 m/s unbiased	[26, 66, 81]
1D GMF MVE	$\hat{\sigma}_0$, DDMA, LES, TES, DDMV & ADDMV	Combines multiple estimates produced using independent GMF's for each observable	RMSE : 1.65 m/s unbiased	[80]
2D GMF MVE	$\hat{\sigma}_0$, DDMA & LES with θ_i	As above, except the individual, GMFs have both the observable and incidence angle as input. Performance evaluated on synthetic data and median RCG result is shown	RMSE : 1.57 m/s unbiased	[61]
2D hybrid GMF MVE	$\hat{\sigma}_0$, DDMA & LES with θ_i	Two sets of GMFs exist for YSLF and FDS conditions. The suitable set provides inputs to the MV estimator	RMSE : 2.0(1.4) m/s when $U_{10} \leq 20$ m/s RMSE: 6.5(5.0) m/s when $U_{10} \geq 20$ m/s	[65]
Bayesian	$\hat{\sigma}_0$, DDMA & LES with θ_i	The SWH is used to condition the estimates in a Bayesian estimator	1.4 to 2.5 m/s std. dev. of a figure of merit. Bias is significant and U_{10} dependent	[89]

(Continues)

Table 20.2 Continued

Class	Input	Note	Reported performance	References
CDF MVE	$\hat{\sigma}_0$, DDMA & LES with θ_i and RCG	Assuming monoticity between the observables and U_{10}, wind speed estimates can be made using their respective empirical CDFs. These estimates are fed into the MVE. Reported performance shown here is against ASCAT	RMSE : 1.56 m/s bias: 0.29 m/s	[93]
MLP NN	$\hat{\sigma}_0$, θ_i, θ_e, $G(S)$, S, R_z^{ECEF} & DDM numerical scaling	One hidden layer, six neurons, undisclosed activation function	RMSE : 2.2 m/s unbiased	[98]
MLP NN	DDMA, LES θ_i & windowed DDM	Undisclosed architecture configuration	RMSE : 1.79 m/s unbiased	[97]
MLP DNN	DDMA, LES θ_i, RCG & S	Two hidden layers, 16 neurons, sigmoid activation function	RMSE : 1.51 m/s unbiased	[99]
Stare Processing	SMN	MSS estimated (not U_{10}) from a time series of SMN samples. Performance evaluated against WW3 at large and small scales	Correlation coefficient of 0.684 and 0.742	[17, 64, 111]

to increase our understanding of fundamental theory of including scattering models and ocean processes. One example of this is the use of multiple satellites in the constellation which make near-coincident observations of the same spot on Earth's surface where the sea state can be considered stationary on such timescales but the observations are made at differing geometries. Limited fetch conditions, coherent scattering regimes and extreme – both high and low – wind speeds have proven to be difficult due to the assumptions made in most models and techniques developed to date. A substantial amount of work remains to be done in these areas where the vast amount of data collected by CyGNSS may prove to be most useful. In addition to this, with GNSS-R now being a proven technique, the integration of GNSS-R-derived data products into numerical weather prediction models is becoming increasingly important.

There are many additional concepts of ML that we have not considered in this chapter such as data pre-processing, normalisation and collation, cost functions, optimisation algorithms, model initialisation, and training schedules and schemes to mention a few; furthermore, we only considered the MLP NN architecture. Many of these concepts draw parallels to techniques that are already in place for training empirical GMFs, and it would be of great benefit to apply the learnings of the ML community to the GNSS-R wind speed estimation problem. This is not just in the context of designing networks for wind speed prediction but in the way projects to build them are designed and executed. Another recurring theme in ML that we should take note of is the public release of standardised datasets and trained models for benchmarking purposes. Such an approach greatly improves the speed and ability of the community to compare the performance of new models and would greatly benefit the GNSS-R community when comparing wind speeds predicted with either empirical GMFs or ML models. As we begin to see diminishing incremental improvements in the predictors produced in the GNSS-R community, the benefit of a standardised benchmark increases as more valid comparisons are required. This is exemplified by Table 20.2, wherein we are not able to make entirely definitive comparisons of the techniques themselves due to the differences in data used to report performance. It is expected that in the near future, more ML techniques will be applied to GNSS-R where, with the aid of a large amount of domain knowledge and physics-based models, significant advancements will be made. Future challenges involve the identification, adaption and application of established ML approaches to GNSS-R, where suitable.

The stare processing technique is appealing due to its ability to combine a significant amount of data in the DDM that is normally discarded without the commensurate loss of spatial resolution associated with more traditional GNSS-R wind speed retrieval techniques that process the specular zone only. Furthermore, it has been shown that wind direction signatures exist in non-specular regions of the DDM [35] and ambiguous stare processing techniques may be able to exploit this [25, 112]. The ambiguous stare processing technique involves extracting multiple SMN profiles from a set of points on the Earth's surface such that a significant portion of the DDM sequence is then used in the inversion process. Stare processing with properly designed constellations of GNSS-R receivers will increase both the number and diversity in the stare profiles. This not only will increase the performance of wind speed estimation based on (ambiguous) stare processing but may provide the best practical solution to obtaining wind direction estimates using space-borne platforms.

References

[1]   Woodward P.M. *Probability and Information Theory, with Applications to Radar*. Elsevier; 1980.

[2]   Blahut R.E. *Theory of remote image formation*. Cambridge University Press; 2004.

[3]   Ulaby F.T., Long D.G., Blackwell W.J. 'Radar Remote Sensing of the Ocean'. *Microwave radar and radiometric remote sensing*. 4. University of Michigan Press Ann Arbor; 2014. pp. 793–8.

[4]   Eustice D., Baylis C., R. J. M. I. 'Woodward's ambiguity function: From foundations to applications'. 2015 Texas Symposium on Wireless and Microwave Circuits and Systems (WMCS); 2015.

[5]   Skolnik M.I. *Radar Handbook*. McGraw-Hill; 2008.

[6]   Kelly E.J., Wishner R.P. 'Matched-filter theory for high-velocity, accelerating targets'. *IEEE Transactions on Military Electronics*. 1965;**9**(1):56–69.

[7]   Voronovich A.G., Zavorotny V.U. 'Bistatic radar equation for signals of opportunity revisited'. *IEEE Transactions on Geoscience and Remote Sensing*. 2018;**56**(4):1959–68.

[8]   Baylis C., Cohen L., Eustice D., Marks R. 'Myths concerning Woodward's ambiguity function: analysis and resolution'. *IEEE Transactions on Aerospace and Electronic Systems*. 2016;**52**(6):2886–95.

[9]   Lowe S.T., Kroger P., Franklin G., *et al.* A delay/Doppler-mapping receiver system for GPS-reflection remote sensing'. *IEEE Transactions on Geoscience and Remote Sensing*. 2002;**40**(5):1150–63.

[10] Glennon E.P., Dempster A.G. 'A review of GPS cross correlation mitigation techniques'. The 2004 International Symposium on GNSS/GPS. Fairfax: The Institute of Navigation; 2004.

[11] Glennon E.P., Dempster A.G. 'Cross correlation mitigation by adaptive orthogonalization using constraints – new results'. Proceedings of the 19th International Technical Meeting of the Satellite Division of The Institute of Navigation (ION GNSS 2006); September 2006. pp. 1811–20.

[12] Jin S., Cardellach E., Xie F. 'GNSS Remote Sensing: Theory, Methods and Applications'. Springer; 2014.

[13] 'Mission Description – GNSS Reflectometry on TDS-1 with the SGR-ReSI'. Surrey Satellite Technology Limited, Technical Report; 2015.

[14] Southwell B.J., Dempster A.G. 'An adaptive window for GNSS-R stare processing'. IEEE International Geoscience and Remote Sensing Symposium (IGARSS). IEEE; 2019. pp. 8733–6.

[15] De Vos Van Steenwijk R., Unwin M., Jales P. 'Introducing the SGR-ReSI: a next generation spaceborne GNSS receiver for navigation and

remotesensing'. 5th ESA Workshop on Satellite Navigation Technologies and European Workshop on GNSS Signals and Signal Processing, NAVITEC2010; 2010.

[16]   'MERRByS product manual – GNSS reflectometry on TDS-1 with the SGR-ReSI'. Surrey Satellite Technology Limited, Technical Report; 2015.

[17]   Jales P. 'Spaceborne receiver design for scatterometric GNSS reflectometry [Ph.D. dissertation]'. University of Surrey; 2012.

[18]   Zavorotny V.U., Voronovich A.G. 'Scattering of GPS signals from the ocean with wind remote sensing application'. *IEEE Transactions on Geoscience and Remote Sensing.* 2000;**38**(2):951–64.

[19]   Garrison J.L., Komjathy A., Zavorotny V.U., Katzberg S.J. 'Wind speed measurement using forward scattered GPS signals'. *IEEE Transactions on Geoscience and Remote Sensing.* 2002;**40**(1):50–65.

[20]   Marchan-Hernandez J.F., Camps A., Rodriguez-Alvarez N., Valencia E., Bosch-Lluis X., Ramos-Perez I. 'An efficient algorithm to the simulation of delay-Doppler maps of reflected global navigation satellite system signals'. *IEEE Transactions on Geoscience and Remote Sensing.* 2009;**47**(8):2733–40.

[21]   Valencia E., Camps A., Marchan-Hernandez J.F., *et al.* Ocean surface's scattering coefficient retrieval by delay-Doppler map inversion'. *IEEE Geoscience and Remote Sensing Letters.* 2011;**8**(4):750–4.

[22]   Park H., Valencia E., Rodriguez-Alvarez N., Bosch-Lluis X., Ramos-Perez I., Camps A. 'New approach to sea surface wind retrieval from GNSS-R measurements'. IEEE International Geoscience and Remote Sensing Symposium. IEEE; 2011. pp. 1469–72.

[23]   Martín-Neira M., D'Addio S., Buck C., Floury N., Prieto-Cerdeira R. 'The Paris ocean altimeter in-orbit demonstrator'. *IEEE Transactions on Geoscience and Remote Sensing.* 2011;**49**(6):2209–37.

[24]   Camps A., Park H., Valencia i Domenech E, Domenech V.I., *et al.* Optimization and performance analysis of interferometric GNSS-R altimeters: application to the Paris IoD mission'. *IEEE Journal of Selected Topics in Applied Earth Observations and Remote Sensing.* 2014;**7**(5):1436–51.

[25]   Southwell B.J. 'Techniques for spaceborne remote sensing of Earth's oceans using reflected GNSS signals,' PhD Thesis, UNSW Sydney, Sydney, Australia. 2019. Available from https://www.unsworks.unsw.edu.au/permalink/f/5gm2j3/unsworks_62774.

[26]   Gleason S. 'Remote sensing of ocean, ice and land surfaces using bistatically reflected GNSS signals from low earth orbit PhD dissertation, University of Surrey, September'. 2006.

[27]   Yan Q., Huang W., Foti G. 'Quantification of the relationship between sea surface roughness and the size of the glistening zone for GNSS-R'. *IEEE Geoscience and Remote Sensing Letters.* 2018;**15**(2):237–41.

[28]   Southwell B.J., Dempster A.G. 'A new approach to determine the specular point of forward reflected GNSS signals'. *IEEE Journal of Selected Topics in Applied Earth Observations and Remote Sensing.* 2018;**11**(2):639–46.

[29] Bass F.G., Fuks I.M. *Wave Scattering from Statistically Rough Surfaces: International Series in Natural Philosophy.* **93**. Elsevier; 1979.

[30] Ulaby F.T., Moore R.K., Fung A.K. *Microwave Remote Sensing: Active and Passive. Vol. Ii. Radar Remote Sensing and Surface Scattering and Emission Theory.* Addison-Wesley Reading, MA; 1982.

[31] Beckmann P., Spizzichino A. *The Scattering of Electromagnetic Waves from Rough Surfaces.* Pergamon Press; 1963.

[32] Ruf C.S., Gleason S., Jelenak Z. 'The CYGNSS nanosatellite constellation Hurricane mission'. IEEE International Geoscience and Remote Sensing Symposium (IGARSS); 2012. pp. 214–16.

[33] Armatys M. 'Estimation of sea surface winds using reflected GPS signals'. *Ph.D. dissertation, University of Colorado*; 2001.

[34] Cardellach E., Rius A. 'A new technique to sense non-Gaussian features of the sea surface from L-band bi-static GNSS reflections'. *Remote Sensing of Environment.* 2008;**112**(6):2927–37.

[35] Guan D., Park H., Camps A., *et al.* Wind direction signatures in GNSS-R observables from space'. *Remote Sensing.* 2018;**10**(2):198.

[36] Cox C., Munk W. 'Measurement of the roughness of the sea surface from photographs of the sun's glitter'. *Journal of the Optical Society of America.* 1954;**44**(11):838–50.

[37] Carreno-Luengo H., Camps A. 'Unified GNSS-R formulation including coherent and incoherent scattering components'. IEEE International Geoscience and Remote Sensing Symposium (IGARSS); 2016. pp. 4815–8.

[38] Zavorotny V.U., Voronovich A.G. 'GNSS-R delay-Doppler maps of ocean surface at weak winds'. IEEE International Geoscience and Remote Sensing Symposium (IGARSS); 2017. pp. 2664–6.

[39] Voronovich A.G., Zavorotny V.U. 'The transition from weak to strong diffuse radar bistatic scattering from rough ocean surface'. *IEEE Transactions on Antennas and Propagation.* 2017;**65**(11):6029–34.

[40] Mashburn J., O'Brien A., Axelrad P. 'A comparison of waveform model re-tracking methods using data from CYGNSS'. IEEE International Geoscience and Remote Sensing Symposium (IGARSS). IEEE; 2018. pp. 4289–92.

[41] Zuffada C., Haines B., Hajj G., *et al.* 'Assessing the altimetric measurement from CYGNSS data'. IEEE International Geoscience and Remote Sensing Symposium (IGARSS); 2018. pp. 8292–5.

[42] Camps A. 'Spatial resolution in GNSS-R under coherent scattering'. *IEEE Geoscience and Remote Sensing Letters.* 2020;**17**(1):32–6.

[43] Elfouhaily T.M., Guérin C.-A. 'A critical survey of approximate scattering wave theories from random rough surfaces'. *Waves in Random Media.* 2004;**14**(4):R1–40.

[44] Voronovich A.G., Zavorotny V.U. 'Full-polarization modeling of monostatic and bistatic radar scattering from a rough sea surface'. *IEEE Transactions on Antennas and Propagation.* 2014;**62**(3):1362–71.

[45] Zuffada C., Fung A., Parker J., Okolicanyi M., Huang E. 'Polarization properties of the GPS signal scattered off a wind-driven ocean'. *IEEE Transactions on Antennas and Propagation.* 2004;**52**(1):172–88.

[46] Clarizia M.P., Gommenginger C., Di Bisceglie M., Galdi C., Srokosz M.A. 'Simulation of L-band bistatic returns from the ocean surface: a facet approach with application to ocean GNSS reflectometry'. *IEEE Transactions on Geoscience and Remote Sensing.* 2012;**50**(3):960–71.

[47] Soriano G., Guérin C.A. 'A cutoff invariant two-scale model in electromagnetic scattering from sea surfaces'. *IEEE Geoscience and Remote Sensing Letters.* 2008;**5**(2):199–203.

[48] Zavorotny V.U., Gleason S., Cardellach E., Camps A. 'Tutorial on remote sensing using GNSS bistatic radar of opportunity'. *IEEE Geoscience and Remote Sensing Magazine.* 2014;**2**(4):8–45.

[49] Masters D., Axelrad P., Katzberg S. 'Initial results of land-reflected GPS bistatic radar measurements in SMEX02'. *Remote Sensing of Environment.* 2004;**92**(4):507–20.

[50] Katzberg S.J., Torres O., Grant M.S., Masters D. 'Utilizing calibrated GPS reflected signals to estimate soil reflectivity and dielectric constant: results from SMEX02'. *Remote Sensing of Environment.* 2006;**100**(1):17–28.

[51] Alonso-Arroyo A., Zavorotny V.U., Camps A. 'Sea ice detection using U.K. TDS-1 GNSS-R data'. *IEEE Transactions on Geoscience and Remote Sensing.* 2017;**55**(9):4989–5001.

[52] Cardellach E., Flato G., Fragner H, Wickert J., Baggen R., *et al.* GNSS transpolar earth reflectometry exploring system (G-TERN): mission concept'. *IEEE Access.* 2018;**6**(980–14):13980–4018.

[53] Loria E., Brien A.O., Gupta I.J. 'Detection & separation of coherent reflections in GNSS-R measurements using CyGNSS data'. IEEE International Geoscience and Remote Sensing Symposium (IGARSS); 2018. pp. 4003–6.

[54] Clarizia M.P., di Bisceglie M., Galdi C., Gommenginger C., Landi L. 'Delay super resolution for GNSS-R'. 2009 IEEE International Geoscience and Remote Sensing Symposium; 2009. pp. 134–7.

[55] Beltramonte T., di Bisceglie M., Galdi C., Russo I., Zuffada C. 'Super resolution in GNSS coherent scattering'. 2019 Specialist Meeting on Reflectometry using GNSS and other Signals of Opportunity, GNSS+R 2019. IEEE; 2019.

[56] Cheong J.W., Kuthenoor P., Dempster A.G. 'Validation of super-resolution GNSS-R using an airborne field trial'. 2020 IEEE International Geoscience and Remote Sensing Symposium. IEEE; 2020.

[57] Chen-Zhang D.D., Ruf C.S., Ardhuin F., Park J. 'GNSS-R nonlocal sea state dependencies: model and empirical verification'. *Journal of Geophysical Research: Oceans.* 2016;**121**(11):8379–94.

[58] Tolman H.L. User manual and system documentation of WAVEWATCH III TM version 3.14, p. 220. 2009. Available from https://polar.ncep.noaa.gov/mmab/papers/tn276/MMAB_276.pdf [Accessed 6 Aug. 2019].

[59] Elfouhaily T., Chapron B., Katsaros K., Vandemark D. 'A unified directional spectrum for long and short wind-driven waves'. *Journal of Geophysical Research: Oceans*. 1997;**102**(C7):15781–96.

[60] Brown G. 'Backscattering from a Gaussian-distributed perfectly conducting rough surface'. *IEEE Transactions on Antennas and Propagation*. 1978;**26**(3):472–82.

[61] Clarizia M.P., Ruf C.S. 'Wind speed retrieval algorithm for the cyclone global navigation satellite system (CYGNSS) mission'. *IEEE Transactions on Geoscience and Remote Sensing*. 2016;**54**(8):4419–32.

[62] 'MAX2769 universal GPS receiver, Maxim Integrated'. *Technical Report*. 2010.

[63] Alonso-Arroyo A., Querol J., Lopez-Martinez C., *et al*. SNR and standard deviation of cGNSS-R and iGNSS-R Scatterometric measurements'. *Sensors*. 2017;**17**(1):183.

[64] Tye J., Jales P., Unwin M., Underwood C. 'The first application of stare processing to retrieve mean square slope using the SGR-ReSI GNSS-R experiment on TDS-1'. *IEEE Journal of Selected Topics in Applied Earth Observations and Remote Sensing*. 2016;**9**(10):4669–77.

[65] Ruf C.S., Balasubramaniam R. 'Development of the CYGNSS geophysical model function for wind speed'. *IEEE Journal of Selected Topics in Applied Earth Observations and Remote Sensing*. 2019;**12**(1):66–77.

[66] Foti G., Gommenginger C., Jales P., *et al*. Spaceborne GNSS reflectometry for ocean winds: first results from the UK TechDemoSat-1 mission'. *Geophysical Research Letters*. 2015;**42**(13):5435–41.

[67] Lungu T. *Seawinds science data product user's manual [online]*. 2006. Available from https://podaac-tools.jpl.nasa.gov/drive/files/allData/sea-winds/L2BE/docs/SWS_SDPUG_V2.0.pdf [Accessed 6 Aug 2019].

[68] ASCAT product guide [online]. 2015. Available from https://www.eu-metsat.int/website/wcm/idc/idcplg?IdcService=GET_FILE&dDoc-Name=PDF_ASCAT_PRODUCT_GUIDE&RevisionSelectionMethod=LatestReleased&Rendition=Web [Accessed 6 Aug 2019].

[69] Physical oceanography distributed active archive centre glossary [online]. Available from https://podaac.jpl.nasa.gov/Glossary [Accessed 28 Feb 2019].

[70] Gleason S., Ruf C.S., Clarizia M.P., O'Brien A.J. 'Calibration and unwrapping of the normalized scattering cross section for the cyclone global navigation satellite system'. *IEEE Transactions on Geoscience and Remote Sensing*. 2016;**54**(5):2495–509.

[71] Gleason S. 'Algorithm theoretical basis document level 1A DDM calibration'. Southwest Research Institute, Technical Report revision 2; 2014.

[72] Gleason S. Algorithm theoretical basis document level 1B DDM calibration. Southwest Research Institute, Technology Report 2018, revision 2; 2018.

[73] Gleason S., Ruf C.S., O Brien A.J., McKague D.S. 'The CYGNSS level 1 calibration algorithm and error analysis based on on-orbit measurements'.

Ieee Journal of Selected Topics in Applied Earth Observations and Remote Sensing. 2018:37–49.

[74] Jales P., Unwin M., Duncan S., *et al.* 'Advancements and improvements to TDS-1 level 1b meta-data and radiometric calibration'. 2017 Specialist Meeting on Reflectometry using GNSS and other Signals of Opportunity, GNSS+R 2017; 2017.

[75] Wang T., Ruf C.S., Block B., McKague D.S., Gleason S. 'Design and performance of a GPS constellation power monitor system for improved CYGNSS L1B calibration'. *IEEE Journal of Selected Topics in Applied Earth Observations and Remote Sensing.* 2019;**12**(1):26–36.

[76] Andersson E., Persson A., Tsonevsky I. 'User guide to ECMWF forecast products'. **1**. ECMWF, v2; 2015. p. 121.

[77] Clarizia Z.V.U., Paola M., Ruf C.S. 'Algorithm theoretical basis document level 1A DDM calibration'. Southwest Research Institute, Technical Report 2018, revision 5; 2018.

[78] Fabra F., Cardellach E., Li W., Rius A. 'WAVPY: A GNSS-R open source software library for data analysis and simulation'. IEEE International Geoscience and Remote Sensing Symposium (IGARSS); 2017. pp. 4125–8.

[79] Gleason S. 'Space-based GNSS scatterometry: ocean wind sensing using an empirically calibrated model'. *IEEE Transactions on Geoscience and Remote Sensing.* 2013;**51**(9):4853–63.

[80] Clarizia M.P., Ruf C.S., Jales P., Gommenginger C. 'Spaceborne GNSS-R minimum variance wind speed estimator'. *IEEE Transactions on Geoscience and Remote Sensing.* 2014;**52**(11):6829–43.

[81] Unwin M., Jales P., Tye J., Gommenginger C., Foti G., Rosello J. 'Spaceborne GNSS-reflectometry on TechDemoSat-1: early mission operations and exploitation'. *IEEE Journal of Selected Topics in Applied Earth Observations and Remote Sensing.* 2016;**9**(10):4525–39.

[82] Steigenberger P., Thölert S., Montenbruck O. 'Flex power on GPS block IIR-M and IIF'. *GPS Solutions.* 2019;**23**(1):8.

[83] Ruf C., Clarizia M.P., Gleason S. 'CYGNSS Handbook'. University of Michigan, Ann Arbor, MI, Technical Report; 2016.

[84] Kanamitsu M. 'Description of the NMC global data assimilation and forecast system'. *Weather and Forecasting.* 1989;**4**(3):335–42.

[85] Ruf C.S., Gleason S., McKague D.S. 'Assessment of CYGNSS wind speed retrieval uncertainty'. *IEEE Journal of Selected Topics in Applied Earth Observations and Remote Sensing.* 2019;**12**(1):87–97.

[86] Soisuvarn S., Jelenak Z., Said F., Chang P.S., Egido A. 'The GNSS reflectometry response to the ocean surface winds and waves'. *IEEE Journal of Selected Topics in Applied Earth Observations and Remote Sensing.* 2016;**9**(10):4678–99.

[87] Foti G., Gommenginger C., Unwin M., Jales P., Tye J., Rosello J. 'An assessment of non-geophysical effects in spaceborne GNSS reflectometry data from the UK TechDemoSat-1 mission'. *IEEE Journal of Selected Topics in Applied Earth Observations and Remote Sensing.* 2017;**10**(7):3418–29.

[88]   Wang T., Zavorotny V.U., Johnson J., Ruf C., Yi Y. 'Modeling of sea state conditions for improvement of CYGNSS L2 wind speed retrievals'. IEEE International Geoscience and Remote Sensing Symposium (IGARSS); 2018. pp. 8297–300.

[89]   Clarizia M.P., Ruf C.S. 'Bayesian wind speed estimation conditioned on significant wave height for GNSS-R Ocean observations'. *Journal of Atmospheric and Oceanic Technology*. 2017;**34**(6):1193–202.

[90]   Wang T., Zavorotny V.U., Johnson J., Yi Y., Ruf C. 'Integration of CYGNSS wind and wave observations with the WAVEWATCH III numerical model'. IGARSS 2019–2019 IEEE International Geoscience and Remote Sensing Symposium. IEEE; 2019. pp. 8350–3.

[91]   Uhlhorn E.W., Black P.G., Franklin J.L., Goodberlet M., Carswell J., Goldstein A.S. 'Hurricane surface wind measurements from an operational stepped frequency microwave radiometer'. *Monthly Weather Review*. 2007;**135**(9):3070–85.

[92]   Freilich M.H., Challenor P.G. 'A new approach for determining fully empirical altimeter wind speed model functions'. *Journal of Geophysical Research*. 1994;**99**(C12):25051.

[93]   Clarizia M.P., Ruf C.S. 'Statistical derivation of wind speeds from CYGNSS data'. *IEEE Transactions on Geoscience and Remote Sensing*. 2020;**58**(6):3955–64.

[94]   Murphy K.P. 'Machine learning: a probabilistic perspective'. MIT Press; 2012.

[95]   Goodfellow I., Bengio Y., Courville A., Bengio Y. 'Deep Feedforward Networks'. *Deep learning*. 2. **1**. MIT Press; 2016. pp. 163–97.

[96]   Wang F., Yang D., Zhang B., Li W. 'Waveform-based spaceborne GNSS-R wind speed observation: demonstration and analysis using UK TechDemoSat-1 data'. *Advances in Space Research*. 2018;**61**(6):1573–87.

[97]   Liu Y., Collett I., Morton Y.J. 'Application of neural network to GNSS-R wind speed retrieval'. *IEEE Transactions on Geoscience and Remote Sensing*. 2019;**57**(12):9756–66.

[98]   Asgarimehr M., Zhelavskaya I., Foti G., Reich S., Wickert J. 'A GNSS-R geophysical model function: machine learning for wind speed retrievals'. *IEEE Geoscience and Remote Sensing Letters*. 2020;**17**(8):1333–7.

[99]   Reynolds J., Clarizia M.P., Santi E. 'Wind speed estimation from CYGNSS using artificial neural networks'. *IEEE Journal of Selected Topics in Applied Earth Observations and Remote Sensing*. 2020;**13**:708–16.

[100]   Bengio Y., Courville A., Vincent P. 'Representation learning: a review and new perspectives'. *IEEE Transactions on Pattern Analysis and Machine Intelligence*. 2013;**35**(8):1798–828.

[101]   Montavon G., Samek W., Müller K.-R. 'Methods for interpreting and understanding deep neural networks'. *Digital Signal Processing*. 2018;**73**(8):1–15.

[102]   Lapuschkin S., Binder A., Montavon G., Muller K.-R., Samek W. 'Analyzing classifiers: Fisher vectors and deep neural networks'. Proceedings of the

IEEE Conference on Computer Vision and Pattern Recognition; 2016. pp. 2912–20.

[103] Ramachandram D., Taylor G.W. 'Deep multimodal learning: a survey on recent advances and trends'. *IEEE Signal Processing Magazine*. 2017;**34**(6):96–108.

[104] Srivastava N., Salakhutdinov R. 'Multimodal learning with deep Boltzmann machines'. *The Journal of Machine Learning Research*. 2014;**15**(1):2949–80.

[105] He H., Garcia E.A. 'Learning from imbalanced data'. *IEEE Transactions on Knowledge and Data Engineering*. 2009;**21**(9):1263–84.

[106] Chawla N.V., Bowyer K.W., Hall L.O., Kegelmeyer W.P. 'SMOTE: synthetic minority over-sampling technique'. *Journal of Artificial Intelligence Research*. 2002;**16**:321–57.

[107] Torgo L., Ribeiro R.P., Pfahringer B., Branco P. *Portuguese Conference on Artificial Intelligence*. Springer; 2013. pp. 378–89.

[108] Branco P., Torgo L., Ribeiro R.P. 'Smogn: a pre-processing approach for imbalanced regression'. First International Workshop on Learning with Imbalanced Domains: Theory and Applications. PMLR; 2017. pp. 36–50.

[109] Marquis W.A., Reigh D.L. 'The GPS block IIR and IIR-M broadcast L-band antenna panel: its pattern and performance'. *Navigation*. 2015;**62**(4):329–47.

[110] Caparrini M., Germain O., Soulat F., Ruffini L., Ruffini G. 'A system for monitoring a feature of a surface with broad swath and high resolution'. European Patent Office, no. EP1279970; 2003.

[111] Jales P. 'GNSS-reflectometry: techniques for scatterometric remote sensing'. Proceedings of the 23rd International Technical Meeting of The Satellite Division of the Institute of Navigation (ION GNSS; 2010. pp. 2761–70.

[112] Southwell B.J. 'Investigating the sensitivity of delay doppler maps to wind direction using ambiguous stare processing'. Proceedings of the 31st International Technical Meeting of The Satellite Division of the Institute of Navigation (ION GNSS+); 2018.

[113] Southwell B.J., Dempster A.G. 'Incoherent range walk compensation for spaceborne GNSS-R imaging'. *IEEE Transactions on Geoscience and Remote Sensing*. 2019;**57**(5):2535–42.

Chapter 21

GNSS-R ocean altimetry

Hugo Carreno-Luengo[1]

Mesoscale ocean altimetry remains a challenging area for satellite observations and yet of great interest for oceanographers trying to validate and derive their ocean circulation models with real measurements. Global navigation satellite systems (GNSS) Earth-reflected signals can be used as sources of opportunity for mesoscale ocean altimetry with improved spatio-temporal resolution as compared to traditional monostatic radar altimetry.

21.1 Historical overview: technical relevant aspects

The potential of c/iGNSS-R for ocean altimetry was first published in 1993 [1]. Over smooth ocean surfaces, the altimetric range as it is obtained from the delay of the peak was first proposed in [2]. In 2002, a new approach was formulated through fitting a theoretical model to the data. The best fit model implicitly indicates the delay location where the specular point lies [3]. In 2010, it was demonstrated that the maximum of the derivative of the waveform's leading edge corresponds to the specular ray-path delay (except for filtering effects of the limited bandwidth) [4]. The delay Doppler map (DDM) multi-look technique was proposed later. It uses the full DDM as a way to perform multi-look altimetry beyond the typical pulse-limited region [5]. Additionally, improved altimetric techniques based on phase observations have been tested from an aircraft [6, 7] and a zeppelin [8]. The results, for low-elevation angles up to ~30°, show altimetric precisions comparable to Nadir-looking peak-derivative methods over open sea waters. These results have been confirmed from space, both in near-Nadir geometry over smooth sea ice surfaces [9] and grazing-angle geometries over relatively calm seas [10] as well as over sea ice and ice sheets [11]. All of the above algorithms are based on cGNSS-R. The peak-derivate method has also been applied to iGNSS-R [12] and to rGNSS-R (Figure 21.1) in 2014 [13].

It is also worth noting that in the space-borne era, the peak-derivative method [14, 15] and the phase observations [10] have also been successfully explored using cGNSS-R. An alternative approach similar to the peak-derivative method was

[1]Climate and Space Sciences and Engineering Department, University of Michigan, United States

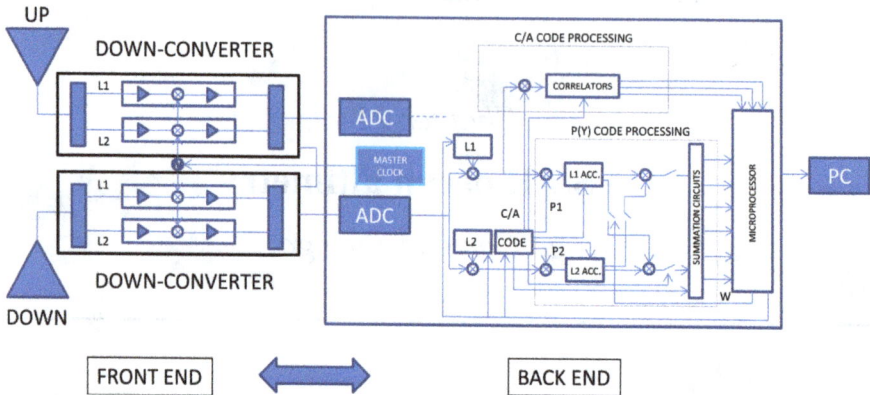

Figure 21.1   *Block diagram of the P(Y) & C/A reflectometer (PYCARO) instrument. The correlation approach used in the down-looking channel by PYCARO providing P-code processing of the encrypted global positioning system (GPS) signals without knowledge of the encrypted code, in addition to the coarse acquisition (C/A) code cGNSS-R, is sketched. Both channels use a similar correlation approach [13].*

proposed in [12]. It is based on the assumption that the specular path delay corresponds to the delay at the 75% [12] and 70% [16] of the peak power. In order to achieve the centimetric accuracy required to track the mean sea level and its spatio-temporal variations[a], one of the challenging errors to be corrected for is the electromagnetic (EM) bias. The EM bias exhibits a dependence on the elevation and azimuth angles [17–19]. Additionally, bandwidth has an important impact on the iGNSS-R waveforms [20]. As the bandwidth is reduced, the autocorrelation function (ACF) becomes wider and the iGNSS-R waveform shape approximates to the cGNSS-R one using the coarse acquisition (C/A) code only. The displacement produced is small (~14 cm) for 20 MHz (40 MHz in radio frequency RF). At 10 MHz (20 MHz in RF), the displacement obtained is ~25 cm, which could start to be relevant for accurate ocean altimetry.

Regarding scattering theories, the first few experiments showed that a specialized receiver should be developed to collect the Earth-reflected signals over a rough surface because the scattering process distorted the signal, and the receiver lost tracking [21, 22]. In 2000, a bistatic model of the ocean scattered GPS signals was developed [23], which provides an analytical expression of the waveforms under the Kirchhoff approximation. Assuming that coherent scattering is negligible, the bistatic scattering coefficient was derived under the geometric optics limit for a sea surface model with Gaussian distribution of the slopes. Then, a final expression

[a]These sea levels variations are related to large-scale circulation, ocean currents and eddies, or El Niño events.

of the waveform was derived. During the last decades, additional experimental [3, 24–26] and theoretical [12, 27–29] works have been performed to investigate the feasibility of this bistatic radar system for accurate ocean altimetry. The most common practice is the use of open-loop receivers because the scattering is mostly diffuse over the ocean. To that end, a model of the scattering geometry is applied to center the delay and Doppler tracking windows.

21.2   Altimetric tracking point

It has been demonstrated [4] that for an ideal case (a receiver with infinite bandwidth), the delay corresponding to the maximum derivative of the reflected waveform is equal to the delay corresponding to the specular reflection (altimetric tracking point).

In a more real scenario, a more complete analysis must be performed by considering a finite sampling interval of the waveforms in the receiver and different sources of noise: thermal noise (additive), speckle noise[b] (multiplicative), and processing noise due to errors in the way that the signals are processed.

For a band-limited receiver, the impulse response is filtered out and the filtering introduces a bias b_{spec} [4, 20, 30] in the position of the specular reflection point. This bias has a systematic instrument-related component and a component dependent on the sea state [27]. The first one can be, in principle, calibrated, while the latter one is a sort of EM bias that is not yet fully understood at L-band and for a bistatic configuration. Additionally, the receiver's bandwidth impacts the ACF shape and thus the shape of the received power waveform. The impact of the receiver's bandwidth on the ACF shape was showed in [20]. As it can be appreciated, the impact obtained for an intermediate frequency (IF) bandwidth of 20 MHz (40 MHz at RF) is practically negligible, since it has been shown in [20] that the bandwidth of the GPS L1 composite signal is about 20 MHz (40 MHz at RF). The impact starts to be noticeable at 10 MHz (20 MHz at RF) and becomes significant at 5 MHz (10 MHz at RF) and at 2.5 MHz (5 MHz at RF). From this bandwidth, the ACF shape tends to become wider, yielding to the loss of the accuracy gain introduced by the iGNSS-R processing with respect to the cGNSS-R case.

Additionally, other biases must be corrected. For example, not all the scattering elements equally contribute to the radar return, since the valleys of the waves tend to reflect GNSS pulses better than crests. Thus, the centroid of the mean reflecting surface is shifted away from the mean sea level toward the valleys of the waves. If the sea state is rougher, this bias will be larger, thus reducing the accuracy of the GNSS-R altimetry products. This shift is the EM bias b_{em} that causes an

---

[b]Speckle is a source of noise that involves diffuse scattering from rough surfaces. If the ocean is reasonably rough with respect to the incoming GPS signal wavelength, the different heights and orientations of the waves within the glistening zone will randomly shift the phases of the GNSS signals. Some of these paths will destructively interfere and others will constructively interfere, as a consequence, the signal power level will fluctuate randomly at the antenna input. From this, it can be derived that the signal-to-noise ratio (*SNR*) of the reflected signal is severely affected by the sea state. To mitigate this effect, incoherent averaging of consecutive uncorrelated signals must be performed [27].

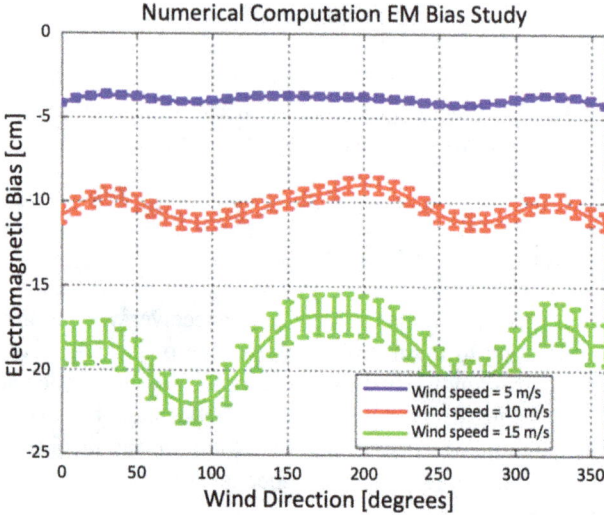

Figure 21.2   EM bias versus wind direction using non-Gaussian sea surface and incidence angle of $\theta_i = 25°$ for wind speed $U_{10} = 10, 15,$ and 20 m/s. Adapted from [18].

overestimation in the sea surface height (SSH) measurement, and it is well understood in classical Nadir-looking altimetry, but little is known in bistatic configurations at L-band. In [18], it was found that the EM bias is almost insensitive to the sea surface spectra selected and increases with increasing wind speed and incidence/scattering angle (up to approximately −20 cm at $\theta_i = 45°$ and $U_{10} = 12$ m/s), and it also exhibits a non-negligible azimuthal dependence, which must be accounted for in the error budgets of upcoming GNSS-R altimetry missions (Figure 21.2).

In order to extract the geometric distance between the reflected and the direct paths ρ_{geo}, several biases have to be corrected, including b_{spec} and b_{em}. Assuming that the direct and reflected propagation directions are parallel and that the surface is flat, the raw experimental height between the up-looking antenna and the sea surface can be calculated as follows:

$$H = \frac{\rho_{geo}}{2\sin\theta_e} + b_{spec} + b_{em} + b_{atm} + b_{ins} + n \tag{21.1}$$

where b_{atm} is the atmospheric-induced delay (ionospheric and tropospheric effects), b_{ins} is the bias induced by instrumental components (clock errors, sub-system delays, and antenna offsets), n is noise, and θ_e is the elevation angle at the specular point.

In an air-borne scenario, the ionosphere affects both direct and reflected signals in the same way. Therefore, the relative delay between them is independent of the bias induced by the ionosphere. In the case of a space-borne platform, such as the proposed passive reflectometry and interferometry system in-orbit-demonstrator (PARIS-IoD) mission [30], the reflected signal is seriously affected by the ionosphere, while the direct signal is only marginally influenced (depending ultimately on the platform's height). Therefore, in this scenario it is required to compensate for this delay but not in an air-borne one.

Finally, in order to extract the SSH, H must be subtracted from the up-looking antenna altitude relative to a mean sea surface model H_{model}. Then, the ultimate accuracy of the GNSS-R technique is determined by the inherent error of the measurement:

$$SSH_{measured} = H_{model} - H \qquad (21.2)$$

21.3 Height precision

Few models have been developed to predict the height precision (Figure 21.3). Each one drives the analysis from different hypotheses. The first code range precision model was proposed in [31]. Later, a new one based on the Cramer–Rao bound (CRB) approach was proposed in [32]. The CRB method allows to predict the optimum behavior in stochastic problems that can be described by a probability density function. A simpler model will be introduced here since it allows to analyze the height precision as a function of different parameters involved in the GNSS-R

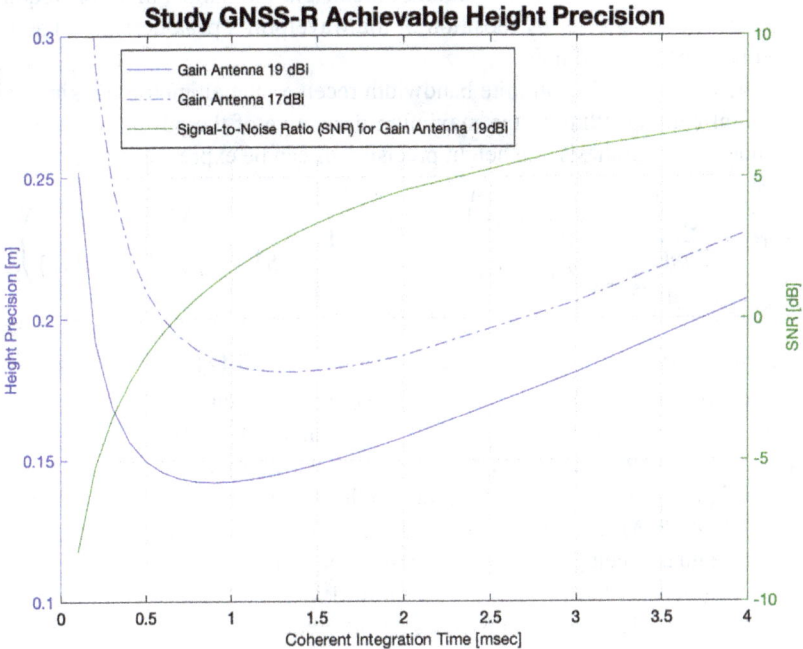

Figure 21.3   *Height precision versus coherent integration time for two values of antenna gain. Along-track spatial resolution is fixed as 50 km, corresponding to a total observation time of 9 s. Adapted from [5].*

scenario such as the signal-to-noise ratio (*SNR*), the speckle noise, the observation geometry, and the autocorrelation properties of GNSS signals [33].

The standard deviation σ_R of the total received waveform power $\left\langle |Z(\tau)|^2 \right\rangle$ is dependent on the *SNR* and as shown in (21.3) decreases as a function of the incoherent averaging N_{inc} ([34], p. 492):

$$\sigma_R = \frac{\left\langle |Z(\tau)|^2 \right\rangle}{\sqrt{N_{inc}}} = \frac{\left\langle |Z_s(\tau)|^2 \right\rangle}{\sqrt{N_{inc}}} \left(1 + \frac{1}{SNR(\tau)} \right) \tag{21.3}$$

where subscript *s* means signal. The noise power standard deviation is

$$\sigma_N = \frac{\left\langle |Z_s(\tau)|^2 \right\rangle}{\sqrt{N_{inc}}} \left(\frac{1}{SNR(\tau)} \right) \tag{21.4}$$

If the signal and the noise are uncorrelated, the standard deviation of the signal power then becomes ([34], p. 493):

$$\sigma_S = \sqrt{\sigma_R^2 + \sigma_N^2} = \frac{\left\langle |Z_s(\tau)|^2 \right\rangle}{\sqrt{N_{inc}}} \sqrt{\left(1 + \frac{1}{SNR(\tau)} \right)^2 + \left(\frac{1}{SNR(\tau)} \right)^2} \tag{21.5}$$

In order to obtain the standard deviation of the height estimation, first, it is required to estimate the delay error σ_d (position of the waveform's peak derivative) associated with the presence of noise.

In the ideal case of an infinite bandwidth receiver, the altimetric tracking point is the point corresponding to the maximum derivative of the power waveform [4, 35]. Under this hypothesis, the height precision σ_h can be expressed as follows:

$$\sigma_h = \frac{c\sigma_d}{2\sin\theta_e} = \frac{c\left\langle |Z_s(\tau)|^2 \right\rangle}{2\left\langle |Z_s(\tau)|^2 \right\rangle' \sin\theta_e} \frac{1}{\sqrt{N_{inc}}} \sqrt{\left(1 + \frac{1}{SNR(\tau)} \right)^2 + \left(\frac{1}{SNR(\tau)} \right)^2} \tag{21.6}$$

where *c* is the speed of light in the vacuum or air and $\left\langle |Z_s(\tau)|^2 \right\rangle'$ is the first derivative of the average signal power waveform. Note that, to achieve a high precision, a high *SNR* is required. The height precision is also inversely proportional to the slope of the power waveform at the tracking point. The waveform slope, in turn, increases with the signal bandwidth. Therefore, due to the wider bandwidth of the composite GPS L1 (C/A, P, M) signal, the ACF is much narrower than that of the C/A code only, hence guaranteeing improved altimetry precision.

This model has been previously employed in [4, 30, 33, 36] to analyze the expected achievable height precision using different techniques. In cGNSS-R, the received signal is cross-correlated with a local replica of the transmitted signal. In iGNSS-R, the reflected signal is cross-correlated with a measured direct signal. iGNSS-R allows exploiting the full power spectral density of the GNSS signals, thus improving the ranging precision with an acceptable degradation in signal amplitude.

All embedded codes in a given GNSS frequency band do contribute to the cross-correlation shape, including the high-chip rate restricted access codes.

In order to evaluate the ranging performance of the altimeter, it is fundamental to analyze the *SNR*. *SNR* is a critical parameter for the interferometric processing configuration, since it is affected by thermal noise at both up-looking and down-looking chains. This, in turn, has a major impact on the sizing of both up- and down-looking antennas. The *SNR* in the iGNSS-R case (SNR_i) is related to the *SNR* in the cGNSS-R case (SNR_c) as follows:

$$SNR_i = \frac{SNR_c}{1 + \frac{1+SNR_R}{SNR_D}} \tag{21.7}$$

where SNR_D and SNR_R are the *SNR*s of the direct and reflected channels of the iGNSS-R case. Due to the wider ACF and the lack of thermal noise in the clean replica signal, SNR_c is higher than SNR_i. Thus, in the iGNSS-R case, the incoherent integration time should be increased and higher antenna gains are required. On the other hand, increasing the incoherent averaging reduces the spatial resolution. The altimetric performance of iGNSS-R was proven to be at least twice as good as the cGNSS-R approach [12, 27]. More recently, the rGNSS-R technique [13] was developed to improve the altimetric performance using the reflectometer PYCARO instrument. The advantages of rGNSS-R rely mainly on the larger bandwidth of the P(Y) codes, as compared to the C/A ones, and the larger *SNR*, despite the losses of the semicodeless approach. The improvement factor was found to be around twice as good as the cGNSS-R approach, using a low-gain down-looking antenna [13].

21.4 Impact of GNSS codes on altimetric performance

There are a wide variety of GNSS codes with different properties [37]. The altimetric performance is affected by the influence of these properties on the *SNR* [38]. There is an impact of the platform's height on the peak of the DDMs through the Woodward ambiguity function (WAF). The different size of the scattering area as a function of the height is translated into different power levels of the reflected signals [39]. Depending on the platform's height, the Earth region contributing to the incoherent component is on the order of several iso-delay ellipses, which are a function of the ACF of the different GNSS codes. On the other hand, the area contributing to the coherent component is approximately limited by the first Fresnel zone [40], which actually depends on the signal wavelength. Therefore, in phase altimetry applications, the spatial resolution does not depend on the GNSS code.

21.5 Experimental field campaigns

21.5.1 Ground-based

The first GNSS-R proof-of-concept was carried out during a ground-based experiment in a controlled environment at the Zeeland Bridge, The Netherlands, in 1997 [2].

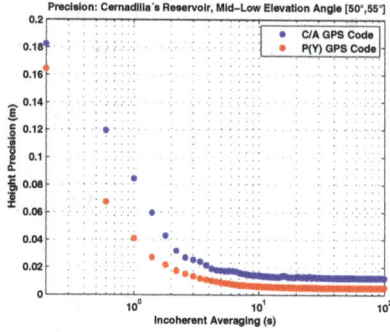

(a)

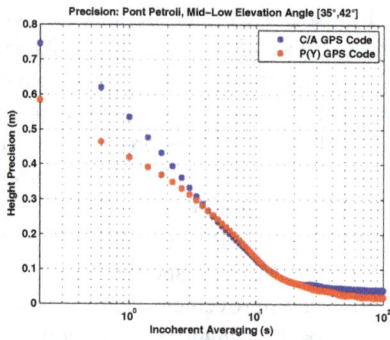

(b)

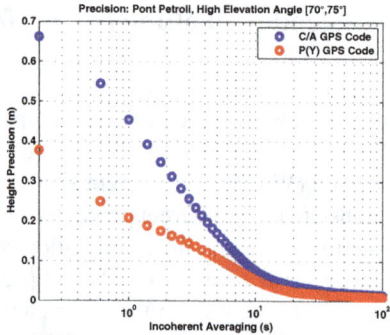

(c)

Figure 21.4   *C/A and P(Y)-derived height precision as a function of the incoherent averaging in case of (a) smooth surface at mid-low satellite's elevation angles (θ_e from 50° to 55°) and a mean platform's height over the surface ~65 m; (b) rough surface at mid-low satellite's elevation angles (θ_e from 35° to 42°) and a mean platform's height over the surface ~4.76 m; and (c) rough surface at high satellite's elevation angles (θ_e from 70° to 75°) and a mean platform's height over the surface ~4.76 m [13].*

cGNSS-R was used in this experiment. The key result of the experiment was the demonstration of a root mean squared (RMS) height precision within 5 s of 1% of the used code chip (3 m for C/A code). In 2001, a ground-based experiment, which also employed cGNSS-R, was performed at the Crater Lake and showed that the surface height was estimated with a 2 cm precision in 1 s [41]. On the other hand, iGNSS-R was first tested during an experiment over the Zealand Bridge [42], showing a height precision as good as 3 cm. More recently, the rGNSS-R [13, 43] was tested in a controlled scenario over the canal d'investigacio i experimentacio maritima (CIEM)/Universitat Politecnica de Catalunya (UPC)-BarcelonaTech wave channel [44] and also over the Mediterranean Sea to evaluate the achievable altimetric precision. The results using the P(Y) code, from this static and low-altitude scenario over the Mediterranean Sea at high satellite's elevation angles, showed a 2 cm height precision for 20 s of incoherent averaging [13] (Figure 21.4).

21.5.2 Air-borne

A wide variety of air-borne experimental field campaigns have also been carried out to develop and test GNSS-R techniques, including e.g. Platform Harvest [3], ocean salinity airborne campaign (COSMOS) [4], and Skyvan [26] flights (Figure 21.5). The air-borne experiment is quite important because of its larger footprint as compared to the ground-based scenario. A comparative analysis of the results obtained during the two campaigns on-board the Skyvan in the Gulf of Finland with Harvest and COSMOS experiments was performed in [26]. cGNSS-R altimetry with the P(Y) code was employed during the aircraft experiment performed in January 2001 over the Platform Harvest off the coast of California. In this experiment, the altimetric point position was extracted by fitting the reflected signal to a model. The mean delay precision obtained from two GPS satellites with elevation angles of 55° and 60° was RMS = 56 cm [T_c = 10 ms and N_{inc} = 200]. The trajectory consisted of 7 paths of 3–4.5 min each, so about 60 min of data were collected at a flight altitude of 3 000 m. It was observed that the scatter was stronger at lower aircraft speeds due to the lesser reduction of the speckle noise: RMS = 46 cm for $v_{ircraft}$ = 288 km/h and RMS = 64 cm for $v_{ircraft}$ = 188 km/h. After incoherent averaging over 7 km, the speckle was reduced and the achieved delay precision was RMS = 5.5 cm. Speckle, calculation, and antenna beam pattern model seemed to be the sources of the largest systematic errors. The proposed methods to improve the altimetric performance were to fly at higher aircraft altitudes to reduce the speckle,[c] more fully populated geometric parameter space, and more detailed wind-vector retrieval or using external wind-vector measurements. Data from a previous GPS open-loop differential real-time receiver (GOLD-RTR) flight experiment were used for a comparative study with the same receiver. These data were acquired during the COSMOS campaign

---

[c]The correlation time between consecutive waveforms depends on the square root of the flight altitude. Therefore, at higher altitudes, speckle is more correlated and averaging is not as effective in reducing the variance of the observations. However, at higher flight altitudes the number of scatterers in the glistening zone is larger than at low flight altitudes, and the signal fluctuation is less important.

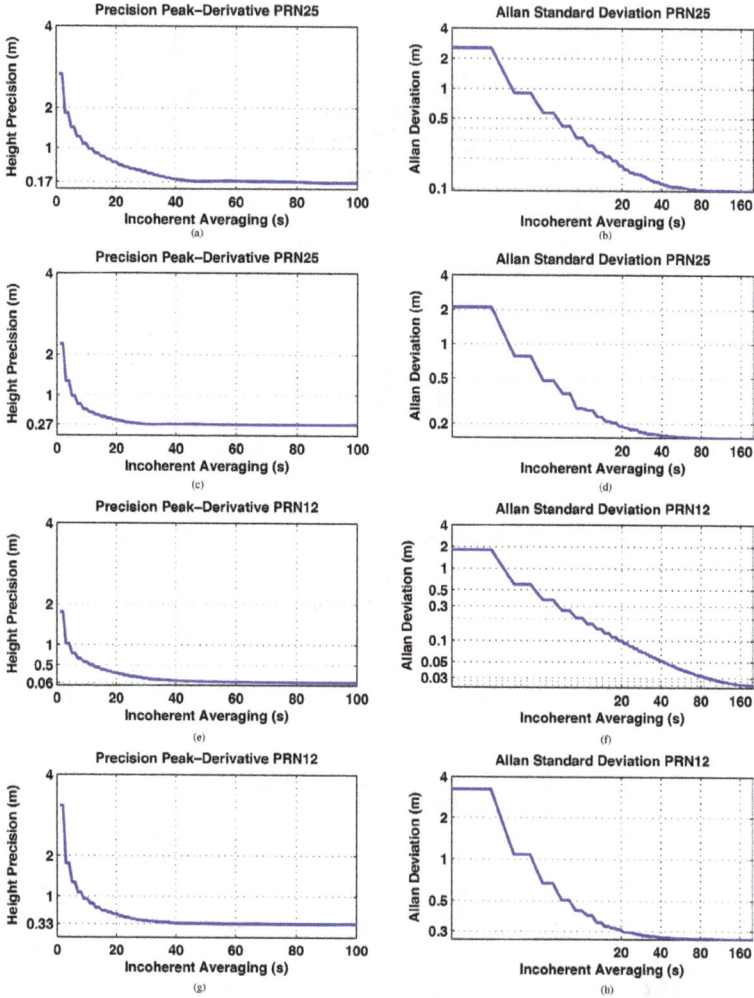

Figure 21.5   (a) Height precision: h =500 m, θ_e from 73° to 76°, GPS time from
16.4 to 16.6 h, $v_{ircraft}$ =302 km/h. (b) Allan's standard deviation:
h=500 m, θ_e from 73° to 76°, GPS time from 16.4 to 16.6 h,
$v_{ircraft}$=302 km/h. (c) Height precision: h=150 m, θ_e from 62° to 66°,
GPS time from 17.5 to 17.7 h, $v_{ircraft}$ =205 km/h. (d) Allan's standard
deviation: h=150 m, θ_e from 62° to 66°, GPS time from 17.5 to 17.7
h, $v_{ircraft}$ =205 km/h. (e) Height precision: h=3 000 m, θ_e from 66° to
85°, GPS time from 5.4 to 6.6 h, $v_{ircraft}$ =237 km/h.
(f) Allan's standard deviation: h=3 000 m, θ_e from 66° to 85°, GPS
time from 5.4 to 6.6 h, $v_{ircraft}$ =237 km/h. (g) Height precision:
h=3 000 m, θ_e from 70° to 77°, GPS time from 5.6 to 6.2 h, $v_{ircraft}$
=237 km/h. (h) Allan's standard deviation: h=3 000 m, θ_e from 70°
to 77°, GPS time from 5.6 to 6.2 h, $v_{ircraft}$ =237 km/h [26].

[4] performed on 15 April 2006. The flight was at 3 000 m altitude with a speed $v_{ircraft}$ = 270 km/h. Data were collected during 1900 s for altimetric purposes corresponding to three different satellites with elevation angle ranges of 70° to 77°, 45° to 55°, and 35° to 50°. In the first elevation angle range, the delay precision achieved was 2.51 m for T_c = 1 ms, N_{inc} = 1 000 and 6 cm for T_c = 1 ms, N_{inc} = 1 751 000. In order to compare with the previous results, a similar elevation angle range from 70° to 77° was selected in [26]. The expected results (Figure 21.5(g)) presented a delay standard deviation of 6 cm for T_c = 1 ms, N_{inc} = 1 550 000, which was consistent with the COSMOS results.

An additional parameter that is important in ocean altimetry is the achievable accuracy. In the Skyvan flights [12, 26], the feasibility of c/iGNSS-R to measure the topography of the sea surface was demonstrated. This was done by superposing the measured sea surface height over the aircraft's ground track with the geoid undulations, which were about 1 m (Figure 21.6).

One key point for accurate space-borne ocean altimetry is the ionospheric delay correction. In order to correct for the effects of the ionosphere, at least two GNSS frequencies are required. Nominally, these should be the GPS L1—Galileo E1 (1 575.42 MHz) and GPS L5—Galileo E5a (1 176.45 MHz) bands since these provide the greatest separation in frequency and, hence, the best accuracy for the ionospheric delay incurred [30].

21.6 Space-borne missions

21.6.1 PARIS IoD

Dr. Manuel Martin-Neira stated in 1993 [1]:

> As was recognized during the Consultative Meeting on Imaging Altimeter Requirements and Techniques held in June 1990 at Mullard Space Science Laboratory (UK), the ability to carry out high precision ocean altimetry over a swath with high spatial resolution would revolutionize many fields of Earth science: Some form of multi-beam altimetry would offer the possibility of achieving satisfactory sampling of the ocean mesoscale flows, and would, in addition, improve the ability to study other spatially and temporally variable oceanographic phenomena such as wave and wind fields, and ocean sea-ice interactions. The PARIS concept is directed toward such a multi-beam altimetry objective.

The PARIS IoD altimeter was proposed as a new type of passive instrument that combines bistatic radar and radiometer techniques, including interferometry (iGNSS-R). The PARIS concept has been developed through several steps, including comparisons with different techniques c/r/piGNSS-R. PARIS was the trigger of the GNSS-R community, which more recently has also used the Earth-reflected GNSS signals for different scientific applications such as wind speed, soil moisture content, biomass mapping, etc.

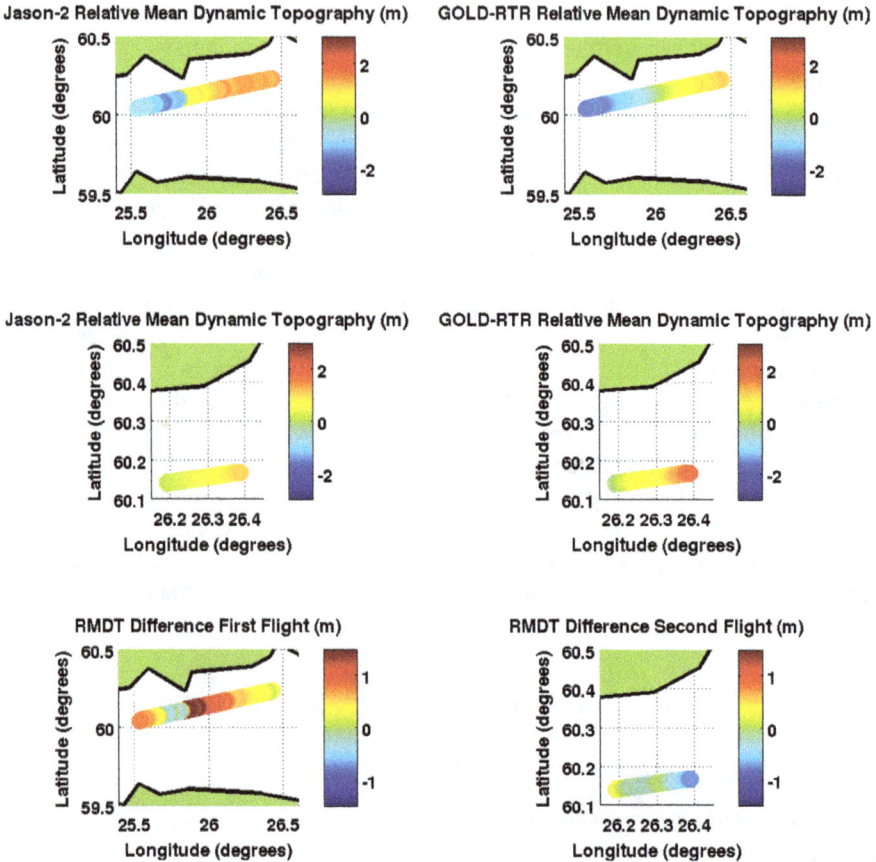

Figure 21.6   Comparison of ocean altimetry performance between conventional ocean altimeters (Jason-2) and GNSS-R using the GOLD-RTR instrument from an air-borne experiment [26].

Following PARIS IoD Phase A (Figure 21.7), several space-borne missions proposals were designed for testing iGNSS-R from space, including GNSS reflectometry, radio occultation, and scatterometry onboard the international space station (GEROS-ISS) Phase A [45], "Cookie" mission concept [46] (Figure 21.8), and the GNSS transpolar Earth reflectometry exploring system (G-TERN) proposal [47]. The structure of the satellite is an important factor to be accounted for in accurate altimetry [48]. Multiple scattering from the ISS structure has revealed that (1) the strength of the received signals is affected (important for scatterometry), (2) the waveform peak is shifted by ~0.06 C/A chips = 18 m (important for altimetry), (3) there is a region where signals are blocked by the ISS itself, and (4) there are other regions in the space where two or up to three "rays" may reach the antenna arrays.

An error budget was theoretically estimated for a space-borne instrument in [27]. This work evaluated the optimization and performance analysis of cGNSS-R

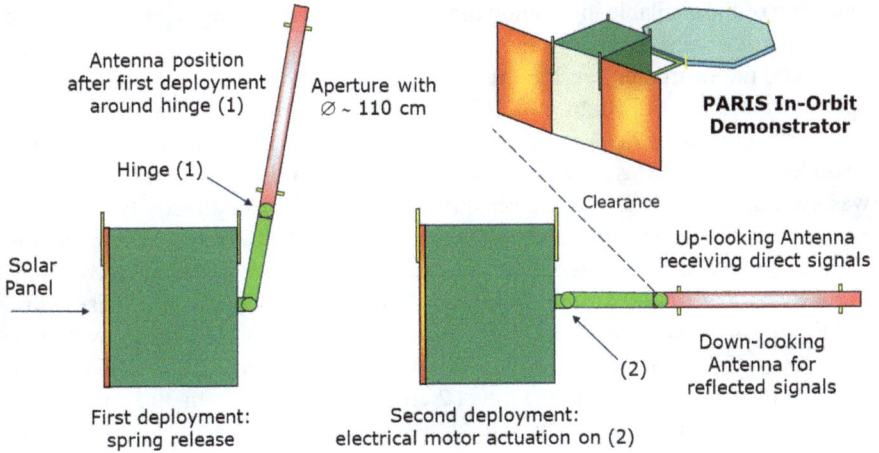

Figure 21.7   *Accommodation and deployment of the PARIS antenna on the small German technologie erprobungs träger (TET) platform. Adapted from [30].*

and iGNSS-R altimeters. First, the pros and cons of both configurations were studied with a special emphasis on the *SNR* degradation of the direct signal in the iGNSS-R case. This degradation leads to the need of very large and steerable antennas and the

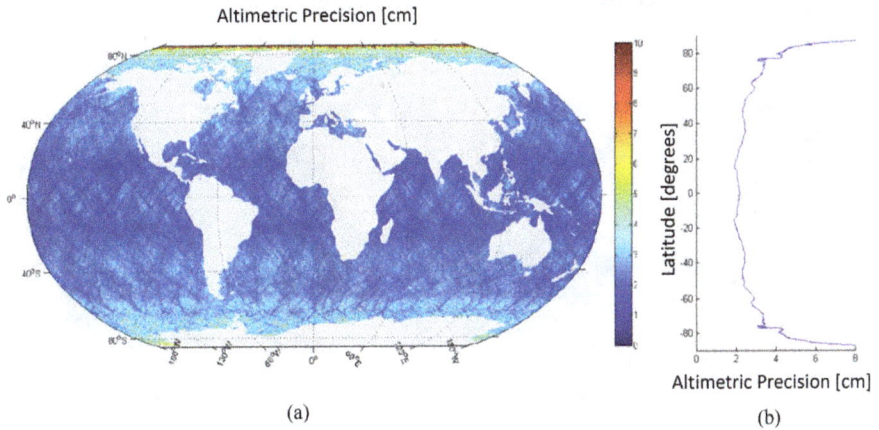

(a)   (b)

Figure 21.8   *(a) Global distribution of the precision in SSH measurements obtained by the three "Cookie" satellite constellation after accumulating near-Nadir and grazing GNSS reflection observations over a 10-day period and a spatial resolution of 20 km × 20 km. (b) Longitudinal average SSH precision as a function of latitude. Adapted from [46].*

reduction of the available integration time in the "slow swapping" technique, which is only partly compensated by adding a second channel.

Then, the instrument performance was revised for the ideal case and for the optimum receiver bandwidths [28]. For a dwell line of 100 km, the main results were $\sigma_{h,L1}$ = 13.9 cm, $\sigma_{h,L5}$ = 23.9 cm, and $\sigma_{h,E1}$ =10.3 cm, $\sigma_{h,E5}$ = 6.3 cm. Finally, instrument errors were classified as antenna, receiver, and correlator errors, and their impact was evaluated by computing the sensitivities versus antenna directivity or receivers' noise temperature. The impact of the instrument errors was assessed, including the error propagation and slight amplification due to the ionospheric corrections. For a dwell line of 100 km, maximum antenna directivity $\leq$20.3 dB, noise figure = 2.92 dB, and the typical transmitted powers, the final Level 1 errors were $\sigma_{h,L1}$ from 16.4 to 37.2 cm, and $\sigma_{h,L5}$ from 29.7 to 56 cm, $\sigma_{h,E1}$ from 12.8 to 26.6 cm, and $\sigma_{h,E5}$ from 8.3 to 15.5 cm, and the Level 2 error were $\sigma_{h,L1\&L5}$ from 30.5 to 60.5 cm, and $\sigma_{h,E1\&E5}$ from 13.7 to 27.7 cm. These results make them suitable for most mesoscale altimetry applications. Better results can be obtained with larger directivity antennas and less noisy receivers.

21.6.2   *Space-borne Imaging Radar-C*

The first space-borne GNSS-R evidence was found in segments of space-borne imaging radar-C (SIR-C) data acquired when radar returns were not present [31]. The main conclusion, which was made concerning altimetry from a ~20 dB generic space-based platform using cGNSS-R, was that the C/A code *SNR* should be expected to be about ~5–7 times of the Y code (known) *SNR* because of the C/A code's greater transmission power and a factor of 10 times greater reflection area, which is enough to compensate for the decreased coherence time.

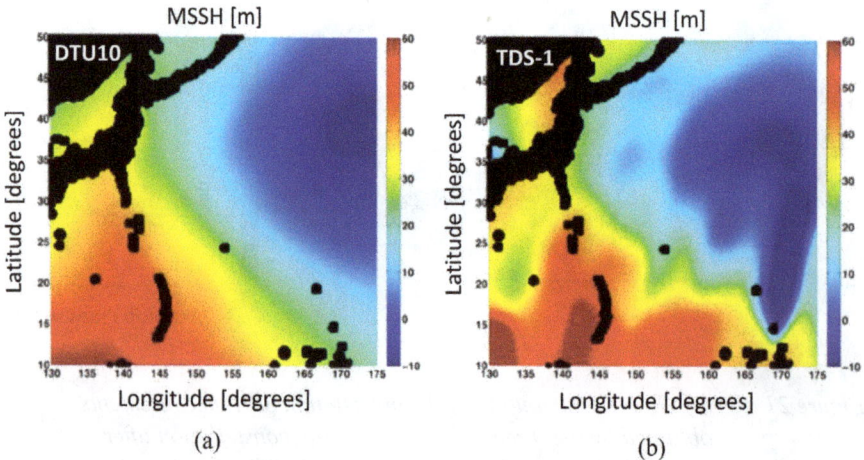

Figure 21.9   (a) Technical University of Denmark (DTU) 10 mean sea surface height (MSSH) over the North Pacific and (b) UK TDS-1-derived MSSH. Adapted from [14].

21.6.3   UK TDS-1

More recently, the ability of space-borne cGNSS-R to provide SSH information was demonstrated using data from the UK techdemosat-1 (TDS-1) satellite [14]. Figure 21.9 shows the comparison of SSH maps for the North Pacific region. An overall agreement between measured and DTU 10 heights was found. One important limitation of this analysis was the sparse space and time sampling of the available data as a consequence of the GNSS-R payload being switched on for only 2 out of every 8 days. However, with the launch of cyclone global navigation satellite system (CYGNSS) in 2016, the space–time coverage became quite dense over ±40° latitude, allowing for an assessment of the SSH retrieval capabilities of space-borne GNSS-R over a wider area.

21.6.4   CYGNSS

The national aeronautics and space administration (NASA) CYGNSS mission (down-looking antenna gain ~14.5 dB, Left Hand Circular Polarization) was proposed by Prof. Christopher Ruf [49, 50] for ocean wind speed estimation over tropical cyclones (TCs) using GPS L1 C/A signals of opportunity (Figure 21.10). The orbital configuration of each CYGNSS satellite is a circular low Earth orbit (LEO) with a height ~520 km and an inclination angle of ~35°. Each delay Doppler mapping instrument (DDMI) aboard the 8 CYGNSS microsatellites collects forward

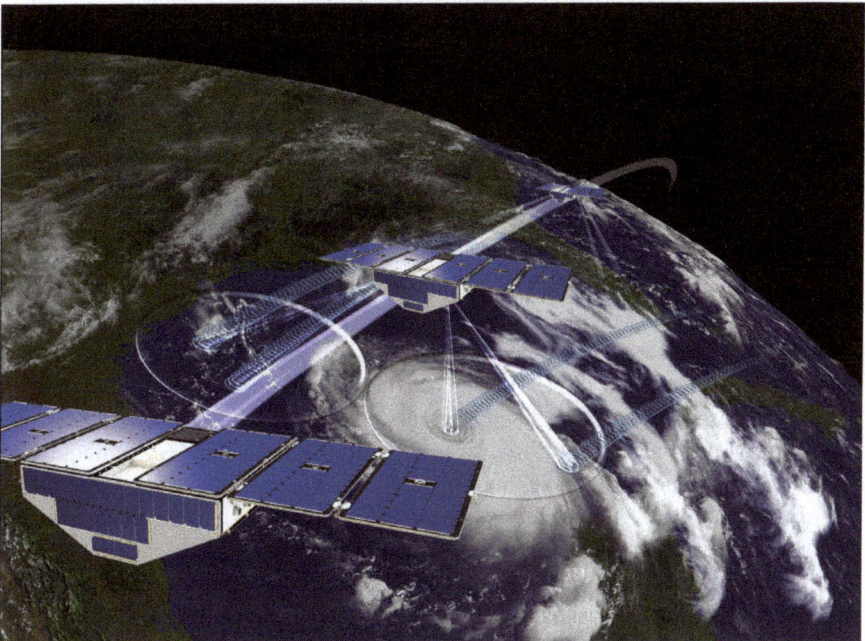

Figure 21.10   Artist's view of the NASA's CYGNSS constellation coordinated by the University of Michigan (UM). Image credits NASA.

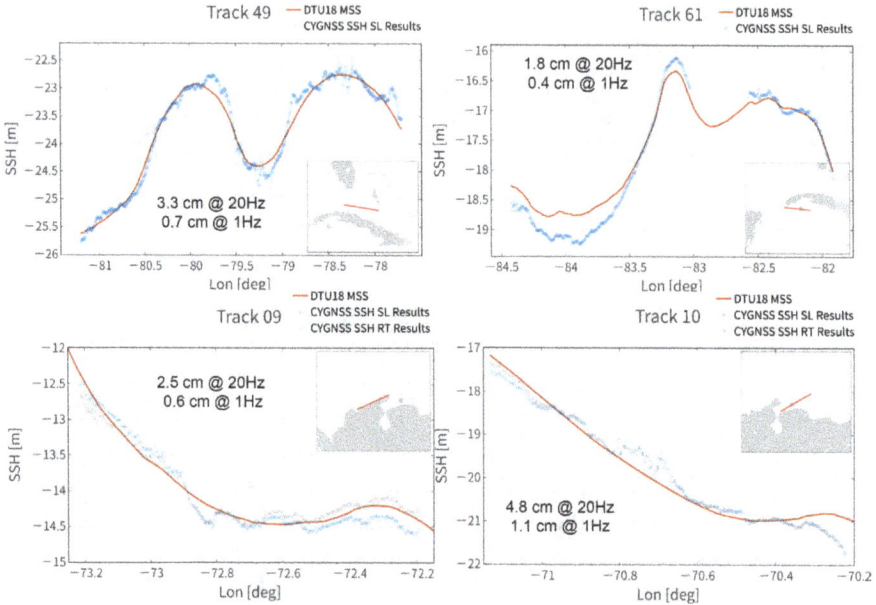

Figure 21.11   *CYGNSS-derived precise phase ocean altimetry results.*
Measurement precision: Track 49 (3.3 cm @ 20 Hz, 0.7 cm @ 1
Hz), track 61 (1.8 cm @ 20 Hz, 0.4 cm @ 1 Hz), track 09 (2.5 cm
@ 20 Hz, 0.6 cm @ 1 Hz), and track 10 (4.8 cm @ 20 Hz, 1.1 cm
@ 1 Hz). Adapted from [10].

scattered signals along four specular directions corresponding to four different transmitting GPS spacecrafts, simultaneously. As such, CYGNSS allows one to sample
the Earth's surface along 32 tracks simultaneously, within a wide range of satellites'
elevation angles $\theta_e \sim [20, 90]°$ over tropical latitudes $\sim[-40, 40]°$.

CYGNSS data have also been explored for ocean altimetry using both cGNSS-R
and phase altimetry (Figure 21.11). The first precise phase altimetry study with
CYGNSS [10] presented phase altimetry retrievals up to $\theta_e \sim 25°$, resulting in a
measurement system precision of $\sim$3 cm at 20 Hz sampling, a centimeter-level with
1 s integration, despite the fact that the signals were received away from the antenna
main beam. The combined precision was between 9 and 39 cm (16 cm median, 20
cm mean) in 50 ms integration time, mostly driven by residual systematic errors
due to simple or missing corrections. This precision level could scale down to a few
centimeter-level after 1 s integration, with typically 7–8 km along-track spatial resolution. The measurement system component of the precision was similar to those
obtained with dedicated monostatic radar altimeters. However, further investigation
is required to test the potential correction of the different error contributions, especially because grazing carrier phase altimetry tracks do not repeat, a fact upon which
some of the radar altimeter corrections rely. This could degrade the final accuracy

achievable by the grazing carrier phase altimetry technique, despite its high precision of measurement system.

21.7 Conclusions

In this chapter, the fundamentals of GNSS-R ocean altimetry have been presented, including key milestones, theoretical elements, a wide variety of examples, and relevant results. The impact of the platform's height on the achievable height precision has been analyzed from ground-based, air-borne, and space-borne platforms.

A historical overview of the development of the GNSS-R concept has been presented, including key experiments, theoretical investigations, and several proposals for space-borne missions. Future research activities should include the use of iGNSS-R and/or rGNSS-R to improve the achievable height precision. The main challenge in iGNSS-R is the large size of the antennas to provide the required directivity for both the up- and the down-looking antennas. This is the reason why PARIS IoD was designed with a deployment mechanism to accommodate the large antenna to the platform, while CYGNSS uses two low-directivity antennas, because the main goal was to further advance extreme weather predictions with a focus on TCs inner core process studies. The larger antenna imposes a significant increment in the cost of the mission. Additionally, iGNSS-R instruments require a dedicated calibration strategy. rGNSS-R is an alternative technique for accurate ocean altimetry, showing an improvement factor ~2 as compared to cGNSS-R.

Acknowledgement

The author would like to thank the following researchers for providing their figures: Dr. Salvatore D'Addio (Figure 21.3), Dr. Manuel Martin-Neira (Figure 21.7), Dr. Christopher Ruf (Figure 21.10), and Dr. Estel Cardellach (Figure 21.11).

References

[1] Martin-Neira M. 'A passive reflectometry and interferometry system (PARIS): application to ocean altimetry'. *ESA Journal*. 1993;**17**(9):331–55.

[2] Martin-Neira M., Caparrini M., Font-Rossello J., Lannelongue S., Vallmitjana C.S. 'The PARIS concept: an experimental demonstration of sea surface altimetry using GPS reflected signals'. *IEEE Transactions on Geoscience and Remote Sensing*. 2001;**39**(1):142–50.

[3] Lowe S.T., Zuffada C., Chao Y., Kroger P., Young L.E., LaBrecque J.L. '5-cm-precision aircraft ocean altimetry using GPS reflections'. *Geophysical Research Letters*. 2002;**29**(10):13-1–13-14.

[4] Rius A., Cardellach E., Martin-Neira M. 'Altimetric analysis of the sea-surface GPS-reflected signals'. *IEEE Transactions on Geoscience and Remote Sensing*. 2010;**48**(4):2119–27.

[5]   D'Addio S., Martin-Neira M., di Bisceglie M., Galdi C., Martin Alemany F. 'GNSS-R altimeter based on Doppler multi-looking'. *IEEE Journal of Selected Topics in Applied Earth Observations and Remote Sensing.* 2014;**7**(5):1452–60.

[6]   Semmling A.M., Schmidt T., Wickert J., *et al.* 'On the retrieval of the specular reflection in GNSS carrier observations for ocean altimetry'. *Radio Science.* 2012;**47**(6):RS6007.

[7]   Semmling A.M., Beckheinrich J., Wickert J., *et al.* 'Sea surface topography retrieved from GNSS reflectometry phase data of the GEOHALO flight mission'. *Geophysical Research Letters.* 2014;**41**(3):954–60.

[8]   Semmling A.M., Wickert J., Schon S., *et al.* 'A zeppelin experiment to study airborne altimetry using specular global navigation satellite system reflections'. *Radio Science.* 2013;**48**(4):427–40.

[9]   Cardellach E., Ao C.O., de la Torre Juarez M., Hajj G.A. 'Carrier phase delay altimetry with GPS-reflection/occultation interferometry from low Earth orbiters'. *Geophysical Research Letters.* 2004;**31**(10):L10402

[10] Cardellach E., Li W., Rius A., *et al.* 'First precise spaceborne sea surface altimetry with GNSS reflected signals'. *IEEE Transactions on Geoscience and Remote Sensing.* 2020;**13**:102–12.

[11] Li W., Cardellach E., Fabra F., Rius A., Ribo S., Martin-Neira M. 'First spaceborne phase altimetry over sea ice using TechDemoSat-1 GNSS-R signals'. *Geophysical Research Letters.* 2017;**44**(16):8369–76.

[12] Cardellach E., Rius A., Martin-Neira M., *et al.* 'Consolidating the precision of interferometric GNSS-R ocean altimetry using airborne experimental data'. *IEEE Transactions on Geoscience and Remote Sensing.* 2013;**52**(8):4992–5004.

[13] Carreno-Luengo H., Camps A., Ramos-Perez I., Rius A. 'Experimental evaluation of GNSS-reflectometry altimetric precision using the P(Y) and C/A signals'. *IEEE Journal of Selected Topics in Applied Earth Observations and Remote Sensing.* 2014;**7**(5):1493–500.

[14] Clarizia M.P., Ruf C., Cipollini P., Zuffada C. 'First spaceborne observation of sea surface height using GPS-reflectometry'. *Geophysical Research Letters.* 2016;**43**(2):767–74.

[15] Li W., Cardellach E., Fabra F., Ribo S., Rius A. 'Assessment of spaceborne GNSS-R ocean altimetry performance using CYGNSS mission raw data'. *IEEE Transactions on Geoscience and Remote Sensing.* 2020;**58**(1):238–50.

[16] Mashburn J., Axelrad P., Zuffada C., Loria E., O'Brien A., Haines B. 'Improved GNSS-R ocean surface altimetry with CYGNSS in the seas of Indonesia'. *IEEE Transactions on Geoscience and Remote Sensing.* 2020;**58**(9):6071–87.

[17] Ghavidel A., Camps A. 'Time-domain statistics of the electromagnetic bias in GNSS-Reflectometry'. *Remote Sensing.* 2015;**7**(9):11151–62.

[18] Ghavidel A., Schiavulli D., Camps A. 'Numerical computation of the electromagnetic bias in GNSS-R altimetry'. *IEEE Transactions on Geoscience and Remote Sensing.* 2016;**54**(1):489–98.

[19] Ghavidel A., Camps A. 'Impact of rain, swell, and surface currents on the electromagnetic bias in GNSS-reflectometry'. *IEEE Journal of Selected Topics in Applied Earth Observations and Remote Sensing.* 2016;**9**(10):4643–49.

[20] Martin F., Camps A., Park H., D'Addio S., Martin-Neira M., Pascual D. 'Cross-correlation waveform analysis for conventional and interferometric GNSS-R approaches'. *IEEE Journal of Selected Topics in Applied Earth Observations and Remote Sensing.* 2014;**7**(5):1560–72.

[21] Auber J.C., Bibaut A., Rigal J.M. 'Characterization of multipath on land and sea at GPS frequencies'. Proceedings of the 7th International Technical Meeting of the Satellite Division of the Institute of Navigation; Salt Lake City, UT, USA, September; 1994. pp. 1155–71.

[22] Garrison J.L., Katzberg S.J. 'Detection of ocean reflected GPS signals: theory and experiment'. Engineering New Century. Proceedings of the 1997 IEEE Southeastcon'97; Blacksburg, VA, USA, April; 1997. pp. 290–4.

[23] Zavorotny V.U., Voronovich A.G. 'Scattering of GPS signals from the ocean with wind remote sensing application'. *IEEE Transactions on Geoscience and Remote Sensing.* 2000;**38**(2):951–64.

[24] Garrison J.L., Komjathy A., Zavorotny V.U., Katzberg S.J. 'Wind speed measurement using forward scattered GPS signals'. *IEEE Transactions on Geoscience and Remote Sensing.* 2002;**40**(1):50–65.

[25] Rius A., Fabra F., Ribo S., *et al.* 'PARIS interferometric technique proof of concept: sea surface altimetry measurements'. Proceedings of the 2012 IEEE International Geoscience and Remote Sensing Symposium; Munich, Germany, July; 2012. pp. 7067–70.

[26] Carreno-Luengo H., Park H., Camps A., Fabra F., Rius A. 'GNSS-R derived centimetric sea topography: an airborne experiment demonstration'. *IEEE Journal of Selected Topics in Applied Earth Observations and Remote Sensing.* 2013;**6**(3):1468–78.

[27] Camps A., Park H., Domènech E.Vi., *et al.* 'Optimization and performance analysis of interferometric GNSS-R altimeters: application to the Paris IoD mission'. *IEEE journal of selected topics in applied earth observations and remote sensing.* 2014;**7**(5):1436–51.

[28] Pascual D., Camps A., Martin F., Park H., Arroyo A.A., Onrubia R. 'Precision bounds in GNSS-R ocean altimetry'. *IEEE Journal of Selected Topics in Applied Earth Observations and Remote Sensing.* 2014;**7**(5):1416–23.

[29] Martin F., D'Addio S., Camps A., Martin-Neira M. 'Modeling and analysis of GNSS-R waveforms sample-to-sample correlation'. *IEEE Journal of Selected Topics in Applied Earth Observations and Remote Sensing.* 2014;**7**(5):1545–59.

[30] Martin-Neira M., D'Addio S., Buck C., Floury N., Prieto-Cerdeira R. 'The PARIS ocean altimeter in-orbit demonstrator'. *IEEE Transactions on Geoscience and Remote Sensing.* 2011;**49**(6):2209–37.

[31] Lowe S.T., LaBrecque J.L., Zuffada C., Romans L.J., Young L.E., Hajj G.A. 'First spaceborne observation of an Earth-reflected GPS signal'. *Radio Science.* 2002;**37**(1):7-1–7-28.

[32] Germain O., Ruffini G. 'A revisit to the GNSS-R code range precision'. Proceedings of the 2006 GNSS-R Workshop; Noordwijk, The Netherlands, June; 2006.

[33] D'Addio S., Buck C., Martin-Neira M. 'PARIS altimetry precision prediction with GALILEO-signals-in-space'. *Proceedings of the 2008 IEEE International Geoscience and Remote Sensing Symposium; Boston, MA, USA, July; 2008. pp. III-63–6.*

[34] Ulaby F.T., Moore R.K., Fung A.K. *Microwave Remote Sensing: Active and Passive, Vol. 2: Radar Remote Sensing and Surface Scattering and Emission Theory*. Reading, MA, USA: Addison-Wesley; 1982.

[35] Chelton D., Walsh E., MacArthur J. 'Pulse compression and sea level tracking in satellite altimetry'. *Journal Atmospheric and Oceanic Technology*. 1998;**6**(3):406–38.

[36] Gleason S., Gommenginger C., Cromwell D. 'Fading statistics and sensing accuracy of ocean scattered GNSS and altimetry signals'. *Advances in Space Research*. 2010;**46**(2):208–20.

[37] Zavorotny V.U., Gleason S., Cardellach E., Camps A. 'Tutorial on remote sensing using GNSS bistatic radar of opportunity'. *IEEE Geoscience and Remote Sensing Magazine*. 2014;**2**(4):8–45.

[38] Carreno-Luengo H., Camps A. 'First dual-band multiconstellation GNSS-R scatterometry experiment over boreal forests from a stratospheric balloon'. *IEEE Journal of Selected Topics in Applied Earth Observations and Remote Sensing*. 2015;**9**(10):4743–51.

[39] D'Addio S., Martin-Neira M. 'Comparison of processing techniques for remote sensing of Earth-exploiting reflected radio-navigation signals'. *Electronics Letters*. 2013;**49**(4):292–3.

[40] Carreno-Luengo H., Ruf C., Warnock A., Brunner K. 'Investigating the impact of coherent and incoherent scattering terms in GNSS-R delay Doppler maps'. Proceedings of the 2020 IEEE International Geoscience and Remote Sensing Symposium; Hawaii, USA, July; 2020. pp. 6202–5.

[41] Treuhaft R.N., Lowe S.T., Zuffada C., Chao Y. '2-cm GPS altimetry over Crater Lake'. *Geophysical Research Letters*. 2001;**28**(23):4343–46.

[42] Nogues-Correig O., Ribo S., Arco J.C., *et al.* 'The proof of concept for 3-cm altimetry using the PARIS interferometric technique'. Proceedings of the 2010 IEEE International Geoscience and Remote Sensing Symposium; Honolulu, Hawaii, USA, July; 2010. pp. 3620–3.

[43] Carreno-Luengo H., Camps A., Via P., *et al.* '3Cat-2—an experimental nanosatellite for GNSS-R Earth observation: mission concept and analysis'. *IEEE Journal of Selected Topics in Applied Earth Observations and Remote Sensing*. 2016;**9**(10):4540–51.

[44] Carreno-Luengo H., Camps A. 'Empirical results of a surface-level GNSS-R experiment in a wave channel'. *Remote Sensing*. 2015;**7**(6):7471–93.

[45] Wickert J., Cardellach E., Martin-Neira N., *et al.* 'GEROS-ISS: GNSS reflectometry, radio occultation, and scatterometry onboard the international space

station'. *IEEE Journal of Selected Topics in Applied Earth Observations and Remote Sensing.* 2016;**9**(10):4552–81.

[46] Martiın-Neira M., Li W., Andres-Beivide A., Ballesteros-Sels X. 'Cookie: a satellite concept for GNSS remote sensing constellations'. *IEEE Transactions on Geoscience and Remote Sensing.* 2016;**9**(10):4593–610.

[47] Cardellach E., Flato G., Fragner H., *et al.* 'GNSS transpolar Earth reflectometry exploring system (G-TERN): mission concept'. *IEEE Access.* 2018;**6**:13980–14018.

[48] Camps A., Park H., Sekulic I., Rius J. 'GNSS-R altimetry performance analysis for the GEROS experiment on board the international space station'. *Sensors.* 2017;**17**(7):1583.

[49] Ruf C., Unwin M., Dickinson J., *et al.* 'CYGNSS: enabling the future of hurricane prediction [remote sensing satellites]'. *IEEE Geoscience and Remote Sensing Magazine.* 2013;**1**(2):52–67.

[50] Ruf C. *CYGNSS Handbook: cyclone global navigation satellite system* [online]. 2021. Available from https://cygnss.engin.umich.edu/wp-content/uploads/sites/534/2021/06/CYGNSS_Handbook_April2016.pdf [Accessed 02 Feb 2021].

Chapter 22

Sea ice sensing using the GNSS-R technique

Qingyun Yan[1], Weimin Huang[2], and Shuanggen Jin[3]

Sea ice condition plays a significant role in offshore oil and gas development, marine transportation, and climate change. Thus, it is important to collect such information. During the past two decades, sea ice sensing using Global Navigation Satellite System-Reflectometry (GNSS-R) has attracted tremendous interest. In this chapter, an overview of the state-of-the-art methods for sea ice remote sensing with the GNSS-R technique is presented. More specifically, recent progress in sea ice sensing including sea ice detection, sea ice concentration (SIC) estimation, sea ice type classification, sea ice thickness (SIT) retrieval, and sea ice altimetry from GNSS-R data is summarized. The fundamentals of these applications and corresponding performance are described.

22.1 Background and overview

Due to global climate change [1], shrinking of ice cover and thinning of sea ice have been reported during the last decades [2–4]. Sea ice has a huge impact on keeping high surface albedo, limiting air-sea interaction [5], and modulating the distribution of freshwater and seawater [6]. Moreover, managing and securing marine operations, such as offshore oil and gas development and global shipping, rely on the information about sea ice conditions [7]. Thus, knowledge of sea ice conditions is critical. However, in situ sea ice observation is cumbersome and expensive with limited coverage. On the contrary, remote sensing techniques provide a more efficient and cost-effective method for sea ice monitoring.

Sea ice observed from space has been widely investigated with intensive remote sensing data [8]. Specifically, the sea ice extent, drift, growth stage, concentration, and thickness can be estimated using different sensors, for example, passive microwave [9, 10], scatterometer [11, 12], radar altimeter [4, 13], and synthetic aperture radar (SAR) [14, 15]. However, passive microwave and scatterometer products are typically

[1]School of Remote Sensing and Geomatics Engineering, Nanjing University of Information Science and Technology, China
[2]Faculty of Engineering and Applied Science, Memorial University, Canada
[3]Shanghai Astronomical Observatory, Chinese Academy of Sciences, China

characterized by coarse resolutions (25–50 km). Although SAR and radar altimeter can collect data with finer resolutions, their costs are generally high. In addition, interpretation of SAR images is usually time-consuming and subjective [15], and empirical retracking using altimeter lacks an appropriate physical model [16].

Since the concept of GNSS-R was proposed in 1988 [17], various remote sensing tasks have been accomplished with the assistance of the GNSS-R technique. Currently, GNSS-R has been successfully utilized to monitor winds and surface roughness over the ocean [18–22]. In addition, snow depth estimation [23–26], soil moisture retrieval [27–30], and tsunami detection [31, 32] have also been explored. GNSS-R employs a forward scattering configuration, in which the transmitter (Tx) and the receiver (Rx) are not collocated. A Tx can be any GNSS satellite such as the global positioning system (GPS), global navigation satellite system (GLONASS), Galileo, and Beidou/COMPASS [33–35]. The transmitted signals are scattered by the Earth's surface, for example, ocean, land, and ice carry information about the surface conditions and they could be captured by one or more GNSS-R Rxs. Among existing GNSSs, GPS is the most widely used and it transmits L-band signals with a wavelength of 19 cm. Due to the continuous and widely covered GNSS signals, a passive GNSS-R Rx is inexpensive with low mass and low power and it is easy and flexible for deployment. More detailed introduction of the GNSS-R technique can be found in Camps's chapter [36]. In general, GNSS-R can be divided into three categories according to the platform of Rxs, that is, spaceborne, airborne, and ground-based. The first type is able to provide large-scale or global surveillance while the latter two are usually for regional and local observations.

The first demonstration of sea ice remote sensing using GNSS-R was reported by Komjathy *et al.* [37] in which an airborne GPS Rx was used. The sensitivity of the reflected GPS signal to the presence of sea ice was observed since the received signal consistently displayed a sharp and narrow waveform shape when the flight was over the ice surface. Recently, some ground-based and airborne tests were also conducted [38–42]. The first spaceborne GNSS-R measurement was accomplished by the UK Disaster Monitoring Constellation (UK-DMC) satellite in 2004 and the capacity of sea ice remote sensing was illustrated by Gleason *et al.* [43–45]. Although only two sets of data were acquired (16-s long in total), the analyses and investigation are valuable. These two measurements were conducted at different locations with varying SICs, which is defined as the ratio of sea ice over seawater within a certain region. The data collected from the area with higher SIC showed stronger peak and less spreading in the delay-Doppler map (DDM), which indicated the sensitivity of GNSS-R to SIC. It was also found that a rough seawater surface could decrease the overall coherent reflection and lead to a larger glistening zone, which results in a DDM with more spreading along the delay and Doppler axes. A comparison of DDMs from regions of varying SICs is shown in Figure 22.1. Gleason also mentioned [44] that for a coherent signal off the ice surface, the carrier phase information can be recovered for accurate altimetry measurement. With the launch of the UK TechDemoSat-1 in 2014, the application of GNSS-R in sea ice sensing has been rapidly augmented [46–50].

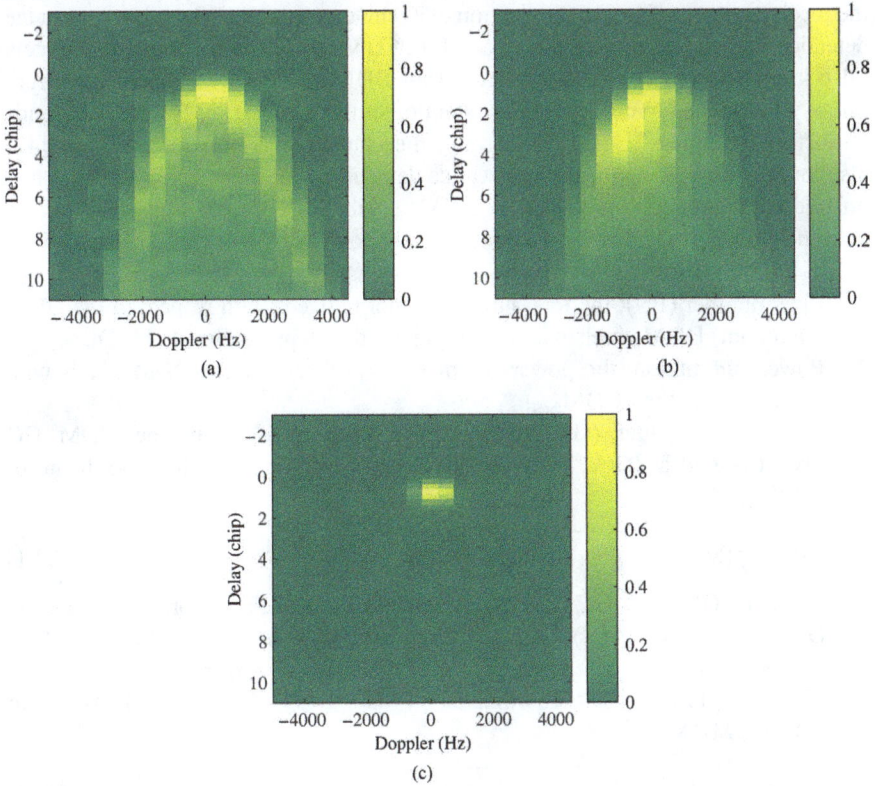

Figure 22.1 *TDS-1 delay-Doppler maps (DDMs) collected over regions with varying sea ice concentrations (SICs): (a) 0%, (b) 50%, and (c) 92%, respectively. Each DDM has been normalized by its peak power value.*

In this chapter, most of which is adapted from the open access article by Yan and Huang [51], methods and remaining challenges of sea ice sensing are introduced according to the specific applications.

22.2 Sea ice detection

The methods for sea ice detection can be classified into four categories: DDM observable-based, scattering coefficient retrieval-based, machine learning based, and empirical model-based.

22.2.1 DDM observable-based method

Although there is no exact theoretical model describing the received GNSS-R signals from sea ice, it has been empirically shown that the coherent component dominates the received power. The first application of TechDemoSat-1 (TDS-1) data for sea ice detection was shown by Yan and Huang [46, 52] via evaluation of some parameters (also known as observables) extracted from DDMs. The DDM observables were initially introduced to determine the extent of spreading for DDMs, which depends on surface roughness [19, 20, 53]. Around the same time, Alonso-Arroyo *et al.* [47] also investigated the application of sea ice detection with different observables and pointed out that the outline of sea ice DDMs exhibits a K-shape.

The typical observables used include the follows:

1. Pixel number (PN): the total number of normalized (with respect to the DDM maximum) DDM pixels with power greater than a preset threshold (DDM_{thres}).
2. Power summation: the power summation I_0 of normalized DDM pixels with power greater than DDM_{thres}.
3. Geometrical center (GC) distance: the distance d_1 from the DDM GC pixel (located at (GC_τ, GC_f)) to the peak power pixel (with a coordinate of (MAX_τ, MAX_f))

$$d_1 = \sqrt{(MAX_\tau - GC_\tau)^2 + (MAX_f - GC_f)^2}, \tag{22.1}$$

   where GC_τ and GC_f are the mean values of delay coordinates (τ) and Doppler coordinates (f) for the DDM pixels with power greater than DDM_{thres}.
4. Center-of-mass (CM) distance: the distance d_2 from the DDM CM pixel (whose coordinate is (CM_τ, CM_f)) to (MAX_τ, MAX_f)

$$d_2 = \sqrt{(MAX_\tau - CM_\tau)^2 + (MAX_f - CM_f)^2}, \tag{22.2}$$

   where CM_τ and CM_f are defined as

$$CM_\tau = I_0^{-1} \sum\sum_{DDM(\tau,f)>DDM_{thres}} \tau \cdot DDM(\tau,f), \tag{22.3}$$

$$CM_f = I_0^{-1} \sum\sum_{DDM(\tau,f)>DDM_{thres}} f \cdot DDM(\tau,f). \tag{22.4}$$

5. CM taxicab distance: the taxicab distance d_3 from (CM_τ, CM_f) to (MAX_τ, MAX_f)

$$d_3 = |MAX_\tau - CM_\tau| + |MAX_f - CM_f|. \tag{22.5}$$

6. Delay-Doppler map average (DDMA): the average value of the normalized DDM around its peak.
7. Trailing edge slope (TES): the slope of Doppler integrated waveform (DIW, summation of all delay waveforms at each Doppler bin) between its maximum and the value at, for example, three bins after the peak.
8. Matched filer (MF): the correlation coefficient between the obtained DIW and a Doppler cut of the so-called Woodward ambiguity function (WAF).

The shapes of DDMs for seawater and sea ice are different. More specifically, DDMs of seawater exhibit more spreading, showing a horseshoe shape. The

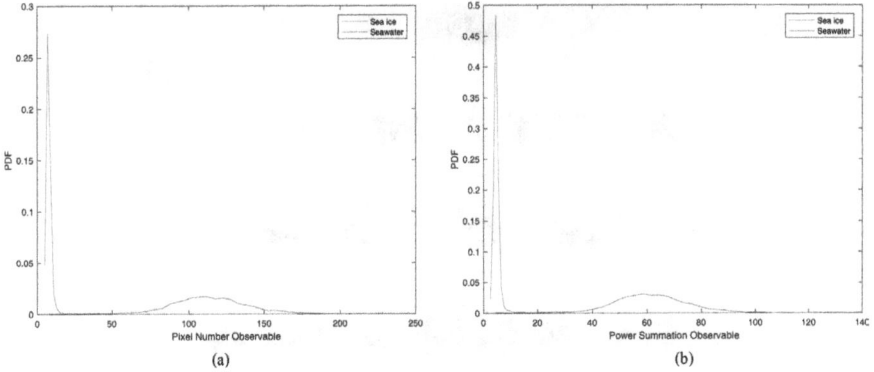

Figure 22.2   Sample probability density functions (PDFs) of observables: (a) pixel number and (b) power summation

above-mentioned observables are capable of reflecting the extension of a DDM. For the first six observables, a larger value indicates a higher degree of spreading. Two examples of sample probability density functions (PDFs) of Observables 1 and 2 for sea ice and seawater are presented in Figure 22.2. Thus, sea ice detection can be implemented based on the value of a DDM observable and corresponding threshold. The flowchart for sea ice detection using these observables is displayed in Figure 22.3. Although for the last two observables, a smaller value indicates more spreading, the associated sea ice detection procedure is quite similar. Observables 1–5 were used in the study by Yan and Huang [46], and observables 6–8 were employed by Alonso-Arroyo *et al.* [47]. There are other applicable observables for sea ice detection [54–57].

22.2.2   Scattering coefficient retrieval-based method

Schiavulli *et al.* and Yan and Huang [48, 58] presented a different scheme for water/ ice transition observation and sea ice detection based on retrieving scattering coefficient (σ^0) in the spatial domain from TDS-1 DDMs.

As illustrated in an earlier chapter, a DDM depicts the power distribution of the scattered GNSS signal in the delay-Doppler domain and it can be matched to surface σ^0 distribution in the spatial domain via the Zavorotny-Voronovich (Z-V) model [59], which can be expressed as a 2-D convolution as shown by Marchan-Hernandez *et al.* [60]

$$\langle |Y(\tau,f)|^2 \rangle = \chi^2(\tau,f) * \Sigma(\tau,f), \tag{22.6}$$

where $\chi^2(\tau,f) \cong \Lambda^2(\tau)S^2(f)$ is the WAF, S is the sinc function and Λ is the triangle function [59], and

$$\Sigma(\tau,f) = T_i^2 \iint\limits_A \frac{D^2(\vec{\rho})\sigma^0(\vec{\rho})\delta(\tau)\delta(f)}{4\pi R_R^2(\vec{\rho})R_T^2(\vec{\rho})} d^2\vec{\rho}, \tag{22.7}$$

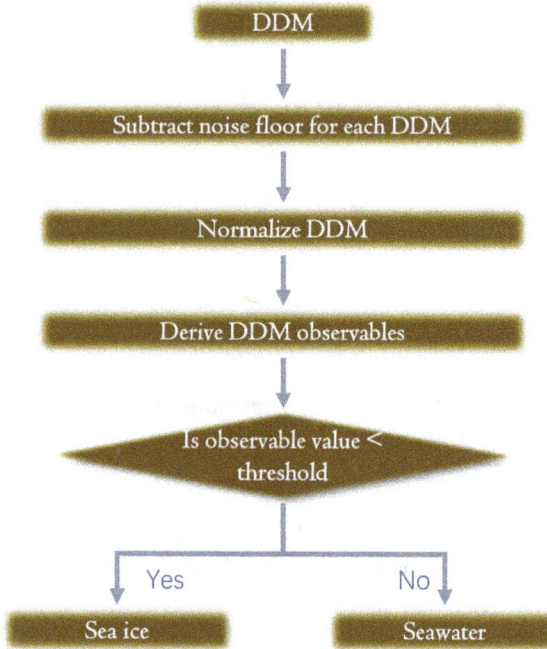

Figure 22.3   A flowchart for sea ice detection based on delay-Doppler map (DDM) observables

where T_i is the coherent integration time, $\vec{\rho}$ is the displacement vector of a surface point, D is the antenna pattern. R_T and R_R are the distances from a point on the ocean surface to the GNSS-R Tx and Rx. A represents the glistening zone. By applying change of variable, the integral in (22.7) can be simplified as [60]

$$\Sigma(\tau,f) = T_i^2 \frac{D^2(\vec{\rho}(\tau,f))\sigma^0(\vec{\rho}(\tau,f))\,|J(\tau,f)|}{4\pi R_R^2(\vec{\rho}(\tau,f))R_T^2(\vec{\rho}(\tau,f))}, \tag{22.8}$$

where $J(\cdot)$ is the Jacobian of the change of variables.

In order to retrieve σ_0, Σ needs to be obtained first. In the study by Schiavulli *et al.* [48], the 2-D-truncated singular value decomposition (TSVD) is adopted. While in the study by Yan and Huang [58], the spatial integration approach (SIA) [61] and multi-scan technique are used. These two methods are described as follows.

22.2.2.1   2-D-truncated singular value decomposition

The reconstruction of Σ from (22.6) is cast in the following form

$$L(X) = B, \tag{22.9}$$

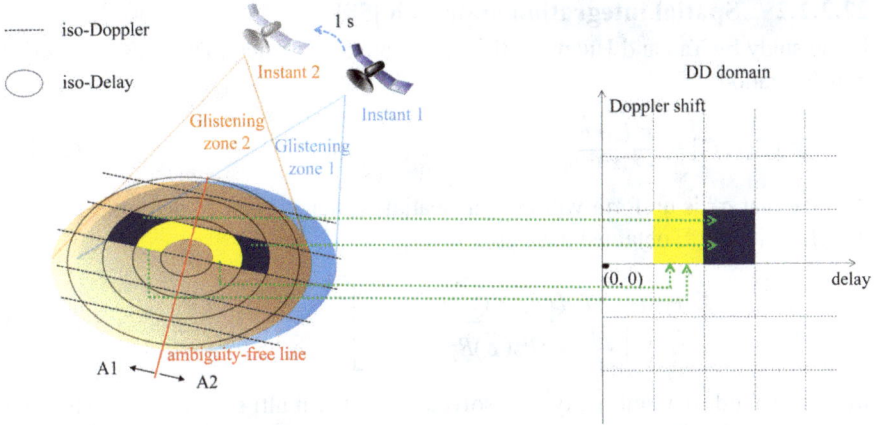

Figure 22.4   Demonstration of the ambiguity problem and its solution by the multi-scan method. The glistening zones of two different delay-Doppler map (DDM) acquisitions consecutively are shown in blue and orange, respectively. A1 and A2 denote the separate regions that are symmetrical to the ambiguity-free line in red.

where X represents the discretized Σ and B is the corresponding DDM, that is, $\langle|Y(\tau,f)|^2\rangle$. $L(\cdot)$ is the convolution operator [48]. Equation (22.9) can be expressed in matrix form by transferring B and X into lexicographically ordered vectors $\mathbf{b}$ and $\mathbf{x}$

$$\mathbf{A}\mathbf{x} = \mathbf{b} \tag{22.10}$$

where $\mathbf{A}$ is an n-by-n matrix and n is the length of $\mathbf{b}$ and $\mathbf{x}$. Equation (22.10) is solved using the regularization methods via minimizing the norm of the residual

$$\min \|\mathbf{A}\mathbf{x} - \mathbf{b}\|, \text{ subject to } \mathbf{x} \in P. \tag{22.11}$$

P represents the space in which the minimization is proceeded. Schiavulli *et al.* [48, 62] selected the Hilbert spaces, and used the 2-D TSVD approach. Based on TSVD, a regularized solution of (22.10) is derived. After removing the noisiest singular values, only the first k singular values are retrained in the TSVD solution

$$\mathbf{x}_k = \sum_{l=1}^{k} \frac{\mathbf{u}_l^T \mathbf{b}}{\varsigma_l} \mathbf{v}_l \tag{22.12}$$

where ς_l represents the selected singular values of $\mathbf{A}$; $\mathbf{u}_l^T \mathbf{b}$ and $\mathbf{v}_l$ are the left and right singular vectors, respectively. Once Σ is obtained, σ^0 can be determined according to (22.8). However, an ambiguity problem exists when converting a DDM pixel into the spatial coordinate because each DDM pixel corresponds to two different spatial clusters that are symmetric about the ambiguity-free line (see Figure 22.4). In the study by Schiavulli *et al.* [48], the ambiguity problem is resolved based on visual observation of consecutive DDMs acquired within the same track.

22.2.2.2   Spatial integration approach [58]

In the study by Yan and Huang [58], Σ is deconvolved from a DDM using Fourier transformation ($\mathscr{F}[\cdot]$)

$$\mathscr{F}[\Sigma(\tau,f)] = \frac{\mathscr{F}\left[\langle|Y(\tau,f)|^2\rangle\right]}{\mathscr{F}\left[\chi^2(\tau,f)\right]}. \tag{22.13}$$

Assume that σ^0 is uniform within each spatial cluster, σ^0 can be determined using the SIA with the Σ obtained through

$$\sigma^0(\vec{\rho}) = \frac{\Sigma(\tau,f)}{T_i^2}\left[\iint\limits_A \frac{D^2(\vec{\rho})\delta(\tau)\delta(f)}{4\pi R_R^2(\vec{\rho})R_T^2(\vec{\rho})}d^2\vec{\rho}\right]^{-1}. \tag{22.14}$$

In this method, the ambiguity is resolved using the multi-scan data, usually two consecutive DDMs. Because the time gap between two consecutive TDS-1 measurements is short (1 s), two adjacent DDMs are generated almost from the same glistening zone by assuming that the illuminated area is about 100 km and its surface scattering property remains unchanged within 1 s. Thus, the Σ for two adjacent DDMs can be expressed as

$$\begin{aligned}\Sigma_1(\tau,f) &= T_i^2\sigma_{A1}^0(\vec{\rho})\iint\limits_{A1} \frac{D_1^2(\vec{\rho})\delta(\tau)\delta(f)}{4\pi R_{R1}^2(\vec{\rho})R_{T1}^2(\vec{\rho})}d^2\vec{\rho} \\ &+ T_i^2\sigma_{A2}^0(\vec{\rho})\iint\limits_{A2} \frac{D_1^2(\vec{\rho})\delta(\tau)\delta(f)}{4\pi R_{R1}^2(\vec{\rho})R_{T1}^2(\vec{\rho})}d^2\vec{\rho}\end{aligned} \tag{22.15}$$

$$\begin{aligned}\Sigma_2(\tau,f) &= T_i^2\sigma_{A1}^0(\vec{\rho})\iint\limits_{A1} \frac{D_2^2(\vec{\rho})\delta(\tau)\delta(f)}{4\pi R_{R2}^2(\vec{\rho})^2 R_{T2}^2(\vec{\rho})}d^2\vec{\rho} \\ &+ T_i^2\sigma_{A2}^0(\vec{\rho})\iint\limits_{A2} \frac{D_2^2(\vec{\rho})\delta(\tau)\delta(f)}{4\pi R_{R2}^2(\vec{\rho})R_{T2}^2(\vec{\rho})}d^2\vec{\rho},\end{aligned} \tag{22.16}$$

where subscripts $A1$ and $A2$ denote two regions that are symmetric about the ambiguity-free line, and subscripts 1 and 2 indicate the two DDMs, respectively. By solving (22.15) and (22.16), σ_{A1}^0 and σ_{A2}^0 for both sides of the ambiguity line can be determined.

Besides the above methods, another efficient methodology for resolving the ambiguity problem can be found in the study by Southwell and Dempster [63].

22.2.3   *Machine learning-based method*

Recently, machine learning has become a rapidly growing technique in various fields because it is able to construct a relationship between the observation data and desired parameters via a data-driven mode. Without exception, machine learn-ing- based methods have also been successfully applied to sea ice remote sensing using, for example, SAR [14], altimeter [64], and passive microwave data [65]. Implementing machine learning typically involves three steps: preprocessing of the

Pre-processed DDM + Reference Data

Training set

Test set

Train model

No

Is stop criteria reached for training?

Yes

Export trained model

Produce sea ice estimation results

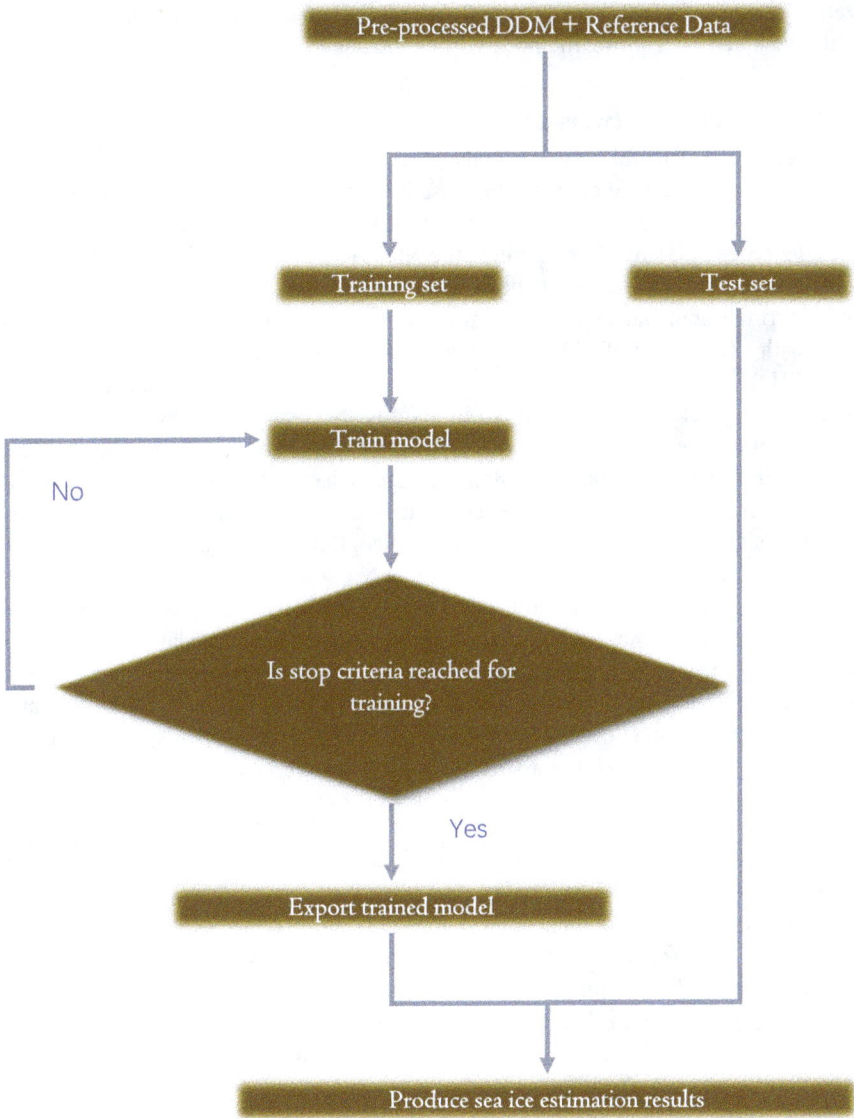

Figure 22.5   Block diagram of machine learning-based sea ice sensing techniques. DDM, delay-Doppler map.

input (data), training, and test. Machine learning approaches can be divided into two types: classification and regression. For sea ice detection, which can be regarded as a binary (sea ice and seawater) classification problem, classification-orientated machine learning methods are usually adopted. The general procedure of a machine learning- based sea ice sensing method is depicted in Figure 22.5. Here, three representative machine learning methods will be introduced.

22.2.3.1   Neural networks

The algorithm of neural networks (NNs) has been well accepted as a robust technique for remote sensing practices because it is able to approximate desired outputs without *a prior* assumption about their distribution and directly from data through a learning process. Before being input to the NN, each DDM should undertake the following four preprocess steps: 1) noise floor subtraction, 2) normalization, 3) signal box determination, and 4) data stretching. An original TDS-1 DDM consists of 128 bins (with a resolution of 244 ns) in delay and 20 bins in Doppler (with a resolution of 500 Hz). To mitigate the noise effect for each DDM, the noise floor is deducted as the average of a signal-free box (containing the first four delay bins along all Doppler bins). Next, each DDM is normalized by its peak power, that is, the local maxima in the DDM. Then, the signal box is selected as a region consisting of 4 rows before and 35 rows after the local maxima in delay and all the Doppler bins. The 2-D signal box is transferred into a lexicographically ordered vector, which contains 800 elements and is the input to the NN.

An NN contains input, hidden, and output layers each of which has a certain number of nodes that are interconnected between layers. Yan and Huang [50] designed a simple but effective NN consisting of 1 input layer (800 units, i.e., $a_l^{(0)}$, $l \in [1, 800]$), 1 hidden layer (3 units, i.e., $a_l^{(1)}$, $l \in [1, 3]$), and 1 output layer (1 unit, i.e., $a_1^{(2)}$ that produces the detection results), as shown in Figure 22.6(a). In general, the associations between the layers can be expressed in matrix form as

$$\mathbf{a}^{(k+1)} = \varphi^{(k+1)}(\mathbf{W}^{(k+1)}\mathbf{a}^{(k)} + \mathbf{c}^{(k+1)}), k \in [0, 1], \tag{22.17}$$

where $\varphi^{(k+1)}(\cdot)$, $\mathbf{a}^{(k+1)}$, $\mathbf{c}^{(k+1)}$, and $\mathbf{W}^{(k+1)}$ represent the activation function, input vector, bias vector, and weight matrix, respectively. The widely used sigmoid function $\varphi^{(1)}(x) = 1/(1 + e^{-x})$ and the softmax function are selected as the activation functions for the hidden and the output layers, respectively.

Training a NN is to determine the optimal weights and bias using a learning algorithm such as, the back-propagation (BP) learning [67] and the Levenberg-Marquardt algorithm [68] based on the difference between predicted results and expected (label) data for given inputs. A detailed description can be found in the paper by Yan *et al.* [50]. After the NN is trained using a set of DDM data and corresponding seawater/ice label data, it can be deployed to determine whether a region is covered by sea ice or not according to the input DDM.

22.2.3.2   Convolutional neural networks

CNN is a class of deep NNs that deploys extra convolutional and pooling layers. Unlike a simple NN, the convolutional layer can take 2-D DDM as input, thus, preserving the spatial information of DDMs. The pooling layer can reduce the redundancy in data and make CNNs less sensitive to the misalignment of the signal box

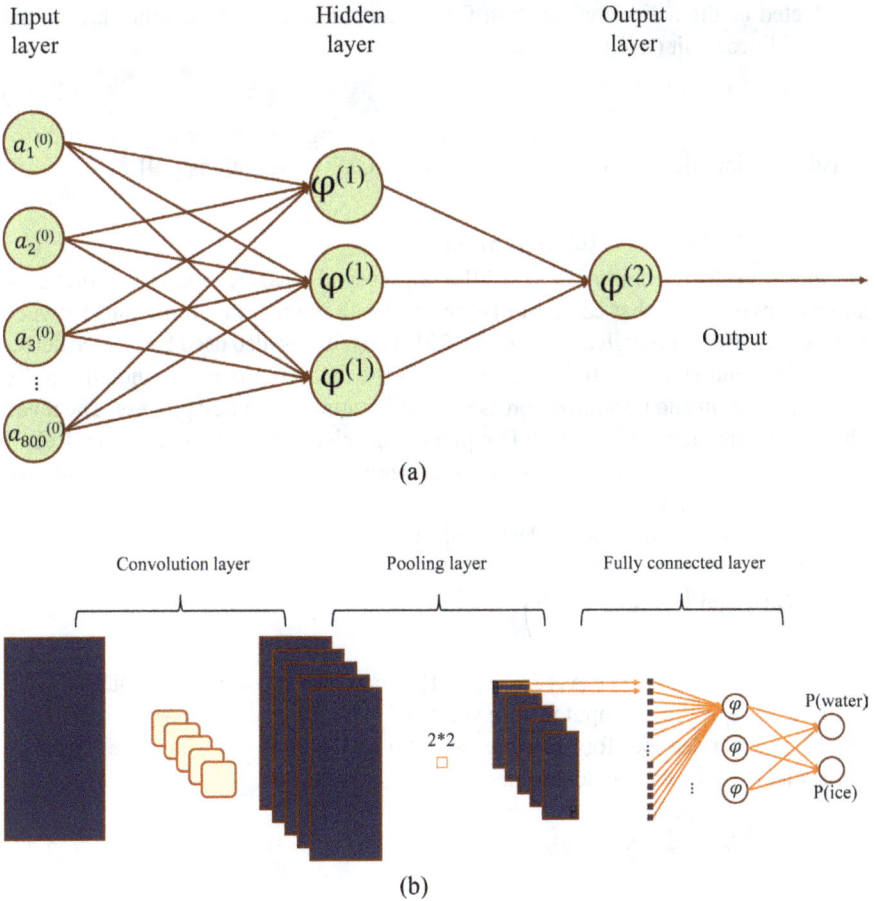

Figure 22.6   Diagram of (a) a neural network (NN) from Yan et al. [50] and (b) a convolutional neural network (CNN) from Yan and Huang [66] for sea ice detection

within a DDM frame. CNNs also require less data preprocessing steps than NNs, that is, noise floor subtraction, normalization, and (optional) signal box determination.

Figure 22.6 (b) depicts the sea ice detection CNN designed by Yan and Huang [66]. It consists of one convolution layer (which is made of five seven-by-seven filters) followed by one two-by-two pooling layer and two fully connected layers (whose functionality is similar to the input and hidden layers for NNs). For the convolution layer,

$$\mathbf{h}^k = \varphi((\mathbf{H}^k * \mathbf{a}) + b) \tag{22.18}$$

where $\mathbf{a}$, b, $\mathbf{h}^k$, and $\mathbf{H}^k$ represent the input, the bias, the kth ($k \in [1, 5]$) convolved image, and filter, respectively. $*$ is the convolution operator. The softmax function is selected as the activation function for the output layer. For the other layers, the rectified linear unit (ReLU) function

$$\varphi(z) = \max(0, z) \tag{22.19}$$

is chosen. Training of the CNN is accomplished using the BP learning [67] and stochastic gradient descendant with momentum (SGDM) algorithms [69].

22.2.3.3   Support vector machines

A support vector machine (SVM) [70] is a powerful classification model that constructs a hyperplane that can best distinguish (with the maximum margin) between different classes with high accuracy [71, 72]. Thus, it was also used for sea ice detection in Yan and Huang's study [72], in which the data preprocessing includes noise floor subtraction and normalization as well as a feature extraction process. The mean value along the delay axis at each Doppler bin is calculated and processed by ReLU. As a result, 20 mean values are obtained and normalized by its maximum, and then used as the feature vector.

The SVM classifier can be described as

$$f(\mathbf{z}) = \text{sgn}\left(\sum_j \alpha_j y_j \mathbf{z}^T \mathbf{z}_j + b\right), \tag{22.20}$$

based on a training set $\{(\mathbf{z}_1, y_1), ..., (\mathbf{z}_n, y_n)\}$, where n is the number of DDMs in the training set, $\mathbf{z}_j$ is the jth input feature vector ($j \in [1, n]$), and $y_j \in \{-1, 1\}$ is the class label (seawater/sea ice) for the corresponding DDM. $\boldsymbol{\alpha} = \{\alpha_1, ..., \alpha_n\}$ are Lagrange multipliers, which can be determined via

$$\max_{\boldsymbol{\alpha}} \left(\sum_j \alpha_j - \tfrac{1}{2}\sum_j\sum_k \alpha_j\alpha_k y_j y_k \mathbf{z}_j^T \mathbf{z}_k\right) \tag{22.21}$$

s.t.

$$0 \le \alpha_j \le C, \text{ and } \sum_j \alpha_j y_j = 0. \tag{22.22}$$

and b can be derived as the mean value of the term $(1/y_i - \sum_i \alpha_i y_i \mathbf{z}_i^T \mathbf{z}_i)$.

22.2.4   *Empirical model-based method*

The methods introduced above are based on spaceborne GNSS-R data. Although limited ground-based experiments have been conducted for ice sensing, the results are promising. In 2012, signal-to-noise ratio (SNR) data of GNSS-R signals were collected at the Onsala Space Observatory, Sweden [42, 73]. It was found that the high-frequency part of the SNR data, referred to as δSNR, can be modeled as

$$\delta\text{SNR} = [C_1 \sin(\tfrac{4\pi m}{\lambda} \sin\theta) + C_2 \cos(\tfrac{4\pi m}{\lambda} \sin\theta)] \times \exp(-4k^2\gamma\sin^2\theta) \tag{22.23}$$

where λ is the signal wavelength, m is the height of Rx above the reflecting surface, θ is the elevation angle, C_1 and C_2 are the amplitudes, and γ is referred to as the damping parameter that depends on the characteristics of scattering surface. C_1, C_2,

Table 22.1   Sea ice detection methods

Method	Source	Platform	Accuracy
Observable	[46, 47, 52, 54–57]	Spaceborne	95.40%–99.73%
σ^0 retrieval	[48, 58, 63]	Spaceborne	–
Machine learning	[50, 66, 72]	Spaceborne	97.17%–98.56%
Empirical model	[42, 73]	Ground-based	–

m, and γ can be determined through non-linear least-squares fitting using (22.23). It was observed that the value of γ is low when the surface is covered by ice, and thus it can be used for ice detection [42].

The methods for sea ice detection are summarized in Table 22.1.

22.3   SIC estimation

It was observed in Gleason's study [44] that spaceborne GNSS-R data collected over regions of varying SIC exhibit different extents of spreading in DDM. According to this observation, SIC information can be extracted from GNSS-R data. By now, most existing SIC estimation methods are machine learning based, including NNs [50], CNNs [66], and support vector regression (SVR) [74]. Unlike sea ice detection which is a classification task, SIC estimation is a regression problem. However, the procedures for both applications are similar. One main difference is that the target output for the latter is SIC value instead of surface type. In addition, the linear function $\varphi(x) = x$ rather than the softmax function is chosen as the activation function in the output layer for NNs and CNNs, and the regression model is employed for SVR. Since the methods for SIC estimation and sea ice detection are similar, they will not be recapped here; interested readers can refer to Yan *et al.* [50, 66, 74]. An example of the SIC results obtained using NN, CNN, and SVR methods is presented in Figure 22.7. The accuracy results are summarized in Table 22.2.

Other than machine learning algorithms, an empirical model was developed for estimating SIC from DDM observable as [75]

$$SIC = A\epsilon^{B(r-\mu)} + C \tag{22.24}$$

where r is the DDM observable that is the summation of a waveform beyond its right edge. A, B, C, ϵ, and μ can be determined through least-squares fitting using a set of training data.

From August 25, 2016 to September 13, 2016, an experiment was conducted during the Fram Strait cruise of the RV Lance, which showed that SIC can be inverted from power ratios using a non-linear least-squares fitting [76]. The cost function between the observed and modeled power ratios ($p^{(i)}$ and h_p)

$$\delta p^2(SIC, \sigma) = \frac{1}{m}\sum_{i=1}^{m}(p^{(i)} - h_p(SIC, \sigma)) \tag{22.25}$$

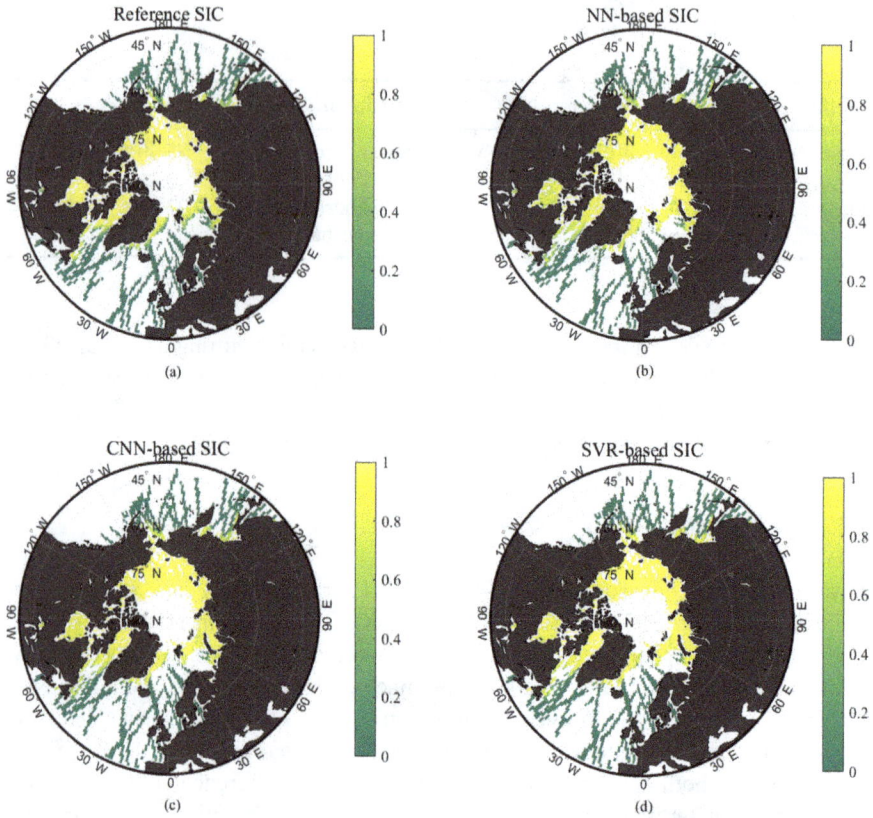

Figure 22.7   Sea ice concentration (SIC) results of (a) reference, (b) neural network (NN)-based, (c) convolutional neural network (CNN)-based, and (d) support vector regression (SVR-based methods

Table 22.2   Sea ice concentration (SIC) estimation accuracy of different machine learning algorithms

	Convolutional neural network (CNN)	Neural network (NN)	Support vector regression (SVR)
r	0.93	0.93	0.94
Root mean square difference (RMSD)	0.16	0.15	0.15

Table 22.3   Approaches of sea ice concentration estimation

Technique	Source	Platform	Accuracy (RMSE)
Machine learning	[50, 66, 74]	Spaceborne	0.15–0.16
Empirical model	[75]	Spaceborne	0.12
Empirical model	[76]	Ground-based	0.25

RMSE, root mean square error.

is minimized for different combinations of SIC and σ (roughness effect) with SIC chosen as {0, 20, 40, 60, 80, 100}% and σ out of {0, 5, 10, 15, 20, 25} cm.

In 2013, ground-based tests were conducted at the Bohai Bay (China), and the ratio between direct and reflected signals (ρ) was found to be sensitive to SIC [40], but no SIC estimation results were obtained.

The methods for SIC estimation are summarized in Table 22.3.

22.4   SIT retrieval

Most recently, SIT information was also successfully obtained from GNSS-R data. Existing methods include three-layer model-based [77], empirical model-based [41], and phase altimetry-based [78] algorithms.

22.4.1   Three-layer model

The potential of spaceborne GNSS-R for SIT retrieval was reported by Li *et al.* [49]. It was argued that the main reflection occurred at the ice-water interface for the case of first-year ice (FYI) with a thickness of 20–60 cm [49]. If only coherent GNSS reflections are considered, the reflectivity (Γ) of an ice-covered ocean surface can be modeled as the product of the propagation loss of sea ice and the reflection coefficient at sea ice-water boundary [77, 79] (see Figure 22.8).

22.4.1.1   Relationship between SIT and Γ

Shown in the paper by Tsang and Newton [80],

$$\Gamma = |R|^2 \cdot \exp\left[-\left(\tfrac{4\pi}{\lambda}\sigma_{rms}\cos\theta\right)^2\right], \tag{22.26}$$

where R is the overall Fresnel reflection coefficient, σ_{rms} is the surface RMS height, and θ is the incidence angle. The exponential term in this equation describes the surface roughness effect, which can be approximated by 1 for smooth surfaces with a very small σ_{rms}. With this approximation, then

$$\Gamma = |R|^2. \tag{22.27}$$

By introducing a three-layer air-ice-seawater model (see Figure 22.8), R can be formulated as shown by Ulaby *et al.* [81],

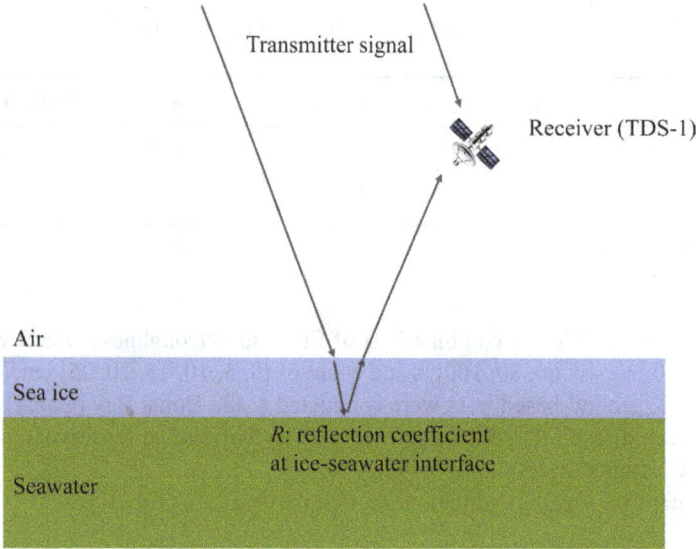

Figure 22.8   *Schematic of a Global Navigation Satellite System-Reflectometry (GNSS-R) signal reflected from a three-layer model of air, sea ice, and seawater*

$$R = \frac{R_1 + R_2 e^{-2ik_{zi}d}}{1 + R_1 R_2 e^{-2ik_{zi}d}},$$ (22.28)

where R_1 and R_2 denote the reflection coefficients at the upper (air-sea ice) and lower (ice-seawater) interfaces, respectively. d is the SIT, and k_{zi} is the z-component of the signal propagation vector in sea ice. The formula of k_{zi} contains a real part (the phase constant β), and an imaginary part (the attenuation coefficient α), that is,

$$k_{zi} = \beta - i\alpha.$$ (22.29)

As noted by Li *et al.* [49], due to a lower dielectric contrast between air and sea ice than that between sea ice and seawater, R_1 is less than R_2. As such, the signals are mainly reflected by the ice-water interface. This is advocated by the analyses in Li *et al.*'s paper [49] that the gap between the estimated surface height and the mean sea surface is consistent with collocated SIT. For this reason, the factor of R_1 in (22.28) can be neglected by setting it as 0 here, which gives

$$R \approx R_2 e^{-2ik_{zi}d},$$ (22.30)

and by combining (22.27), (22.29), and (22.30), thus,

$$\Gamma = \left| R_2 \right|^2 e^{-4\alpha d}.$$ (22.31)

The expression for α is (see e.g., Kaleschke *et al.* [82]),

$$\alpha = \frac{2\pi}{\lambda} \cos\theta \left| \mathrm{Im}\left\{ \sqrt{\varepsilon_i} \right\} \right|,$$ (22.32)

with ε_i being the relative permittivity of sea ice.

In summary, in this model GNSS-R signals are first attenuated by the ice layer and then reflected at the ice-water interface. Therefore, SIT d can be calculated as

$$d = \frac{-1}{4\alpha} \ln \frac{\Gamma}{|R_2|^2},$$

(22.33)

with Γ and the values of R_2 and α, whose derivation will be described in the following section.

22.4.1.2 Dielectric models

From (22.32), it can be seen that α is dependent on ε_i. The model of Vant *et al.* [83] can be employed here for deriving ε_i based on the relative brine volume (V_b, in %, or per 1000),

$$\varepsilon_i = 3.1 + 0.0084V_b + i(a_1 + a_2V_b),$$

(22.34)

with $a_1 = 0.037, a_2 = 0.00445$ for FYI, or $a_1 = 0.003, a_2 = 0.00435$ for multi-year ice (MYI). To derive V_b, the following empirical formula, which was presented by Ulaby *et al.* [84], can be used

$$V_b = 10^{-3}S\left(-\frac{49.185}{T} + 0.532\right),$$

(22.35)

where S and T are sea ice salinity (in) and temperature (in °C), respectively.

R_2 can be obtained by differentiating the horizontal and vertical polarization components (i.e., R_{hh} and R_{vv}) [59], as

$$R_2 = \frac{1}{2}\left(R_{vv} - R_{hh}\right),$$

(22.36)

and

$$R_{vv} = \frac{\varepsilon_r \cos\theta_i - \sqrt{\varepsilon_r - \sin^2\theta_i}}{\varepsilon_r \cos\theta_i + \sqrt{\varepsilon_r - \sin^2\theta_i}},$$

(22.37)

$$R_{hh} = \frac{\cos\theta_i - \sqrt{\varepsilon_r - \sin^2\theta_i}}{\cos\theta_i + \sqrt{\varepsilon_r - \sin^2\theta_i}},$$

(22.38)

$$\theta_i = \arcsin\frac{\sin\theta}{\sqrt{\varepsilon_i}}, \varepsilon_r = \varepsilon_w/\varepsilon_i,$$

(22.39)

and ε_w is the relative permittivity of seawater, which can be calculated using the model of Klein and Swift [85].

Therefore, SIT can be estimated based on (22.33) with TDS-1 Γ, R_2 obtained from (22.34)–(22.39), and α from (22.32), (22.34), and (22.35). Two examples of retrieved results are displayed in Figure 22.9 for March 2017 and April 2018. Overall, a good agreement between the TDS-1 and soil moisture and ocean salinity (SMOS) results is achieved, with a correlation coefficient (r) of 0.84 and a root mean square difference (RMSD) of 9.39 cm [77].

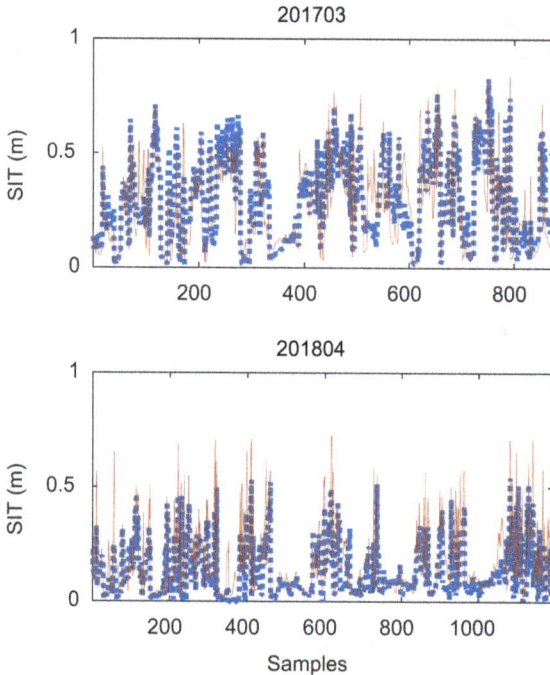

Figure 22.9   Comparison between sea ice thickness (SIT) estimation results from TDS-1 and SMOS data

22.4.2   Empirical SIT estimation model

The investigation of SIT estimation from ground-based GNSS-R data was shown by Gao *et al.* [41], in which an empirical method was used. Based on the field data collected at Liaodong Bay (China) in 2016 [41], an empirical relationship between SIT and the power ratio ρ of the direct signal to reflected signal was obtained through a fitting process as

$$SIT = 2.086\rho^{-0.021} - 2.697. \tag{22.40}$$

This model was verified by Gao *et al.* [41] for sea ice with a thickness of 10–20 cm.

22.4.3   Phase altimetry-based SIT retrieval

A coastal experiment was performed from January 22, 2016 to February 19, 2016, in Dashentang, Tianjin, China [78]. The sea ice surface height and seawater surface height are calculated using the phase altimetry technique (see Section 22.5.2) and their difference is known as the freeboard h_f. Consequently, SIT can be estimated through

$$SIT = \frac{\rho_w}{\rho_w - \rho_i}h_f, \tag{22.41}$$

Table 22.4   *Applications of sea ice thickness retrieval*

Technique	Source	Platform
Three-layer model	[77]	Spaceborne
Empirical model	[41]	Ground-based
Phase altimetry	[78]	Ground-based

where ρ_w and ρ_i are, respectively, the densities of seawater and sea ice that are assumed to be 1.0257 g/cm³ and 0.851 g/cm³ [78]. It is worth noting that the coastal BeiDou-reflected signals were employed for SIT estimation for the first time.

The applications of SIT retrieval are summarized in Table 22.4.

22.5   Ice altimetry techniques

Another argument made by Gleason [44] was that the coherent reflection from ice surfaces allows accurate altimetry measurements, which was later verified bu Hu *et al.* [86] through investigating delay waveforms and via processing phase measurements of raw data in the study by Li *et al.* [49]. These two methods are discussed as follows.

22.5.1   Waveform-based method

The waveforms are generated from raw data by cross-correlating received signals with local replicas. Based on the direct and reflected waveforms, the path delay $\Delta\rho$ between these two signals can be calculated as

$$\Delta\rho = c \cdot (t^R - t^D) + V, \tag{22.42}$$

where c is the speed of light, t^R and t^D are the arrival times of reflected and direct signals, respectively. V is an error term that accounts for the ionospheric and tropospheric effects, and noise. t^R and t^D are determined based on the peak of their corresponding waveforms. The height of the surface under investigation can be determined from the obtained path delay using the method described by, for example, Clarizia *et al.* [87]. The estimated precision is about 1 m with a spatial resolution of 3.8 km.

22.5.2   Phase-based method

Unlike the waveform-based method, the phase delay is directly extracted from raw data through an open-loop tracking of the reflected signal for surface height h determination via

$$h = -\frac{(\phi_o - \phi_m) - \varepsilon_{\text{fit}}}{2\sin\theta}, \tag{22.43}$$

where ϕ_o and ϕ_m represent, respectively, the observed and modeled phase delays, ε_{fit} is the estimated error that is determined through least-squares fitting, and θ is the Tx

Table 22.5   Comparison between sea ice altimetry methods

Technique	Source	Platform	Accuracy
Waveform-based	[86]	Spaceborne	4.4 m (root mean square difference)
Phase-based	[49]	Spaceborne	4.7 cm (root mean square difference)

elevation angle at the specular point. By applying this method to a 40-s long dataset collected in the Hudson Bay [49], the RMSD between the estimated and reference results is 4.7 cm with an along-track sampling distance of about 140 m.

A comparison between these two methods is presented in Table 22.5.

22.6   Other applications

22.6.1   Sea ice classification

A strong sensitivity of the GNSS-R signals to the surface roughness of several primary ice types, that is, FYI, MYI, and young ice, was found in the study by Rodriguez-Alvarez *et al.* [88]. This leads to different DDM shapes for various ice types. Based on this perspective, several DDM observables were developed for sea ice type classification. The observables used include DDMA, TES, leading edge slope (LES), generalized linear observable (see Rodriguez-Alvarez and Garrison [89]), and waveform width observable (the width of a waveform when its power reaches two-thirds of the peak value). The first step for sea ice classification is to identify the DDMs of sea ice using the methods similar to those described in Section 22.2.1. Next, the observables for classification are calculated from the identified data. Since no intuitive relationship exists between ice type and these observables, the standard classification and regression tree (CART) model (see Breiman *et al.* [90]) is used and trained as shown by Rodriguez-Alvarez *et al.* [88] for ice classification. The algorithm details can be found in the model by Rodriguez-Alvarez *et al.* [88]. This algorithm achieves an accuracy of 54.5%, 94%, and 69.7% for classifying FYI, MYI, and young ice, respectively.

22.6.2   Sea ice permittivity and roughness retrieval

The possibility of sea ice permittivity and roughness extraction from a reflected GPS signal was investigated by Rivas *et al.* [38]. The received waveform is modeled as the product of the reflection coefficient and the GPS Coarse/Acquisition (C/A) code autocorrelation function in the limit of small surface slopes for Kirchhoff approximation. For this reason, the power strength and shape of a waveform can be used to derive the permittivity and roughness, respectively. The surface permittivity is derived from the peak power, and the surface roughness (parameterized by surface mean square slope) is obtained based on a least-squares fitting between the modeled and measured waveforms.

22.7 Conclusions

In this chapter, the application of GNSS-R signals for sea ice detection, SIC and SIT estimation, altimetry, classification, and so on, are introduced. Existing experiment results showed that both the DDM observables-based and machine learning-based methods can provide satisfactory sea ice detection and SIC estimation. The difference between DDM observables-based and machine learningbased methods lies in that the latter accomplish the classification through a learning process exclusively depending on the data, while the former are based on the interpretation by researchers. The difference between DDMs of sea ice and seawater is distinct, which can be well represented by the derived observables, and thus the accuracy is plausible. However, it is challenging to deal with complex problems such as SIC estimation and sea ice type classification by solely using observables, and in such cases machine learning methods can be adopted. Nonetheless, empirical models have also proved to be effective. For sea ice altimetry, the phase-based method shows better performance than the waveform-based one, which is due to 1) an extra error mitigation processed by the former, 2) the restriction on delay resolution of the waveform by the latter, and 3) shorter integration time adopted by the former that results in better spatial resolution. The available accuracy of each application has been provided in Tables 22.1–22.3 and 22.5. However, it should be noted that these results were achieved based on the different datasets, locations, or experiment setups, so that a direct comparison may be biased and unfair.

Although existing studies have demonstrated high accuracy and efficiency in sea ice detection, SIC estimation and type classification, it is necessary to derive other sea ice parameters from the GNSS-R signals in the future. Existing researches on sea ice sensing are mainly based on data collected by TDS-1 satellite, whose service was ceased in December 2018. Although the Cyclone GNSS (CyGNSS) mission is in orbit, it is unable to cover high-latitude sea ice infested areas. Thus, the launch of new GNSS-R satellites for sea ice monitoring is urgent.

Moreover, limited airborne and ground-based tests have been conducted for sea ice sensing so far. Although the area covered by an airborne or a ground-based Rx is not large, such systems are relatively cheap and easy for deployment and can provide long-term or short-term observations for specific regions of interests. Increasing ground GNSS stations and airborne tests is necessary for the development of sea ice sensing schemes in the future.

References

[1] Hartman D.L., Klein Tank A.M.G., Rusicucci M. 'Observations: Atmosphere and Surface'. Cambridge University Press; 2013.

[2] Rothrock D.A., Yu Y., Maykut G.A. 'Thinning of the Arctic sea-ice cover'. *Geophys Research Letters*. 1999;**26**(23):3469–72.

[3] Comiso J.C., Parkinson C.L., Gersten R., *et al.* 'Accelerated decline in the Arctic sea ice cover'. *Geophysical Research Letters*. 2008;**35**(1):L01703.

[4]   Laxon S.W., Giles K.A., Ridout A.L., *et al.* 'CryoSat-2 estimates of Arctic sea ice thickness and volume'. *Geophysical Research Letters*. 2013;**40**(4):732–7.

[5]   Barry R.G., Serreze M.C., Maslanik J.A., *et al.* 'The Arctic sea ice-climate system: observations and modeling'. *Reviews of Geophysics*. 1993;**31**(4):397.

[6]   McPhee M.G., Proshutinsky A., Morison J.H., *et al.* 'Rapid change in freshwater content of the Arctic ocean'. *Geophysical Research Letters*. 2009;**36**(10):L10602.

[7]   Galley R.J., Else B.G.T., Prinsenberg S.J., *et al.* 'Summer sea ice concentration, motion, and thickness near areas of proposed offshore oil and gas development in the Canadian Beaufort Sea – 2009'. *ARCTIC*. 2013;**66**(1):105–16.

[8]   Sandven S., Johannessen O.M., Kloster K. *Sea Ice Monitoring by Remote Sensing*. Chichester, UK: John Wiley & Sons, Ltd; 2006.

[9]   Petrou Z.I., Tian Y. 'Prediction of sea ice motion with convolutional long short-term memory networks'. *IEEE Transactions on Geoscience and Remote Sensing*. 2019;**57**(9):6865–76.

[10] Zhang Z., Yu Y., Li X., *et al.* 'Arctic sea ice classification using microwave scatterometer and radiometer data during 2002–2017'. *IEEE Transactions on Geoscience and Remote Sensing*. 2019;**57**(8):5319–28.

[11] Otosaka I., Belmonte Rivas M., Stoffelen A. 'Bayesian sea ice detection with the ERS scatterometer and sea ice Backscatter model at C-band'. *IEEE Transactions on Geoscience and Remote Sensing*. 2018;**56**(4):2248–54.

[12] Belmonte Rivas M., Otosaka I., Stoffelen A., *et al.* 'A scatterometer record of sea ice extents and backscatter: 1992–2016'. *The Cryosphere*. 2018;**12**(9):2941–53.

[13] Rose S.K., Andersen O.B., Passaro M., *et al.* 'Arctic ocean sea level record from the complete radar altimetry era: 1991–2018'. *Remote Sensing*. 2019;**11**(14):1672.

[14] Wang L., Scott K.A., Xu L., Clausi D.A. 'Sea ice concentration estimation during melt from dual-pol SAR scenes using deep convolutional neural networks: a case study'. *IEEE Transactions on Geoscience and Remote Sensing*. 2016;**54**(8):4524–33.

[15] Gao F., Wang X., Gao Y., Dong J., Wang S. 'Sea ice change detection in SAR images based on convolutional-wavelet neural networks'. *IEEE Geoscience and Remote Sensing Letters*. 2019;**16**(8):1240–4.

[16] Longepe N., Thibaut P., Vadaine R., *et al.* 'Comparative evaluation of sea ice lead detection based on SAR imagery and altimeter data'. *IEEE Transactions on Geoscience and Remote Sensing*. 2019;**57**(6):4050–61.

[17] Hall C.D., Cordey R.A. 'Multistatic scatterometry'. IEEE International Geoscience and Remote Sensing Symposium; 1988. pp. 561–2.

[18] Katzberg S.J., Torres O., Ganoe G. 'Calibration of reflected GPS for tropical storm wind speed retrievals'. *Geophysical Research Letters*. 2006;**33**(18):n/a.

[19] Marchan-Hernandez J.F., Rodriguez-Alvarez N., Camps A., Bosch-Lluis X., Ramos-Perez I., Valencia E. 'Correction of the sea state impact in the L-band brightness temperature by means of delay-Doppler maps of global

navigation satellite signals reflected over the sea surface'. *IEEE Transactions on Geoscience and Remote Sensing*. 2008;**46**(10):2914–23.

[20] Rodriguez-Alvarez N., Akos D.M., Zavorotny V.U., Smith J.A., Camps A., Fairall C.W. 'Airborne GNSS-R wind retrievals using delay Doppler maps'. *IEEE Transactions on Geoscience and Remote Sensing*. 2013;**51**(1):626–41.

[21] Foti G., Gommenginger C., Jales P., *et al.* 'Spaceborne GNSS reflectometry for ocean winds: first results from the UK TechDemoSat-1 mission'. *Geophysical Research Letters*. 2015;**42**(13):5435–41.

[22] Yan Q., Huang W., Foti G. 'Quantification of the relationship between sea surface roughness and the size of the Glistening zone for GNSS-R'. *IEEE Geoscience and Remote Sensing Letters*. 2018;**15**(2):237–41.

[23] Najibi N., Jin S., Najibi N., *et al.* 'Physical reflectivity and polarization characteristics for snow and ice-covered surfaces interacting with GPS signals'. *Remote Sensing*. 2013;**5**(8):4006–30.

[24] Jin S., Najibi N. 'Sensing snow height and surface temperature variations in Greenland from GPS reflected signals'. *Advances in Space Research*. 2014;**53**(11):1623–33.

[25] Najibi N., Jin S., Wu X. 'Validating the variability of snow accumulation and melting from GPS-reflected signals: forward modeling'. *IEEE Transactions on Antennas and Propagation*. 2015;**63**(6):2646–54.

[26] Jin S., Qian X., Kutoglu H. 'Snow depth variations estimated from GPS-reflectometry: a case study in Alaska from L2P SNR data'. *Remote Sensing*. 2016;**8**(1):63.

[27] Katzberg S.J., Torres O., Grant M.S., Masters D. 'Utilizing calibrated GPS reflected signals to estimate soil reflectivity and dielectric constant: results from SMEX02'. *Remote Sensing of Environment*. 2006;**100**(1):17–28.

[28] Egido A., Paloscia S., Motte E., *et al.* 'Airborne GNSS-R polarimetric measurements for soil moisture and above-ground biomass estimation'. *IEEE Journal of Selected Topics in Applied Earth Observations and Remote Sensing*. 2014;**7**(5):1522–32.

[29] Camps A., Park H., Pablos M., *et al.* 'Sensitivity of GNSS-R spaceborne observations to soil moisture and vegetation'. *IEEE Journal of Selected Topics in Applied Earth Observations and Remote Sensing*. 2016;**9**(10):4730–42.

[30] Yan Q., Huang W., Jin S., Jia Y. 'Pan-tropical soil moisture mapping based on a three-layer model from CYGNSS GNSS-R data'. *Remote Sensing of Environment*. 2020;**247**:111944.

[31] Yan Q., Huang W. 'Tsunami detection and parameter estimation from GNSS-R delay-Doppler map'. *IEEE Journal of Selected Topics in Applied Earth Observations and Remote Sensing*. 2016;**9**(10):4650–9.

[32] Yan Q., Huang W. 'GNSS-R delay-Doppler map simulation based on the 2004 Sumatra-Andaman tsunami event'. *Journal of Sensors*. 2016;**2016**(14):1–14.

[33] Jin S., Komjathy A. 'GNSS reflectometry and remote sensing: new objectives and results'. *Advances in Space Research*. 2010;**46**(2):111–7.

[34] Jin S., Feng G.P., Gleason S. 'Remote sensing using GNSS signals: current status and future directions'. *Advances in Space Research*. 2011;**47**(10):1645–53.

[35]   Zavorotny V.U., Gleason S., Cardellach E., Camps A. 'Tutorial on remote sensing using GNSS bistatic radar of opportunity'. *IEEE Geoscience and Remote Sensing Magazine*. 2014;**2**(4):8–45.

[36]   Camps A. 'Introduction to remote sensing using GNSS signals of opportunity' in Huang W., Gill E.W. (eds.). *Remote Sensing Technologies: High Frequency, Marine and GNSS-Based Radar*. IET; 2021

[37]   Komjathy A., Maslanik J., Zavorotny V.U., *et al.* 'Sea ice remote sensing using surface reflected GPS signals'. *IEEE International Geoscience and Remote Sensing Symposium*. 2000;**7**:2855–7.

[38]   Rivas M.B., Maslanik J.A., Axelrad P. 'Bistatic scattering of GPS signals off Arctic sea ice'. *IEEE Transactions on Geoscience and Remote Sensing*. 2010;**48**(3):1548–53.

[39]   Fabra F., Cardellach E., Rius A., *et al.* 'Phase altimetry with dual polarization GNSS-R over sea ice'. *IEEE Transactions on Geoscience and Remote Sensing*. 2012;**50**(6):2112–21.

[40]   Zhang Y., Meng W., Gu Q., *et al.* 'Detection of Bohai Bay sea ice using GPS-reflected signals'. *IEEE Journal of Selected Topics in Applied Earth Observations and Remote Sensing*. 2015;**8**(1):39–46.

[41]   Gao H., Yang D., Zhang B., *et al.* 'Remote sensing of sea ice thickness with GNSS reflected signal'. *Journal of Electronics and Information Technology*. 2017;**39**(5):1096–100.

[42]   Strandberg J., Hobiger T., Haas R. 'Coastal sea ice detection using ground-based GNSS-R'. *IEEE Geoscience and Remote Sensing Letters*. 2017;**14**(9):1552–6.

[43]   Gleason S., Adjrad M., Unwin M. 'Sensing ocean ice and land reflected signals from space: Results from the UK-DMC GPS Reflectometry experiment'. ION GNSS 18th International Technical Meeting of theSatellite Division; 13–16 September 2005; 2005. pp. 1679–85.

[44]   Gleason S. 'Remote sensing of ocean, ice and land surfaces using bistatically scattered GNSS signals from low earth orbit'. University of Surrey; 2006.

[45]   Gleason S. 'Towards sea ice remote sensing with space detected GPS signals: demonstration of technical feasibility and initial consistency check using low resolution sea ice information'. *Remote Sensing*. 2010;**2**(8):2017–39.

[46]   Yan Q., Huang W. 'Spaceborne GNSS-R sea ice detection using delay-Doppler maps: first results from the U.K. TechDemoSat-1 mission'. *IEEE Journal of Selected Topics in Applied Earth Observations and Remote Sensing*. 2016;**9**(10):4795–801.

[47]   Alonso-Arroyo A., Zavorotny V.U., Camps A. 'Sea ice detection using U.K. TDS-1 GNSS-R data'. *IEEE Transactions on Geoscience and Remote Sensing*. 2017;**55**(9):4989–5001.

[48]   Schiavulli D., Frappart F., Ramillien G., *et al.* 'Observing sea/ice transition using radar images generated from TechDemoSat-1 delay Doppler maps'. *IEEE Geoscience and Remote Sensing Letters*. 2017;**14**(5):734–8.

[49]   Li W., Cardellach E., Fabra F., Rius A., Ribó S., Martín-Neira M. 'First spaceborne phase altimetry over sea ice using TechDemoSat-1 GNSS-R signals'. *Geophysical Research Letters*. 2017;**44**(16):8369–76.

[50]   Yan Q., Huang W., Moloney C. 'Neural networks based sea ice detection and concentration retrieval from GNSS-R delay-Doppler maps'. *IEEE Journal of Selected Topics in Applied Earth Observations and Remote Sensing*. 2017;**10**(8):3789–98.

[51]   Yan Q., Huang W. 'Sea ice remote sensing using GNSS-R: a review'. *Remote Sensing*. 2019;**11**(21):2565.

[52]   Yan Q., Huang W. 'Sea ice detection from GNSS-R delay-Doppler map'. 17th International Symposium on Antenna Technology and Applied Electromagnetics; 2016. pp. 1–2.

[53]   Clarizia M.P., Ruf C.S., Jales P., Gommenginger C. 'Spaceborne GNSS-R minimum variance wind speed estimator'. *IEEE Transactions on Geoscience and Remote Sensing*. 2014;**52**(11):6829–43.

[54]   Zhu Y., Yu K., Zou J., Wickert J. 'Sea ice detection based on differential delay-Doppler maps from UK TechDemoSat-1'. *Sensors*. 2017;**17**(7):1614.

[55]   Zhang G., Guo J., Yang D., *et al.* 'Sea ice edge detection using spaceborne GNSS-R signal'. *Geomatics and Information Science of Wuhan University*. 2019;**44**(5):668–74.

[56]   Zhu Y., Tao T., Yu K., *et al.* 'Sensing sea ice based on Doppler spread analysis of spaceborne GNSS-R data'. *IEEE Journal of Selected Topics in Applied Earth Observations and Remote Sensing*. 2020;**13**:217–26.

[57]   Cartwright J., Banks C.J., Srokosz M. 'Sea ice detection using GNSS-R data from TechDemoSat-1'. *Journal of Geophysical Research: Oceans*. 2019;**124**(8):C015327

[58]   Yan Q., Huang W. 'Sea ice detection based on unambiguous retrieval of scattering coefficient from GNSS-R delay-Doppler maps'. MTS/IEEE Oceans; 2018. pp. 1–5.

[59]   Zavorotny V.U., Voronovich A.G. 'Scattering of GPS signals from the ocean with wind remote sensing application'. *IEEE Transactions on Geoscience and Remote Sensing*. 2000;**38**(2):951–64.

[60]   Marchan-Hernandez J.F., Camps A., Rodriguez-Alvarez N., Valencia E., Bosch-Lluis X., Ramos-Perez I. 'An efficient algorithm to the simulation of delay-Doppler maps of reflected global navigation satellite system signals'. *IEEE Transactions on Geoscience and Remote Sensing*. 2009;**47**(8):2733–40.

[61]   Valencia E., Camps A., Marchan-Hernandez J.F., *et al.* 'Ocean surface's scattering coefficient retrieval by delay-Doppler map inversion'. *IEEE Geoscience and Remote Sensing Letters*. 2011;**8**(4):750–4.

[62]   Schiavulli D., Nunziata F., Migliaccio M., Frappart F., Ramilien G., Darrozes J. 'Reconstruction of the radar image from actual DDMs collected by TechDemoSat-1 GNSS-R mission'. *IEEE Journal of Selected Topics in Applied Earth Observations and Remote Sensing*. 2016;**9**(10):4700–8.

[63]   Southwell B.J., Dempster A.G. 'Sea ice transition detection using incoherent integration and deconvolution'. *IEEE Journal of Selected Topics in Applied Earth Observations and Remote Sensing*. 2020;**13**:14–20.

[64]   Shen X., Zhang J., Zhang X., Meng J., Ke C. 'Sea ice classification using Cryosat-2 altimeter data by optimal classifier-feature assembly'. *IEEE Geoscience and Remote Sensing Letters*. 2017;**14**(11):1948–52.

[65]   Bobylev L.P., Zabolotskikh E.V., Mitnik L.M. Neural-Network based algorithm for ice concentration retrievals from satellite passive microwave data. Microwave Radiometry and Remote Sensing of the Environment Specialist Meeting; 2008. pp. 1–4.

[66]   Yan Q., Huang W. 'Sea ice sensing from GNSS-R data using convolutional neural networks'. *IEEE Geoscience and Remote Sensing Letters*. 2018;**15**(10):1510–4.

[67]   Werbos P. *Beyond regression : New tools for prediction and analysis in the behavioral sciences*. Harvard University; 1974.

[68]   Marquardt D.W. 'An algorithm for least-squares estimation of nonlinear parameters'. *Journal of the Society for Industrial and Applied Mathematics*. 1963;**11**(2):431–41.

[69]   LeCun Y.A., Bottou L., Orr G.B., *et al. Efficient BackProp*. Berlin, Heidelberg: Springer; 2012.

[70]   Cortes C., Vapnik V. 'Support-vector networks'. *Machine Learning*. 1995;**20**(3):273–97.

[71]   Pal M., Mather P.M. 'Support vector machines for classification in remote sensing'. *International Journal of Remote Sensing*. 2005;**26**(5):1007–11.

[72]   Yan Q., Huang W. 'Detecting sea ice from TechDemoSat-1 data using support vector machines with feature selection'. *IEEE Journal of Selected Topics in Applied Earth Observations and Remote Sensing*. 2019;**12**(5):1409–16.

[73]   Hobiger T., Strandberg J., Haas R. 'Inverse modeling of ground-based GNSS-R – results and new possibilities'. *IEEE International Geoscience and Remote Sensing Symposium*. 2017:2671–81.

[74]   Yan Q., Huang W. 'Sea ice concentration estimation from TechDemoSat-1 data using support vector regression'. IEEE Radar Conference2019. pp. 1–6.

[75]   Zhu Y., Tao T., Zou J., Yu K., Wickert J., Semmling M. 'Spaceborne GNSS reflectometry for retrieving sea ice concentration using TDS-1 data'. *IEEE Geoscience and Remote Sensing Letters*. 2020;**18**(4):612–6.

[76]   Semmling A.M., Rosel A., Divine D.V., *et al.* 'Sea-ice concentration derived from GNSS reflection measurements in fram strait'. *IEEE Transactions on Geoscience and Remote Sensing*. 2019;**57**(12):10350–61.

[77]   Yan Q., Huang W. 'Sea ice thickness measurement using spaceborne GNSS-R: first results with TechDemoSat-1 data'. *IEEE Journal of Selected Topics in Applied Earth Observations and Remote Sensing*. 2020;**13**:577–87.

[78]   Zhang Y., Hang S., Han Y., *et al.*Sea ice thickness detection using coastal BeiDou reflection setup in Bohai Bay'. *IEEE Geoscience and Remote Sensing Letters*. 2021;**18**(3):381–5.

[79] Yan Q., Huang W. 'Sea ice thickness estimation from TechDemoSat-1 data. ocean. 2019 – Marseille'. IEEE Oceans; 2019. pp. 1–4.

[80] Tsang L., Newton R.W. 'Microwave emissions from soils with rough surfaces'. *Journal of Geophysical Research.* 1982;**87**(C11):9017.

[81] Ulaby F.T., Moore R.K., Fung A.K. 'Microwave remote sensing: active and passive. Vol. 1'. *Addison-Wesley Pub. Co., Advanced Book Program/World Science Division.* 1982.

[82] Kaleschke L., Maaß N., Haas C., Hendricks S., Heygster G., Tonboe R.T. 'A sea-ice thickness retrieval model for 1.4 GHz radiometry and application to airborne measurements over low salinity sea-ice'. *The Cryosphere.* 2010;**4**(4):583–92.

[83] Vant M.R., Ramseier R.O., Makios V. 'The complex-dielectric constant of sea ice at frequencies in the range 0.1–40 GHz'. *Journal of Applied Physics.* 1978;**49**(3):1264–80.

[84] Ulaby F.T., Moore R.K., Fung A.K. 'Microwave remote sensing : Active and passive'. Addison-Wesley Pub. Co., Advanced Book Program/World Science Division; 1986. p. 2.

[85] Klein L., Swift C. 'An improved model for the dielectric constant of sea water at microwave frequencies'. *IEEE Transactions on Antennas and Propagation.* 1977;**25**(1):104–11.

[86] Hu C., Benson C., Rizos C., Qiao L. 'Single-Pass sub-meter space-based GNSS-R ice altimetry: results from TDS-1'. *IEEE Journal of Selected Topics in Applied Earth Observations and Remote Sensing.* 2017;**10**(8):3782–8.

[87] Clarizia M.P., Ruf C., Cipollini P., Zuffada C. 'First spaceborne observation of sea surface height using GPS-reflectometry'. *Geophysical Research Letters.* 2016;**43**(2):767–74.

[88] Rodriguez-Alvarez N., Holt B., Jaruwatanadilok S., Podest E., Cavanaugh K.C. 'An Arctic sea ice multi-step classification based on GNSS-R data from the TDS-1 mission'. *Remote Sensing of Environment.* 2019;**230**(4):111202.

[89] Rodriguez-Alvarez N., Garrison J.L. 'Generalized linear observables for ocean wind retrieval from calibrated GNSS-R delay–Doppler maps'. *IEEE Transactions on Geoscience and Remote Sensing.* 2016;**54**(2):1142–55.

[90] Breiman L., Friedman J.H., Olshen R.A. 'Classification and regression trees'. Routledge; 1984.

Chapter 23

Triton – GNSS reflectometry mission in Taiwan

Yung-Fu Tsai[1], Chen-Tsung Lin[1], Jyh-Ching Juang[2], Wen-Hao Yeh[1], Shih-Hung Lo[1], Lin Zhang[3], and Hwa Chien[3]

23.1 Introduction

The goal of a global navigation satellite system (GNSS) is to provide accurate position, velocity and time (PVT) information through the reception of the GNSS signal. As of 2020, four fully operational GNSS systems are available: the American Global Positioning System (GPS) [1], Russian Global Navigation Satellite System (GLONASS) [2], Chinese BeiDou Navigation Satellite System [3] and European Galileo system [4]. Because of the omnipresence of GNSS signals and well-maintained infrastructure, many GNSS applications have been explored. Two typical applications for space-borne GNSS receivers are GNSS radio occultation (GNSS-RO) and GNSS reflectometry (GNSS-R) [5]. The GNSS-RO technique involves probing atmospheric and ionospheric profiles, with the bending angle and total electron content calculated from the GNSS signal propagating through the atmosphere and ionosphere. The GNSS-R technique involves remote sensing of the earth's surface through reception of reflected GNSS signals. A comprehensive introduction to the GNSS-R technique is presented in Chapter 18. The aim of this chapter is to introduce the recent development of a Taiwanese GNSS-R system, named Triton.

The first GNSS space application in Taiwan was the FORMOSAT-3/COSMIC (FS-3) programme (launched in 2006), which was a GNSS-RO programme with international collaboration between the National Space Organization (NSPO) of Taiwan and the University Corporation for Atmospheric Research of the United States. FORMOSAT-7/COSMIC-2 (FS-7), the subsequent programme launched in 2019, was a collaboration between the NSPO and the US National Oceanic and Atmospheric Administration [6]. Because the FS-7 GNSS-RO payload is capable of

[1]National Space Organization, Taiwan
[2]Department of Electrical Engineering, National Cheng Kung University, Taiwan
[3]Institute of Hydrological and Oceanic Sciences, National Central University, Taiwan

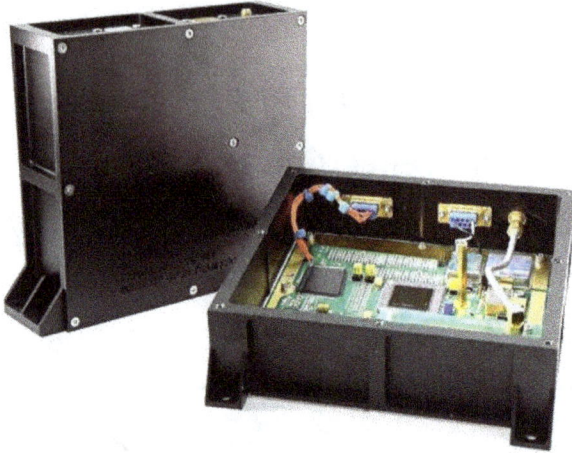

Figure 23.1   First space-grade GPS receiver in Taiwan

receiving not only GPS signals but also GLONASS and Galileo signals, the generated data volume of the FS-7 constellation is three times that of the FS-3. Hence,
the FS-7 mission dramatically increases the amount of atmospheric and ionospheric
observation data at low latitudes. Consequently, on the basis of satellite development and the GNSS-RO data processing experience, the NSPO is developing a satellite programme, named Triton, for the GNSS-R mission [7]. Triton is designated
to be a research-oriented programme aimed at the integration and testing of payload
modules, retrieval algorithms, product capability and operational applications.

The GNSS-R payload of the Triton satellite is designed on the basis of the
NSPO's development experience with its in-house-built space-grade GPS receiver
(GPSR). Being equipped with a space-grade GPSR is regarded as indispensable for
low earth orbit (LEO) missions because of the convenience of autonomous onboard
positioning and timing. Taiwanese companies, such as MediaTek and SkyTraq
Technology, have established strong capabilities in designing and manufacturing
civil GPSR modules and chipsets. However, such civil GPSR modules or chipsets
cannot be applied to satellite missions because of the International Traffic in Arms
Regulations [8]. To overcome this export control, the NSPO, which plays an essential role in promoting autonomous technologies in the space industry, developed
the first space-grade GPSR in Taiwan (Figure 23.1). The space-borne GNSS-R
receiver was designed for the Triton programme to collect earth-surface-reflected
GNSS signals. The collected data will be used for studies in several remote sensing
domains [9, 10]; for example, for retrieval algorithm studies and air–sea flux estimation, which contribute to the estimation of sea surface wind speed in severe sea
states. The retrieved sea surface wind speed data are expected to be used in the data
assimilation system of weather forecasting models to improve forecast accuracy,
especially for areas that lack observations.

The Triton satellite mission is described in Section 23.2. The development of the GPSR and GNSS-R mission payloads is described in Sections 23.3 and 23.4, respectively. Subsequently, the validation of the GNSS-R payload is described in Section 23.5. The development of the wind speed retrieval algorithm is discussed in Section 23.6. Finally, the Triton programme is summarised.

23.2   Triton satellite mission

The Triton satellite is an experimental microsatellite designed and manufactured by the NSPO and planned along with the FS-7 programme. Hence, it initially used the FORMOSAT serial number, with the addition of an 'R' for identification (FS-7R). In late 2018, the name 'Triton' was officially given to the GNSS-R mission. In addition to the GNSS-R mission, the Triton satellite mission is conceived as a technology demonstration mission to verify certain key satellite bus technologies, including an electrical power subsystem, attitude determination and control subsystems, an onboard data handling subsystem, a communication subsystem and some associated components to ultimately provide the baseline design for a satellite platform with self-reliant components and technologies for future NSPO missions [11]. Notably, the pointing accuracy of the satellite must be higher than 0.1° (3σ). However, because of the change in the constellation size of the FS-7 programme (from 12 + 1 satellites to 6 satellites), the Triton programme became less dependent on the FS-7 programme. The Triton satellite shares some common components with FS-7 satellites, such as the solar array, propulsion system, S-band transceiver, reaction wheel, magnetometer, coarse sun sensor, magnetic torquers and battery (Figure 23.2).

The Triton satellite's wet mass is approximately 285 kg, and its deployed and stowed configurations are depicted in Figure 23.3. The dimensions of the stowed configuration are approximately $100 \times 120 \times 125$ cm^3. In the deployed configuration, the solar panel will be stretched into space to generate an average power of 162 W. The subsystems, including onboard computer, power control unit, H_2O_2 propulsion, fibre optical gyro, space-grade GPSR and solar array control mechanism, are designed and developed indigenously as part of the Triton mission.

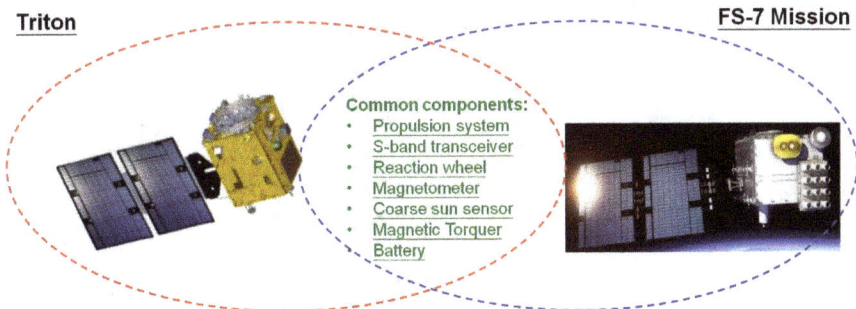

Figure 23.2   Triton and FS-7 satellites

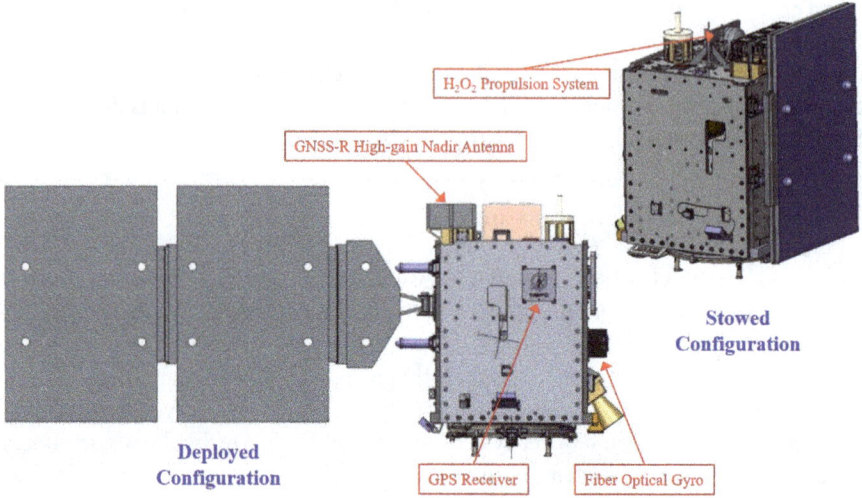

Figure 23.3   Triton satellite configurations

23.3   GPSR development

The hardware structure of the NSPO in-house-built space-grade GPSR consists of a radio frequency (RF) module, field-programmable gate array (FPGA) unit, digital signal processor (DSP), erasable programmable read-only memory and power supply system, as shown in Figure 23.4. The MAX2769 IC [12] is the main chip of the RF module and is designed for GPS L1 signal conversion and digitalisation. In addition to providing the interface between the DSP and other onboard devices and generating time pulses, the unpacking and filtering of the digitalised RF signal are implemented in the FPGA unit to reduce the computational load of the DSP. The main processor, the DSP, is responsible for the fast search algorithm engine, signal

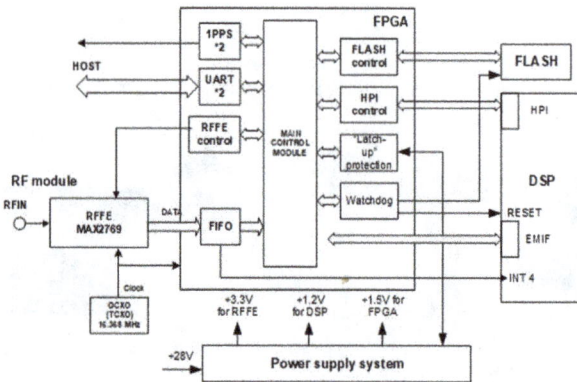

Figure 23.4   NSPO GPSR system hardware structure

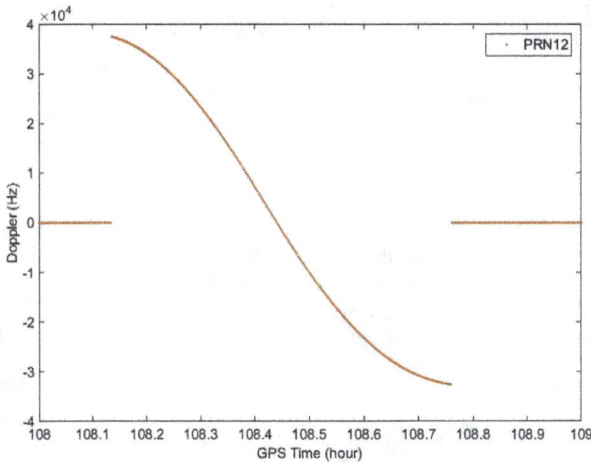

Figure 23.5   Doppler shift simulation of LEO missions

tracking, ephemeris decoding, navigation solution, orbit propagation, and host command and message decoding, among other tasks. One key feature of the space-grade GPSR is that it can overcome the high Doppler shifts and Doppler shift rates of the highly dynamic environment of LEO missions. The Doppler shift caused by an LEO satellite at an altitude of 550 km is ±40 kHz and the contact period is only approximately 40 minutes (Figure 23.5). Because of the implementation of the fast search algorithm, the NSPO in-house-built GPSR cold starts within 90 s, which is faster than most space-grade GPS receivers.

Another challenge for a space-grade GPSR is the harsh operating environment, characterised by challenges such as the extreme vibration during the launch phase, wide temperature range and the high-radiation conditions of space. The key specifications of the NSPO in-house-built GPSR are listed in Table 23.1. Two types of performance tests were conducted for this GPSR: the terrestrial test and the high dynamic environment test. The terrestrial test duration was 4 days. The GPS antenna for receiving real GPS signals is located on the rooftop of the NSPO's Assembly Integration and Test Building. The 3D position and velocity errors of the static GPSR performance were [4.9576, 7.9232, 5.3995] (m) and [0.0466, 0.0602, 0.0396] (m/s), respectively.

Because the NSPO GPSR is capable of flying onboard an LEO mission, the end-to-end test platform must reproduce the RF signal that the receiver will experience in highly dynamic environments. Two approaches can be employed to reproduce the designated GPS signal through the GPS simulator of Spirent Communications Public Limited Company. One approach involves using preloaded motion data files or predetermined scenarios in the Spirent GPS simulator. The other approach is to construct an attitude control and LEO mission dynamic environment simulation platform, which is implemented on the PCI (Peripheral Component Interconnect) eXtensions for Instrumentation real-time system of the National Instruments

Table 23.1   NSPO in-house-built GPSR specifications

Input power	28 V/3.5 W
Electrical interface	RS422/UART@115200 bps & 1 PPS output
Tracking channels	GPS L1 C/A*12
Tracking threshold	33 dB-Hz
Max. Doppler shift	±65 kHz
Position accuracy	Better than 8 m (1 σ)
Velocity accuracy	Better than 0.1 m/s (1σ)
Time accuracy	Better than 1 microsecond
Cold start time	Within 90 seconds
Trajectory dynamic	Up to 12 g
Mass	1.8 kg
Dimension	160 × 120 × 40 mm
Mission life	5 years
Total dose	35 krad

Corporation to generate the corresponding satellite attitude that is fed into the Spirent GPS simulator in real time [13]. Both approaches were applied for the high dynamic tests. The first high dynamic test was conducted under the predetermined scenario of orbiting at an altitude of 780 km and a slope of 45°. The position and velocity errors of the nine-orbit navigation performance test were [1.7637, 1.6696, 1.3000] (m) and [0.0283, 0.0277, 0.0239] (m/s), respectively. The average time to first fix of the NSPO GPSR over 30 cold starts with the same high dynamic scenario was 65.82 s. The other high dynamic test was conducted with the satellite attitude changing at a rate of 1°/s to validate the NSPO GPSR's tracking stability under tumbling conditions. The NSPO GPSR maintained the navigation solution even when the satellite was in a tumbling state (after 150 s; Figure 23.6) [14].

23.4   GNSS reflectometry mission payload

The principal task of the GNSS-R receiver is to receive and process the reflected GNSS signals from the earth; subsequently, the features of the earth surface can be retrieved. Because the FS-7 mission is designed to probe atmospheric data, the Triton GNSS-R mission will employ FS-7 GNSS-RO data to enhance its capability in weather prediction and severe weather research. Therefore, the retrieval of the wind speed over the ocean is regarded as a vital data product of the Triton mission. The design of the NSPO self-developed GNSS-R payload is based on several applicable GNSS-R processing options, which are discussed in [5, 15]. To meet relevant needs and with reference to the SGR-ReSI (existing GNSS-R receiver) design [16], the GNSS-R payload should satisfy the following requirements [17].

- The GNSS-R payload should process GPS L1 reflection signals and then generate an associated delay–Doppler map (DDM) either autonomously or according to a schedule.

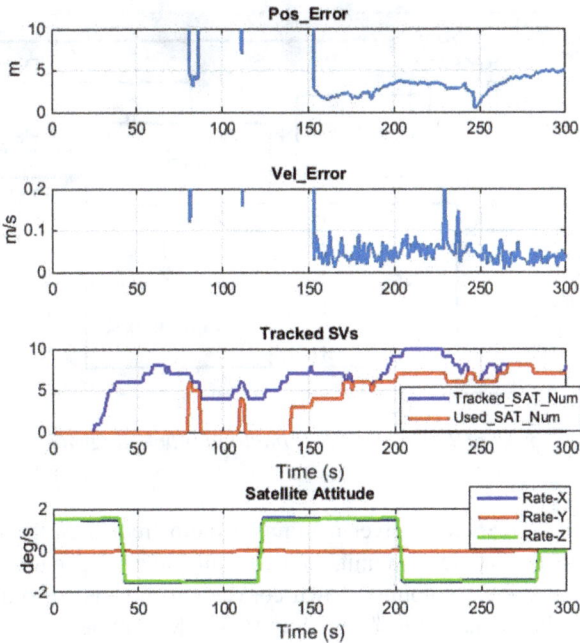

Figure 23.6   NSPO GPSR tracking stability

- The generated DDM resolution should be at least 128 (in code phase) × 64 (in frequency bin) and each entry of the DDM should be ≥16 bits.
- The GNSS-R payload should be capable of processing at least four DDMs simultaneously.
- The DDM update rate should be at least 1 s.
- The GNSS-R payload should potentially be extended to process reflected quasi-Zenith Satellite System (QZSS) and Galileo L1 signals and generate DDMs of the aforementioned space, magnitude and time resolution in real time.
- The GNSS-R payload should record direct line-of-sight and reflected GPS L1/L2 band signals in raw data format at the intermediate frequency for ground postprocessing purposes.
- The GNSS-R payload should include at least 512 MB of random-access memory to store the raw data.
- The GNSS-R payload should be designed with a process-and-record mode for ground debugging purposes.
- The GNSS-R payload should facilitate the retrieval of metadata to support calibration and retrieval.

To meet the aforementioned design requirements, several design configurations were proposed and examined in terms of complexity, resource utilisation and technical maturity. The GNSS-R payload block diagram, including the zenith antenna, nadir antenna, low-noise amplifiers (LNAs) and GNSS-R receiver, is depicted in

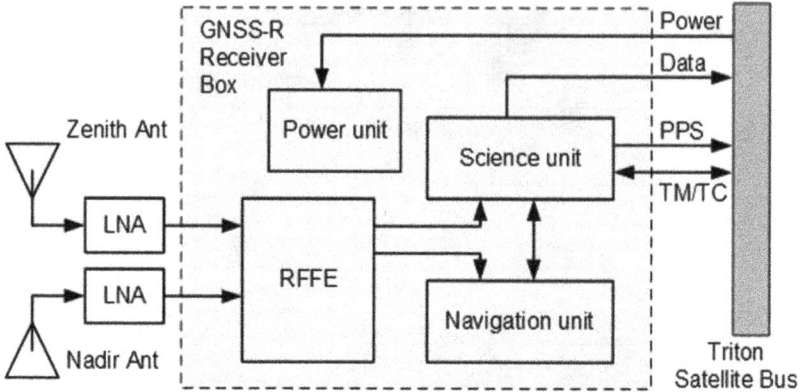

Figure 23.7   GNSS-R payload block diagram

Figure 23.7. The GNSS-R receiver contains a radio frequency front end (RFFE) with two RF inputs, a navigation unit, a power unit and a science unit. The RFFE is responsible for signal conditioning, down-conversion and analog-to-digital conversion. The incoming signals are sampled at 16.368 M samples/s, and each sample is represented as 4-bit data (i.e. a 2-bit in-phase component and 2-bit quadrature component). The navigation unit processes the line-of-sight signal to provide PVT information and GNSS satellite position and velocity information. The power unit regulates the input power from the Triton satellite bus for the distribution of regulated voltage to other units. The core technologies of the GNSS-R navigation unit and power unit are based on the NSPO space-grade GPSR. The interface of the telemetry/telecommand for the GNSS-R payload operation and status monitoring is the RS422. The science unit output data, including the generated DDM, stored raw data or both, are transmitted to the Triton satellite bus through the SpaceWire interface. The transmitted pulse-per-second signal is used for synchronisation.

Photos of the GNSS-R payload system are displayed in Figure 23.8. The zenith antenna is a dual-frequency (L1 and L2) right-hand circularly polarised antenna that is used to receive direct line-of-sight signals from GNSS satellites. The nadir antenna is a dual-frequency left-hand circularly polarised antenna to receive reflected or scattered signals from GNSS satellites. The LNAs, which are installed close to the antenna, are used to provide appropriate amplification (~33 dB) and filtering to enhance the signal-to-noise ratio. Both the nadir antenna and LNA are designed and developed by the NSPO with domestic organisations.

The GNSS-R receiver's kernel and science unit is responsible for the reception and processing of reflected GNSS signals. Hence, a reflection management routine is used to determine the specular reflection point with respect to each GNSS satellite on the basis of the information provided by the navigation unit. The relative code shift and Doppler shift are also computed to correctly assign the bins for DDM processing. The relative code shift is the code delay from the specular reflection point to the receiver. The relative Doppler shift changes the frequency of the GNSS signal

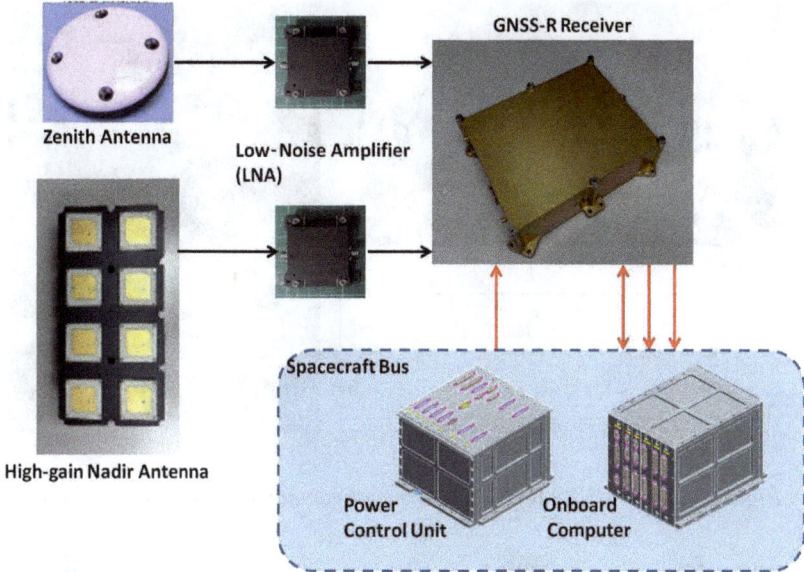

Figure 23.8   Photo of GNSS-R payload system

for the receiver's movement relative to the specular reflection point and the GNSS satellite's movement relative to the specular reflection point. The science unit relies on the reflected scenario to adjust the code phase and Doppler frequency during generation. Moreover, special code and carrier generation schemes were developed to process the data in parallel to the produced DDMs (Figure 23.9). The science unit of

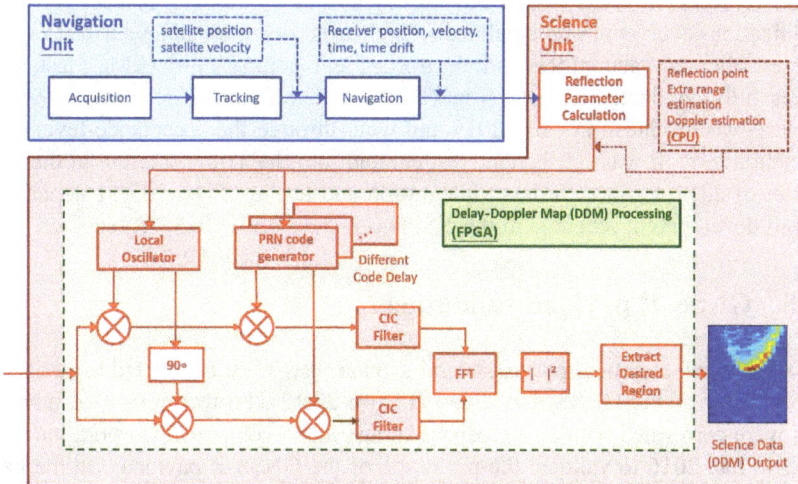

Figure 23.9   GNSS-R receiver function block

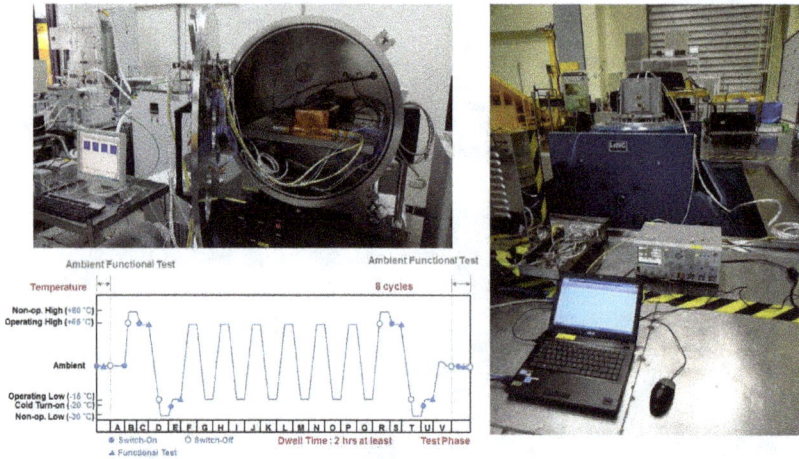

Figure 23.10   GNSS-R receiver thermal vacuum test and vibration test

the NSPO self-developed GNSS-R receiver is implemented by using a reprogrammable FPGA coprocessor, Zynq-7045, to meet the demand for real-time processing. The design of the reflectometry receiver is similar, in terms of major functions, to that of the SGR-ReSI used in the TechDemoSat-1 and Cyclone Global Navigation Satellite System (CYGNSS) missions [18]. However, the NSPO self-developed GNSS-R payload has some distinguishing features, such as reflected QZSS signal processing. The key improvements of the NSPO GNSS-R receiver are the capability to generate more DDMs (eight DDMs/s) and on-orbit GNSS-R receiver software modification.

The prototype and engineering qualification model of the NSPO self-developed GNSS-R receiver were developed in 2016 and mid-2017, respectively. All qualification-level environmental tests, including the thermal cycle test, vibration test, thermal vacuum test and electromagnetic compatibility (EMC) test, were successfully conducted in 2018 (Figure 23.10). The flight model of the GNSS-R receiver was produced in early 2019 and went through the acceptance-level environmental tests; it was subsequently integrated into the Triton satellite in the third quarter of 2019 for satellite-level integration and testing. Figure 23.11 depicts the overall development schedule of the NSPO self-developed GNSS-R receiver.

23.5   GNSS-R payload validation

In the early development phase, several ground tests were conducted to verify the functionality of the GNSS-R payload (Figure 23.12). Furthermore, two airborne tests were performed with an Aerospace Industrial Development Corporation aeroplane in late 2016 to validate the prototype of the GNSS-R payload. On the basis of the flight path plan, the link budget analysis was conducted before both airborne experiments. Figure 23.13 depicts one of the actual flight test paths as well as one

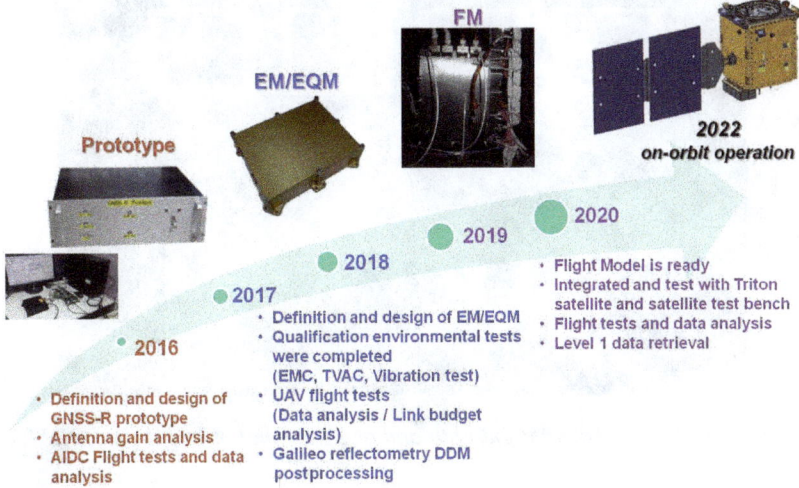

Figure 23.11   GNSS-R receiver development schedule

snapshot of a real-time DDM. Data collection during each airborne experiment lasted for approximately 30 minutes, and the collected data included DDM and raw data. Quality assessments were performed with the collected raw data [19]; for example, the direct and reflected signals were compared as were the reflected delay waveforms. The reflected signals were occasionally stronger than the direct signals,

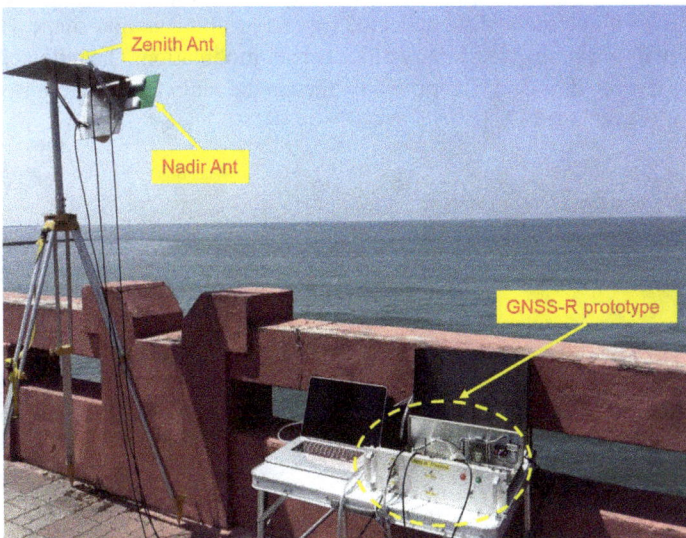

Figure 23.12   Ground test with the GNSS-R prototype

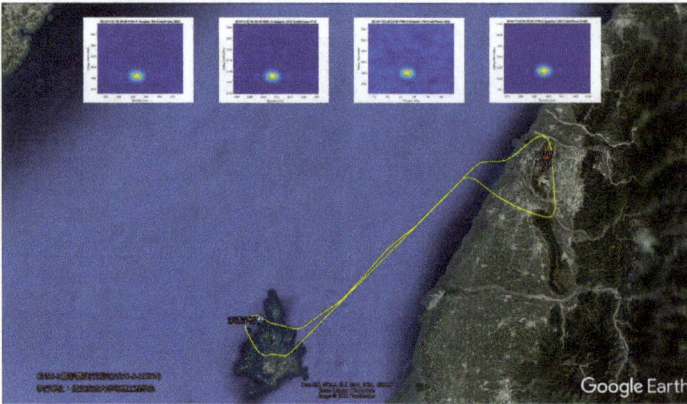

Figure 23.13   Actual flight path and one snapshot of a real-time DDM

as shown in Figure 23.14. Notably, the leading edges and trailing edges exhibited distinct shapes, consistent with expectations.

Figure 23.15 reveals the consistent delay waveforms that were collected, especially under high-elevation angles (above 45°). The consistency of the waveforms indicates that the quality of the GNSS-R payload data is satisfactory and can be used for wind data retrieval [20]. In addition to GPS signals, the QZSS, Galileo and Satellite-Based Augmentation System (SBAS) signals were processed to demonstrate the GNSS-R payload capability. Figure 23.16 illustrates the processing results of the QZSS, Galileo and SBAS (Multi-functional Satellite Augmentation System (MSAS) and GPS-aided GEO augmented navigation (GAGAN)) satellites. Because QZSS and SBAS signals are modulated in the same manner as GPS signals, the resulting delay waveforms of the reflected signals exhibit a similar shape. However, Galileo signals are subject to a binary offset carrier modulation. The reflected Galileo signals are expected to be different, which means the main lobe and the two humps (sidelobes) are included [21].

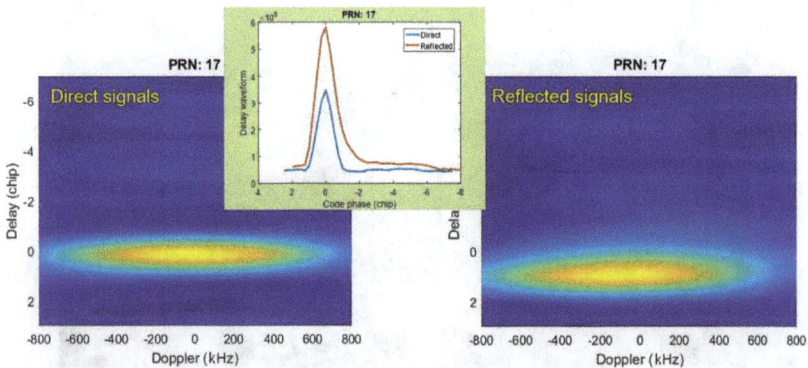

Figure 23.14   Postprocessing of the direct line-of-sight and reflected signals

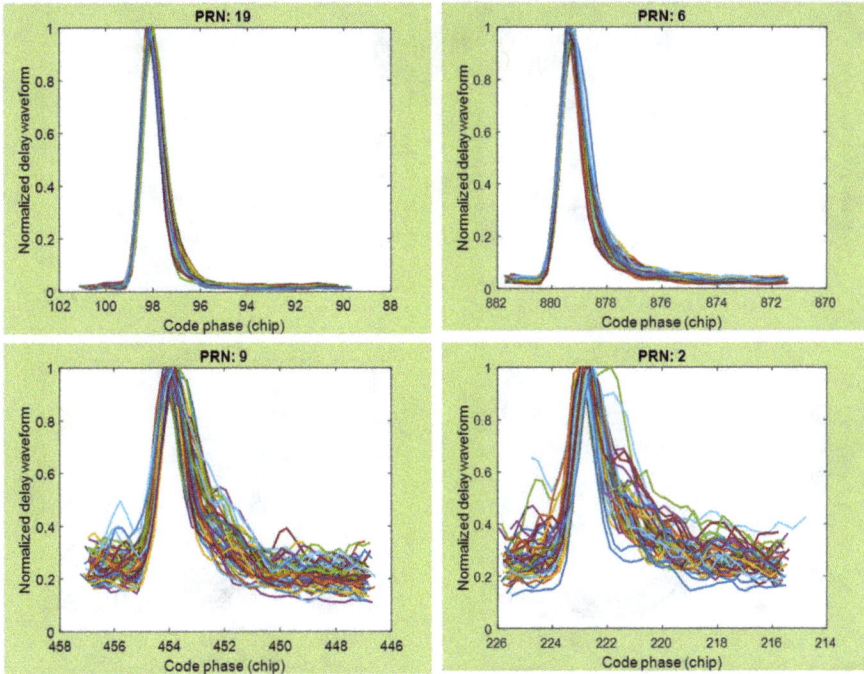

Figure 23.15   Delay waveforms of the scattered signals at various epochs with respect to the corresponding GPS satellites

To support the Triton mission, a science team has been formed to perform tasks related to data calibration, data archiving, data retrieval and data utilisation. Figure 23.17 depicts the science data processing of Triton's GNSS-R payload, which is under development by domestic universities and research centres. Triton's GNSS-R process system will be established in the Taiwan Radio Occultation Process System, which is developed for processing a large amount of FS-7 RO data [22].

23.6   Wind speed retrieval algorithm

A detailed discussion of wind estimation from GNSS-R signals is presented in Chapter 20. The wind retrieval module of the Triton system contains two parts. One is the level 1b (L1b) algorithm, which generates the DDM observable from the normalised bistatic radar cross section (NBRCS), and the other one is the level 2 (L2) algorithm, which retrieves the wind speed with a given DDM observable through the geophysical model function (GMF).

The inputs of the L1b system are the PVT of the GNSS (transmitter) and Triton (receiver) as well as the measured and calibrated digital–analog in the DDM from the Triton L1a product. In total, 128 and 64 DDMs are included in the Doppler and delay bins, respectively. The L1b system contains several modules. Specific

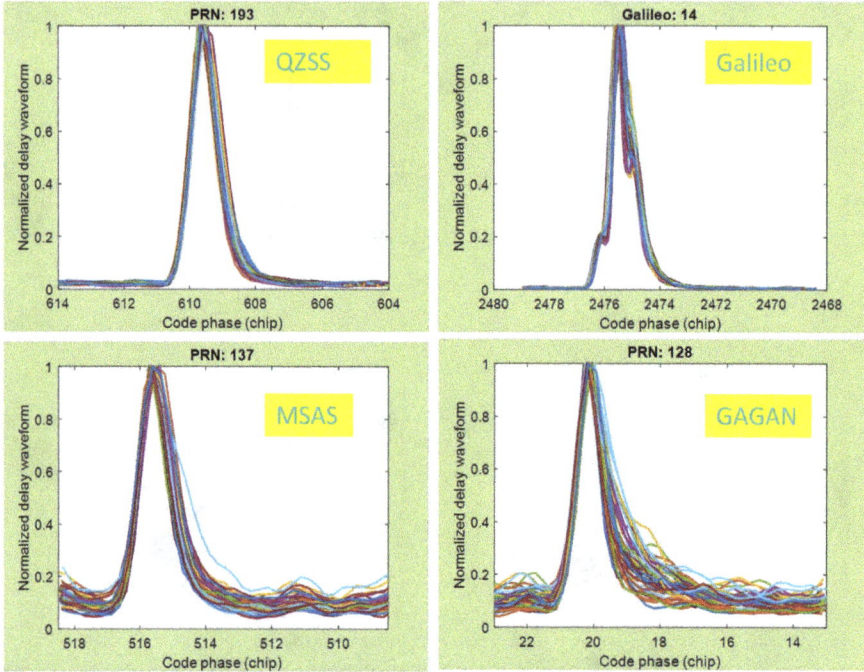

Figure 23.16   Delay waveforms of the scattered signals at various epochs with respect to the corresponding QZSS/Galileo/MSAS/GAGAN satellites

modules are responsible for computing the specular point position, the delay and Doppler pattern in the glistening zone, and the physical scattering area and effective scattering area (ESA). The ESA is used to remove the ambiguity effect from the bistatic radar cross section to generate the NBRCS. The L1b products consist of two DDM observables, the delay–Doppler map average (DDMA), a window-averaged NBRCS magnitude, and the leading edge slope of the waveform obtained from integrating the NBRCS along the delay axis. The strategy is to separate the procedure of wind speed retrieval from the DDM observable.

In the first stage, the L2a GMF, which is the relationship between the DDM observables and mean square slope (MSS), sea-truth data is established. The L2a GMF is then used to infer the MSS from Triton's DDM observables. In the second stage, the correlation of the MSS sea truth with the sea surface wind speed under various conditions is investigated and later used for wind speed retrieval. The advantage of this strategy is the mitigation of the confusion caused by the mixed sources of uncertainties that are involved in the two processes of the algorithm; in this manner, the uncertainties can be identified and addressed, especially in the case of extreme wind conditions.

MSS plays a pivotal role in the algorithm to link the DDM observables and the wind speed. In the geometric optic limit of the Kirchhoff approximation, the NBRCS

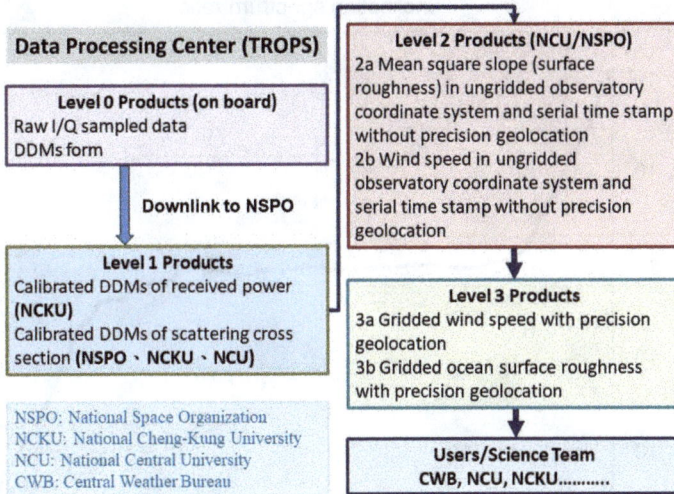

Figure 23.17   Science data processing chain

is a function of the Fresnel reflectance coefficient, scattering vector and the large-scale 'smoothed' surface slopes. The probability distribution of the surface slope is typically approximated by the anisotropic bivariate Gaussian distribution [23, 24], the variance of which is the MSS. MSS is the dominant factor influencing air–sea gas fluxes, momentum, and heat as well as radar scattering processes [25]. Studies have investigated the relationship between sea surface roughness and wind speed under a clean sea surface condition or over sea surfaces with oil slicks by using model results [26], satellite remote sensing data [25] and in situ measurements.

Obtaining MSS data is challenging [27–29]. By using existing remote sensing techniques, MSS can be measured using a Ku-band altimeter or computed from soil moisture and ocean salinity (SMOS) mission passive radiometer products. Collecting sea truth is highly challenging and expensive. The only possible

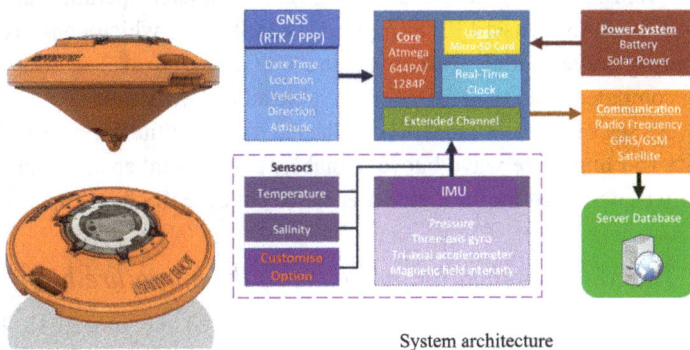

System architecture

Figure 23.18   System architecture of the miniature buoy

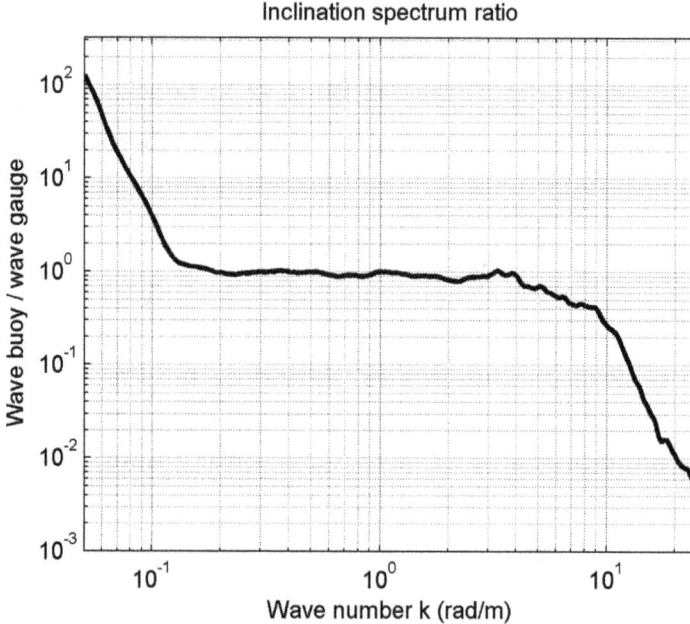

Figure 23.19   Transfer function and ratio of the slope spectrum measured by a miniature buoy to the slope spectrum measured by a wave gauge

approaches are direct measurements using a miniature data buoy or through an indirect airborne wide swath radar altimeter (WSRA).

MSS remote sensing data can be obtained from the SMOS mission (https://smos-diss.eo.esa.int/oads/access/) and WSRA Level 4 dataset (https://www.pro-sensing.com/wsra-level-4-data/). The payload for SMOS is a passive microwave radiometer (1.4 GHz). The algorithm for retrieving the MSS from the radar back-scattered data can be found in the SMOS L2 OS Algorithm Theoretical Baseline Document. WSRA is a digital beamforming radar altimeter operating at 16 GHz. The ECMWF-ERA5 dataset [30] also provides MSS data, which comprise results obtained using wave models. Moreover, the CYGNSS Level 2 dataset provides MSS data computed from their DDM observable and DDMA.

The MSS sea truth for Triton will be collected from satellite remote sensing data, model data and miniature wave buoys. Notably, a vast number of miniature buoy clusters will be deployed for the Triton mission to collect MSS direct measurements.

23.6.1   The MSS observation principle of the miniature buoy

The system architecture of the miniature buoy is displayed in Figure 23.18. The miniature buoy is equipped with a three-axis gyro, accelerometer and a digital compass. The sampling rate for rotation and acceleration measurements is 10 Hz.

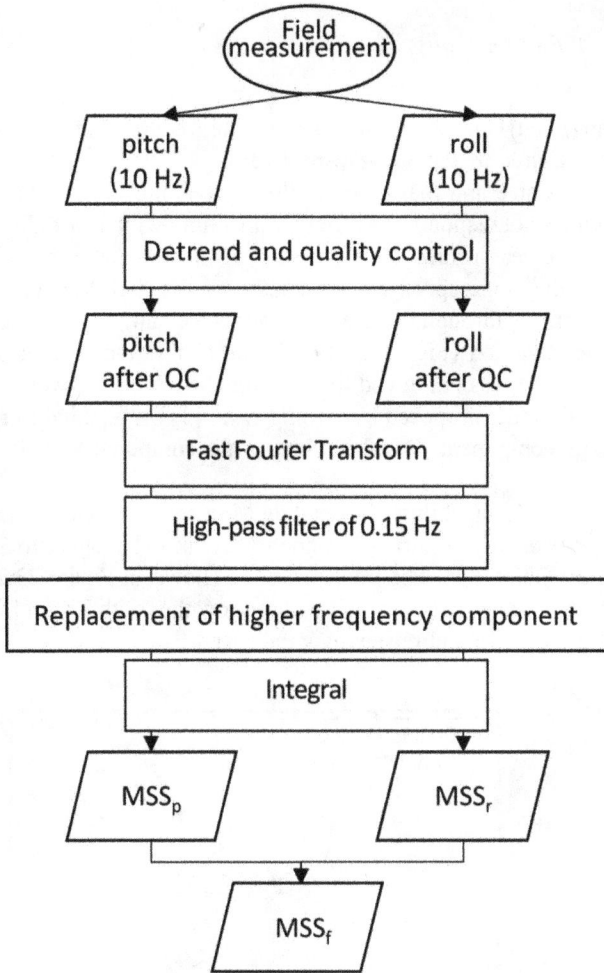

Figure 23.20   Flow chart of MSS measurement

Because the MSS of the sea surface is defined as the variance of the surface slope, the calculated MSS is obtained using the pitch and roll of the surface slope measured directly at sea by a miniature buoy as follows:

$$MSS = \sqrt{\left(MSS_p\right)^2 + \left(MSS_r\right)^2} \tag{23.1}$$

where

$$MSS_p = \int_0^\infty F\left(f, \tan\left(pitch\right)\right) \, df \tag{23.2}$$

$$MSS_r = \int\limits_0^\infty F\left(f, \tan\left(roll\right)\right) \mathrm{d}f \tag{23.3}$$

where $F\left(f, \tan\left(pitch\right)\right)$ and $F\left(f, \tan\left(roll\right)\right)$ are the frequency spectra of $\tan\left(pitch\right)$ and $\tan\left(roll\right)$ measured by the buoy, respectively.

Because of its size and mass inertia, the buoy is not ideal in terms of wave-following and does not respond adequately or in a timely manner to high-frequency fluctuations on the sea surface. The accuracy of the calculated MSS depends on the accuracy of the slope spectrum, especially for the high-frequency band. Lab calibration, performed through the use of a giant wave tank, was undertaken to provide the transfer function (Figure 23.19). In the calibration, a high-pass filter of 0.15 Hz (0.1 rad m^{-1}) was employed to eliminate the effects of swells. The empirical wave spectral model proposed by Hwang *et al.* [31] is adopted to represent the higher-frequency component. The flow chart of the computed MSS is presented in Figure 23.20.

To ensure Triton's capability of providing MSS sea-truth data, a cluster of miniature buoys were deployed during Typhoon Maysak (31 August to 3 September 2020) in the East China Sea. The trajectories are shown in Figure 23.21. The data collected from the miniature buoys will be used together with Triton DDM observables for generating GMF and retrieving wind speed.

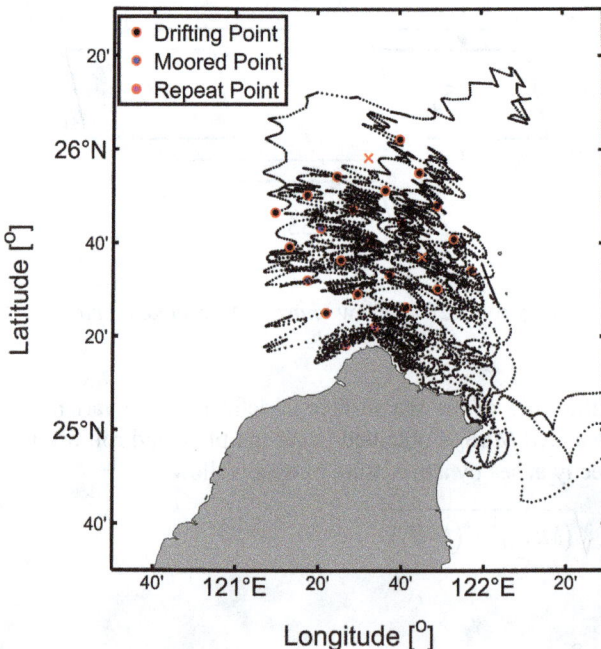

Figure 23.21   Example of the testing trajectories of miniature data buoys deployed in the East China Sea during Typhoon Maysak

23.7   Summary

The Triton programme started in 2012 and passed several milestones, such as system design review, preliminary design review, critical design review and integration and test review, before 2018. The NSPO team completed the installation of most components, harness routing and integration in 2019. The satellite-level EMC test, preliminary end-to-end test and mission test were completed in 2020. All environmental tests are scheduled for completion in 2021 to enable the launch of the Triton satellite in 2022. Once the data are available, its application in ocean observation will be evaluated.

References

[1] GPS Information [online]. Available from https://www.gps.gov/systems/gps/space/ [Accessed 17 Sept 2021].

[2] GLONASS Information [online]. Available from https://www.glonass-iac.ru/en/index.php [Accessed 17 Sept 2021].

[3] BeiDou Information [online]. Available from http://en.beidou.gov.cn/SYSTEMS/System/ [Accessed 17 Sept 2021].

[4] Galileo Information [online]. Available from https://www.gsc-europa.eu/ [Accessed 17 Sept 2021].

[5] Zavorotny V.U., Gleason S., Cardellach E., Camps A. 'Tutorial on remote sensing using GNSS bistatic radar of opportunity'. *IEEE Geoscience and Remote Sensing Magazine.* 2014;**2**(4):8–45.

[6] FORMOSAT-7 Program [online]. Available from https://www.nspo.org.tw/inprogress.php?c=20022301&ln=en [Accessed 17 Sept 2021].

[7] TRITON Program [online]. Available from https://www.nspo.org.tw/inprogress.php?c=20030305&ln=en [Accessed 17 Sept 2021].

[8] Juang J.C., Lin C.F., Hu C.M., Chang C.C., Tsai Y.F., Lin C.T. 'Implementation and test of a space-borne GPS receiver payload of university microsatellite'. *Journal of Aeronautics, Astronautics and Aviation, Series A.* 2012;**44**:141–8.

[9] Rodriguez-Alvarez N., Camps A., Vall-llossera M., *et al.* 'Land geophysical parameters retrieval using the interference pattern GNSS-R technique'. *IEEE Transactions on Geoscience and Remote Sensing.* 2010;**49**(1):71–84.

[10] Cardellach E., Fabra F., Nogués-Correig O., Oliveras S., Ribó S., Rius A. 'GNSS-R ground-based and airborne campaigns for ocean, land, ice, and snow techniques: application to the GOLD-RTR data sets'. *Radio Science.* 2011;**46**(6):1–16.

[11] Tsai Y.F., Lin C.T., Juang J.C. 'Taiwan's GNSS Reflectometry Mission – The FORMOSAT-7 Reflectometry (FS-7R) Mission'. *Journal of Aeronautics, Astronautics and Aviation.* 2018;**50**(4):391–404.

[12] Datasheet of MAX2769 Universal GPS Receiver [online]. Available from https://www.maximintegrated.com/en/products/comms/wireless-rf/MAX2769.html [Accessed 17 Sept 2021].

[13]   Chang H.-Y., Chang H.-C., Lin C.-T. 'Performance demonstration of NSPO space-borne GPS receiver'. *Proceedings of the 26th International Technical Meeting of the Satellite Division of the Institute of Navigation (ION GNSS+ 2013)*; Nashville, TN, September; 2013. pp. 3368–77.

[14]   Chang H.Y., Chiang W.L., KL W. 'A space-borne GNSS receiver for evaluation of the LEO navigation based on real-time platform' in Dołęga B., Głębocki R., Kordos D., Żugaj M. (eds.). *Advances in Aerospace Guidance, Navigation and Control*. Cham: Springer; 2018. pp. 529–50.

[15]   Tsai Y.-F., Yeh W.-H., Juang J.-C., Yang D.-S., Lin C.-T. 'From GPs receiver to GNSS reflectometry payload development for the triton satellite mission'. *Remote Sensing*. 2021;**13**(5):999.

[16]   Unwin M., de Vos Van Steenwijk R., Curiel A.D.S., Cutter M., Abbott B., Gommenginger C. 'Remote sensing using GPS signals – the SGR-ReSI instrument'. Presented at the 25th Annual AIAA/USU Conference on Small Satellites; Logan, UT, USA; 2011.

[17]   Juang J.-C., Ma S.-H., Lin C.-T. 'Study of GNSS-R techniques for FORMOSAT mission'. *IEEE Journal of Selected Topics in Applied Earth Observations and Remote Sensing*. 2016;**9**(10):4582–92.

[18]   CYGNSS Mission [online]. Available from https://cygnss.engin.umich.edu/ [Accessed 17 Sept 2021].

[19]   Juang J.C., Lin C.T. 'Recent development in GNSS-reflected signal processing and receiver research in Taiwan'. Presented on COSMIC Data User Workshop and IROWG-6 Meeting; Colorado USA; 2017.

[20]   Juang J.C., Ma S.H., Lin C.T. 'Data analysis of Galileo reflected signals in an airborne experiment'. Presented on Speciallist Meeting on Reflectometry using GNSS and other Signals of Opportunity. MI USA; Ann Arbor; 2017.

[21]   Juang J.C., Tsai Y.F., Lin C.T. 'FORMOSAT-7R Mission for GNSS reflectometry'. *IEEE International Geoscience and Remote Sensing Symposium, IGARSS 2019*; 2019.

[22]   Yeh W.-H., Huang C.-Y., Tseng T.-P., *et al.* 'Taiwan/TriG radio occultation process system (TROPS): brief introduction'. Presented on 26th IUGG General Assembly; Prague, Czech Republic, Jun. 22–Jul. 2; 2015.

[23]   Elfouhaily T., Chapron B., Katsaros K., Vandemark D. 'A unified directional spectrum for long and short wind-driven waves'. *Journal of Geophysical Research: Oceans*. 1997;**102**(C7):15781–96.

[24]   Zavorotny V.U., Voronovich A.G. 'Scattering of GPs signals from the ocean with wind remote sensing application'. *IEEE Transactions on Geoscience and Remote Sensing*. 2000;**38**(2):951–64.

[25]   Li S., Zhao D., Zhou L., Liu B. 'Dependence of mean square slope on wave state and its application in altimeter wind speed retrieval'. *International Journal of Remote Sensing*. 2013;**34**(1):264–75.

[26]   Zhang H., Yang K., Lou X., *et al.* 'Observation of sea surface roughness at a pixel scale using multi-angle sun glitter images acquired by the aster sensor'. *Remote Sensing of Environment*. 2018;**208**(C6):97–108.

[27] Gleason S., Zavorotny V.U., Akos D.M., *et al.* 'Study of surface wind and mean square slope correlation in Hurricane Ike with multiple sensors'. *IEEE Journal of Selected Topics in Applied Earth Observations and Remote Sensing.* 2018;**11**(6):1975–88.

[28] Vandemark D., Chapron B., Sun J., Crescenti G.H., Graber H.C. 'Ocean wave slope observations using radar backscatter and laser altimeters'. *Journal of Physical Oceanography.* 2004;**34**(12):2825–42.

[29] Walsh E.J., Vandemark D.C., Friehe C.A., *et al.* 'Measuring sea surface mean square slope with a 36-GHz scanning radar altimeter'. *Journal of Geophysical Research: Oceans.* 1998;**103**(C6):12587–601.

[30] *ERA5: fifth generation of ECMWF atmospheric re-analyses of the global climate* [online]. Copernicus Climate Change Service Climate Data Store (CDS). 2017. Available from https://cds.climate.copernicus.eu/cdsapp#!/home [Accessed 17 Sept 2021].

[31] Hwang P.A., Fois F. 'Surface roughness and breaking wave properties retrieved from polarimetric microwave radar backscattering'. *Journal of Geophysical Research: Oceans.* 2015;**120**(5):3640–57.

Appendix: List of Reviewers

Stuart J. Anderson	University of Adelaide (Australia)
Adriano Camps	Universitat Politècnica de Catalunya (Spain)
Zhongbiao Chen	Nanjing University of Information Science and Technology (China)
Brian M. Emery	University of California, Santa Barbara (USA)
Eric W. Gill	Memorial University (Canada)
Erin E. Hackett	Coastal Carolina University (USA)
Malcolm L. Heron	James Cook University (Australia)
Yukiharu Hisaki	University of the Ryukyus (Japan)
Weimin Huang	Memorial University (Canada)
Giovanni Ludeno	National Research Council (Italy)
Bjoern Lund	University of Miami (USA)
Jeffrey D Paduan	Naval Postgraduate School (USA)
Ryan Riddolls	Defence Research and Development Canada (Canada)
Francesco Serafino	National Research Council (Italy)
Wei Shen	Nanjing University of Science and Technology (China)
Benjamin J. Southwell	Dolby Laboratories (Australia)
Lucy R. Wyatt	University of Sheffield (UK)
Qingyun Yan	Nanjing University of Information Science and Technology (China)
Valery Zavorotny	University of Colorado Boulder (USA)

Index